Biology of Marijuana

Biology of Marijuana

From gene to behavior

Edited by

Emmanuel S. Onaivi

Department of Psychiatry and Pharmacology
Vanderbilt University
Nashville, Tennessee, USA, and
NIDA–NIH, MNB,
5500 Nathan Shock
Baltimore, MD 21224, and
Department of Biology
William Paterson University
Wayne, NJ 07508, USA

CRC Press
Taylor & Francis Group
Boca Raton London New York

CRC Press is an imprint of the
Taylor & Francis Group, an **informa** business

A TAYLOR & FRANCIS BOOK

CRC Press
Taylor & Francis Group
6000 Broken Sound Parkway NW, Suite 300
Boca Raton, FL 33487-2742

First issued in paperback 2019

© 2002 by Taylor & Francis Group, LLC
CRC Press is an imprint of Taylor & Francis Group, an Informa business

No claim to original U.S. Government works

ISBN-13: 978-0-415-27348-0 (hbk)
ISBN-13: 978-0-367-39619-0 (pbk)

Typeset in Baskerville BT by
Integra Software Services Pvt. Ltd, Pondicherry, India

Every effort has been made to ensure that the advice and information in this book
is true and accurate at the time of going to press. However, neither the publisher
nor the authors can accept any legal responsibility or liability for any errors or
omissions that may be made. In the case of drug administration, any medical procedure
or the use of technical equipment mentioned within this book, you are strongly
advised to consult the manufacturer's guidelines.

British Library Cataloguing in Publication Data
A catalogue record for this book is available
from the British Library

Library of Congress Cataloging in Publication Data
A catalog record for this book has been requested

Visit the Taylor & Francis Web site at
http://www.taylorandfrancis.com

and the CRC Press Web site at
http://www.crcpress.com

Contents

Figures

Plates

Color plates appear between pages 204 and 205.

Contributors

Dr Babatunde E Akinshola
Dept Pharmacology
Howard University
Washington DC 20059, USA

Francis Barth
Sanofi Recherche Center de labege
Innopole, B. P. 137, 31676
Labege Cedex, France

Alvin Berger
Nestle Research Center Lausanne
Nestec LTD
Vers-Chez-Les-Blanc
1000 Lausanne 26, Switzerland

Tiziana Bisogno
Endocannabinoid Research Group
Instituto per la Chimica di Molecole
de Intersse Biologico
Consiglo Nazionale delle Ricerche
Via Campi Flegrei 34
Comprensorio Olivetti
Fabbricato 70, 80078
Pozzuli (Napoli), Italy

Monsif Bouaboula
Sanofi Recherche Center de labege
Innopole, B. P. 137, 31676
Labege Cedex, France

Dr Sumner H Burstein
Department of Biochemistry
and Molecular Biology
University of Massachusetts
Medical School
Worcester, MA 01655–0103, USA

Dr Guy A Cabral
Department of Microbiology
and Immunology
Medical College of Virginia of Virginia
Commonwealth University
1101 E. Marshall St., Room 7065
Richmond, VA 23298–0678

Bernard Calandra
Sanofi Recherche Center de labege
Innopole, B. P. 137, 31676
Labege Cedex, France

R Cancro
New York University
Medical Center
550 First Avenue
New York
NY 10016, USA

Pierre Casellas
Sanofi Recherche Center de labege
Innopole, B. P. 137, 31676
Labege Cedex, France

Dr Nissar A Darmani
Kirksville College of Osteopathic
Medicine
Department of Pharmacology
800 West Jefferson Street
Kirksville, Missouri 63501, USA

Dr Sam A Deadwyler
Dept Physiology and Pharmacology
Bowman Gray School of Medicine
Wake Forest University
Medical Center Boulevard
Winston-Salem NC 27157–1083
USA

Dr Vincenzo Di Marzo
Endocannabinoid Research Group
Instituto per la Chimica
di Molecole de Interesse
Biologico, C.N.R.
Via Toiano 6, 80072
Arco Felice, Naples, Italy

Luciano De Petrocellis
Endocannabinoid Research Group
Instituto per la Chimica di Molecole
de Intersse Biologico
Consiglo Nazionale delle Ricerche
Via Campi Flegrei 34
Comprensorio Olivetti
Fabbricato 70, 80078
Pozzuli (Napoli), Italy

Dr S K Dey
University of Kansas Medical Center
Kansas City, 66160
USA

Pascual Ferrara
Sanofi Recherche Center de labege
Innopole, B. P. 137, 31676
Labege Cedex, France

Prof Ester Fride
Department of Behavioral Sciences
College of Judea and Samaria
Ariel 44837, Israel

Dr Eliot L Gardner
NIDA Intramural Research Program
National Institutes of Health
5500 Nathan Shock Drive
Baltimore, MD 21224, USA

Dr Michelle Glass
Dept Pharmacology
University of Auckland
New Zealand

Paul Gouldson
Sanofi Recherche Center de labege
Innopole, B. P. 137, 31676
Labege Cedex,
France

Dr Elena Grigorenko
Department of Physiology
Wake Forest University School of
Medicine
Winston-Salem, NC 27157, USA

Dr Robert E Hampson
Department of Physiology and
Pharmacology
Bowman Gray School of Medicine
Wake Forest University Medical
Center Boulevard
Winston-Salem, NC 27157, USA

David Harris
Physiology and Pharmacology
University of Nottingham
Medical School
Nottingham, NG7 2UH

D Harvey
New York University
Medical Center
550 First Avenue
New York
NY 10016, USA

Dr Stephen J Heishman
Clinical Pharmacology and
Therapeutics Branch
National Institute on Drug Abuse
National Institutes of Health
5500 Nathan Shock Drive
Baltimore MD 21224
USA

Dr John R Hubbard
2016 Baxter lane
Franklin
TN 37069
USA

Susan M Huang
Schrier Research Laboratory
Department of Psychology
Brown University
P. O. Box 1853
Providence, RI 02912, USA

Hiroki Ishiguro
Molecular Neurobiology Branch
NIDA Intramural Program
National Institutes of Health
Baltimore, MD 21224, USA

Dr David A Kendall
Physiology and Pharmacology
University of Nottingham Medical
    School
Nottingham, NG7 2UH

Atmaram D Khanolkar
University of Connecticut
School of Pharmacy
372 Fairfield Road
Box 2092, Storrs
CT 06269, USA

Gerard Le Fur
Sanofi Recherche Center de labege
Innopole, B. P. 137, 31676
Labege Cedex, France

Zhicheng Lin
Molecular Neurobiology Branch
NIDA Intramural Program
National Institutes of Health
Baltimore, MD 21224, USA

Dr Alexandros Makriyannis
University of Connecticut
School of Pharmacy

372 Fairfield Road
Box 2092
Storrs, CT 06269
USA

Dr Roy J Mathew
Department of Psychiatry
Duke University Medical Center
Durham, NC 27710
USA

Sean D McAllister
Forbes Norris MDA/ALS Research
    Center
2351 Clay Street
Suite 416
California Pacific Medical Center
San Francisco, CA 94115
USA

Dr Raphael Mechoulam
Department of Medicinal Chemistry
    and Natural Products
Medical Faculty, Hebrew University
Jerusalem 91120
Israel

Dr Laura L Murphy
Department of Physiology
Southern Illinois University
School of Medicine
Carbondale, IL 62901–6512
USA

Dr Richard E Musty
Department of Psychology
John Dewey Hall
The University of Vermont
Burlington
Vermont 05405, USA

Dr Gabriel G Nahas
New York University
Medical Center
550 First Avenue
New York, NY 10016
USA

Dr Emmanuel S Onaivi
NIDA Intramural Research Program
National Institutes of Health
5500 Nathan Shock Drive
Baltimore, MD 21224
Department of Biology
William Paterson University
Wayne, NJ 07508
USA

Sonya L Palmer
University of Connecticut
School of Pharmacy
372 Fairfield Road
Box 2092, Storrs
CT 06269, USA

Dr B C Paria
Dept Molecular and Integrative
   Physiology
Ralph L Smith Research Center
University of Kansas Medical Center
Kansas City 66160–7338
USA

Marielle Portier
Sanofi Recherche Center de labege
Innopole, B. P. 137, 31676
Labege Cedex
France

Dr Michael D Randall
University of Nottingham Medical
   School
Nottingham, NG7 2UH, UK

Murielle Rinaldi-Carmona
Sanofi Recherche Center de labege
Innopole, B. P. 137, 31676
Labege Cedex
France

Dr Patricia H Reggio
Professor of Chemistry
Kennesaw State University
Kennesaw
GA 30122, USA

Dr M Clara Sañudo-Peña
Schrier Research Laboratory
Department of Psychology
Brown University
P. O. Box 1853
Providence, RI 02912, USA

Dr Paul Schweitzer
Neuropharmacology CVN-12
The Scripps research institute
3545 John Hopkins Court
San-Diego, CA 92121, USA

Mr David Shire
Sanofi Recherche Center de Labege
Innopole, B. P. 137, 31676
Labege Cedex
France

Dr Nadia Solowij
National Drug and Alcohol
   Research Center
University of New South Wales
Sydney
NSW 2052, Australia

Nicole M Strangman
Schrier Research Laboratory
Department of Psychology
Brown University
P. O. Box 1853
Providence, RI 02912
USA

K M Sutin
New York University
Medical Center
550 First Avenue
New York
NY 10016, USA

H Turndorf
New York University
Medical Center
550 First Avenue
New York
NY 10016, USA

Dr George R Uhl
Chief, Molecular Neurobiology
Branch, NIDA Intramural Research
    Program
National Institutes of Health
5500 Nathan Shock Drive
Baltimore, MD 21224, USA

Dr J Michael Walker
Schrier Research Laboratory
Department of Psychology
Brown University
P. O. Box 1853

Providence, RI 02912
USA

Dr William H. Wilson
Duke University Medical
    Center
Durham, NC 27710, USA

Ping-Wu Zhang
Molecular Neurobiology Branch
NIDA Intramural Program
National Institutes of Health
Baltimore, MD 21224, USA

Preface

Marijuana is the prototypical cannabinoid, and is one of the most widely used drugs in the world. However, until the last decade, cannabinoid research lagged behind comparable work on other intoxicating natural products such as opiates and cocaine. This decade has seen an explosion of work and rapid advances in knowledge about cannabinoid receptors and the endogenous compounds that act on them.

The advances described here can be seen as a pharmacological success story, analogous to the prior successes in understanding opiates. Opiates were isolated as crystalline salts in the 19th century, while the active constituents in cannabis, delta-9-tetrahydrocannabinol (Δ^9-THC) remained elusive until 1964. Research on morphine led to the discovery of the naturally occurring family of morphine-like neurotransmitter peptides, the enkephalins and endorphins. Research on the active constituents of cannabis, (Δ^9-THC), also led to the discovery of cannabinoid ligands, which are derivatives of arachidonic acid. The advances described herein can also be viewed as a triumph, and possibly a paradigmatic triumph, for the role of molecular biological approaches to neuropharmacology which is likely to become even more widespread over the next decade. A major breakthrough in the field occured when an orphan G-protein coupled heptahelical receptor cDNA was cloned and then identified as the major brain cannabinoid CB_1 receptor. Identification of this receptor cDNA allowed rapid identification of the related CB_2 receptor and their genes. Cells expressing this receptor could then be made available for large scale screening, such as the efforts that identified the first selective CB_1 and CB_2 antagonists. Genomic characterization allowed production and elucidation of the features of knockout mice that confirmed major roles of cannabinoid systems in brain systems controlling locomotion, motor coordination, drug reward, and other features as well as roles for CB_2 systems in immune modulation. It could be argued that the CB_1 receptor might be one of the shining examples of the role that characterizing orphan receptors could play in the coming decade.

From both of these perspectives, it is exciting to view the prospect for cannabinoid research that is displayed so nicely in this book. It describes the identification of the members of this gene family in several species. The CB_1 knockout mice were shown to survive and develop; yet some of them suddenly die. From the knockout animals, the involvement of the cannabinoid system in the physiology of pain, motor and reward pathways were demonstrated. There are discussions of the acute and chronic actions that occur in animals and humans with much more insight due to the availability of CB_1 and CB_2 knockout mice and specific cannabinoid receptor antagonists useful both in animals and in humans. In addition, chapters describe the second messenger pathways that the surprisingly high-copy number receptor impacts, its possible role in selective sequestration of significant numbers of selected G proteins due to this high number, and a novel mechanism for receptor–receptor interaction in brain.

Exciting and tantalizing information and possibilities about endocannabinoids, the brain compounds that act at these receptors, have accompanied advances in cannabinoid research. Finding the endocannabinoids, e.g. anandamide, noladin-ether and 2-arachidonyl glycerol (2-AG) complements identification of the CB_1 and CB_2 receptors, and opens up a previously unknown endocannabinoid system of potentially central importance in biology and therapeutics.

The overarching goal for this book is thus to provide a reference text on this exciting new biology of marijuana and a launch pad for future discoveries in this area. Because of the ubiquity of the cannabinoid system in the body and brain, these observations should inform our understanding of a number of biological functions from movement, memory, learning, appetite stimulation, pain and emotions, to blood pressure, eye pressure, sexual function etc. We believe that this work will be of interest to undergraduate and graduate student in the basic science and biomedical sciences, researchers, medical practitioners, pharmaceutical scientists and science journalists.

George R. Uhl and Emmanuel S. Onaivi

Chapter 1

Cannabinoid receptor genetics and behavior

Emmanuel S. Onaivi, Hiroki Ishiguro, Zhicheng Lin,
Babatunde E. Akinshola, Ping-Wu Zhang and George R. Uhl

ABSTRACT

The last decade has seen more rapid progress in marijuana research than any time in the thousands of years that marijuana has been used by humans. cDNA and genomic sequences encoding G-protein coupled cannabinoid receptors (*Cnrs*) from several species are now cloned. Endogenous cannabinoid (endocannabinoid) ligands for these receptors, synthetic and hydrolyzing enzymes and transporters that define cannabinoid neurochemically-specific brain pathways have been identified. Endocannabinoid lipid signaling molecules alter activity at G-protein coupled receptors and possibly even anandamide-gated ion channels, such as vanilloid receptors. Availability of increasingly-specific CB_1 and CB_2 antagonists and of CB_1 and CB_2 receptor knockout mice increases our understanding of these cannabinoid systems and provides tantalizing evidence for even more GPCR-*Cnrs*. Initial studies of *Cnr* gene structure, regulation and polymorphisms whet our appetite for more data about these interesting genes, their variants, and roles in vulnerabilities to addictions and other neuropsychiatric disorders. Behavioral studies of cannabinoids document the complex interactions between rewarding and adverse effects of these drugs. Pursuing cannabinoid-related molecular, pharmacological and behavioral leads will add greatly to our understanding of endogenous brain neuromodulator systems, abused substances and potential therapeutics. The studies of CB_1 and CB_2 receptor genes reviewed in this chapter provide a basis for many of these studies.

Key Words: marijuana, genes, cannabinoids, endocannabinoids, mRNA, behavior

INTRODUCTION

Cannabinoids are the constituents of the marijuana plant (*Cannabis sativa*) of which the principal psychoactive ingredient is Δ^9-tetrahydrocannabinol (Δ^9-THC). Marijuana has remained one of the most widely used and abused drugs in the world. Although research on the molecular and neurobiological bases of the physiological and neurobehavioral effects of marijuana use was slowed by the lack of specific tools and technology for many decades, much progress has been achieved in cannabinoid research in the last decade. A central feature of this progress has been the elucidation of the cDNAs and genes that encode G-protein coupled (GPCR) cannabinoid receptors (*Cnrs*). This has facilitated discoveries of endogenous ligands, (endocannabinoids) which has led in turn to use of these endocannabinoids to help define other potential GPCR and even ligand-gated channel cannabinoid receptors. Even understanding the currently-understood CB_1 and CB_2 GPCR *Cnrs* has documented their importance for mediating most of the psychoactive effects of marijuana, other neurobehavioral alterations, and the bulk of the cellular, biochemical and physiological effects of cannabinoids (Martin, 1986).

Two cannabinoid receptor GPCR subtypes have been cloned to date. These are designated as CNR1 and CNR2 or CB_1 and CB_2. They belong to the large superfamily of receptors that couple to guanine-nucleotide-nucleotide-binding proteins and that thread through cell membranes seven times (heptahelical receptors). The CB_1 *Cnr* is predominantly expressed in brain and spinal cord, and thus, is often referred to as the brain *Cnr*. The CB_2 *Cnr* is, at times, referred to as the peripheral *Cnr* because of its largely-peripheral expression in immune cells. cDNA sequences encoding the rat (Matsuda *et al.*, 1990), human (Gerard *et al.*, 1991; Munro *et al.*, 1993), *murine* (Chakrabarti *et al.*, 1995; Abood *et al.*, 1997), bovine (Wessner, Genebank submission, 1997), feline (Gebremedhin *et al.*, Genebank submission, 1997), puffer fish (Yamaguchi *et al.*, 1996), leech (Stefano *et al.*, 1997), and newt (Soderstrom *et al.*, Genebank submission, 1999), CB_1 or CB_2 like receptors have been reported. The CB_1 *Cnr* is highly conserved across species, whereas the CB_2 receptor shows more cross-species variation. Human CB_1 and CB_2 receptors share 44% overall amino acid identity. Although this might suggest significant overall evolutionary divergence, the receptors' amino acid identities range from 35% to as high as 82% in different CB_1 transmembrane regions (Shire *et al.*, 1999).

CB_1 and CB_2 receptor gene products are expressed in relative abundance in specific tissues and cell types (Herkenham *et al.*, 1991; Bouaboula *et al.*, 1993; Matsuda *et al.*, 1993). The CB_1 *Cnr* is expressed at relatively high levels in brain regions such as hippocampus and cerebellum, and expressed at low levels in peripheral tissues including spleen, testis, and leucocytes. Das *et al.* (1995) demonstrated that the CB_1 mRNA but not CB_2 mRNA was expressed in the mouse uterus, where endocannabinoids can also be synthesized. CB_2 is not expressed in even moderate abundance in any brain region, but is expressed in peripheral tissues including white blood cells (Munro *et al.*, 1993; Facci *et al.*, 1995).

The identification of endogenous ligands for *Cnrs* has focused on modified eicosanoid-like fatty acids. Devane *et al.* (1992) named anandamide after the Sanskrit word for "bliss". The second principal endocannabanoid ligand 2-arachidonylglycerol (2-AG) was identified by Mechoulam *et al.*, 1995 and Sugiura *et al.*, 1995. Di Marzo *et al.* (this volume) review the biosynthesis, pharmacological, physiological functions of anandamide and other endocannabinoids. Potent synthetic cannabinoid agonists and antagonists have also been developed (Shire *et al.*, 1999).

This chapter discusses the current state of description of the genes encoding *Cnrs*, from their identification by chance to their at least partial elucidation in many species (Table 1.1). The chapter also defines some of the limitations of current knowledge. The pharmacology of the *Cnrs* is still less well understood than that of many GPCRs. Only scant information describes how these genes are regulated. A moderate store of data describes some of the signal transduction pathways engaged by cannabinoid receptor activation (see for example Hillard

Table 1.1 Molecular biological characteristics of G-protein coupled cannabinoid receptors

Receptor	Second messenger [a]Inbition of adenylate cyclase [b]Inbition of Ca^{2+} channels	Species cloned	Chromosomal location	Amino acid sequence	Genebank accession	Primary reference
CB_1	a and b	Human	6q14–15	472	X54937	Gerard et al., 1991
	a and b	Rat	–	473	X55813	Matsuda et al., 1990
	a and b	Mouse	Prox. 4	473	U17985	Chakrabarti et al., 1995
		Amphibian (Newt)	–	473	AF181894	Soderstrom et al., 2000
		Cat	–	472	U94342	Gebremedlin et al., 1997
		Bovine		Partial	U77348	Wessner et al., 1997
		Leech	–	480	–	Stefano et al., 1997
		Fish	–	470	X94402	Yamaguchi et al., 1996
		Fish		468	X94401	Yamaguchi et al., 1996
CB_1A		Human	1p36	360	X74328	Munro et al., 1993
CB_2		Rat	–	360	AF1763550	Griffin et al., 1999
		Mouse	Distal 4	347	NM009924	Shire et al., 1996

and Auchampach, 1994; Glass and McAllister, in this volume). However, many topics including the ways in which the abundant CB_1 receptors could alter activities mediated through other coexpressed GPCRs by sequestering G-proteins and other means are still in their infancy. As we discuss these GPCR receptors, we also need to be aware of the possible ligand gated ion channels influenced by cannabinoids. We need to bear in mind the data suggesting that cannabinoids can exert receptor-independent effects on biological (Hillard *et al.*, 1985; Makriyannis *et al.*, 1989) and enzyme systems such as protein kinase C. Despite these caveats, it is impressive to review the large amount of data about GPCR cannabinoid receptors amassed over the last decade.

Cannabinoid research and the use of cannabis products continue to attract significant attention. The current dramatic advances in molecular biology and technology, which increased scientific knowledge in cannabinoid research, will certainly contribute to a better policy on the medical use of marijuana. For example, preliminary studies with *Cnr* antagonists have contributed to resolving the long-standing debate about addiction to marijuana. Specifically, the controversial question of physical dependence on psychoactive cannabinoids has now been addressed, using the antagonist SR 141716A [N-(piperidin-1-yl)5-(4-chlorophenyl)-(2,4-dichlorophenyl)-4-methyl-1H-pyrazole-3carboxamidehydrochloride, to precipitate withdrawal reactions in rats injected with increasing doses of Δ^9-THC. Aceto *et al.* (1995) reported a precipitated withdrawal syndrome that was absent in control animals, providing evidence that Δ^9-THC could produce physical dependence. In general, it had been claimed that the psychoactivity and euphoria induced by cannabinoids limit their use in the clinic for numerous therapeutic applications for which they are currently being evaluated. Potential therapeutic applications include anti-emetic, appetite stimulant, glaucoma, epilepsy, multiple sclerosis, hepatitis C, Tourett's syndrome, migraine etc. In some cases, the issue of euphoria induced by cannabinoids does not outweigh the overall quality of life in the terminally ill patients. In many cases, there are however, a number of medical uses that can certainly benefit from the dissociation of the psychoactivity induced by marijuana and cannabinoids from their therapeutic actions. This chapter discusses the evidence for the existence of genes encoding *Cnrs* in the mammalian systems and the cloning of the rat CB_1 *Cnr* cDNA followed by the cloning of the genes from other species (Table 1.1). The synthesis of cannabinoid agonist and antagonist along with the discovery of endocannabinoids were pivotal to these current advances. With the availability of these genes, gene products and other *Cnr* research tools, it is speculated that the properties of these genes and regulation will be intensely studied as to reveal, how the psychoactivity can be dissociated from the therapeutic properties of marijuana and cannabinoids, or could it be that certain therapeutic actions of marijuana and cannabinoids cannot be separated from their psychoactivity?

Cannabinoid receptor (*Cnr*) genes

In this section, the expression of *Cnr* genes in different species and in many tissues of the mammalian system is reviewed along with the functional implication of selective *Cnr* subtype gene knockout by homologous recombination. Even with the significant advancements in cannabinoid research and the availability of molecular probes the cloning of the first *Cnr* gene was fortunate.

Genes encoding rat cannabinoid receptors

The rat CB_1 cDNA was identified from brain distribution data concerning an orphan G-protein coupled receptor cDNA. Matsuda *et al.* reported the identification in 1990. They identified the orphan GPCR cDNA in a rat cerebral cortex cDNA library, probed with a 56-base pair oligonucleotide probe, complementary to sequences encoding the second transmembrane domain of the bovine GPCR substance-P receptor. When *in situ* hybridization data paralleled the distribution of CB_1 receptor autoradiograms developed in an adjacent laboratory, the identity of the orphan cDNA was suspected and rapidly confirmed pharmacologically. Like other GPCRs, the *Cnrs* contain an N-terminal extracellular domain that possesses glycosylation sites, seven transmembrane segments and a C-terminal intracellular domain that may be coupled to a G-protein complex. One distinguishing feature of the CB_1 receptor is its long N-terminal putative extracellular segment (Shire *et al.*, 1995).

The rat CB_2 cDNA was described following description of the *murine* CB_2 gene (Shire *et al.*, 1996; see below). Griffin *et al.* (1999) used primers homologous to the predicted translation initiation and termination sites of the mouse CB_2 gene and PCR amplifications to generate a ~ 1.1 kb fragment from rat genomic DNA that allowed subsequent identification of the rat CB_2 receptor. Table 1.1 shows that sequence analysis of the coding region of the rat CB_2 receptor clone indicate 90% nucleic acid identity (93% amino acid identity) between rat and mouse and 81% nucleic acid identity (81% amino acid identity) between rat and human (Griffin *et al.*, 1999).

Genes encoding human cannabinoid receptors

The human CB_1 cDNA was isolated by Gerard *et al.* (1991) from a human brain stem cDNA library using a 600 bp DNA probe and polymerase chain reaction. The deduced amino acid sequences of the rat and human receptors showed that they encode protein residues of 473 and 472 amino acids respectively with 97.3% homology. These proteins share the seven hydrophobic transmembrane domains and residues common among the family of G-protein receptors (Matsuda *et al.*, 1991; Gerard *et al.*, 1991; Shire *et al.*, 1995). The human and rat CB_1 receptors also share pharmacological characteristics including the inhibition of adenylate cyclase activity via Gi/o in a stereoselective and pertusis sensitive manner following activation by cannabinoids (Devane *et al.*, 1988; Matsuda *et al.*, 1990; Gerard *et al.*, 1991). The CB_1 receptors alter potassium channel conductance (Hampson *et al.*, 1995) and decrease calcium channel conductance (Mackie and Hille, 1992).

The CB_2 *Cnr* clone was isolated from myeloid cells by PCR and degenerate primers using a cDNA template from the human promyelocytic leukaemic line HL60 (Munro *et al.*, 1993). This clone was shown to be related to the rat CB_1 and was therefore used to screen the HL60 library. The primary structure of the CB_1 was essential in the identification of this subtype of *Cnr* gene. Most importantly, there were the similar hydrophobic domains 1, 2, 5, 6 and 7, in which 50% or greater of the amino acids were identical between CB_1 and CB_2 *Cnrs*. The extracellular domain also contained sequence motifs that were common to both clones (Matsuda, 1997). The protein encoded by the CB_2 shows 44% identity with the human CB_1. A number of functional and expression studies have been performed with the CB_2 gene and the results indicate that the CB_2 is the predominant *Cnr* in the immune system, where it is expressed in B and T cells (Munro *et al.*, 1993; Schatz *et al.*, 1997; Gurwiz and Kloog, 1998). Presence of *Cnrs* in invertebrate immunocytes attests to their primordial role in immune regulation (Gurwiz and Kloog, 1998).

It has been proposed that dysfunction of normal endocannabinoid systems, especially those derived from macrophages, might participate in the immune system dysfunctions in HIV-infected individuals. Activation of B and T cell CB_2 receptors by cannabinoids leads to inhibition of adenylate cyclase in these cells and a subsequent reduced response to immune challenge. In *in vitro murine* T cell lines, inhibition of signal transduction via the adenylate cyclase/cAMP pathway can induce T cell dysfunction and reduced interleukin 2 (IL-2) gene transcription, for example (Condie *et al.*, 1998).

Genes encoding mouse cannabinoid receptors

The mouse *Cnr* CB_1 cDNA was cloned and compared to the rat and human sequences by Chakrabarti *et al.* (1995) and Onaivi *et al.* (1996). C57BL/6 mouse cDNA libraries were screened using a rat CB_1-specific probe. Clones were purified to homogeneity by repeated screening, inserts were sub-cloned and sequenced, and CB_1 cDNA sequences submitted to GENEBANK by Chakrabarti *et al.* (1995) (#U17985). Later reports of other mouse CB_1 cDNA sequences confirmed our initially submitted sequences (Abood *et al.*, 1997). The mouse genomic clone data displays 95% nucleic acid and 99.5% amino acid identity to rat. There was 90% nucleic acid identity and 97% amino acid identity between mouse and human CB_1 gene. Examination of the 5' untranslated sequence of the mouse CB_1 genomic clone revealed a splice junction site approximately 60 bp upstream from the translational start site. Binding studies using CB_1 gene coding regions stably transfected and expressed in 293 cells revealed substantial specific [^{3}H]SR 141716A and [^{3}H]CP-55940 binding with B_{max} and K_D values similar to those of native mouse CB_1 receptors in brain (Abood *et al.*, 1997).

The molecular cloning, expression and function of the *mouse* CB_2 peripheral *Cnr* was reported by Shire *et al.*, 1996, from a mouse splenocyte cDNA library. They used a radiolabelled human CB_2 cDNA to screen a *mouse* spleen cDNA library and detected four positive clones. The sequence was found to contain a short 5' untranslated region (UTR), a 1041 bp open reading frame (ORF) and a 3'-UTR of 2.5 kb containing three canonical polyadenylation signal motifs. The ORF is 81% identical to the human CB_2 and the 3'-UTR 1.7 kb longer than that of the human CB_2. Unlike the human CB_1 and rat CB_1, which differ in only 13 residues, the deduced mouse CB_2 protein sequence of 347 residues differs from human CB_2 in 60 residues with 82% identity. It is also shorter by 13 residues at its carboxy terminus. Shire *et al.* (1996), thus observed that the differences between the two receptors were scattered throughout their sequences, but with a relatively high concentration in N-terminal and other extracellular regions. Amino-acids amenable to potential post-translational modification are to be found in both human and mouse CB_2. These include some serines/threonine that are potential phosphoacceptor sites and sites for asparagines N-linked glycosylation.

Other vertebrate and invertebrate Cnr genes

Other vertebrate and non-vertebrate *Cnr* genes have been cloned, sequenced and in some cases partially characterized (Table 1.1). Since *Cnr* genes and signaling exists in monkey, cat, dog, cattle, goat, pig, sheep, zebra finch, frog and many lower organisms described below, this signaling system has clearly been conserved during evolution (Akinshola *et al.*, 1999a; Martin *et al.*, 2000).

Genes encoding puffer fish cannabinoid receptors

The puffer fish, *Fugu rubribes* (Fugu), has been proposed as a model primitive vertebrate genome. Yamaguchi *et al.* (1996), characterized two putative G-protein coupled receptor genes, FCB_1A and FCB_1B by degenerate PCR and low-stringency hybridization of a Fugu genomic library. These two genes show high homology to the human CB_1 and less homology with CB_2 *Cnr* subtype. The amino acid sequences of the FCB_1A and FCB_1B genes are 66.2% identical. They display 72.2% and 59.0% homologies to the human CB_1 sequences. The overall homology between FCB_1A or FCB_1B and the human CB_2 is only 34.9% and 31.7%. Transcripts of both the FCB_1A and FCB_1B receptors are abundant in brain. Analysis of the receptor structure by Yamaguchi *et al.* (1996) showed that in contrast to CB_2 *Cnr* subtype, both the two Fugu receptors and the mammalian CB_1 *Cnrs* contain a longer third intracellular and a longer cytoplasmic tail than the CB_2 *Cnr* subtype. The tissue distribution of these *Cnr* genes in the puffer fish appear to be high in the brain and expressed more moderately in spleen, ovary and testis. Yamaguchi *et al.* (1996), concluded that this distribution was similar to human CB_1 than the CB_2 and that both the FCB_1A and FCB_1B are more similar to the human CB_1 than to the human CB_2 *Cnr*.

Genes encoding leech cannabinoid receptors

In the leech CNS, the CB_1 *Cnr* gene has been cloned, sequenced and partially characterized (Stefano *et al.*, 1997). The deduced amino acid sequence analysis from a 480 bp amplified RT-PCR fragment cDNA exhibits a 49.3% and 47.2% sequence identity with human and rat CB_1 *Cnrs* respectively. Thus, the leech *Cnr* shares similar properties with CB_1 with highly conserved regions particularly in the putative transmembrane domains 1 and 2. As parts of the sequence of the leech *Cnr* are highly conserved and has 61% sequence homology with the human CB_1, it would appear that the presence of *Cnrs* in these lower organisms indicate that this signaling system is highly conserved during evolution. Interestingly, anandamide is also the endocannabinoid in leech CNS. It displays a monophasic high affinity binding site that is coupled to nitric oxide release (Stefano *et al.*, 1997). Analyses of G-protein coupled receptors isolated from the leech CNS have also revealed evidence for a chimeric cannabinoid/melanocortin receptor (Elphick, 1998). Two regions of the leech sequence display high levels of amino acid identity with mammalian *Cnrs*, while a third region is 98% identical to part of the bovine adrencorticotropic hormone receptor. Therefore, the leech receptor may resemble a putative ancestor of mammalian cannabinoid and melanocortin receptors.

Other invertebrate *Cnr* genes

Sequence from the *Caenorhabdities elegans* (*C. elegans*) genome share weak similarity to *Cnrs* (Wilson *et al.*, 1994). *Cnr* binding activity has been reported in the marine cyanobacterium, *Lyngbya majuscula* (Sitachitta and Gerwick, 1998). Cannabinoids can inhibit the growth of pathogenic amoeboflagellate *Naegleria fowleri* and prevents enflagellation and encystment, but do not impair amoeboid movements of this organism (Pringle *et al.*, 1979). Cannabinoids influence the social behavior of ants (*Formica pratensis*) (Frischknecht and Waser, 1980), movement and cellular growth in *Tetrahymena* species (McClean and Zimmerman, 1976), nerve cell excitability in *Aplysia* (Acosta-Urquidi and Chase, 1975), electrophysiological activity of the squid and lobster giant axons (Brady and Carbone, 1973; Turkanis and Karler, 1988), regeneration in the planarian *Dugesia tigrina*, (Lenicque *et al.*, 1972), fertility in sea urchin sperm (Schuel *et al.*, 1987, 1994), feeding in *Hydra* (Cndaria) (De Petrocellis *et al.*, 1999), and physiological actions in molluscs (Sepe *et al.*, 1998). The effects of cannabinoids in the vertebrate and invertebrate systems may indicate the existence of *Cnr* genes and subtypes that have yet to be cloned. Recent studies have shown the existence of elaborate invertebrate cannabinoid signaling cascades complete with enzymatic activities responsible for the biosynthesis and degradation of endocannabinoids. Such endocannabinoid signaling systems have been described in sea urchin and hydra (Bisogno *et al.*, 1997; De Petrocellis *et al.*, 1999).

Cannabinoid receptor gene expression

The expression of CB_1 *Cnr* in the CNS has been extensively studied (Onaivi *et al.*, 1996). The CB_2 receptor gene has been detected particularly in the immune system and the expression of its transcripts found in spleen, tonsils, thymus, mast cells and blood cells. CB_1 and CB_2 *Cnrs* can be co-expressed in some of the same cells, in which cannabimimetic effects can be mediated by their combination. The relative abundance of the endocannabinoids and the relatively large numbers of expressed cannabinoid receptors may allow these systems to influence many biochemical systems. It thus may not be surprising that, intimate links between cannabinoid systems and dopaminergic, glutamatergic, serotonergic, opioidergic and other important neurotransmitters can be readily identified.

The expression of the CB_1 *Cnr* genes has been detected in the brains of many species including, human, monkey, pig, dog, cat, cattle, guinea pig, rat, mouse, frog, fish, leech etc. CB_1 expression can be detected in regions that influence a number of key functions including mood, motor coordination, autonomic function, memory, sensation and cognition. Expression is more abundant in hippocampus, cerebral cortex, some olfactory regions, caudate, putamen, nucleus accumbens and the horizontal limb of the diagonal band. In concert with the localization and distribution pattern of the CB_1 *Cnr* gene, the radioligand binding sites to CB_1 protein reported in the rat brain, mirror the expression of mRNA products of the CB_1 *Cnr* gene. The highest densities of the expression of the CB_1 receptors are consistent with the marked effects that cannabinoids exert on motor function tests, like spontaneous locomotor activity and catalepsy in rodents (Onaivi *et al.*, 1996; Chakrabarti *et al.*, 1998). There is however, a relatively low number of CB_1 receptors in the human cerebellum in comparison with rodents. The low abundance of CB_1 receptors in the human cerebellum is consistent with the more subtle defects noted in human gross motor functioning following marijuana use. Soderstrom and Johnson (2000) reported on CB_1 *Cnr* expression in brain regions associated with zebra finch song control. They demonstrated using *in situ* hybridization that, CB_1 mRNA is expressed at high levels in the higher vocal center (HVC) and the robust nucleus of the archstriatum (RA), areas that are involved in song learning and production.

Interaction between these receptors and alterations in mental and neurological disorders has been reviewed by Musty in this book. While the specific effects of *Cnr* gene expression in mental and neurological function is incompletely understood, Tourette syndrome (GTS), obsessive compulsive disorder (OCD), Parkinson's disease, Alzheimer's disease and other neuropsychiatric or neurological disturbance are candidates to be influenced by possible variants in the *Cnr*, CB_1 receptor gene (Gadzicki *et al.*, 1999). Altered CB_1 expression has been reported and clinical trials began on the use of cannabinoids to treat a number of mental disorders as well as brain injury. The expression of the CB_1 and to a lesser extent CB_2 *Cnr* genes has been studied at different stages in development using brain tissues and preimplantation embryo and in the aging brain. CB_1 expression can be detected in tissue from newborn infants (Mailleux *et al.*, 1992). The ontogeny of rat *Cnr* expression allows the receptor to be detected at postnatal day 2 and at an even earlier stage in rat embryonic brains (de Fonesca *et al.*, 1993; McLaughlin and Abood, 1993). Pre-implantation mouse embryos express both CB_1 and CB_2 *Cnr* genes (Belue *et al.*, 1995). While the CB_2 *Cnr* gene has been detected from single-cell embryo through the blastocyst stages, the CB_1 gene is expressed in the four-cell embryo.

There appears to be a general decline in the expression of CB_1 *Cnr* genes with age in the human and rodent brain (Westlake *et al.*, 1994), although conflicting data (decline or no change in the expression of CB_1 *Cnr* gene) are reported in aged rats.

Both CB_1 and CB_2 *Cnr* genes are differentially expressed although in the immune cells, spleen and bone marrow the CB_2 is more abundant (Galiegue *et al.*, 1995). In many of these cells, in peripheral tissues and some cell lines, the CB_1 message is detected only after PCR amplification. CB_1 receptor-specific PCR products can be obtained from human polymorphonuclear (PMN) monocytes, T4 and T8 cells, B cells, natural killer cells (NK), T leukemia cells, lymphoma cells, B lymphoblasts and lymphocytes, immortalized monocytes, mouse NL-like cells and T cells. By contrast, abundant levels of CB_2 message have been reported in human lung, uterus, pancreas, tonsils, thymus, peripheral blood mononuclear cells, NK cells, B cells, macrophages, PMN cells, mast cells, basophilic leukemia cells (RBL-2H3 cells), T4 and T8 cells (Galiegue *et al.*, 1995 and Daaka *et al.*, 1996). The expression of both CB_1 and CB_2 *Cnr* genes in the human placenta was demonstrated by Kenney *et al.* (1999) and found to play a role in the regulation of serotonin transporter activity.

Since the human placenta is a direct target for cannabinoids, use of marijuana during pregnancy could affect the placental clearance of serotonin. Cannabinoids negatively regulate AP-1 activity through inhibition of c-fos and c-jun proteins (Faubert and Kaminski, 2000). They inhibit interleukin-2 gene transcription (Yea et al., 2000).

The emergence of novel research tools has accelerated cannabis research in the last decade, more than at any time in the thousands of years of marijuana use in human history. Although it is not yet known why the marijuana (cannabinoid) system is so abundant in the nervous system, the analysis of the receptor proteins and genes encoding these *Cnrs* may shed some light on the mode of action of cannabinoids, and the biological role of these genes in the nervous system. In our studies, we have analyzed both CB_1 and CB_2 *Cnr* genes in normal humans who do not use marijuana. The expression of the *Cnr* proteins in different human population according to gender and ethnic background in Asians, blacks and whites were compared. The finding that the expression of *Cnrs* in humans varies according to gender and ethnic differences among whites, blacks and Asians population, while a significant observation, should be confirmed in a larger sample size. The implications and physiological relevance of this finding are only speculative and premature if unconfirmed. However, this is not surprising as numerous studies have linked genetic determinants and differences to the neurobehavioral responses of abused drugs in man and animals, (Le et al., 1994; Harada et al., 1996). For example, genetic differences in alcohol and compulsive drug taking behavior have been demonstrated in animals and man (Le et al., 1994; Harada et al., 1996). Genetic variation in some receptor and enzyme systems e.g. cholecystkinin and serotonin 1A receptors and liver enzymes, alcohol and aldehyde dehydrogenase, may be associated with alcohol dependence due to the modified function in physiological and behavioral responses, (Thomasson et al., 1991; Harada et al., 1996). Thus, the implication and relevance of the differential expression of *Cnrs* in humans according to ethnic background remains to be determined. If it turns out that these levels are relevant to the psychoactivity, toxicity and perhaps therapeutic efficacy then determination of the expression of cannabinoids in human blood may be used to predict the outcome of their actions. Endocannabinoids may also play important roles in the regulation and activation of *Cnr* and genes *in vivo*. With the identification of genes associated with human diseases and approaches to regulating gene transcription, modulation of gene activity for treating disease may be applied to the development of cannabinoid therapeutics.

Large scale gene expression changes during long-term exposure to Δ^9-THC in rats have been analyzed by Kittler et al. (1999). They used cDNA microarrays to assess changes in expression levels of very large numbers of genes. They randomly selected and arrayed at high density 24, 456 rat brain cDNA clones to investigate differential gene expression profiles following acute (24 h), short-term (7 days) and chronic (21 days) treatment with Δ^9-THC. They found a total of 64 different genes altered by Δ^9-THC, of these 43 were known, 10 had transcripts to homologous ESTs and 11 transcripts had no homology to known sequences in the Genebank database. In addition, they found that a slightly higher percentage of altered genes were down-regulated (58%) than up-regulated (42%), while some genes showed both up and down-regulation at different times during chronic Δ^9-THC treatment. The study indicated that utilizing large scale screening demonstrated that different sets of genes were altered at different times during chronic exposure to Δ^9-THC. The complete identity of these altered genes when known, may throw more light on the mechanism of action of cannabinoids. The same group investigated the effects of long-term exposure to Δ^9-THC on expression of $CB_1 Cnr$ mRNA in different rat brain regions (Zhuang *et al.*, 1998). They found that, in the striatum, the levels of CB_1 transcripts were significantly reduced from days 2–14 and returned to control levels by day 21.

Molecular characteristics of cannabinoid receptors (*Cnrs*)

Marijuana, *Cnr* gene had been elusive to clone but evidence for the existence of the receptor had been demonstrated since the 1980s (Howlett *et al.*, 1988; Devane *et al.*, 1988). It has now been shown and recognized that cannabinoids have specific receptors with endogenous ligands and inhibit adenylate cyclase. The CB_1 receptors also modulate the activities of calcium and potassium channels. Although a number of approaches are now available for the cloning of genes encoding different receptors, the most common methods previously available, which involved the purification to homogeneity of the gene protein product, did not work for the cannabinoid receptors.

Despite the wealth of information and major advances that have transformed cannabinoid research into mainstream science, little information is available at the molecular level about *Cnr* gene structure, regulation and polymorphisms. Therefore, much research remains to be conducted at the molecular level about the 5′ untranslated regions, particularly cannabinoid promoter structure and regulation and the 3′ untranslated regions, which apart from containing several polyadenylation signals may also play important regulatory roles (Shire *et al.*, 1999). In order to start to characterize the genomic structure of the *Cnrs*, we have cloned, sequenced, (Chakrabarti *et al.*, 1995) constructed the 3D model (Onaivi *et al.*, 1998) and localized the mouse CB_1 *Cnr* gene to chromosome 4 (Stubbs *et al.*, 1996). In addition, a EMBL3SP6/T7 library of C57BL/6N genomic DNA was screened by a full length (2.2 kb) CB_1 cDNA and obtained some clones. From these clones λAC21 was used and restriction enzyme digestion produced three bands from the CB_1 genomic insert. The bands were 9 kb, 6.5 kb and 1.5 kb in size. Southern hybridization of these bands with CB_1 cDNA lights up the 9 kb and 1.5 kb bands. Thus, the 9 kb and 1.5 kb bands are linked and host the cDNA. The 6.5 kb band is flanking either the *C*-terminal or in the *N*-terminal. The 9 kb band showed stronger hybridization than the 1.5 kb band. In the 2.2 kb band the long sfi I site is 0.8 kb apart from the *N*-terminal fragment of the cDNA. Compilation of these data result in the following two possibilities for the structure of the CB_1 genomic DNA insert (Figure 1.1). To characterize the insert in further detail long PCR (XLPCR) amplification was performed with *N*-terminal GSP and universal primers which amplified a band of about 7 kb. This indicates that the 6.5 kb band flanks the structural part of the gene at its *N*-terminal. Hence the map of the genomic DNA insert in λAC21 is as shown in (Figure 1.1). The currently available information on the genomic structure of CB_1 and CB_2 is sketchy and the regulation of these genes is poorly

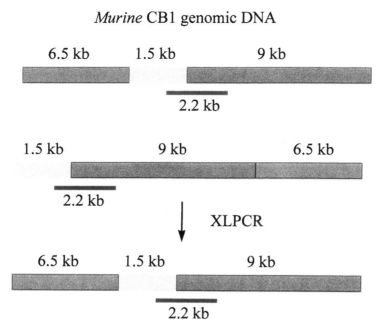

Figure 1.1 Initial characterization of the *murine* CB₁ genomic DNA using CB₁ cDNA. It was determined that the 6.5 kb band flanks the structural part of the gene at its N-terminal.

understood, the emerging putative structure for the CB_1 gene is depicted in Figure 1.2. As discussed below, the CB_1 *Cnr* gene structure is polymorphic with implication, not only for substance abuse but also in other neuropsychiatric disorders.

In order to determine whether there are introns in the coding sequences of human, rat and mouse CB_1 and CB_2 *Cnr* genes, primer pairs spanning the cDNA sequences of these genes were generated to test whether the DNA fragments amplified by these primer pairs were identical with both genomic DNA and cDNA templates. The hypothesis we tested was that if the DNA fragment sizes were identical with both templates, then the gene was intronless, if not, the intron location, size and structure can be determined. To determine whether the structure of the testis CB_1 receptor gene or its transcripts is different from those in the brain, we isolated DNA and RNA from rat testes and brain and used them as templates for PCR with primer pairs. There was no difference in sizes of the PCR amplified DNA bands between brain DNA and RNA, between testes DNA and RNA or between the brain and testis DNA or RNA. Specified amplified DNA bands were identified by Southern hybridization with human CB_1 cDNA as probe, and data conform with those expected. Cannabinoids are known to have effects on male fertility by their ability to lower testosterone levels in testis (Iversen, 1993), CB_1 gene was also found to be expressed in testis (Gerard *et al.*, 1991). Northern analysis of CB_1 mRNA levels in rat brain and testis expresses this gene about 20–25 fold

Structure of Cannabinoid receptor 1 gene

*It has several alternative splicing forms.

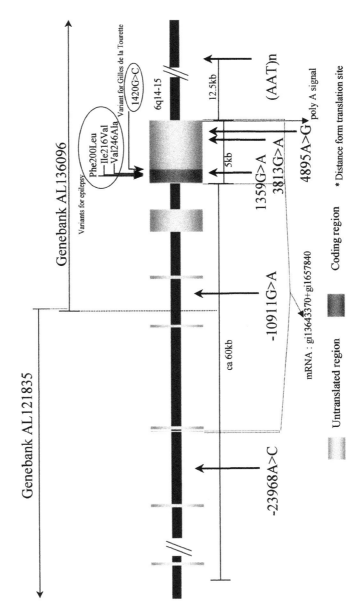

Figure 1.2 Structure of CB₁ *Cnr* gene. The emerging structure of the CB₁ *Cnr* gene, containing the single coding exon and 112 kb published sequence from the 3' region on chromosome 6q14-q15. A number of polymorphisms indicated have been identified in the CB₁ *Cnr* gene.

less than the brain tissue (Gerard *et al.*, 1991). The data obtained indicated that there was no apparent CB_1 rearrangement in testis and there are no size variants of this gene in rat brain and testis. Such variations are known for dopamine D2 receptor transcripts, which arise from alternate splicing (O'Dowd, 1993). If there is any testis-specific subtype of the *Cnr*, it is likely to be coded by a largely dissimilar gene. CB_2 and an unknown *Cnr* subtype may be a good candidate for a *Cnr* subtype in the testis. In our study, primer pairs specific for human CB_2 cDNA failed to amplify specific DNA molecules from rat testis DNA or RNA templates under conditions when human DNA or RNA yielded positive results. Similar experiments were performed with DNA and RNA isolated from brains of three mouse strains, the C57BL/6, DBA/2 and ICR. These mouse strains show similarities and marked differences in cannabinoid-induced neurobehavioral patterns (Onaivi *et al.*, 1995). We have investigated, whether the mouse CB_1 *Cnr* gene is also intronless, whether the mouse strains differ from each other in the *Cnr* gene (CB_1) structure or whether there are any size variants in the CB_1 transcripts in their brains. The DNA PCR data shows that the CB_1 *Cnr* gene in the three strains appears to be identical and intronless. We also tested the CB_1 and CB_2 *Cnr* gene structures in human blood cells. The data indicated that the human CB_1 *Cnr* gene might also be intronless, at least in its coding region. Similar observations indicated that the CB_2 *Cnr* gene in human cells might also be intronless at least in the coding region. Furthermore, the rat and human CB_1 cDNA sequences are very similar (Matsuda *et al.*, 1990; Gerard *et al.*, 1991). Unlike the CB_1 *Cnr*, which is highly conserved across the human, rat and mouse species, the CB_2 *Cnr* is much more divergent (Griffin *et al.*, 1999). This divergence in mouse, rat and human CB_2 *Cnr* leading to differences in functional assays may be related to species specificity.

Although many of the GPCRs are found to be intronless (O'Dowd, 1993), there are exceptions, such as some dopamine receptors. In common with many of the genes encoding members of the GPCRs, the genes encoding the dopamine, D1 and D5 lack introns in their coding regions (O'Dowd, 1993). But the dopamine D1 receptor gene is reported to have an intron at the 5′ non-coding region. However, the dopamine, D2, D3 and D4 receptor genes which have large introns, are distinguishable from many members of the GPCR family, where the entire protein is encoded by just one exon. It is interesting to find that both of the subtypes of the *Cnrs* may be coded by single-exon genes. There is of course the possibility of these genes having intron(s) at the upstream or downstream non-coding regions. We have not tested the possibility, but Shire *et al.* (1995) have found the presence of two introns in the CB_1 gene, one in the 5′ UTR and the second in the coding region of the receptor. The advantages or disadvantages of being intronless are subject to speculation (Lambowitz and Belfort, 1993). One obvious advantage is that the expression of these genes has a major RNA processing event to skip, thus making the conditions of their expression relatively quick and simple. This advantage may have implications related to the biological functions of these *Cnr* proteins. This issue may seem to be complex at the moment, because the structural features of the *Cnr* genes are currently incompletely understood.

The existence of a subtype of CB_1 *Cnr* gene, originally designated as CB_1A (now designated CB_1B and described by Shire *et al.* (1995)), has not been detected in any species *in vivo*. Therefore, while it is unlikely and doubtful that CB_1B exists in the form described by Shire *et al.* (1995), this does not mean that other *Cnr* subtypes may not exist. For example, the molecular cloning of two cannabinoid type 1-like receptor genes from the puffer fish has been characterized by Yamaguchi *et al.* (1996). They characterized two putative G-protein coupled receptor encoding genes, FCB_1A and FCB_1B, obtained by degenerate PCR and low-stringency hybridization of a Fugu genomic library. It was found that these two genes showed high homology to the human CB_1, but very low homology to the CB_2 *Cnr* gene. The amino acid sequences of the FCB_1A and FCB_1B genes are 66.2% identical, and the homology of each gene to human CB_1 is 72.2% and 59%, respectively. The transcripts of both the FCB_1A and FCB_1B receptors are abundant in the brain. No CB_2 *Cnr* gene could be cloned from the puffer fish. Therefore, the puffer fish has two subtypes of CB_1 *Cnrs*, which is distinct from the CB_2 subtype. The primary structure of the CB_1 and CB_2 *Cnrs* are similar to those of other G-protein coupled receptors with the characteristic features of a typical seven hydrophobic domains with some highly conserved amino acid residues. A detailed comparison of the molecular properties of the human, rat and mouse CB_1 and where applicable, CB_2 *Cnrs* had been previously reviewed (Onaivi *et al.*, 1996; Matsuda, 1997). These receptors mediate their intracellular actions by a pathway that involves activation of one or more guanine nucleotide-binding regulatory proteins, which responds to cannabinoids including the endocannabinoids. The conservatism of the CB_1 *Cnr* sequence contrasts with the variability seen with the CB_2 *Cnr* as discussed by Shire in this volume. The composition and amino acid sequence alignments of CB_1 and CB_2 *Cnrs* show considerable structural homology and distribution in the CNS between species with substantial amino acid conservation but with significant differences with the CB_2 *Cnr* whose presence in the CNS is controversial. Like other GPCRs, the primary structures of the *Cnr* are characterized by the seven hydrophobic stretches of 20–25 amino acids predicted to form transmembrane α helices, connected by alternating extracellular and intracellular loops. In comparing the composition of the *N*-terminal 28 amino acids between human CB_1 and CB_2 *Cnrs* and also between human, rat and mouse, it has been reported by Onaivi *et al.* (1996), that: (1) the human and rat *N*-terminal 28 amino acids in the CB_1 *Cnrs* were similar in the total number of non-polar, polar, acidic and basic amino acids; (2) the mouse *N*-terminal 28 amino acids differed from the rat and human CB_1 *Cnrs* in number and composition of the total non-polar and polar amino acids; (3) there are significant differences in the total non-polar, polar, acidic and basic amino acid composition of the *N*-terminal 28 amino acids between human CB_1 and CB_2 *Cnrs*; and (4) the molecular weights of human, rat and mouse CB_1 *Cnrs* are similar. Therefore, the amino acid composition of the mammalian CB_1 *Cnrs* shows strong conservatism in contrast to molecular weights and amino acid composition of CB_2 *Cnrs*.

The three dimensional (3D) model, helix bundle arrangement of human, rat and mouse CB_1 and CB_2 receptors have been constructed and compared (Bramblett *et al.*, 1995; Onaivi *et al.*, 1996) and extensively reviewed in this book by Reggio. The transmembrane helix bundle arrangement obtained for the CB_1 *Cnrs* is consistent with that obtained for other GPCRs. Potential sites for *N*-glycosylation, and the action of protein kinase C, cAMP-dependent protein kinase and Ca-calmodulin-dependent protein kinase II in the derived amino acid sequence of the *Cnr* proteins have been identified (see Figure 1.3) (Onaivi *et al.*, 1996). Most but not all GPCRs are glycoproteins and consensus sites for *N*-glycosylation are mainly concentrated at the *N*-terminus of the protein. There are three potential *N*-glycosylation sites highly conserved in human, rat and mouse. The rodent CB_1 *Cnr* protein has an additional potential *N*-glycosylation site at the *C*-terminal segment that is absent in the human CB_1 *Cnr* protein. One potential *N*-glycosylation site is present in human and rat CB_1 *Cnr* protein but that site is missing in mouse CB_1 *Cnr*. Whether all of these potential *N*-glycosylation sites are naturally glycosylated in CB_1 *Cnr* proteins or whether these *N*-glycosylation are essential for *Cnr* function and whether additional *N*-glycosylation in the CB_1 *Cnr* of different mammalian species imparts differential activity of this protein are yet to be determined. However, mutation of *N*-glycosylation sites in similar GPCRs, e.g. β-adrenergic receptors and muscarinic receptors, abolishes glycosylation, but has essentially no effect on receptor expression and function (Dohlman *et al.*, 1991). The human *Cnr* subtypes CB_1, and CB_2 with some similarities and differences in their receptor function, appear to differ in the number and distribution of their potential *N*-glycosylation sites. Due to the modification of the *N*-terminal region, CB_2 has only one potential *N*-glycosylation site whereas CB_1 *Cnr* has five. There is no potential *N*-glycosylation site at the *C*-terminal segment of CB_2 *Cnr*. The biological significance (if any) of these differences is yet to be determined.

The *C*-terminal regions and the third intracellular loop of GPCRs are known to be rich in serine and threonine residues. In the case of rhodopsin, β-adrenergic and some muscarinic receptors, some of these residues are targets of cAMP-dependent protein kinase and other protein kinases (Strada *et al.*, 1994). These phosphorylations are often agonist dependent and result in desensitization and coupling of the receptor from the G-protein. There are four clusters of potential cAMP-dependent protein kinase and Ca-calmodulin-dependent-protein kinase sites in CB_1 *Cnr* that are conserved in human, rat and mouse proteins. There is a single potential protein kinase C site that is also conserved in all these CB_1 *Cnrs* whereas CB_2 has no such site. The *N*-terminal potential cAMP clusters present in CB_1 appears to be conserved and the CB_2 *Cnr* has two such potential sites. None of the *Cnrs* have any potential protein kinase site at the *C*-terminal regions. The biological significance of these potential protein phosphorylation sites in these receptor molecules is yet to be determined. In addition, many members of GPCRs are known to contain conserved cysteine residues that appear to stabilize the tertiary structure of the receptor because of their involvement in an intra-molecular disulfide bridge. In most receptors, these cysteines occur in the extracellular domains that lie between hydrophobic domains two and three, and hydrophobic domains four and five (second and third extra-cellular domains, on the assumption that *N*-terminal domain is also extra-cellular). In CB_1 and CB_2 *Cnrs*, no cysteines are found within the second extra-cellular domain, but the third extra-cellular domain contains two or more cysteines. One other deviation from most other GPCRs is that CB_1 and CB_2 *Cnrs* lack a highly conserved proline residue in the fifth hydrophobic domain (Matsuda, 1997). The structural features of these proteins that is critical for ligand binding and functional properties have been evaluated in *in vivo* and *in vitro* models (Akinshola *et al.*, 1999). Heterologously, *Cnrs* bind tritiated synthetic ligands like WIN 55,212-2 in a saturable and competitive manner. The

A. Potential N-glycosylation and protein kinase sites in human CB1.

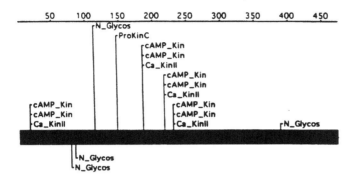

B. Potential N-glycosylation and protein kinase sites in rat CB1

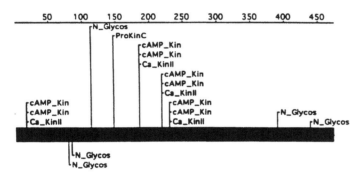

C. Potential N-glycosylation and protein kinase sites in mouse CB1

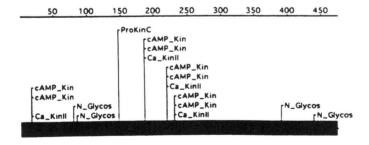

Figure 1.3 Potential modification sites of the CB$_1$ *Cnr* protein. Comparison of potential *N*-glycosylation (*N*-glycos) and protein kinase (PKC: Protein kinase C; camp-Kin: camp-dependent protein kinase; Ca-KinII: Ca-dependent protein kinase II) sites in human (A), rat (B) and mouse CB$_1$ *Cnr* proteins obtained using the MacVector sequence analysis software.

binding activities of CB_1 and CB_2 *Cnrs* have been determined after transfection into CHO cells, COS-7 cells, and mouse AtT20 cells (Felder *et al.*, 1995; Slipetz *et al.*, 1995). In other expression systems, the CB_1 *Cnrs* have been examined in insect Sfi cells, (Pettit *et al.*, 1994), *Xenopus* oocytes (Henry and Chavkin, 1995), mouse L cells, human embryonic kidney 293 cells, (Song and Bonner, 1996) and dissociated rat superior cervical ganglion neurons (Pan *et al.*, 1996). Furthermore, the performance of mutated *Cnrs* with that of wild-type receptors have been compared in a number of functional assays and reviewed by Matsuda, 1997. In those studies, mutant *Cnrs* containing point mutation (a single amino acid substitution) or ones that have been modified by replacement of a series of amino acids from one receptor with that of another (chimeric receptor) have been expressed and studied (Matsuda, 1997).

Cannabinoid receptor gene knockout mice

There are two major experimental approaches that can be used to elucidate the biological role(s) of cannabinoids. One is the traditional pharmacological manipulation, including the use of the highly selective *Cnr* antagonist to determine the involvement of cannabinoids in any biological processes or systems. The second approach is targeted gene disruption and manipulation of *Cnr* genes. Targeting a specific gene also referred to as homologous recombination enables the study of the physiological consequences of invalidating the function of a specific gene. These approaches are being used to study the physiological role of cannabinoids. An important question is whether all of the numerous central effects of cannabinoids are mediated by the CB_1 *Cnr*, which is currently the only known *Cnr* that is expressed in the brain. Thus, the synthesis of the selective and specific *Cnr* receptor antagonists and the development of genetically modified strains of mice in which the expression of the CB_1 and/or the CB_2 has been eliminated, has revealed significant data on the physiological role of the cannabinoid system. The CB_1 and CB_2 *Cnr* knockout mice have been generated. The development of CB_2 *Cnr* mutant mice was reported by Buckley *et al.* (1997). Because the CB_2 *Cnrs* are predominantly in immune cells (B cells, T cells and macrophages), a CB_2 *Cnr* knockout mouse was generated in order to study the effects of cannabinoids on immune cells and immunomodulation. The CB_2 *Cnr* gene was invalidated by using homologous recombination in embryonic stem cells. They replaced the 3' region of the CB_2 coding exon with phosphoglycerate kinase (PGK)-neomycin sequences through homologous recombination in the embryonic stem cells. This mutation eliminated part of intracellular CB_2 *Cnr* for the homologous recombination in 129 embryonic stem (ES) cells and injected into C57BL/6 blastocysts and placed in foster mothers. The mice deficient in CB_2 *Cnr* gene were generally healthy, fertile and care for their offspring and *in situ* hybridization histochemistry demonstrated the absence of the CB_2 mRNA in the knockout mice (Buckley *et al.*, 1997, 2000). Binding studies on intact spleens and splenic membranes using the highly specific [^{3}H]CP-55940 indicate significant binding to spleens derived from wild type but absent from the CB_2 *Cnr* mutant mice (Buckley *et al.*, 1997, 2000). Fluorescence activated cell sorting (FACS) analysis showed no differences in immune cell populations between cannabinoid CB_2 *Cnr* knockout and wild type mice. In addition, the role of the CB_2 *Cnr* on Δ^9-THC inhibition of macrophage co-stimulatory activity was determined. Buckley *et al.* (1997) reported that Δ^9-THC inhibits helper T cell activation through macrophages derived from wild type, but not from the CB_2 mutant mice, indicating cannabinoids inhibit macrophage co-stimulatory activity and T cell-activation via the CB_2 *Cnr*. While these studies continue, mice deficient for the CB_2 *Cnr* gene demonstrated that the CB_2 *Cnr* is involved in cannabinoid-induced immunomodulation and not involved in the CNS effects of cannabinoids. These investigators suggested that it might be possible in the future to separate central and peripheral effects of cannabinoids with selective agents acting at the CB_2 *Cnr* sites that may be devoid of the psychoactive effects known to be mediated via CB_1 activity in the CNS.

CB_1 *Cnr* gene knockout mice have been independently generated by two groups. These two groups produced mutant mice with disrupted CB_1 *Cnr* genes by standard homologous recombination techniques similar to those used in the production of the CB_2 mutant mice. The first report by Ledent *et al.* (1999) showed that the spontaneous locomotor activity of the mutant mice was increased and since they did not respond to cannabinoid drugs, they suggested that the CB_1 *Cnr* was responsible for mediating the analgesic, reinforcement, hypothermic, hypo-locomotive and hypotensive effects of cannabinoids. In their CB_1 mutant mice, Ledent *et al.* (1999), also showed that the acute effects of opiates were unaffected, but that the reinforcing properties of morphine and the severity of the withdrawal syndrome were strongly reduced. The second report was by Zimmer *et al.* (1999) and Steiner *et al.* (1999), who showed that the CB_1 mutant mice appeared healthy and fertile, but had significantly high mortality rates and reduced spontaneous locomotor activity, increased immobility, and were hypoalgesic when compared to the wild type litter mates. In the CB_1 mutant mice Δ^9-THC-induced catalepsy, hypomobility and hypothermia were absent but reported that Δ^9-THC induced analgesia in the tail-flick test and other behavioral (licking of the abdomen) and physiological (diarrhea) responses to THC was still present. Thus, there were similarities and differences in reports of the second group by Zimmer *et al.* (1999) and Steiner *et al.* (1999) with those of Ledent *et al.* (1999). It appears that the differences in responses by the mutant mice from the two groups might be related to methodological differences from different laboratory techniques. The groups, however, differed in their findings on the baseline motility of the CB_1 mutants. Ledent *et al.* (1999), found that the CB_1 mutant mice exhibited higher levels of spontaneous locomotion, even when placed in fear inducing novel environments (like in elevated plus-maze and open field). In contrast, Zimmer *et al.* (1999), found that the CB_1 mutant mice displayed reduced activity in the open-field test and an increased tendency to be cataleptic. In the basal ganglia, a brain structure with high levels of CB_1 *Cnr* and important for sensorimotor and motivational aspects of behavior, it was shown by Steiner *et al.* (1999), that these mutant mice display significantly increased levels of substance P, dynorphin, enkephalin and GAD67 gene expression that may account for the alterations in spontaneous activity observed in the CB_1 mutant mice. These data, however, remain at variance with those of the apparently similar strain of mice tested by Ledent *et al.* (1999). Overall, however, these findings provide many valuable insights into cannabinoid mechanisms despite some differences between reports on the CB_1 *Cnr* gene knockout mice. There is therefore a general agreement that the CB_1 *Cnr* plays a key role in mediating most but not all CNS effects of cannabinoids.

The biological consequences of inactivating the CB_1 and/or CB_2 *Cnr* genes has continued to be studied intensively. The availability of the cannabinoid knockout mice provides an excellent opportunity to study the biological roles of these genes. In a hippocampal model for synaptic changes that are believed to underlie memory at the cellular level, Bohme *et al.* (2000), examined the physiological properties of the Schaffer collateral – CA_1 synapses in mutant mice lacking the CB_1 *Cnr* gene and found that these mice exhibit a half-larger long-term potentiation than wild-type controls, with other properties of these synapse, such as paired-pulse facilitation, remaining unchanged. They concluded that disrupting the CB_1 *Cnr* – mediated neurotransmission at the genome level produces mutant mice with an enhanced capacity to strengthen synaptic connections in a brain region crucial for memory formation (Bohme *et al.*, 2000). Reibaud *et al.* (2000) have also used the CB_1 *Cnr* knockout mice in a two-trial object recognition test to assess the role of *Cnrs* in memory. They showed that the CB_1 knockout mice were able to retain memory for at least 48 h after first trial, whereas the wild-type controls lose their capacity to retain memory after 24 h. This data along with previous findings of other investigators suggest that the endogenous cannabinoid systems play a crucial role in the process of memory storage and retrieval. This finding is supported by previous data indicating enhanced long-term potentiation in mice lacking CB_1 *Cnr* gene (Bohme *et al.*, 2000). These rapid advances in cannabinoid research have continued to add to our knowledge about the biology of marijuana (cannabinoids) in the vertebrate and invertebrate systems. The mice lacking *Cnr* genes have also enabled scientists to investigate the interaction of cannabinoids with other neurochemical networks. The interaction between the cannabinoid and opioid systems was examined by Valverde *et al.* (2000). They demonstrated that the absence of the CB_1 *Cnr* did not modify the antinociceptive effects induced by mu, delta and kappa opioid agonists but the mice exhibited a reduction in stress-induced analgesia. These results therefore indicate that the CB_1 *Cnrs* are not involved in the antinociceptive responses to exogenous opioids, but that a physiological interaction between the opioid and cannabinoid systems is necessary to allow the development of opioid mediated responses to stress. In a different study, Mascia *et al.* (1999), showed that morphine did not modify dopamine release in the nucleus accumbens of CB_1 *Cnr* knockout mice under conditions where it dose dependently stimulates the release of dopamine in the corresponding wild-type mice, indicating that the CB_1 *Cnrs* regulate mesolimbic dopaminergic transmission in brain areas known to be involved in the reinforcing effects of morphine (Mascia *et al.*, 1999).

Other cannabinoid receptor transgenic models

Rapid advances in designing genetically engineered laboratory animals are producing not only research models, but also models that are more effective research tools. As a result, the need for precision genetic characterization and definition of laboratory animals is of primary concern. However, the use of transgenic mouse models for the over expression and other forms of modification of *Cnr* genes (except for the *Cnr* gene inactivation described above) to study the regulation and site specific mechanisms of action of cannabinoids are currently unexplored. For example, the use of cannabinoid transgenes in which genomic regulatory sequences of interest are coupled to a reporter gene, can be used to probe further the mechanism of regulation of the *Cnr* genes. But the current lack of information about the promoters, 5′ and 3′ untranslated regions and other regulatory elements of *Cnr* genes makes it difficult to make necessary *Cnr* gene construct modification for generating such rodent *Cnr* transgenic models. Obviously the use of *Cnr* transgenic animals will provide new *in vivo* systems for studying genetic regulation, development, normal function and dysfunction associated with the cannabinoid system.

Polymorphic structure of cannabinoid receptor genes

Improved information about *Cnr* and its allelic variants in humans and mice can add to our understanding of vulnerabilities to addictions and other neuropsychiatric disorders. However, little information is available at the molecular level about *Cnr* gene structure, regulation and polymorphisms. Different human *Cnr* gene polymorphisms have been reported. A silent mutation of a substitution from G to A, at nucleotide position 1359 in codon 453 (Thr) that turned out to be a common polymorphism in the German population (Gadzicki *et al.*, 1999). In this study, allelic frequencies of 1359(G/A) in genomic DNA samples from German Gilles de la Tourette sydrome (GTS) patients and controls were determined by screening the coding exon of the CB_1 *Cnr* gene using PCR single-stranded conformation polymorphism (PCR-SSCP) analysis (Gadzicki *et al.*, 1999). This was accomplished by the use of a PCR based assay by artificial creation of a *MSP1* restriction site in amplified wild-type DNA (G-allele), which is destroyed by A-allele (Gadzicki *et al.*, 1999). They found no significant differences in allelic distributions between GTS patients and controls within the coding region of the CB_1 *Cnr*. In our studies, the frequencies of this polymorphism are significantly different between Caucasian, African-American and Japanese population (Ishiguro *et al.*, unpublished observation). A HindIII restriction fragment length polymorphism (RFLP) located in an intron approximately 14 kb in 5′ region of the initiation codon of the CB_1 *Cnr* gene has been reported. Caenazzo *et al.* (1991), therefore genotyped 96 unrelated Caucasians using hybridization of human DNA digested with HindIII and identified two allele with bands at 5.5 (A1) and 3.3 kb (A2). The frequencies of these alleles were 0.23 and 0.77 respectively. Another polymorphism is a triplet repeat marker for CB_1 *Cnr* gene. This is a simple sequence repeat polymorphism (SSRP) consisting of nine alleles containing (AAT) 12–20 repeat sequences that was identified by Dawson (1995). This polymorphism has been used in linkage and association studies of the CB_1 *Cnr* gene with mental illness in different population. This CB_1 *Cnr* gene triplet repeat marker was used to test for linkage with schizophrenia using 23 multiplex schizophrenia pedigrees (Dawson, 1995) and association with heroin abuse in a Chinese population (Li *et al.*, 2000) and also for association with intravenous (IV) drug use in Caucasians (Comings *et al.*, 1997). There was no linkage and association of the marker with schizophrenia indicating that CB_1 *Cnr* gene is not a gene of major aetiological effect for schizophrenia but might be a susceptibility locus in certain individuals with schizophrenia, particularly those whose symptoms are apparently precipitated or exacerbated by cannabis use (Dawson, 1995). Comings *et al.* (1997), hypothesized that genetic variants of CB_1 *Cnr* gene might be associated with susceptibility to alcohol or drug dependence and analyzed the triplet repeat marker in the CB_1 *Cnr* gene. They found a significant association of the CB_1 *Cnr* gene with a number of different types of drug dependence (cocaine, amphetamine, cannabis), and intravenous drug use but no significant association with variables related to alcohol abuse/dependence in non-Hispanic Caucasians. In addition, this group also reported that a significant association of the triplet repeat marker in the CB_1 *Cnr* gene alleles with the P300 event related potential that has been implicated in substance abuse (Johnson *et al.*, 1997). Li *et al.* (2000) attempted to replicate the finding of Comings *et al.* (1997), in a sample of Chinese heroin addicts and did not find any evidence that the CB_1 *Cnr* gene AAT repeat polymorphism confers susceptibility to heroin abuse. CB_1 *Cnr* gene is located in human chromosome 6q14-q15 and it is interesting that previous reports showed evidence for suggestive linkage to schizophrenia with chromosome 6q markers (Martinez *et al.*, 1999) and also suggestive evidence for a schizophrenia susceptibility locus on chromosome 6q (Cao *et al.*, 1997). Although there was no linkage and association of the CB_1 *Cnr* triplet marker with schizophrenia, it remains to be determined if linkage and association to schizophrenia might exist with other unknown polymorphisms that might exist in the CB_1 *Cnr* gene structure that is currently poorly characterized. Three other rare variants have been reported in the CB_1 *Cnr* gene of an epilepsy patient (Kathmann *et al.*, 2000). This was obtained from PCR assay with cDNA from hippocampal tissue taken from patients undergoing neurosurgery for intractable epilepsy. They detected four mutations in the coding region of the CB_1 *Cnr* gene, with the first three mutations yielding amino acid substitutions as shown in Figure 1.2.

We have initiated a series of studies to analyze CB_1 *Cnr* gene structure, regulation and expression in the mouse and human models to determine genotypic and/or haplotypic associations of CB_1 *Cnr* gene with addictions and other neuropsychiatric disturbances. Genotypes at markers near the *murine* Chr 4, 13.9 cM CB_1/*Cnr* locus in 9 mouse strains reveal apparent haplotypes that extend from at least D4Mit213 to D4Mit90 (Figure 1.4) These haplotypes can be correlated with strain differences in cannabinoid effects.

The human Chr 6, 91.8~96.1 cM CB_1/*Cnr1* locus encodes at least six exons which account for 24–28 kb of sequence (Figure 1.2). Examination of CB_1 *Cnr* gene sequence variations in distinct populations has revealed a G/A single nucleotide polymorphism (SNP) in CB_1 3' and 5' flanking sequences. The initial values for linkage disequilibrium between these markers and genotypic frequencies of the markers in drug abusing and control populations were calculated. The A allele of the SNP polymorphism was present in fewer African-Americans and Asians than in Caucasians. However, in the Caucasian and African-American samples used, no association between drug abuse and the 1359(G/A) polymorphism could be found (Ishiguro *et al.*, unpublished observation). We searched for mutations in the coding exon of the gene, as well as calculating the linkage disequilibrium between the

Markers	AKR	NOD	AJ	C3H	BALBC	DBA	NON	LP	C57			
					Mouse markers around *cnr1*							
D4MT213	117	117	133	133	133	133	133	133	133			
D4MT237	126	126	116	116	116	116	116	124	124			
Pmv30	a		a	a		a			b	a=3.3kb, b absent		
Cnr1												
D4Nds3J1084	B	B	C	C					C	3 Alleles, A Largest, B Larger, C, Smaller		
D4Nds3J459	c	c				c	c		b	3 Alleles, b-second largest, c 3rd largest, No a		
D4Lgm2J3223			b	b		b			b	a=10.0 & 4.2 kb, b=3.3, 1.0kb		
D4Lgm2J15235						d				c=4.2 kb, d=3.3 & 1.0 kb		
D4MT24	132	132	134	134	134	134	134	134	134			
D4MT23	214	214	188	188	188	188	188	188	188			
D4MT257	134	−1	132	132	132	132	132	132	132			
D4MT90	160	160	160	160	160	160	160	160	160			
D4MT109	149	149	149	149	149	145	145	149	149			
D4MT182	131	131	131	131	131	131	133	135	135			
D4MT286	76	76	76	76	76	76	76	76	96			

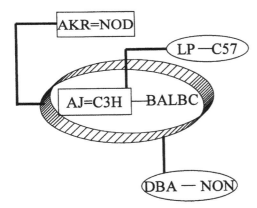

Figure 1.4 Murine haplotypic markers around CB$_1$ *Cnr*. Genotypes at markers near the *murine* chr 4 13.9 cM CB$_1$ *Cnr* locus in nine mouse strains revealed haplotypes that extend from at least D4Mit213 to D4Mit90. These haplotypes can be correlated with strain differences in cannabinoid effects.

polymorphisms to estimate the truly functional genetic variations associated with neuropsychiatric disorders. We found two novel single nucleotide polymorphism, 3813A>G and 4895G>A, in 3′ untranslated region of the gene, while no missense variation was detected. We also found one new polymorphism located at − 10911G>A in 5′ intron, upstream from the coding region. There was no linkage disequilibrium between other SNPs and this polymorpism even in the caucasian population examined. There were significantly big differences of genetic distributions between each race. In addition, the linkage disequilibrium between these physically close markers exists in Caucasians, but does not exist in African-American populations. However, in the Japanese, Caucasian and African-American

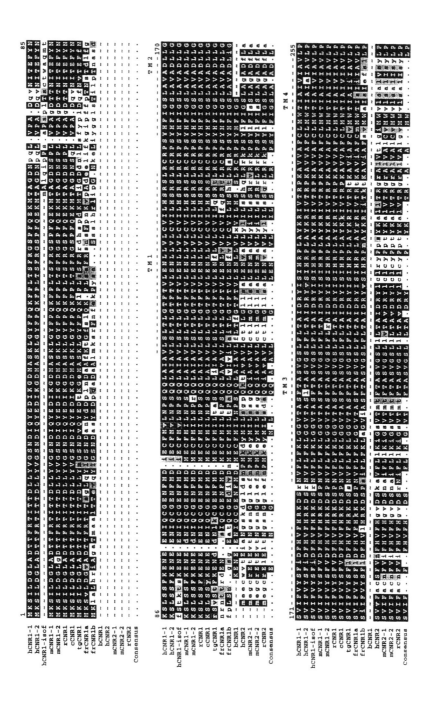

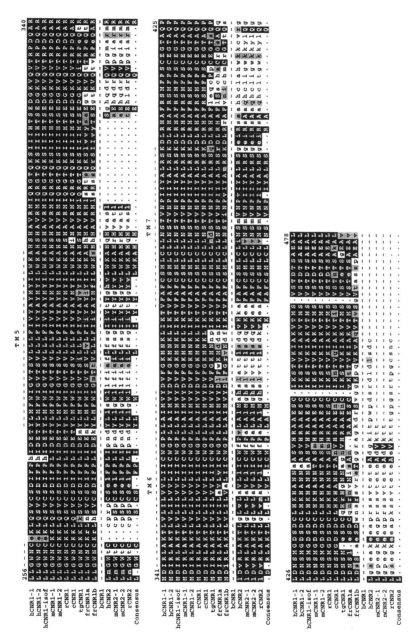

Figure 1.5 Amino acid sequence alignments of CB₁ and CB₂ Cnrs. hCNR1, mCNR1 and mCNR2 has different entries in the Genebank. The plurality used is 4 and the consensus sequence is shown at the bottom of the alignments. Identical amino acid residues are in black, closely related ones in dark gray, less-closely related ones in light gray and unrelated in white. The seven putative transmembrane domains ᵀᴹ are indicated at the top of the alignments. The Genebank accession numbers are U73304 (hCNR1, human), AF107262 (hCNR1–2, human), X81121, U22948 (mCNR1–1, mouse), U17985 (mCNR1–2, mouse), X55812 (rCNR1, rat), U94342 (cCNR1, cat), AF181894 (tgCNR1, Taricha granulosa), X94401 (frCNR1a, type a from F. rubripes), X94402 (frCNR1b, type b), U77348 (bCNR1, partial bovine sequence), X74328 (hCNR2), X93168 (mCNR2–1), X86405 (mCNR2–2), AF176350 (rCNR2).

populations, no association between substance abuse and the SNPs could be found (Ishiguro *et al.*, unpublished observation). Furthermore, the 1359A>G polymorphism has been genotyped in healthy control male Japanese subjects (aged 20–30) from whom personality traits were measured with temperament and character inventory (TCI) test, although the statistical power is weak because of lower frequency of allelic distribution. No association between TCI scores and the polymorphism could be found (Ishiguro *et al.*, unpublished observation).

While studies continue, these findings add to the characterization of CB_1 *Cnr* genes in species in which they can be tested for impact on substance abuse and other neuropsychiatric disorders. The amino acid (AA) sequence alignments (Figure 1.5) and construction of a phylogenetic tree of known *Cnrs* (Figure 1.6) indicate some similarities and significant divergence between CB_1 and CB_2 *Cnrs*.

Chromosomal mapping of the *Cnr* genes

Using genetic linkage mapping and chromosomal *in situ* hybridization, Hoehe *et al.* (1991) have determined the genomic location of the human *Cnr* gene. With *in situ* hybridization using a biotinylated cosmid probe the *Cnr* gene was localized at 6q14-q15, thus confirming the linkage analysis and defining a precise alignment of the genetic and cytogenetic maps (Hoehe *et al.*, 1991). These investigators found that the location of the human CB_1 *Cnr* gene is very near the gene encoding the alpha subunit of chorionic gonadotropin (CGA). After we cloned and sequenced the *murine* CB_1 *Cnr* gene (Chakrabarti *et al.*, 1995), we collaborated with Stubbs *et al.* (1996) and determined that the *murine* CB_1 and CB_2 *Cnr* gene is located in proximal chromosome 4. This location is within a region to which other homologs of human 6q genes are located. In order to localize the *murine* CB_1 and CB_2 *Cnr* gene in the mouse genome, Stubbs *et al.* (1996), traced the inheritance of species-specific variants of the gene in 160 progeny of an interspecific backcross. Therefore, using the interspecific and four DNA probes we mapped the *murine* CB_1 and CB_2 *Cnr* genes to chromosome 4 with map positions calculated for Mos, Cntfr, Pax5 and Cd72, in excellent agreement with previously published results that clearly established linkage between the CB_1 *Cnr* gene and other genes known to be located on mouse chromsome 4. The CB_1 *Cnr* gene, GABRR1, GABRR2 and Cga are linked together both in the mouse and on human chromosome 6q. The genes encoding the peripheral CB_2 *Cnr* and α-L-fucosidase have been shown to be located near a newly identified common virus integration site, Ev11 (Valk *et al.*, 1997). They showed that Ev11 is located at the distal end of mouse chromosome 4, in a region that is synthenic with human 1p36, in agreement with our report (Onaivi *et al.*, 1998), that the mouse CB_2 *Cnr* gene is also located at the distal end of mouse chromosome 4. The results of the chromosomal location of the human (Hoehe *et al.*, 1991) and the mouse (Stubbs *et al.*, 1996) *Cnr* genes add a new marker to this region of the mouse-human homology, and confirms the close linkage of *Cnr* genes in both species (Figure 1.7). The location of the rat CB_1 *Cnr* gene in the rat genome has not been determined, but may be expected to fit the rodent-human homology as the CB_1 *Cnr* genes are highly conserved in the mammalian species. The physical and genetic localization of the bovine CB_1 *Cnr* gene has been mapped to chromosome

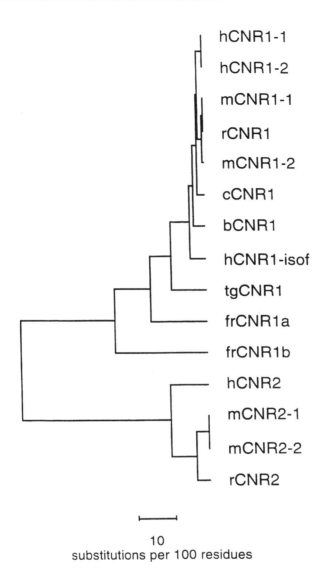

Figure 1.6 Phylogenetic tree of *Cnrs*. The graphical representation of *Cnrs* phylogenetic tree was generated by using GCG programs and GrowTree (UPGMA). Information for each protein is listed in figure legend 5.

9q22 using fluorescence *in situ* hybridization (FISH) and R-banding to identify the chromosome (Pfister-Genskow *et al.*, 1997). The genetic mapping of the CB_1 *Cnr* gene on bovine chromosome 9q22 by *in situ* localization and linkage mapping of a dinucleotide repeat, D9S32, also adds to coverage of the bovine genome map and contributes to the mammalian comparative gene map (Pfister-Genskow *et al.*, 1997). As the neurobiological effects of marijuana and other cannabinoids suggest

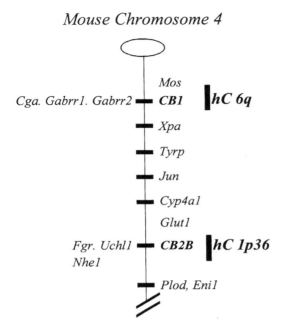

Figure 1.7 Chromosomal localization of CB_1 and CB_2 *Cnr* genes. The chromosomal location of CB_1 and CB_2 *Cnr* genes on mouse chromosome 4 and human chromosome 6 and 1P36 respectively.

the involvement of the *Cnr* genes in mental and neurological disturbances, the mapping of the genes will undoubtedly enhance our understanding of the linkage and possible cannabinoid genetic abnormalities. In the case of cattle, research into the role of *Cnrs* in mediating responses to natural and production-induced stressors could lead to improvement of production inefficiencies that exist in meat and milk animal systems (Pfister-Genskow *et al.*, 1997). The chromosomal location and genomic structure of human and mouse fatty acid amide hydrolase (FAAH) genes have been mapped to chromosomes 1p34-p35 and 4 respectively (Wan *et al.*, 1998). The localization of FAAH and CB_2 *Cnr* genes, in same chromosomal regions in mice and humans, again adds a new marker to this region of the mouse-human homology, and confirms the close linkage of FAAH and *Cnr* genes in both species.

Genes encoding endocannabinoid transporter(s)

The mechanism(s) involved in the inactivation of endocannabinoids *in vivo* is not completely understood. However, functional studies indicate that the biological actions of endocannabinoids are probably terminated by a two-step inactivation process consisting of carrier-mediated uptake and intracellular hydrolysis by FAAH (Di Marzo *et al.*, 1999; Piomelli *et al.*, 1999; Hillard, 2000). Although there is evidence from these functional studies for the existence of some form of cannabinoid transporter(s), their identity, sequence information and biological characteristics at the molecular level are unknown. On the other hand, FAAH has been purified, cloned, sequenced from mouse, rat and human and thus, fairly well characterized. It is a single-copy gene with 579 amino acids and highly conserved primary structure and homologous in mouse, rat and human species (Giang and Cravatt, 1997). This membrane-associated enzyme is 63 kDa and possesses the ability to hydrolyze a range of fatty acid amides including anandamide, 2-arachidonylglycerol and oleamide. The distribution of FAAH and CB_1 Cnr in rat brain is similar, with FAAH often occurring in neuronal somata that are post synaptic CB_1 Cnr expressing axons and therefore consistent with a potential role in the regulation of endocannabinoids (Egertova $et\ al.$, 2000). The biosynthesis and inactivation of endocannabinoids and other cannabimimetic fatty acid derivatives have been extensively reviewed in this volume by Di Marzo $et\ al.$ An acid amidase hydrolyzing anandamide and other N-acylethanolamines distinct from FAAH that can hydrolyze N-acylethanolamines was reported by Ueda $et\ al.$ (2000). Further research will continue to unravel the biochemical pathways associated with endocannabinoids.

There is evidence that the transport of endocannabinoids, like anandamide and 2-AG from one side of a biological membrane to another is accomplished via a protein carrier (Di Marzo $et\ al.$, 1999; Piomelli $et\ al.$, 1999; Hillard, 2000). Evidence has been shown for this carrier-mediated, transmembrane transport of anandamide in human neuroblastoma and lympoma cells (Maccarone $et\ al.$, 1998), in mouse macrophages and RBL-2H3 cells (Bisogno $et\ al.$, 1998) and in neurons (Di Marzo $et\ al.$, 1994; Hilliard $et\ al.$, 1997). This transport process fulfills several criteria of a carrier-mediated process including saturability, temperature dependence, high affinity, substrate selectivity, facilitated diffusion and Na^+-independence (Piomelli $et\ al.$, 1999; Di Marzo $et\ al.$, 1999; Hillard, 2000). Some of these features make this process fundamentally different from the other known transport carriers like the catecholamine and amino acid transporters. Using a relatively potent uptake inhibitor AM404, N-(4 hydroxyphenyl)arachidonylamide, these investigators have demonstrated that a high affinity transport system present in neurons and astrocytes has a role in anandamide uptake and subsequent inactivation by FAAH (Di Marzo $et\ al.$, 1999; Piomelli $et\ al.$, 1999; Hillard, 2000). While there is ample scientific evidence to support the concept that anandamide transport across membranes is protein-mediated, definitive evidence depends on its molecular characterization. However, the differential uptake of the different endocannabinoids, for example anandamide and 2-AG' in different cell types may indicate the possibility of different cannabinoid transporters for the different endocannabinoids (Di Marzo $et\ al.$, 1999; Hillard, 2000). This would not be unprecedented as the monoamines have different transporters for dopamine, serotonin and norepinephrine.

The mammalian vanilloid subtype 1 capsaicin receptor (VR1) has been cloned and shown to be activated by plant derived agonists such as capsaicin (the pungent ingredient in hot chilli pepper) and resiniferatoxin (Caterina $et\ al.$, 1997). In mammals, an endogenous ligand for VR1 has not been described, but the endocannabinoid anandamide has been shown to activate the receptor, resulting in physiological responses that are capsazepine-sensitive and Cnr-insensitive (Zygmunt $et\ al.$, 1999; Szolcsanyi, 2000). It has also been demonstrated that anandamide acts as full agonist at the human VR1 vanilloid receptor (Smart *et al.*, 2000). The interaction between synthetic vanilloids and the endogenous cannabinoid system has been studied because of the structural similarity between some vanilloid agonists eg. olvanil, and endocannabinoid, e.g. anandamide by Di Marzo *et al.*, 1998. They reported that olvanil is a potent inhibitor of the anandamide transporter. While impaired nociception and pain sensation and elimination of capsaicin sensitivity in mice lacking the capsaicin (vanilloid) VR1 receptor has been demonstrated (Caterina *et al.*, 2000), it can be hypothesized that these mice might be sensitive to the antinociceptive effects of *Cnr* activation.

Neurobiology of cannabinoid modulation of other receptor systems

There are numerous reports of *Cnr* agonist interaction with CNS neurotransmitter systems such as GABA, DA, 5-HT, NE, Ach, opiates and the glutamate receptors to mention a few. We do not attempt to provide a comprehensive account of the numerous effects of cannabinoids on other receptor systems. This section, as an example, will only deal with the glutamatergic receptor system that consists of α-amino-3-hydroxy-5-methyl-4-isoxazole-propionic acid (AMPA), *N*-methyl-D-aspartate (NMDA) and kainate (KA) subtypes and how they are influenced by cannabinoids in heterologous expression systems. Glutamate is the major excitatory neurotransmitter in the central nervous system, where the majority of fast synaptic transmissions are mediated by AMPA receptors (Jonas and Sakmann, 1992). However, there have been very few studies of cannabinoid influence on glutamatergic neurotransmission. Reports have shown that *Cnr* agonists modulate NMDA neurotransmission in cultures of brain regions such as the basal ganglia (Glass *et al.*, 1997), hippocampus (Terranova *et al.*, 1995), cerebellum (Hampson *et al.*, 1998) and the forebrain (Nadler *et al.*, 1993). In all the studies, NMDA receptor functions were inhibited. It has been suggested that cannabinoids inhibit glutamatergic neurotransmission presynaptically (Shen *et al.*, 1996) in rat hippocampal neurons. The mechanism of cannabinoid action has been attributed to blockade of N, P and Q type calcium channels (Mackie *et al.*, 1992, 1995). Thus, cannabinoid interactions may alter calcium currents in NMDA neurotransmission if *Cnrs* are directly activated. Evidence in support of a direct inhibitory but neuroprotective effect of cannabinoids on the NMDA receptor via Ca^{2+} dependent mechanisms (Nadler *et al.*, 1993; Striem *et al.*, 1997), have been reported. A novel enhancement of Ca^{2+} signals in cultured cerebellar granule neurons has recently been shown by (Netzeband *et al.*, 1999). Similarly, Ca^{2+} independent mechanisms have been reported for cannabinoid modulation of the NMDA receptor in rat brain slices (Hampson *et al.*, 1998) and in the rat periaqueductal gray neurons (Vaughan *et al.*, 1999). Most of the experiments on the interactions between the cannabinoid and glutamatergic systems were done in neuronal cultures, brain slices and cell lines. It became imperative that other *in vitro* systems that are free of both the *Cnr* and the glutamate receptor be tested, so that definitive conclusions can be reached on the nature of cannabinoid modulation of the glutamatergic system.

Heterologous expression systems such as the *Xenopus laevis* oocyte and cell transfection systems for recombinant receptor expression are good models for conducting these studies. Studies on recombinant *Cnr* (McAllister *et al.*, 1999) and cannabinoid agonist interaction with recombinant NMDA in oocytes (Hampson *et al.*, 1998) are very few. However, results from the latter study show a dual novel effect for anandamide modulation of the NMDA receptor. The discovery of the putative endogenous cannabinoid neurotransmitter, anandamide and the wide range of its CNS effects such as the inhibition of long-term potentiation in brain slices (Terranova *et al.*, 1995), and its *in vivo* and *in vitro* properties, increased the pace of studies on cannabinoid and non-*Cnr* actions. However, there is a big gap in the literature on effects of cannabinoid agonists on non-NMDA receptors. Two studies recently reported the direct inhibition of recombinant AMPA receptor currents in oocytes (Akinshola *et al.*, 1999a, 1999b). The modulation of AMPA receptor currents by anandamide in oocytes was reported to be independent of the *Cnr* system, but dependent on the cAMP signal transduction mechanism and similar for all receptor subunits tested (Akinshola *et al.*, 1999b). Figure 1.8, summarizes the effects of anandamide on recombinant AMPA receptors in the oocytes.

Neurobehavioral and *in vitro* actions of cannabinoids

Despite the decades of extensive investigations and recent developments in cannabinoid research, the identification of specific mechanisms for the actions of cannabinoids have been slow to emerge. We therefore do not attempt to provide a comprehensive account of the numerous *in vivo* and *in vitro* effects of cannabinoids but a few examples from our studies and those of others. The discovery of endocannabinoids such as anandamide and 2-arachidonyl glycerol and the widespread localization of *Cnrs* in the brain and peripheral tissues, suggests that the cannabinoid neurochemical system represents a previously unrecognized ubiquitous network in the nervous system, whose biology and function is unfolding. We have tested the hypothesis that some of the actions of anandamide are independent of a *Cnr* mechanism (Akinshola *et al.*, 1999b). In the first series of experiments, the effects of anandamide or methanandamide on behavior and CB_1 *Cnr* gene expression in three mouse strains was determined. This was accomplished by the use of cannabinoid agonist and antagonist interaction in *in vitro* and *in vivo* test systems. The effects of acute administration of anandamide to C57BL/6, DBA/2 and ICR mice were evaluated in motor function and emotionality tests. The C57BL/6 and ICR mouse strains were more sensitive than the DBA/2 strain to the depression of locomotor and stereotyped behavior caused by anandamide. Although anandamide produced catalepsy in all three strains, anandamide induced ataxia in the minus-maze test only in the C57BL/6 animals and at the lowest dose used. In the plus-maze test, anandamide produced a mild averse response, which became intense aversion to the open arms of the plus-maze following repeated daily treatment. Northern analysis data using the CB_1 cDNA as a probe indicated that there was greater expression of the CB_1 gene in the whole brain of the ICR mouse than in the brains of the C57BL/6 and DBA/2 strains with or without pretreatment with anandamide. Since the anandamide induced neurobehavioral changes that did not correspond to the CB_1 *Cnr* gene expression in the mouse strains, it is unlikely that the CB_1 *Cnr* mediates all the cannabimimetic effects of anandamide in the brain. *In vitro*, we

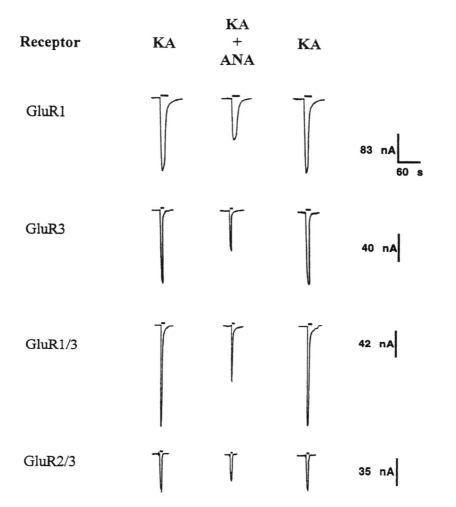

Figure 1.8 Current traces of anandamide action on AMPA receptor subunits. Current traces of anandamide inhibition of kainate-activated currents in *Xenopus* oocytes expressing homomeric and heteromeric AMPA receptor subunit combinations. Each set of traces shows the effect of 100 μM anandamide on currents activated by 200 μM kainic acid in oocytes expressing the GluR subunits indicated. The bars above current records represent the time duration of kainic acid and/or anandamide application.

used *Xenopus laevis* oocytes and two-voltage clamp technique (as described above) in combination with differential display polymerase chain reaction to determine whether the differential display genes following treatment with anandamide may be linked to the AMPA glutamate receptor. The differential expression of genes *in vivo* after the sub-acute administration of anandamide could not be directly linked with AMPA glutamate receptor. For the antagonist studies *in vivo*, SR141716A, the CB_1 antagonist, induced anxiolysis that was dependent on the mouse strain used in the anxiety model and blocked the anxiogenic effects of anandamide or methanandamide whereas, SR141716A had no effect on the anandamide inhibition of kainite activated currents *in vitro*.

We tested another hypothesis that there might exist in the central nervous system a multiplicity of *Cnrs*. The basis for the hypothesis has been the myriad neurobehavioral effects produced after smoking marijuana or the administration of cannabinoids to humans and animals. We therefore studied the neurobehavioral specificity of CB_1 *Cnr* gene expression and whether Δ^9-THC induced neurobehavioral changes are attributable to genetic differences (Onaivi *et al.*, 1996). We also examined whether some of these neurobehavioral changes were mediated by specific brain regions in the mouse model. We found that the differential sensitivity following the administration Δ^9-THC to three mouse strains, C57BL/6, DBA/2 and ICR mice indicated that some of the neurobehavioral changes might be attributable to genetic differences. The objective of the study was to determine the extent to which the CB_1 *Cnr* is involved in the behavioral changes following Δ^9-THC administration. This objective was addressed by experiments using the following strategies: DNA-PCR and reverse PCR; systemic administration of Δ^9-THC and intracerebral microinjection of Δ^9-THC. The site specificity of action of Δ^9-THC in the brain was determined using stereotaxic surgical approaches. The intracerebral microinjection of Δ^9-THC into the nucleus accumbens (ACB) was found to induce catalepsy, while injection of Δ^9-THC into the central nucleus of amygdala resulted in the production of an anxiogenic – like response. Although the DNA-PCR data indicated that the CB_1 *Cnr* gene appeared to be identical and intronless in all the three mouse strains, the reverse PCR data showed two additional distinct CB_1 mRNAs in the C57BL/6 mouse which also differed in pain sensitivity and rectal temperature changes following the administration of Δ^9-THC (Onaivi *et al.*, 1996). We therefore suggested that the diverse neurobehavioral alterations induced by Δ^9-THC may not be mediated by CB_1 *Cnrs* in the brain and that the CB_1 *Cnr* genes may not be uniform in the mouse strains used. The potential promise of antisense oligonucleotides as research tools and therapeutic agents has been the subject of close scrutiny and attention, particularly the application of gene therapy in the clinic. Thus, a number of problems have been identified with their use as research tools. Our use of CB_2 antisense oligonucleotide indicated that CB_2 may be present in the brain to influence behavior. The ICV administration of the CB_2 antisense induced a significant anti-aversive response in the elevated plus-maze test of anxiety. A response similar to that following the administration of the CB_1 *Cnr* antagonist SR141716A. Knowing that this might be a non-specific effect, it is interesting and thought provoking that CB_2 *Cnr* or CB_2-like *Cnrs* might be in the brain. While a number of laboratories have not been able to detect CB_2 expression in the brain, a demonstration of CB_2 expression in the rat microglial cells (Kearn and Hilliard, 1997) in cerebral granule cells (Skaper *et al.*, 1996) and mast cells (Facci *et al.*, 1995) have been reported. We utilized a CB_2 antisense, 5′-TGTCTCCCGGCATC-CCTC-3′, CB_2 sense was 5′-GAGGGATGCCGGGAGACA-3′. The use of the antagonists that are selective for the CB_1 and CB_2 *Cnrs* in these behavioral tests will also contribute to a further understanding of the role of these *Cnr* subtypes in the behavioral effects of cannabinoids.

Other behavioral effects of cannabinoid agents in animal models has been reviewed by Chaperon and Thiebot (1999) and extensively reviewed in this book. Briefly, cannabinomimetics produce complex behavioral and pharmacological effects that probably involve numerous neuronal substrates. Interactions with acetylcholine, dopamine, serotonin, adrenergic, opiate, glutamatergic and GABAergic systems have been demonstrated in several brain structures. In animals, cannabinoid agonists such as, WIN 55,212-2 and CP 55,940 produce a characteristic combination of four prototypic profiles, sometimes referred to as response to the tetrad tests, including, catalepsy, analgesia, hypoactivity and hypothermia. These effects are reversed by the selective CB_1 *Cnr* antagonist, SR141716A, providing evidence for the involvement of CB_1 *Cnr*-related mechanisms. Accumulating evidence indicates that endocannabinoids have cannabinoid and non-*Cnr* mediated effects in these classical cannabinomimetic actions. *Cnr*-related processes seem also involved in cognition, memory, anxiety, control of appetite, emesis, inflammatory and immune response covered in various sections in this volume. Cannabinoid agonist may induce biphasic effects, for example, hyperactivity at low doses and severe motor deficits at larger doses have been documented.

The conditioned place preference (CPP) paradigm has been used extensively to study brain mechanisms of reward and reinforcement. Marijuana (cannabinoid) interactions with the brain substrates for reward and reinforcement were excellently reviewed by Gardner in this book and also by Tzschentke (1998). Although the paradigm has been criticized because of some inherent methodological problems, it is clear that the place preference conditioning has become a valuable and firmly established and a widely used tool in addiction research, Tzschentke (1998). The rewarding properties of cannabinoids and Δ^9-THC are difficult to demonstrate in rodents using standard place preference procedures (Tzschentke, 1998; Valjent and Maldonado, 2000). Furthermore, only few studies have examined the effects of marijuana and hashish, and inconsistent results have been reported. Sanudo-Pena *et al.* (1997) found no CPP at a low dose of THC (1.5 mg/kg) and a conditioned place aversion (CPA) at a high dose (15 mg/kg) while the *Cnr* antagonist SR141716A induced a CPP at a low and high dose (0.5 and 5 mg/kg). In contrast, Mallet and Beninger (1998) found a significant CPA for the same low doses of THC (1.0 and 1.5 mg/kg), and neither CPP or CPA was found for anandamide. Lepore *et al.* (1995) however, reported THC-induced CPP for 2 and 4 mg/kg, but not for 1 mg/kg, when animals received one conditioning session per day. But, when conditioning took place only every other day (to allow for a 24 h washout period for THC), the dose of 1 mg/kg THC was sufficient to produce CPP, whereas the higher doses (2 and 4 mg/kg) produced CPA. The synthetic cannabinoid CP 55940 was reported to produce CPA, (McGregor *et al.*, 1996). The synthetic *Cnr* agonist WIN 55212-2 produced a robust CPA while the CB_1 antagonist SR141716A produced neither CPP nor CPA (Chaperon *et al.*, 1998). The emerging consensus appear to be that cannabinoid antagonism produces CPP while cannabinoid agonism induces place aversion, (Sanudo-Pena *et al.*, 1997). Taken together the effects of cannabinoids in the CPP paradigm suggest that the effects of cannabinoids and perhaps marijuana may be complex and conclusions about their rewarding and adverse actions deserve further intensive study.

Implication for the medical use of marijuana

The use of cannabis for both recreational and medicinal purposes dates back to thousands of years. In recent times, there has been an increase in calls for marijuana to be legalized for medicinal use in AIDS, cancer, multiple sclerosis and other medical conditions where patients might benefit from the biological effects of cannabis. Synthetic cannabinoids such as dronabinol, marinol and nabilone already have an established use as antiemetics in nausea and vomiting associated with cancer chemotherapy. The reported beneficial effects in cancer and AIDS patients might reflect improved weight gain, owing to the well-documented anti-emetic and appetite stimulating effects of cannabinoids. This might be a major advantage for cancer patients undergoing rigorous chemotherapy, or advanced AIDS patients. Interestingly, although cannabis is widely used as a recreational drug in humans, only a few studies revealed an appetitive potential of cannabinoids in animals. However, evidence for the adverse effects of Δ^9-THC, WIN 55,212-2 and CP 55,940 is more readily obtained in a variety of tests. The selective blockade of CB1 *Cnrs* by SR141716A impaired the perception of the appetitive value of positive reinforcers (cocaine, morphine and food) and reduced the motivation for sucrose, beer and alcohol consumption, indicating that positive incentive and/or motivational processes could be under a permissive control of *Cnr*-related mechanisms. In addition, endocannabinoids and other endogenous fatty acid ethanolamides with their roles in sleep and inflammation are emerging as new important biological signaling molecules (Boger *et al.*, 1998). Other targets for potential development of therapeutic agents are cannabinoid uptake inhibitors. When the physiological role of *Cnrs* and endocannabinoids is established, it is likely that other therapeutic targets may be uncovered. Based on what we currently know about the anatomical distribution of *Cnrs* and endocannabinoids and from marijuana users, disorders associated with memory and motor coordination may benefit from novel cannabinoids. Similarly, *Cnrs* and endocannabinoids in the periphery can be targeted in immune disorders and blood pressure regulation (Mechoulam, 1999 and Varga *et al.*, 1998). This is to be expected as many novel drugs are based on chemical modifications of transmitters. Therefore, it seems reasonable to expect that, as with other receptor–transmitter systems, excess or lack of *Cnr* or endogenous ligands may be the cause for disorders in the CNS or those associated with the immune (Mechoulam, 1999) and other peripheral organ systems.

CONCLUDING REMARKS AND FUTURE DIRECTIONS

Faster progress has been achieved in marijuana research within the last decade than in the thousands of years that marijuana has been used in human history. For many decades, research on the molecular and neurobiological bases of the physiological and neurobehavioral effects of marijuana was hampered by the lack of specific research tools and technology. The situation started to change with the availability of molecular probes and other recombinant molecules that have led to the major advances. Thus, we can look forward to a bright future of discoveries concerning the role of endocannabinoids in brain function, immune function, reproductive function and emotional behavior, as well as the discovery of potential medicines to improve the health of many who suffer from various disorders. One notable future direction is in the brain injury and stroke field, for which there are currently no major drugs and one cannabinoid based drug, HU-211 (dexana-binol) has successfully completed phase II trials. Marijuana (cannabinoid) research appears to have a solid scientific background for significant contribution in understanding human biology and in the development of cannabinoid based therapeutics.

ACKNOWLEDGMENTS

We thank Drs. Scott Hall, Fei Xu and Bruce Hope at the NIDA intramural program for critical dialog about some aspects of this chapter. We also thank Dr. Gautam Chaudhuri from Meharry Medical College, Nashville, TN., Dr. Lisa Stubbs at the Livermore National Laboratory, Livermore, CA., Drs. Amitabha Chakrabati and Fridolin Sulser from Vanderbilt University in Nashville, TN, for collaborative research. We thank many other colleagues in NIDA including Dr. Heishman, Axelrad who helped with the review of some of the chapters.

REFERENCES

Abood, M. E., Ditto, K. E., Noel, M. A., Showalter, V. M. and Tao, Q. (1997) "Isolation and expression of a CB$_1$ cannabinoid receptor gene. Comparison of binding properties with those of native CB$_1$ receptors in mouse brain and N18TG2 neuroblastoma cells," *Biochemical Pharmacology* **24**: 207–214.

Aceto, M. D., Martin, B. R., Scates, S. M. and Loew, J. (1995) "Cannabinoid precipitated withdrawal: Induction by the antagonist SR141716A," *FASEB Journal* **9**(3): 2320.

Acousta-Urquidi, J. and Chase, R. (1975) "The effects of delta-9-tetrahydrocannabinol on action potentials in the mollusk Aplysia," *Canadian Journal of Physiology and Pharmacology* **53**: 793–798.

Akinshola, B. E., Chakrabarti, A. and Onaivi, E. S. (1999a) "In vitro and in vivo action of cannabinoids," *Neurochemical Research* **24**: 1233–1240.

Akinshola, B. E., Taylor, R. E., Ogunseitan, A. B. and Onaivi, E. S. (1999b) "Anandamide inhibition of recombinant AMPA receptor subunits in *Xenopus* oocytes is increased by forskolin and 8-bromo-cyclic AMP," *Naunyn-Schmiedeberg's Archives of Pharmacology* **360**: 242–248.

Belue, R. C., Howlett, A. C., Westlake, T. M. and Hutchings, D. E. (1995) "The ontogeny of cannabinoid receptors in the brain of postnatal and aging rats," *Neurotoxicology and Teratology* **17**: 25–30.

Bisogno, T., Katayama, K., Melck, D., Ueda, N., De Petrocellis, l., Yamamoto, S. and Di Marzo, V. (1998) "Biosynthesis and degradation of bioactive fatty acid amides in human breast cancer and rat pheochromocytoma cells – implication for cell proliferation and differentiation," *European Journal of Biochemistry* **254**: 634–642.

Bisogno, T., Ventriglia, M., Milone, A., Mosca, M., Cimino, G. and Di Mazo, V. (1997) "Occurrence and metabolism of anandamide and related acyl-ethanolamides in ovaries of the sea urchin Paracentrotus lividus," *Biochimica et Biophysica Acta* **1345**: 338–348.

Boger, D. L., Henriksen S. J. and Cravatt B. F. (1998) Oleamide: "An endogenous sleep inducing lipid and prototypical member of a new class of biological signaling molecules," *Current Pharmaceutical Design* **4**: 303–314.

Bohme, G. A., Laville, M., Ledent, C., Parmentier, M. and Imperato, A. (2000) "Enhanced long-term potentiation in mice lacking cannabinoid CB$_1$ receptors," *Neuroscience* **95**: 5–7.

Brady, R. O. and Carbone, E. (1973) "Comparison of the effects of 9-tetrahydrocannabinol, 11-hydroxy-9-tetrahydrocannabinol, and ethanol on the electrophysiological activity of the giant axon of the squid," *Neuropharmacology* **12**: 601–605.

Bouaboula, M., Rinaldi, M., Carayon, P., Carrilon, C., Delpech, B., Shire, D. *et al.* (1993) "Cannabinoid receptor expression in human leucocytes," *European Journal of Biochemistry* **214**: 173–180.

Bramblett, R. D., Panu, A. M., Ballesteros, J. A. and Reggio, P. H. (1995) "Construction of a 3D model of the cannabinoids CB$_1$ receptor: Determination of the helix ends and helix orientation," *Life Sciences* **56**: 1971–1982.

Breivogel, C. S., Di Marzo, V., Zimmer, A. M., Zimmer, A. and Martin, B. R. (2000) "Characterization of an unknown cannabinoid receptor in CB$_1$ knockout mouse brain membranes," *2000 Symposium on the cannabinoids, Burlington, Vermont, ICRS*. p. 6.

Buckley, N. E., McCoy, K. L., Mezey, E., Bonner, T., Zimmer, A., Felder, C. C. *et al.* (2000) "Immunomodulation by cannabinoids is absent in mice deficient for the cannabinoid CB$_2$ receptor," *European Journal of Pharmacology* **396**, 141–149.

Buckley, N. E., Mezey, E., Bonner, T., Zimmer, A., Felder, C. C., Glass, M. *et al.* (1997) "Development of a CB$_2$ knockout mouse," (1997) *Symposium on the Cannabinoids, Burlington, Vermont, International Cannabinoid Research Society* **57**.

Caenazzo, L., Hoehe, M. R., Hsieh, W.- T., Berrettini, W. H., Bonner, T. I. and Gershon, E. S. (1991) "HindIII identifies a two allele DNA polymorphism of the human cannabinoid receptor gene (CNR)," *Nucleic Acids Research* **19**: 4798.

Cao, Q., Martinez, M., Zhang, J., Sanders, A. R., Badner, J. A., Cravchik, A. *et al.* (1997) "Suggestive evidence for a schizophrenia susceptibility locus on chromosome 6q and a confirmation in an independent series of pedigrees," *Genomics* **43**: 1–8.

Caterina, M. J., Leffler, A., Malmberg, A. B., Martin, W. J., Trafton, J., Petersen-Zeitz, K. R. *et al.* (2000) "Impaired nociception and pain sensation in mice lacking the capsaicin receptor," *Science* **288**: 306–313.

Caterina, M. J., Schumacher, M. A., Tominaga, M., Rosen, T. A., Levine, J. D. and Julius, D. (1997) "The capsaicin receptor : a heat-activated ion channel in the pain pathway," *Nature* **389**: 816–824.

Chakrabarti, A., Ekuta, J. E. and Onaivi, E. S. (1998) "Neurobehavioral effects of anandamide and cannabinoid receptor gene expression in mice," *Brain Research Bulletin* **45**: 67–74.

Chakrabarti, A., Onaivi, E. S. and Chaudhuri, G. (1995) "Cloning and sequencing of a cDNA encoding the mouse brain-type cannabinoid receptor protein," *DNA Sequence* **5**: 385–388.

Chaperon, F. and Thiebot, M.-H.(1999) "Behavioral effects of cannabinoid agents in animals," *Critical Reviews in Neurobiology* **13**: 243–281.

Chaperon, F., Soubrie, P., Puech, A. J. and Thiebot, M. H. (1998) "Involvement of central cannabinoid (CB$_1$) in the establishment of place conditioning in rats," *Psychopharmacology* **135**: 324–332.

Comings, D. E., Muhleman, D., Gade, R., Johnson, P., Verde, R., Saucier, G. *et al.* (1997) "Cannabinoid receptor gene (CNR1): association with IV drug use," *Molecular Psychiatry* **2**: 161–168.

Condie, R., Herring, A., Koh, W. S., Lee, M. and Kaminski, N. E. (1998) "Cannabinoid inhibition of adenylate cyclase-mediated signal transduction and interleukin 2 (IL-2) expression in the murine T-cell line," *EL4.IL-2. Journal of Biological Chemistry* **271**: 13175–13183.

Daaka, Y., Friedman, H. and Klein T. W. (1996) "Cannabinoid receptor proteins are increased in Jurkat human-T cell line after mitogen activation," *Journal of Pharmacology and Therapeutics* **276**: 776–783.

Das, S. K., Paria, B. C., Chakraborty, I. and Dey, S. K. (1995) "Cannabinoid ligand-receptor signaling in the mouse uterus," *Proceedings of the National Academy of Sciences of the USA* **92**: 4332–4336.

Dawson, E. (1995) "Identification of a polymorphic triplet marker for the brain cannabinoid receptor gene: use in linkage and association studies of schizophrenia," *Psychiatric Genetics* **5**: S50.

de Fonseca, F. R., Ramos, J. A., Bonnin, A. and Fernandez-Ruiz, J. J. (1993) "Presence of cannabinoid binding sites in the brain from early postnatal ages," *Neuroreport* **4**: 135–138.

De Petrocellis, L., Melck, D., Bisogno, T., Milone, A. and Di Marzo, V. (1999) "Finding of the endocannabinoid signaling system in *Hydra*, a very primitive organism: possible role in the feeding response," *Neuroscience* **92**: 377–387.

Devane, W. A., Dysarz, F. A., Johnson, M. R., Melvin, L. S. and Howlett, A. C. (1988) "Determination and characterization of cannabinoid receptor in rat brain," *Molecular Pharmacology* **34**: 605–613.

Devane, W. A., Hanus, L., Breuer, A., Pertwee, R. G., Stevenson, L. A., Griffin, G. *et al.* (1992) "Isolation and structure of a brain constituent that binds to the cannabinoid receptor," *Science* **258**: 1946–1949.

Di Marzo, V., Bisogno, T., De Petrocellis, L., Melck, D. and Martin, B. R. (1999) "Cannabimimetic fatty acid derivatives: The anandamide family and other endocannabinoids," *Current Medicinal Chemistry* **6**: 721–744.

Di Marzo, V., Bisogno, T., Melck, D., Ross, R., Brockie, H., Stevenson, L. *et al.* (1998) "Interactions between synthetic vanilloids and the endogenous cannabinoid system," *FEBS Letters* **436**: 449–454.

Di Marzo, V., Breivogel, C. S., Bisogno, T., Tao, Q., Bridgen, D. T., De Petrocellis, L., Razdan, R. K. *et al.* (2000) "Evidence for non-CB_1, non-CB_2 anandamide receptors in mouse brain," *2000 Symposium on the cannabinoids, Burlington, Vermont, ICRS.* p. 18.

Di Marzo, V., Fontana, A., Cadas, H., Schinelli, S., Cimino, G., Schwartz, J. C. and Piomelli, D. (1994) "Formation and inactivation of endogenous cannabinoid anandamide in central neurons," *Nature* **372**: 686–691.

Dohlman, H. G., Thorner, J., Caron, M. G., Lefkowitz, R. J. (1991) "Model systems for the study of seven-transmembrane-segment receptors," *Annual Review in Biochemistry* **60**: 653–688.

Egertova, M., Cravatt, B. F. and Elphick, M. R. (2000) "Fatty acid amide hydrolase expression in rat choroids plexus: possible role in regulation of the sleep-inducing action of oleamide," *Neuroscience Letters* **282**: 13–16.

Elphick, M. R. (1998) "An invertebrate G-protein coupled receptor is a chimeric cannabinoid/melanocortin receptor," *Brain Research* **780**: 170–173.

Facci, L., Dal Toso, R., Romanello, S., Buriani, A., Skaper, S. D. and Leon, A. (1995) "Mast cells express a peripheral cannabinoid receptor with differential sensitivity to anandamide and palmitoylethanolamide," *Proceedings of the National Academy of Sciences of the USA* **92**: 3376–3380.

Faubert, B. L. and Kaminski, N. E. (2000) "AP-1 activity is negatively regulated by cannabinol through inhibition of its protein components, c-fos and c-jun," *Journal of Leukocyte Biology* **67**: 259–266.

Felder, C. C., Veluz, J. S., Williams, H. L., Briley, E. M. and Matsuda, L. A. (1995) "Cannabinoid agonists stimulate both receptor- and non-receptor-mediated signal transduction pathways in cells transfected with and expressing cannabinoid receptor clones," *Molecular Pharmacology* **42**: 838–845.

Frischknecht, H. R. and Waser, P. G. (1980) "Actions of hallucinogens on ants (Formica pratensis)—III. Social behavior under the influence of LSD and tetrahydrocannabinol," *General Pharmacology* **11**: 97–106.

Gadzicki, D., Muller-Vahl, K. and Stuhrmann (1999) "A frequent polymorphism in the coding exon of the human cannabinoid receptor (CNR1) gene," *Molecular and Cellular Probes* **13**: 321–323.

Galiegue, S., Mary S., Marchand, J., Dussossoy, D., Carriere, D., Carayon, P. *et al.* (1995) "Expression of central and peripheral cannabinoid receptors in human immune tissues and leukocyte subpopulations," *European Journal of Biochemistry* **232**: 54–61.

Gerard, C. M., Mollereau, C., Vassart, G. and Parmentier, M. (1991) "Molecular cloning of a human cannabinoid receptor which is also expressed in the testis," *The Biochemical Journal* **279**: 129–134.

Giang, D. K. and Cravatt, B. F. (1997) "Molecular characterization of human and mouse fatty acid amide hydrolase," *Proceedings of the National Academy of Sciences of the USA* **94**: 2238–2242.

Glass, M., Brotchie, J. M. and Maneuf, Y. P. (1997) "Modulation of neurotransmission by cannabinoids in the basal ganglia," *European Journal of Neuroscience* **9**: 199–203.

Griffin, G., Tao, Q. and Abood, M. E. (1999) "Cloning and pharmacological characterization of the rat CB$_2$ cannabinoid receptor," *The Journal of Pharmacology and Experimental Therapeutics* **292**: 886–894.

Gurwitz, D. and Kloog, Y. (1998) "Do endogenous cannabinoids contribute to HIV-mediated immune failure," *Molecular Medicine Today* **4**: 196–200.

Hampson, A. J., Bornheim, L. M., Scanziani, M., Yost, C. S., Gray, A. T., Hansen, B. M. *et al.* (1998) "Dual effects of anandamide on NMDA receptor-mediated responses and neurotransmission," *Journal of Neurochemistry* **70**: 671–676.

Hampson, R. E., Evans, G. J., Mu, J., Zhuang, S. Y., King, V. C., Childers, S. R. *et al.* (1995) "Role of cyclic AMP dependent protein kinase in cannabinoid receptor modulation of potassium "A-current" in cultured rat hippocampal neurons," *Life Science* **56**: 2081–2088.

Harada, S., Okubo, T., Tsutsumi, M., Takase, S. and Muramatsu, T. (1996) "Investigation of genetic risk factors associated with alcoholism," *Alcoholism, Clinical and Experimental Research* **20**: 293A–296A.

Henry, D. J. and Chavkin, C. (1995) "Activation of inwardly rectifying potassium channels (GIRK1) by co-expressed rat brain cannabinoid receptors in *Xenopus* oocytes," *Neuroscience Letters* **186**: 91–94.

Herkenham, M., Lynn, A. B., Johnson, M. R., Melvin, L. S., de Costa, B. R. and Rice, K. C. (1991) "Characterization and localization of cannabinoid receptors in rat brain: A quantitative in vitro autoradiography study," *Journal of Neuroscience* **11**: 563–583.

Hillard, C. J. (2000) "Biochemistry and pharmacology of the endocannabinoids arachidonylethanolamide and 2-arachidonylglycerol," *Prostaglandins & other Lipid Mediators* **61**: 3–18.

Hillard, C. J. and Auchampach, J. A. (1994) "In vitro inactivation of brain protein kinase C by the cannabinoids," *Biochimica et Biophysica Acta* **1220**: 163–170.

Hillard, C. J., Edgemond, W. S., Jarrhian, A. and Campbell, W. B. (1997) "Accumulation of N-arachidonoyethanolamine (anandamide) into cerebral granule cells occurs via facilitated diffusion," *Journal of Neurochemistry* **69**: 631–638.

Hillard, C. J., Harris, R. A. and Bloom, A. S. (1985) "Effects of the cannabinoids on physical properties of brain membranes and phospholipid vesicles: Fluorescence studies," *The Journal of Pharmacology and Experimental Therepeutics* **232**: 579–588.

Hoehe, M. R., Caenazzo, L., Martinez, M. M., Hsieh, W. -T., Modi, W. S., Gershon, E. S. *et al.* (1991) "Genetic and physical mapping of the human cannabinoid receptor gene to chromosome 6q14-q15," *The New Biologist* **3**: 880–885.

Howlett, A. C., Johnson, M. R., Melvin, L. S., Milne, G. M. (1988) "Nonclassical cannabinoid analgetics inhibit adenylate cyclase: development of a cannabinoid receptor model," *Molecular Pharmacology* **33**: 297–302.

Iversen, L. L. (1993) "Medical uses of marijuana?" *Nature* **365**: 12–13.

Johnson, J. P., Muhleman, D., MacMurray, J., Gade, R., Verde, R., Ask, M. *et al.* (1997) "Association between the cannabinoid gene (CNR1) and the P300 event-related potential," *Molecular Psychiatry* **2**: 169–171.

Jonas, P. and Sakmann, B. (1992) "Glutamate receptor channels in isolated patches from CA1 and CA3 pyramidal cells of rat hippocampal slices," *Journal of Physiology* **455**: 143–171.

Kathmann, M., Haug, K., Heils, A., Nothen, M. M. and Schlicker, E. (2000) "Exchange of three amino acids in the cannabinoid CB_1 receptor (CNR1) of an epilepsy patient," *2000 Symposium on the cannabinoids, Burlington, Vermont. ICRS*. p. 18.

Kearn, C. S. and Hilliard, C. J. (1997) "Rat microglial cell express the peripheral-type cannabinoid receptor (CB_2) which is negatively coupled to adenylyl cyclase," (1997) *Symposium on the Cannabinoids, Burlington, Vermont, International Cannabinoid Research Society* **57**.

Kenney, S. P., Kekuda, R., Prasad P. D., Leibach, L. D., Devoe, L. D. and Ganapathy, V. (1999) "Cannabinoid receptors and their role in the regulation of the serotonin transporter in human placenta," *American Journal Obstetrics Gynecology* **181**: 491–497.

Kittler J. J., Clayton, C., Zhuang S.-Y., Trower, M. M., Wallace, D., Hampson, R. *et al.* (1999) "Large scale gene expression changes during long-term exposure to Δ^9-THC in rats," *Symposium on the Cannabinoids, Burlington, Vermont, ICRS*, 79pp.

Lambowitz, A. M. and Belfort, M. (1993) "Introns as mobile genetic elements," *Annual Review of Biochemistry* **62**: 587–622.

Le, A. D., Ko, J., Chow, S. and Quan, B. (1994) "Alcohol consumption by C57BL/6 Balb/c and DBA/2 mice in a limited access paradigm," *Pharmacology, Biochemistry and Behavior* **47**: 375–378.

Ledent, C., Valverde, O., Cossu, G., Petitet, F., Aubert, J.-F., Beslot, F. *et al.* (1999) "Unresponsiveness to cannabinoids and reduced addictive effects of opiates in CB_1 receptor knockout mice," *Science* **283**: 401–404.

Lenicque, P. M., Paris, M. R. and Poulot, M. (1972) "Effects of some components of *Cannabis sativa* on the regenerating planarian worm *Dugesia tigrina*," *Experientia* **28**: 1399–1400.

Lepore, M., Vorel, S. R., Lowinson, J. and Gardner, E. L. (1995) "Conditioned place preference induced by delta 9-tetrahydrocannabinol: comparison with cocaine, morphine, and food reward," *Life Sciences* **56**: 2073–2080.

Li, T., Liu, X., Zhu, Z. -H., Zhao, J., Hu, X., Ball, D. M. *et al.* (2000) "No association between (AAT)n repeats in the cannabinoid receptor gene (CNR1) and heroin abuse in a Chinese population," *Molecular Psychiatry* **5**: 128–130.

Maccarone, M., Van der Stelt, M., Rossi, A., Veldink, G. A. and Vliegenthart Agro, A. F. (1998) "Anandamide hydrolysis by human cells in culture and brain," *Journal of Biological Chemistry* **273**: 32332–32339.

Mackie, K. and Hille, B. (1992) "Cannabinoids inhibit N-type calcium channels in neuroblastoma-glioma cells," *Proceedings of the National Academy of Sciences of the USA* **89**: 3825–3829.

Mackie, K., Lai, Y., Westenbroek, R. and Mitchell, R. (1995) "Cannabinoids activate an inwardly rectifying potassium conductance and inhibit Q-type calcium currents in AtT20 cells transfected with rat brain cannabinoid receptor," *Journal of Neuroscience* **10**: 6552–6561.

Mailleux, P., Parmentier, M. and Vanderhaeghen, J. (1992) "Distribution of cannabinoid receptor messenger RNA in the human brain: an in situ hybridization histochemistry with oligonucleotides," *Neuroscience Letters* **143**: 200–203.

Makriyannis, A., Banijamali, A., Jarrell, H. C. and Yang, D.-P. (1989) "The orientation of (–)-delta-9-tetrahydrocannabinol in DPPC bilayers as determined by solid state 2H-NMR," *Biochimica et Biophysica Acta* **986**: 141–145.

Mallet, P. E. and Beninger, R. J. (1998) "Delta-9-tetrahydrocannabinol, but not the endogenous cannabinoid receptor ligand anandamide, produces conditioned place avoidance," *Life Sciences* **62**: 2431–2439.

Martin, R. S., Luong, L. A., Welsh, N. J., Eglen, R. M., Martin, G. R. and Maclennan, J. (2000) "Effects of cannabinoid receptor agonists on neuronally-evoked contractions of urinary bladder tissues isolated from rat, mouse, pig, dog, monkey and human," *British Journal of Pharmacology* **219**: 1707–1775.

Martin, B. R. (1986) "Cellular effects of cannabinoids," *Pharmacology Reviews* **38**: 45–74.

Martinez, M., Goldin, L. R., Cao, Q., Zhang, J., Sanders, A. R., Nancarrow, D. J. *et al.* (1999) "Follow-up study on a susceptibility locus for schizophrenia on chromosome 6q," *American Journal of Medical Genetics* **88**: 337–343.

Mascia, M. S., Obinu, M. C., Ledent, C., Parmentier, M., Bohme, G. A., Imperato, A. *et al.* (1999) "Lack of morphine-induced dopamine release in the nucleus accumbens of cannabinoid CB_1 receptor knockout mice," *European Journal of Pharmacology* **383**: R1-R2.

Matsuda, L. A. (1997) "Molecular aspects of cannabinoid receptors," *Critical Review in Neurobiology* **11**: 143–166.

Matsuda, L. A., Bonner, T. I. and Lolait, S. J. (1993) "Localization of cannabinoid mRNA in rat brain," *Journal of Comparative neurology* **327**: 535–550.

Matsuda, L. A., Lolait, T. I., Brownstein, M. J., Young, A. C. and Bonner, T. I. (1990) "Structure of a cannabinoid receptor and functional expression of the cloned cDNA," *Nature* **346**: 561–564.

McAllister, S. D., Griffin, G., Satin, L. S. and Abood, M. E. (1999) "Cannabinoid receptors can activate and inhibit G protein-coupled inwardly rectifying potassium channels in a *xenopus* oocyte expression system," *The Journal of Pharmacology, and Experimental Therapeutics* **291**: 618–626.

McClean, D. K. and Zimmerman, A. M. (1976) "Action of delta 9-tetrahydrocannabinol on cell division and macromolecular synthesis in division-synchronized protozoa," *Pharmacology* **14**: 307–321.

McGregor, I. S., Issakidis, C. N. and Prior, G. (1996) "Aversive effects of the synthetic cannabinoid receptor agonist CP 55, 940 in rats," *Pharmocology, Biochemistry and Behavior* **53**: 657–664.

McLaughlin, C. R. and Abood, M. E. (1993) "Developmental expression of cannabinoid receptor mRNA," *Developmental Brain Research* **76**: 75–81.

Mechoulam, R. (1999) "Recent advances in cannabinoid research," *Forschende Komplementarmedizin und Klassische Naturheilkunde* **6**: 16–20.

Mechoulam, R., Ben-Shabat, S., Hanus, L., Ligumsky, M., Kaminski, N. E., Schatz, N. E. *et al.* (1995) "Identification of an endogenous 2-monoglyceride, present in canine gut, that binds to cannabinoid receptors," *Biochemical Pharmacology* **50**: 83–90.

Munro, S., Thomas, K. L. and Abu-Shaar, M. (1993) "Molecular characterization of a peripheral cannabinoid receptor," *Nature* **365**: 61–65.

Nadler, V., Mechoulam, R. and Sokolovsky, M. (1993) "Blockade of 45Ca^{2+} influx through the N methyl-D-aspartate receptor ion channel by the non-psychoactive cannabinoid HU-211," *Brain Research* **622**: 79–85.

Netzeband, J. G., Conroy, S. M., Parsons, K. L. and Gruol, D. L. (1999) "Cannabinoids enhance NMDA-elicited Ca^{2+} signals in cerebellar granule neurons in culture," *Journal of Neuroscience* **19**: 8765–8777.

O'Dowd, B. F. (1993) "Structures of dopamine receptors," *Journal of Neurochemistry* **60**: 804–816.

Onaivi, E. S., Chakrabarti, A. and Chaudhuri, G. (1996) "Cannabinoid receptor genes," *Progress in Neurobiology* **48**: 275–305.

Onaivi, E. S., Chakrabarti, A., Gwebu, E. T. and Chaudhuri, G. (1995) "Neurobehavioral effects of delta-9-THC and cannabinoid receptor gene expression in mice," *Behavioural and Brain Research* **72**: 115–125.

Onaivi, E. S., Stubbs, L., Chakrabarti, A., Chittenden, L., Hurst, D. P., Akinshola, B. E. *et al.* (1998) "Murine Cannabinoid receptor genetics," *The FASEB Journal* **12**(4): A194.

Pan, X., Ikeda, S. R. and Lewis, D. L. (1996) "Rat brain cannabinoid receptor modulates N-type Ca2+ channels in a neuronal expression system," *Molecular Pharmacology* **49**: 707–714.

Pettit, D. A., Showalter, V. M., Abood, M. E. and Cabral, G. A. (1994) "Expression of a cannabinoid receptor in baculovirus-infected insect cells," *Biochemical Pharmacology* **48**: 1231–1243.

Pfister-Genskow, M., Weesner, G. D., Hayes, H., Eggen, A. and Bishop, M. D. (1997) "Physical and genetic localization of the bovine cannabinoid receptor (CNR1) gene to bovine chromosome 9," *Mammalian Genome* **8**: 301–302.

Piomelli, D., Beltramo, M., Glasnapp, S., Lin, S. Y., Goutopoulos, A., Xie, X.-Q. *et al.* (1999) "Structural determinants for recognition and translocation by the anandamide transporter," *Proceedings of the National Academy of Sciences of the USA* **96**: 5802–5807.

Pringle, H. L., Bradley, S. G. and Harris, L. S. (1979) "Susceptibility of *Naegleria fowleri* to delta-9-tetrahydrocannabinol," *Antimicrobial agents and Chemotherapy* **16**: 674–679.

Reibaud, M., Obinu, M. C., Ledent, C., Parmentier, M., Bohme, G. A. and Imperato, A. (1999) "Enhancement of memory in cannabinoid CB_1 receptor knockout mice," *European Journal of Pharmacology* **379**: R1–R2.

Sanudo-Pena, M. C., Tsou, K., Delay, E. R., Hohman, A. G., Force, M. and Walker, J. M. (1997) "Endogenous cannabinoids as an aversive or counter-rewarding system in the rat," *Neuroscience Letters* **223**: 125–128.

Schatz, A. R., Lee, M., Condie, R. B., Pulaski, J. T. and Kamiski, N. E. (1997) "Cannabinoid receptors CB_1 and CB_2: A characterization of expression and adenylate cyclase modulation within the immune system," *Toxicology and Applied Pharmacology* **142**: 278–287.

Schuel, H., Goldstein, E., Mechoulam, R., Zimmerman, A. M. and Zimmerman, S. (1994) "Anandamide (arachidonylethanolamide), a brain cannabinoid receptor agonist, reduces sperm fertilizing capacity in sea urchins by inhibiting the acrosome reaction," *Proceedings of the National Academy of Sciences of the USA* **91**: 7678–7682.

Schuel, H., Schuel, R., Zimmerman, A. M. and Zimmerman, S. (1987) "Cannabinoids reduce fertility of sea urchin sperm," *Biochemistry and Cell Biology* **65**: 130–136.

Sepe, N., De Petrocellis, L., Montanaro, F., Cimino, G. and Di Marzo, V. (1998) "Bioactive long chain N-acylethanolamines in five species of edible bivalve mollusks. Possible implications for mollusk physiology and sea food industry," *Biochimica et Biophysica Acta* **1389**: 101–111.

Shen, M., Piser, T. M., Seybold, V. S. and Thayer, S. A. (1996) "Cannabinoid receptor agonists inhibit glutamatergic synaptic transmission in rat hippocampal cultures," *Journal of Neuroscience* **16**: 4322–4334.

Shire, D., Calandra, B., Bouaboula, M., Barth, F., Rinaldi-Carmona, M., Casellas, P. *et al.* (1999) "Cannabinoid receptor interactions with the antagonists SR 141716A and SR 144528," *Life Sciences* **65**: 627–635.

Shire, D., Calendra, B., Rinaldi-Carmona, M., Oustric, D., Pessegue, B., Bonnin-Cabanne, O. *et al.* (1996) "Molecular cloning, expression and function of the murine CB_2 peripheral cannabinoid receptor," *Biochimica et Biophysica Acta* **1307**: 132–136.

Shire, D., Carillon, C., Kaghad, M., Rinaldi-Carmona, M., Le Fur, G., Caput, D. *et al.* (1995) "An amino terminal variant of the central cannabinoid receptor resulting from alternative splicing," *Journal of Biological Chemistry* **270**: 3726–3731.

Sitachitta, N. and Gerwick, W. H. (1998) "Grenadadiene and grenadamide, cyclopropyl-containing fatty acid metabolites from the marine cyanobacterium *Lyngbya majuscula*," *Journal of Natural Products* **61**: 681–684.

Skaper, S. D., Buriani, A., Dal Toso, R., Petrelli, L., Romanello, S., Facci, L. *et al.* (1996) "The ALIAmide palmitoylethanolamide and cannabinoids, but not anandamide, are protective in a delayed postglutamate paradigm of excitotoxic death in cerebellar granule neurons," *Proceedings of the National Academy of Sciences of the USA* **93**: 3984–3989.

Slipetz, D. M., O'Neill, G. P., Favreau, L., Dufresne, C., Gallant, M., Gareau, Y. *et al.* (1995) "Activation of the human peripheral cannabinoid receptor results in inhibition of adenylyl cyclase," *Molecular Pharmacology* **48**: 352–361.

Smart, D., Gunthorpe, M. J., Jerman, J. C., Nasir, S., Gray, J., Muir, A. I. *et al.* (2000) "The endogenous lipid anandamide is a full agonist at the human vanilloid receptor (hVR1)," *British Journal of Pharmacology* **129**: 227–230.

Soderstrom, K. and Johnson, F. (2000) "CB_1 cannabinoid receptor expression in brain regions associated with zebra finch song control," *Brain Research* **857**: 151–157.

Song, Z. H. and Bonner, T. I. (1996) "A lysine residue of the cannabinoid receptor is critical for receptor recognition by several agonists but not WIN55212-2," *Molecular Pharmacology* **49**: 891–896.

Stefano, G. B., Salzet, B. and Salzet, M. (1997) "Identification and characterization of the leech CNS cannabinoid receptor: coupling to the nitric oxide release," *Brain Research* **753**: 219–224.

Steiner, H., Bonner, T. I., Zimmer, A. M., Kitai, S. T., Zimmer, A. (1999) "Altered gene expression in striatal projection neurons in CB_1 cannabinoid receptor knockout mice," *Proceedings Naionational Academy Sciences* **96**: 5786–5790.

Strada, C. D., Fong, T. M., Tota, M. R., Underwood, D. and Dixon, R. A. F. (1994) "Structure and function of G-coupled receptors," *Annual Review of Biochemistry* **63**: 101–132.

Striem, S., Bar-Joseph, A., Berkovitch, Y. and Biegon, A. (1997) "Interaction of dexanabinol (HU-211), a novel NMDA receptor antagonist, with the dopaminergic system," *European Journal of Pharmacology* **338**: 205–213.

Stubbs, L., Chittenden, L., Chakrabarti, A. and Onaivi, E. S. (1996) "The mouse cannabinoid receptor gene is located in proximal chromosome 4," *Mammalian Genome* **7**: 165–166.

Sugiura, T., Kondo, S., Sukagawa, A., Nakane, A., Shinoda, A., Itoh, K. *et al.* (1995) "2-Arachidonoylglycerol: A possible endogenous cannabinoid ligand in brain," *Biochemical and Biophysical Research Communications* **215**: 89–97.

Szolcsanyi, J. (2000) "Are cannabinoids endogenous ligands for the VR1 capsaicin receptor," *TIPS* **21**: 41–42.

Terranova, J. P., Michaud, J. C., Le Fur, G. and Soubrie, P. (1995) "Inhibition of long-term potentiation in rat hippocampal slices by anandamide and WIN55212-2: reversal by SR141716 A, a selective antagonist of CB_1 cannabinoid receptors," *Naunyn Schmiedebergs Archives of Pharmacology* **352**: 576–579.

Thomasson, H. R., Edenberg, H. J., Crabb, D. W., Mai, X. L., Jerome, R. E., Li, T. K. *et al.* (1991) "Alcohol and alcohol dehydrogenase genotypes and alcoholism in Chinese men," *Americam Journal of Human genetics* **48**: 677–681.

Turkanis, S. A. and Karler, R. (1988) "Changes in neurotransmitter release at a neuromuscular junction of the lobster caused by cannabinoids," *Neuropharmacology* **27**: 737–742.

Tzschentke, T. M. (1998) "Measuring reward with the conditioned place preference paradigm: A comprehensive review of drug effects, recent progress and new issues," *Progress in Neurobiology* **56**: 613–672.

Ueda, N., Yamanaka, K., Terasawa, Y. and Yamamoto, S. (2000) "An acid amidase hydrolyzing anandamide and other N-acylethanolamines," *2000 Symposium on the cannabinoids*, Burlington, Vermont, ICRS. p. 13.

Valjent, E. and Maldonado (2000) "A behavioral model to reveal place preference to Δ^9-tetrahydrocannabinol in mice," *Psychopharmacology* **147**: 436–438.

Valk, P. J. M., Hol, S., Vankin, Y., Ihle, J. N., Askew, D., Jenkins, N. A., Gilbert, D. J., Copeland, N. G., DE Both, N. J., Lowenberg, B. and Delwel, R. (1997) "The genes encoding the peripheral cannabinoid receptor and a-L-fucosidase are located near a newly identified common virus integration site, Evi11," *Journal of Virology* **71**: 6796–6804.

Valverde, O., Ledent, C., Beslot, F., Parmentier, M. and Roques, B. P. (2000) "Reduction of stress-induced analgesia but not of exogenous opioid effects in mice lacking CB_1 receptors," *European Journal of Neuroscience* **12**: 533–539.

Varga, K., Wagner, J. A., Bridgen, D. T. and Kunos, G. (1998) "Platelet- and macrophage-derived endogenous cannabinoids are involved in endotoxin-induced hypotension," *FASEB Journal* **12**: 1035–1044.

Vaughan, C. W., McGregor, I. S. and Christie, M. J. (1999) "Cannabinoid receptor activation inhibits GABAergic neurotransmission in rostral ventromedial medulla neurons in vitro," *British Journal of Pharmacology* **127**: 935–940.

Wan, M., Cravatt, B. F., Ring, H. Z., Zhang, X. and Franke, U. (1998) "Conserved chromosomal location and genomic structure of human and mouse fatty-acid amide hydrolase genes and evaluation of clasper as a candidate neurological mutation," *Genomics* **54**: 408–414.

Westlake, T. M., Howlett, A. C., Bonner, T. L., Matsuda, L. A. and Herkenham, M. (1994) "Cannabinoid receptor binding and messenger RNA expression in human brain: an in vitro receptor autoradiography and in situ hybridization histochemistry study of normal aged and Alzheimer's brains," *Neuroscience* **63**: 637–652.

Wilson, R., Ainscough, R., Anderson, K., Baynes, C., Berks, M., Bonfield, J. *et al.* (1994) "2.2 Mb of contiguous nucleotide sequence from chromosome III of C. elegans," *Nature* **368**: 32–38.

Yamaguchi, F., Macrae, A. D. and Brenner, S. (1996) "Molecular cloning of two cannabinoid type1-like receptor genes from the Puffer Fish *Fugu rubripes*," *Genomics* **35**: 603–605.

Yea, S. S., Yang, K. H. and Kaminski, N. E. (2000) "Role of nuclear factor of activated T-cells and activator protein-1 in the inhibition of interleukin-2 gene transcription by cannabinol in EL4 T-cells," *The Journal of Pharmacology and Experimental Therapeutics* **292**: 597–605.

Zhuang, S.-Y., Kittler, J., Grigorenko, E. V., Kirby, M. T., Sim, L. J., Hampson, R. E. *et al.* (1998) "Effects of long-term exposure to Δ^9-THC on expression of cannabinoid (CB_1) mRNA in different rat brain regions," *Molecular Brain Research* **62**: 141–149.

Zimmer, A., Zimmer, A. M., Hohmann, A. G., Herkenham, M., Bonner, T. I. (1999) "Increased mortality, hypoactivity, and hypoalgesia in cannabinoid CB_1 receptor knockout mice," *Proceedings of the National Academy of Sciences of the USA* **96**: 5780–5785.

Zygmunt, P. M., Petersson, J., Andersson, D. A., Chuang, H.-H., Sorgard, M., Di Marzo, V. *et al.* (1999) "Vanilloid receptors on sensory nerves mediate the vasodilator action of anandamide," *Nature* **400**: 452–457.

Chapter 2

Cannabinoid therapeutic potential in motivational processes, psychological disorders and central nervous system disorders

Richard E. Musty

ABSTRACT

Cannabis has been used for medical purposes for centuries. With the discovery of cannabinoid receptors, there has been an explosion of research on both natural and synthetic cannabinoids. This chapter reviews both animal and human research demonstrating the potential role of cannabinoids in motivational processes and their associated disorders (hunger, appetite, pain), psychological disorders (anxiety, depression, bipolar disorder, schizophrenia, alcohol dependence) and central nervous system disorders (vomiting and nausea, spasticity, dystonia, brain damage, epilepsy). The most likely applications for cannabinoid agonists are for the treatment of loss of appetite, pain, anxiety, vomiting, nausea and epilepsy. The most likely applications for cannabinoid antagonists may be for anxiety, schizophrenia, spasticity, and dystonia. It is difficult to formulate an hypothesis concerning the potential treatment of depression, bipolar disorder and alcohol dependence since very little work has been done with these disorders at this point in time. In addition, one synthetic cannabinoid may be helpful in the treatment of brain damage. Basic research through clinical trials is needed to understand the potential application for psychological and central nervous system disorders.

Key Words: cannabis, cannabinoid, therapeutic, psychological disorders, CNS disorders

Cannabis preparations were used for central nervous system disorders for many years prior to the 1930s, when these preparations were removed from pharmacopeias throughout the world. *The Dispensatory of the United States of America*, published in 1892 (Wood *et al.*, 1892), listed many indications for medicinal use, including neuralgia, convulsions, spasms, hysteria (anxiety), "nervous disquietude", mental depression, delirium tremens, insanity, pain, and insomnia. Preparation of an alcoholic extract for prescription is described in the Dispensatory, as well as dosage recommendations.

The present chapter reviews both animal and human research demonstrating the potential role of cannabinoids in motivational processes and their associated disorders (hunger, appetite, pain), psychological disorders (anxiety, depression, bipolar disorder, schizophrenia, alcohol dependence) and central nervous system disorders (vomiting and nausea, spasticity, dystonia, brain damage, epilepsy).

HUNGER AND APPETITE

Animal studies

McLauglin *et al.* (1979) tested the effects of cannabinoids on food intake in sheep. Drugs were administered i.v. Δ^9-tetrahydrocannabinol (Δ^9-THC), d- and l-isomers of tetrahydrocannabinol and with 9-aza-cannabinol (9AC). At 30 min post-injection, Δ^9-THC and the d-isomer did not increase food intake, but the l-isomer and 9-AC increased intake.

Graceffo and Robinson (1998) administered THC (1.0 or 2.0 mg/kg) 30, 60, or 120 min or saline 30 min prior to being placed in the presence of a highly palatable food (Nilla Wafers soaked in water). Food intake did not differ between doses, but spontaneous exploration decreased at the 2.0 mg/kg dose of THC at the 60 min time point. Since motor depression was observed, it is possible that these doses were too high, thereby not enhancing hunger.

Human studies

Appetite stimulation has been reported anecdotally for years. For patients with wasting syndrome (e.g. HIV/AIDS, cancer-chemotherapy induced anorexia, anorexia nervosa), cannabinoid agonists may have clinical usefulness.

Table 2.1 summarizes the effects of THC on anorexia. In two of the three studies, patients gained weight. In the Gross study rather large doses were given, which may have produced sedation as suggested by the fact that patients experienced dysphoria, thus not increasing hunger. Furthermore, the patients suffered from anorexia nervosa, a condition quite different from weight loss in individuals associated with HIV/AIDS.

Two additional studies deserve mention. First, in a prospective open label study, Nelson *et al.* (1994), administered oral THC (2.5 mg/kg 2 times per day, 1 hour after eating) to terminal cancer patients. Appetite was improved. Second, in a double blind, placebo controlled study of normal volunteers, Mattes *et al.* (1994) found that whether THC, administered by oral capsules, rectal suppositories or smoking, induced stimulation of appetite as well an increase of calories consumed. Rectal suppositories and oral THC were given at a dosage of 2.5 mg twice daily. Patients assigned to smoked marijuana had to inhale for 3 sec and hold the smoke deeply in their lungs for 12 sec; this process was continued until the cigarette was smoked to a stub. The plasma THC levels peaked more quickly with the inhaled THC but also decreased more quickly; in contrast, the levels achieved with suppository THC were more sustained. Since the volunteers were experienced marijuana users, drug acceptance was good. Efficacy of the three treatment conditions were not different. Smoked marijuana was no more effective than suppository THC in stimulating appetite, as measured by calorie intake.

Appetite-stimulating effects of THC may be beneficial for patients with wasting related to the acquired immunodeficiency syndrome (AIDS) and those with severe cancer-related anorexia. The literature contains few studies with objective data on the use of either pure THC or crude marijuana for appetite stimulation. Future

Table 2.1 Human studies of THC on hunger

Reference	Subjects	Drug and dose	Type of study	Results
Gross et al. (1983)	11 patients with primary anorexia nervosa	Oral THC 7.5 mg–30 mg/day Diazepam 3–15 mg/day for 2 weeks	Double blind, placebo controlled	No significant difference in weight gain between THC and diazepam. More side effects (dysphoria) with THC
Plasse et al. (1991)	10 patients with AIDS-related diseases receiving anti-viral therapy	Dronabinol (THC) 2.5 mg tds "as needed" for 5 months	Open trial	Significant weight gain or reduction of weight loss compared to pre-treatment
Beale et al. (1995)	72 patients with AIDS-related illness	Dronabinol (THC) 2.5 mg bd Placebo for 6 weeks	Double blind, placebo controlled	Significant reduction in nausea and weight loss increased appetite and improved mood with THC compared to placebo

British Medical Association (1997) © Overseas Publishers Association N.V., with permission from Gordon and Breach Publishers.

research should examine improved appetite, caloric intake and weight gain, all necessary to establish the efficacy of the cannabinoids being tested.

PAIN

Analgesia induced by cannabinoid agonists has been repeatedly reported (Segal, 1986). In 1851, Christison reviewed therapeutic applications of cannabis as pain medication. At that time, tinctures of cannabis were used and he reported its usefulness in the treatment of rheumatic pain, sciatic nerve pain and tooth pain (Christison, 1851).

Animal studies

Bicher and Mechoulam (1968) found Δ^9-THC and Δ^8-THC (i.p.) were about 1/2 as effective as morphine (s.c.). On three tests of analgesia: the hot plate test, the acetic-acid writing test and the tail flick test, Sofia and colleagues (1975) conducted a comparison of the pain relieving effects of Δ^9-THC, a crude marijuana extract (CME), cannabinol (CBN), cannabidiol (CBD), morphine SO-4 and aspirin (all p.o). They used acetic-induced writhing, hot plate tests and the Randall-Selitto paw pressure tests in rats. Δ^9-THC and morphine were equipotent in all tests except that morphine was significantly more potent in elevating pain threshold in the uninflamed rat hind paw. In terms of Δ^9-THC content, CME was nearly equipotent in the hot plate and Randall-Selitto tests, but was 3 times more potent in the acetic acid writhing test. On the other hand, CBN, like aspirin, was only effective in reducing writhing frequency in mice (3 times more potent than aspirin) and raising pain threshold of the inflamed hind paw of the rat (equipotent with aspirin). CBD did not display a significant analgesic effect in any of the test systems used. The results of this investigation seem to suggest that both Δ^9-THC and CME possess analgesic activity similar to morphine, while CBN appears to be non-analgesic at the doses used.

In a recent report, Chichewizc and Welch (1999) found that Δ^9-THC (20 mg/kg) and morphine (20 mg/kg) induced analgesia in both vehicle treated and morphine tolerant mice. In both groups, analgesia was equally effective "indicating that analgesia produced by the combination is not hampered by existing morphine treatment (no cross tolerance to the combination)". Mice were tested with Δ^9-THC (20 mg/kg) and morphine (20 mg/kg) twice daily for 6.5 days and tested for tolerance and on day 8, Δ^9-THC tolerance was observed, but morphine tolerance did not occur. These results suggest low-dose combinations of Δ^9-THC and morphine might prevent morphine tolerance. The authors conclude that combinations of these drugs may be useful in chronic pain patients over morphine administration alone.

Ko and Woods (1999) applied capsaicin (100 mg) locally in the tail of rhesus monkeys which decreased pain threshold to the tail when exposed to 46° water causing a tail withdrawal reflex. Which does not occur in the absence of capsaicin. Δ^9-THC (10–320 mg) was coadministered with capsaicin in the tail to determine if pain thresholds were decreased. THC reduced pain responses in a dose dependent fashion. When SR141716 (10–100 mg) was applied to the tail reversal of pain blockade did not occur, suggesting that topical THC might block pain at the level of skin receptors. Overall, animal research is very consistent in regard to the analgesic effects of the natural cannabinoids.

Synthetic cannabinoid receptor agonists also have analgesic properties. Wilson and May (1975) and Wilson *et al.* (1976) found the following ED50s for synthetic 11-hydroxy- and 9-nor-THC derivatives in mice using the hot plate test. These were 9-nor-9β-OH-THC(1.6),11-hydroxy-Δ^9-THC(1.9) 11-hydroxy-Δ^8-THC (1.9), Δ^9-THC (8.8) and Δ^8-THC (9.6). Johnson and Melvin (1986) reviewed the discovery of non-classical cannabinoids synthesized by Pfizer. A large number of variations were tested and the reader is referred to this article for details. The work culminated in nantrodol as their initial effort to find a potent non-opioid analgesic. After studies of the stereospecificity of nantrodol, the isomer, levonantrodol was found to be the most potent analgesic. Welch and Stevens (1992) administered cannabinoid agonists intrathecally in mice and found the following rank order of antinociceptive potencies: using the tail-flick test was levonantradol >CP-55,940=CP-56,667 > 11-hydroxy-Δ^9-THC > Δ^9-THC > Δ^8-THC; dextronantradol was inactive at the tested dose ED50s were 0.4, 12.3, 4.2, 15, 45 and 72 micrograms/mouse in the tail-flick test, respectively. Cannabinoid agonist analgesia was not blocked by naloxone, suggesting these compounds do not act on opioid receptors. Martin *et al.* (1999) tested intrathecal administration of the CB$_1$ agonist, WIN 55,212-2 using a model of persistent pain (injection of Freund's adjuvant in the plantar surface of the rat hindpaw). Withdrawal thresholds were determined using Von Frey hairs. WIN 55,212-2 decreased paw withdrawal which was reversed by co-administration of SR141716. Further, Houser *et al.* (2000) measured the release of endogenous opioids, dynorphins in the rat spinal cord, after administration of anandamide (AEA), Δ^9-THC and CP-55,940. At peak analgesia, both Δ^9-THC and CP-55,940 induced release of dynorphin A, but AEA did not induce release of dynorphin suggesting a different mechanism of action for the analgesic effects of AEA. Edsall *et al.* (1996) using an antisense oligodeoxynucleotide 'knock-down' technique found that the analgesic effect of CP-55,940 was not present after the administration of the antisense oligodeoxynucleotide. Administration of oligodeoxynucleotides which had no effect on the CB$_1$ receptor had no effect on CP-55,940 induced analgesia, suggesting the CB$_1$ receptor is necessary for antinociception induced by cannabinoid agonists.

Endogenous cannabinoids also produce analgesia. Fride and Mechoulam (1993) found that anandamide (ANA) induced analgesia. Barg *et al.* (1995) found that two additional endogenous anandamides, docosatetraenylethanolamide and homo-gamma-linolenylethanol-amide produced analgesia using the hot plate test. Walker *et al.* (1999) stimulated both the dorsal and lateral periaquaductal gray which was followed by an increase in anandamide release suggesting anandamide mediates cannabinoid analgesia. Taken together, these studies suggest cannabinoid agonists act on brain cannabinoid receptors. At least at the level of the spinal cord, exogenous cannabinoids contribute to the release of endogenous opioids. Teasing out the mechanisms of cannabinoid action in antinociception continues to be an area for further research.

Dajani *et al.* (1999) found that 1',1'-Dimethylheptyl Δ^8-tetrahydrocannabinol-11-oic acid (CT-3) induced analgesia using the tail-clip and hot-plate tests. The potency of this cannabinoid was similar to morphine sulfate but has a longer duration of action. Using two routes of administration (i.p. and i.g.) the median effective dose (ED(50)) was 5 mg/kg. In addition, CT-3 does not induce gastric ulcers acutely at 100 mg/kg or after 30 days of administration at 30 mg/kg. The authors suggest that clinical testing of CT-3 is warranted based on these findings.

Human clinical studies

Recently, Mikuriya (1999) reported on interviews of 1800 patients who used marijuana for various medical conditions. Of these patients, he reported that 41% experienced analgesia following traumatic inflammation induced pain, autoimmune disorder-induced pain and ideopathic pain. Similarly, Consroe *et al.* (1997) found self-reported reductions in pain in patients with multiple sclerosis. Consroe *et al.* (1998) found similar self-reported pain reduction in patients with spinal cord injury. Schnelle *et al.* (1999) used an anonymous standardized survey of the medical use of cannabis and cannabis products of patients in Germany, Austria and Switzerland. Data from 128/170 patients were usable. Of these, 5.4% used cannabis for back pain and 3.6% for headache. Table 2.2 lists human studies of cannabinoid effects on pain.

Of the anecdotal reports listed, these are consistent with the more recent reports discussed above. Among the 4 double blind placebo controlled studies, the CB_1 receptor agonists Δ^9-THC or levonantradol, lead to significant pain relief, while the CB_1 receptor antagonist (Petitet *et al.*, 1998), CBD had no effect. While these studies had less number of patients, the results are fairly consistent. Taken together with the results from animal research, it seems clear that more clinical trials are needed with cannabinoids in order to determine which may have clinical potential.

ANXIETY

Animal studies

Musty *et al.* (1985) found CBD increased licking for water in the lick suppression test in a dose related fashion (mg/kg). Equivalent effects were found with the classic anxiolytic drug diazepam. In an effort to find more potent effects, they tested two analogs, 2-pinyl-5-dimethylheptyl resorcinol (PR-DMH) and Mono-methyl cannabidiol (ME-CBD-2). ME-CBD-2 had anxiolytic activity, but was less potent than CBD, while PR-DMH had no anxiolytic properties. Of the two active compounds, both were less potent than diazepam.

In another series of experiments, Musty (1984) found that CBD inhibited the development of stress-induced ulcers in rats as compared with diazepam, which produced equivalent reduction in the number of stress-induced ulcers. Guimaraes *et al.* (1990) tested rats in the elevated plus maze. In the first test, rats are placed in a plus shaped maze which is elevated off the floor. Two of the maze arms are enclosed with walls and two are not. Time spent in the enclosed arms is taken as

Table 2.2 Human studies on the analgesic effects of cannabis and cannabinoids

Reference	Subjects	Drug and dose	Type of study	Results
Noyes et al. (1975a)	10 patients with cancer pain	Oral THC 5, 10, 15, 20 mg in random order	Double blind, placebo controlled	Significant pain relief with 15 and 20 mg THC compared to placebo. Drowsiness and mental clouding common
Noyes et al. (1975b)	36 patients with cancer pain	Oral THC 10 and 20 mg, oral codeine 60 and 120 mg	Double blind, placebo controlled	THC 20 mg and codeine 120 mg gave equivalent and significant pain relief compared with placebo. THC caused sedation and mental clouding
Raft et al. (1977)	10 patients undergoing extraction of impacted molar teeth	IV THC 0.22 mg/kg and 0.44 mg/kv, IV diazepam 0.157 mg/kg	Double blind, placebo controlled	No analgesic effects of THC detected. Higher dose of THC was rated as least effective, diazepam most effective. 6 subjects preferred placebo to THC, 4 preferred low dose THC to placebo
Petro (1980)	2 patients with painful muscle spasms (1 spinal cord injury, 1 MS)	Cannabis	Open clinical report	Relief from pain and muscle spasms
Jain et al. (1981)	56 patients with postoperative pain	IM levonantradol 1.5–3 mg	Double blind, placebo controlled	Significant pain relief with both doses of levonantradol compared with placebo Drowsiness common with levonantradol
Lindstrom et al. (1987) Cited by Consroe and Sandyk (1992)	10 patients with chronic neuropathic pain	Oral cannabidiol 450 mg/day in divided doses	Double blind, placebo controlled	No analgesic effect of cannabidiol compared to placebo. Sedation with cannabidiol in 7 patients
Maurer et al. (1990)	1 patient with spinal cord injury	Oral THC 5 mg, oral codeine 50 mg, each given 18 times over 5 months	Anecdotal reports	THC and codeine alleviated pain to a similar degree: THC also relieved spasticity
Grinspoon and Bakalar (1993)	3 patients with various severe acute/chronic pain not controlled with opiates. 1 patient with migraine	Smoking cannabis		Pain relief reported in all cases allowing reduction in other analgesics; no "high" reported

British Medical Association (1997) © Overseas Publishers Association N.V., with permission from Gordon and Breach Publishers.

a measure of anxiety or fear. Both CBD and diazepam decreased the amount of time spent in the enclosed arms. Since these studies were conducted, Petitet *et al.* (1998) reported CBD is an antagonist of the CB_1 receptor in the micromolar range suggesting that CBD may have pharmacological effects as an antagonist of the CB_1 receptor.

Since the discovery of the synthetic, highly potent CB_1 receptor antagonist, SR 141716, by Rinaldi-Carmona, Barth, Heaulme *et al.*, several other studies seem to support the hypothesis that CB_1 receptor antagonists have anxiolytic properties. Onaivi *et al.* (1998) used two tests of anxiety in mice, the elevated plus maze (as discussed above) and the two compartment black and white box test. When administered SR141716, in the elevated plus maze, mice spent more time in the open arms indicating a reduction of anxiety. In the second test, mice are allowed to choose to spend time in a two compartment box, one which is white and brightly lit, the other is black and dimly lit. Time spent in the dark compartment is taken as an index of anxiety. When administered SR141716A, mice spent more time in the white, brightly lit compartment indicating a reduction in anxiety. Taken together, these studies suggest that CB_1 receptor antagonists have anxiolytic properties.

Human studies

Consroe *et al.* (1997) found that anxiety was reduced in 85% of patients, using cannabis, with multiple sclerosis in a self-report questionnaire. In another self-report study (Consroe *et al.*, 1998) patients with spinal cord injuries, who used cannabis, reported similar reductions in anxiety.

In a laboratory setting, when subjects were instructed to smoke marijuana until they reached their "usual level of intoxication", regression analysis of a visual analog scale of the word "anxious" predicted decreased scores on this scale. These data support the hypothesis that THC has anxiolytic properties at low doses (Musty, 1988).

In normal volunteers, Zuardi *et al.* (1982) tested the hypothesis that CBD would antagonize anxiety induced by THC. They used a dose of 0.5 mg/kg of THC for a 150 lb (68 kg)/subject which is a rather large dose, exceeding the dose a person would take for the intoxicating effect of the drug. Subjects report a pleasant high at 0.25 mg/kg using the same route of administration without an increase in anxiety. A second study (Zuardi *et al.*, 1993) induced anxiety in normal subjects by having them prepare a 4 minute speech about a topic from a course they had taken during the year. They were told the speech would be videotaped for later analysis by a psychologist. The subject began the speech while viewing his/her image on a video monitor. Anxiety measures were taken using a visual-analog scale of mood (anxiety, physical sedation, mental sedation and other feelings, e.g. interested) at five time points: baseline, immediately before instructions, immediately before the speech in the middle speech, and after the speech. Heart rate and blood pressure measures were also taken. The subjects were randomly assigned to one of four drug conditions: CBD (300 mg), isapirone (5 mg), diazepam (10 mg) or placebo. CBD, diazepam and isapirone decreased anxiety and systolic blood pressure. Neither CBD nor isapirone had effects on physical sedation, mental sedation, or other feelings, but diazepam induced feelings of physical sedation. Based on both the evidence from animal and human studies, it seems worthwhile to continue research on the anxiolytic properties of CB_1 receptor antagonists.

DEPRESSION

In a study of normal subjects, Musty (1988) found a positive correlation on the depression scale of the MMPI with feelings of euphoria after smoking marijuana, while there was no correlation between anxiety (Hysteria Scale) and somatic concerns (Hypochondriasis Scale) with feeling euphoric, suggesting an antidepressive effect from marijuana use. Schnelle, Grotenhermen, Reif *et al.* (1999), in a survey of 128 patients in Germany, 12% reported marijuana use for relief of depression. Consroe *et al.* (1997) found that depression was reduced in patients with multiple sclerosis (who smoked cannabis) using a self-report questionnaire. In another self-report study (Consroe *et al.*, 1998) of patients with spinal cord injuries similar reductions in depression were reported. In cancer patients, Regelson (1976) found THC relieved depression in advanced cancer patients. Finally, Warner and colleagues (1994) found reported relief from depression, in a survey of 79 mental patients. At present, there are very little data supporting the hypothesis that cannabinoids might relieve depression, but tests of both agonists and antagonists of the CB_1 receptor are clearly indicated to test this hypothesis.

BIPOLAR DISORDER

Grinspoon and Bakalar (1993, 1997) presented 6 case studies of people with bipolar disorder, who used cannabis to treat their symptoms:

> Some used it to treat mania, depression, or both. They stated that it was more effective than conventional drugs, or helped relieve the side effects of those drugs. One woman found that cannabis curbed her manic rages.... Others described the use of cannabis as a supplement to lithium (allowing reduced consumption) or for relief of lithium's side effects.

These clinical observations are important leads to the potential use of cannabinoids for treating manic depressive disorder and suggest that clinical trials should be conducted.

SCHIZOPHRENIA

Animal studies

Zuardi *et al.* (1991) tested the effects of CBD and haloperidol in a model which predicts anti-psychotic activity in rats. Apomorphine induces stereotyped sniffing and biting. Both drugs decreased the frequency of these behaviors. CBD did not induce catalepsy, even at very high doses, although haloperidol induced catalepsy. The authors conclude that CBD has a pharmacological profile similar to the atypical antipsychotic drugs.

Musty *et al.* (2000) tested the effects of the of the CB_1 receptor antagonist SR141716 in two animal models of schizophrenia. In the first model, ibotenic acid lesions of the hippocampus were made in neonatal rats, which results in a brain degeneration pattern similar to that observed in schizophrenics, as well as abnormal play behavior in an anxiety-provoking environment. In the second model, ketamine-induced enhancement of pre-pulse inhibition was tested. In both of these tests, SR141716 reversed the abnormal behavior. These findings in animal models are consistent with the hypothesis that CB_1 receptor antagonists have antipsychotic activity.

Human studies

Use of cannabis has been associated with exacerbation of symptoms of schizophrenia (Negrete *et al.*, 1986), but other reports suggest use of cannabis helped patients manage their symptoms of schizophrenia. Several studies have shown potential symptom-relieving effects of cannabis use.

Peralta and Cuesta (1992) studied 95 schizophrenics who had used cannabis in the last year. They found lower scores in the schizophrenics on delusions and alogia scales of Andreasen's Scales for the Assessment of Positive and Negative Symptoms, suggesting cannabis may affect the negative symptoms of schizophrenia. In a sample of community-based mentally ill patients (Warner *et al.*, 1994), reported fewer hospital admissions and fewer symptoms of anxiety, depression and insomnia among users of marijuana.

Zuardi *et al.* (1995) reported an experiment in a single case, in which the patient was being treated with haloperidol. The medication was stopped due to side effects followed by a return of symptoms, leading to hospitalization. At this point, the patient was given placebo medication for four days, after which she was administered CBD (two-doses per day) on an increasing dose schedule to 750 mg/dose until the 26th day. This was followed by 4 days of placebo and finally by a return to haloperidol for 4 weeks. Interviews were conducted and videotaped, which were followed by rating of interviews using the Brief Psychiatric Rating Scale (BPRS) and the Interactive Observation Scale for Psychiatric Patients (IOSPP). A psychiatrist rated the patient, blind to treatment conditions on the BPRS and nurse assistants independently and blind to treatment conditions rated the patient on the IOSPP. Comparing placebo to the CBD condition, Hostilty-Suspiciousness dropped by 50% of the BPRS maximum scale score (mss), Thought–Disturbance decreased by 37.5% of the mss, Anxiety–Depression by 43.7% of the mss, Activation by 41.6% of the mss, and Anergia decreased by 31.3% of the mss. During 4 days of placebo, which followed all four scale scores increased somewhat. The patient was returned to haloperidol treatment and the subsequent scores were close to those with CBD treatment. This N of 1 experiment demonstrates that antagonists of the CB_1 receptor are candidates for testing in human schizophrenia.

ALCOHOL DEPENDENCE

Musty (1984) found CBD, Δ^9-THC and clonidine reduced body tremor and audiogenic seizures during alcohol withdrawal in C57Bl6J mice that were forced to become alcohol tolerant on a liquid diet containing alcohol. Equivalent reductions in tremors and seizures were found with clonidine. Grinspoon and Bakalar (1997) reported two cases of individuals who used marijuana to deal with alcohol dependence.

VOMITING AND NAUSEA

Table 2.3 shows the results of controlled studies of the effects of THC on nausea and vomiting (emesis).

A mixed picture emerges from these studies. Of the 10 studies cited, 7/10 demonstrated that THC was equal to or better than placebo or alternate drug therapies in suppression of nausea and vomiting. Six states in the USA conducted several studies of cannabinoid anti-nausea and antiemetic effects in the 1980s. Musty and Rossi (in press) reviewed these studies using smoked marijuana compared with THC and other antiemetics or smoked marijuana alone. They found smoked marijuana was at least as effective as older antiemetics. In Phase III study, 100% efficacy was reported. In addition, smoked marijuana was preferred by patients who could tolerate smoking. Finally, Abrahamov and Mechoulam (1995), in an open-label trial, gave sublingual Δ^8-THC to child and adolescent cancer patients and found this drug to be highly effective against emesis and nausea with almost no side effects.

Table 2.4 shows the effects of oral Nablione on nausea and vomiting. This drug produced significant decreases in nausea and vomiting over placebo, phencyclidine, or domperidone in all 13 studies cited.

Johnson and Melvin (1986) reviewed the efficacy of levonantradol (LVN) as an antiemetic in humans. In 12 studies they cited, 343 patients were administered LVN in a variety of research designs ranging from placebo controlled, randomized, double-blind studies to open phase II trials. Relief was reported in all of the studies ranging from partial to complete control of emesis. Some patients withdrew due to ineffectiveness or adverse reactions. In the phase II trials, LVN was given p.o. (0.5 mg or 1.0 mg) with 57% and 72% of patients having partial to complete relief respectively. The most prominent side effects were somnolence and dysphoria.

Overall, these data suggest cannabinoid agonists can be effective as antiemetic and anti-nausea drugs. Since the inhalation route gives the patient more control over the therapeutic effect of the drug, it is clear that alternate routes of administration, which would accelerate absorption should be exploited (i.e., sublingual, rectal or inhaled without smoke). In addition, a variety of natural and synthetic cannabinoids should be tested.

SPASTICITY

Animal studies

The cannabinoid agonist WIN 55,212-2 relieves symptoms of dystonia in mutant dystonic hamsters (Richter and Losher, 1994). Cannabidiol attenuates dystonia and torticollus in mutant rats (Consroe et al., 1988), apomorphine induced turning

Table 2.3 Selected well controlled studies on the antiemetic effects of THC in patients on cancer chemotherapy

Reference	Subjects	Drug and dose	Type of study	Results
Sallan et al. (1975)	22 patients on cancer chemotherapy, resistant to standard drugs	THC 10 mg/m^2	db, pc, r, x	THC significantly more effective than placebo for nausea and vomiting (patients' self-report); sedation and euphoria on THC
Chang et al. (1979)	15 patients on high dose methotrexate	THC 10 mg/m^2 oral, 3 hrly 17 mg smoking*	db, pc, r, x	14 of the 15 patients had decreased nausea and vomiting on the THC compared with placebo
Frytak et al. (1979)	116 Patients with gastrointestinal carcinoma on combined 5-fluouracil and semustine	THC 15 mg tds PCP 10 mg tds	db, pc	THC and PCP equally effective, both better than placebo. Side effects of THC sometimes intolerable sedation, "high", dysopria, hypotension, tachycardia
Orr and McKernan (1981)	55 patients on various cancer chemotherapy	THC 7 mg/m^2 qds PCP 7 mg/m^2 qds	db, pc, r, x	THC better than PCP, both better than placebo. THC produced "high" in 82%
Lucas and Laszlo (1980)	53 patients on various cancer chemotherapy	THC 15 mg THC 5 mg × 2 standard antiemetics	pc, r	THC more effective than standard regimes and placebo

Study	Patients	Drugs/Dose	Design	Results
Chang et al. (1981)	8 patients on adriamycin and cyclophosphamide	THC 10 mg/m² oral, 17.4 mg smoking*, 3 hrly	db, pc, r	THC ineffective compared with placebo
Niedhart et al. (1981)	52 patients on various cancer chemotherapy	THC haloperidol	db, pc, r, x	No difference between THC and haloperidol in nausea and vomiting
Gralla et al. (1982)	27 patients on cisplatin	THC 10 mg/m² MCP 2 mg/kg IV	db, pc, r	MCP better than THC; both better than placebo for nausea and vomiting
Ungerleider et al. (1982)	214 patients on various cancer chemotherapy	THC 7.5–12.5 mg 4 hrly PCP 10 mg 4 hrly	db, r, x	No significant difference between PCP and THC in control of nausea and vomiting; more side effects on THC but preferred by more patients
Lane et al. (1991)	62 patients on various cancer chemotherapy	THC (dronabinol) 10 mg qds PCP 10 mg qds Both drugs together	db, r	No nausea and vomiting 51% for THC, 83% for PCP; PCP and THC combined better than either drug alone

db – double blind; pc – placebo controlled; r – randomized; x – crossover; PCP – prochlorperazine; MCP – metoclopramide; tds – 3 times daily; qds – 4 times daily; * – THC smoked rather than taken orally if vomiting occurred

British Medical Association (1997) © Overseas Publishers Association N.V., with permission from Gordon and Breach Publishers.

Table 2.4 Selected well-controlled studies on the antiemetic effects of Nabilone™ in patients on cancer chemotherapy

Reference	Subjects	Drug and dose	Type of study	Results
Nagy et al. (1978)	47 patients on cisplatin combination therapy	Nabilone 2 mg 6 hrly PCP 10 mg 6 hrly	db, pc, r, x	Nabilone more effective than PCP or placebo for nausea and vomiting
Herman et al. (1979)	113 patients on cisplatin, doxorubicin/cyclophosphamide, mustine	Nabilone 2 mg 6 or 8 hrly PCP 10 mg 6 or 8 hrly	db, pc, r, x	Nabilone significantly more effective than PCP or placebo for nausea and vomiting
Einhorn et al. (1981)	100 patients on various cancer chemotherapy	Nabilone 2 mg 6 hrly PCP 10 mg 6 hrly	db, r, x	Nabilone significantly more effective than PCP for nausea and vomiting, and preferred by 75% of patients. Lethargy and hypotension with Nabilone
Jones et al. (1982)	54 patients on various chemotherapy	Nabilone 2 mg 6 hrly Placebo	db, pc, r, x	Significant reduction of nausea and vomiting with Nabilone compared with placebo. Side effects common but acceptable (dizziness 65%; drowsiness 51%)
Wada et al. (1982)	114 patients on various cancer chemotherapy	Nabilone 2 mg bd Placebo	db, pc, r, x	Nabilone superior to placebo for nausea and vomiting. Side effects frequent with nabilone but preferred by more patients
Levitt (1982)	36 patients on various cancer chemotherapy	Nabilone (dose not stated) Placebo	db, pc, r, x	Nabilone superior to placebo for nausea and vomiting. Side effects frequent with nabilone but preferred by more patients
Johnannson et al. (1982)	18 patients on various cancer chemotherapy	Nabilone 2 mg bd PCP 10 mg bd	db, r, x	Nabilone superior to PCP for nausea and vomiting. Nabilone preferred to PCP by patients, despite side effects
Ahmedizal et al. (1983)	26 patients with lung cancer on cyclophosphmadide, adrimycin, and etoposide	Nabilone 2 mg bd PCP 10 mg tds	db, r, x	Nabilone significantly more effective than PCP for nausea and vomiting. More side effects (drowsiness, dizziness) with Nabilone but preferred by patients

Niiranan and Mattson (1985)	24 patients on various chemotherapy	Nabilone 2 mg 6 hrly PCP 15 mg 6 hrly	db, r, x	Nabilone significantly better than PCP for nausea and vomiting and preferred by most patients though more side effects.
Niederle et al. (1986)	20 patients on cisplatin	Nabilone 2 mg 6 hrly Alizapride 150 mg tds	db, r, x	Nabilone more effective than alizapride for nausea and vomiting, though more side effects
Pomeroy et al. (1986)	38 patients on various cancer chemotherapy	Nabilone 1 mg 6 hrly Domperidone 20 mg 6 hrly	db, r	Nabilone significantly superior to domperidone for vomiting episodes
Dalzell et al. (1986)	23 children on various cancer chemotherapy	Nabilone 0.5 mg bd or tds Domperidone 5–15 mg tds	db, r, x	Nabilone more effective than domperidone for nausea and vomiting though more side effects. Two-third of children preferred Nabilone
Chan et al. (1987)	30 children on various cancer chemotherapy	Nabilone PCP	db, r, x	Nabilone was superior to PCP in nausea and vomiting

db – double-blind; pc – placebo controlled; r – randomized; x – crossover; PCP – prochlorperazine; bd – 2 times daily; tds – 3 times daily

British Medical Association (1997) © Overseas Publishers Association N.V., with permission from Gordon and Breach Publishers.

in rats with unilateral nigrostriatal lesions and aggressive behavior in an l-pyro-glutamate-induced model of Huntington's disease (Conti *et al.*, 1988).

In a recent review of Cannabinoids and Spasticity, Musty and Consroe (in press) reported:

> Δ^9-THC (or Δ^8-THC) was found to reduce histological and clinical features of experimental autoimmune encephalomyelitis (EAE) and experimental auto-immune neuritis (EAN) in experimental animals (26, 27, 54). EAE is the commonly used animal model for MS, while EAN is a commonly used animal model for Guillain-Barre Syndrome (Lyman, 1991). Δ^9-THC has immunosup-pressant effects, presumably by acting on primarily CB_2 receptors located in the immune system (Pertwee, 1997). These findings suggest that cannabinoid receptor agonists might affect the underlying disease process of MS.
>
> Lastly, recent findings (Baker, Pryce, Croxford *et al.*, 2000) have clearly established that cannabinoid CB_1 agonists can ameliorate specifically the signs of spasticity and tremor in an MS animal model (Biozzi ABH mice). Using the mouse model of chronic relapsing encephalomyelitis, the CB_1 agonists, WIN 55121, Δ9-THC, JW-133 and methanandamide each blocked spasticity and tremor in the mutant mice. Further, the CB_1 antagonist drug, SR 144528, reversed the effect of the CB_1 agonists and when given alone the SR 144528 caused an exacerbation of these 2 clinical signs. In addition, treat-ment with the CB_2 receptor agonists, palmitoylethanolamide reduced spasti-city. Since this compound has no actions on the CB_1 receptor, it is possible that CB_2 receptor agonists may have therapeutic potential without intoxicat-ing effects. These results provide direct evidence that: (a) cannabinoid CB_1 agonists can obtund two major clinical signs of MS; (b) endogenously present anandamide may be tonically active in the control of spasticity and tremor of MS; and (c) by implication of the results with Δ9-THC, patient claims of reduction of spasticity and tremor from marijuana smoking are genuine pharmacological effects.

HU-211, a non-psychoactive cannabinoid that has anti-inflammatory properties, was administered to rats with experimental autoimmune EAE. The symptoms EAE were reduced after administration. Histological examination of the brain and spinal cord revealed a decrease in inflammation in this model (Achiron *et al.*, 2000).

Human studies

Reviews and critiques of the clinical literature on spasticity effects have been published (Consroe and Snider, 1986; Consroe and Sandyk, 1992; Pertwee, 1987). Scott Imler, who managed one of the cannabis buyers clubs in California illus-trates:

> I have a videotape of a quadriplegic whose legs are bucking and bouncing. A couple of puffs and the spasms stop cold.... Our quad members tell us this happens every morning: They wake up with knees in their mouths, stiff as a board, and two or three puffs later, their muscles relax... (Musty *et al.*, 1998).

Table 2.5 summarizes studies which have been conducted with smoked marijuana, Δ^9-THC and nabilone in patients with Multiple Sclerosis.

Among anecdotal reports and the questionnaire study, patients report improvement of symptoms of their MS. In the only study in which patients smoked marijuana, posture and balance was impaired in all. Of the 3 studies using oral THC in patients with MS, most patients reported subjective relief of symptoms, but a smaller number of patients showed objective improvement in symptoms. In the n of 1 study with Nabilone™, improvement was observed. Due to the very small number of patients in these studies, it is difficult to draw any final conclusion concerning the efficacy of oral THC in MS patients. Given the fact that animal studies have shown specific pharmacological effects in combination with these human studies, further testing of cannabinoids is certainly warranted.

Table 2.6 summarizes studies of cannabinoids in patients with spinal cord of injury. While these studies are rather small, ranging from n of 1, to open label and double blind the results are consistent suggesting further research on the effects of symptoms associated with spinal cord injury are needed.

EPILEPSY

Animal studies

There are many studies which show that Δ^9-THC can induce seizures at high doses. This literature has been reviewed by Consroe and Snider (1986). Δ^9-THC, Δ^8-THC, Δ^9-THC acids, cannabinol, and CBD raise electroshock induced seizure thresholds in mice (Consroe and Snider, 1986). Further, metabolites of Δ^9-THC, also reduce seizure thresholds (11-hydroxy-Δ^9-THC, 8-hydroxy-Δ^9-THC and 8-dihydroxy-Δ^9-THC). Synthetic analogs also reduce thresholds (9-nor-Δ^8-THC, 1,2 dimethylheptyl isomers of $^{6-10}$THC, 9-nor-9 hydroxy-hexahydro-cannabinol and 9-nor-9 hydroxy-hexahydro-cannabinol. Several CBD analogs are also active, i.e., those with a 1,2 dimethylheptyl side chain of the resorcinal moiety and the + isomer of CBD. These later CBD analogs have somewhat greater potency than the parent homologs. Further, CBD analogs have the greatest protective index[PI] (Toxic Dose$_{50}$/Effective Dose$_{50}$). Interestingly, CBD has a PI comparable to classic anti-epileptic drugs (Phenytoin, phenobarbital and carbamazpine) which are effective against grand mal and partial seizures. Seizures are commonly induced chemically by pentylentetrazol (PTZ). In mice, Δ^9-THC, Δ^8-THC and the THC metabolites 11-hydroxy-Δ^9-THC and 11-oxo-Δ^8-THC reduce the incidence of tonic convulsions. Lethal outcomes were also reduced. In genetically susceptible animals, the literature is mixed, but CBD has been shown to be effective in more strains than it has not. Cannabidiol has been shown to decrease convulsions in mice (Consroe and Snider, 1986). In addition, cannabinoids have been shown to have anti-seizure and anti-convulsant effects in a variety of species and paradigms (Consroe and Snider, 1986). Leite *et al.* (1982) also demonstrated (−) and (+) isomers of CBD and CBD dimethylheptyl homologs have anti-convulsant activity.

Table 2.5 Studies on the effects of cannabis and cannabinoids in multiple sclerosis (MS)

Reference	Subjects	Drug and dose	Type of study	Results
Petro and Ellenberg (1981)	9 patients with MS	Oral THC 5, 10 mg, single doses	Double blind, placebo controlled	Significant reduction in objective spasticity scores overall after THC; one patient improved objectively after placebo, three patients subjectively improved, two of these objectively improved. Minimal psychoactive effects
Clifford (1983)	8 Patients with MS	Oral THC 5–15 mg 6 hrly, up to 18 h	Single blind, placebo controlled	5 patients showed mild subjective but not objective improvement in tremor and well being after THC. 2 patients showed subjective and objective improve-ment in tremor, but not inataxia or other symptoms. All experienced a "high" after THC; 2 became dysphoric
Ungerleider et al. (1987)	13 patients with MS	Oral THC 2.5 mg once or twice daily for 5 days	Double blind, placebo controlled	Significant subjective improvement in spasticity at doses of 7.5 mg THC and above, but no change in objective measurements of weakness, spasticity, coordination, gait or reflexes. Side effects in 12 patients on THC and 5 patients on placebo
Meinck et al. (1989)	1 patient with MS	Smoking cannabis, dose not stated, single dose	Open trial	Improvement in tremor, spasticity and ataxia
Greenberg et al. (1994)	10 patients with MS 10 normal controls	Smoking cannabis (1.54% THC) single dose	Double blind, placebo controlled	Cannabis impaired posture and balance in all subjects, causing greater impairment in MS patients. No other objective neurological changes but subjective improvement in some patients

Reference	Subjects	Treatment	Method	Results
Martyn et al. (1995)	1 patient with MS	Oral nabilone 1mg every second day for 2 periods of four weeks	Double blind, placebo controlled	Improvement in general well-being, muscle spasms and frequency of nocturia with nabilone. Mild sedation but no euphoria with nabilone
Consroe et al. (1997)	112 patients with MS in UK and USA	Smoking cannabis, dosage unknown	Questionnaire survey (48% response rate from 223 questionnaires)	Improvement in muscle spasms and pain, depression, tremor, anxiety, paraesthesiae, weakness, balance, constipation. No information on adverse effects
Grinspoon and Bakalar (1993), Davies (1992), Doyle (1992), Ferriman (1993), Handscombe (1993), Hodges (1992, 1993), James (1993)	~10 patients with MS	Smoking or oral cannabis, dosage not known	Anecdotal reports	Increased well being, improvement of walking, appetite, breathing, bladder control, and relief of muscle spasms

Table 2.6 Studies on the effects of cannabis and cannabinoids in spinal cord injury and movement disorders

Reference	Subjects	Drug and dose	Type of study	Results
Dunn and Davis (1974)	10 patients with a range of problems arising from spinal cord injury	Cannabis	Patient survey of perceived effects	5/8 noted improvement in spasticity, 4/9 noted improvements in phantom limb pain, 1/9 noted worsening of bladder spasms and 2/10 worsening of urinary retention
Petro (1980)	2 patients; 1 with spinal cord injury; 1 with MS	Cannabis	Open clinical report	Relief from pain and muscle spasms
Malec et al. (1982)	24 patients with spinal cord injuries	Cannabis	Questionnaire survey (53% response rate from 48 questionnaires)	21 of 24 who had used cannabis reported decrease in spasticity
Maurer et al. (1990)	1 patient with spinal cord injury	Oral THC 5 mg, oral codeine 50 mg, placebo, each given 18 times over 5 months	Double blind, placebo controlled	THC and codeine alleviated pain to a similar degree; THC had a additional beneficial effect on spasticity
Consroe et al. (1997)	5 patients with various dystonias	Oral cannabidiol, 100–600 mg/day over 6 weeks	Open trial	20–50% improvement in dystonia, in all cases, but exacerbation of tremor and hypokinaesia in 2 patients with co-existing Parkinsonism
Sandyk and Awerbach (1988)	3 patients with Tourette's syndrome	Cannabis smoking	Case reports	Alleviation of motor tics reported by patients; authors suggest that effects due to anxiolytic action of cannabis
Frankel et al. (1990)	5 patients with Parkinson's disease	Cannabis cigarette (2.9% THC) diazepam 5 mg oral, levodopa/carbi-dopa, (25 mg/25 mg) oral, apomorphine 1–5 mg s.c. on consecutive days	Open study	Improvement in tremor with levodopa and apomorphine; no improvement in any disability with cannabis or diazepam
Consroe et al. (1991)	15 patients with Huntington's disease	Oral cannabidiol 10 mg/kg/day for 6 weeks	Double blind, placebo controlled	No beneficial effects

British Medical Association (1997) © Overseas Publishers Association N.V., with permission from Gordon and Breach Publishers.

Human studies

Grinspoon and Bakalar (1997, 1998) presented 3 case studies of people with epilepsy, who used cannabis to treat their symptoms. Schnelle *et al.* (1999) found, in a survey of 128 patients in Germany, 3.6% reported marijuana use for relief of epilepsy.

Table 2.7 presents a summary of cannabinoids in epilepsy. Of the 3 controlled studies, mixed results were obtained, but it seems that CBD has some potential to control epilepsy. One study warrants a detailed discussion. Cunha *et al.* (1980) studied 15 patients with secondary generalized epilepsy with temporal lobe foci. They were tested in a double-blind two group procedure with CBD. Patients remained on the anti-epileptic drug which they were receiving for the duration of the study. Doses given were 200–300 mg of CBD in gelatin capsules. Patients were in the study from a minimum of 6 weeks to 18 weeks. One week prior to beginning administration, the placebo group had a median of 4.5 focal convulsions and a median of 1 (range 1–3) generalized convulsion(s) per week and the CBD group had a median of 6.5 focal convulsions and a median of 2.5 (range 1–4) generalized convulsion(s) per week. Each week patients were rated on a clinical efficacy scale, where 0 = total absence of convulsions and self-reported improvement; 2 = subjective self improvement; 3 = no reduction in convulsions and no self-reported improvement. Over the total drug phase for each subject the placebo group had a median rating of 3 (6 rated 3 and 1 rated 0) and the CBD treated group had a median rating of 1.6 (4 rated 0, 2 rated 2 and 1 rated 3). No side effects occurred in any of the patients. These data, at least support the hypothesis that CBD could be used as an adjunctive drug to current drug treatment. Due to the small sample size, further clinical trials are warranted.

BRAIN DAMAGE FROM HEAD INJURY, EXCITOTOXINS, ISCHEMIA, INFECTION AND POISON

A nonpsychotropic cannabinoid, HU-211, has cerebroprotective effects. HU-211 (Dronabinol) is a synthetic, nonpsychotropic cannabinoid, which has been shown to act as a noncompetitive *N*-methyl-D-aspartate (NMDA) receptor antagonist (Shohami *et al.*, 1993). These authors administered HU-211 1, 2 and 3 h after closed head injury in anesthetized rats. The drug reduced blood–brain barrier breakdown and improved motor functions on beam walking and balancing tasks. Raising NMDA and glutamate levels in the brain leads to neuronal death. Nadler *et al.* (1993) exposed cultured neurons to these neuromodulators and treated them with HU-211 which reduced cell death in a dose-related manner. The authors suggest HU-211 may be neuroprotective against the toxicity of these neuromodulators. In a similar study, the excitotoxins NMDA, quisqualate, AMPA or kainate were administered to neuronal cultures (Eshhar *et al.*, 1993). HU-211 was found to protect neurons from NMDA and quisqualate, but not from after administration of HU-211. Occlusion of the middle cerebral artery in rats was used to produce cerebral ischemia in rats by Leker *et al.* (1999). When HU-211 was administered, infarct volumes were reduced compared with control rats. Gallily *et al.* (1997) demonstrated HU-211 provided protection against septic shock after *Escherichia coli*

Table 2.7 Human Studies on the effects of cannabis and cannabinoids in epilepsy

Reference	Subjects	Drug and dose	Type of study	Results
Keeler and Reifler (1967)	1 patient with generalized epilepsy	Smoking cannabis (7 occasions over 3 weeks)	Case report	Cannabis appeared to precipitate convulsions after a 6-month fit-free period without medication. Causal relationship not clear
Consroe et al. (1975)	1 patient with generalized epilepsy poorly controlled on standard drugs	Smoking cannabis 2–5 times daily while continuing standard medication	Case report	Unsatisfactory control on standard medication; no further fits while also smoking cannabis
Cunha et al. (1980)	15 patients with generalized epilepsy poorly controlled on standard drugs; 16 healthy volunteers	Oral cannabidiol 200–300 mg/day to standard therapy in patients. Oral cannabidiol 3 mg/kg/day vs. placebo in controls. Treatment for 4.5 months in both patients and controls	Double blind, placebo controlled	Cannabidiol improved control in 7 of 8 patients; 1 of 7 patients improved on placebo. Somnolence reported by 4 patients on cannabidiol. No psychotropic or neurological effects of cannabidiol noted in controls
Ames and Cridland (1986)	12 epileptic patients not controlled on standard drugs	Oral cannabidiol 200–300 mg/day for 4 weeks	Double blind, placebo controlled	No significant effect of cannabidiol on seizure frequency
Trembly and Sherman (1990) (Conference report cited by Consroe and Sandyk 1992)	10 patients with generalized, focal or complex partial epilepsy poorly controlled on standard drugs. 1 patient with epilepsy	Oral cannabidiol 300 mg/day for 6 months in addition to standard drugs. Oral cannabidiol 900–1200 mg/day for 10 months	Double blind, placebo controlled. Open trial	No effect of cannabidiol on seizure pattern or frequency. Reduction of seizure frequency while on cannabidiol
Grinspoon and Bakalar (1993)	1 patient with complex partial seizures; 1 patient with generalized absence attacks	Smoking cannabis	Anecdotal reports	Cannabis appeared to alleviate seizures in both patients

British Medical Association (1997) © Overseas Publishers Association N.V., with permission from Gordon and Breach Publishers.

055:B5 inoculation by decreasing lethality in mice. *In vitro*, HU-211 induced suppression of tumor necrosis factor alpha and nitric oxide production by both mouse and rat macrophages after *Escherichia coli* 055:B5 exposure. Brain damage may be affected by reactive oxygen species (ROS). Shohami *et al.* (1997) reported HU-211 acts as an antioxidant and may reduce brain damage through this mechanism. In a study of soman induced seizures in the rat, HU-211 had no effect, but reduced median lesion volume 86% when administered 5 min and 81.5% when administered 40 min postsoman. Histological examination showed less pyriform cortex damage (Filbert *et al.*, 1999). In sum, these animal studies suggest the HU-211 has properties which seem to reduce brain inflammation from closed head injury, endogenous excitotoxins, cerebral ischemia, bacterial infection and neurotoxic poisons.

Finally, Striem *et al.* (1997) found HU-211 interacts with the dopaminergic system, in addition to its NMDA antagonist effects and anti-oxidant effects. In vitro HU-211, enhanced the conversion of [3H]adenine to cyclic AMP. In catalepsy tests, HU-211 decreased the catalepsy induced by D1, D2 and non-selective dopamine receptor antagonists. Further research on the interactions with the dopaminergic system are needed to establish what role HU-211 might have.

Human studies

Pharmos (Huggett, 2001) has completed phase I and Phase II clinical trials of HU-211 in patients with traumatic head injury. The phase II trial was a double blind, placebo controlled, randomized study of 101 patients. Although they reported data analysis is not final, they found significant reduction in intracranial pressure and maintenance of systolic blood pressure, as well as a greater number of patients able to resume normal life.

CONCLUSION

This review has suggested that cannabinoids may be useful in the treatment of many disorders. The most likely applications for cannabinoid agonists are for the treatment of loss of appetite, pain, anxiety, vomiting, nausea and epilepsy. The most likely applications for cannabinoid antagonists may be for anxiety, schizophrenia, spasticity, and dystonia. It is difficult to formulate an hypothesis concerning the potential treatment of depression, bipolar disorder and alcohol dependence since very little work has been done with these disorders at this point of time.

Another possibility is that mixtures of agonists and antagonists might be useful since it has been demonstrated that CBD attenuates the psychoactivity of THC (Karniol and Carlini, 1973; Karniol *et al.*, 1974). At present, GW Pharmacueticals has launched a series of animal and human research studies, using extracts from cloned cannabis plants which have different amounts of cannabinoids in them to test this hypothesis using a sublingual route of administration.

In addition, the recent Institute of Medicine report (Watson *et al.*, 2000) recommended:

> Clinical trials of marijuana use for medical purposes should be conducted under the following limited circumstances: trials should involve only short-term use

(less than six months), should be conducted in patients with conditions for which there is reasonable expectation of efficacy, should be approved by institutional review boards, and should collect data about efficacy.

Research with smoked marijuana could reveal differential effects of the multiple cannabinoids in the plant as compared with the effects of testing pure compounds. At present, however, it is unlikely this type of research will be done, since a supply of cannabis with different ratios of cannabinoids is not readily available to the research community even though cloned plant extracts have been developed.

Since the discovery of the CB_1 and CB_2 receptors and endogenous cannabinoid ligands, the potential for development of synthetic drugs seems possible. Several drugs seem to have some potential and are under intensive study, as discussed above: the CB_1 and CB_2 receptor anatagonist SR141716 (Sanofi-Synthelabo), HU-211 (Pharmos) and CT-3 (Atlantic Pharmaceuticals).

ACKNOWLEDGEMENT

This work was supported by an individual project fellowship from the Open Society Institute.

REFERENCES

Abrahamov, A. and Mechoulam, R. (1995) "An efficient new cannabinoid antiemetic in pediatric onocology," in *The 1994 International Symposium on Cannabis and the Cannabinoids. Life Sciences*, edited by R. E. Musty, P. Reggi, P. Consroe and A. Makriynnis, **56**: 22–24, 1931–1932.

Achiron, A., Miron, S., Lavie, V., Margalit, R. and Biegon, A. (2000) "Dexanabinol (HU-211) effect on experimental autoimmune encephalomyelitis: implications for the treatment of acute relapses of multiple sclerosis," *Journal of Neuroimmunology* **102**: 26–31.

Ahmedzai, S., Carlyle, D. L., Calder, I. T. and Moran, F. (1983) "Anti-emetic efficacy and toxicity of nabilone, a synthetic cannabinoid, in lung cancer chemotherapy," *British Journal of Cancer* **48**: 657–663.

Ames, F. R. and Cridland, S. (1986) "Anticonvulsant effect of cannabidiol," *South African Medical Journal* **69**: 14.

Baker, D., Pryce, G., Croxford, J. L., Brown, P., Pertwee, R. G., Huffmann, J. W. *et al.* (2000) "Cannabinoids control spasticity and tremor in a multiple sclerosis model," *Nature* **404**: 84–87.

Barg, J., Fride, E., Hanus, L., Levy, R., Matus-Leibovitch, N., Heldman, E. *et al.* (1995) "Cannabinomimetic behavioral effects of and adenylate cyclase inhibition by two new endogenous anandamides," *European Journal of Pharmacology* **287**: 145–52.

Beal, J. A., Olson, R., Laubenstein, L., Morales, J. O., Bellman, P., Yangco, B. *et al.* (1995) "Dronabinol as a treatment for anorexia associated with weight loss in patients with AIDS," *Journal of Pain and Symptom Management* **10**: 89–97.

Bicher, H. I. and Mechoulam, R. (1968) "Pharmacological effects of two active constituents of marijuana constituents," *Archieves International Pharmacodynamics* **172**: 24–31.

Chan, H. S. L., Correia, J. A. and MacLeod, S. M. (1987) "Nabilone versus prochlorperazine for control of cancer chemotherapy-induced emeisis in children: a double-blind crossover trial," *Pediatrics* **70**: 946–952.

Chang, A. E., Schiling, D. J., Stillman, R. C., Godberg, N. H., Seipp, C. A., Barofsky, I. *et al.* (1979) "Delta-9-THC as an antiemetic in cancer patients receiving high dose methotrexate," *Annals of Internal Medicine* **91**: 819–830.

Chang, A. E., Shiling, D. J., Stillman, R. C., Goldberg, N. H., Seipp, C. A., Barofsky, I. *et al.* (1981) "A prospective evaluation of delta-9-tetrahydrocannabinol as an antiemetic in patients receiving adriamycin and cytoxan chemotherapy," *Cancer* **47**: 1746–1751.

Chichewizc, D. L. and Welch, S. (1999) *Symposium on the Cannabinoids*, Burlington, VT: International Cannabinoid Research Society, 66.

Christison (1851) "On the physical and medicinal qualities of Indian hemp," *Mon Journal of Medicinal Sciences* **13**, 26 & 177.

Clifford, D. B. (1983) "Tetrahydrocannabinol for tremor in multiple sclerosis," *Annals of Neurology* **13**: 669–671.

Consroe, P. and Snider, S. R. (1986) "Therapeutic potential of cannabinoids in neurological disorders," in *Cannabinoids as Therapeutic Agents*, edited by R. Mechoulam, pp. 21–49. Boca Raton, FL: CRC Press Inc.

Consroe, P., Laguna, J., Allender, J., Snider, S., Stern, L., Sandyk, R. *et al.* (1991) "Controlled clinical trial of cannabidiol in Huntington's Disease," *Pharmacology, Biochemistry & Behaviour* **40**: 701–708.

Consroe, P., Musty, R. and Conti, L. H. (1988) "Effects of cannabidiol in animal models of neurologic dysfunction," in *Marihuana: An international research report*, edited by G. Chesher, P. Consroe and R. E. Musty, pp. 141–146. Canberra: Australian Government Publishing Service.

Consroe, P., Musty, R., Rein, J., Tillery, W. and Pertwee, R. (1997) "Perceived effects of cannabis smoking in patients with multiple sclerosis," *European Neurology* **38**: 44–48.

Consroe, P. and Sandyk, R. (1992) "Potential role of cannabinoids for therapy of neurological disorders," in *Marijuana/Cannabinoids: Neurobiology and Neurophysiology*, edited by A. Bartke and L. Murphy, pp. 459–524. Boca Raton, FL: CRC Press Inc.

Consroe, P., Tillery, W., Rein, J. and Musty, R. E. (1998) "Reported marijuana effects in patients with spinal cord injury," *1998 Symposium on the Cannabinoids*, Burlington, VT: International Cannabinoid Research Society, 64.

Consroe, P. F., Wood, G. C. and Buchsbaum, H. (1975) "Anticonvulsant nature of marijuana smoking," *JAMA* **235**: 306–307.

Conti, L. H., Johanessen, J., Musty, R. E. and Consroe, P. (1988) "Anti-dyskinetic effects of cannabidiol," in *Marihuana: An international research report*, edited by G. Chesher, P. Consroe, and R. E. Musty, pp. 153–156. Cannberra: Australian Government Publishing Service.

Cunha, J. M., Carlini, E. A., Pereira, A. E., Ramos, O. L., Pimentel, C., Gagliardi, R. *et al.* (1980) "Chronic administration of cannabidiol to healthy volunteers and epileptic patients," *Pharmacology* **21(3)**: 175–185.

Dalzell, A. M., Bartlett, H. and Lilleyman, J. S. (1986) "Nabilone: an alternative antiemetic for cancer chemotherapy," *Archives of Diseases of Childhood* **B9**: 1314–1319.

Dajani, E. Z., Larsen, K. R., Taylor, J., Dajani, N. E., Shahwan, T. G., Neeleman, S. D. *et al.* (1999) "A novel, orally effective cannabinoid with analgesic and anti-inflammatory properties," *The Journal of Pharmacology and Experimental Therapeutics* **291**: 31–38.

Davies, C. (1992) "Drug dealers saved my wife from her MS hell," *The Mail on Sunday*, 15th November.

Doyle, C. (1992) "High, dry and happier," *The Daily Telegraph*, 24th November.

Dunn, M. and Davis, R. (1974) "The perceived effects of marijuana on spinal cord injured males," *Paraplegia* **12**: 175.

Einhorn, L. H., Nagy, C., Furnas, B. and Williams, S. D. (1981) "Nabilone: an effective antiemetic in patients receiving cancer chemotherapy," *Journal of Clinical Pharmacology* **21**: 377s–382s.

Edsall, S. A., Knapp, R. J., Vanderah, T. W., Roeske, W. R., Consroe, P. and Yamamura, H. (1996) "Antisense oligodeoxynucleotide treatment to the brain cannabinoid receptor inhibits antinociception," *Neuroreport* **7**: 593–596.

Eshhar, N., Striem, S. and Biegon, A. (1993) "HU-211, a non-psychotropic cannabinoid, rescues cortical neurones from excitatory amino acid toxicity in culture," *Neuroreport* **13**: 237–240.

Ferriman, A. (1993) "Marihuana: the best medicine?" *The Times*, 4th May 1993.

Filbert, M. G., Forster, J. S., Smith, C. D. and Ballough, G. P. (1999) "Neuroprotective effects of HU-211 on brain damage resulting from soman-induced seizures," *Annuals of the New York Academy of Sciences* **890**: 505–514.

Frankel, J. P., Hughes, A., Lees, A. J. and Stern, G. M. (1990) "Marijuana for Parkinsonian tremor," *Journal of Neurology, Neurosurgery and Psychiatry* **53**: 436.

Fride, E. and Mechoulam, R. (1993) "Pharmacological activity of the cannabinoid receptor agonist, anandamide, a brain constituent," *European Journal of Pharmacology* **231**: 313–314.

Frytak, S., Moertel, C. G., O'Fallon, J. R., Rubin, J., Creagan, E. T., O'Connell, M. J. *et al.* (1979) "Delta-9-tetrahydrocannabinol as an antiemetic for patients receiving cancer chemotherapy. A comparison with prochlorperazine and a placebo," *Annals of Internal Medicine* **91**: 825–830.

Gallily, R., Yamin, A., Waksmann, Y., Ovadia, H., Weidenfeld, J., Bar-Joseph, A. *et al.* (1997) "Protection against septic shock and suppression of tumor necrosis factor alpha and nitric oxide production by dexanabinol (HU-211), a nonpsychotropic cannabinoid," *The Journal of Pharmacology and Experimental Therapeutics* **283**: 918–924.

Graceffo, T. J. and Robinson, J. K. (1998) "Delta-9-tetrahydrocannabinol (THC) fails to stimulate consumption of a highly palatable food in the rat," *Life Sciences* **62**: 85–88.

Gralla, R. J., Tyson, L. B., Clark, R. A., Bordin, L. A., Kelsen, D. P. and Kalman, L. B. (1982) "Antiemetic trials with high dose metoclopramide: superiority over THC, and preservartion of efficacy in subsequent chemotherapy courses," *Proceedings of ASCO Meeting*, C-222.

Greenberg, H. S., Werness, S. A., Pugh, J. E., Andrus, R. O., Anderson, D. J. and Domino, E. F. (1994) "Short-term effects of smoking marijuana on balance in patients with multiple sclerosis and normal volunteers," *Clinical Pharmacology & Therapeutics* **55**: 324–328.

Grinspoon, L. and Bakalar, J. B. (1997) *Marijuana the Forbidden Medicine*, pp. 150–162. New Haven: Yale University Press.

Grinspoon, L. and Bakalar, J. B. (1993) *Marihuana, the Forbidden Medicine*. New Haven and London: Yale University Press.

Grinspoon, L. and Bakalar, J. B. (1998) "The use of cannabis as a mood stabilizer in bipolar disorder: anecdotal evidence and the need for clinical research," *Psychoactive Drugs* **30(2)**: 171–177.

Gross, H., Egbert, M. H., Faden, V. B., Goldberg, S. C., Kaye, W. H., Caine, E. D. *et al.* (1983) "A double-blind trial of delta-9-THC in primary anorexia nervosa," *Journal of Clinical Psychopharmacology* **3**: 165–171.

Guimaraes, F. S., Chiaretti, T. M., Graeff, F. G. and Zuardi, A. W. (1990) "Antianxiety effect of cannabidiol in the elevated plus-maze," *Psychopharmacology* **100**: 558–559.

Handscombe, M. (1993) "Cannabis: why doctors want it to be legal," *The Independent*, 23rd February.

Herman, T. S., Einhorn, L. H., Jones, S. E., Nagy, C., Chester, A. B., Dean, J. C. *et al.* (1979) "Superiority of nabilone over prochlorperazine as an antiemetic in patients receiving cancer chemotherapy," *New England Journal of Medicine* **300**: 1295–1297.

Hodges, C. (1992) *Very alternative medicine*. The Spectator, 1st August.

Hodges, C. (1993) *I wish I could get it at the chemist's*. The Independent, 23rd February.

Houser, S. J., Eads, M., Embrey, J. P. and Welch, S. P. (2000) "Dynorphin B and spinal analgesia: induction of antinociception by the cannabinoids CP55,940, Delta(9)-THC and anandamide," *Brain Research* **857**: 337–342.

Huggett (2001) "Pharmos starts phase III study of dexanabind in brain injury," *Bioworld Today* **12**.

Jain, A. K., Ryan, J. R., McMahon, F. G. and Smith, G. (1981) "Evaluation of intramuscular levonantradol and placebo in acute postoperative pain," *Journal of Clinical Pharmacology* **21**: 320S–326S.

James, T. (1993) "Breaking the law to beat MS," *The Yorkshire Post*, 27th September.

Johansson, R., Kikku, P. and Groenroos, M. (1982) "A double-blind, controlled trial of nabilone vs prochlorperazine for refractory emesis induced by cancer chemotherapy," *Cancer Treatment Reviews* **9** (Supplement B): 25–33.

Johnson, M. R. and Melvin, L. S. (1986) "The discovery of nonclassical cannabinoid analgetics," in *Cannabinoids as Therapeutic Agents*, edited by R. Mechoulam, pp. 121–145. Boca Raton, FL: CRC Press.

Jones, S. E., Durant, J. R., Greco, F. A., and Robertone, A. (1982) "A multi-institutional phase III study of nabilone vs placebo in chemotherapy-induced nausea and vomiting," *Cancer Treatment Review* **9**: 45s–48s.

Karniol, I. G. and Carlini, E. A. (1973) "Pharmacological interaction between cannabidiol and delta 9-tetrahydrocannabinol," *Psychopharmacologia* **33**: 53–70.

Karniol, I. G., Shirakawa, I., Kasinski, N. *et al.* (1974) "Cannabidiol interfers with the effects of delta 9-tetrahydrocannabinol in man," *European Journal of Pharmacology* **28**: 172–177.

Keeler, M. H. and Reifler, C. B. (1967) "Grand mal convulsions subsequent to marihuana use," *Diseases of the Nervous System* **18**: 474–475.

Ko, M. C. and Woods, J. H. (1999) "Local administration of delta-9-tetrahydrocannabinol attenuates capsaicin-induced thermal nociception in rhesus monkeys: A peripheral cannabinoid action," *Psychopharmacology (Berliner)* **143(3)**: 322–326.

Kotin, J., Post, R. M. and Goodwin, F. K. (1973) "Delta-9-tetrahydrocannabinol in depressed patients," *Archives of General Psychiatry* **3**: 345–348.

Lane, M., Vogel, C. L. and Ferguson, J. (1991) "Dronabinol and prochlorperazine in combination are better than either agent alone for treatment of chemotherapy-induced nausea and vomiting," *Proceedings of the American Society of Clinical Oncologists* **8**: 326.

Leite, J. R., Carlini, E. A., Lander, N. and Mechoulam, R. (1982) "Anti-convulsant effects of the (–) and + isomers of cannabdiol and their dimethylheptyl homologs," *Pharmacology* **24**: 141–146.

Leker, R. R., Shohami, E., Abramsky, O. and Ovadia, H. (1999) "Dexanabinol; a novel neuroprotective drug in experimental focal cerebral ischemia," *Journal of Neurology and Sciences* **162**: 114–119.

Levitt, M. (1982) "Nabilone vs placebo in the treatment of chemotherapy-induced nausea and vomiting in cancer patients," *Cancer Treatment Reviews* **9** (Supplement B): 49–53.

Lindstrom, P., Lindblom, U. and Boreus, L. (1987) "Lack of effect of cannabidiol in sustained neuropathia," Paper presented at *Marijuana '87 International Conference on Cannabis*, Melbourne, September 2 to 4: 1987 (cited by Consroe & Sandyk, 1992).

Lucas, V. S. Jr. and Lazlo, J. (1980) "9-THC for refractory vomiting induced by cancer chemotherapy," *Journal of the American Medical Association* **243**: 1241–1243.

Lyman, W. D. (1991) "Drugs of abuse and experimental autoimmune diseases," in *Drugs of Abuse, Immunity and Immunodeficiency*, edited by H. Friedman, S. Specter, and T. W. Klein, pp. 81–92. New York, NY: Plenum Press.

Malec, J., Harvey, R. F. and Cayner, J. J. (1982) "Cannabis effect on spasticity in spinal cord injury," *Archives of Physical and Medical Rehabilitation* **63**: 116–118.

Martyn, C. N., Illis, L. S. and Thom, J. (1995) "Nabilone in the treatment of multiple sclerosis," *Lancet* **345**: 579.

Martin, W. J., Loo, C. M. and Basbaum, A. I. (1999) "Spinal cannabinoids are anti-allodynic in rats with persistent inflammation," *Pain* **82**: 199–205.

Mattes, R. D., Engelman, K., Shaw, L. M. and Elsohly, M. A. (1994) "Cannabinoids and appetite stimulation," *Pharmacology Biochemistry and Behavior* **49**: 187–195.

Maurer, M., Henn, V., Dittrich, A. and Hoffmann, A. (1990) "Delta-9-tetrahydrocannabinol shows antispastic and analgesic effects in a single case double-blind trial," *European Archives of Psychiatry and Clinical Neuroscience* **240**: 1–4.

McLaughlin, C. L., Baile, C. A. and Bender, P. E. (1979) "Cannabinols and feeding in sheep," *Psychopharmacology (Berliner)*, **64(3)**: 321–323.

Meinck, H. M., Schonle, P. W. and Conrad, B. (1989) "Effects of cannabinoids on spasticity and ataxia in multiple sclerosis," *Journal of Neurology* **236**: 120–122.

Mikuriya, T. H. (1999) "Clinical Report: Medical Uses of Cannabis in California," *1999 Symposium on the Cannabinoids*, Burlington, VT, International Cannabinoid Research Society 89.

Musty, R. E. (1984) "Possible anxiolytic effects of cannabidiol," in *The Cannabinoids*, edited by S. Agurell, W. Dewey and R.Willette, pp. 829–844. New York: Academic Press.

Musty, R. E. (1988) "Individual differences as predictors of marihuana phenomenology," in *Marihuana: An International Research Report*, edited by G. Chesher, P. Consroe, and R. E. Musty, pp. 201–207. Canberra: Australian Government Publishing Service.

Musty, R. E. (moderator), MacDonald, (Mac) J., Christie, N. E., Buckley, S. I., Tod, M., Roger, G. *et al.* (1998) "Medical uses of marijuana: a debate," in *HMS Beagle: The BioMed-Net Magazine* (http://hmsbeagle.com/1997/01/cutedge/overview.htm), May 15, Issue 30.

Musty, R. E., Conti, L. H. and Mechoulam, R. (1985) "Anxiolytic properties of cannabidiol," In *Marihuana* **84**: edited by D. Harvey, pp. 713–719.

Musty, R. E., Deyo, R. A., Baer, J. L., Darrow, S. M. and Coleman, B. (2000) "Effects of SR141716 on animal models of schizophrenia," *Symposium on the Cannabinoids*, Burlington, VT: International Cannabinoid Research Society (in press).

Nadler, V., Mechoulam, R. and Sokolovsky, M. (1993) "The non-psychotropic cannabinoid (+)-(3S,4S)- 7-hydroxy-delta 6-tetrahydrocannabinol 1,1-dimethylheptyl (HU-211) attenuates N-methyl-D-aspartate receptor-mediated neurotoxicity in primary cultures of rat forebrain," *Neuroscience Letters* **162**: 43–45.

Nagy, C. M., Furnas, B. E., Einhorn, L. H. and Bond, W. H. (1978) "Nabilone: antiemetic crossover study in cancer chemotherapy patients," *Proceedings of the American Society for Cancer research* **19**: 30.

Negrete, J. C., Knapp, W. P., Douglas, D. E. and Smith, W. B. (1986) "Cannabis affects the severity of schizophrenic symptoms: results of a clinical survey," *Psychological Medicine*, **16(3)**: 515–520.

Nelson, K., Walsh, D., Deeter, P. and Sheehan, F. (1994) "A phase II study of delta-9-tetrahydrocannabinol for appetite stimulation in cancer-associated anorexia," *Journal of Palliative Care* **10**: 14–18.

Niederle, N., Schutte, J. and Schmidt, C. G. (1986) "Cross over comparison of the antiemetic efficacy of nabilone an alizapride in patients with nonseminomatous testicular cancer receiving cisplatin therapy," *Klinische Wochenschrigt* **64**: 362–365.

Niedhart, J., Gagen, M., Wilson, H., and Young, D. (1981) "Comparative trial of the antiemetic effects of THC and haloperidol," *Journal of Clinical Pharmacology* **21**: 38s.

Niiranen, A. and Mattson, K. (1985) "A cross comparison of nabilone and prochlorperazine for emesis induced by cancer chemotherapy," *American Journal of Clinical Oncology* **8**: 336–340.

Noyes, R., Brunk, S. F., Baram, D. A. and Baram, A. (1975a) "Analgesic effect of delta-9-tetrahydrocannabinol," *Journal of Clinical Pharmacology* **15**: 139–143.

Noyes, R., Brunk, S. F., Baram, D. A. and Canter, A. (1975b) "The analgesic properties of delta-9-THC and codeine," *Clinical Pharmacology & Therapeutics* **18**: 84–89.

Onaivi, E. S., Babatunde, E. A. and Chakrabarti, A. (1998) "Cannabinoid (CB$_1$) receptor antaginism induces anxiolysis," *1998 Symposium on the Cannabinoids*, Burlington, VT: International Cannabinoid Research Society, 58.

Orr, L. E. and McKernan, J. F. (1981) "Antiemetic effect of delta-9-THC in chemotherapy-associated nausea and emiesis as compared to placebo and compazine," *Journal of Clinical Pharmacology* **21**: 76–80.

Peralta, V. and Cuesta, M. J. (1992) "Influence of cannabis abuse on schizophrenic psychopathology," *Acta Psychiatrica Scandinavica* **85(2)**: 127–130.

Pertwee, R. G. (1987) "The central neuropharmacology of psychotropic cannabinoids," *Pharmacology Therapeutics* **36**: 189–261.

Pertwee, R. G. (1997) "Pharmacology of cannabinoid CB$_1$ and CB$_2$ receptors," *Pharmacology Therapeutics* **74**: 129–180.

Petitet, F., Jeantaud, B., Reibaud, M., Imperato, A. and Dubroeucq, M. C. (1998) "Complex pharmacology of natural cannabinoids: evidence for partial agonist activity of delta-9-tetrahydrocannabinol and antagonist activity of cannabidiol on rat brain cannabinoid receptors," *Life Science* **63(1)**: PL1–6.

Petro, D. J. (1980) "Marijuana as a therapeutic agent for mucsle spasm or spasticity," *Psychosomatics* **21**: 81–85.

Petro, D. J. and Ellenberger, C. (1981) "Treatment of human spasticity with Δ^9-tetrahydrocannabinol," *Journal of Clinical Pharmacology* **21**: 413S–416S.

Plasse, T. F., Gorter, R. W., Krasnow, S. H., Lane, M., Shepard, K. V. and Wadleigh, R. G. (1991) "Recent clinical experience with dronabinol," *Pharmacology Biochemistry & Behavior* **40**: 695–700.

Pomeroy, M., Fennelly, J. J. and Towers, M. (1986) "Prospective randomized double-blind trial of nabilone versus domperidone in the treatment of cytotoxic-induced emesis," *Cancer Chemotherapy and Pharmacology* **17**: 285–288.

Raft, D., Gregg, J., Ghia, J. and Harris, L. (1977) "Effects of intravenous tetrahydrocannabinol on experimental and surgical pain," Psychological correlates of the analgesic response. *Clinical Pharmacology and Therapeutics* **21**: 26–33.

Regelson, W., Butler, J. R., Schultz, J., Kirk, T., Peek, L., Green, M. L. and Zakis, O. (1976) "Δ^9-tetrahydrocannabinol as an effective antidepressant and appetite stimulating agent in advanced cancer patients," in *Pharmacology of Marijuana*, edited by M. C. Braude and S. Szara, pp. 763–776, New York: Raven Press.

Richter, A. and Loscher, W. (1994) "(+)-WIN 55, 212–2, a novel cannabinoid receptor agonist, exerts antidystonic effects in mutant dystonic hamsters," *European Journal Pharmacology* **264**: 371–377.

Sallan, S., Zinberg, N. and Frei, E. III. (1975) "Antiemetic effect of delta-9-tetrahydrocannabinol in patients receiving cancer chemotherapy," *New England Journal of Medicine* **293**: 795–797.

Sandyk, R. and Awerbuch, G. (1988) "Marijuana and Tourette's syndrome," *Journal of Clinical Psychopharmacology* **8**: 444.

Schnelle, M., Grotenhermen, F., Reif, M. and Gorter, R. W. (1999) "Ergebnisse einer standardisierten Umfrage zur medizinischen Verwendung von Cannabisprodukten im deutschen Sprachraum, [Results of a standardized survey on the medical use of cannabis products in the German-speaking area]," *Forschende Komplementarmedizin und Klassische Naturheikunde* **3**: 28–36.

Shohami, E., Novikov, M. and Mechoulam, R. (1993) "A nonpsychotropic cannabinoid, HU-211, has cerebroprotective effects after closed head injury in the rat," *Journal of Neurotrauma* **10**: 109–119.

Shohami, E., Beit-Yannai, E., Horowitz, M. and Kohen, R. (1997) "Oxidative stress in closed-head injury: brain antioxidant capacity as an indicator of functional outcome," *Journal of Cerebral Blood Flow and Metabiolism* **10**: 1007–1019.

Segal, M. (1986) "Cannabinoids and analgesia," in *Cannabinoids as Therapeutic Agents*, edited by R. Mechoulam, pp. 106–120. Boca Raton, FL: CRC Press Inc.

Sofia, R .D., Vassar, H. B. and Knobloch, L. C. (1975) "Comparative analgesic activity of various naturally occurring cannabinoids in mice and rats," *Psychopharmacologia* **40(4)**: 285–295.

Striem, S., Bar-Joseph, A., Berkovitch, Y. and Biegon, A. (1997) "Interaction of dexanabinol (HU-211), a novel NMDA receptor antagonist, with the dopaminergic system," *European Journal of Pharmacology* **338**: 205–213.

Trembley, B. and Sherman, M. (1990) "Double-blind clinical study of cannabidiol as a secondary anticonvulsant," Paper presented at *Marijuana '90 International Conference on Cannabis and Cannabiniods*, Kolympari (Crete), July 8 to 11 (cited by Consroe & Sandyk, 1992).

Ungerleider, J. T., Andrysiak, T., Fairbanks, L., Goonight, J., Sarna, G., and Jamison, K. (1982) "Cannabis and cancer chemotherapy. A comparison of oral delta-9-THC and prochlorperazine," *Cancer* **50**: 636–645.

Ungerleider, J. T., Andrysiak, T., Fairbanks, L., Ellison, G. W. and Myers, L. W. (1988) "Delta-9-tetrahydrocannabinol in the treatment of spasticity associated with multiple sclerosis," *Advances in Alcoholism and Substance Abuse* **7**: 39–50.

Wada, J. K., Bogdon, D. L., Gunnell, J. C., Hum, G. J., Gota, C. H. and Rieth, T. E. (1982) "Double-blind, randomized, crossover trial of nabilone vs placebo in cancer chemotherapy," *Cancer Treatment Review* **9** (Supplement B): 39–44.

Walker, J. M., Huang, S. M., Strangman, N. M., Tsou, K. and Sanudo-Pena, M. C. (1999) Pain modulation by release of the endogenous cannabinoid anandamide. *Proceedings of the National Academy of Sciences of the USA* **96** 12198–12203.

Warner, R., Taylor, D., Wright, J., Sloat, A., Springett, G., Arnold, S. *et al.* (1994) "Substance use among the mentally ill: prevalence, reasons for use, and effects on illness," *American Journal of Orthopsychiatry* **64(1)**: 30–39.

Watson, S. J., Benson, J. A. and Joy, J. E. (2000) "Marijuana and medicine: assessing the science base – a summary of the 1999 Institute of Medicine Report," *Archieves of General Psychiatry* **57**: 5786–5790.

Welch, S. P. and Stevens, D. L. (1992) "Antinociceptive activity of intrathecally administered cannabinoids alone, and in combination with morphine, in mice," *Journal of Pharmacology and Experimental Therapeutics* **262**: 10–18.

Willette, R. (1984) *The Cannabinoids*, pp. 829–844. New York: Academic Press.

Wilson, R. S. and May, E. L. (1975) "9-nor-Δ^8-tetrahydrocannabinol, a cannabinoid of metabolic interest," *Journal of Medicinal Chemistry* **17**: 475–476.

Wilson, R. S., May, E. L., Martin, B. R. and Dewey, W. L. (1976) "Analgesic properties of the tetrahydocannabinols. Synthesis, some behavioral and analgesic comparisons with the tetrahyrocannabinols," *Journal of Medicinal Chemistry* **19**: 1165.

Wood, H.C., Remington, J. P. and Satdler, S. P. (1892) *The dispensatory of the United States of America*, pp. 348–351. Philadelphia: Lippincott.

Zuardi, A. W., Shirakawa, I., Finkelfarb, E. and Karniol, I. G. (1982) "Action of cannabidiol on the anxiety and other effects produced by delta 9-THC in normal subjects," *Psychopharmacology (Berliner)* **76(3)**: 245–250.

Zuardi, A. W., Rodrigues, J. A. and Cunha, J. M. (1991) "Effects of cannabidiol in animal models predictive of antipsychotic activity," *Psychopharmacology* **104**: 260–264.

Zuardi, A. W., Cosme, R. A., Graeff, F. G. and Guimaraes, F. S. (1993) "Effects of ipsapirone and cannabidiol on human experimental anxiety," *Journal of Psychopharmacology* **7**: 82–88.

Zuardi, A. W., Morais, S. L., Guimaraes, F. S. and Mechoulam, R. (1995) "Antipsychotic effect of cannabidiol," *Journal of Clinical Psychiatry* **56**: 485–486.

Chapter 3

Marijuana addiction and CNS reward-related events

Eliot L. Gardner

ABSTRACT

The reward substrates of the central nervous system (CNS) consist of: (1) a core dopaminergic/enkephalinergic neural system synaptically interconnecting the ventral tegmental area, nucleus accumbens, and ventral pallidum, and which appears to mediate reinforcement; (2) a glutamatergic neural network originating in the frontal cortex and deep temporal lobe, which feeds into the core dopaminergic/enkephalinergic system and which appears to mediate aspects of reward-related incentive motivation; and (3) additional neural inputs – which use many different neurotransmitters, including 5-hydroxytryptamine (serotonin), gamma-aminobutyric acid (GABA), and dynorphin – into the core dopaminergic/enkephalinergic system, which appear to regulate additional aspects of reward. These complex and inter-related systems are strongly implicated in drug addiction, and in such addiction-related phenomena as withdrawal dysphoria and craving. These systems are also implicated in the pleasures produced by such natural rewards as food and sex. On the basis of more than 15 years of work, cannabinoids are now known to activate these CNS substrates and influence reward-related behaviors. From these actions, presumably, derive both the addictive potential of cannabinoids and possible clinical benefit in mood disorders such as depression.

Key Words: addiction, brain stimulation, cannabinoid, cannabis, central nervous system, CNS, dependence, dopamine, drug abuse, ICSS, marijuana, medical forebrain bundle, microdialysis, nucleus accumbens, place preference, reinforcement, reward, self-administration, ventral tegmental area

INTRODUCTION

At present, the field of cannabinoid pharmacology is witnessing extraordinary advances. None of these advances are more dramatic than those relating to our understandings of cannabinoid action on the central nervous system (CNS) and behavior (Murphy and Bartke, 1992; Pertwee, 1995; Mechoulam *et al.*, 1996; Felder and Glass, 1998; Piomelli *et al.*, 1998). In the CNS, cannabinoids act through a specific G protein-coupled cannabinoid receptor (the CB_1 receptor) (Howlett *et al.*, 1990; Martin *et al.*, 1994) which has been characterized at the molecular level (Matsuda *et al.*, 1990). Like most other neurotransmitter or neuromodulator receptors, this receptor shows a heterogeneous distribution in the CNS, but appears to be especially dense in the basal ganglia, hippocampus, and cerebellum (Herkenham *et al.*, 1991a). Of relevance to CNS reward substrates (see below), the CB_1 receptor is found in moderate density in the ventral striatum, including the nucleus accumbens (Moldrich and Wenger, 2000), and in moderate-to-high density in the ventral mesencephalon. As noted by Childers and Breivogel (1998), cannabinoid receptor density is extremely high compared to other G protein-coupled receptors in the CNS, so even areas with relatively low levels of cannabinoid receptors may have relatively high levels compared to other G protein-coupled receptors. A large number of CB_1 receptor agonists have been developed (Duane Sofia, 1978; Makriyannis, 1993; Melvin *et al.*, 1993, 1995; Eissenstat *et al.*, 1995; Ryan *et al.*, 1995; Tius *et al.*, 1995; Xie *et al.*, 1995, 1996; Bloom *et al.*, 1997), as well as at least one highly selective and potent CB_1 receptor antagonist (Rinaldi-Carmona *et al.*, 1995). Other compounds with CB_1 antagonist activity may also exist (see, e.g. Fernando and Pertwee, 1997; Felder *et al.*, 1998), together with at least one compound that appears to display inverse agonist properties at the CB_1 receptor (Landsman *et al.*, 1998). At least two endogenous ligands bind to cannabinoid receptors and fulfill classical requirements for identification as neurotransmitters or neuromodulators – arachidonyl ethanolamide (anandamide) (Devane *et al.*, 1992; Mechoulam *et al.*, 1996; Sugiura *et al.*, 1996) and 2-arachidonylglycerol (2-AG) (Mechoulam *et al.*, 1995, 1996; Sugiura *et al.*, 1995; Stella *et al.*, 1997). Additional anandamide-like endogenous ligands may also exist in the CNS, creating an entire class of "endocannabinoid" neurotransmitters or neuromodulators (Mechoulam *et al.*, 1994; Barg *et al.*, 1995). Non-anandamide-like endogenous cannabinoid ligands are also thought to exist in the CNS (Evans *et al.*, 1994; Childers and Breivogel, 1998).

As pointed out by Mechoulam (1986), cannabinoids constitute one of humanity's most ancient classes of psychoactive substance. Among the reasons for this long and sustained use is the fact that cannabinoids produce a dose-dependent euphoric high, which in natural cannabinoid preparations such as marijuana and hashish appears to derive from the psychoactive constituent Δ^9-tetrahydrocannabinol (Δ^9-THC). Other subjective effects may also contribute to the addictive liability of cannabinoids, but the euphorigenic property is the single pharmacological effect which cannabinoids share with other addictive drugs (Gardner, 1992, 1997, 1999, 2000; Gardner and Lowinson, 1991; Gardner and Vorel, 1998; Gardner and David, 1999). Upwards of 20 million people are current marijuana users in the United States, and approximately 50% of American teenagers have used marijuana by the time they complete 12th grade (MacCoun and Reuter, 1997). These figures would not be troubling were it not for credible evidence that, among teenagers (especially those with conduct disorder, attention deficit/hyperactivity disorder, or major depression), progression from first marijuana use to dependent marijuana use is as rapid as the progression from first tobacco use to tobacco dependence, and significantly more rapid than that for alcohol (Crowley *et al.*, 1998). Approximately, 10% of regular marijuana and hashish users, and 25% of heavy users, meet strict diagnostic criteria for drug dependence (Anthony *et al.*, 1994; Hall *et al.*, 1994), and actual physical dependence on cannabinoids is well documented (Crowley *et al.*, 1998). Most troublingly, a significant percentage of heavy cannabis users suffer lingering adverse health consequences (Hall *et al.*, 1994; Pope and Yurgelun-Todd, 1996).

ESSENTIAL COMMONALITIES OF ADDICTIVE DRUGS

Millions of chemicals are listed in such standard compendia as *Chemical Abstracts*, yet only a few score of these have addictive liability. Those with addictive liability have neither chemical nor classical pharmacological commonalities. For example, the chemical structures of opiates (e.g. heroin) do not in the least resemble those of the psychostimulants (e.g. cocaine, amphetamines), and the classical pharmacological actions of opiates (e.g. analgesia, sedation) do not in the least resemble those of the psychostimulants (e.g. arousal, locomotor activation, anxiety). In fact, for decades it was not clear that any commonalities existed amongst drugs with addictive potential. However, in recent years it has become evident that the essential commonality is a drug-induced enhancement of CNS reward functions (for reviews, see Gardner, 1997, 2000; Gardner and David, 1999), which appears to have face validity in view of the fact that most human drug addicts report that their first drug use was "to get high." The evidence for this enhancement of CNS reward substrates is several-fold: (1) almost without exception, drugs with addictive liability enhance electrical brain-stimulation reward or lower CNS reward thresholds (Wise, 1980, 1984; Kornetsky, 1985; Gardner, 1997; Wise, 1998); (2) almost without exception, drugs with addictive liability enhance basal neuronal firing and/or basal neurotransmitter release in CNS reward circuits (Di Chiara and Imperato, 1988; Wise and Rompré, 1989; Gardner, 1997; Wise, 1998); (3) laboratory animals work for microinjections of addicting drugs into CNS reward loci, but not into other CNS loci (Phillips and LePiane, 1980; Bozarth and Wise, 1981a; Goeders and Smith, 1983; Hoebel *et al.*, 1983; Goeders *et al.*, 1984; Gardner, 1997); (4) neuropharmacological blockade of CNS reward circuits markedly inhibits the rewarding properties of self-administered addictive drugs (Jönsson, 1972; Johanson *et al.*, 1976; Yokel and Wise, 1976; de Wit and Wise, 1977; Woolverton, 1986; Gardner, 2000); (5) lesions of CNS reward circuits markedly inhibit the rewarding properties of self-administered addictive drugs (Lyness *et al.*, 1979; Bozarth and Wise, 1981b; Roberts and Koob, 1982; Spyraki *et al.*, 1983; Gardner, 2000). Thus, CNS reward enhancement is the single essential commonality of drugs possessing addictive liability, and such drugs act on CNS reward substrates to produce the subjective "high" that drug users seek (for reviews, see Wise and Rompré, 1989; Gardner, 1997, 2000; Wise, 1996a; Wise, 1998; Gardner and David, 1999). Furthermore, aberrations within these CNS reward circuits appear to confer vulnerability to drug addiction and dependence (Nestler, 1993; Self and Nestler, 1995; Blum *et al.*, 1996; Nestler *et al.*, 1996; Koob and Le Moal, 1997; Kreek and Koob, 1998).

CNS REWARD SUBSTRATES

The reward substrates of the CNS consist of: (1) a core dopaminergic/enkephalinergic neural system associated with the medial forebrain bundle – which synaptically interconnects the ventral tegmental area, nucleus accumbens, and ventral pallidum, and which appears to directly mediate reinforcement; (2) a glutamatergic neural network originating in the frontal cortex and deep temporal lobe (especially the amygdala), which feeds into the core dopaminergic/enkephalinergic system and which appears to mediate aspects of reward learning and reward-related incentive motivation; and (3) additional neural inputs – which use a variety of neurotransmitters, including 5-hydroxytryptamine (serotonin), gamma-aminobutyric acid (GABA), and dynorphin – into the core dopaminergic/enkephalinergic system, which appear to regulate additional aspects of reward. These three component systems will be briefly discussed in turn. However, it must be emphasized that these are exceedingly complex neural systems, and a detailed analysis is beyond the present scope. Interested readers are referred to recent reviews by Gardner (1997, 2000) and Wise (1998).

The core dopaminergic/enkephalinergic reward system

The core reward system of the CNS consists of an "in-series" (Wise and Bozarth, 1984) set of neural circuits, interconnected with one another and running for a major portion of its length within the medial forebrain bundle (Gardner, 1997). "First-stage" reward neurons originate in the anterior bed nuclei of the medial forebrain bundle, a diffuse set of anterior ventral limbic forebrain nuclei. "First-stage" reward neurons run posteriorly within the medial forebrain bundle in a myelinated moderately fast-conducting pathway of unknown neurotransmitter type, and synapse on "second-stage" dopamine (DA) neurons in the ventral tegmental area of the ventral midbrain. "Second-stage" DA neurons run anteriorly within the medial forebrain bundle, and synapse on "third-stage" enkephalinergic neurons in the nucleus accumbens of the anterior limbic forebrain. "Third-stage" neurons run a comparatively short distance – carrying the reward signal one link farther – to the ventral pallidum. The "second-stage" DA neurons relevant to reward functions appear to constitute only a portion of the total mesoaccumbens DA projection from ventral tegmental area to nucleus accumbens. The "third-stage" pathway comprises a large set of output neurons of the nucleus accumbens. The "first-stage" reward neurons appear preferentially activated by electrical brain-stimulation reward. The "second-stage" neurons appear preferentially activated by drugs with addictive potential – enhancing CNS reward functions and producing the pleasurable "high" sought by drug addicts. The "third-stage" neurons appear critical for the expression of reward-related behaviors. The "first-stage," "second-stage," and "third-stage" components of the system are interlinked by extensive reciprocal neural interconnections. In addition to the "third-stage" enkephalinergic neurons, another nucleus accumbens output pathway – the medium spiny output neurons which use gamma-aminobutyric acid (GABA) as their neurotransmitter – may constitute yet another CNS reward final common output path, in which the critical reward event is inhibition of the GABAergic medium spiny neurons (see, e.g. Carlezon and Wise, 1996).

The glutamatergic synaptic inputs to the core reward system

The ventral tegmental area receives substantial glutamatergic inputs from both the medial prefrontal cortex (Sesack and Pickel, 1992) and amygdala (Wallace *et al.*, 1992). In quite parallel fashion, the nucleus accumbens also receives substantial glutamatergic inputs from both the medial prefrontal cortex (Tarazi and Baldessarini, 1999) and amygdala (Kelley *et al.*, 1982; Groenewegen *et al.*, 1991).

These glutamatergic inputs modulate DA within the core reward system (e.g. Floresco *et al.*, 1998) and appear crucial to the mediation of a variety of reward and reward-related functions. The glutamatergic feeds from medial prefrontal cortex appear particularly implicated in mediating the development of drug-induced sensitization (Cador *et al.*, 1999; Li *et al.*, 1999), a neural process believed by some workers to underlie the development of addictive patterns of drug use. The glutamatergic feeds from the amygdala (especially the basolateral amygdala) appear particularly implicated in mediating aspects of reward learning and the incentive motivational processes that may underlie drug craving and relapse (Cador *et al.*, 1989; Everitt *et al.*, 1989; Everitt *et al.*, 1991; Robbins and Everitt, 1992; Robbins and Everitt, 1996; Hitchcott *et al.*, 1997; Hitchcott and Phillips, 1997, 1998a, 1998b). The present author and his colleagues have recently shown that electrical stimulation of the basolateral amygdala triggers relapse to drug-taking behavior (Hayes *et al.*, 1999), and Grimm and See (2000) have elegantly shown that temporary inactivation of the basolateral amygdala by intra-amygdaloid microinjections of tetrodotoxin eliminates the relapse to drug-taking behavior that is normally triggered by environmental cues formerly associated with the drug-taking habit.

Additional synaptic inputs to the core reward system

There are several additional neural inputs to the core reward system that may modulate drug reward by modulating DA function within the core system. GABAergic efferents from the nucleus accumbens form a feedback loop to the ventral tegmental area, and nucleus accumbens medium spiny GABAergic neurons also project to other GABAergic neurons synaptically linked to both the accumbens and ventral tegmental area (Alexander and Crutcher, 1990; Kalivas *et al.*, 1993; Van Bockstaele and Pickel, 1995). Endogenous opioid peptidergic neurons also provide synaptic regulation of core mesoaccumbens DA function and of the accumbens-ventral pallidal projection (Alexander and Crutcher, 1990; Heimer and Alheid, 1991; Zahm and Brog, 1992; Kalivas *et al.*, 1993; McGinty, 1999). Both the ventral tegmental area and the nucleus accumbens also receive serotonergic inputs, and manipulation of these serotonergic inputs appears to modulate reward functions. In fact, serotonergic lesions appear to make cocaine more rewarding (Loh and Roberts, 1990). Cocaine is also rendered more rewarding by fluoxetine-induced acute enhancement of forebrain serotonergic levels, which – in turn – may inhibit serotonergic cell firing by stimulation of serotonergic autoreceptors (Chen *et al.*, 1996). Congruent with this is the observation that microinfusion of the serotonin agonist 8-OH-DPAT into the dorsal raphé nucleus – which produces autoreceptor-mediated inhibition of serotonergic cell firing – potentiates the rewarding effects of electrical brain-stimulation reward (Fletcher *et al.*, 1995). It is also now recognized that the core mesoaccumbens reward system receives a substantial modulatory cholinergic input (Oakman *et al.*, 1995), which would appear to have relevance for nicotine-induced reward.

It would be a serious misstatement to convey the impression that these CNS reward substrates are either anatomically or functionally simple. Indeed, a very large body of experimental evidence (for reviews, see Gardner, 1997, 2000) suggests that these reward-related systems are functionally heterogeneous – with some neurons encoding reward magnitude *per se* while others encode expectancy of reward, errors in reward-prediction, prioritized reward, and other more complex aspects of reward-driven learning and reward-related incentive motivation. However, it seems equally clear that one of the primary functions of those reward substrates is to compute hedonic tone and neural "payoffs," that this computation takes place in large measure within the circuits delineated above, that the "second-stage" DA component is the common site of action for addictive drugs (and is crucial to their addictive features), that drug reward *per se* and drug potentiation of electrical brain-stimulation reward have common mechanisms, and that electrical brain-stimulation reward and the pharmacological rewards of addictive drugs are habit-forming because they act in the CNS circuits that subserve more natural, biologically significant rewards (Wise, 1996b; Shizgal, 1997).

Very importantly, the "second-stage" DA component of these CNS reward substrates appears to be the crucial convergence upon which drugs with euphorigenic properties and/or addictive potential (regardless of chemical structure or pharmacological category) act to enhance neural reward functions, subjective experience of reward, and reward-related behaviors.

CANNABINOID EFFECTS ON CNS REWARD SUBSTRATES

Although cannabinoids have been claimed to be devoid of interaction with CNS reward substrates, that position is not tenable. In fact, cannabinoids interact with CNS reward substrates in a manner strikingly analogous to that of other addictive drugs.

Cannabinoid effects on electrically-induced CNS reward

Self-delivered electrical stimulation of CNS reward circuits (through surgically implanted electrodes deep in the brain) provides a very direct *in vivo* assay of drug effects on reward substrates. More than a full decade ago, the author's research group showed that Δ^9-THC enhances brain-stimulation reward in laboratory rats (Gardner *et al.*, 1988a). These experiments were carried out using a two-lever "auto-titration" threshold-measuring quantitative electrophysiological brain-stimulation technique, in which experimental animals indicate their threshold for brain-stimulation reward in terms of microamperes of current delivered to the tip of the implanted electrode. Low doses of Δ^9-THC (1.5 mg/kg intraperitoneally) significantly enhanced brain-stimulation reward (lowered reward thresholds) in the medial forebrain bundle (Gardner *et al.*, 1988a; Gardner and Lowinson, 1991). More recently, we repeated these experiments using a different quantitative electrophysiological brain-stimulation reward threshold paradigm – one based on rate-frequency curve-shift electrophysiological trade-off functions (Gardner *et al.*, 1995; Lepore *et al.*, 1996). Again, low Δ^9-THC doses (1.0 mg/kg i.p.) significantly enhanced electrical brain-stimulation reward (lowered reward thresholds), especially in Lewis strain rats, a strain long-recognized as sensitive to the reward-enhancing effects of addictive drugs (George and Goldberg, 1989; Guitart *et al.*, 1992; Kosten *et al.*, 1994).

Cannabinoid effects on neuronal activity in CNS reward loci

It is well established that some addictive drugs (e.g. opioids, nicotine) enhance mesoaccumbens DA reward substrates by enhancing the firing rate of the "second-stage" DA neurons within these reward substrates (see, e.g. Gysling and Wang, 1983; Grenhoff *et al.*, 1986). To assess whether cannabinoids act similarly, several research groups have combined cannabinoid administration in laboratory animals with *in vivo* single-neuron electrophysiological recording techniques. With one exception (Gifford *et al.*, 1997), the findings are that Δ^9-THC and the potent synthetic cannabinoids WIN-55212-2 and CP-55940 enhance neuronal firing of DA neurons in forebrain reward substrates (French, 1997; French *et al.*, 1997; Gessa *et al.*, 1998). The effect is seen in mesoaccumbens DA neurons, and also in the adjacent nigrostriatal and mesoprefrontal DA neurons (French *et al.*, 1997; Diana *et al.*, 1998; Gessa *et al.*, 1998). The enhanced DA firing is more pronounced in "second-stage" DA reward neurons than in other forebrain DA neurons (French *et al.*, 1997), which is congruent with the known preferential action of addictive drugs for the "second-stage" DA reward neurons as compared to other CNS DA neurons (Di Chiara, 1995; Pontieri *et al.*, 1995; Di Chiara and Imperato, 1986; Gardner, 1997). The effect is blocked by the selective CB_1 cannabinoid receptor antagonist SR-141716A (French, 1997; Diana *et al.*, 1998; Gessa *et al.*, 1998). In a small number of mesoprefrontal DA neurons, cannabinoid administration did not increase basal firing rate. Instead, cannabinoid administration to those neurons increased neuronal burst firing (Diana *et al.*, 1998), a firing pattern which dramatically increases DA release at axon terminals (Overton and Clark, 1997).

Two other research teams (Miller and Walker, 1995; Tersigni and Rosenberg, 1996) have used single-neuron recording techniques to study cannabinoid effects on the neostriatal outflow neurons – extraordinarily enriched with cannabinoid receptors (Herkenham *et al.*, 1991a; Herkenham *et al.*, 1991b; Jansen *et al.*, 1992; Mailleux and Vanderhaeghen, 1992; Matsuda *et al.*, 1993) – that synapse into the ventral mesencephalon near the "second-stage" DA reward cell fields. Both teams found that systemic injections or local microinjections of the cannabinoid agonists WIN-55212-2 or CP-55940 increased firing rates in substantia nigra pars reticulata neurons. Tersigni and Rosenberg (1996) also found that local microinjections of the selective CB_1 antagonist SR-141716A decreased basal firing of those neurons, suggesting that they are under tonic cannabinoid regulatory control. Miller and Walker (1995) found that the cannabinoid agonist WIN-55212-2 inhibited electrical activation of substantia nigra pars reticulata neurons, and that this effect was reversed by GABA antagonism. An inferential extrapolation from these data would be the suggestion that cannabinoid receptors on axon terminals in the ventral tegmental area may produce neuronal disinhibition by inhibiting GABA release – a mechanism similar to that subserving opioid and nicotine activation of the "second-stage" DA reward neurons.

Cannabinoid effects on synaptic dopamine in CNS reward loci

In vitro *effects*

A large number of published studies have addressed cannabinoid effects on DA function in forebrain reward loci using *in vitro* biochemical measurements. This is a large corpus of published work spanning more than 20 years, and a detailed analysis is beyond the present scope. In summary, though, a remarkable consistency emerges from this work: (1) cannabinoids (at doses relevant to human use) enhance DA synthesis, release, and turnover (Hattendorf *et al.*, 1977; Bloom and Dewey, 1978; Bloom and Kiernan, 1980; Bloom, 1982; Kumar *et al.*, 1984; Patel *et al.*, 1985; Bowers and Hoffman, 1986; Sakurai-Yamashita *et al.*, 1989; Rodríguez de Fonseca *et al.*, 1992; Bonnin *et al.*, 1993; Navarro *et al.*, 1993a, 1993b; Jentsch *et al.*, 1997, 1998); and (2) cannabinoids inhibit DA reuptake in CNS reward loci (Banerjee *et al.*, 1975; Hershkowitz and Szechtman, 1979; Poddar and Dewey, 1980).

In vivo *effects*

Almost 15 years ago, the author's research group – using *in vivo* brain microdialysis – reported that Δ^9-THC enhances extracellular DA overflow in forebrain reward loci (Ng Cheong Ton and Gardner, 1986). Subsequent work, both by the author's group (Ng Cheong Ton *et al.*, 1988; Chen *et al.*, 1989, 1990a, 1990b; Gardner and Lowinson, 1991; Gardner, 1992) and others (e.g. Taylor *et al.*, 1988; Tanda *et al.*, 1997), has confirmed those original reports. The DA-enhancing effect is tetrodotoxin-sensitive, calcium-dependent, and is blocked both by the opiate antagonist naloxone and the cannabinoid CB_1 antagonist SR-141716A (Chen *et al.*, 1990b; Gardner and Lowinson, 1991; Gardner, 1992; Tanda *et al.*, 1997). The *in vivo* DA-enhancing effect of cannabinoids is seen not only in the nucleus accumbens but also in other reward-relevant forebrain DA terminal projection loci, including medial prefrontal cortex (Chen *et al.*, 1990a) and neostriatum (Ng Cheong Ton *et al.*, 1988; Taylor *et al.*, 1988). Within the nucleus accumbens, the DA-enhancing effect occurs preferentially within the "shell" anatomical sub-domain (Tanda *et al.*, 1997), an important finding in view of the accumbens shell's specialization for mediating drug-enhanced CNS reward functions (Pontieri *et al.*, 1995; Gardner, 1997). The author's research group has also used an additional technique for obtaining *in vivo* measurements of extracellular DA in forebrain reward loci – voltammetric electrochemistry – and seen significant cannabinoid enhancement of extracellular DA (Ng Cheong Ton *et al.*, 1988). Thus, despite one negative report (Castañeda *et al.*, 1991), the preponderance of evidence from several different laboratories and using two different *in vivo* neurochemical techniques is that cannabinoids enhance extracellular DA in forebrain reward loci.

Genetic variation in cannabinoid effects on CNS reward substrates

As reviewed by George and Goldberg (1989), genetic influences have long been known to be significant determinants of drug self-administration at both the human and animal levels.

At the human level, evidence from family, twin, and adoption studies all support a substantial genetic component in both initial vulnerability to drug addiction and in continued drug dependence (for review, see Uhl *et al.*, 1995). As vulnerability to drug addiction at the human level does not follow clear Mendelian patterns of inheritance, most genetic studies in the field have been association studies – statistical correlations between an inherited condition and polymorphisms occurring in strong candidate genes (Lander and Schork, 1994; Elston, 1995). Given the wealth of animal research data on the importance of DA CNS reward mechanisms in drug addiction, polymorphisms in genes that regulate DA neurotransmission have been considered prime candidates as vulnerability factors for addiction (Koob and Bloom, 1988). A meta-analysis of published studies supports a positive association between drug and alcohol addiction and DA D_2 receptor polymorphisms (Uhl *et al.*, 1994).

At the animal level, genetic influences contribute heavily to both drug preference and propensity for drug self-administration (George, 1987; Suzuki *et al.*, 1988; George and Goldberg, 1989; Guitart *et al.*, 1992; Kosten *et al.*, 1994). In fact, some inbred animal strains generalize their increased vulnerability from one addictive drug class to others, supporting the concept that generalized *poly*-drug addiction has a genetic component (see, e.g. George and Meisch, 1984; Khodzhagel'diev, 1986; George, 1987) and suggesting, in turn, that some genetically inbred animal strains may have a generalized vulnerability to drug-induced reward. The Lewis strain rat is notable in this regard. Lewis rats appear to be inherently drug-seeking and drug-preferring as compared to rats of other strains. Lewis rats work harder for psychostimulant and opiate self-administration, place-condition more readily to opiates and cocaine, and voluntarily drink ethanol more readily – all in comparison to rats of other strains (Suzuki *et al.*, 1988; George and Goldberg, 1989; Guitart *et al.*, 1992). To see if cannabinoid effects on CNS reward substrates are subject to similar genetic variation, the author's research group compared Δ^9-THC's effects on electrical brain-stimulation reward thresholds in drug-preferring Lewis rats, drug-neutral Sprague-Dawley rats, and drug-resistant Fischer 344 rats. We found that Δ^9-THC produces robust enhancement of brain-stimulation reward in drug-preferring Lewis rats, moderate enhancement of brain-stimulation reward in drug-neutral Sprague-Dawley rats, and no enhancement of brain-stimulation reward in drug-resistant Fischer 344 rats (Gardner *et al.*, 1988b; Gardner *et al.*, 1989a; Gardner *et al.*, 1995; Lepore *et al.*, 1996). We also compared Δ^9-THC's effect on nucleus accumbens DA – using *in vivo* brain microdialysis – in Lewis, Sprague-Dawley, and Fischer 344 rats. We found that Δ^9-THC produces robust enhancement of nucleus accumbens DA in drug-preferring Lewis rats, moderate enhancement in drug-neutral Sprague-Dawley rats, and no enhancement in drug-resistant Fischer 344 rats (Gardner *et al.*, 1989a; Chen *et al.*, 1991). We believe that these data – from two different laboratory paradigms, one electrophysiological and one neurochemical – strongly support the concept that cannabinoid effects on CNS reward substrates are strongly influenced by genetic variables. We have further suggested (Lepore *et al.*, 1996) that significant genetic variations may also exist for other cannabinoid-induced pharmacological actions.

CANNABINOID WITHDRAWAL EFFECTS ON CNS REWARD SUBSTRATES

Whereas administration of addictive drugs produces enhancement of electrical brain-stimulation reward and mesoaccumbens DA, *withdrawal* from such drugs produces *inhibition* of electrical brain-stimulation reward and *depletion* of DA in CNS reward loci (see, e.g. Kokkinidis *et al.*, 1980; Cassens *et al.*, 1981; Schaefer and Michael, 1986; Frank *et al.*, 1988; Kokkinidis and McCarter, 1990; Parsons *et al.*, 1991; Robertson *et al.*, 1991; Pothos *et al.*, 1991; Rossetti *et al.*, 1992; Schulteis *et al.*, 1994; Spanagel *et al.*, 1994; Wise and Munn, 1995). Based on such findings, elevations in brain-stimulation reward thresholds and DA depletion in CNS reward substrates have been proposed as the underlying neural basis for post-drug-use anhedonia and drug craving (Dackis and Gold, 1985; Koob *et al.*, 1989; Markou and Koob, 1991). As noted by Wise and Munn (1995), "dopamine depletion and... attendant subsensitivity of the reward system offers a withdrawal symptom that may be more significant for drug self-administration than classic [physical withdrawal]... symptoms" and "subsensitivity of the reward system... is more obviously linked to the habit-forming property of drugs rather than to correlated side effects." Importantly, since elevations in brain-stimulation reward thresholds and correlated DA depletion in CNS reward substrates – unlike *physical* withdrawal symptoms – offer a set of withdrawal symptoms *common* to opiates, psychostimulants, and ethanol, they may constitute the long-sought common denominator for addiction.

Koob and colleagues have proposed yet another common denominator of withdrawal from addictive drugs – elevations of corticotropin-releasing factor (CRF) in the central nucleus of the amygdala (Merlo Pich *et al.*, 1995; Koob, 1996). This is a provocative hypothesis, since the amygdala has been suggested to mediate neural substrates of fear and anxiety (Le Doux *et al.*, 1988; Davis, 1992) and – as noted above – to mediate neural substrates of an emotional memory system that facilitates stimulus-reward learning (Cador *et al.*, 1989; Everitt *et al.*, 1989; Gaffan, 1992) and drug-seeking behavior (Hiroi and White, 1991; White and Hiroi, 1993).

Cannabinoids appear to interact with these CNS substrates of drug withdrawal in a fashion strikingly similar to that shown by other addictive drugs. The present author has reported that significant elevations in brain-stimulation reward thresholds (i.e., inhibition of CNS reward substrates) are seen during acute withdrawal from low doses of Δ^9-THC (Gardner and Lepore, 1996; Gardner and Vorel, 1998). And Rodríguez de Fonseca and colleagues (1997) have shown that acute cannabinoid withdrawal is accompanied by marked CRF elevations in the central nucleus of the amygdala, with maximal CRF elevations correlated with maximal cannabinoid withdrawal signs. These data suggest that cannabinoid withdrawal – at least with respect to effects on CNS reward substrates – is strikingly similar to that seen with other addictive drugs.

ENDOGENOUS CNS OPIOID INVOLVEMENT IN CANNABINOID EFFECTS ON CNS REWARD SUBSTRATES

As noted earlier in this review, endogenous CNS opioid peptide systems constitute an integral component of the reward substrates of the CNS. Indeed, some would argue that the opioid peptide components are as important as the DA components in the overall neural computation of reward. As to cannabinoid–opioid linkages, cannabinoid CB_1 receptors are co-localized with mu opioid receptors in the nucleus accumbens (Navarro *et al.*, 1998). Furthermore, cannabinoids and opioids appear to interact at the level of their signal transduction mechanisms (Thorat and Bhargava, 1994), with both cannabinoid receptors and opioid receptors being coupled to similar postsynaptic intracellular signaling mechanisms involving activation of G_i proteins and consequent inhibition of adenylyl cyclase and cAMP production (Childers *et al.*, 1992). Endogenous CNS opioid involvement in cannabinoid effects on CNS reward substrates is thus an exciting and rapidly developing domain of investigation.

Cannabinoid effects on endogenous CNS opioid systems

Effects on endogenous CNS opioid receptors

A full 15 years ago, the present author's research group reported that cannabinoids alter *in vitro* opioid receptor binding in both membrane-bound and solubilized partially purified preparations of rat CNS opioid receptors (Vaysse *et al.*, 1985, 1987). We found that Δ^9-THC produces a dose dependent inhibition of mu and delta opioid receptor binding, but does not alter kappa opioid, sigma, DA, or muscarinic acetylcholine receptors (Vaysse *et al.*, 1985, 1987; Gardner and Lowinson, 1991; Gardner, 1992). When subjected to Scatchard analyses, the mu opioid receptor binding data indicated that Δ^9-THC produces a significant decrease in receptor density with no change in receptor affinity, consistent with a noncompetitive mechanism for cannabinoid inhibition of mu opioid receptors. Studies of Δ^9-THC on solubilized, partially purified opioid receptors revealed similar findings (Vaysse *et al.*, 1987; Gardner and Lowinson, 1991; Gardner, 1992). On the basis of these findings, we suggested as early as 1985 that Δ^9-THC produces a direct allosteric modulation of the opioid receptor complex rather than a nonspecific perturbation of the membrane lipid bilayer (Vaysse *et al.*, 1985). We also compared the potencies of a large number of cannabinoids to inhibit mu opioid receptor binding, and found a good correlation with psychoactive potency at the human level (except for a lower-than-expected potency for 11-hydroxy-Δ^9-THC and a greater-than-expected potency for cannabidiol) (Vaysse *et al.*, 1987; Gardner and Lowinson, 1991; Gardner, 1992). A caveat is warranted. Although we repeated our experiments, replicated our findings, and are confident of our results, Ali and colleagues (1989) were unable to find similar effects in a somewhat analogous experiment.

Effects on endogenous CNS opioid neurotransmitters

Acute Δ^9-THC administration in adult rats produces increased CNS methionine-enkephalin levels (Kumar and Chen, 1983). Chronic Δ^9-THC administration in adult rats produces increased methionine-enkephalin-like immunoreactivity and increased beta-endorphin-like immunoreactivity in the preoptic area and mediobasal hypothalamus (Kumar *et al.*, 1984). Subchronic Δ^9-THC administration in newborn rat pups produces elevated methionine-enkephalin and beta-endorphin levels in the anterior hypothalamus/preoptic area and in the mediobasal hypothalamus (Kumar *et al.*, 1990). Subchronic Δ^9-THC administration in adult rats produces increased mRNA levels for pro-opiomelanocortin, the precursor for the opioid neurotransmitter beta-endorphin (Corchero *et al.*, 1997a), and also for pro-enkephalin and pro-dynorphin, the precursors for the opioid enkephalins and dynorphins, respectively (Corchero *et al.*, 1997b), although none of these findings was in CNS reward areas. More relevant to CNS reward substrates, pro-enkephalin mRNA levels in the nucleus accumbens are elevated in animals given subchronic administration of the potent synthetic cannabinoid CP-55940 (Manzanares *et al.*, 1998). On the other hand, nucleus accumbens pro-enkephalin mRNA was unchanged following subchronic Δ^9-THC or R-methanandamide (although the latter two cannabinoids did increase pro-enkephalin mRNA in other CNS areas) (Manzanares *et al.*, 1998). Other inconsistencies include findings by Kumar *et al.* (1986) that perinatal Δ^9-THC *decreases* methionine-enkephalin-like and beta-endorphin-like immunoreactivity in the anterior hypothalamus/preoptic area while *increasing* them in the mediobasal hypothalamus, and a report that Δ^9-THC has no effect on endogenous CNS opioid levels (Ali *et al.*, 1989).

ENDOGENOUS OPIOID MEDIATION OF CANNABINOID EFFECTS ON CNS REWARD SUBSTRATES

As noted above, acute enhancement of CNS reward substrates appears to be the single essential commonality of drugs with addictive potential. Strikingly, this drug-induced enhancement of CNS reward substrates is blocked or attenuated by such highly specific and selective opiate antagonists as naloxone and naltrexone. This holds not only for addictive drugs of the opiate class but also for non-opiates such as ethanol, amphetamines, cocaine, barbiturates, benzodiazepines, and phencyclidine (for review, see Gardner, 1997). Such findings – from dozens of laboratories over a span of more than 20 years – clearly implicate endogenous opioid mechanisms in mediating the rewarding actions of such drugs. To determine whether cannabinoid-induced enhancement of CNS reward substrates might be similarly blocked or attenuated by opiate antagonists, the author's research group carried out a series of experiments using both electrical brain-stimulation reward and *in vivo* brain microdialysis (Chen *et al.*, 1989; Gardner *et al.*, 1989b; Chen *et al.*, 1990b; Gardner *et al.*, 1990a; Gardner and Lowinson, 1991; Gardner, 1992). We found that the selective opiate antagonist naloxone, at doses low enough to preclude nonspecific action on other neurotransmitter systems, significantly attenuated Δ^9-THC's enhancement of electrical brain-stimulation reward (Chen *et al.*, 1989; Gardner *et al.*, 1989b; Gardner and Lowinson, 1991; Gardner, 1992). We also found that naloxone, again at doses low enough to preclude nonspecific action on other neurotransmitter systems, significantly attenuated Δ^9-THC's enhancement of nucleus accumbens DA (Chen *et al.*, 1990b; Gardner *et al.*, 1990a; Gardner and Lowinson, 1991; Gardner, 1992). More recently, Tanda *et al.* (1997) independently confirmed our *in vivo* brain microdialysis findings. Importantly, though, Tanda *et al.* (1997) also showed that the selective μ_1 opiate receptor antagonist naloxonazine duplicates the naloxone blockade – implicating the μ_1 opiate receptor subtype in mediating cannabinoid effects on CNS reward substrates. These findings by our group and by

Tanda and colleagues are congruent with an older report that naloxone attenuates Δ^9-THC's augmentation of CNS DA synthesis measured *in vitro* (Bloom and Dewey, 1978). On the other hand, French (1997) did not observe any naloxone attenuation of Δ^9-THC's augmentation of ventral tegmental area DA neuronal firing rates. The reason for this discrepancy is not clear.

ENDOGENOUS CNS CANNABINOID INVOLVEMENT IN OPIOID EFFECTS ON CNS REWARD SUBSTRATES

Since endogenous CNS opioid mechanisms are so clearly involved in mediating at least some cannabinoid effects on CNS reward substrates, it might logically be asked whether endogenous CNS cannabinoid systems are involved in mediating opioid effects on CNS reward substrates. There is little work addressing this possibility, but the few published studies are highly provocative to this reviewer's mind, notably the facts that morphine's rewarding properties and its ability to enhance nucleus accumbens DA are both markedly reduced in knockout mice lacking the CB_1 cannabinoid receptor (Ledent *et al.*, 1999; Mascia, 1999; Cossu *et al.*, 2001). In contrast, CB_1 receptor knockout mice *retain* their normal responses to the rewarding effects of cocaine, d-amphetamine, and nicotine – as assessed by intravenous self-administration (Cossu *et al.*, 2001). Furthermore, CB_1 receptor knockout mice also retain their normal responses to the DA-enhancing effects of cocaine, as assessed by *in vivo* brain microdialysis (Mascia *et al.*, 1999).

NEURAL AND SYNAPTIC MODELS OF CANNABINOID ACTION ON CNS REWARD SUBSTRATES

From the evidence reviewed above, and from other data, it is possible to generate hypotheses concerning: (1) *where* cannabinoids act to alter CNS reward substrates; (2) *how* cannabinoids act to alter CNS reward substrates; and (3) integrated models of cannabinoid action on CNS reward substrates.

Sites of cannabinoid action on CNS reward substrates

As noted above, different addictive drugs enhance reward by acting at different sites within the reward substrates of the CNS. Nicotine, ethanol, benzodiazepines, and barbiturates appear to act – transsynaptically – within somatic and dendritic regions of the "second-stage" DA neurons in the ventral tegmental area; cocaine, amphetamines, and dissociative anesthetics appear to act primarily on the axon terminal projections of the "second-stage" DA neurons within the nucleus accumbens. Opiates act on reward substrates within the ventral tegmental area, nucleus accumbens, and ventral pallidum (Gardner, 1997). The site(s) of cannabinoid action on CNS reward substrates has been addressed in several ways – some direct and some inferential. One of the direct ways used in the present author's laboratory has been to study the effects of local intracranial cannabinoid microinjections on DA levels in the nucleus accumbens as measured by *in vivo* brain microdialysis

(Chen *et al.*, 1993). In those studies, we found that direct microinfusions of Δ^9-THC into the nucleus accumbens dose-dependently enhanced accumbens DA levels. We also found that direct microinjections of Δ^9-THC into the ventral tegmental area dose-dependently enhanced *local* somatodendritic DA release within the ventral tegmental area, but *did not* enhance nucleus accumbens DA levels (Chen *et al.*, 1993). This suggests that *locally-applied* ventral tegmental area Δ^9-THC does not alter local DA neuronal firing, and further suggests that the elevated nucleus accumbens DA levels and enhanced brain-stimulation reward produced by systemic cannabinoid administration result from local pharmacological action at or near the "second-stage" DA axon terminals in the nucleus accumbens. However, as noted above, *systemic* cannabinoid administration *does* enhance the neuronal firing of the "second-stage" DA reward neurons (French, 1997; French *et al.*, 1997; Gessa *et al.*, 1998). Furthermore, intracranial microinjection of the μ_1 opioid antagonist naloxonazine *directly into* the ventral tegmental area attenuates cannabinoid-induced enhancement of nucleus accumbens DA levels (Tanda *et al.*, 1997). In addition – and puzzlingly – *systemic* naloxone (at doses high enough to block endogenous CNS opioid mechanisms) does *not* inhibit Δ^9-THC's enhancement of "second-stage" DA neuronal firing in the ventral tegmental area-nucleus accumbens axis (French, 1997). To this reviewer, this combination of findings is frankly puzzling, as it is by no means clear why local CNS microinjections of an opioid antagonist should attenuate the cannabinoid effects, while systemic administration of an opioid antagonist (at doses clearly high enough to enter the CNS and affect local CNS circuits) fails to do so. However, trusting that all of these findings are correct, one is forced to surmise that cannabinoids enhance DA in the nucleus accumbens "second-stage" DA terminal projection area by acting at a combination of CNS loci: (1) within the nucleus accumbens – by acting on a neuronal mechanism closely linked to axon terminal DA release; (2) within the ventral tegmental area – by acting on an endogenous opioid mechanism *not* linked to activation of neuronal firing, but rather linked to mechanisms of DA synthesis, transport, and/or release; and (3) also within the ventral tegmental area – by acting on a *non*-endogenous-opioid mechanism linked to activation of "second-stage" DA neuronal firing. While this combination of putative cannabinoid sites of action is somewhat complex, it is hardly beyond possibility. Drugs acting on the CNS often act at multiple sites; indeed, more often than not. That cannabinoids should share such complexity is not surprising, at least to this reviewer.

Mechanisms of cannabinoid action on CNS reward substrates

Just as different addictive drugs act at different sites of action within the CNS to enhance reward substrates, so too do different addictive drugs enhance reward substrates by acting through different *mechanisms*. Amphetamines (and probably some phencyclidine-like dissociative anesthetics) act as presynaptic DA releasers, cocaine as a presynaptic DA reuptake blocker, opiates and nicotine as transsynaptic enhancers of DA neuronal firing, and other addictive drugs by yet other mechanisms (for review, see Gardner, 1997). The mechanisms of action by which cannabinoids alter reward have been addressed using a variety of experimental approaches.

As noted in the previous section on CNS *sites* of cannabinoid action, studies using local intracerebral Δ^9-THC microinjections have led to the conclusion that a major cannabinoid site of action on CNS reward substrates is proximal to the "second-stage" DA axon terminals in the nucleus accumbens (Chen *et al.*, 1993). As also noted above, Δ^9-THC's enhancement of DA in the nucleus accumbens is calcium-dependent and tetrodotoxin-sensitive (Chen *et al.*, 1990b; Gardner and Lowinson, 1991; Gardner, 1992), implicating an action potential-dependent mechanism. Additional studies carried out in the present author's laboratory – using *in vivo* voltammetric electrochemistry to study the electrochemical profile of Δ^9-THC's enhancement of DA overflow in forebrain reward DA axon terminal loci – indicate that Δ^9-THC's electrochemical "signature" resembles that of a DA reuptake blocker rather than a presynaptic DA releaser (Ng Cheong Ton *et al.*, 1988). Additional experiments have also shed light on the question of neural mechanism(s). First, the author's research group has studied the effects of various combinations of Δ^9-THC and the DA antagonist haloperidol on nucleus accumbens DA using *in vivo* brain microdialysis (Gardner *et al.*, 1990b). The rationale for these experiments was the following – impulse-induced facilitation of DA release is known to underlie a synergy between DA antagonists and DA reuptake blockers (Westerink *et al.*, 1987). We found that acute pretreatment with the DA antagonist haloperidol has a synergistic effect on acute Δ^9-THC's enhancement of DA in the nucleus accumbens, and acute Δ^9-THC pretreatment before haloperidol has a similar synergistic effect on acute haloperidol's enhancement of DA in the nucleus accumbens (Gardner *et al.*, 1990b). Tetrodotoxin perfused locally into the nucleus accumbens abolished this synergy between Δ^9-THC and haloperidol (Gardner *et al.*, 1990b). Since this type of synergistic effect on DA typifies the effect produced by co-administration of a DA antagonist and a DA reuptake blocker such as GBR-12909 (Shore *et al.*, 1979; Westerink *et al.*, 1987), our findings of such a synergistic effect with the Δ^9-THC-haloperidol combinations suggest that Δ^9-THC's enhancing action on nucleus accumbens DA derives from DA reuptake blockade at nucleus accumbens DA terminals (albeit perhaps indirectly mediated) (Gardner *et al.*, 1990b; Gardner and Lowinson, 1991; Gardner, 1992). We further explored this possibility using *in vivo* brain microdialysis of the DA metabolite 3-methoxytyramine (3-MT) (Chen *et al.*, 1994). While only a small portion of released DA is metabolized to it, 3-MT is believed to be a sensitive index of extracellular DA (Wood and Altar, 1988) and – critically for our experiments – a sensitive marker for distinguishing DA releasing agents from DA reuptake blockers since DA releasers such as amphetamine and methamphetamine increase 3-MT levels while DA reuptake blockers such as bupropion and nomifensine do not (Heal *et al.*, 1990). We therefore carried out experiments on the effects of amphetamine, cocaine, nomifensine, and Δ^9-THC on extracellular nucleus accumbens 3-MT levels using *in vivo* brain microdialysis. We found that the DA releaser amphetamine significantly increased both DA and 3-MT in the nucleus accumbens, while the DA reuptake blockers cocaine and nomifensine increased only DA (Chen *et al.*, 1994). Δ^9-THC increased only DA, resembling the DA reuptake blockers (Chen *et al.*, 1994). These *in vivo* findings are congruent with older *in vitro* studies showing that cannabinoids have DA reuptake blockade actions in the CNS, as noted previously in this review (e.g. Banerjee *et al.*, 1975; Hershkowitz and Szechtman, 1979; Poddar and Dewey, 1980). We have suggested that such a mechanism may underlie cannabinoid action on DA reward substrates locally within the nucleus accumbens. Cannabinoid action within the ventral tegmental area – acting on endogenous opioid mechanisms *not* linked to activation of neuronal firing, but rather linked to mechanisms of DA synthesis, transport, and/or release – may be mediated by a cannabinoid-receptor-mediated inhibition of an inhibitory endogenous opioid peptidergic synaptic link to ventral tegmental area DA neurons that regulates their synthesis, transport, and/or release of DA rather than cell firing. *Additional* cannabinoid action within the ventral tegmental area – acting on *non*-endogenous-opioid mechanisms linked to activation of neuronal firing – could conceivably be mediated by a cannabinoid-receptor-mediated inhibition of feedback neurons from the nucleus accumbens to the ventral tegmental area which normally exert inhibitory synaptic tone on ventral tegmental area DA cells. This suggestion is congruent with the known co-localization of cannabinoid receptors with DA D_1 receptors on striatonigral inhibitory projection neurons (Herkenham *et al.*, 1991b).

Hypothetical models of cannabinoid action on CNS reward substrates

From many of the considerations reviewed above, hypothetical models of cannabinoid action on CNS reward substrates can be developed.

One model (Gardner and Lowinson, 1991; Gardner, 1992) takes as its starting point the observations that: (1) intracranial cannabinoid microinfusion into the nucleus accumbens enhances accumbens DA while cannabinoid microinjection into the ventral tegmental area does not; (2) the electrochemical "signature" of cannabinoid-induced enhancement of forebrain DA resembles that of a DA reuptake blocker rather than that of a DA releaser; (3) the synergistic interaction between Δ^9-THC and haloperidol resembles that of a DA reuptake blocker rather than that of a DA releaser; (4) cannabinoid action on 3-MT resembles that of a DA reuptake blocker rather than that of a DA releaser; and (5) cannabinoids allosterically modulate endogenous CNS opioid receptors. This model posits – from this evidence – that the principal site of cannabinoid action on CNS reward substrates lies within the DA axon terminal region of the nucleus accumbens and associated cell fields of the ventral striatum. This model further accepts that the "second-stage" DA reward neurons synapse on endogenous opioid peptide neurons which carry the reward signal to the ventral pallidum. It also accepts that other modulatory endogenous opioid peptidergic neurons synapse, in axo-axonic fashion, onto the "second-stage" DA axon terminals in the nucleus accumbens. It presumes that these axo-axonic synapses modulate the flow of reward-relevant neural signals through the DA circuitry, by either: (1) classical presynaptic excitation and inhibition; or (2) a modulatory interaction between opioid receptors located on the DA axon terminals and the DA reuptake mechanism within those terminals. This last possibility is admittedly not a classically described form of axo-axonic functional interaction, but evidence for such functional coupling between presynaptic receptor activation and neurotransmitter uptake within nerve terminals does exist (e.g. Galzin *et al.*, 1982, 1985; Langer and Moret, 1982; Cubeddu *et al.*, 1983; Göthert *et al.*, 1983; Moret and Briley, 1988). This model further presumes the existence of opioid autoreceptors within this synaptic complex, for which suggestive evidence exists (e.g. Gintzler and Xu, 1991). The model hypothesizes that: (1) cannabinoid CB_1 receptors are located on opioid peptide axon terminals forming axo-axonic connections with the "second-stage" DA axon terminals; and (2) these cannabinoid receptors allosterically modulate opioid autoreceptors on the opioid peptide axon terminals.

A second model, based on recent findings by Hoffman and Lupica (2001), posits a presynaptic locus of action for cannabinoids within the nucleus accumbens – similar to the just-cited model – but on GABAergic neurons instead of opioid peptide neurons. According to this model, cannabinoids produce inhibition of GABAergic inhibitory inputs to the "second-stage" DA reward neurons, resulting in an enhancement of DA neural tone in the nucleus accumbens. However, this model does not account for the robust opioid antagonist-induced inhibition of cannabinoid effects on reward substrates.

A third model, based largely on the work of Tanda *et al.* (1997), posits that the principal neural site for cannabinoid action on CNS reward substrates is in the ventral tegmental area, presumably on the feedback neurons from the accumbens to the ventral tegmental area – which are known to have CB_1 receptors on them. This model further posits that activation of these CB_1 receptors inhibits – in turn – opioid peptide/GABA-mediated inhibition of the "second-stage" DA reward neurons in the ventral tegmental area, resulting in an enhancement of DA neural tone in the "second-stage" DA reward neurons. This second model is certainly attractive in its simplicity. It is also congruent with data showing that cannabinoids inhibit GABAergic afferent inputs into the ventral mesencephalon (e.g. Szabo *et al.*, 2000). However, such a model is difficult to reconcile with findings that: (1) local cannabinoid microinjections into the ventral tegmental area do not enhance DA release in the nucleus accumbens (Chen *et al.*, 1993); and (2) naloxone does not alter cannabinoid effects on ventral tegmental DA neurons (French, 1997).

It should be stressed that these three models are *not* mutually incompatible, in as much as it is common for CNS-active drugs to have multiple sites and mechanisms of action. At the same time, it is much to be hoped that further studies may yield additional insights that will permit the development of a unified model which fits all relevant data.

CANNABINOID EFFECTS ON REWARD-RELATED BEHAVIORS

Cannabinoid effects on conditioned place preference

One of the most widely used behavioral techniques for inferring appetitiveness or aversiveness of pharmacological agents in laboratory animals is conditioned place/ cue preference (or aversion). Conditioned place or cue preference is the learned approach to a previously neutral set of environmental stimuli which have been paired with administration of a rewarding treatment (for review, see van der Kooy, 1987). When used to assess the rewarding properties of drugs, animals are given drug injections while confined in one of two cue-distinctive chambers and vehicle injections while confined in the other. Prior to injections, both chambers are tested to ensure their neutrality (i.e., animals voluntarily spending equal time in each cue-distinctive chamber). After drug injections, if the animal shows a marked preference for the drug-paired chamber, the drug is considered to have been rewarding. If the animal avoids the drug-paired chamber, the drug is considered to have been aversive. This technique has been used by several research groups to assess the rewarding or aversive properties of cannabinoids. Three research groups have reported that cannabinoids produce conditioned place *aversions* (Parker and Gillies, 1995; McGregor *et al.*, 1996; Sañudo-Peña *et al.*, 1997). Another group – using cue preference/aversion rather than place preference/aversion – also reported cannabinoid aversion (Goett and Kay, 1981). In contrast, the author's research group found robust cannabinoid-induced conditioned place *preferences* (Lepore *et al.*, 1995). The crucial differences appear to be related to cannabinoid dose and timing. When the drug-place pairing interval was 24 h (within the post-cannabinoid dysphoric rebound – see above section on cannabinoid withdrawal), we found that 1.0 mg/kg Δ^9-THC produced no preference, while 2.0 or 4.0 mg/kg produced robust place preferences. When the drug-place pairing interval was 48 h (past the post-drug dysphoric rebound) 1.0 mg/kg Δ^9-THC produced robust place preferences, while 2.0 or 4.0 mg/kg produced robust place *aversions* (Lepore *et al.*, 1995). We believe that at the shorter pairing interval, the post-cannabinoid rebound dysphoria (Gardner and Lepore, 1996) attenuated Δ^9-THC's rewarding effect, eliminating the reward of the 1.0 mg/kg dose and lowering the 2.0 and 4.0 mg/kg doses into a rewarding range. We believe that at the longer pairing interval, the post-cannabinoid dysphoric rebound had passed, accentuating the effects of all doses – allowing the 1.0 mg/kg dose to become rewarding, and pushing the 2.0 and 4.0 mg/kg doses into an aversive dose-response zone. This dose-dependent switch from reward to aversion (low dose-reward; high dose-aversion) has been long recognized for other addictive drugs (Fudala *et al.*, 1985; Jorenby *et al.*, 1990; Gardner, 1992), and it has also been long recognized that timing of drug administration during place conditioning is as strong a determinant of place preference or aversion as the drug itself (Fudala and Iwamoto, 1990). At the human level, a parallel finding has long been noted, i.e., that low-to-moderate doses of Δ^9-THC produce a subjective "high" but higher doses are aversive (Noyes *et al.*, 1975; Raft *et al.*, 1977; Laszlo *et al.*, 1981). Very recently, our findings of cannabinoid-induced conditioned place preferences (and of the crucial importance of cannabinoid dose to either preference or aversion) have been independently confirmed (Sala and Braida, 2000; Valjent and Maldonado, 2000; Braida *et al.*, 2001a).

Cannabinoid effects on naturally rewarding behaviors

There exists an extensive published literature dealing with cannabinoid effects on naturally rewarding behaviors. This literature can be divided into two areas: (1) cannabinoid effects on sexual behavior; and (2) cannabinoid effects on consumption of highly rewarding (e.g. sweet) foods and liquids.

The literature dealing with cannabinoid effects on sexual behavior is extensive, and beyond the present scope. The interested reader is referred to that literature (e.g. Bloch *et al.*, 1978; Sieber *et al.*, 1981; Turley and Floody, 1981; Cutler and Mackintosh, 1984; Navarro *et al.*, 1993b, 1995, 1996).

The literature dealing with cannabinoid effects on voluntary consumption of sweet foods and liquids is smaller. And, in the view of this reviewer, it is more germane to the considerations of the present review, in light of the following facts: (1) animals genetically selected for high rates of electrical brain-stimulation reward (Lieblich *et al.*, 1978; Gross-Isseroff *et al.*, 1992) show significantly enhanced consummatory responses to sweet stimuli (Ganchrow *et al.*, 1981); and (2) a number of other, non-cannabinoid, addictive drugs share the common feature of augmenting consumption and increasing the reward value of sweet stimuli (for review, see Milano *et al.*, 1988). With respect to cannabinoids, a number of studies have found them to augment choice and consumption of sweet foods and solutions (e.g. sucrose) (Sofia and Knoblock, 1976; Brown *et al.*, 1977; Milano *et al.*, 1988) and to increase their reward value as assessed by increased progressive-ratio break-points (Gallate *et al.*, 1999). In the latter study, carried out by McGregor and colleagues, the cannabinoid effect was reversed by the CB_1 receptor antagonist SR-141716A, confirming the specificity of the cannabinoid-induced effects (Gallate *et al.*, 1999). Provocatively, it has also been reported that sucrose consumption is *inhibited* by SR-141716A (Arnone *et al.*, 1997), suggesting that CNS endocannabinoid neural tone is important for the CNS reward substrates activated by such natural rewards, and may act to mediate their reward value. Trojniar and Wise (1991) reported that Δ^9-THC augments food consumption elicited by electrical stimulation of the lateral hypothalamus, a CNS site long known to be involved in the regulation of food intake and incentive-reward. The McGregor research group has reported cannabinoid-induced augmentation of beer consumption and increased progressive-ratio break-points for beer by laboratory rodents (Gallate *et al.*, 1999), and *decreased* progressive-ratio break-points for beer reinforcement produced by the CB_1 antagonist SR-141716A (Gallate and McGregor, 1999). Provocatively, the opioid antagonist naloxone also produced a *decrease* in progressive-ratio break-points for beer reinforcement (Gallate and McGregor, 1999), which speaks – presumably – to the equal importance of endogenous opioid peptidergic neural tone in mediating the incentive-reward value of beer. Finally, Freedland *et al.*, (2000) have reported that the CB_1 antagonist SR-141716A decreases both the appetitive phase and the consummatory phase of a complex sucrose-rewarded operant paradigm, again arguing for the importance of CNS endocannabinoid neural tone in mediating the CNS reward substrates activated by natural rewards.

Cannabinoid self-administration in animals

Given all the data reviewed above, it is quite to be expected that cannabinoids would be reliably self-administered by animals. Until recently, that did not seem to be the case. A number of older studies failed to demonstrate reliable cannabinoid self-administration in laboratory animals (Kaymakçalan, 1972; Corcoran and Amit, 1974; Harris *et al.*, 1974; Leite and Carlini, 1974; Carney *et al.*, 1977; Takahashi and Singer, 1981). In those older studies, the few instances of cannabinoid self-administration reported were unfortunately contaminated by methodological flaws, such as: (1) observing intravenous cannabinoid self-stimulation in monkeys only after self-administration of other highly reinforcing drugs (Pickens *et al.*,

1973); (2) observing intravenous cannabinoid self-administration only in monkeys made physically dependent on cannabinoids (Deneau and Kaymakçalan, 1971; Kaymakçalan, 1972); or (3) observing cannabinoid self-administration in rats only under food deprivation conditions (Takahashi and Singer, 1979, 1980).

Recently, however, a number of laboratories have independently reported successful cannabinoid self-administration in laboratory animals under methodologically cleaner conditions. Thus, Fratta, Martellotta, Ledent and colleagues have reported obtaining robust and consistent intravenous self-administration of the potent synthetic cannabinoid agonist WIN-55212-2 in drug-naive mice, and showed that this effect was blocked by the selective CB_1 antagonist SR-141716A (Fratta *et al.*, 1997; Martellotta *et al.*, 1998; Ledent *et al.*, 1999). Also, Sala and colleagues have recently reported obtaining successful intracerebroventricular self-administration of the potent synthetic cannabinoid agonist WIN-55212-2 in laboratory rats, and reported that this was blocked by the selective CB_1 antagonist SR-141716A (Sala and Braida, 2000; Braida *et al.*, 2001b). Perhaps most compellingly of all, Tanda, Munzar and Goldberg (2000) have recently reported persistent high rates of intravenous Δ^9-THC self-administration in squirrel monkeys, and reported that this was blocked by the selective CB_1 antagonist SR-141716A. What makes this last report so compelling is that the self-administered cannabinoid is the same one used by humans (Δ^9-THC) and that the self-administration paradigm used was classically straightforward (intravenous administration and a simple schedule of reinforcement).

SUMMARY

On the basis of more than 15 years of consistent laboratory findings using a wide variety of electrophysiological, biochemical, and behavioral techniques, and with independent evidence from many separate research groups, it seems clear that cannabinoids enhance CNS reward substrates in a manner not unlike that of more traditionally-studied addictive drugs. Also on the basis of electrophysiological and biochemical evidence from many separate research groups, cannabinoid withdrawal appears to activate the same CNS withdrawal substrates as activated by withdrawal from other addictive drugs. Cannabinoids are euphorigenic and have addictive liability at the human level, and are self-administered by laboratory animals. Cannabinoid action on CNS reward substrates appears to explain these behavioral properties. Indeed, cannabinoids appear to share common final neural actions with other addictive drugs: (1) activation of CNS reward substrates during acute administration; and (2) inhibition of CNS reward substrates (as well as activation of amygdaloid CRF systems) during withdrawal. While these neuropharmacological properties appear to confer addictive potential on cannabinoids, they may also open up another intriguing possibility – that cannabinoids may have clinical utility in dysphoric states. This latter possibility – that some cannabinoids or cannabinoid derivatives may have antidepressant efficacy – should not be discounted, as enhancement of CNS reward substrates is recognized as an underlying mechanism of action for some antidepressant medications. If the reward enhancing properties of at least some cannabinoids prove to translate into clinical usefulness, this may open up entirely new areas of therapeutics for an ancient class of psychoactive substances and their modern synthetic analogs.

ACKNOWLEDGMENTS

Cannabinoid research cited in this paper from the author's laboratory was supported by research grants from the U.S. Public Health Service, National Institutes of Health (grants AA09547, DA02089, DA03622, and RR05397); the U.S. National Science Foundation (grant BNS-86-09351); the Natural Sciences and Engineering Research Council of Canada; the New York State Office of Alcoholism and Substance Abuse Services; the New York State Office of Mental Health; the Aaron Diamond Foundation; and the Julia Sullivan Medical Research Fund. The work of the following former and present students, postdoctoral fellows, and laboratory technicians is gratefully acknowledged: Thomas Seeger, Jules Nazzaro, Pierre Vaysse, Jean Max Ng Cheong Ton, Cassandra Milling, Amy Donner, David Cohen, David Morrison, Jianping Chen, Jin Li, Diane Smith, Ronen Marmur, Eric Rosenbaum, Addy Pulles, Virginia Savage, Daniel Matalon, Michael Froehler, Jordan Spector, Marino Lepore, Xinhe Liu, Robert Hayes, Stanislav Robert Vorel, and William Paredes. Preparation of this manuscript was supported by the Intramural Research Program, National Institute on Drug Abuse, U.S. Public Health Service.

REFERENCES

Alexander, G. E. and Crutcher, M. D. (1990) "Functional architecture of basal ganglia circuits: neural substrates of parallel processing," *Trends in Neurosciences* **276**: 186–188.

Ali, S. F., Newport, G. D., Scallet, A. C., Gee, K. W., Paule, M. G., Brown, R.M. *et al.* (1989) "Effects of chronic delta-9-tetrahydrocannabinol (THC) administration on neurotransmitter concentrations and receptor binding in the rat brain," *Neurotoxicology* **10**: 491–500.

Anthony, J. C., Warner, L. A. and Kessler, R. C. (1994) "Comparative epidemiology of dependence on tobacco, alcohol, controlled substances and inhalants: basic findings from National Comorbidity Study," *Experimental and Clinical Psychopharmacology* **2**: 244–268.

Arnone, M., Maruani, J., Chaperon, F., Thiébot, M.-H., Poncelot, M., Soubrié, P. *et al.* (1997) "Selective inhibition of sucrose and ethanol intake by SR 141716, an antagonist of central cannabinoid (CB1) receptors," *Psychopharmacology* **132**: 104–106.

Banerjee, S. P., Snyder, S. H. and Mechoulam, R. (1975) "Cannabinoids: influence on neurotransmitter uptake in rat brain synaptosomes," *Journal of Pharmacology and Experimental Therapeutics* **194**: 74–81.

Barg, J., Fride, E., Hanus, L., Levy, R., Matus-Leibovitch, N., Heldman, E. *et al.* (1995) "Cannabimimetic behavioral effects of and adenylate cyclase inhibition by two new endogenous anandamides," *European Journal of Pharmacology* **287**: 145–152.

Bloch, E., Thysen, B., Morrill, G. A., Gardner, E. and Fujimoto, G. (1978) "Effects of cannabinoids on reproduction and development," *Vitamins and Hormones* **36**: 203–258.

Bloom, A. S. (1982) "Effect of Δ^9-tetrahydrocannabinol on the synthesis of dopamine and norepinephrine in mouse brain synaptosomes," *Journal of Pharmacology and Experimental Therapeutics* **221**: 97–103.

Bloom, A. S. and Dewey, W. L. (1978) "A comparison of some pharmacological actions of morphine and Δ^9-tetrahydrocannabinol in the mouse," *Psychopharmacology* **57**: 243–248.

Bloom, A. S. and Kiernan, C. J. (1980) "Interaction of ambient temperature with the effects of Δ^9-tetrahydrocannabinol on brain catecholamine synthesis and plasma corticosterone levels," *Psychopharmacology* **67**: 215–219.

Bloom, A. S., Edgemond, W. S. and Moldvan, J. C. (1997) "Nonclassical and endogenous cannabinoids: effects on the ordering of brain membrane," *Neurochemical Research* **22**: 563–568.

Blum, K., Cull, J. G., Braverman, E. R. and Comings, D. E. (1996) "Reward deficiency syndrome," *American Scientist* **84**: 132–145.

Bonnin, A., Fernández-Ruiz, J. J., Martin, M., Rodríguez de Fonseca, F., Hernández, M. L. and Ramos, J. A. (1993) "δ^9-Tetrahydrocannabinol affects mesolimbic dopaminergic activity in the female rat brain: interactions with estrogens," *Journal of Neural Transmission* **92**: 81–95.

Bowers, M. B. Jr. and Hoffman, F. J. Jr. (1986) "Regional brain homovanillic acid following Δ^9-tetrahydrocannabinol and cocaine," *Brain Research* **366**: 405–407.

Bozarth, M. A. and Wise, R. A. (1981a) "Intracranial self-administration of morphine into the ventral tegmental area of rats," *Life Sciences* **28**: 551–555.

Bozarth, M. A. and Wise, R. A. (1981b) "Heroin reward is dependent on a dopaminergic substrate," *Life Sciences* **29**: 1881–1886.

Braida, D., Pozzi, M., Cavallini, R. and Sala, M. (2001a) "Conditioned place preference induced by the cannabinoid agonist CP 55,940: interaction with the opioid system," *Neuroscience* **104**: 923–926.

Braida, D., Pozzi, M., Parolaro, D. and Sala, M. (2001b) "Intracerebral self-administration of the cannabinoid receptor agonist CP 55,940 in the rat: interaction with the opioid system," *European Journal of Pharmacology* **413**: 227–234.

Brown, J. E., Kassouny, M. and Cross, J. K. (1977) "Kinetic studies of food intake and sucrose solution preference by rats treated with low doses of delta-9-tetrahydrocannabinol," *Behavioral Biology* **20**: 104–110.

Cador, M., Robbins, T. W. and Everitt, B. J. (1989) "Involvement of the amygdala in stimulus reward associations: interaction with the ventral striatum," *Neuroscience* **30**: 77–86.

Cador, M., Bjijou, Y., Cailhol, S. and Stinus, L. (1999) "D-amphetamine-induced behavioral sensitization: implication of a glutamatergic medial prefrontal cortex-ventral tegmental area innervation," *Neuroscience* **94**: 705–721.

Carlezon, W. A. Jr. and Wise, R. A. (1996) "Rewarding actions of phencyclidine and related drugs in nucleus accumbens shell and frontal cortex," *Journal of Neuroscience* **16**: 3112–3122.

Carney, J. M., Uwaydah, I. M. and Balster, R. L. (1977) "Evaluation of a suspension system for intravenous self-administration of water insoluble substances in the rhesus monkey," *Pharmacology Biochemistry and Behavior* **7**: 357–364.

Cassens, G., Actor, C., Kling, M. and Schildkraut, J. J. (1981) "Amphetamine withdrawal: effects on threshold of intracranial reinforcement," *Psychopharmacology* **73**: 318–322.

Castañeda, E., Moss, D. E., Oddie, S. D. and Whishaw, I. Q. (1991) "THC does not affect striatal dopamine release: microdialysis in freely moving rats," *Pharmacology Biochemistry and Behavior* **40**: 587–591.

Chen, J., Paredes, W., Li, J., Smith, D. and Gardner, E. L. (1989) "In vivo brain microdialysis studies of Δ^9-tetrahydrocannabinol on presynaptic dopamine efflux in nucleus accumbens of the Lewis rat," *Society for Neuroscience Abstracts* **15**: 1096.

Chen, J., Paredes, W., Lowinson, J. H. and Gardner, E. L. (1990a) "Δ^9-Tetrahydrocannabinol enhances presynaptic dopamine efflux in medial prefrontal cortex," *European Journal of Pharmacology* **190**: 259–262.

Chen, J., Paredes, W., Li, J., Smith, D., Lowinson, J. and Gardner, E. L. (1990b) "Δ^9-Tetrahydrocannabinol produces naloxone-blockable enhancement of presynaptic basal dopamine efflux in nucleus accumbens of conscious, freely-moving rats as measured by intracerebral microdialysis," *Psychopharmacology* **102**: 156–162.

Chen, J., Paredes, W., Lowinson, J. H. and Gardner, E. L. (1991) "Strain-specific facilitation of dopamine efflux by Δ^9-tetrahydrocannabinol in the nucleus accumbens of rat: an *in vivo* microdialysis study," *Neuroscience Letters* **129**: 136–140.

Chen, J., Marmur, R., Pulles, A., Paredes, W. and Gardner, E. L. (1993) "Ventral tegmental microinjection of Δ^9-tetrahydrocannabinol enhances ventral tegmental somatodendritic dopamine levels but not forebrain dopamine levels: evidence for local neural action by marijuana's psychoactive ingredient," *Brain Research* **621**: 65–70.

Chen, J., Paredes, W. and Gardner, E. L. (1994) "Δ^9-Tetrahydrocannabinol's enhancement of nucleus accumbens dopamine resembles that of reuptake blockers rather than releasers – evidence from *in vivo* microdialysis experiments with 3-methoxytyramine," *National Institute on Drug Abuse Research Monograph Series* **141**: 312.

Chen, J., Liu, X., Paredes, W. and Gardner, E. L. (1996) "Acute fluoxetine enhances cocaine's action on brain reward," *Biological Psychiatry* **39**: 537.

Childers, S. R. and Breivogel, C. S. (1998) "Cannabis and endogenous cannabinoid systems," *Drug and Alcohol Dependence* **51**: 173–187.

Childers, S. R., Fleming, L., Konkoy, C., Marckel, D., Pacheco, M., Sexton, T. *et al.* (1992) "Opioid and cannabinoid receptor inhibition of adenylyl cyclase in brain," *Annals of the New York Academy of Sciences* **654**: 33–51.

Corchero, J., Fuentes, J. A. and Manzanares, J. (1997a) "Δ^9-Tetrahydrocannabinol increases proopiomelanocortin gene expression in the arcuate nucleus of the rat hypothalamus," *European Journal of Pharmacology* **323**: 193–195.

Corchero, J., Avila, M. A., Fuentes, J. A. and Manzanares, J. (1997b) "Δ^9-Tetrahydrocannabinol increases prodynorphin and proenkephalin gene expression in the spinal cord of the rat," *Life Sciences [Pharmacology Letters]* **61**: PL39–PL43.

Corcoran, M. E. and Amit, Z. (1974) "Reluctance of rats to drink hashish suspensions: free-choice and forced consumption, and the effects of hypothalamic stimulation," *Psychopharmacologia* **35**: 129–147.

Cossu, G., Ledent, C., Fattore, L., Imperato, A., Böhme, G. A., Parmentier, M. and Fratta, W. (2001) "Cannabinoid CB_1 receptor knockout mice fail to self-administer morphine but not other drugs of abuse," *Behavioural Brain Research* **118**: 61–65.

Crowley, T. J., Macdonald, M. J., Whitmore, E. A. and Mikulich, S. K. (1998) "Cannabis dependence, withdrawal, and reinforcing effects among adolescents with conduct symptoms and substance use disorders," *Drug and Alcohol Dependence* **50**: 27–37.

Cubeddu, L. X., Hoffmann, I. S. and James, M. K. (1983) "Frequency-dependent effects of neuronal uptake inhibitors on the autoreceptor-mediated modulation of dopamine and acetylcholine release from the rabbit striatum," *Journal of Pharmacology and Experimental Therapeutics* **226**: 88–94.

Cutler, M. G. and Mackintosh, J. H. (1984) "Cannabis and delta-9-tetrahydrocannabinol. Effects on elements of social behaviour in mice," *Neuropharmacology* **23**: 1091–1097.

Dackis, C. A. and Gold, M. S. (1985) "New concepts in cocaine addiction: the dopamine depletion hypothesis," *Neuroscience and Biobehavioral Reviews* **9**: 469–477.

Davis, M. (1992) "The role of amygdala in fear and anxiety," *Annual Review of Neuroscience* **15**: 353–375.

Deneau, G.A. and Kaymakçalan, S. (1971) "Physiological and psychological dependence to synthetic Δ^9-tetrahydrocannabinol (THC) in rhesus monkeys," *Pharmacologist* **13**: 246.

Devane, W. A., Hanus, L., Breuer, A., Pertwee, R. G., Stevenson, L. A., Griffin, G. *et al.* (1992) "Isolation and structure of a brain constituent that binds to the cannabinoid receptor," *Science* **258**: 1946–1949.

de Wit, H. and Wise, R. A. (1977) "Blockade of cocaine reinforcement in rats with the dopamine receptor blocker pimozide, but not with the noradrenergic blockers phentolamine or phenoxybenzamine," *Canadian Journal of Psychology* **31**: 195–203.

Diana, M., Melis, M. and Gessa, G. L. (1998) "Increase in meso-prefrontal dopaminergic activity after stimulation of CB1 receptors by cannabinoids," *European Journal of Neuroscience* **10**: 2825–2830.

Di Chiara, G. (1995) "The role of dopamine in drug abuse viewed from the perspective of its role in motivation," *Drug and Alcohol Dependence* **38**: 95–137.

Di Chiara, G. and Imperato, A. (1986) "Preferential stimulation of dopamine release in the nucleus accumbens by opiates, alcohol, and barbiturates: studies with transcerebral dialysis in freely moving rats," *Annals of the New York Academy of Sciences* **473**: 367–381.

Di Chiara, G. and Imperato, A. (1988) "Drugs abused by humans preferentially increase synaptic dopamine concentrations in the mesolimbic system of freely moving rats," *Proceedings of the National Academy of Sciences of the United States of America* **85**: 5274–5278.

Duane Sofia, R. (1978) "Cannabis: structure–activity relationships," in *Handbook of Psychopharmacology*, Vol. 12, edited by L .L. Iversen, S. D. Iversen and S. H. Snyder, pp. 319–371. New York: Plenum Press.

Eissenstat, M. A., Bell, M. R., D'Ambra, T. E., Alexander, E. J., Daum, S. J., Ackerman, J. H. *et al.* (1995) "Aminoalkylindoles: structure–activity relationships of novel cannabinoid mimetics," *Journal of Medicinal Chemistry* **38**: 3094–3105.

Elston, R. C. (1995) "Linkage and association to genetic markers," *Experimental and Clinical Immunogenetics* **12**: 129–140.

Evans, D. M., Lake, T. J., Johnson, M. R. and Howlett, A. C. (1994) "Endogenous cannabinoid receptor binding activity released from rat brain slices by depolarization," *Journal of Pharmacology and Experimental Therapeutics* **268**: 1271–1277.

Everitt, B. J., Cador, M. and Robbins, T. W. (1989) "Interactions between the amygdala and ventral striatum in stimulus-reward association: studies using a second-order schedule of sexual reinforcement," *Neuroscience* **30**: 63–75.

Everitt, B. J., Morris, K. A., O'Brien, A., Burns, L. and Robbins, T. W. (1991) "The basolateral amygdala-ventral striatal system and conditioned place preference: further evidence of limbic-striatal interactions underlying reward-related processes," *Neuroscience* **41**: 1–18.

Felder, C. C. and Glass, M. (1998) "Cannabinoid receptors and their endogenous agonists," *Annual Review of Pharmacology and Toxicology* **38**: 179–200.

Felder, C. C., Joyce, K. E., Briley, E. M., Glass, M., Mackie, K. P., Fahey, K. J. *et al.* (1998) "LY320135, a novel cannabinoid CB1 receptor antagonist, unmasks coupling of the CB1 receptor to stimulation of cAMP accumulation," *Journal of Pharmacology and Experimental Therapeutics* **284**: 291–297.

Fernando, S. R. and Pertwee, R. G. (1997) "Evidence that methyl arachidonyl fluorophosphonate is an irreversible cannabinoid receptor antagonist," *British Journal of Pharmacology* **121**: 1716–1720.

Fletcher, P. J., Pampakeras, M. and Yeomans, J. S. (1995) "Median raphe injections of 8-OH-DPAT lower frequency thresholds for lateral hypothalamic self-stimulation," *Pharmacology Biochemistry and Behavior* **52**: 65–71.

Floresco, S. B., Yang, C. R., Phillips, A. G. and Blaha, C. D. (1998) "Basolateral amygdala stimulation evokes glutamate receptor-dependent dopamine efflux in the nucleus accumbens of the anesthetized rat," *European Journal of Neuroscience* **10**: 1241–1251.

Frank, R. A., Martz, S. and Pommering, T. (1988) "The effect of chronic cocaine on self-stimulation train-duration thresholds," *Pharmacology Biochemistry and Behavior* **29**: 755–758.

Fratta, W., Martellotta, M. C., Cossu, G. and Fattore, L. (1997) "WIN 55,212-2 induces intravenous self-administration in drug-naive mice," *Society for Neuroscience Abstracts* **23**: 1869.

Freedland, C. S., Sharpe, A. L., Samson, H. H. and Porrino, L. J. (2001) "Effects of SR141716A on ethanol and sucrose self-administration," *Alcoholism: Clinical and Experimental Research* **25**: 277–282.

French, E. D. (1997) "Δ^9-Tetrahydrocannabinol excites rat VTA dopamine neurons through activation of cannabinoid CB1 but not opioid receptors," *Neuroscience Letters* **226**: 159–162.

French, E. D., Dillon, K. and Wu, X. (1997) "Cannabinoids excite dopamine neurons in the ventral tegmentum and substantia nigra," *NeuroReport* **8**: 649–652.

Fudala, P. J. and Iwamoto, E. T. (1990) "Conditioned aversion after delay place conditioning with amphetamine," *Pharmacology Biochemistry and Behavior* **35**: 89–92.

Fudala, P. J., Teoh, K. W. and Iwamoto, E. T. (1985) "Pharmacologic characterization of nicotine-induced conditioned place preference," *Pharmacology Biochemistry and Behavior* **22**: 237–241.

Gaffan, D. (1992) "Amygdala and the memory of reward," in *The Amygdala: Neurobiological Aspects of Emotion*, edited by J. P. Aggleton, pp. 471–483. New York: Wiley.

Gallate, J. E. and McGregor, I. S. (1999) "The motivation for beer in rats: effects of ritanserin, naloxone and SR 141716," *Psychopharmacology* **142**: 302–308.

Gallate, J. E., Saharov, T., Mallet, P. E. and McGregor, I. S. (1999) "Increased motivation for beer in rats following administration of a cannabinoid CB_1 receptor agonist," *European Journal of Pharmacology* **370**: 233–240.

Galzin, A. M., Dubocovich, M. L. and Langer, S. Z. (1982) "Presynaptic inhibition of dopamine receptor agonists of noradrenergic neurotransmission in the rabbit hypothalamus," *Journal of Pharmacology and Experimental Therapeutics* **221**: 461–471.

Galzin, A. M., Moret, C., Verzier, B. and Langer, S. Z. (1985) "Interaction between tricyclic and nontricyclic 5-hydroxytryptamine uptake inhibitor and the presynaptic 5-hydroxytryptamine inhibitory autoreceptors in the rat hypothalamus," *Journal of Pharmacology and Experimental Therapeutics* **235**: 200–211.

Ganchrow, J. R., Lieblich, I. and Cohen, E. (1981) "Consummatory responses to taste stimuli in rats selected for high and low rates of self-stimulation," *Physiology and Behavior* **27**: 971–976.

Gardner, E. L. (1992) "Cannabinoid interaction with brain reward systems – the neurobiological basis of cannabinoid abuse," in *Marijuana/Cannabinoids: Neurobiology and Neurophysiology*, edited by L. L. Murphy and A. Bartke, pp. 275–335. New York: CRC Press.

Gardner, E. L. (1997) "Brain reward mechanisms," in *Substance Abuse: A Comprehensive Textbook*, 3rd edn, edited by J. H. Lowinson, P. Ruiz, R. B. Millman and J. G. Langrod, pp. 51–85. Baltimore: Williams & Wilkins.

Gardner, E. L. (1999) "Cannabinoid interaction with brain reward systems," in *Marihuana and Medicine*, edited by G. G. Nahas, S. Agurell, K. M. Sutin and D. J. Harvey, pp. 187–205. Totowa, New Jersey: Humana Press.

Gardner, E. L. (2000) "What we have learned about addiction from animal models of drug self-administration," *American Journal on Addictions* **9**: 285–313.

Gardner, E. L. and David, J. (1999) "The neurobiology of chemical addiction," in *Getting Hooked: Rationality and the Addictions*, edited by J. Elster and O.-J. Skog, pp. 93–136. Cambridge, England: Cambridge University Press.

Gardner, E. L. and Lepore, M. (1996) "Withdrawal from a single dose of marijuana elevates baseline brain-stimulation reward thresholds in rats," Paper presented at meetings of the Winter Conference on *Brain Research*, Aspen, Colorado.

Gardner, E. L. and Lowinson, J. H. (1991) "Marijuana's interaction with brain reward systems: update 1991," *Pharmacology Biochemistry and Behavior* **40**: 571–580.

Gardner, E. L. and Vorel, S. R. (1998) "Cannabinoid transmission and reward-related events," *Neurobiology of Disease* **5**: 502–533.

Gardner, E. L., Paredes, W., Smith, D., Donner, A., Milling, C., Cohen, D. *et al.* (1988a) "Facilitation of brain stimulation reward by Δ^9-tetrahydrocannabinol," *Psychopharmacology* **96**: 142–144.

Gardner, E. L., Paredes, W., Smith, D., Seeger, T., Donner, A., Milling, C. *et al.* (1988b) "Strain-specific sensitization of brain stimulation reward by Δ^9-tetrahydrocannabinol in laboratory rats," *Psychopharmacology* **96**(suppl.): 365.

Gardner, E. L., Chen, J., Paredes, W., Li, J. and Smith, D. (1989a) "Strain-specific facilitation of brain stimulation reward by Δ^9-tetrahydrocannabinol in laboratory rats is mirrored by strain-specific facilitation of presynaptic dopamine efflux in nucleus accumbens," *Society for Neuroscience Abstracts* **15**: 638.

Gardner, E. L., Paredes, W., Smith, D. and Zukin, R. S. (1989b) "Facilitation of brain stimulation reward by Δ^9-tetrahydrocannabinol is mediated by an endogenous opioid mechanism," *Advances in the Biosciences* **75**: 671–674.

Gardner, E. L., Chen, J., Paredes, W., Smith, D., Li, J. and Lowinson, J. (1990a) "Enhancement of presynaptic dopamine efflux in brain by Δ^9-tetrahydrocannabinol is mediated by an endogenous opioid mechanism," in *New Leads in Opioid Research*, edited by J. M. van Ree, A. H. Mulder, V. M. Wiegant and T. B. van Wimersma Greidanus, pp. 243–245. Amsterdam: Elsevier Science Publishers.

Gardner, E. L., Paredes, W. and Chen, J. (1990b) "Further evidence for Δ^9-tetrahydrocannabinol as a dopamine reuptake blocker: brain microdialysis studies," *Society for Neuroscience Abstracts* **16**: 1100.

Gardner, E. L., Liu, X., Paredes, W., Savage, V., Lowinson, J. and Lepore, M. (1995) "Strain-specific differences in Δ^9-tetrahydrocannabinol (THC)-induced facilitation of electrical brain stimulation reward (BSR)," *Society for Neuroscience Abstracts* **21**: 177.

George, F. R. (1987) "Genetic and environmental factors in ethanol self-administration," *Pharmacology Biochemistry and Behavior* **27**: 379–384.

George, F. R. and Goldberg, S. R. (1989) "Genetic approaches to the analysis of addiction processes," *Trends in Pharmacological Sciences* **10**: 78–83.

George, F. R. and Meisch, R. A. (1984) "Oral narcotic intake as a reinforcer: genotype × environment interaction," *Behavior Genetics* **14**: 603.

Gessa, G. L., Melis, M., Muntoni, A. L. and Diana, M. (1998) "Cannabinoids activate mesolimbic dopamine neurons by an action on cannabinoid CB1 receptors," *European Journal of Pharmacology* **341**: 39–44.

Gifford, A. N., Gardner, E. L. and Ashby, C. R. Jr. (1997) "The effect of intravenous administration of delta-9-tetrahydrocannabinol on the activity of A10 dopamine neurons recorded in vivo in anesthetized rats," *Neuropsychobiology* **36**: 96–99.

Gintzler, A. R. and Xu, H. (1991) "Different G proteins mediate the opioid inhibition or enhancement of evoked [5-methionine]enkephalin release," *Proceedings of the National Academy of Sciences of the United States of America* **88**: 4741–4745.

Goeders, N. E. and Smith, J. E. (1983) "Cortical dopaminergic involvement in cocaine reinforcement," *Science* **221**: 773–775.

Goeders, N. E., Lane, J. D. and Smith, J. E. (1984) "Self-administration of methionine enkephalin into the nucleus accumbens," *Pharmacology Biochemistry and Behavior* **20**: 451–455.

Goett, J. M. and Kay, E. J. (1981) "Lithium chloride and delta-9-THC lead to conditioned aversions in the pigeon," *Psychopharmacology* **72**: 215–216.

Göthert, M., Schliker, E. and Kostermann, F. (1983) "Relationship between transmitter uptake inhibition and effects of α-adrenoreceptor agonists on serotonin and noradrenaline release in the rat brain cortex," *Naunyn-Schmiedeberg's Archives of Pharmacology* **322**: 121–128.

Grenhoff, J., Aston-Jones, G. and Svensson, T. H. (1986) "Nicotinic effects on the firing pattern of midbrain dopamine neurons," *Acta Physiologica Scandinavica* **128**: 351–358.

Grimm, J. W. and See, R. E. (2000) "Dissociation of primary and secondary reward-relevant limbic nuclei in an animal model of relapse," *Neuropsychopharmacology* **22**: 473–479.

Groenewegen, H. J., Berendse, H. W., Meredith, G. E., Haber, S. N., Voorn, P., Wolters, J. G. *et al.* (1991) "Functional anatomy of the ventral limbic system innervated striatum," in *The*

Mesolimbic Dopamine System: From Motivation to Action, edited by P. Willner and J. Scheel-Kruger, pp. 19–60. New York: John Wiley and Sons.

Gross-Isseroff, R., Cohen, E. and Shavit, Y. (1992) "Comparison of mu opioid receptors in brains of rats bred for high or low rate of self-stimulation," *Physiology and Behavior* **51**: 1093–1096.

Guitart, X., Beitner-Johnson, D., Marby, D. W., Kosten, T. A. and Nestler, E. J. (1992) "Fischer and Lewis rat strains differ in basal levels of neurofilament proteins and their regulation by chronic morphine in the mesolimbic dopamine system," *Synapse* **12**: 242–253.

Gysling, K. and Wang, R. Y. (1983) "Morphine-induced activation of A10 dopamine neurons in the rat," *Brain Research* **277**: 119–127.

Hall, W., Solowij, N. and Lemon, J. (1994) *The Health and Psychological Consequences of Cannabis Use* (National Drug Strategy Monograph Series No. 25). Canberra: Australian Government Publishing Service.

Harris, R. T., Waters, W. and McLendon, D. (1974) "Evaluation of reinforcing capability of Δ^9-tetrahydrocannabinol in monkeys," *Psychopharmacologia* **37**: 23–29.

Hattendorf, C., Hattendorf, M., Coper, H. and Fernandes, M. (1977) "Interaction between Δ^9-tetrahydrocannabinol and *d*-amphetamine," *Psychopharmacology* **54**: 177–182.

Hayes, R. J., Vorel, S. R., Liu, X., Spector, J., Lachman, H. and Gardner, E. L. (1999) "Electrical stimulation of the basolateral nucleus of the amygdala reinstates cocaine-seeking behavior," *Society for Neuroscience Abstracts* **25**: 559.

Heal, D. J., Frankland, A. T. J. and Buckett, W. R. (1990) "A new and highly sensitive method for measuring 3-methoxytyramine using HPLC with electrochemical detection: studies with drugs which alter dopamine metabolism in the brain," *Neuropharmacology* **29**: 1141–1150.

Heimer, L. and Alheid, G. F. (1991) "Piecing together the puzzle of basal forebrain anatomy," *Advances in Experimental Medicine and Biology* **295**: 1–42.

Herkenham, M., Lynn, A. B., Johnson, M. R., Melvin, L. S., de Costa, B. R. and Rice, K. C. (1991a) "Characterization and localization of cannabinoid receptors in rat brain: a quantitative in vitro autoradiographic study," *Journal of Neuroscience* **11**: 563–583.

Herkenham, M., Lynn, A. B., De Costa, B. R. and Richfield, E. K. (1991b) "Neuronal localization of cannabinoid receptors in the basal ganglia of the rat," *Brain Research* **547**: 267–274.

Hershkowitz, M. and Szechtman, H. (1979) "Pretreatment with Δ^1-tetrahydrocannabinol and psychoactive drugs: effects on uptake of biogenic amines and on behavior," *European Journal of Pharmacology* **59**: 267–276.

Hiroi, N. and White, N. M. (1991) "The lateral nucleus of the amygdala mediates expression of the amphetamine conditioned place preference," *Journal of Neuroscience* **11**: 2107–2116.

Hitchcott, P. K. and Phillips, G. D. (1997) "Amygdala and hippocampus control dissociable aspects of drug-associated conditioned rewards," *Psychopharmacology* **131**: 187–195.

Hitchcott, P. K. and Phillips, G. D. (1998a) "Effects of intra-amygdala R(+)7-OH-DPAT on intra-accumbens d-amphetamine-associated learning. I. Pavlovian conditioning," *Psychopharmacology* **140**: 300–309.

Hitchcott, P. K. and Phillips, G. D. (1998b) "Effects of intra-amygdala R(+)7-OH-DPAT on intra-accumbens d-amphetamine-associated learning. II. Instrumental conditioning," *Psychopharmacology* **140**: 310–318.

Hitchcott, P. K., Bonardi, C. M. and Phillips, G. D. (1997) "Enhanced stimulus-reward learning by intra-amygdala administration of a D3 dopamine receptor agonist," *Psychopharmacology* **133**: 240–248.

Hoebel, B. G., Monaco, A. P., Hernandez, L., Aulisi, E. F., Stanley, B. G. and Lenard, L. (1983) "Self-injection of amphetamine directly into the brain," *Psychopharmacology* **81**: 158–163.

Hoffman, A. F. and Lupica, C. R. (2001) "Direct actions of cannabinoids on synaptic transmission in the nucleus accumbens: a comparison with opioids," *Journal of Neurophysiology* **85**: 72–83.

Howlett, A. C., Bidaut-Russell, M., Devane, W. A., Melvin, L. S., Johnson, M. R. and Herken-ham, M. (1990) "The cannabinoid receptor: biochemical, anatomical, and behavioral characterization," *Trends in Neurosciences* **13**: 420–423.

Jansen, E. M., Haycock, D. A., Ward, S. A. and Seybold, V. S. (1992) "Distribution of can-nabinoid receptors in rat brain determined with aminoalkylindoles," *Brain Research* **575**: 93–102.

Jentsch, J. D., Andrusiak, E., Tran, A., Bowers, M. B. Jr. and Roth, R. H. (1997) "Δ^9-Tetra-hydrocannabinol increases prefrontal cortical catecholaminergic utilization and impairs spatial working memory in the rat: blockade of dopaminergic effects with HA966," *Neuropsychopharmacology* **16**: 426–432.

Jentsch, J. D., Wise, A., Katz, Z. and Roth, R. H. (1998) "α-Noradrenergic receptor modulation of the phencyclidine- and Δ^9-tetrahydrocannabinol-induced increases in dopamine utilization in rat prefrontal cortex," *Synapse* **28**: 21–26.

Johanson, C. E., Kandel, D. A. and Bonese, K. (1976) "The effects of perphenazine on self-administration behavior," *Pharmacology Biochemistry and Behavior* **4**: 427–433.

Jönsson, L. E. (1972) "Pharmacological blockade of amphetamine effects in amphetamine dependent subjects," *European Journal of Clinical Pharmacology* **4**: 206–211.

Jorenby, D. E., Steinpreis, R. E., Sherman, J. E. and Baker, T. B. (1990) "Aversion instead of preference learning indicated by nicotine place conditioning in rats," *Psychopharmacology* **101**: 533–538.

Kalivas, P. W., Churchill, L. and Klitenick, M. A. (1993) "GABA and enkephalin projection from the nucleus accumbens and ventral pallidum to the ventral tegmental area," *Neuro-science* **57**: 1047–1060.

Kaymakçalan, S. (1972) "Physiology and psychological dependence on THC in Rhesus mon-keys," in *Cannabis and its Derivatives*, edited by W. D. M. Paton and J. Crown, pp. 142–149. London: Oxford University Press.

Kelley, A. E., Domesick, V. B. and Nauta, W. J .H. (1982) "The amygdalostriatal projection in the rat: an anatomical study by anterograde and retrograde tracing methods," *Neuroscience* **7**: 615–630.

Khodzhagel'diev, T. (1986) "Formirovanie vlecheniia k nikotinu u myshei linii C57Bl/6 i CBA [Development of nicotine preference in C57Bl/6 and CBA mice]," *Biulleten Eksperi-mentalnoi Biologii i Meditsiny* **101**: 48–50.

Kokkinidis, L. and McCarter, B. D. (1990) "Postcocaine depression and sensitization of brain-stimulation reward: analysis of reinforcement and performance effects," *Pharmacology Biochemistry and Behavior* **36**: 463–471.

Kokkinidis, L., Zacharko, R. M. and Predy, P. A. (1980) "Post-amphetamine depression of self-stimulation responding from the substantia nigra: reversal by tricyclic antidepres-sants," *Pharmacology Biochemistry and Behavior* **13**: 379–383.

Koob, G. F. (1996) "Drug addiction: the yin and yang of hedonic homeostasis," *Neuron* **16**: 893–896.

Koob, G. F. and Bloom, F. E. (1988) "Cellular and molecular mechanisms of drug depend-ence," *Science* **242**: 715–723.

Koob, G. F. and Le Moal, M. (1997) "Drug abuse: hedonic homeostatic dysregulation," *Science* **278**: 52–58.

Koob, G. F., Stinus, L., Le Moal, M. and Bloom, F. E. (1989) "Opponent process theory of motivation: neurobiological evidence from studies of opiate dependence," *Neuroscience and Biobehavioral Reviews* **13**: 135–140.

Kornetsky, C. (1985) "Brain-stimulation reward: a model for the neuronal bases for drug-induced euphoria," *National Institute on Drug Abuse Research Monograph Series* **62**: 30–50.

Kosten, T. A., Miserendino, M. J., Chi, S. and Nestler, E. J. (1994) "Fischer and Lewis rat strains show differential cocaine effects in conditioned place preference and behavioral sensitization but not in locomotor activity or conditioned taste aversion," *Journal of Pharmacology and Experimental Therapeutics* **269**: 137–144.

Kreek, M. J. and Koob, G. F. (1998) "Drug dependence: stress and dysregulation of brain reward pathways," *Drug and Alcohol Dependence* **51**: 23–47.

Kumar, M. S. and Chen, C. L. (1983) "Effect of an acute dose of delta-9-THC on hypothalamic luteinizing hormone releasing hormone and met-enkephalin content and serum levels of testosterone and corticosterone in rats," *Substance and Alcohol Actions/Misuse* **4**: 37–43.

Kumar, M. S., Patel, V. and Millard, W. J. (1984) "Effect of chronic administration of Δ^9-tetrahydrocannabinol on the endogenous opioid peptide and catecholamine levels in the diencephalon and plasma of the rat," *Substance and Alcohol Actions/Misuse* **5**: 201–210.

Kumar, A. M., Solomon, J., Patel, V., Kream, R. M., Drieze, J. M. and Millard, W. J. (1986) "Early exposure to Δ^9-tetrahydrocannabinol influences neuroendocrine and reproductive functions in female rats," *Neuroendocrinology* **44**: 260–264.

Kumar, A. M., Haney, M., Becker, T., Thompson, M. L., Kream, R. M. and Miczek, K. (1990) "Effect of early exposure to Δ^9-tetrahydrocannabinol on the levels of opioid peptides, gonadotropin-releasing hormone and substance P in the adult male rat brain," *Brain Research* **525**: 78–83.

Lander, E. S. and Schork, N. J. (1994) "Genetic dissection of complex traits," *Science* **265**: 2037–2048.

Landsman, R. S., Makryannis, A., Deng, H., Consroe, P., Roeske, W. R. and Yamamura, H. I. (1998) "AM630 is an inverse agonist at the human CB_1 receptor," *Life Sciences [Pharmacology Letters]* **62**: PL109–PL113.

Langer, S. Z. and Moret, C. (1982) "Citalopram antagonizes the stimulation by lysergic acid diethylamide of presynaptic inhibitory serotonin autoreceptors in the rat hypothalamus," *Journal of Pharmacology and Experimental Therapeutics* **222**: 220–226.

Laszlo, J., Lucas, V. S., Hanson, D. C., Cronin, C. M. and Sallan, S. E. (1981) "Levonantradol for chemotherapy-induced emesis: phase I–II oral administration," *Journal of Clinical Pharmacology* **21**: 51S–56S.

Ledent, C., Valverde, O., Cossu, G., Petitet, F., Aubert, J. F., Beslot, F. *et al.* (1999) "Unresponsiveness to cannabinoids and reduced addictive effects of opiates in CB1 receptor knockout mice," *Science* **283**: 401–404.

Le Doux, J. A., Iwata, J., Cicchetti, P. and Reis, D. J. (1988) "Different projections of the central amygdaloid nucleus mediate autonomic and behavioral correlates of conditioned fear," *Journal of Neuroscience* **8**: 2517–2529.

Leite, J. R. and Carlini, E. A. (1974) "Failure to obtain 'cannabis directed behavior' and abstinence syndrome in rats chronically treated with cannabis sativa extracts," *Psychopharmacologia* **36**: 133–145.

Lepore, M., Vorel, S. R., Lowinson, J. and Gardner, E. L. (1995) "Conditioned place preference induced by Δ^9-tetrahydrocannabinol: comparison with cocaine, morphine, and food reward," *Life Sciences* **56**: 2073–2080.

Lepore, M., Liu, X., Savage, V., Matalon, D. and Gardner, E. L. (1996) "Genetic differences in Δ^9-tetrahydrocannabinol-induced facilitation of brain stimulation reward as measured by a rate–frequency curve-shift electrical brain stimulation paradigm in three different rat strains," *Life Sciences [Pharmacology Letters]* **58**: PL365–PL372.

Li, Y., Hu, X. T., Berney, T. G., Vartanian, A. J., Stine, C. D., Wolf, M. E. *et al.* (1999) "Both glutamate receptor antagonists and prefrontal cortex lesions prevent induction of cocaine sensitization and associated neuroadaptations," *Synapse* **34**: 169–180.

Lieblich, I., Cohen, E., Beiles, A. and Beiles, A. (1978) "Selection for high and for low rates of self-stimulation in rats," *Physiology and Behavior* **21**: 843–849.

Loh, E. A. and Roberts, D. C. S. (1990) "Break-points on a progressive ratio schedule reinforced by intravenous cocaine increase following depletion of forebrain serotonin," *Psychopharmacology* **101**: 262–266.

Lyness, W. H., Friedle, N. M. and Moore, K. E. (1979) "Destruction of dopaminergic nerve terminals in nucleus accumbens: effect on d-amphetamine self-administration," *Pharmacology, Biochemistry and Behavior* **11**: 553–556.

MacCoun, R. and Reuter, P. (1997) "Interpreting Dutch cannabis policy: reasoning by analogy in the legalization debate," *Science* **278**: 47–52.

Mailleux, P. and Vanderhaeghen, J. J. (1992) "Distribution of neuronal cannabinoid receptor in the adult rat brain: a comparative receptor binding radioautography and in situ hybridization histochemistry," *Neuroscience* **48**: 655–668.

Manzanares, J., Corchero, J., Romero, J., Fernandez-Ruiz, J. J., Ramos, J. A. and Fuentes, J. A. (1998) "Chronic administration of cannabinoids regulates proenkephalin mRNA levels in selected regions of the rat brain," *Molecular Brain Research* **55**: 126–132.

Makriyannis, A. (1993) "Probes for the cannabinoid sites of action," *National Institute on Drug Abuse Research Monograph Series* **134**: 253–267.

Markou, A. and Koob, G. F. (1991) "Postcocaine anhedonia: an animal model of cocaine withdrawal," *Neuropsychopharmacology* **4**: 17–26.

Martellotta, M. C., Cossu, G., Fattore, L., Gessa, G. L. and Fratta, W. (1998) "Self-administration of the cannabinoid receptor agonist WIN 55,212-2 in drug-naive mice," *Neuroscience* **85**: 327–330.

Martin, B. R., Welch, S. P. and Abood, M. (1994) "Progress toward understanding the cannabinoid receptor and its second messenger system," *Advances in Pharmacology* **25**: 341–397.

Mascia, M. S., Obinu, M. C., Ledent, C., Parmentier, M., Böhme, G. A., Imperato, A. *et al.* (1999) "Lack of morphine-induced dopamine release in the nucleus accumbens of cannabinoid CB_1 receptor knockout mice," *European Journal of Pharmacology* **383**: R1–R2.

Matsuda, L. A., Lolait, S. J., Brownstein, M. J., Young, A. C. and Bonner, T. I. (1990) "Structure of a cannabinoid receptor and functional expression of the cloned cDNA," *Nature* **346**: 561–564.

Matsuda, L . A., Bonner, T. I. and Lolait, S. J. (1993) "Localization of cannabinoid receptor mRNA in rat brain," *Journal of Comparative Neurology* **327**: 535–550.

McGinty, J. F. (1999) "Regulation of neurotransmitter interactions in the ventral striatum," *Annals of the New York Academy of Sciences* **877**: 129–139.

McGregor, I. S., Issakidis, C. N. and Prior, G. (1996) "Aversive effects of the synthetic cannabinoid CP 55,940 in rats," *Pharmacology Biochemistry and Behavior* **53**: 657–664.

Mechoulam, R. (1986) "The pharmacohistory of cannabis sativa," in *Cannabinoids as Therapeutic Agents*, edited by R. Mechoulam, pp. 1–16. Boca Raton, Florida: CRC Press.

Mechoulam, R., Hanus, L. and Martin, B. R. (1994) "Search for endogenous ligands of the cannabinoid receptor," *Biochemical Pharmacology* **48**: 1537–1544.

Mechoulam, R., Ben Shabat, S., Hanus, L., Ligumsky, M., Kaminski, N. E., Schatz, A. R. *et al.* (1995) "Identification of an endogenous 2-monoglyceride, present in canine gut, that binds to cannabinoid receptors," *Biochemical Pharmacology* **50**: 83–90.

Mechoulam, R., Ben Shabat, S., Hanus, L., Fride, E., Vogel, Z., Bayewitch, M. *et al.* (1996) "Endogenous cannabinoid ligands – chemical and biological studies," *Journal of Lipid Mediators and Cell Signalling* **14**: 45–49.

Melvin, L. S., Milne, G. M., Johnson, M. R., Subramaniam, B., Wilken, G. H. and Howlett, A. C. (1993) "Structure–activity relationships for receptor-binding and analgesic activity: studies of bicyclic cannabinoid analogs," *Molecular Pharmacology* **44**: 1008–1015.

Melvin, L. S., Milne, G. M., Johnson, M. R., Wilken, G. H. and Howlett, A. C. (1995) "Structure–activity relationships defining the ACD-tricyclic cannabinoids: cannabinoid receptor binding and analgesic activity," *Drug Design and Discovery* **13**: 155–166.

Merlo Pich, E., Lorang, M., Yeganeh, M., Rodríguez de Fonseca, F., Raber, J., Koob, G. F. *et al.* (1995) "Increase in extracellular corticotropin-releasing factor-like immunoreactivity levels in the amygdala of awake rats during restraint stress and ethanol withdrawal as measured by microdialysis," *Journal of Neuroscience* **15**: 5439–5447.

Milano, W. C., Wild, K. D., Hui, Y. Z., Hubbell, C. L. and Reid, L. D. (1988) "PCP, THC, ethanol, and morphine and consumption of palatable solutions," *Pharmacology Biochemistry and Behavior* **31**: 893–897.

Miller, A. S. and Walker, J. M. (1995) "Effects of a cannabinoid on spontaneous and evoked neuronal activity in the substantia nigra pars reticulata," *European Journal of Pharmacology* **279**: 179–185.

Moldrich, G. and Wenger, T. (2000) "Localization of the CB_1 cannabinoid receptor in the rat brain. An immunohistochemistry study," *Peptides* **21**: 1735–1742.

Moret, C. and Briley, M. (1988) "Sensitivity of the response of 5-HT autoreceptors to drugs modifying synaptic availability of 5-HT," *Neuropharmacology* **27**: 43–49.

Murphy, L. and Bartke, A. (eds) (1992) *Marijuana/Cannabinoids: Neurobiology and Neurophysiology.* Boca Raton, Florida: CRC Press.

Navarro, M., Fernández-Ruiz, J. J., de Miguel, R., Hernández, M. L., Cebeira, M. and Ramos, J. A. (1993a) "Motor disturbances induced by an acute dose of Δ^9-tetrahydrocannabinol: possible involvement of nigrostriatal dopaminergic alterations," *Pharmacology, Biochemistry and Behavior* **45**: 291–298.

Navarro, M., Fernández-Ruiz, J. J., de Miguel, R., Hernández, M. L., Cebeira, M. and Ramos, J. A. (1993b) "An acute dose of δ^9-tetrahydrocannabinol affects behavioral and neurochemical indices of mesolimbic dopaminergic activity," *Behavioural Brain Research* **57**: 37–46.

Navarro, M., Rubio, P. and Rodríguez de Fonseca, F. (1995) "Behavioural consequences of maternal exposure to natural cannabinoids in rats," *Psychopharmacology* **122**: 1–14.

Navarro, M., de Miguel, R., Rodríguez de Fonseca, F., Ramos, J. A. and Fernández-Ruiz, J. J. (1996) "Prenatal cannabinoid exposure modifies the sociosexual approach behavior and the mesolimbic dopaminergic activity of adult male rats," *Behavioural Brain Research* **75**: 91–98.

Navarro, M., Chowen, J., Carrera, M. R. A., del Arco, I., Villanúa, M. A., Martin, Y. *et al.* (1998) "CB_1 cannabinoid receptor antagonist-induced opiate withdrawal in morphine-dependent rats," *NeuroReport* **9**: 3397–3402.

Nestler, E. J. (1993) "Molecular mechanisms of drug addiction in the mesolimbic dopamine pathway," *Seminars in the Neurosciences* **5**: 369–376.

Nestler, E. J., Berhow, M. T. and Brodkin, E. S. (1996) "Molecular mechanisms of drug addiction: adaptations in signal transduction pathways," *Molecular Psychiatry* **1**: 190–199.

Ng Cheong Ton, J. M. and Gardner, E. L. (1986) "Effects of delta-9-tetrahydrocannabinol on dopamine release in the brain: intracranial dialysis experiments," *Society for Neuroscience Abstracts* **12**: 135.

Ng Cheong Ton, J. M., Gerhardt, G. A., Friedemann, M., Etgen, A. M., Rose, G. M., Sharpless, N. S. *et al.* (1988) "The effects of Δ^9-tetrahydrocannabinol on potassium-evoked release of dopamine in the rat caudate nucleus: an in vivo electrochemical and in vivo microdialysis study," *Brain Research* **451**: 59–68.

Noyes, J. R., Brunk, S. F., Avery, D. H. and Canter, A. (1975) "The analgesic properties of delta-9-tetrahydrocannabinol and codeine," *Clinical Pharmacology and Therapeutics* **18**: 84–89.

Oakman, S. A., Faris, P. L., Kerr, P. E., Cozzari, C. and Hartman, B. K. (1995) "Distribution of pontomesencephalic cholinergic neurons projecting to substantia nigra differs significantly from those projecting to ventral tegmental area," *Journal of Neuroscience* **15**: 5859–5869.

Overton, P. G. and Clark, D. (1997) "Burst firing in midbrain dopaminergic neurons," *Brain Research Reviews* **25**: 312–334.

Parker, L. A. and Gillies, T. (1995) "THC-induced place and taste aversions in Lewis and Sprague-Dawley rats," *Behavioral Neuroscience* **109**: 71–78.

Parsons, L. H., Smith, A. D. and Justice, J. B. Jr. (1991) "Basal extracellular dopamine is decreased in the rat nucleus accumbens during abstinence from chronic cocaine," *Synapse* **9**: 60–65.

Patel, V., Borysenko, M. and Kumar, M. S. (1985) "Effect of Δ^9-THC on brain and plasma catecholamine levels as measured by HPLC," *Brain Research Bulletin* **14**: 85–90.

Poddar, M. K. and Dewey, W. L. (1980) "Effects of cannabinoids on catecholamine uptake and release in hypothalamic and striatal synaptosomes," *Journal of Pharmacology and Experimental Therapeutics* **214**: 63–67.

Pertwee, R. (ed.) (1995) *Cannabinoid Receptors*. London: Academic Press.

Phillips, A. G. and LePiane, F. G. (1980) "Reinforcing effects of morphine microinjection into the ventral tegmental area," *Pharmacology Biochemistry and Behavior* **12**: 965–968.

Pickens, R., Thompson, T. and Muchow, D. C. (1973) "Cannabis and phencyclidine self-administered by animals," in *Psychic Dependence* [Bayer-Symposium IV], edited by L. Goldfarb and F. Hoffmeister, pp. 78–86. Berlin: Springer-Verlag.

Piomelli, D., Baltramo, M., Giufridda, A. and Stella, N. (1998) "Endogenous cannabinoid signalling," *Neurobiology of Disease* **5**: 462–473.

Pontieri, F. E., Tanda, G. and Di Chiara, G. (1995) "Intravenous cocaine, morphine, and amphetamine preferentially increase extracellular dopamine in the 'shell' as compared with the 'core' of the rat nucleus accumbens," *Proceedings of the National Academy of Sciences of the United States of America* **92**: 12304–12308.

Pope, H. G. Jr. and Yurgelun-Todd, D. (1996) "The residual cognitive effects of heavy marijuana use in college students," *Journal of the American Medical Association* **275**: 521–527.

Pothos, E., Rada, P., Mark, G. P. and Hoebel, B. G. (1991) "Dopamine microdialysis in the nucleus accumbens during acute and chronic morphine, naloxone-precipitated withdrawal and clonidine treatment," *Brain Research* **566**: 348–350.

Raft, D., Gregg, J., Ghia, J. and Harris, L. (1977) "Effects of intravenous tetrahydrocannabinol on experimental and surgical pain. Psychological correlate of the analgesic response," *Clinical Pharmacology and Therapeutics* **21**: 26–33.

Rinaldi-Carmona, M., Barth, F., Heaulme, M., Alonso, R., Shire, D., Congy, C. *et al.* (1995) "Biochemical and pharmacological characterisation of SR141716A, the first potent and selective brain cannabinoid receptor antagonist," *Life Sciences* **56**: 1941–1947.

Roberts, D. C. S. and Koob, G. F. (1982) "Disruption of cocaine self-administration following 6-hydroxydopamine lesions of the ventral tegmental area in rats," *Pharmacology Biochemistry and Behavior* **17**: 901–904.

Robertson, M. W., Leslie, C. A. and Bennett, J. P. (1991) "Apparent synaptic dopamine deficiency induced by withdrawal from chronic cocaine treatment," *Brain Research* **538**: 337–339.

Robbins, T. W. and Everitt, B. J. (1992) "Amygdala-ventral striatal interactions and reward related processes," in *The Amygdala: Neurobiological Aspects of Emotions, Memory and Mental Dysfunction*, edited by J. P. Aggleton, pp. 401–429. New York: Wiley-Liss.

Robbins, T. W. and Everitt, B. J. (1996) "Neurobehavioral mechanisms of reward and motivation," *Current Opinion in Neurobiology* **6**: 228–236.

Rodríguez de Fonseca, F., Fernández-Ruiz, J. J., Murphy, L. L., Cebeira, M., Steger, R. W., Bartke, A. *et al.* (1992) "Acute effects of Δ-9-tetrahydrocannabinol on dopaminergic activity in several rat brain areas," *Pharmacology Biochemistry and Behavior* **42**: 269–275.

Rodríguez de Fonseca, Carrera, M. R. A., Navarro, M., Koob, G. F. and Weiss, F. (1997) "Activation of corticotropin-releasing factor in the limbic system during cannabinoid withdrawal," *Science* **276**: 2050–2054.

Rossetti, Z. L., Hmaidan, Y. and Gessa, G. L. (1992) "Marked inhibition of mesolimbic dopamine release: a common feature of ethanol, morphine, cocaine and amphetamine abstinence in rats," *European Journal of Pharmacology* **221**: 227–234.

Ryan, W., Singer, M., Razdan, R. K., Compton, D. R. and Martin, B. R. (1995) "A novel class of potent tetrahydrocannabinols (THCs): 2′-yne-Δ^8 and Δ^9 THCs," *Life Sciences* **56**: 2013–2020.

Sakurai-Yamashita, Y., Kataoka, Y., Fujiwara, M., Mine, K. and Ueki, S. (1989) "Δ^9-Tetra-hydrocannabinol facilitates striatal dopaminergic transmission," *Pharmacology Biochemistry and Behavior* **33**: 397–400.

Sala, M. and Braida, D. (2000) "Interaction between cannabinoids and opiates on rewarding effects: place preference and i.c.v. self-administration studies in rats," Paper presented at meetings of the International Cannabinoid Research Society, Hunt Valley, Maryland.

Sañudo-Peña, M. C., Tsou, K., Delay, E. R., Hohman, A. G., Force, M. and Walker, J. M. (1997) "Endogenous cannabinoids as an aversive or counter-rewarding system in the rat," *Neuroscience Letters* **223**: 125–128.

Schaefer, G. J. and Michael, R. P. (1986) "Changes in response rates and reinforcement thresholds for intracranial self-stimulation during morphine withdrawal," *Pharmacology Biochemistry and Behavior* **25**: 1263–1269.

Schulteis, G., Markou, A., Gold, L. H., Stinus, L. and Koob, G. F. (1994) "Relative sensitivity of multiple indices of opiate withdrawal: a quantitative dose–response analysis," *Journal of Pharmacology and Experimental Therapeutics* **271**: 1391–1398.

Self, D. W. and Nestler, E. J. (1995) "Molecular mechanisms of drug reinforcement and addiction," *Annual Review of Neuroscience* **18**: 463–495.

Sesack, S. R. and Pickel, V. M. (1992) "Prefrontal cortical efferents in the rat synapse on unlabeled neuronal targets of catecholamine terminals in the nucleus accumbens septi and on dopamine neurons in the ventral tegmental area," *Journal of Comparative Neurology* **320**: 145–160.

Shizgal, P. (1997) "Neural basis of utility estimation," *Current Opinion in Neurobiology* **7**: 198–208.

Shore, P. A., McMillen, B. A., Miller, H. H., Sanghera, M. K., Kiserand, R. S. and German, D. C. (1979) "The dopamine neuronal storage system and non-amphetamine psychotogenic stimulants: a model for psychosis," in *Catecholamines: Basic and Clinical Frontiers*, edited by E. Usdin, I. J. Kopin and J. Barchas, pp. 722–735. New York: Pergamon Press.

Sieber, B., Frischknecht, H. R. and Waser, P. G. (1981) "Behavioral effects of hashish in mice. IV. Social dominance, food dominance, and sexual behavior within a group of males," *Psychopharmacology* **73**: 142–146.

Sofia, R. D. and Knoblock, L. C. (1976) "Comparative effects of various naturally occurring cannabinoids on food, sucrose and water consumption by rats," *Pharmacology Biochemistry and Behavior* **4**: 591–599.

Spanagel, R., Almeida, O. F., Bartl, C. and Shippenberg, T. S. (1994) "Endogenous kappa-opioid systems in opiate withdrawal: role in aversion and accompanying changes in mesolimbic dopamine release," *Psychopharmacology* **115**: 121–127.

Spyraki, C., Fibiger, H. C. and Phillips, A. G. (1983) "Attenuation of heroin reward in rats by disruption of the mesolimbic dopamine system," *Psychopharmacology* **79**: 278–283.

Stella, N., Schweitzer, P. and Piomelli, D. (1997) "A second endogenous cannabinoid that modulates long-term potentiation," *Nature* **388**: 773–778.

Sugiura, T., Kondo, S., Sukagawa, A., Nakane, S., Shinoda, A., Itoh, K. *et al.* (1995) "2-Arachidonoylglycerol: a possible endogenous cannabinoid receptor ligand in brain," *Biochemical and Biophysical Research Communications* **215**: 89–97.

Sugiura, T., Kondo, S., Sukagawa, A., Tonegawa, T., Nakane, S., Yamashita, A. *et al.* (1996) "N-arachidonoylethanolamide (anandamide), an endogenous cannabinoid ligand, and related lipid molecules in the nervous system," *Journal of Lipid Mediators and Cell Signalling* **14**: 51–56.

Suzuki, T., George, F. R. and Meisch, R. A. (1988) "Differential establishment and maintenance of oral ethanol reinforced behavior in Lewis and Fisher 344 inbred rat strains," *Journal of Pharmacology and Experimental Therapeutics* **245**: 164–170.

Szabo, B., Wallmichrath, I., Mathonia, P. and Pfreundtner, C. (2000) "Cannabinoids inhibit excitatory neurotransmission in the substantia nigra pars reticulata," *Neuroscience* **97**: 89–97.

Takahashi, R. N. and Singer, G. (1979) "Self-administration of Δ^9-tetrahydrocannabinol by rats," *Pharmacology Biochemistry and Behavior* **11**: 737–740.

Takahashi, R. N. and Singer, G. (1980) "Effects of body weight levels on cannabis self-administration," *Pharmacology Biochemistry and Behavior* **13**: 877–881.

Takahashi, R .N. and Singer, G. (1981) "Cross self-administration of delta 9-tetrahydrocannabinol and D-amphetamine in rats," *Brazilian Journal of Medical and Biological Research* **14**: 395–400.

Tanda, G., Pontieri, F. E. and Di Chiara, G. (1997) "Cannabinoid and heroin activation of mesolimbic dopamine transmission by a common μ_1 opioid receptor mechanism," *Science* **276**: 2048–2050.

Tanda, G., Munzar, P. and Goldberg, S. R. (2000) "Self-administration behavior is maintained by the psychoactive ingredient of marijuana in squirrel monkeys," *Nature Neuroscience* **3**: 1073–1074.

Tarazi, F. I. and Baldessarini, R. J. (1999) "Regional localization of dopamine and ionotropic glutamate receptor subtypes in striatolimbic brain regions," *Journal of Neuroscience Research* **55**: 401–410.

Taylor, D. A., Sitaram, B. R. and Elliot-Baker, S. (1988) "Effect of Δ-9-tetrahydrocannabinol on release of dopamine in the corpus striatum of the rat," in *Marijuana: An International Research Report*, edited by G. Chesher, P. Consroe and R. Musty, pp. 405–408. Canberra: Australian Government Publishing Service.

Tersigni, T. J. and Rosenberg, H. C. (1996) "Local pressure application of cannabinoid agonists increases spontaneous activity of rat substantia nigra pars reticulata neurons without affecting response to iontophoretically-applied GABA," *Brain Research* **733**: 184–192.

Thorat, S. N. and Bhargava, H. N. (1994) "Evidence for a bidirectional cross-tolerance between morphine and Δ^9-tetrahydrocannabinol in mice," *European Journal of Pharmacology* **260**: 5–13.

Tius, M. A., Hill, W. A., Zou, X. L., Busch-Petersen, J., Kawatami, J. K., Fernandez-Garcia, M. C. *et al.* (1995) "Classical/non-classical cannabinoid hybrids: stereochemical requirements for the southern hydroxylalkyl chain," *Life Sciences* **56**: 2007–2012.

Trojniar, W. and Wise, R. A. (1991) "Facilitatory effect of Δ^9-tetrahydrocannabinol on hypothalamically induced feeding," *Psychopharmacology* **103**: 172–176.

Turley, W. A. Jr. and Floody, A. R. (1981) "Δ-9-Tetrahydrocannabinol stimulates receptive and proceptive sexual behaviors in female hamsters," *Pharmacology Biochemistry and Behavior* **14**: 745–747.

Uhl, G., Blum, K., Noble, E. and Smith, S. (1994) "Substance abuse vulnerability and D2 receptor gene," *Trends in Neurosciences* **16**: 83–88.

Uhl, G. R., Elmer, G. I., Labuda, M. C. and Pickens, R. W. (1995) "Genetic influences in drug abuse," in *Psychopharmacology: The Fourth Generation of Progress*, edited by F. E. Bloom and D. J. Kupfer, pp. 1793–1806. New York: Raven Press.

Valjent, E. and Maldonado, R. (2000) "A behavioural model to reveal place preference to Δ^9-tetrahydrocannabinol in mice," *Psychopharmacology* **147**: 436–438.

Van Bockstaele, E. J. and Pickel, V. M. (1995) "GABA-containing neurons in the ventral tegmental area project to the nucleus accumbens in rat brain," *Brain Research* **682**: 215–221.

van der Kooy, D. (1987) "Place conditioning: a simple and effective method for assessing the motivational properties of drugs," in *Methods for Assessing the Reinforcing Properties of Abused Drugs*, edited by M. A. Bozarth, pp. 229–240. New York: Springer-Verlag.

Vaysse, P. J.-J., Gardner, E. L. and Zukin, R. S. (1985) "Modulation of rat brain opiate receptors by cannabinoids," Paper presented at meetings of the International Narcotics Research Conference, Falmouth, Massachusetts.

Vaysse, P. J.-J., Gardner, E. L. and Zukin, R. S. (1987) "Modulation of rat brain opioid receptors by cannabinoids," *Journal of Pharmacology and Experimental Therapeutics* **241**: 534–539.

Wallace, D. M., Magnuson, D. J. and Gray, T. S. (1992) "Organization of amygdaloid projections to brainstem dopaminergic, noradrenergic, and adrenergic cell groups in the rat," *Brain Research Bulletin* **28**: 447–454.

Westerink, B. H., Tuntler, J., Damsma, G., Rollema, H. and de Vries, J. B. (1987) "The use of tetrodotoxin for the characterization of drug-enhanced dopamine release in conscious rats studied by brain dialysis," *Naunyn Schmiedeberg's Archives of Pharmacology* **336**: 502–507.

White, N. M. and Hiroi, N. (1993) "Amphetamine conditioned cue preference and the neurobiology of drug-seeking," *Seminars in the Neurosciences* **5**: 329–336.

Wise, R. A. (1980) "Action of drugs of abuse on brain reward systems," *Pharmacology Biochemistry and Behavior* **13**(suppl.1): 213–223.

Wise, R. A. (1984) "Neural mechanisms of the reinforcing action of cocaine," *National Institute on Drug Abuse Research Monograph Series* **50**: 15–33.

Wise, R. A. (1996a) "Neurobiology of addiction," *Current Opinion in Neurobiology* **6**: 243–251.

Wise, R. A. (1996b) "Addictive drugs and brain stimulation reward," *Annual Review of Neuroscience* **19**: 319–340.

Wise, R. A. (1998) "Drug-activation of brain reward pathways," *Drug and Alcohol Dependence* **51**: 13–22.

Wise, R. A. and Bozarth, M. A. (1984) "Brain reward circuitry: four circuit elements 'wired' in apparent series," *Brain Research Bulletin* **12**: 203–208.

Wise, R. A. and Munn, E. (1995) "Withdrawal from chronic amphetamine elevates baseline intracranial self-stimulation thresholds," *Psychopharmacology* **117**: 130–136.

Wise, R. A. and Rompré, P.-P. (1989) "Brain dopamine and reward," *Annual Review of Psychology* **40**: 191–225.

Wood, P. L. and Altar, C. A. (1988) "Dopamine release in vivo from nigrostriatal, mesolimbic, and mesocortical neurons: utility of 3-methoxytyramine measurements," *Pharmacological Reviews* **40**: 163–187.

Woolverton, W. L. (1986) "Effects of a D_1 and D_2 dopamine antagonist on the self-administration of cocaine and piribedil by rhesus monkeys," *Pharmacology Biochemistry and Behavior* **24**: 531–535.

Xie, X. Q., Eissenstat, M. and Makriyannis, A. (1995) "Common cannabimimetic pharmacophoric requirements between aminoalkyl indoles and classical cannabinoids," *Life Sciences* **56**: 1963–1970.

Xie, X. Q., Melvin, L. S. and Makriyannis, A. (1996) "The conformational properties of the highly selective cannabinoid receptor ligand CP-55,940," *Journal of Biological Chemistry* **271**: 10640–10647.

Yokel, R. A. and Wise, R. A. (1976) "Attenuation of intravenous amphetamine reinforcement by central dopamine blockade in rats," *Psychopharmacology* **48**: 311–318.

Zahm, D. S. and Brog, J. S. (1992) "On the significance of subterritories in the 'accumbens' part of the rat ventral striatum," *Neuroscience* **50**: 751–767.

Chapter 4

Effects of marijuana on human performance and assessment of driving impairment

Stephen J. Heishman

ABSTRACT

Marijuana is the most widely used illicit drug in the United States with 72 million people reporting having used marijuana in their life. Among individuals who reported driving after drug use, 70% indicated they had smoked marijuana within 2 h of driving. Laboratory studies in which subjects smoked marijuana under controlled conditons have documented that marijuana reliably impaired sensory-perceptual abilities, gross motor coordination, psychomotor abilities, divided and sustained attention, and cognitive functioning, including learning and memory. Other laboratory research has focused on the applied question of whether marijuana impairs driving abilities as measured by standardized field sobriety tests used by police to detect impaired drivers. This research has shown that marijuana impaired balance and psychomotor coordination in a dose-dependent manner. Such behavioral impairment may underlie marijuana-induced decrements in tests of on-road driving. Plasma Δ^9-tetrahydrocannabinol (Δ^9-THC) concentration in the range of 15–30 ng/ml was associated with impaired balance and motor coordination, suggesting that driving abilities would be impaired at the same Δ^9-THC concentration. Future research should continue to explore the relationship between plasma Δ^9-THC concentration and human performance, and more studies are needed on the effects of marijuana combined with other commonly used drugs, such as alcohol, on driving abilities.

Key Words: marijuana, THC, performance, cognition, field sobriety tests, driving

INTRODUCTION

Marijuana is by far the most widely used illicit drug in the United States, accounting for 81% of all reported illicit drug use in 1998. Recent surveys indicated that 33% of the United States population over age 12 (72 million people) reported use of marijuana in their lifetime, and 5% (11 million) were current users, reporting use in the past month (Substance Abuse and Mental Health Services Administration, 1999).

Driving under the influence of drugs has become a major public health concern. A recent study found that 28% of drivers in the United States (46.8 million people) reported having driven within 2 h of drinking alcohol and/or using drugs in the past year. Of these drivers, 70% reported using marijuana before driving, and the majority reported heavy or weekly marijuana use (Townsend *et al.*, 1998). The critical question is, of course, whether driving ability is impaired as a result of recent marijuana use. In the study, 56% of drivers indicated that marijuana did not adversely affect their driving ability. However, many studies have documented that acute marijuana administration can impair cognitive and psychomotor functioning. Thus, driving under the influence of marijuana may play a causative role in traffic accidents and injuries (Bates and Blakely, 1999).

The effects of smoked marijuana and Δ^9-tetrahydrocannabinol (Δ^9-THC), the primary psychoactive constituent of marijuana, have been investigated in numerous studies over the past several decades (Chait and Pierri, 1992; Beardsley and Kelly, 1999). The purpose of this chapter is to provide a brief overview of the *acute* effects of marijuana on human performance as assessed in the laboratory. The chapter by Solowij (this volume) provides a review of the effects of *chronic* marijuana use on cognitive functioning. Following this overview is a discussion of two studies in which my colleagues and I investigated the validity of a standardized assessment procedure used by police to determine whether someone is impaired as the result of drug use and whether they can safely operate a motor vehicle.

MARIJUANA AND HUMAN PERFORMANCE

Marijuana consists of the dried and crushed leaves and stems of the plant, *Cannabis sativa*, which grows worldwide. In the United States, marijuana is typically rolled in cigarettes (joints) and smoked. Unless otherwise noted, the studies reviewed in this chapter were conducted with experienced marijuana users who smoked standard marijuana cigarettes provided by the National Institute on Drug Abuse. These marijuana cigarettes resemble an unfiltered tobacco cigarette in size, weigh 700–900 mg, and are assayed to determine the percentage of Δ^9-THC by weight. Doses are typically manipulated by using cigarettes that differ in Δ^9-THC content or by varying the number of puffs administered to subjects. Placebo marijuana cigarettes are also available, which have had the Δ^9-THC removed chemically from the plant material. When burned, placebo cigarettes smell identical to active marijuana cigarettes.

The overview of the acute effects of marijuana is divided into sensory, motor, attentional, and cognitive abilities, which provides a focus on the behaviors being assessed, rather than emphasizing the individual performance tests. This section concludes with a brief discussion of two behavioral effects of marijuana that have received much research attention, next-day or hangover effects and an amotivational syndrome.

Sensory abilities

A frequently used measure of central nervous system (CNS) functioning is critical flicker frequency (CFF) threshold. The task requires subjects to view a light stimulus and to note the point (CFF threshold) at which the steady light begins to flicker (and vice versa), as the frequency of the light is manipulated. An increase in CFF threshold indicates increased cortical and behavioral arousal, whereas a decrease suggests lowered CNS arousal (Smith and Misiak, 1976). Surprisingly, few studies have investigated the effect of marijuana on CFF threshold. Block *et al.* (1992) reported that one marijuana cigarette (2.6% Δ^9-THC) decreased CFF threshold compared with placebo. However, Liguori *et al.* (1998) reported that one marijuana cigarette (1.8 and 4.0% Δ^9-THC) did not affect CFF threshold. Methodological difference between the studies may account for the conflicting results. More studies are needed to clarify the effect of marijuana on CFF threshold.

Although more a perceptual process than a sensory ability, a commonly reported effect of marijuana is to increase the subjective passage of time relative to clock time. This typically results in subjects either overestimating an experimenter-generated time interval (Chait *et al.*, 1985) or underproducing a subject-generated interval (Chait and Perry, 1994). However, Heishman *et al.* (1997) found that 3.6% Δ^9-THC marijuana (4, 8, or 16 puffs) had no effect on either time estimation or production.

Motor abilities

In their review, Chait and Pierri (1992) indicated that marijuana moderately impaired balance (increased body sway) and hand steadiness. Consistent with this motor impairment, one marijuana cigarette (1.5 or 4.0% Δ^9-THC) was found to decrease postural balance as subjects attempted to maintain balance while standing on a moving platform (Greenberg *et al.*, 1994; Liguori *et al.*, 1998). The circular lights test is another measure of gross motor coordination in which subjects extinguish lights by pressing buttons that are arranged in a circle on a wall-mounted panel. Cone *et al.* (1986) found that two marijuana cigarettes (2.8% Δ^9-THC) impaired performance on the circular lights task, whereas Heishman *et al.* (1988) reported no effect of marijuana (1.3 and 2.7% Δ^9-THC, 2 cigarettes) on circular lights performance.

The effect of marijuana has also been examined on fine motor control. The time taken to sort a deck of playing cards was increased after smoking one 2.9% Δ^9-THC marijuana cigarette (Chait *et al.*, 1985). In contrast, several studies have shown that marijuana did not influence finger tapping rate (Chait and Pierri, 1992), which is considered to be a measure of pure motor activity.

Attentional abilities

Attention is a broad psychological construct encompassing behaviors such as searching, scanning, and detecting visual and auditory stimuli for brief or long periods of time (Kinchla, 1992). In nearly all tests assessing attention, responding is measured in some temporal form, such as reaction or response time, time off target, or response rate. Response accuracy may also be reported. Because of differential drug effects, a distinction between focused, selective, divided, and sustained attention can be instructive (Heishman *et al.*, 1994).

Focused attention can be defined as attending to one task for a brief period of time, usually 5 min or less. A relatively large number of studies have investigated the effects of marijuana on focused attention, including reaction time tests. Marijuana (1.8 and 3.6% Δ^9-THC) was shown to slow responding on a simple, visual reaction time task (Wilson *et al.*, 1994), whereas others have not found marijuana to impair simple reaction time performance (Chait and Pierri, 1992; Foltin *et al.*, 1993; Heishman *et al.*, 1997). In contrast, marijuana has been shown to impair complex or choice reaction time tasks in a consistent manner (Block *et al.*, 1992; Chait and Pierri, 1992).

Another commonly used test of psychomotor skills and focused attention is the digit symbol substitution test (DSST), which requires subjects to type a pattern associated with each numeral 1–9 (McLeod *et al.*, 1982). In general, marijuana impairs performance on the DSST. In concentrations ranging from 1.8 to 3.6% Δ^9-THC, marijuana has been shown to decrease number of attempted responses (speed) and/or decrease number of correct responses (accuracy) on the DSST (Heishman *et al.*, 1988, 1989, 1997; Azorlosa *et al.*, 1992; Kelly *et al.*, 1993b; Chait and Perry, 1994). Oral Δ^9-THC (10 and 20 mg) also impaired DSST performance (Kamien *et al.*, 1994). However, other studies have reported no effect of marijuana (1.3–3.6% Δ^9-THC) on the DSST (Chait *et al.*, 1985; Foltin *et al.*, 1993; Azorlosa *et al.*, 1995). The reasons for a lack of effect in these latter studies is unclear because doses of marijuana were comparable across studies, and, in one study (Azorlosa *et al.*, 1995), task presentation was identical to those studies reporting impairment. Marijuana (1.2% Δ^9-THC) also impaired selective attention as evidenced by slower responding and greater interference scores in the Stroop color naming test (Hooker and Jones, 1987).

Divided attention has generally been shown to be impaired by marijuana. Most divided attention tests consist of a central or primary task and a secondary or peripheral task. Several studies have shown that marijuana impaired detection accuracy and/or stimulus reaction time in one or both test components (Chait *et al.*, 1988; Perez-Reyes *et al.*, 1988; Marks and MacAvoy, 1989; Azorlosa *et al.*, 1992; Chait and Perry, 1994). Kelly *et al.* (1993a) used a complex, 5-minute divided attention test, in which an arithmetic task (addition and subtraction of three-digit numbers) was presented in the center of the video monitor and three other stimulus detection tasks were presented in the corners of the monitor. Performance was impaired in a dose-related manner after smoking one marijuana cigarette (2.0 or 3.5% Δ^9-THC). This finding illustrates that marijuana disrupts performance in complex tasks requiring the ability to shift attention rapidly between various stimuli. These same abilities are required when operating a motor vehicle. Not surprisingly, laboratory tests that model various components of driving (Moskowitz, 1985; Liguori *et al.*, 1998) and tests of on-road driving (Robbe, 1994) have been shown to be impaired by marijuana (see Marijuana and Driving below).

Marijuana also impairs sustained attention. In a 30-minute vigilance task, hashish users exhibited more false alarms than non-using control subjects (Bahri and Amir, 1994). This finding is consistent with the observation that the impairing effects of marijuana on sustained attention are most evident in tests that last 30–60 min; tests with durations of 10 min are typically not adversely affected by marijuana (Chait and Pierri, 1992).

Cognitive abilities

Marijuana has been shown to impair learning in the repeated acquisition and performance of response sequences task, which comprises separate acquisition (learning) and performance components (Bickel *et al.*, 1990). This task allows independent assessment of drug effects on acquisition of new information and on performance of previously learned information. Increased errors in the acquisition phase were reported after smoked marijuana (Kelly *et al.*, 1993b) and oral Δ^9-THC (Kamien *et al.*, 1994). However, other studies have found no effect of smoked marijuana on this test (Foltin and Fischman, 1990; Foltin *et al.*, 1993). Block *et al.* (1992) reported that one 2.6% Δ^9-THC marijuana cigarette impaired paired-associative learning.

One of the most reliable behavioral effects of marijuana is the impairment of memory processes. Numerous studies have found that smoked marijuana decreased the number of words or digits recalled and/or increased the number of intrusion errors in immediate or delayed tests of free recall after presentation of information to be remembered (Chait *et al.*, 1985; Hooker and Jones, 1987; Heishman *et al.*, 1989, 1990, 1997; Azorlosa *et al.*, 1992; Block *et al.*, 1992; Kelly *et al.*, 1993a). Using an extensive battery of cognitive tests, Block *et al.* (1992) reported that marijuana (2.6% Δ^9-THC) slowed response time for producing word associations, slowed reading of prose, and impaired tests of reading comprehension, verbal expression, and mathematics. Heishman *et al.* (1990) also found that simple addition and subtraction skills were impaired by smoking one, two, or four marijuana cigarettes (2.6% Δ^9-THC). Finally, Kelly *et al.* (1993a) reported that marijuana (2.0 and 3.5% Δ^9-THC) slowed response time in a spatial orientation test requiring subjects to determine whether numbers and letters were displayed normally or as a mirror image when they were rotated between 90 and 270 degrees.

Next-day or hangover effects

Over the years, an intriguing research question with important practical implications has been whether marijuana impairs performance beyond the period of acute intoxication, which typically lasts 2–6 h after smoking one or two cigarettes. Recently, studies have documented performance decrements 12–24 h after smoking marijuana (Pope *et al.*, 1995). One series of studies reported that 24 h after smoking one marijuana cigarette (2.2% Δ^9-THC), experienced aircraft pilots were impaired attempting to land a plane in a flight simulator (Yesavage *et al.*, 1985; Leirer *et al.*, 1991); however, a third study failed to replicate this next-day effect (Leirer *et al.*, 1989). In another series of studies, a comprehensive battery of tests revealed that only time estimation (Chait *et al.*, 1985) and memory (Chait, 1990) were impaired 9–17 h after smoking two marijuana cigarettes (2.1–2.9% Δ^9-THC), leading the authors to conclude that evidence for next-day performance effects of marijuana was weak. Yet another series of studies found next-day impairment on tests of memory and mental arithmetic after smoking two or four marijuana cigarettes (2.6% Δ^9-THC) over a 4 h period (Heishman *et al.*, 1990), but not after smoking one marijuana cigarette (Heishman *et al.*, 1990; Fant *et al.*, 1998). Thus, residual impairment appears to be a dose-related phenomenon, with effects more likely to be observed at higher marijuana doses.

Amotivational syndrome

A long-standing, controversial issue has been the amotivational syndrome allegedly caused by heavy, chronic marijuana use. This syndrome has been characterized by feelings of lethargy and apathy and an absence of goal-directed behavior (McGlothin and West, 1968; Kupfer *et al.*, 1973). However, studies conducted in countries where segments of the population use marijuana heavily (Comitas, 1976; Stefanis *et al.*, 1977; Page, 1983) and laboratory studies in the United States (Mendelson *et al.*, 1976; Kelly *et al.*, 1990) have not found empirical support for an amotivational syndrome. Foltin and colleagues (Foltin *et al.*, 1989, 1990a,b) have conducted several inpatient studies lasting 15–18 days with subjects reporting weekly marijuana use. Subjects were required to perform low-probability tasks such as the DSST, word-sorting, and vigilance to gain access to more highly desired (high-probability) work and recreational activities. On days that subjects smoked active marijuana, the amount of time spent on low-probability tasks increased, which is inconsistent with an amotivational syndrome. Additionally, the effect of marijuana on time spent on low- versus high-probability activities differed for work and recreational activities, indicating that the behavioral effects of marijuana are context-dependent and not readily predicted by a simplistic amotivational hypothesis (Beardsley and Kelly, 1999).

Summary of performance effects

Laboratory studies in which subjects smoked marijuana documented that marijuana impaired sensory-perceptual abilities by increasing the subjective passage of time relative to clock time. Marijuana impaired gross motor coordination as measured by body sway and postural balance. However, inconsistent findings have been reported for fine motor control; hand steadiness was impaired, whereas several studies have shown no effect of marijuana on finger tapping. Marijuana has been shown to impair complex, but not simple, reaction time tests. A majority of studies have found that marijuana disrupted performance on the DSST. Complex divided attention tests, including driving a vehicle, were readily impaired by marijuana, as were tests requiring sustained attention for more than 30 min. Numerous studies have documented that smoked marijuana and oral Δ^9-THC impaired learning, memory, and other cognitive processes.

MARIJUANA AND DRIVING

Background

Motor vehicle accidents are the leading cause of death in the United States for people aged 1 to 34 (Morbidity and Mortality Weekly Report, 1994). Studies investigating the prevalence rate of drugs other than alcohol in fatally-injured drivers have reported varied results, ranging from 6% to 37% (Williams *et al.*, 1985; Soderstrom *et al.*, 1988; Budd *et al.*, 1989; Marzuk *et al.*, 1990; Terhune *et al.*, 1992). Among individuals stopped for reckless driving who were judged to be clinically intoxicated, urine drug testing indicated 85% were positive for cannabinoids, cocaine metabolites, or both (Brookoff *et al.*, 1994). These relatively high prevalence rates reinforce the general assumption that psychoactive drugs are capable of impairing driving (O'Hanlon and de Gier, 1986; Marowitz, 1995). Drug prevalence rates do not imply impaired driving (Consensus Development Panel, 1985; Bates and Blakely, 1999). However, because certain drugs, including marijuana (see above), reliably degrade psychomotor and cognitive performance in the laboratory (Heishman, 1998), many drug-related vehicular accidents and DUI/DWI arrests probably involve impaired behaviors critical for safe driving.

Currently, the only standardized procedure for detecting drug-induced impairment is the Drug Evaluation and Classification (DEC) program (National Highway Traffic Safety Administration, 1991), which is used by police departments throughout the nation. The DEC program was developed by the Los Angeles Police Department during the 1970s because of a need to document legally whether a driver was impaired due to recent drug ingestion. The DEC program consists of a standardized evaluation conducted by a trained police officer (Drug Recognition Examiner, DRE) and the toxicological analysis of a biological specimen. The evaluation involves a breath alcohol test, examination of the suspect's appearance, behavior, eye movement and nystagmus, field sobriety tests, vital signs, and questioning of the suspect. From the evaluation, the DRE concludes (1) if the suspect is behaviorally impaired such that he or she is unable to operate a motor vehicle safely; (2) if the impairment is drug-related; and (3) the drug class(es) likely causing the impairment. The toxicological analysis either confirms or refutes the DRE's drug class opinion.

Several field studies have indicated that DREs' opinions were confirmed by toxicological analysis in 74–92% of cases when DREs concluded suspects were impaired (Compton, 1986; Preusser et al., 1992; Adler and Burns, 1994; Tomaszewski et al., 1996; Kunsman et al., 1997). These studies attest to the validity of the DEC program as a measurement of drug-induced behavioral impairment in the field. However, the validity of the DEC evaluation has not been rigorously examined under controlled laboratory conditions. We have recently conducted two studies to determine the validity of the individual measures of the DEC evaluation in predicting whether research volunteers were administered marijuana and other drugs of abuse (Heishman et al., 1996, 1998).

Laboratory study

Method

Eighteen research volunteers who reported a history of past and current marijuana use were recruited from the community. Before the study, participants were given psychological and physical examinations to determine whether they were healthy and capable of participating in the study. On three experimental days, which were separated by at least 48 h, participants smoked two marijuana cigarettes that contained either 0%, 1.77%, or 3.55% Δ^9-THC under double-blind conditions. There was a 5–7 min break between cigarettes. The order of marijuana dose conditions was randomized across participants. The DEC evaluation began 10 min after smoking ended and lasted about 25 min. During each experimental session, eight blood samples were obtained for analysis of plasma Δ^9-THC concentration (Foltz et al., 1983). Plasma concentration from the blood sample collected half-way through the DEC evaluation (16 min after smoking ended) will be reported here.

Of particular relevance to the issue of marijuana and driving safety is the part of the DEC evaluation in which performance of four standardized field sobriety tests (SFST) are evaluated. The SFST were Romberg Balance, Walk and Turn, One Leg Stand, and Finger to Nose. The Romberg Balance assesses body sway and tremor while subjects stand for 30 sec with feet together, arms at sides, head tilted back, and eyes closed. The Walk and Turn test requires participants to take nine heel-to-toe steps along a straight line marked on the floor, turn, and return with nine heel-to-toe steps. The One Leg Stand assesses balance by having participants stand on one leg, with the other leg elevated in a stiff-leg manner 6 inches off the floor for 30 sec. There was a brief rest period between testing of each leg. In the Finger to Nose test, participants stand as in the Romberg Balance and bring the tip of their index finger of the left or right hand (as instructed) directly to the tip of the nose.

Results

The 76 variables derived from the DEC evaluation were first analyzed using step-wise discriminant analysis to determine the variables that best predicted the presence or absence of marijuana. This subset of best-predictor variables was then subjected to a discriminant function analysis that predicted and classified whether subjects were dosed or not dosed with marijuana. The resulting data were classified as true positive, true negative, false positive, or false negative. These parameters were then used to calculate several measures of predictive accuracy of the DEC evaluation, including sensitivity, specificity, and efficiency.

The stepwise discriminant analysis resulted in a subset of 28 variables that best predicted the presence or absence of marijuana. The 28 best-predictor variables in descending order of predictive weight were: (1) increased pulse; (2) droopy eyelids; (3) low oral temperature; (4) abnormal speech; (5) lack of rebound dilation of the pupils under direct illumination; (6) sum of the pupillary diameter measures during four illumination conditions; (7) increased systolic and diastolic blood pressure; (8) low volume speech; (9) increased body sway during Romberg Balance test; (10) incoherent speech; (11) abnormal pupillary reaction to light; (12) eyes that did not appear normal; (13) bloodshot eyes; (14) abnormal muscle tone; (15) abnormal appearance of eyes; (16) abnormal facial appearance; (17) increased eye or body tremors during Finger to Nose test; (18) increased errors on executing the turn on Walk and Turn test; (19) less than the complete number of steps in Walk and Turn test; (20) decreased errors on Finger to Nose test; (21) increased errors on Walk and Turn test; (22) marijuana breath odor; (23) abnormal breath odor; (24) lack of hippus (alternating dilation and constriction) of the pupils under direct illumination; (25) failure of eyes to converge; (26) stale breath odor; (27) cigarette breath odor; and (28) slurred speech.

The discriminant function comprising these 28 variables predicted the presence (sensitivity = 100%) and absence (specificity = 98.1%) of marijuana with extremely high accuracy, resulting in minimal false positive (1.9%) and false negative (0%) rates. Overall predictive efficiency was 98.8%. A second discriminant function

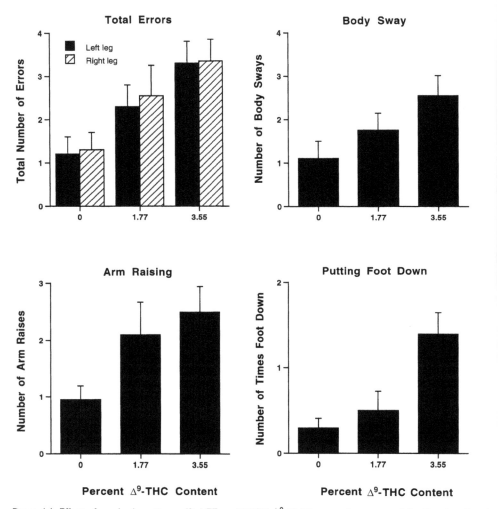

Figure 4.1   Effect of smoked marijuana (0, 1.77, and 3.55% Δ^9-THC) on performance of the One Leg Stand test. Subjects were required to stand on one leg with arms at sides and the other leg raised 6 inches off the ground for 30 sec. Marijuana significantly ($p < 0.05$) increased total errors similarly for the left and right legs compared with placebo. Data for both legs were combined to show that marijuana significantly ($p < 0.05$) increased body sway, arm raising, and putting the foot down to maintain balance during the test.

using only the five best predictive variables resulted in only slightly less sensitivity (90.6%) and specificity (92.6%) compared with the 28-variable model. False positive (7.4%) and false negative (9.4%) rates were slightly higher, and predictive efficiency was 91.9% with the 5-variable model.

Performance on the four SFST was analyzed using analysis of variance. Marijuana significantly impaired performance on two of the tests, One Leg Stand and Finger to Nose. Figure 4.1 shows that total number of errors for each leg in the One

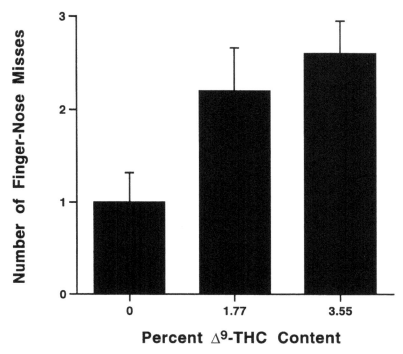

Figure 4.2 Effect of smoked marijuana (0, 1.77, and 3.55% Δ^9-THC) on performance of the Finger to Nose test. Subjects were required to stand with arms at sides, head slightly tilted back, eyes closed, and when instructed, bring the tip of their left or right index finger to the tip of their nose. Both active doses of marijuana significantly ($p < 0.05$) increased number of misses compared with placebo.

Leg Stand test was significantly increased in a dose-dependent manner by marijuana. The component measures of the One Leg Stand test (body sway, arm raising, and putting foot down) were all significantly affected by marijuana (Figure 4.1). Figure 4.2 indicates that both active doses of marijuana increased the number of times participants missed the tip of their nose in the Finger to Nose test.

Marijuana produced peak plasma Δ^9-THC concentration immediately after smoking, which had declined to 15.4 ± 3.0 and 28.2 ± 4.2 ng/ml for 1.75% and 3.55% Δ^9-THC, respectively, at the time of the DEC evaluation. Concentration of 11-nor-9-carboxy-THC, the primary inactive metabolite of Δ^9-THC, reached a maximum during the DEC evaluation of 18.4 ± 3.6 and 27.7 ± 4.2 ng/ml after low and high marijuana doses, respectively.

Conclusion

The validity of the DEC evaluation was examined by developing mathematical models based on discriminant functions that identified which subsets of variables best predicted whether subjects were dosed with placebo or active marijuana.

The data clearly indicated that the variables of the DEC evaluation accurately predicted acute administration of marijuana. This predictive validity was optimal when predictions were made using 28 variables, but the model using the five best variables was also highly accurate.

Although the SFST used in this study were validated for use by police to detect alcohol-impaired driving, we found that they were sensitive to the impairing effects of marijuana as well. The two SFST significantly affected by marijuana were the One Leg Stand and Finger to Nose. The nature of the observed effect in the One Leg Stand was clearly impaired balance. Marijuana caused participants to sway, raise their arms, and put their foot down to maintain balance. Increased errors in the Finger to Nose test indicated impaired psychomotor coordination. The marijuana-induced impairment in balance and motor coordination is consistent with the data reviewed in the first section of this chapter. Impaired balance and coordination may contribute to the deficits in driving observed after marijuana use (Liguori et al., 1998). Such impairment has been characterized by increased lateral movement of a vehicle within the driving lane in a test of highway driving (Robbe, 1994). Lastly, the data indicated that a plasma Δ^9-THC concentration range of 15–30 ng/ml was associated with impaired SFST performance, suggesting that driving abilities would be impaired at the same plasma Δ^9-THC concentration.

CONCLUSION

Laboratory studies in which subjects smoked marijuana under controlled conditons have documented that marijuana reliably impaired sensory-perceptual abilities (time estimation); gross motor coordination (body sway, balance); psychomotor abilities (complex reaction time, DSST); divided and sustained attention; and cognitive functioning, including learning and memory. Other laboratory research has focused on the applied question of whether marijuana impairs driving abilities as measured by field sobriety tests used by police to detect impaired drivers. This research has shown that marijuana impaired balance and psychomotor coordination in a dose-dependent manner. Such behavioral impairment may underlie marijuana-induced decrements in tests of on-road driving.

Two areas for future research include the simultaneous measurement of plasma Δ^9-THC concentration with assessment of performance and studies of the interaction of marijuana with other drugs. Very few of the studies reviewed in this chapter provided data on the dose of Δ^9-THC actually delivered to subjects. This is especially critical in studies with marijuana because the large variability in smoking topography (e.g. duration and volume of puffs and depth of inhalation) (Herning et al., 1981; Heishman et al., 1989) and the low bio-availability of smoked drugs (Ohlsson et al., 1980; Heishman et al., 1994) result in highly variable delivered Δ^9-THC doses (Huestis et al., 1992). Such data are necessary to relate performance impairment with a known plasma Δ^9-THC concentration. Relatively few studies have investigated the interactive effects of marijuana and other drugs on human behavior. Such basic information is critically needed because the simultaneous use of drugs with different pharmacological effects (e.g. alcohol and marijuana) is a common practice today. The combined effect of two or more drugs can be very different from that of each drug alone.

Much progress has been made over the past decade on the neurophysiology and neuropharmacology of the cannabinoid system. Cannabinoid receptors have been cloned and characterized, endogenous ligands have been discovered, and receptor antagonists have been synthesized and tested. The goal of future cannabinoid research will be to explore the functional significance of this neuroscientific knowledge with respect to human behavior. The end result may be the development of novel therapeutic agents for the treatment of conditions known to have cannabinoid involvement, such as memory loss, movement disorders, and mood disturbances.

REFERENCES

Adler, E. V. and Burns, M. (1994) Drug recognition expert (DRE) validation study. Final Report to Governor's Office of Highway Safety, State of Arizona.

Azorlosa, J. L., Greenwald, M. K. and Stitzer, M. L. (1995) "Marijuana smoking: Effects of varying puff volume and breathhold duration," *Journal of Pharmacology and Experimental Therapeutics* **272**: 560–569.

Azorlosa, J. L., Heishman, S. J., Stitzer, M. L. and Mahaffey, J. M. (1992) "Marijuana smoking: Effect of varying Δ^9-tetrahydrocannabinol content and number of puffs," *Journal of Pharmacology and Experimental Therapeutics* **261**: 114–122.

Bahri, T. and Amir, T. (1994) "Effect of hashish on vigilance performance," *Perceptual and Motor Skills* **78**: 11–16.

Bates, M. N. and Blakely, T. A. (1999) "Role of cannabis in motor vehicle crashes," *Epidemiologic Reviews* **21**: 222–232.

Beardsley, P. M. and Kelly, T. H. (1999) "Acute effects of cannabis on human behavior and central nervous system function," in *The Health Effects of Cannabis*, edited by H. Kalant, W. Corrigal, W. Hall and R. Smart, pp. 127–169. Toronto: Addiction Research Foundation.

Bickel, W. K., Hughes, J. R. and Higgins, S. T. (1990) "Human behavioral pharmacology of benzodiazepines: Effects on repeated acquisition and performance of response chains," *Drug Development Research* **20**: 53–65.

Block, R. I., Farinpour, R. and Braverman, K. (1992) "Acute effects of marijuana on cognition: Relationships to chronic effects and smoking techniques," *Pharmacology Biochemistry and Behavior* **43**: 907–917.

Brookoff, D., Cook, C. S., Williams, C. and Mann, C. S. (1994) "Testing reckless drivers for cocaine and marijuana," *New England Journal of Medicine* **331**: 518–522.

Budd, R. D., Muto, J. J. and Wong, J. K. (1989) "Drugs of abuse found in fatally injured drivers in Los Angeles County," *Drug and Alcohol Dependence* **23**: 153–158.

Chait, L. D. (1990) "Subjective and behavioral effects of marijuana the morning after smoking," *Psychopharmacology* **100**: 328–333.

Chait, L. D., Corwin, R. L. and Johanson, C. E. (1988) "A cumulative dosing procedure for administering marijuana smoke to humans," *Pharmacology Biochemistry and Behavior* **29**: 553–557.

Chait, L.D., Fischman, M. W. and Schuster, C.R. (1985) "'Hangover' effects the morning after marijuana smoking," *Drug and Alcohol Dependence* **15**: 229–238.

Chait, L. D. and Perry, J. L. (1994) "Acute and residual effects of alcohol and marijuana, alone and in combination, on mood and performance," *Psychopharmacology* **115**: 340–349.

Chait, L. D. and Pierri, J. (1992) "Effects of smoked marijuana on human performance: A critical review," in *Marijuana/Cannabinoids: Neurobiology and Neurophysiology*, edited by L. Murphy and A. Bartke, pp. 387–423. Boca Raton: CRC Press.

Comitas, L. (1976) "Cannabis and work in Jamaica: A refutation of the amotivational syndrome," *Annals of the New York Academy of Science* **282**: 24–32.

Compton, R. P. (1986) Field evaluation of the Los Angeles Police Department drug detection procedure. National Highway Traffic Safety Administration, Report No. DOT HS 807 012. Washington, DC: U.S. Department of Transportation.

Cone, E. J., Johnson, R. E., Moore, J. D. and Roache, J. D. (1986) "Acute effects of smoking marijuana on hormones, subjective effects and performance in male human subjects," *Pharmacology Biochemistry and Behavior* **24**: 1749–1754.

Consensus Development Panel (1985) "Drug concentrations and driving impairment," *Journal of the American Medical Association* **254**: 2618–2621.

Fant, R. V., Heishman, S. J., Bunker, E. B. and Pickworth, W. B. (1998) "Acute and residual effects of marijuana in humans," *Pharmacology Biochemistry and Behavior* **60**: 777–784.

Foltin, R. W. and Fischman, M. W. (1990) "The effects of combinations of intranasal cocaine, smoked marijuana, and task performance on heart rate and blood pressure," *Pharmacology Biochemistry and Behavior* **36**: 311–315.

Foltin, R. W., Fischman, M. W., Brady, J. V., Bernstein, D. J., Capriotti, R. M., Nellis, M. J. and Kelly, T. H. (1990a) "Motivational effects of smoked marijuana: Behavioral contingencies and low-probability activities," *Journal of the Experimental Analysis of Behavior* **53**: 5–19.

Foltin, R. W., Fischman, M. W., Brady, J. V., Bernstein, D. J., Nellis, M. J. and Kelly, T. H. (1990b) "Marijuana and behavioral contingencies," *Drug Development Research* **20**: 67–80.

Foltin, R. W., Fischman, M. W., Brady, J. V., Kelly, T. H., Bernstein, D. J. and Nellis, M. J. (1989) "Motivational effects of smoked marijuana: Behavioral contingencies and high-probability recreational activities," *Pharmacology Biochemistry and Behavior* **34**: 871–877.

Foltin, R. W., Fischman, M. W., Pippen, P. A. and Kelly, T. H. (1993) "Behavioral effects of cocaine alone and in combination with ethanol or marijuana in humans," *Drug and Alcohol Dependence* **32**: 93–106.

Foltz, R. L., McGinnis, K. M. and Chinn, D. M. (1983) "Quantitative measurement of Δ^9-tetrahydrocannabinol and two major metabolites in physiological specimens using capillary column gas chromatography negative ion chemical ionization mass spectrometry," *Biomedical Mass Spectrometry* **10**: 316–323.

Greenberg, H. S., Werness, S. A. S., Pugh, J. E., Andrus, R. O., Anderson, D. J. and Domino, E. F. (1994) "Short-term effects of smoking marijuana on balance in patients with mulitple sclerosis and normal volunteers," *Clinical Pharmacology and Therapeutics* **55**: 324–328.

Heishman, S. J. (1998) "Effects of abused drugs on human performance: Laboratory assessment," in *Drug Abuse Handbook*, edited by S. B. Karch, pp. 206–235. Boca Raton, FL: CRC.

Heishman, S. J., Arasteh, K. and Stitzer, M. L. (1997) "Comparative effects of alcohol and marijuana on mood, memory, and performance," *Pharmacology Biochemistry and Behavior* **58**: 93–101.

Heishman, S. J., Huestis, M. A., Henningfield, J. E. and Cone, E. J. (1990) "Acute and residual effects of marijuana: Profiles of plasma Δ^9-THC levels, physiological, subjective, and performance measures," *Pharmacology Biochemistry and Behavior* **37**: 561–565.

Heishman, S. J., Singleton, E. G. and Crouch, D. J. (1996) "Laboratory validation study of Drug Evlauation and Classification program: Ethanol, cocaine, and marijuana," *Journal of Analytical Toxicology* **20**: 468–483.

Heishman, S. J., Singleton, E. G. and Crouch, D. J. (1998) "Laboratory validation study of Drug Evlauation and Classification program: Alprazolam, *d*-amphetamine, codeine, and marijuana," *Journal of Analytical Toxicology* **22**: 503–514.

Heishman, S. J., Stitzer, M. L. and Bigelow, G. E. (1988) "Alcohol and marijuana: Comparative dose effect profiles in humans," *Pharmacology Biochemistry and Behavior* **31**: 649–655.

Heishman, S. J., Stitzer, M. L. and Yingling, J. E. (1989) "Effects of tetrahydrocannabinol content on marijuana smoking behavior, subjective reports, and performance," *Pharmacology Biochemistry and Behavior* **34**: 173–179.

Heishman, S. J., Taylor, R. C. and Henningfield, J. E. (1994) "Nicotine and smoking: A review of effects on human performance," *Experimental and Clinical Psychopharmacology* **2**: 345–395.

Herning, R. I., Jones, R. T., Bachman, J. and Mines, A. H. (1981) "Puff volume increases when low-nicotine cigarettes are smoked," *British Medical Journal* **283**: 1–7.

Hooker, W. D. and Jones, R. T. (1987) "Increased susceptibility to memory intrusions and the Stroop interference effect during acute marijuana intoxication," *Psychopharmacology* **91**: 20–24.

Huestis, M. A., Henningfield, J. E. and Cone, E. J. (1992) "Blood cannabinoids. I. Absorption of Δ^9-THC and formation of 11-OH-THC and THCCOOH during and after smoking marijuana," *Journal of Analytical Toxicology* **16**: 276–282.

Kamien, J. B., Bickel, W. K., Higgins, S. T. and Hughes, J. R. (1994) "The effects of Δ^9-tetrahydrocannabinol on repeated acquisition and performance of response sequences and on self-reports in humans," *Behavioural Pharmacology* **5**: 71–78.

Kelly, T. H., Foltin, R. W., Emurian, C. S. and Fischman, M. W. (1990) "Multidimensional behavioral effects of marijuana," *Progress in Neuro-Psychopharmacology and Biological Psychiatry* **14**: 885–902.

Kelly, T. H., Foltin, R. W., Emurian, C. S. and Fischman, M. W. (1993a) "Performance-based testing for drugs of abuse: Dose and time profiles of marijuana, amphetamine, alcohol, and diazepam," *Journal of Analytical Toxicology* **17**: 264–272.

Kelly, T. H., Foltin, R. W. and Fischman, M. W. (1993b) "Effects of smoked marijuana on heart rate, drug ratings and task performance by humans," *Behavioural Pharmacology* **4**: 167–178.

Kinchla, R. A. (1992) "Attention," *Annual Review of Psychology* **43**: 711–742.

Kunsman, G. W., Levine, B., Costantino, A. and Smith, M. L. (1997) "Phencyclidine blood concentrations in DRE cases," *Journal of Analytical Toxicology* **21**: 498–502.

Kupfer, D. J., Detre, T., Koral, J. and Fajans, P. (1973) "A comment on the 'amotivational syndrome' in marijuana smokers," *American Journal of Psychiatry* **130**: 1319–1322.

Leirer, V. O., Yesavage, J. A. and Morrow, D. G. (1989) "Marijuana, aging, and task difficulty effects on pilot performance," *Aviation Space and Environmental Medicine* **60**: 1145–1152.

Leirer, V. O., Yesavage, J. A. and Morrow, D. G. (1991) "Marijuana carry-over effects on aircraft pilot performance," *Aviation Space and Environmental Medicine* **62**: 221–227.

Liguori, A., Gatto, C. P. and Robinson, J. H. (1998) "Effects of marijuana on equilibrium, psychomotor performance, and simulated driving," *Behavioral Pharmacology* **9**: 599–609.

Marks, D. F. and MacAvoy, M. G. (1989) "Divided attention performance in cannabis users and non-users following alcohol and cannabis separately and in combination," *Psychopharmacology* **99**: 397–401.

Marowitz, L. A. (1995) "Drug arrests and drunk driving," *Alcohol, Drugs and Driving* **11**: 1–22.

Marzuk, P. M., Tardiff, K., Leon, A. C., Stajic, M., Morgan, E. B. and Mann, J. J. (1990) "Prevalence of recent cocaine use among motor vehicle fatalities in New York City," *Journal of the American Medical Association* **263**: 250–256.

McGlothin, W. H. and West, L. J. (1968) "The marihuana problem: An overview," *American Journal of Psychiatry* **125**: 370–378.

McLeod, D. R., Griffiths, R. R., Bigelow, G. E. and Yingling, J. (1982) "An automated version of the digit symbol substitution task (DSST)," *Behavioral Research Methods and Instrumentation* **14**: 463–466.

Mendelson, J. H., Kuehnle, J. C., Greengerg, I. and Mello, N. K. (1976) "Operant acquisition of marihuana in man," *Journal of Pharmacology and Experimental Therapeutics* **198**: 42–53.

Morbidity and Mortality Weekly Report (1994) "Update: Alcohol-related traffic fatalities-United States, 1982–1993," *Morbidity and Mortality Weekly Report* **43**: 861–867.

Moskowitz, H. (1985) "Marihuana and driving," *Accident Analysis and Prevention* **17**: 323–345.

National Highway Traffic Safety Administration (1991) Drug evaluation and classification program, briefing paper. Washington, DC: U.S. Department of Transportation.

O'Hanlon, J. F. and de Gier, J. J. (1986) *Drugs and Driving*. London: Taylor and Francis.

Ohlsson, A., Lindgren, J. E., Wahlen, A., Agurell, S., Hollister, L. E. and Gillespie, H. K. (1980) "Plasma delta-9-tetrahydrocannabinol concentrations and clinical effects after oral and intravenous administration and smoking," *Clinical Pharmacology and Therapeutics* **28**: 409–416.

Page, J. B. (1983) "The amotivational syndrome hypothesis and the Costa Rica study: Relationship between methods and results," *Journal of Psychoactive Drugs* **15**: 261–267.

Perez-Reyes, M., Hicks, R. E., Bumberry, J., Jeffcoat, A. R. and Cook, C. E. (1988) "Interaction between marihuana and ethanol: Effects on psychomotor performance," *Alcoholism: Clinical and Experimental Research* **12**: 268–276.

Pope, H. G., Gruber, A. J. and Yurgelun-Todd, D. (1995) "The residual neuropsychological effects of cannabis: The current status of research," *Drug and Alcohol Depdendence* **38**: 25–34.

Preusser, D. F., Ulmer, R. G. and Preusser, C. W. (1992) Evaluation of the impact of the drug evaluation and classification program on enforcement and adjudication. National Highway Traffic Safety Administration, Report No. DOT HS 808 058. Washington, DC: U.S. Department of Transportation.

Robbe, H. W. J. (1994) *Influence of Marijuana on Driving*. Maastricht: University of Limburg.

Smith, J. M. and Misiak, H. (1976) "Critical Flicker Frequency (CFF) and psychotropic drugs in normal human subjects – a review," *Psychopharmacology* **47**: 175–182.

Soderstrom, C. A., Trifillis, A. L., Shankar, B. S., Clark, W. E. and Crowley, R. A. (1988) "Marijuana and alcohol use among 1023 trauma patients," *Archives of Surgery* **123**: 733–737.

Stefanis, C., Dornbush, R. and Fink, M. (1977) *Hashish: Studies of Long-term Use*. New York: Raven.

Substance Abuse and Mental Health Services Administration (1999) National Household Survey on Drug Abuse: Population Estimates, 1998. DHHS Pub. No. (SMA) 99–3327. Washington, DC: U.S. Government Printing Office.

Terhune, K. W., Ippolito, C. A., Hendricks, D. L., Michalovic, J. G., Bogema, S. C., Santinga, P., Blomberg, R. and Preusser, D. F. (1992) The incidence and role of drugs in fatally injured drivers. National Highway Traffic Safety Administration, Report No. DOT HS 808 065. Washington, DC: U.S. Department of Transportation.

Tomaszewski, C., Kirk, M., Bingham, E., Saltzman, B., Cook, R. and Kulig, K. (1996) "Urine toxicology screens in drivers suspected of driving while impaired from drugs," *Clinical Toxicology* **34**: 37–44.

Townsend, T. N., Lane, J., Dewa, C. S. and Brittingham, A. M. (1998) Driving after drug or alcohol use: Findings from the 1996 National Household Survey on Drug Abuse. Washington, DC: U.S. Department of Transportation.

Williams, A. F., Peat, M. A., Crouch, D. J., Wells, J. N. and Finkle, B. S. (1985) "Drugs in fatally injured young male drivers," *Public Health Report* **100**: 19–25.

Wilson, W. H., Ellinwood, E. H., Mathew, R. J. and Johnson, K. (1994) "Effects of marijuana on performance of a computerized cognitive-neuromotor test battery," *Psychiatry Research* **51**: 115–125.

Yesavage, J. A., Leirer, V. O., Denari, M. and Hollister, L. E. (1985) "Carry-over effects of marijuana intoxication on aircraft pilot performance: A preliminary report," *American Journal of Psychiatry* **142**: 1325–1329.

Chapter 5

Biology of endocannabinoids

Vincenzo Di Marzo, Luciano De Petrocellis,
Tiziana Bisogno, Alvin Berger and
Raphael Mechoulam

ABSTRACT

The discovery of endocannabinoids, endogenous ligands of cannabinoid receptors, opened a new age in research not only on the biology of marijuana but also on the natural role of fatty acid derivatives like the *N*-acylethanolamines and the monoacylglycerols. In fact, the endocannabinoids discovered so far all happen to be polyunsaturated homologues belonging to these two classes of lipids. Certainly the most studied endocannabinoids are anandamide (*N*-arachidonoylethanolamine) and 2-arachidonoylglycerol. Concomitantly to the discovery of the properties as cannabinoid receptor ligands of these two compounds, two other fatty acid derivatives, *N*-palmitoylethanolamine and oleamide (*cis*-9-octadecenoamide), were also found to exhibit cannabimimetic activity in some tests despite their very low affinity for the two cannabinoid receptor subtypes known to date. In this chapter, we review the landmarks in the history of endocannabinoids, starting with the background leading to their discovery and the analytical techniques developed for their analysis and quantification in tissues and biological fluids, and ending with a description of the mechanisms responsible for the regulation of their tissue levels, and of their possible physiological and pathological role. We also briefly describe the possible biological and evolutionary relevance of the finding of these lipids in simple animal organisms and plants, and discuss the possible implications of the presence of cannabimimetic fatty acid derivatives in foods.

Key Words: cannabinoid, anandamide, 2-arachidonoylglycerol, oleamide, palmitoylethanolamide, *N*-acylethanolamines

INTRODUCTION

Huang Ti, a Chinese emperor in around 2600 BC, probably did not suspect that the *Cannabis sativa* preparations that he advised for the treatment of cramps, malaria, rheumatic pains and "female disorders" (Mechoulam, 1986) had, in his own body, endogenous counterparts potentially capable of exerting similar beneficial effects. In fact, only about 4,500 years later, in the 1990s, did the finding of specific binding sites for *Cannabis* psychoactive element, Δ^9-tetrahydrocannabinol (THC) (Gaoni and Mechoulam, 1964), i.e., the "cannabinoid receptors" (Devane *et al.*, 1988; Matsuda *et al.*, 1990; Munro *et al.*, 1993), open the way to the discovery of endogenous cannabinoid-like molecules, or "endocannabinoids" (Devane *et al.*, 1992b; Hanus *et al.*, 1993; Mechoulam *et al.*, 1995; Sugiura *et al.*, 1995; Di Marzo and Fontana, 1995). These molecules are synthesized and metabolized in several tissues from a whole range of animal organisms, from lower invertebrates to mammals, and not only bind with high affinity to cannabinoid receptors but also reproduce in laboratory animals several of the pharmacological actions of THC. In this chapter, we provide an overview of our still limited knowledge of the multi-faceted biology of endocannabinoids, an ever growing source of surprises of which, over the last 9 years, scientists have gained probably only a glimpse. In particular, we discuss the history of the isolation and characterization of the endocannabinoids, a process that many do not yet consider a closed chapter, and review the techniques developed to date for analysis of these compounds in biological samples. Next, we describe the pathways for the biosynthesis and inactivation of these compounds in animal organisms. As the knowledge of the mechanisms for the regulation of the levels of endocannabinoids may provide important indications of their possible physiopathological role only if accompanied by data on their pharmacological activity, these latter data will be also critically analyzed. Finally, we conclude this chapter with a description of the occurrence of endocannabinoids and their analogs in non-mammalian organisms, plants and, subsequently, in foods, exploring the possible clinical, pharmacological and evolutionary importance of the endocannabinoid signaling system. In addition to the endocannabinoids, defined as endogenous molecules capable of binding to and efficiently activating the two cannabinoid receptor subtypes (Di Marzo, 1998), we examine other cannabimimetic fatty acid derivatives, which exhibit THC-like activity in some tests but do not bind with high affinity to these receptors.

DISCOVERY OF ENDOCANNABINOIDS AND OTHER CANNABIMIMETIC FATTY ACID DERIVATIVES

Discovery of cannabinoid receptors

Thousands of papers have been published on THC since its identification as the active constituent of *Cannabis* (Gaoni and Mechoulam, 1964). In fact, there are few natural products that have been investigated so thoroughly. Over the years, we have learned much about its metabolism, physiological and pharmacological effects and numerous clinical trials have been undertaken. However, until the mid 1980s we knew very little about its mechanism of action. Conceptual problems hampered work in this direction. One of these was the presumed lack of stereospecific activity by THC. Compounds acting through a biomolecule – a receptor for example – generally show a very high degree of stereoselectivity. However, synthetic $(+)$-Δ^9-THC showed some cannabimimetic activity, although considerably lower compared with that of natural $(-)$-Δ^9-THC. This observation was not compatible with the existence of a specific cannabinoid receptor and of cannabinoid mediators. Only in the mid 1980s was it established that cannabinoid activity is in fact highly stereoselective and that the previous observations resulted from problems occurring during enantiomer separation (Mechoulam *et al.*, 1988; Mechoulam *et al.*, 1992). A further hurdle towards the elucidation of the mechanism of THC action was that THC was considered to belong to the group of biologically active lipophiles, which act on and through biological membranes.

The action of cannabinoids could be explained, apparently, without postulating the existence of a specific cannabinoid receptor (Paton, 1975). However, by the mid 1980s, it became clear that the membrane perturbation theory of *Cannabis* action represents at best only part of the picture. Makriyannis (1995) has recently reviewed the evidence and has concluded that "although the cellular membrane may not be the principal target for cannabinoid activity, it nevertheless plays a role in the mechanism of action." Following the important observation that cannabinoids inhibit adenylate cyclase (Howlett and Fleming, 1984) a specific binding site with high affinity for THC was discovered (Devane *et al.*, 1988). Its distribution was consistent with the pharmacological properties of psychotropic cannabinoids. This receptor was shortly thereafter cloned (Matsuda *et al.*, 1990) and designated CB_1. A peripheral receptor (CB_2) was later identified in the spleen and immune cells (Munro *et al.*, 1993). THC was found to bind to both CB_1 and CB_2 receptors.

Discovery of the endocannabinoids: anandamide and 2-arachidonoylglycerol

Goldstein (1976), on the subject of opiates, commented: "it seemed unlikely, *a priori*, that such highly stereospecific receptors should have developed in nature to interact with the alkaloids from the opium poppy." The same conclusion seemed to apply for cannabis. It certainly seemed unlikely that the brain should synthesize receptors for a plant constituent. It was assumed that the presence of a specific cannabinoid receptor indicates the existence of endogenous, specific cannabinoid ligands that activate these receptors and a search for such ligands was started. As all plant or synthetic cannabinoids are lipid-soluble compounds, the isolation procedures employed were based on the assumption that the endogenous ligands are lipids – an assumption that ultimately proved to be correct. First, a highly potent radioactively labeled probe, $[^3H]$-HU-243 (Devane *et al.*, 1992a) was prepared and used for CB_1 receptor binding assays. Porcine brains and later canine gut were extracted with organic solvents, and the extracts were purified by chromatography numerous times according to standard protocols for the separation of lipids. The chromatographic fractions were tested in binding assays by using the above probe. Some fractions obtained from brain displaced the labeled probe, which indicated the presence of an active endocannabinoid. Two major endocannabinoids were ultimately isolated – arachidonoylethanolamide (AEA, also known as *N*-arachidonoyl-ethanolamine) (Devane *et al.*, 1992b) and 2-arachidonylglycerol (2-AG, Mechoulam *et al.*, 1995; Sugiura *et al.*, 1995). AEA, the first endocannabinoid ever isolated, was named anandamide from the Sanskrit word *ananda* for "inner bliss" (Devane *et al.*, 1992b). Several additional active anandamide-type compounds were also isolated (Hanus *et al.*, 1993). These compounds were present in tissues in miniscule amounts, and their chemical structures (Figure 5.1) were determined by physical measurements – e.g. NMR and mass spectrometry. Detailed descriptions of the isolation and structural elucidation of AEA and 2-AG have been published (Devane *et al.*, 1992b; Mechoulam *et al.*, 1995).

The identification of fatty acid derivatives as endogenous cannabinoids was unexpected. Although fatty acid ethanolamides (named also *N*-acylethanolamines, NAEs) were known as natural products (Schmid *et al.*, 1990), there was no indication

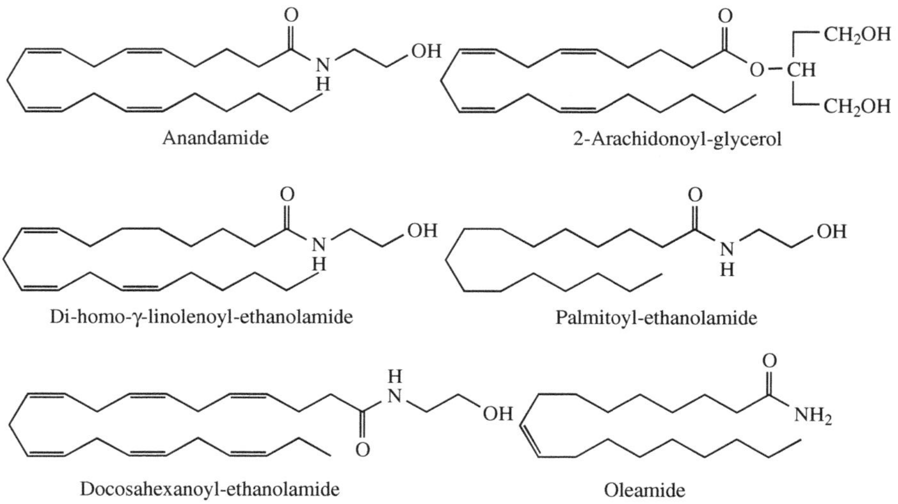

Figure 5.1   Chemical structures of endocannabinoids and other cannabimimetic fatty acid derivatives. Endocannabinoids (2-arachidonoylglycerol and some polyunsaturated C20-C22 *N*-acylethanolamines) are defined as endogenous metabolites capable of binding and activating either CB_1 or CB_2 cannabinoid receptors, or both (Di Marzo, 1998). Other fatty acid derivatives, such as palmitoylethanolamide and oleamide, have THC-like activity in some tests, but do not exhibit high affinity for the two known cannabinoid receptor subtypes.

that such compounds had any relationship to cannabinoids. In retrospect, one can state that *N*-arachidonoyl-ethanolamine and 2-arachidonylglycerol should have been investigated in the past. Arachidonic acid derivatives are major components in numerous animal biological systems – see prostaglandins and leukotrienes, for example. Ethanolamides of fatty acids, as mentioned above, are also well known and some have been shown to have many biological activities (see below). Monoacylglycerols are metabolic intermediates of important constituents, i.e., triacylglycerols, diacylglycerols and phosphoglycerides. It is indeed strange that these facts were not put together and that arachidonic acid amides and esters were not investigated earlier.

Other cannabimimetic fatty acid amides: before and after 1992

The physiological activities of several long chain fatty acid ethanolamides have been examined in the past. At low–medium μM concentrations, they have been shown to have membrane-stabilizing effects, to stimulate the rate of Ca^{2+} sequestration and Ca^{2+} and Mg^{2+} ATPase activity, and to increase the retention time of Ca^{2+} by vesicles (Epps *et al.*, 1982). At high concentration these amides have inhibitory effects. It is of interest that AEA has also been found to have biphasic effects (Sulcova *et al.*, 1998). Palmitoylethanolamide (PEA) was initially isolated from egg yolk, peanut meal and soybean lecithin and found to have anti-inflammatory activity (Kuehl *et al.*, 1957), and was later found also in animal tissues (see below). PEA has a rather checkered history as an anti-inflammatory agent. In a guinea pig joint anaphylaxis assay it was active at $0.3\,\mu g/kg$ (Ganley *et al.*, 1958). It also suppresses passive anaphylaxis in mice (5 mg/kg) (Ganley and Robinson, 1959) and inhibits release of histamine from rat peritoneal cells induced by Russel Viper venom (Goth and Knoohuizen, 1962). It had no activity on primary lesions of adjuvant arthritis in the rat, but inhibited the tuberculin reaction in guinea pigs (Perlik *et al.*, 1971). In a clinical trial, PEA did not prevent rheumatic fever in children (Coburn and Rich, 1960). PEA has also been reported to lower the alcohol-induced impairment of psychomotor tasks in man (Krsiak *et al.*, 1972), to induce non-specific resistance to viral and bacterial infection (Raskova *et al.*, 1972) and to have preventive therapeutic effects in respiratory tract infections (Kahlich *et al.*, 1979). In double blind field trials in army units PEA limited clinical infections significantly. It was introduced in clinical practice in Eastern Europe, under the name "Impulsin," but it is apparently no longer used.

PEA has also been shown to inhibit ion transport through the veratridine-activated fast sodium channels (Gulaya *et al.*, 1993). Several other fatty acid ethanolamides have been found to have positive ionotropic effects on muscle strips from guinea pig heart (Epps *et al.*, 1983). The ethanolamide of oleic acid was found to protect guinea pig atria from the reduction of contractile force caused by hypoxia (Epps *et al.*, 1983). It has been suggested that long chain fatty acid ethanolamides have "protective effects in ischemic myocardium and [are] produced as the result of a defense mechanism against ischemic injury" (Schmid *et al.*, 1990). Surprisingly, no further work in this important area has been reported. The pharmacological effects described above have been well summarized in considerable detail in a review by Schmid *et al.* (1990).

More recent pharmacological and biochemical work on PEA is difficult to rationalize. Most researchers have found that PEA does not activate either CB_1 or CB_2 receptors (Sheskin *et al.*, 1997; Lambert *et al.*, 1999; Martin *et al.*, 1999; Sugiura *et al.*, 2000). Yet, PEA induces in some cases biological responses that are similar to those caused by cannabinoids (see Lambert and Di Marzo, 1999 for review), such as the inhibition of the release of inflammatory mediators and cytokines by mast cells and macrophages (Facci *et al.*, 1995; Berdyshev *et al.*, 1997). Furthemore, PEA was found to reduce peripheral pain, an effect that was antagonized by a CB_2 receptor antagonist (Calignano *et al.*, 1998). When AEA and PEA were administered together they acted synergistically by reducing pain responses 100-fold more potently than each compound alone. Interestingly, Facci *et al.* (1995) have reported that PEA is capable of displacing a cannabinoid receptor ligand, WIN55,212-2, from its specific binding sites in mast cells and rat basophilic leukemia (RBL) cells, although this finding has not been confirmed by using a slightly different experimental protocol (Lambert *et al.*, 1999). So, does PEA act on a non-CB_1, non-CB_2 cannabinoid receptor, or as an endogenous "enhancer" of AEA actions (see below)? More recently, two other fatty acid ethanolamides that do not bind to CB_1 and CB_2 receptors, the oleic and linoleic acid homologues, have also been found to induce some THC-like effects in vivo at high doses (Watanabe *et al.*, 1999).

The effect of AEA on sleep has not been directly investigated, although cannabis and THC are well known for their hypnotic properties. However, Santucci *et al.* (1996) have found that the CB_1 antagonist SR-141716A increases the time spent awake at the expense of both slow-wave and rapid eye movement (REM) sleep. They suggested that "....An endogenous cannabimimetic (anandamidergic?) system may regulate the organization of the sleep-waking cycle." Indeed a fatty acid amide, the primary amide of oleic acid (oleamide, OA), has received considerable attention recently as a potential sleep factor (Boger *et al.*, 1998). It was first identified in the cerebrospinal fluid of sleep-deprived cats (Lerner *et al.*, 1994) and was shown to induce sleep in rodents and cats (Cravatt *et al.*, 1995). An increase of mean duration of slow wave sleep at the expense of waking was noted; distribution of REM sleep was not altered (Boger *et al.*, 1998). OA modulates serotonin receptor responses, with potentiation of several subtypes of these receptors (Huidobro-Toro and Harris, 1996; Boger *et al.*, 1998), including the $5-HT_7$ receptor, which has been associated with sleep (Thomas *et al.*, 1997). Although OA does not bind to either CB_1 or CB_2, it does parallel the effect of AEA in its action on the tetrad of pharmacological effects, which are considered specific for cannabinoids (Mechoulam *et al.*, 1997). This discrepancy has been explained by the inhibitory action of OA on the enzyme FAAH (which hydrolyses AEA), which possibly results in an enhancement of AEA action. On the same basis it has been suggested that OA acts on sleep induction through AEA (Mechoulam *et al.*, 1997). Boger *et al.* (1998) have however suggested that oleamide and AEA "exert their effects through a common mechanism albeit not exclusively through the studied cannabinoid receptors." At present, neither the mechanism of the hypnotic action of AEA nor that of OA is fully established.

QUANTITATIVE ANALYSIS OF ENDOCANNABINOIDS AND OTHER CANNABIMIMETIC FATTY ACID DERIVATIVES

The study of the biological significance of endocannabinoids would not be possible without methods for determining their concentration in tissues and biological fluids. Analytical techniques allowing the accurate measurement of the levels of AEA, 2-AG and their natural analogs have been developed and are necessary, for example, to study the possible correlation between some physiopathological conditions and the levels of these compounds. Today it is possible to measure with confidence AEA, PEA, OA and 2-AG in picomol amounts in samples ranging from small brain regions to blood, CSF and brain microdialysates. Essentially, three types of analytical techniques have been described for these compounds. First, gas chromatography coupled to electron impact mass spectrometric detection (GC-EIMS) was developed in order to allow the separation of the homologs of the NAE family of lipids to which AEA and PEA belong, after a chemical derivatization step required to render these compounds more volatile and less susceptible to pyrolysis (Di Marzo *et al.*, 1994; Fontana *et al.*, 1995, Schmid *et al.*, 1995). Mass spectrometric detection of fragment ions, selected depending on the type of derivatizing agent used, normally allows considerable improvements in sensitivity. Since MS is not a quantitative method *per se*, and calibration curves with external standards are not accurate, isotope-dilution GC-MS procedures have been developed (see Kempe *et al.*, 1996, for example) and used once that deuterated AEA, NAEs and, more recently, 2-AG, became commercially available. Later, GC-EIMS was also applied to the identification and quantification of 2-AG in neuronal cells (Bisogno *et al.*, 1997c) and rat brain (Stella *et al.*, 1997). Simultaneous determination of several components of both the monoacylglycerol and NAE family of lipids is also possible (Schmid *et al.*, 2000). Variations of this method were used for: (1) the quantification of endocannabinoids in mammalian brain at different stages of development (Berrendero *et al.*, 1999), or in different rat brain regions (Bisogno *et al.*, 1999a); (2) the identification of AEA or 2-AG in blood cells (Bisogno *et al.*, 1997a,c; Wagner *et al.*, 1997; Varga *et al.*, 1998; Kuwae *et al.*, 1999; Di Marzo *et al.*, 1999b), rat serum (Giuffrida and Piomelli, 1998) and various rat organs (Kondo *et al.*, 1998); (3) the measurement of NAEs in several invertebrate species (Bisogno *et al.*, 1997d; Sepe *et al.*, 1998; De Petrocellis *et al.*, 1999), rat renal tissues (Deutsch *et al.*, 1997), some rat and human tumor cells (Bisogno *et al.*, 1998) and in mouse uterus at different stages of embryo implantation (Schmid *et al*, 1997); and (4) the detection of minute amounts of these compounds in microdialysates from rat striatum (Giuffrida *et al.*, 1999). A particularly interesting variation in the derivatization protocol was recently exploited to directly analyze not only AEA but also its biosynthetic precursor, *N*-arachidonoyl-phosphatidylethanolamine (NArPE), in rat brain regions (Yang *et al.*, 1999). Furthermore, GC-EIMS was also used to measure OA in some tumor cells (Bisogno *et al.*, 1998), foods (Di Marzo *et al.*, 1998e) and in CSF and plasma samples (Hanus *et al.*, 1999b). Finally, by using GC-EIMS we have recently detected AEA and 2-AG in human blood (V. Di Marzo, L. De Petrocellis and T. Bisogno, unpublished observations), where, unlike most tissues analyzed so far, the amounts of AEA are higher than those of 2-AG (10.3 ± 4 pmol/ml and 2.1 ± 0.5 pmol/ml, means $\pm$ SD, $n = 3$).

Mass spectrometry has also been interfaced to high-pressure liquid chromatography (HPLC) in order to avoid derivatization and to skip the previous purification of the sample by HPLC (Koga *et al.*, 1997). Also in this case, the compounds are quantified by comparison to deuterated internal standards. However, this method, although more sensitive than GC-MS, does not provide structural information on the analyzed compounds since the ionization method used in the MS source (normally atmospheric pressure chemical ionization [APCI]) is such that no other fragment ion except for the quasi-molecular ion is detected. Therefore, some authors have used a second MS analyzer (LC-MS-MS) in order to identify AEA in tissues with more accuracy (Felder *et al.*, 1996). At any rate, the development of LC-MS techniques led to the detection of AEA in the low fmol range, or in even lower amounts, to allow the measurement of the endocannabinoid in a few microliters of microdialysates from small rat brain regions such as the periaqueductal grey (Walker *et al.*, 1999b), or the analysis of both AEA and 2-AG in small amounts (0.04–0.10 g) of tissue such as those derived from mouse brain areas (Di Marzo *et al.*, 2000d, Figure 5.2).

Finally, HPLC coupled to fluorimetric or UV detection of the column eluate has also been exploited for the quantification of AEA and 2-AG (Sugiura *et al.*, 1996b,c; Koga *et al.*, 1995; Kondo *et al.*, 1998; Arai *et al.*, 2000; Yagen and Burstein, 2000). This method still requires the derivatization of the compounds with fluorescent

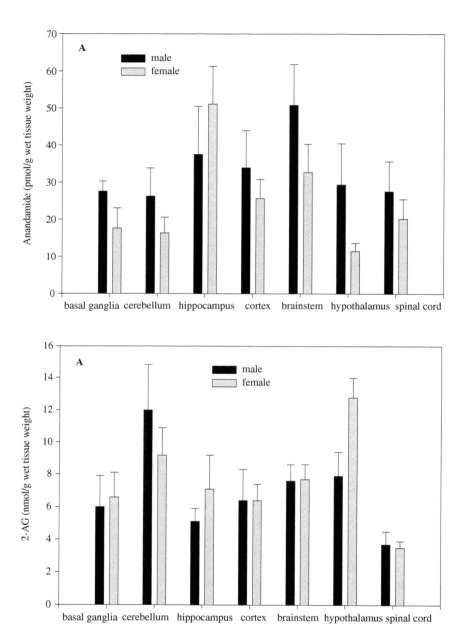

Figure 5.2 Levels of 2-arachidonoylglycerol (2-AG) and anandamide in brain regions and spinal cord from male and female C57/BL6 mice. The amounts of the two endocannabinoids were determined by athmospheric pressare chemical ionization liquid chromatography-mass spectrometry as described elsewhere (Di Marzo *et al.*, 2000d). "Basal ganglia" denotes the striatum, as other extra-pramidal ares of motor control (e.g. the substantia nigra and the gloibus pallidus) yielded amounts too little of tissue to be used for endocannabinoid determination.

tags, but has the advantage that the internal deuterated standard is not needed and calibration curves with external standards can be constructed. However, little chemical information on the analyzed compound is provided.

In conclusion, methods for the accurate, albeit time consuming, quantitative determination of endocannabinoids in tissues and body fluids are straightforward and now accessible to all. However, the lack of selective immunological assays for these compounds prevents the screening of several samples at a time, which is detrimental to the thorough characterization of possible correlations between endocannabinoid levels and physiopathological conditions, particularly as several replicates per sample need to be analyzed in order to minimize the usually large variations with which AEA, PEA and 2-AG are normally detected.

BIOSYNTHESIS AND INACTIVATION OF ENDOCANNABINOIDS AND OTHER CANNABIMIMETIC FATTY ACID DERIVATIVES

For a molecule to be considered as an endogenous mediator of physiological and pathological responses, mechanisms for its synthesis and inactivation need to be present in cells. It is commonly accepted that the endocannabinoids, unlike other mediators, are not produced by 'resting' cells and are, on the contrary, synthesized and released 'on demand' by adequately stimulated cells, following the Ca^{2+}-dependent remodeling of phospholipid precursors. Hence, the release of these compounds is not mediated by pre-stored vescicles. The termination of the action of the endocannabinoids is achieved through their diffusion into cells (which can be facilitated by membrane transporters), usually followed by their hydrolysis or esterification into membrane phospholipids, or both. Enzymatic oxidation of anandamide has also been described, leading to ethanolamides of either hydroxy-eicosatetraenoic acids or prostaglandins (Di Marzo *et al.*, 1999a; Burstein *et al.*, 2000, for reviews). The former compounds are still active at cannabinoid receptors and the meaning of this metabolic reaction is still not known. Although several points of contact exist between the metabolic pathways described so far for AEA, 2-AG and other cannabimimetic fatty acid derivatives, for the sake of clarity we discuss the biosynthesis and inactivation of these compounds one by one.

Biosynthesis of anandamide and congeners

The metabolic pathways of the NAE class of compounds, including PEA, have been investigated since the 1960s, but only after the discovery of AEA in 1992 was it possible to extend the results of those studies to this endocannabinoid. In fact, the metabolism of NAEs in animals and plants had been just reviewed by Schmid and co-workers in 1990, when Devane *et al.* (1992b) found the arachidonoyl homolog in pig brain. Subsequent experiments were aimed at assessing whether AEA could be biosynthesized and degraded in cells through the same mechanisms. The work performed by Schmid's group in the 1970s and 1980s (see Schmid *et al.*, 1990, 1996; Hansen *et al.*, 1998, for reviews) had shown that, rather than through

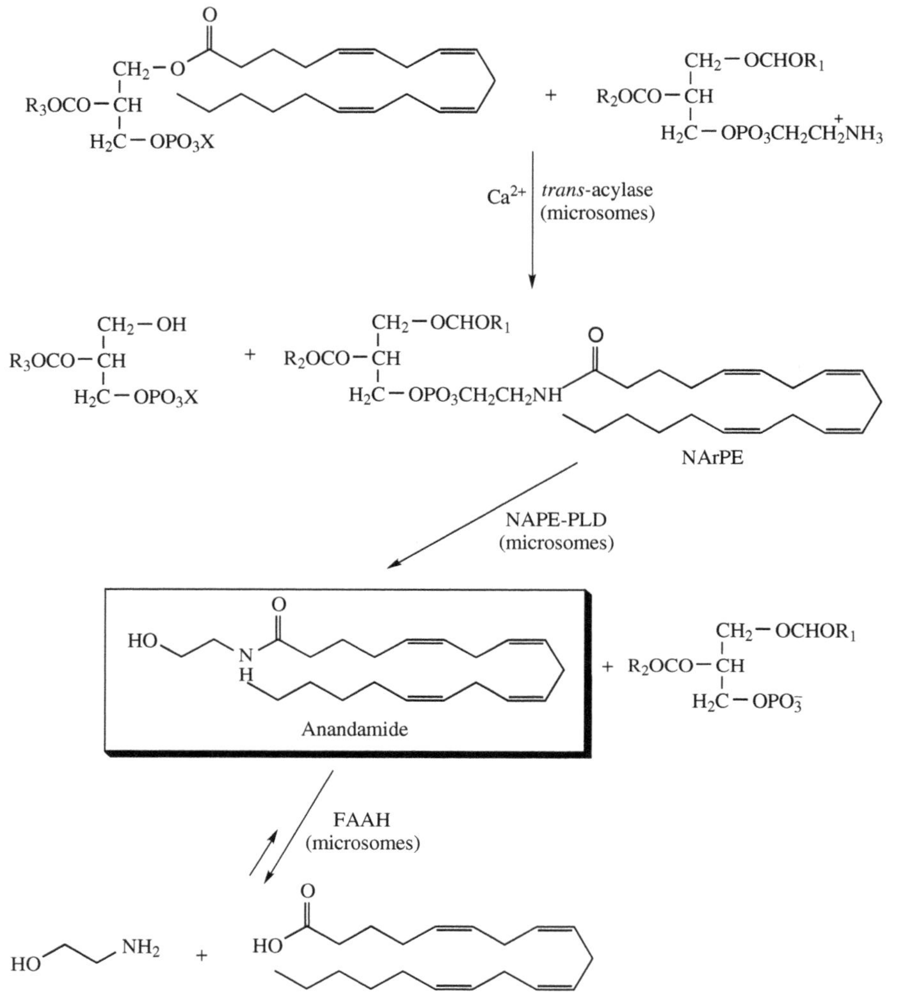

Figure 5.3 Biosynthesis of anandamide. The "phospholipid-mediated pathway" is described. All the enzymes involved in this pathway have been found in microsomal preparations. NarPE, N-arachidonoyl-phosphatidylethanolamine; NAPE-PLD, phospholipase D selective for the N-acyl-phosphatidylethanolamines; FAAH, fatty acid amide hydrolase. Under particular, and possibly non-physiological, conditions FAAH has also been shown to act "in reverse" and catalyze the energy-free condensation of arachidonate and ethanolamine to yield anandamide.

the energy-free condensation of fatty acids and ethanolamine, as previously suggested by Udenfriend and colleagues in the 1960s (Bachur and Udenfriend, 1966), NAEs were biosynthesized via a phospholipid-dependent pathway. According to this pathway (Figure 5.3) the enzymatic hydrolysis of N-acyl-phosphatidylethanolamine (NAPE) precursors, a minor class of membrane phosphoglycerides, leads to the release of NAEs. The hydrolytic enzyme catalyzing this reaction was identified as a phospholipase D selective for NAPEs (NAPE-PLD) and with low affinity for other membrane phospholipids. NAPEs, in turn, are produced through the acyl transfer from the *sn*-1 position of phospholipids to the *N*-position of phosphatidylethanolamine (PE), catalyzed by a Ca^{2+}-dependent *trans*-acylase. Neither the PLD nor the *trans*-acylase appeared to exhibit any selectivity for a particular fatty acid moiety. Thus, according to this biosynthetic scheme, the fatty acid composition of the NAE family of compounds would uniquely depend on the fatty acid composition on the *sn*-1 position of the last phospholipid precursor. This observation made it seem difficult that AEA could be produced through this same pathway *selectively* over other NAEs, since very little arachidonic acid is normally present on the *sn*-1 position of phospholipids. In fact, the studies carried out by Schmid's group in the 1980s did not detect any AEA in the NAEs produced, for example, during ischemic conditions, in the brain or the heart (Epps *et al.*, 1980; Natarajan *et al.*, 1986). After the discovery of AEA, however, it became obvious that most cells are capable of producing this compound in minor amounts as compared to other NAEs, thus supporting the hypothesis, proposed by Di Marzo and colleagues (1994), that the endocannabinoid was also biosynthesized according to the phospholipid-mediated pathway. Subsequent studies (Sugiura *et al.*, 1996b,c; Di Marzo *et al.*, 1996a,b; Cadas *et al.*, 1997) confirmed the occurrence of NArPE, the NAPE precursor of AEA, in *murine* brain, testes and leukocytes, and showed that cell-free preparations could convert NArPE into AEA, as well as biosynthesize NArPE from phospholipid precursors, namely phosphatidylcholine (PC) and PE, used as *sn*-1 donor and phospholipid acceptor of arachidonate, respectively. The enzymes catalyzing these reactions were partially characterized, and their lack of selectivity for any particular NAPE was confirmed. Furthermore, the precursor–product relationship between NArPE and AEA, at least in nervous cells, was confirmed by the finding of a parallel distribution of the two compounds in nine different brain areas (Bisogno *et al.*, 1999a), as well as of similarly increasing levels of NArPE and AEA in the developing brain (Berrendero *et al.*, 1999). At the same time, the "condensation pathway" (Figure 5.3), which had been shown for AEA only in cell free homogenates and in the presence of very high concentrations of arachidonic acid and ethanolamine (Deutsch and Chin, 1993; Devane and Axelrod, 1994; Kruszka and Gross, 1994; Sugiura *et al.*, 1996c), was being found to be due to the enzyme that under physiological conditions catalyzes the hydrolysis of AEA, rather than its synthesis (Ueda *et al.*, 1995; Arreaza *et al.*, 1997; Kurahashi *et al.*, 1997) (see below). Nevertheless, the lack of selectivity of the phospholipid-mediated pathway poses a series of questions that have not yet found an answer. Why do tissues, or stimulated neurons and blood cells, produce AEA as a minor component of a large family of lipids, whose physiological importance has not been clarified yet? Is there a way to enhance the amount of AEA vs. other NAEs in stimulated cells? Although Ca^{2+}-dependent remodeling of phospholipids may, at least in principle, lead to the enrichment of arachidonic acid on the *sn*-1 position of phosphoglycerides (see for example Kuwae *et al.*, 1997), a possible reason for the non-selective synthesis of AEA has been proposed. It is possible, in fact, that the other NAEs are used by cells to prevent the degradation of the endocannabinoid before its release into the extracellular milieu, thereby acting as "entourage compounds" (Mechoulam et al., 1998b). This explanation is based on several experimental observations. First, non-endocannabinoid NAEs, i.e., those AEA congeners that do not significantly bind and activate the two cannabinoid receptor subtypes known to date, are substrates and competitive inhibitors of the intracellular hydrolytic enzymes responsible for AEA degradation (Di Marzo et al., 1994; Désarnaud et al., 1995; Maurelli et al., 1995). Second, not all these compounds are released at the same rate from the cell, since the facilitated transport mechanism likely to be responsible for AEA release (Hillard et al., 1997) does not transport most of the NAEs (Piomelli et al., 1999). Third, some NAEs have cannabimimetic activity in some assays in vivo and in vitro (for example see Keefer et al., 1999; Watanabe et al., 1999), although only at doses much higher than those required for AEA to exert the same effects, thus suggesting that they may act via interference of tonic degradation of endogenous AEA. Analogous "entourage effects" have been demonstrated also for some 2-AG congeners (see below).

Biosynthesis of 2-arachidonoyl glycerol

In comparison with AEA, whose basal levels are low because of the minute amounts of sn-1-arachidonoyl phosphoglyceride precursors, the levels of 2-AG in unstimulated tissues and cells are usually two orders of magnitude higher than those of AEA, and sufficient in principle to permanently activate both cannabinoid receptor subtypes. In whole rat brain, for example, the reported levels of monoarachidonoylglycerol, which comprises a mixture of the 1-, 2- and 3-isomers all of which were shown to variedly activate CB_1 receptors, are around 4 nmol/g wet weight tissue (Sugiura et al., 1995; Stella et al., 1997). This is likely to yield a tissue concentration close to 4 µM. Therefore, it is clear that only a minor part of 2-AG found in tissues and cells is used to activate CB_1 and CB_2 cannabinoid receptors. Indeed, 2-AG is at the crossroad of several metabolic pathways, and is an important precursor and degradation product of phospho-, di- and tri-glycerides. Furthermore, it looks as if only 10–20% of the 2-AG produced de novo from stimulated cells is able to diffuse through the plasma membrane and be released outside the cell to interact with extracellular targets such as the cannabinoid receptors (Bisogno et al., 1997c). It is likely that, if 2-AG acts as an endogenous agonist of cannabinoid receptors, as suggested by recent observations by Sugiura and co-workers (1999, 2000), a particular pool of this metabolite is produced only for this purpose and inactivated through a special route. Therefore, of the several possible biosynthetic pathways already known for 2-AG formation (see Di Marzo, 1998, for review), it is necessary to concentrate on those routes that result in the release of this compound outside the cell (Figure 5.4). These pathways should be Ca^{2+}-dependent since the de novo formation of 2-AG, like for AEA, is presumably induced by plasma membrane depolarization. To date, the only reports that 2-AG biosynthesized de novo following Ca^{2+}-influx is released outside the cells have been obtained in mouse neuroblastoma N18TG2 cells stimulated with ionomycin (Bisogno et al., 1997c) and in human umbilical vein endothelial cells stimulated with the calcium ionophore A23187 or thrombin (Sugiura et al., 1998). However, only in the first case was the pathway for 2-AG biosynthesis investigated (Bisogno et al., 1999c) again suggesting a phospholipid-mediated route. According to this

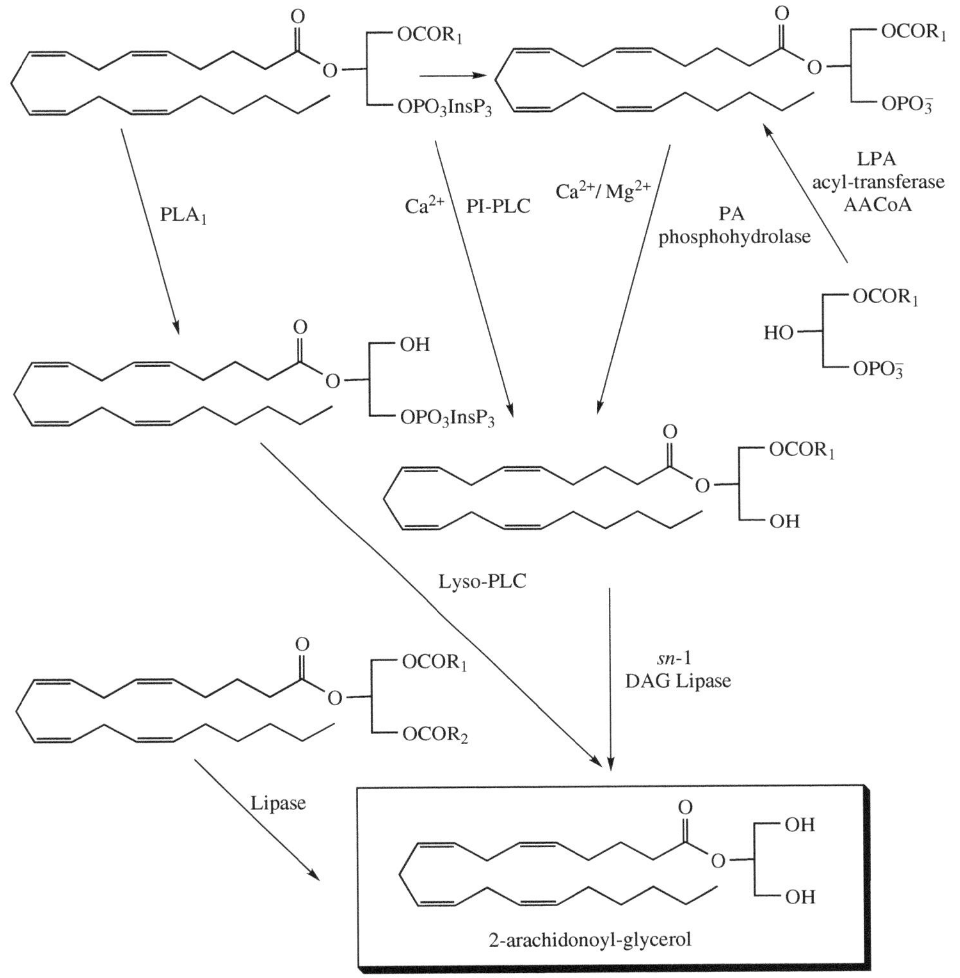

Figure 5.4 Pathways for the biosynthesis of 2-arachidonoylglycerol. Two major routes have been found in intact cells (Allen *et al.*, 1992; Bisogno *et al.*, 1997c, 1999b; Stella *et al.*, 1997), and a third one in rat brain homogenates (Ueda *et al.*, 1995). In all these pathways phospholipid-selective phosphodiesterases play a major role. In the gut, the endocannabinoid can also be formed from the lipase-catalyzed hydrolysis of triacylglycerols. PLA$_1$, phospholipase A$_1$; PI-PLC, phospholipase C selective for phosphatidylinositols; PA, phosphatidic acid; LPA, lysophosphatidic acid; AA, arachidonate; Lyso-PLC, lysophospholipase C; DAG, diacylglycerol; CoA, coenzyme A.

pathway, 2-AG is produced from the hydrolysis of *sn*-2-arachidonate-containing diacylglycerols (DAGs) catalyzed by a *sn*-1 selective DAG lipase. DAGs are in turn produced from the hydrolysis of *sn*-2-arachidonate-containing phosphatidic acid (PA), catalyzed by a PA phosphohydrolase. Finally, *sn*-2-arachidonoyl-PA is produced from the remodeling of phosphatidic acid, possibly involving secretory phospho-lipase A_2 and lysophosphatidic acid *trans*-acylase (Figure 5.4). Clearly, other possible Ca^{2+}-dependent pathways for the biosynthesis of DAGs serving as precursors for "endocannabinoid 2-AG," such as the hydrolysis of phosphoinositides catalyzed by selective phospholipase C (PI-PLC), should not be overlooked. This route was shown to lead to 2-AG formation in platelets (Prescott and Majerus, 1983), dorsal root ganglia (Allen *et al.*, 1992) and cortical neurons (Stella *et al.*, 1997), but little evidence exists showing that it is actually followed by 2-AG release from cells. Finally, several congeners, such as mono-oleoyl-glycerol, can, in principle, be co-synthesized with 2-AG. This has been shown so far to occur in N18TG2 cells (Di Marzo *et al.*, 1998c) but not in hippocampal slices (Stella *et al.*, 1997). Since arachi-donic acid is usually a major substituent on the *sn*-2 position of phosphoglycerides, it is likely that 2-AG, unlike AEA, is the most abundant component of its family of lipids, the 2-acyl-glycerols, as shown, for example, for several rat tissues (Kondo *et al.*, 1998). Nevertheless, 2-AG is likely to co-exist with several mono-acylglycerols, most of which are almost totally inactive at cannabinoid receptors but may retard its degradation by hydrolytic enzymes, as shown in RBL-2H3 and N18TG2 cells and mouse J774 macrophages (Ben-Shabat *et al.*, 1998; Di Marzo *et al.*, 1998b). This phenomenon may explain in part why these "entourage" mono-acylglycerols enhance 2-AG pharmacological activity *in vitro* and *in vivo* (Ben-Shabat *et al.*, 1998).

Biosynthesis of oleamide and of fatty acyl glycines

Although not an endocannabinoid, according to the definition given by Di Marzo (1998), OA produces typical cannabimimetic responses in the mouse 'tetrad' (Mechoulam *et al.*, 1997; Di Marzo *et al.*, 1998e), and potentiates AEA pharmaco-logical effects seemingly by inhibiting its degradation (Maurelli *et al.*, 1995; Mechou-lam *et al.*, 1997; Bisogno *et al.*, 1998). Furthermore, the typical hypnotic effects of this putative sleep-inducing factor are reduced by blockade of cannabinoid CB_1 receptors (Mendelson and Basile, 1999). Since OA binds to these receptors only at concentrations $\geq 10\,\mu M$ (Cheer *et al.*, 1999), it has been suggested that this metabol-ite can also behave as an "entourage congener" for the effects of AEA and act, at least in part, through enhanced levels of *endogenous* AEA (Mechoulam *et al.*, 1997 and Lambert and Di Marzo, 1999 for a recent review). This suggestion is confirmed by the observation that N18TG2 neuroblastoma cells contain and bio-synthesize OA in higher amounts than AEA (Bisogno *et al.*, 1997b). However, the pathways for the production of OA in neuronal cells have not been clarified yet. An early report showing that the compound could be produced by rat brain membranes from the condensation of oleic acid and ammonia (Sugiura *et al.*, 1996a) was subsequently explained with the finding that the enzyme responsible for the hydrolysis of both AEA and OA can also catalyze the reverse reactions start-ing from an excess of fatty acids and ethanolamine or ammonia (Figure 5.5; Arreaza *et al.*, 1997; Bisogno *et al.*, 1997b; Kurahashi *et al.*, 1997). The possibility that OA as well as other primary fatty acid amides are produced from the amidation of fatty acyl glycines using the same enzyme required for the C-terminal amidation of peptides, the peptide amidating enzyme (PAM), has also been put forward (Merkler

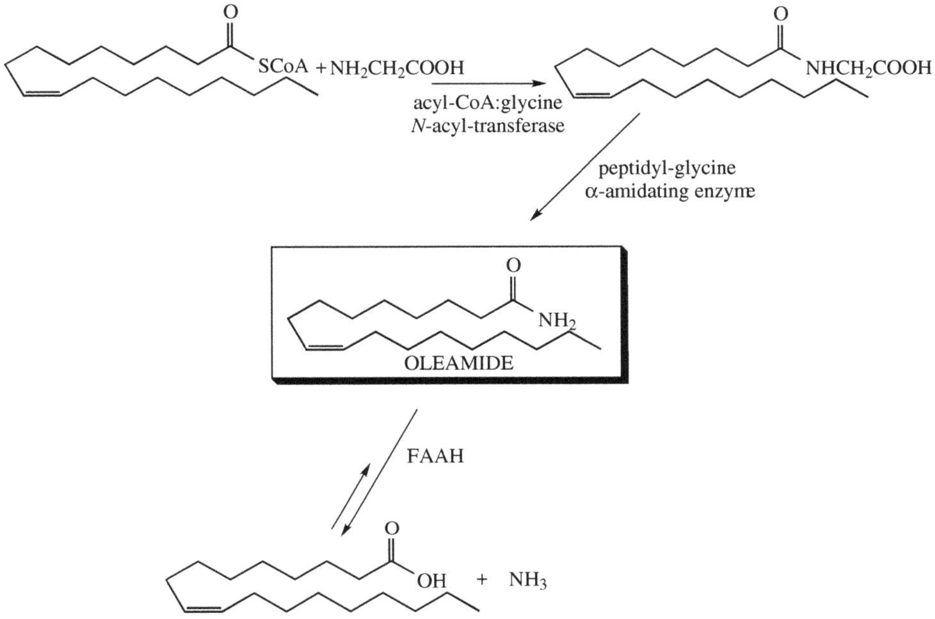

Figure 5.5 Two possible pathways for the biosynthesis and degradation of oleamide. None of these two pathways has been shown to occur in intact cells. The enzyme for the synthesis of oleoylglycine is not present in the CNS. Therefore, in order to generate oleamide thorough the upper pathway, oleoylglycine must cross the blood brain barrier. A third, phospholipid-mediated pathway has also been proposed (Bisogno *et al.*, 1997), and awaits experimental support. FAAH, fatty acid amide hydrolase; CoA, coenzyme A.

et al., 1996). This possibility is supported by the capability of PAM to recognize oleoylglycine (Figure 5.5) as a substrate (Wilcox *et al.*, 1999), and by the finding that this enzyme is expressed in N18TG2 cells (Ritenour-Rodgers *et al.*, 2000), where high amounts of OA were detected (Bisogno *et al.*, 1997b). However, no evidence exists in the presence, in nervous tissues, of oleoylglycine. This compound could be biosynthesized through the action on oleic acid of an acyl glycine *N*-acyl transferase (ACGNAT), an enzyme present in the kidneys and liver, but not in the CNS. Therefore, oleoylglycine, in order to act as a precursor for OA, should be produced in the periphery and then cross the blood–brain barrier (which, given the likely negative charge of oleoylglycine, would require a specific transporter) to become the substrate of PAM in the brain. This biosynthetic mechanism, although reasonable, contrasts with experimental evidence showing that intact isolated N18TG2 cells can biosynthesize OA starting from oleic acid (Bisogno *et al.*, 1997b). Nevertheless, it is possible that neurons utilize two pathways for OA formation, only one of which starts from *exogenous* oleoylglycine. Interestingly, the enzyme ACGNAT could also lead to the formation of arachidonoylglycine, a synthetic molecule that does not bind to cannabinoid receptors and has been found in several rat tissues (Huang *et al.*, 2001). This putative metabolite may be the endogenous correspondent of the non-psychotropic cannabinoid acids (Burstein, 1999, 2000).

Inactivation of endocannabinoids

AEA and 2-AG are two lipophilic molecules capable, in principle, of diffusing through the plasma membrane and, therefore, of being taken up by cells if their amount in the extracellular milieu is higher than their intracellular concentration. However, if these compounds play a physiological role as extracellular chemical signals, their concentration near their molecular sites of action, i.e., the cannabinoid receptors, needs to be rapidly decreased once these receptors have been activated. Thus, simple passive diffusion may be too slow a process to ensure the rapid inactivation of the endocannabinoid signal. Therefore, in order to be rapid, the diffusion of AEA and 2-AG inside the cell needs to be driven by controllable and selective mechanisms, such as a membrane transporter protein and/or intracellular enzymatic process capable of rapidly reducing the intracellular concentration of the two compounds by metabolizing them into different compounds, (Figure 5.6). These arguments hold, of course, if one assumes that the inactivation of endocannabinoids is not effected via extracellular processes, as in fact seems to be the case (see below). Both membrane carrier(s) and intracellular metabolic processes can inactivate AEA and 2-AG. In the case of AEA, its enzymatic hydrolysis to arachi-

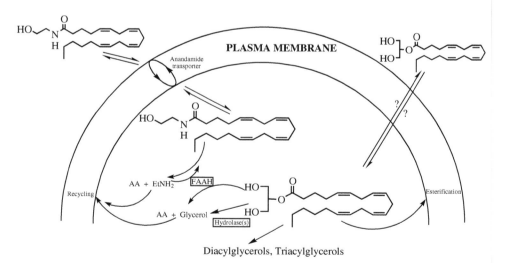

Figure 5.6 Inactivation of endocannabinoids by intact cells. Either facilitated or passive diffusion of anandamide and 2-arachidonoylglycerol, respectively, through the plasma membrane allow the two endocannabinoids to interact with their metabolic enzymes. Hydrolysis is catalyzed by fatty acid amide hydrolase (FAAH) or other hydrolases. The products of the hydrolyses are rapidly recycled into membrane phospholipids. 2-Arachidonoylglycerol can also be directly esterified into membrane phospholipids or diacyl- and triacylglycerols, and its facilitated diffusion into the cell via the anandamide transporter has been proposed and is still being debated.

donic acid and ethanolamine occurs only after its accumulation into intact cells. Accordingly, the compound is taken up by numerous cell types via selective, saturable, temperature-dependent and Na^+-independent facilitated transport mechanisms (Di Marzo *et al.*, 1994; Bisogno *et al.*, 1997a; Beltramo *et al.*, 1997; Hillard *et al.*, 1997; Maccarrone *et al.*, 1998, 1999, 2000a,b; for a comprehensive review, Hillard and Jarrahian, 2000). However, this mechanism is not followed by the efficient hydrolysis of the compound in all cell types examined so far (see for example Bisogno *et al.*, 1997a; Hillard *et al.*, 1997; Piomelli *et al.*, 1999). Since facilitated diffusion must be driven by a gradient of concentration across the cell membrane, this observation may suggest that other ways of reducing the concentration of the AEA taken up by cells exist at the intracellular level. On the other hand, in those cell lines where robust enzymatic hydrolysis of AEA is observed, the possibility exists that AEA diffusion through the plasma membrane may be simply induced by its rapid intracellular hydrolysis. In this latter case, selective inhibitors of AEA facilitated transport and AEA hydrolysis can be used to show that these two processes are indeed distinct events. For example, in rat basophilic RBL-2H3 cells it is possible to inhibit AEA uptake with *N*-arachidonoyl-vanillyl-amides at concentrations 10-fold lower than those required to inhibit AEA enzymatic hydrolysis (Di Marzo *et al.*, 1998a; Melck *et al.*, 1999). Likewise, in the same cells, typical inhibitors of AEA hydrolysis affected the accumulation of AEA into cells only at high concentrations (Rakhshan *et al.*, 2000). These data suggest that, although AEA intracellular metabolism does influence AEA facilitated diffusion, the uptake process is mediated by a mechanism distinct from the one catalyzing AEA hydrolysis.

The picture is more complicated for 2-AG inactivation. 2-AG is capable of competitively inhibiting the uptake of AEA, especially when the concentration of labeled AEA used in the assay is several-fold lower than the apparent K_m of the transporter for AEA (K_m values range between 0.1 and 41 µM depending on the cell type, Hillard and Jarrahian, 2000). This observation suggests that 2-AG can be recognized by the AEA transporter. However, no conclusive evidence exists for the accumulation of 2-AG into cells being mediated by this transporter. In fact, intracellular 2-AG can be metabolized through several rapid enzymatic reactions, including hydrolysis to arachidonic acid and glycerol – which can be catalyzed by more than one enzyme (see below) – or direct re-esterification into membrane phospholipids (Di Marzo *et al.*, 1998b, 1999b). Therefore, it is possible that these several processes render unnecessary the presence of a facilitated diffusion process for 2-AG since they are capable of efficiently driving 2-AG passive diffusion through the plasma membrane (Figure 5.6). Evidence against the possible existence of a 2-AG transporter comes from the observation that this compound, as compared to AEA, is poorly or not at all released by cells, or into rat dorsal striatum microdialysates and human CSF (Bisogno *et al.*, 1997c; Sugiura *et al.*, 1998; Giuffrida *et al.*, 1999; Leweke *et al.*, 1999). In fact, the AEA transporter was proposed to mediate also AEA release from neurons (Hillard *et al.*, 1997). If it were a substrate for this protein, based on the Ki values reported for its inhibition of AEA facilitated transport (Piomelli *et al.*, 1999; Jarrahian *et al.*, 2000), 2-AG should interact with the AEA transporter with a similar affinity as AEA, and, being biosynthesized in much higher amounts than AEA, should be released into the extracellular space, CSF or microdialysates more efficiently than the latter mediator. However, studies with selective inhibitors of the AEA transporter are necessary in order to understand whether this mechanism also limits 2-AG pharmacological activity.

Fatty acid amide hydrolase – an "enzyme for all seasons"

One enzyme has been identified that is capable of catalyzing the hydrolysis of all the cannabimimetic fatty acid derivatives discovered so far. This enzyme, originally known as "anandamide amidohydrolase" (Désarnaud et al., 1995; Hillard et al., 1995; Maurelli et al., 1995; Ueda et al., 1995; Tiger et al., 2000), was purified and cloned from rat liver microsomes (Cravatt et al., 1996) and named "fatty acid amide hydrolase" (FAAH) due to its wide substrate selectivity. The structural and kinetic properties of this enzyme have been widely reviewed in the literature (Di Marzo et al., 1999a; Ueda and Yamamoto, 2000), and several more or less selective FAAH inhibitors have also been developed (reviewed by Di Marzo and Deutsch, 1998). Briefly, FAAH is a membrane protein containing an highly conserved amidase consensus amino acid sequence, and capable of catalyzing the hydrolysis of long chain fatty acid amides, including OA and PEA, and esters, including 2-AG and its 1- and 3-isomers. The enzyme has an alkaline optimal pH, is present only in intracellular membranes, and its localization as well as quaternary structure are regulated by the presence of particular domains. In fact, the putative transmembrane hydrophobic domain of the protein seems to be important for the formation of active oligomers rather than for the localization of the enzyme on intracellular membranes, which instead is regulated by an SH-3 consensus proline-rich sequence necessary for enzymatic activity. The catalytic amino acid of the enzyme has also been identified as Ser241, and two other residues of the amidase consensus sequence, Ser217 and Cys249, have been shown to participate in the enzymatic activity. Structure activity relationship studies carried out with AEA, OA and 2-AG congeners and analogs, as well as with various inhibitors, suggest that the fatty acyl moiety is important for these compounds only to assume the correct conformation for an optimal interaction of the carbonyl group with the catatytically active serine group of the enzyme. The presence of electronegative substituents on the molecule 'head' renders the carbonyl group more electrophilic to the point that, in the presence of a strong electrophilic group in the substrate, the fatty acyl chain becomes less important for interaction with the binding site.

FAAH has also been cloned from mouse, human and pig tissues, and sequence analyses have shown high homology between species, even though differences in the substrate specificity of the recombinant enzymes have been pointed out (Giang and Cravatt, 1997; Goparaju et al., 1999a). The rat and pig enzymes are capable, however, of also catalyzing the hydrolysis 'reverse reaction', i.e., the condensation of fatty acids and ethanolamine or ammonia in the presence of high concentrations (high µM and mM, respectively) of the two substrates (Kurahashi et al., 1997; Arreaza et al., 1997; Goparaju et al., 1999a). Thus, this enzyme may be at the basis of the 'condensation pathway' for AEA (and OA) biosynthesis described above (Figure 5.3), which is unlikely to occur in intact cells under physiological conditions. The identity between FAAH and 'anandamide synthase' is suggested also by the identical tissue distribution, sensitivity to inhibitors and catalytic properties of the two enzymes. However, as purified FAAH is also capable of catalyzing the synthesis of very small amounts of AEA in the presence of low concentrations of substrates, and high concentration of AEA (Katayama *et al.*, 1999), the possibility that, under particular conditions, this enzyme is also used for the selective biosynthesis of AEA cannot be completely ruled out.

Despite the flexibility of FAAH and its wide tissue distribution, other additional enzymes may catalyze the hydrolysis of AEA, 2-AG and OA. A hydrolytic enzyme for AEA, with high affinity also for PEA, OA and 2-AG and an acidic optimal pH, was recently identified in a human megakaryoblastic cell line as well as in rat lungs (Ueda *et al.*, 1999; Ueda and Yamamoto, 2000). The presence of this likely lysosomal enzyme in those cells expressing little FAAH activity and an active AEA transporter has not been determined, nor has its purification and structural characterization reported to date. As to 2-AG, at least two enzymatic activities distinct from FAAH have been partially characterized so far in porcine brain and rat circulating platelets and macrophages, as well as mouse J774 macrophages (Di Marzo *et al.*, 1999b; Goparaju *et al.*, 1999b). These enzymes also recognize other unsaturated monoacylglycerols such as, for example, mono-oleoyl-glycerol, which may act as a competitive inhibitor for 2-AG inactivation by intact cells (Ben-Shabat *et al.*, 1998; Di Marzo *et al.*, 1998b), but little is known on whether they also recognize fatty acid amides. All these enzymes were found to exhibit a sensitivity to some typical inhibitors of AEA hydrolysis different from that observed with FAAH. One may wonder why so many enzymes exist for endocannabinoid hydrolysis. In rat brain, immunocytochemical studies showed that FAAH is expressed in neurons likely to receive synapses from CB_1-receptor expressing neurons (Egertova *et al.*, 1998). Therefore, it is possible that FAAH concurs to the inactivation of AEA and 2-AG when these molecules act as neuromodulators via pre-synaptic CB_1 cannabinoid receptors (see Di Marzo *et al.*, 1998d for review). The lysosomal enzyme may be used by cells to drive AEA facilitated diffusion into those cells where FAAH is not abundantly expressed, or to inactivate the NAEs produced following cell damage and death (see below). As to the various '2-AG lipases' found so far, it is possible that they reflect the fact that 2-AG is not used by cells only as an endocannabinoid. It is possible that one or more of these enzymes catalyze the hydrolysis of 2-AG when this reaction represents a mechanism for the production of free arachidonate and the synthesis of its metabolites (the eicosanoids), rather than for the inactivation of an extracellular agonist of cannabinoid receptors.

Regulation of endocannabinoid metabolism and activity

The regulation of the signal mediated by an endogenous molecule can be effected either by modulating the potency or duration of its activity, or by influencing its concentration near the receptor(s) that mediate(s) its actions. Three possibilities exist for the modulation of the levels of an endogenous substance. Firstly, its production can be regulated by enhancing or reducing the activity and/or levels of the enzyme(s) which catalyze(s) the rate-limiting step(s) of its biosynthetic pathway. Secondly, the activity of the proteins mediating the release of the substance into the extracellular milieu can also be regulated. Finally, the inactivation of the substance can be subject to regulation. In the case of endocannabinoids, whose release from cells is almost uniquely dependent on their *de novo* formation, possible regulative mechanisms have been reported for both their biosynthesis and inactivation. As the major biosynthetic pathways proposed for both AEA and 2-AG depend on the phosphodiesterase-mediated remodeling of phosphoglycerides (Figures 5.3 and 5.4), it is reasonable to expect that some of the stimuli triggering phosphodiesterase and *trans*-acylase activity also regulate the biosynthesis of these two substances. In fact, Ca^{2+} influx into cells, arachidonate mobilization and protein kinase A-mediated protein phosphorylation have been suggested to up-regulate 2-AG and/or AEA formation in neurons and macrophages (Di Marzo *et al.*, 1994; Cadas *et al.*, 1997; Bisogno *et al.*, 1997c; Pestonjamasp and Burstein, 1998). Stimulation of cells with lipopolysaccharide (LPS), which is also known to stimulate phospholipid remodeling, also induced AEA and 2-AG synthesis in macrophages (Wagner *et al.*, 1997; Pestonjamasp and Burstein, 1998; Varga *et al.*, 1998; Di Marzo *et al.*, 1999b).

The two-step inactivation mechanism of AEA was also recently shown to be subject to regulation. Nitric oxide (NO) donors stimulate AEA re-uptake by human neuroblastoma and lymphoma cells, as well as by platelets, mast cells and endothelial cells (Maccarone *et al.*, 1998, 1999, 2000b,c). In fact, peroxynitrite, formed by the reaction of NO and superoxide (O_2^-), is an even more potent stimulant of AEA facilitated transport, and O_2^- scavengers such as glutathione reduce this effect. Since AEA and psychotropic cannabinoids, by activating CB_1 receptors, can enhance NO release, for example in endothelial cells, it is possible that the endocannabinoid exerts a control on its own levels by enhancing its reuptake. Accordingly, HU-210 was found to enhance the facilitated transport of AEA in human endothelial cells (Maccarone *et al.*, 2000b). In a preliminary study, activation of protein kinase C was also shown to inhibit AEA uptake in RBL-2H3 cells, seemingly by decreasing the affinity of the transporter for AEA (Rakhshan *et al.*, 1999). NO has no effect on FAAH, but this enzyme seems to be down-regulated by LPS in human lymphocytes (Maccarone and Di Marzo, unpublished observations). LPS also decreases the activity of '2-AG lipase' enzymes in rat platelets and macrophages (Di Marzo *et al.*, 1999b), and this effect may account in part for the raised levels of 2-AG observed upon treatment of these blood cells with LPS.

Another possible way of enhancing the levels of AEA and 2-AG has been mentioned above and consists of the "entourage" effects by NAEs and monoacylglycerols that are poorly active or inactive at CB_1 and CB_2 receptors but are capable of competing with either the transporters or the enzymes mediating the inactivation of endocannabinoids (Mechoulam *et al.*, 1997; Ben-Shabat *et al.*, 1998). However these "entourage" effects are probably not uniquely due to inhibition of the inactivation of AEA and 2-AG. In fact, saturated analogs of these two compounds, such as PEA and 2-palmitoyl glycerol, do not efficiently inhibit the inactivation of AEA and 2-AG, respectively (see Bisogno *et al.*, 1998; Ben-Shabat *et al.*, 1998, for examples). Yet, these compounds do enhance the effects of the endocannabinoids. It is possible that these saturated analogs enhance the binding of AEA and PEA to cannabinoid receptors and facilitate the coupling of activated cannabinoid receptors to G-proteins or the stimulation of intracellular signaling events by endocannabinoids. In fact, PEA was recently shown to induce G-protein activity at concentrations (1–$10 \mu M$) that do not bind to either CB_1 or CB_2 receptors (Griffin *et al.*, 2000).

POSSIBLE PHYSIOLOGICAL AND PATHOLOGICAL FUNCTIONS OF ENDOCANNABINOIDS IN MAMMALS

Although endocannabinoid levels in tissues may reflect tonic cell stimulation, it is the finding of changes in anandamide and 2-AG levels during physiological or pathological conditions that, together with observations of their pharmacological activity *in vivo* and *in vitro*, can provide useful information on the possible physiopathological role of endocannabinoids. Furthermore, although the cannabinoid receptor antagonists developed so far do not appear to behave as "pure antagonists" but also exhibit, under certain conditions, "reverse agonist" properties (Landsman *et al.*, 1997; Portier *et al.*, 1999), pharmacological studies carried out with these compounds administered alone can reveal a possible "tone" of endocannabinoid-induced CB_1 and CB_2 receptor activation during certain conditions. Finally, accurate and specific behavioral observations carried out in CB_1 and CB_2 receptor "knockout" mice are also being used to understand whether the endocannabinoid system is tonically activated in the control of some CNS functions. A few studies in these directions are being performed and will be briefly mentioned here. For a more detailed discussion of the possible physiological significance of the endocannabinoid system in mammals the reader is referred to other more specific chapters in this book.

Central nervous functions and the control of behavior

AEA and 2-AG levels in whole rat brain were found to vary in a different way during development from gestation to birth and until adulthood. The amounts of AEA, NArPE, and of CB_1 cannabinoid receptors progressively increased during development, whereas 2-AG levels remained constant except for a peak at postnatal day 1 (Berrendero *et al.*, 1999). Furthermore, increasingly higher levels of AEA in the hippocampus and cerebellum were found in rats when passing from 2 to 12 weeks of age, before declining after 12 weeks (Koga *et al.*, 1997). These findings may suggest that AEA plays a role as a CB_1 ligand in adult as opposed to newly born animals. Conversely, 2-AG in brain may be more important at the early stages of development, for example in the regulation of early learning processes or other vital functions such as feeding (see below). The pharmacological bases for a possible role of endocannabinoids in brain development have been extensively reviewed (Fernandez-Ruiz *et al.*, 2000).

Both AEA and 2-AG were shown to reproduce in rodents the inhibition of spontaneous activity and induction of catalepsy typical of psychoactive cannabinoids and CB_1 receptor agonists (Fride and Mechoulam, 1993; Mechoulam *et al.*, 1995). On the other hand, CB_1 receptor knockout mice seem to have different baseline locomotor activity than wild-type mice, although it is not clear yet whether deletion of the CB_1 receptor gene in these transgenic animals leads to hypo- or hyper-motility (Ledent et al., 1999; Zimmer et al., 1999). An endogenous cannabinoid tone controlling spontaneous activity and motor behavior was suggested by the finding of AEA, but not 2-AG, in microdialysates from the dorsal striatum of freely moving rats (Giuffrida et al., 1999). The levels of AEA were increased by stimulation of D2 dopamine receptors, whereas possible antagonism of AEA action at CB_1 receptors with the selective antagonist SR141716A enhanced the motor-stimulatory effect of the D2 agonist quinpirole. Another study carried out in rats (Di Marzo et al., 2000e) showed that the amounts of AEA are also very high in the substantia nigra and external layer of the globus pallidus. Indeed, intrapallidal injection of either AEA or synthetic cannabinoids induces catalepsy (Wickens and Pertwee, 1993), possibly by enhancing the action of GABA via reduction of its reuptake (see Glass et al., 1997, for review). It was found that the levels of 2-AG in the globus pallidus, but not in other brain areas, were significantly increased after treatment of rats with reserpine, which, by causing catecholamine (and dopamine) depletion from the striatum, induces immobility and Parkinson's disease symptoms. Treatment of reserpine-treated rats with D1 and D2 receptor agonists led to partial restoration of locomotor ability which was accompanied by a significant reduction of both 2-AG and AEA levels in the globus pallidus. Finally, co-administration of the D2 receptor agonist, quinpirole, and SR141716A to reserpine-treated rats almost fully restored locomotion (Di Marzo et al., 2000e). These findings show an inverse correlation between spontaneous activity and the levels and action of endocannabinoids in the external layer of the globus pallidus, and suggest that these lipids may contribute to the generation of symptoms of Parkinson's disease. Furthermore, these data suggest that the endocannabinoid system may act in the basal ganglia to put a 'brake' on the motor stimulatory actions of the dopaminergic system.

The role of endocannabinoids in the control of locomotion and, possibly, in reward mechanisms was suggested by a recent study (Di Marzo et al., 2000b) where tolerance to THC was induced in rats and the levels of AEA and 2-AG in different brain regions were compared to those in vehicle-treated rats. Lower levels of both 2-AG and AEA, as well as of cannabinoid binding and coupling to G-proteins, were detected in the striatum of THC-tolerant rats. Interestingly, studies recently reviewed by Rodriguez de Fonseca et al. (1998) showed that THC-tolerant rats are more responsive to drugs that enhance dopamine actions such as amphetamine. These rats also exhibit increased motor behavior after interruption of THC treatment. Therefore, the finding of reduced 2-AG and AEA levels in the striatum of THC-tolerant rats establishes again a correlation between increased spontaneous activity and low levels of endocannabinoid signaling in the basal ganglia. In the study by Di Marzo et al. (2000b) it was also found that the only area where chronic THC treatment did not induce down-regulation of cannabinoid receptor levels and trans-membrane signaling was the limbic forebrain, where an almost 4-fold increase of AEA levels was also observed. In this brain area, cannabinoids, by enhancing the release of dopamine from dopaminergic terminals from the ventral tegmental area, may exert reinforcing actions on the effects of other drugs of abuse or, under more physiological conditions, may participate in the regulation of reward pathways (Gardner and Vorel, 1998, for a review). Di Marzo et al. (2000b) speculated that dopamine released in the nucleus accumbens (a part of the limbic forebrain) upon chronic treatment with THC (Tanda *et al.*, 1997) triggers in turn AEA formation, as previously shown for the dorsal striatum (Giuffrida *et al.*, 1999). That a tonic activation of CB_1 receptors is involved in reinforcing, for example, the actions of opiates, is suggested by recent work carried out with transgenic mice. These studies showed reduced addictive effects of opiates in CB_1 receptor knockout mice (Ledent *et al.*, 1999) as well as lack of morphine-induced dopamine release in the nucleus accumbens of these transgenic animals (Mascia *et al.*, 1999). Thus, contrary to the basal ganglia, endocannabinoids released in the nucleus accumbens may act to *enhance* the action of dopamine, thereby participating in reward mechanisms and reinforcement of drug of abuse effects.

Several observations, such as the finding of CB_1 receptors in some hypothalamic nuclei, e.g. the arcuate nucleus and the medial preoptic area (Fernandez-Ruiz *et al.* 1997), suggest that endocannabinoids modulate hypothalamic functions, including not only the control of body temperature (Fride and Mechoulam, 1993; Mechoulam *et al.*, 1995), but also of food intake, appetite and pituitary hormone release. The CB_1 receptor selective antagonist, SR141716A, inhibits palatable food-intake in rodents (Arnone *et al.*, 1997; Simiand *et al.*, 1998; Colombo *et al.*, 1998). Although this effect could be due to the inverse agonist properties of SR141716A (Landsman *et al.*, 1997), the possibility that SR141716A acts by reverting a food-intake stimulating tone of endocannabinoids also exists, particularly in view of the two following findings: (1) AEA induces hyperphagia in rats in a fashion sensitive to SR141716A (Williams and Kirkham, 1999) and stimulates food-intake at very low concentrations (Hao *et al.*, 2000); and (2) endocannabinoids have been found in the hypothalamus (Gonzalez *et al.*, 1999). Interestingly, the hypothalamic levels of AEA in female rats depend on the phase of the ovarian cycle (Gonzalez *et al.*, 2000), and craving for palatable food has also been correlated with the various phases of the ovarian cycle (Dye and Blundell, 1997). AEA levels in female rat anterior pituitary also change during the estrus cycles, and in a fashion complementary to that observed in the hypothalamus (Gonzalez *et al.*, 2000), thus suggesting the existence of possible feedback mechanisms between the two regions, as previously shown for other mediators. In fact, AEA and THC have been reported to modulate the release of pituitary hormones, e.g. prolactin, corticotropin releasing factor and luteinizing hormone (see Wenger *et al.*, 1999, for a recent review).

The notion that endocannabinoids are involved in the control of learning and memory processes, particularly at the level of the hippocamopus, is supported by at least three different types of observations. First, both AEA and 2-AG inhibit hippocampal long-term potentiation (LTP) and/or transformation (Collin *et al.*, 1995; Terranova *et al.*, 1996; Stella *et al.*, 1997) and modulate the release of glutamate or acetylcholine from hippocampal slices (see Di Marzo *et al.*, 1998d for a review). Also, AEA modulates both short-term and long-term memory (Mallet and Beninger, 1998; Castellano *et al.*, 1997). Second, blockade of CB_1 receptors with SR141716A enhances LTP (Terranova *et al.*, 1996), whereas CB_1 receptor knockout mice exhibit enhancement of memory as well as of LTP (Reibaud *et al.*, 1999; Bohme *et al.*, 2000). Third, both AEA and FAAH (but not always 2-AG) are most abundant in the hippocampus of humans, rats and mice (Felder *et al.*, 1996; Bisogno *et al.*, 1999a; Di Marzo *et al.*, 2000b; Di Marzo *et al.*, 2000d). Interestingly, the levels of AEA (but not 2-AG or FAAH) in the hippocampus of CB_1 receptor knockout mice were significantly lower than in wild-type mice, thus suggesting that tonic activation of CB_1 receptors in this region leads to stimulation of AEA formation (Di Marzo et al., 2000d). All these reports, taken together, point to the presence of an AEA/CB_1 tone regulating learning and memory processes at the hippocampal level.

Several studies (recently reviewed by Martin and Lichtman, 1998, and Walker et al., 1999a) have used a pharmacological approach to assess whether endocannabinoids are involved in the control of nociception and, in particular, inflammatory pain. On the other hand, experiments carried out with cannabinoid receptor antagonists have reported contrasting results (see Di Marzo et al., 2000a for a review), thus indicating the need to perform analytical investigations that could correlate the tissue levels of endocannabinoids to various nociceptive responses. Two such studies have been carried out in rats thus far. In one case, electrical stimulation of the periaqueductal gray (PAG) was shown to induce CB_1-mediated analgesia while leading to the release of AEA in microdialysates from this region of the brainstem (Walker et al., 1999b). Also the injection of the chemical irritant formalin into the paw induced a nociceptive response concomitantly to the release of AEA from the PAG. Although the amounts of AEA found in the microdialysates were probably too little to activate CB_1 receptors, it must be noted that, in both rats and mice, the brainstem is one of the regions with the highest levels of this endocannabinoid and the lowest density of CB_1 receptors (Bisogno et al., 1999; Di Marzo et al., 2000d). Indeed, an earlier investigation had suggested that an endocannabinoid tone may down-modulate pain perception via CB_1 receptors in another region of the brainstem, the rostral ventromedial medulla, through the same circuit previously shown to contribute to the pain-suppressing effects of morphine (Meng et al., 1998). In order to assess whether 2-AG, AEA and PEA participate in inflammatory pain also as local mediators, as suggested by pharmacological studies carried out by Calignano et al. (1998), the amounts of these compounds in the hind paw of rats were measured after injection of formalin, and during the maximal nociceptive response. No statistically significant difference with vehicle-treated rat paws was found for either 2-AG, AEA or PEA (Beaulieu et al., 2000). These studies suggest that endocannabinoids do not necessarily modulate perception of inflammatory pain at the site of inflammation. Further studies have shown that blockade of spinal CB_1 receptors by either SR141716A or CB_1 receptor anti-sense oligonucleotides leads to hyperalgesia (Richardson et al., 1997), thus suggesting the existence of an endocannabinoid tone down-modulating nociceptive response also at the spinal level. Finally, in agreement with a report by Calignano et al. (1998), the involvement of CB_2 receptors in inflammatory pain was recently shown by using a selective and potent agonist of these receptors (Hanus et al., 1999a).

Immune and cardiovascular systems

The regulation of endocannabinoid formation and inactivation by blood and endothelial cells has been studied (see Wagner et al., 1998, for review). There is evidence for the occurrence of anabolic and catabolic reactions for both AEA and

2-AG in *murine* macrophages, mast cells/basophils, neutrophils, lymphocytes, platelets and endothelial cells. Lipopolysaccharides (LPS) of bacterial origin stimulate AEA and 2-AG formation in macrophages (Varga *et al.*, 1998; Pestonjamasp and Burstein, 1998; Di Marzo *et al.*, 1999b). This finding suggests that bacterial infections may trigger the release of endocannabinoids, for which several immune-modulatory actions have been reported (see Parolaro, 1999, for review). LPS-induced hypotension in rats, an animal model for the study of septic shock, was attenuated by the CB_1 antagonist SR141716A (Varga *et al.*, 1998). It was found that platelets contribute to mean arterial pressure through either the formation of 2-AG selectively over anandamide (Varga *et al.*, 1998), or the disposal of endocannabinoids via re-uptake and hydrolysis processes (Maccarrone *et al.*, 1999; Di Marzo *et al.*, 1999b). Accordingly, 2-AG, and even more its non-hydrolyzable analog 2-AG ether, were shown to induce vasodilation both *in vivo* and *in vitro* (Mechoulam *et al.*, 1998a; Varga *et al.*, 1998; Járai *et al.*, 2000). Endothelial cells from the vascular bed also produce and inactivate endocannabinoids (Mechoulam *et al.*, 1998a; Sugiura *et al.*, 1998; Maccarrone *et al.*, 2000b). On the other hand, endocannabinoids derived from macrophages, but not platelets, are the cause of the hypotensive state arising during hemorrhagic shock (Wagner *et al.*, 1997). A possible role of endocannabinoids in immune hyper-reactivity is suggested by the finding that an immortalized mast cell-like line, the rat basophilic leukemia (RBL-2H3) cells, responds to IgE challenge (a stimulus leading to cell degranulation and histamine/serotonin release) by producing anandamide and its congeners (Bisogno *et al.*, 1997a). RBL-2H3 cells and human mast cells also inactivate anandamide and/or 2-AG via several mechanisms (Bisogno *et al.*, 1997a; Di Marzo *et al.*, 1998b; Maccarrone *et al.*, 2000c). These reports, together with the observation that endocannabinoids variedly affect immune cell development, proliferation and functionality (Parolaro, 1999), as well as vascular tone and cardiac output (see Kunos *et al.*, 2000, for a recent review), suggest that these compounds may participate in the control of immune and vascular functions, particularly during uncontrolled immune reactions (allergies, septic shock, auto-immune diseases) or hemorrhagy.

Reproduction

AEA, 2-AG and cannabinoids inhibit mouse embryo development and blastocyst implantation in the uterus through CB_1, but not CB_2, receptors (Paria *et al.*, 1998). This finding, together with the observation that the amounts of AEA or the expression of FAAH are correlated with uterine receptivity to embryo implantation (Schmid *et al.*, 1997; Paria *et al.*, 1999), strongly suggest that endocannabinoids play a role in pregnancy. In particular, it was found that uterine AEA levels are lower when the uterus is receptive – and higher when the organ is refractory – to embryo implantation. FAAH, on the contrary, is highly expressed in the peri-implantation period, which may possibly explain the low levels of AEA found in the uterus at this stage. On the basis of these studies it was suggested for AEA a role in directing the timing and site of embryo implantation, and it is possible to speculate that disorders such as spontaneous termination of pregnancy derive from a malfunctioning uterine endocannabinoid system. Indeed, very recent evidence showed that decreased FAAH levels and activity in lymphocytes, a condition that is likely to determine high levels of endocannabinoids in the uterus, predicts spontaneous abortion in women (Maccarrone *et al.*, 2000a).

Based on studies carried out in sea urchin sperm (Schuel *et al.*, 1994; Berdyshev *et al.*, 1999), endocannabinoids may also interfere with sperm cell fertilizing capability. These findings have been recently extended to human sperm, for which preliminary evidence of the expression of cannabinoid receptors has been provided (Schuel *et al.*, 1998).

Finally, a role for the endocannabinoid system in reproduction is also suggested by the capability of both AEA and 2-AG to inhibit the electrically stimulated contractions of the mouse *vas deferens* (Devane *et al.*, 1992b; Mechoulam *et al.*, 1995). This effect was later shown to be due to activation of CB_1 receptors (Pertwee *et al.*, 1996).

Cell protection

The suggestion that endocannabinoids may have a cell-protecting function during cell injury stems from the fact that a similar role was proposed also for other NAEs, and the corresponding NAPEs (Schmid *et al.*, 1990; Hansen *et al.*, 1998, for reviews). This hypothesis found support in more recent reports that stimuli leading to high intracellular Ca^{2+} concentrations, such as glutamate receptor stimulation, noxious agents like ethanol and sodium azide, and exposure to UVB radiations, all lead to increased synthesis of AEA and other NAEs (Hansen *et al.*, 1995, 1997; Basavarajappa and Hunglund, 1999; Berdyshev *et al.*, 2000). However, in the nervous system, direct evidence for a neuroprotective action for AEA and 2-AG was reported only very recently, and in this case, cannabinoid receptors did not appear to be involved (Sinor *et al.*, 2000). This was quite surprising since AEA, via CB_1 receptors, inhibits Ca^{2+} influx into neurons through voltage-operated calcium channels (Mackie *et al.*, 1993), and was shown also to counteract membrane permeability to Ca^{2+} through NMDA receptor-coupled channels (Hampson *et al.*, 1998a). Therefore, AEA (as well as 2-AG) should be able to inhibit, for example, glutamate-induced excitotoxicity (or other pathological conditions arising from high intracellular Ca^{2+} concentrations) by acting at CB_1 receptors. Furthermore, AEA does not share with other cannabinoids their anti-oxidant effects (Hampson *et al.*, 1998b). Therefore, further studies are necessary in order to assess whether and how endocannabinoids prevent cell damage.

NON-MAMMALIAN CANNABIMIMETIC FATTY ACID DERIVATIVES

NAEs and 2-acylglycerols have been found, in varied amounts, in all animal and plant organisms analyzed so far, and also in several microorganisms (see Schmid *et al.*, 1990). The fatty acid composition of these two families of lipids, however, very much depends on the fatty acid composition of membrane phospholipids, which in turn may greatly change within animal *phyla* and generally undergo dramatic variations when passing from the animal to the plant kingdom. Endocannabinoids are derived from C20 polyunsaturated fatty acids (PUFAs). Thus it is to be expected that tissues particularly rich in these fatty acids will also contain high levels of these compounds. For example, mammalian retinas which contain high concentrations of esterified docosahexaenoic acid ($C_{22:6, n-3}$) also contain the corresponding NAE and 2-acylglycerol, along with AEA and 2-AG (Bisogno *et al.*, 1999b). Conversely, higher plants, which do not synthesize arachidonic acid, should not contain AEA, even though they do produce other 2-acylglycerols, NAEs and NAPEs (Table 5.1 and Chapman *et al.*, 1998 for review). In fact, the finding of AEA in cocoa (di Tomaso *et al.*, 1996) was recently suggested to be due to artifacts occuring during cocoa manufacturing (Di Marzo *et al.*, 1998e). Therefore, not all living organisms and not all tissues may contain those NAEs and 2-acylglycerols that are capable of directly activating the cannabinoid receptors. Conversely, it is likely that non-PUFA NAEs and 2-acylglycerols, which are potentially capable of enhancing the activity/levels of AEA and 2-AG via the "entourage" effects described above and previously (Ben-Shabat *et al.*, 1998), occur in most organic tissues. However, it must be remembered that specific binding sites for, and/or biological responses to, THC and psychotropic cannabinoids have been described so far in animal organisms belonging to most *phyla*, including very simple invertebrates like coelenterates, molluscs, echinoderms and anellids, but not in plants. Therefore, it is obvious that the presence of cannabimimetic fatty acid derivatives in plant seeds (Chapman *et al.*, 1999) has a physiological meaning that has nothing to do with the activation of cannabinoid receptors (see below).

Endocannabinoids in invertebrates

The first example of an effect by cannabinoids on invertebrates was reported by Acosta-Urquidi and Chase (1975). They showed that, in isolated buccal and parieto-visceral ganglia of *Aplysia californica*, THC causes a depression in nerve cell excitability that is consistent with its reported effects in mammals. Later, it was shown that *Aplysia* ganglia contain measurable levels of AEA, NArPE and 2-AG (Di Marzo *et al.*, 1999c). Evidence for a functional endogenous cannabinoid system was recently obtained in the first animal organism to have evolved a neural network, i.e. the coelenterate *Hydra vulgaris* (De Petrocellis *et al.*, 1999). This primitive invertebrate contains AEA, 2-AG, NArPE and a FAAH-like activity. Moreover, in *Hydra* cell membranes, specific binding sites for the CB_1 ligand and antagonist SR141716A were found that could be displaced by increasing concentrations of AEA. These cannabinoid receptors are probably involved in mediating AEA inhibition of *Hydra* feeding behavior, an effect exerted by low doses of the endocannabinoid ($IC_{50} = 7$–10 nM) and blocked by SR141716A. It was hypothesized that endocannabinoids may play a role in the modulation of *Hydra* feeding response, a simple behavior consisting in prey-induced and chemoreceptor-mediated opening and subsequent closure of the mouth.

In the sea urchin *Strongylocentrotus purpuratus*, the cannabinoids THC and cannabidiol (CBD) reduce fertilizing capacity of sperm cells (Schuel *et al.*, 1991). Pretreatment of sperm with THC prevents, in a dose and time dependent manner, the triggering of the acrosome reaction by solubilized egg jelly. THC, CBD and cannabinol (CBN) block the membrane fusion reaction between the sperm plasma membrane and the acrosomal membrane that normally is elicited in response to stimulation by the egg jelly. Schuel *et al.* (1994) provided additional evidence that a cannabinoid receptor and AEA in sperm play a role in blocking the acrosome reaction. Binding of AEA to the cannabinoid receptor modulates stimulus-secretion-coupling in sperm by affecting an event prior to ion channel opening. Further evidence for the existence of an endocannabinoid system in sea urchins was obtained by Bisogno *et al.* (1997d), who showed the presence of AEA, PEA and *N*-stearoylethanolamine in ovaries of *Paracentrotus lividus*. The lipid extract of this tissue also contains the corresponding NAPEs, and whole homogenates from *P. lividus* are capable of converting synthetic [^{3}H]NArPE into [^{3}H]AEA. Moreover, mature ovaries of *P. lividus* also express an amidohydrolase activity which catalyses the hydrolysis of AEA and PEA to ethanolamine and the corresponding fatty acids. This enzyme displayed subcellular distribution, pH/temperature dependency profiles and sensitivity to inhibitors similar but not identical to those of FAAH. These data support the hypothesis that AEA or related substances may be oocyte-derived cannabimimetic regulators of sea urchin fertility. Further support for this hypothesis was recently reported by Berdyshev (1999) who found that also *N*-oleoyl- and *N*-linoleoylethanolamine, but not PEA, inhibit sea urchin sperm fertilizing capacity. The effect was also exerted by THC and other cannabinoid agonists but could not be blocked by the CB_1 antagonist SR141716A, suggesting that the sea urchin sperm cannabinoid receptor is different from CB_1.

Several long chain NAEs, including AEA and PEA (as well as some of the corresponding NAPEs), were found in lipid extracts of bivalve molluscs (Sepe *et al.*, 1998). Analogous to observations in mammalian brain, the amounts of these metabolites, the most abundant being PEA and *N*-stearoyl-ethanolamine, appeared to increase considerably when mussels were extracted 24 h post-mortem. In particulate fractions of homogenates from *Mytilus* a highly selective cannabinoid receptor with an immunomodulatory function is also present. In fact, a CB_1-like cannabinoid receptor is present on immunocytes and microglia from mussels and is coupled to NO release (Stefano *et al.*, 1996). An enzymatic activity capable of catalyzing the hydrolysis of AEA, and displaying pH dependency and inhibitor sensitivity profiles similar to those of mammalian FAAH was also described for *Mytilus* (Sepe *et al.*, 1998). Interestingly, in *Mytilus*, usually inactive concentrations of AEA (10^{-8} and 10^{-7} M) became effective in releasing substantial amounts of NO from immunocytes in the presence of a specific FAAH inhibitor (Stefano *et al.*, 1998). These data indicate that preventing the breakdown of AEA prolongs its activity in these tissues, and, therefore, that FAAH may serve to terminate AEA action under physiological conditions also in invertebrates.

A gene for a cannabinoid receptor has been partially cloned and sequenced in the leech *Hirudo medicinalis*. The cDNA sequence is similar to those obtained from human (49%) and rat (47%) CB_1 receptors (Stefano *et al.*, 1997a). More strikingly, within the sequence, there are two highly conserved motifs – between amino acids 1–97 and 128–153 – which show 80% and 58% homology to human CB_1. Moreover, a third region is 98% identical to part of the bovine adrenocorticotropic hormone receptor. According to Elphick (1998) this protein may resemble the putative ancestor of mammalian cannabinoid and melanocortin receptors. This hypothesis is supported by comparing the sequence of mammalian CB_1 and melanocortin receptors, which contain a fragment in the central part of the two receptors with more than 80% sequence identity. At any rate, also in the leech the cannabinoid receptor is coupled to NO release, and this even leads to modulation of dopamine release from ganglia (Stefano *et al.*, 1997b). Finally, we have found that leech ganglia also contain measurable amounts of AEA, PEA (as well as their corresponding NAPEs) and 2-AG (V. Di Marzo, I. Matias, T. Bisogno and M. Salzet, unpublished observations).

Endocannabinoids in lower vertebrates

Although evidence for the presence and biosynthesis of saturated and mono-unsaturated long chain NAEs and NAPEs has been reported in the nervous tissue of the carp (reviewed by Schmid *et al.*, 1990), no evidence exists as yet for the occurrence of AEA and 2-AG in non-mammalian vertebrates. Immediate targets for the search of these two compounds should be those species which were shown to respond to THC and psychotropic cannabinoids and/or to contain cannabinoid receptors, i.e.: (1) the puffer fish *Fugu rubripes*, where two CB_1-like receptor genes have been cloned (Yamaguchi *et al.*, 1996); (2) the Siamese fighting fish *Betta splendens*, where cannabinoids influence the fighting behavior (Gonzalez *et al.*, 1971); (3) the common frog, where anticonvulsant cannabinoids modulate the post-tetanic neuronal potentiation (Turkanis and Karler, 1975), THC influences neuromuscular transmission (Turkanis and Karler, 1986), and AEA blocks adenylate cyclase at the neuromuscular junction (Van der Kloot *et al.*, 1994); (4) pigeons, where psychotropic cannabinoids variedly affect behavior (Jarbe *et al.*, 1993; Mansbach *et al.*, 1996; Ferrari *et al.*, 1999), seemingly in a stereoselective fashion (Jarbe *et al.*, 1989); (5) the zebra finch songbird, where expression of CB_1 mRNA within the caudal telencephalon appears to change over the course of vocal development, and high-level CB_1 expression in brain regions controlling singing behavior suggests a potential role for cannabinoid signaling in vocal development (Soderstrom and Johnson, 2000); and (6) chicken, goldfish, and tiger salamander, where CB_1-immunoreactivity was recently found in retinas by using a subtype-specific polyclonal antibody (Straiker *et al.*, 1999). In general, we should expect to find in fish, as well as in the vertebrate retinas mentioned above, also the C22:6 n-3 homologs of AEA and 2-AG, since the C22:6 n-3 fatty acid (docosahexaenoic acid) is abundant in these tissues.

Cannabimimetic fatty acid derivatives in plants

Several studies have been carried out on the presence, metabolism and possible biological significance of NAEs in higher plants (Chapman *et al.*, 1998, for review). These compounds, and particularly the 18:2 n-6 homolog, seem to be particularly abundant (with total amounts in the low µg/g tissue range) in plant seeds such as cottonseed, soybean, castor beans, peas, peanuts, hazelnuts, coffee, cocoa, millet, oatmeal, etc. (Chapman *et al.*, 1995, 1998; Di Marzo *et al.*, 1998e, Table 5.1). This was to be expected since the 18:2 n-6 fatty acid (linoleic acid) is the major PUFA in the majority of higher plants. As their amounts decrease upon imbibition of seeds, a role was suggested for NAEs in the regulation of seed germination (Chapman *et al.*, 1999). Fatty acid primary amides, namely OA, were also found in some of

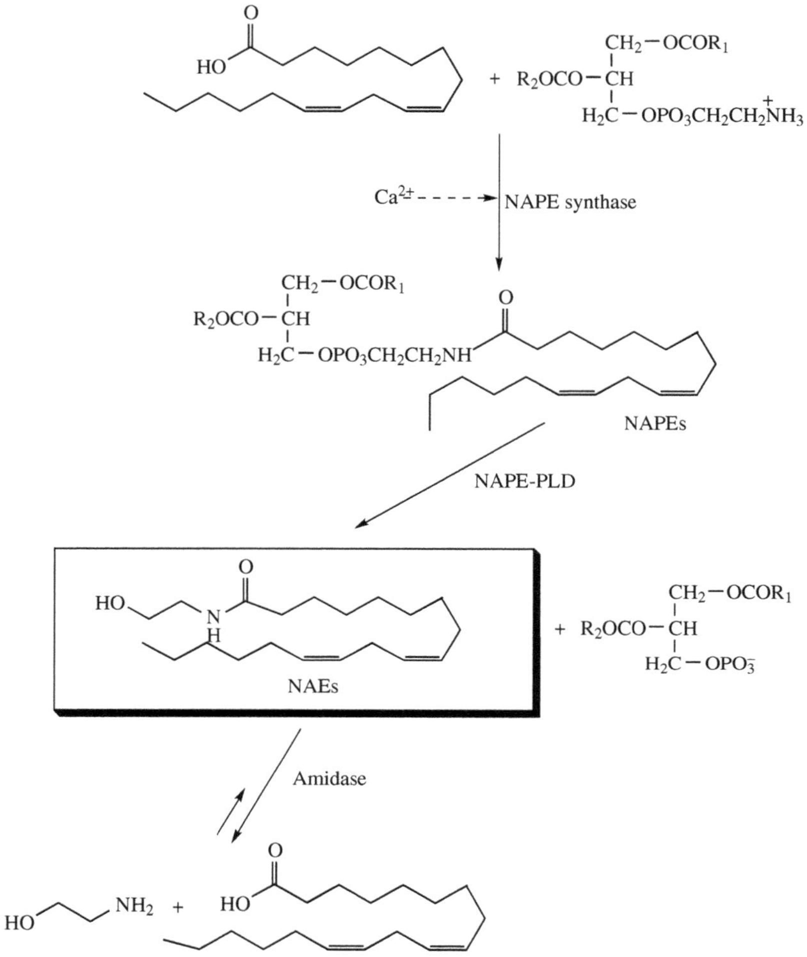

Figure 5.7 Biosynthesis and inactivation of *N*-acylethanolamines (NAEs) in plants. *N*-linoleoyl-ethanol-amine is shown as an example. Unlike animals, in plants the *N*-fatty acid composition of NAEs, as well as of their biosynthetic precursors, the *N*-acyl-phosphatidylethanol-amines (NAPEs) directly reflects the free fatty acid composition of plant tissue. In fact, plant NAPEs are formed from the direct, energy-free, condensation of free fatty acids and phosphatidylethanolamine, catalyzed by "NAPE synthase", an enzyme inhibited by Ca^{2+} (Chapman *et al.*, 1999). NAE hydrolysis is catalyzed by an amidase whose kinetic properties are different from the mammalian fatty acid amide hydrolase.

the above plant materials (Di Marzo *et al.*, 1998e, Table 5.1). Increased release and accumulation of NAEs was found also in cells from the tobacco plant *Nicotiana tabacum* upon elicitation of defense response induced by xylanase (Chapman *et al.*, 1998). More recently, the leaves of the tobacco plant were also found to produce NAEs after treatment with xylanase, and exogenous NAEs were shown to affect the typical elicitor-induced short- and long-term responses in suspended cells and

Table 5.1 N-acylethanolamines (NAEs), oleamide and 2-arachidonoylglycerol in foods. Data are expressed as ng/g starting material (in plant foods) or as ng/g total extracted lipid (in milks, where lipid concentration is about 36g/L), and are means ± SD of at least three separate determinations

Material	NAEs				Oleamide	2-AG
	ng/g Starting Material					
Fatty acyl moiety	16:0	18:1n-9	18:2n-6	20:4n-6		
Coffee						
Caturra coffee cherries with skin, Ecuador		54±35	228±75		720±207	
Coffee green beans, Arabica, Colombia		63±22	17±6		42±10	
Cocoa						
Cocoa beans, unfermented, unroasted, unhulled, Amelonado, Ivory Coast	10±4	148.8±87	108±35		170±43	
Cocoa beans, fermented, unroasted, unhulled		121±53			268±77	
Cocoa roast, fermented, roasted, hulled	20±18	214±144	41±9		5781±1633	
Cocoa powder	1464±401	2172±695	5844±1515	3±2 (or ND)	3687±1237	
Dark Chocolate, 70% cocoa	14±3.9	435±147	224±125		5990±4035	
Nuts, soy, grains, olives						
Peanuts, with salt	77±27	273±31	260±98		620±489	
Hazelnuts	58±22	1055±399	247±102		1476±828	
Walnuts	13±4	21±4	76±27		90±31	
Soybeans, white, whole, dehulled, dried	126±43	302±101	805±249		2289±987	
Oatmeal large flakes		890±298	2750±977		170±93	
Millet	44±12	9±4	431±133		221±29	
Olives, green, with salt, spice water, acidifiant	11±4	95±27	47±20		33±9	
Milks	**ng/g total extracted lipid**					
Bovine milk, early, fresh frozen	112±37	111±41		4.2±0.6	24200±15600	1100±300
Bovine milk, mature, fresh frozen	271±53	65±39	405±198	ND	8500±6100	2400±1000
Bovine milk, pasteurized, Italy	140±29	452±210	117±31	94±6	400±30	1800±300
Human milk early, pooled, frozen				ND	4000±1000	6400±1900
Human milk, mature, pooled, frozen	156±117	227±7	38±15	11±6	1500±300	8700±2800
Goat milk, Italy, commercial	528±7	57±5		9±4	34500±17500	8300±300

leaves from this plant (Tripathy et al., 1999). In particular, the C14:0 NAE inhibited elicitor-induced medium alkalinization by activating ammonia lyase expression. These data suggest that NAEs play a role in the cellular defense responses occurring in plant leaves after fungal infection.

As to NAE metabolic pathways in plants, they seem to be similar but not identical to those described in animals (Figure 5.7). NAEs are produced from the phospholipase D-catalyzed hydrolysis of the corresponding NAPEs (Chapman and Moore, 1993). The latter compounds, however, unlike mammalian NAPEs, originate from the acylCoA-independent and *direct* N-acylation of PE, catalyzed by a 64 kD enzyme which has been partially characterized and purified from cottonseed microsomes (Chapman and Moore, 1994) and named N-acylphosphatidylethanolamine synthase (McAndrew et al., 1995). Finally, NAE inactivation occurs in plants through the hydrolysis of the amide bond catalyzed by an amidohydrolase which, unlike FAAH, is optimally active at pH 6.5, and, like FAAH, is inhibited by the serine protease inhibitor phenyl methyl sulphonyl fluoride (Chapman et al., 1999).

CANNABIMIMETIC FATTY ACID DERIVATIVES IN FOODS

The finding of NAEs and OA in foods of plant origin, such as cocoa, coffee, oatmeal, soybeans and nuts (Table 5.1), raises the question of whether these compounds, if consumed with a meal, may reach the brain in sufficiently high amounts to inhibit the inactivation of *endogenous* AEA, thereby eliciting a cannabimimetic response (di Tomaso et al., 1996). To answer this question, first the activity of compounds capable of *directly* activating the cannabinoid receptors had to be determined after oral administration. Therefore, AEA and 2-AG were administered *per os* to mice, and their pharmacological effects in the mouse 'tetrad' of tests of cannabimimetic activity were assessed (Di Marzo et al., 1998e). It was found that these compounds do cause THC-like central effects in the mouse 'tetrad' (but, interestingly, no effect on defecation during the time frame of the 'tetrad' tests), but only at doses (≥ 300 mg/kg body weight) that cannot possibly be achieved with the consumption of NAE-containing foods. The large differences observed between the doses necessary to induce cannabimimetic effects upon i.p./i.v. or *per os* administration suggested that less than 5% of the oral endocannabinoids reaches the bloodstream, and was explained with the presence of high levels of FAAH in murine digestive system (Katayama et al., 1999), which may cause the degradation of AEA, 2-AG and OA prior to their passage into the blood. Furthermore, these observations, together with the finding that non-endocannabinoid NAEs, such as N-oleoyl- and N-linoleoyl-ethanolamine, and OA are much less potent than AEA and 2-AG when administered i.p. or i.v. (Mechoulam et al., 1997; Watanabe et al., 1999), suggest that oral administration of mono- or di-unsaturated NAEs, at doses similar to those that are normally associated with consumption of plant-derived food, cannot possibly cause cannabimimetic responses such as those measured in the mouse 'tetrad' of tests. In agreement with this hypothesis, OA, which like N-oleoyl- and N-linoleoyl-ethanolamine, also inhibits AEA degradation (Maurelli et al., 1995), and may potentially enhance endogenous AEA levels, was active in these tests only at doses that again cannot possibly be reached with food consumption (Di Marzo *et al.*, 1998e). However, the presence of AEA, PEA and, particularly, 2-AG in foods of animal origin may have some relevance (see for example Sepe *et al.*, 1998). Although the amounts of AEA in animal tissues (0.35–1.75 µg/kg tissue) are well below those necessary to elicit a cannabimimetic response after oral administration, PEA and 2-AG amounts in these tissues can be 1 or 2–3 orders of magnitude higher than those of AEA, respectively. For example, edible mussels may contain up to 55 µg/kg of PEA (Sepe *et al.*, 1998), whereas milk, which may contain up to 330 µg/L of 2-AG (Di Marzo *et al.*, 1998e). Furthermore, while AEA and 2-AG may be degraded in the gastrointestinal (G.I.) tract of mammals, high amounts of 2-AG could be produced by partial gut lipolysis of dietary triacylglycerols, which, in animal tissues, contain usually high amounts of arachidonic acid on the *sn*-2 position. Therefore, it is possible that 2-AG assumed with, or produced after partial digestion of, animal-derived foods may play a role both locally, at the level of the gastrointestinal system (see for example Izzo *et al.*, 1999), and centrally, if adsorbed by the bloodstream, particularly if the co-presence of high amounts of "entourage" 2-acylglycerols is taken into consideration (Ben-Shabat *et al.*, 1998). The possibility that 2-AG contained in milk may play a role in some of the properties of this valuable as well as widely consumed food is currently under investigation in our laboratories.

Other digestive processes starting from fatty foods could, in principle, also lead to the formation of cannabimimetic fatty acids in the gut of animals and humans. The amount of free arachidonic acid in the diet before digestion would be minute, because all would normally be esterified to phospholipids (Carrie *et al.*, 2000) or, more importantly, triacylglycerols. During the digestive process, however, phospholipids containing arachidonate are converted via gut phospholipase A_2 to 1-acyl lysophospholipids and free arachidonic acid. Triacylglycerols containing arachidonate (derived from fish, meats, fungal, and algal sources, all of which are common foods today) can contain the fatty acid on all 3 positions. During digestion by pancreatic, lingual, and other lipases this leads to 2-AG and the free fatty acids formerly on the *sn*-1 and *sn*-3 positions, including arachidonate. The 2-acylglycerols are better absorbed than free fatty acids because the latter metabolites can form insoluble calcium soaps that are poorly absorbed and, consequently, are likely to stay in the gut for a longer period of time. If a man consumes 600 mg of arachidonic acid per day (high levels of this fatty acid are present in foods of animal origin), this means about 200 mg maximum per serving, and up to two thirds of this (120 mg, formerly on the 1- and 3-positions of triacylglycerols) could be converted to free arachidonate in the gut. Thus, it is possible that the arachidonic acid and other free fatty acids are converted in the gut to the corresponding NAEs by FAAH, since Katayama *et al.* (1999) have shown that, due to the presence of lipid inhibiting the hydrolysis reaction in the G.I. tract, the enzyme can catalyze the condensation reaction (i.e., the formation of NAEs and AEA) under these conditions. This hypothesis needs to be tested experimentally by feeding animals diets with and without arachidonic acid, then aspirating the stomach and gut contents, and measuring the levels of NAEs in the different gut regions.

Finally another possible source of NAEs could be lecithin, a widely used additive in most foods. Although one is used to thinking of soy lecithin as a crude source of

PC, it actually contains NAPEs and lysoNAPEs up to 20% by weight. Soybean lecithin also contains PEA (Ganley et al., 1958), but it is not clear if this is artefactually generated from the corresponding NAPE during the preparation of the lecithin. Plants have very active phospholipase D enzymes, which, depending on how the lecithin is obtained, could also be present in lecithin crude preparations.

WHAT'S NEXT?

For a few years after the initial reports on the existence of the endocannabinoids, in 1992 and 1995, the discovery of, and publication on, new actions by these substances was expected to be rather simple: if THC was known to cause an effect, AEA and 2-AG were supposed to do the same. With several thousand papers published on THC, research on the endocannabinoids seemed to have a secure, though somewhat unexciting, future. Gratifyingly, new effects are being found, and even new sites of action other than CB_1 and CB_2 receptors, such as the vanilloid VR1 receptor, are being proposed (see Di Marzo et al., 1999a, 2000c for updated reviews, and Zygmunt et al., 1999). Evidence for the presence also of *novel* cannabinoid/AEA receptors is rapidly accumulating (see Járai et al., 1999; Sagan et al., 1999; Di Marzo et al., 2000d, for examples), and, if confirmed by the isolation, full characterization and cloning of these putative proteins, will potentially lead to new drugs with a different spectrum of pharmacological properties. The finding of endocannabinoids and cannabinoid receptors in animals with a very ancient evolutionary history is telling us that this signaling system has been conserved for millions of years and must be associated to basic functions. Can we now predict the pathways of future research?

Chemistry: There are only few indications at present that novel endocannabinoids are to be found. However, most lipid soluble endogenous ligands – be they steroids, prostaglandins, leukotrienes etc. – exist as groups of related compounds. Hence, one can perhaps expect the discovery of new endocannabinoids, based on unsaturated fatty acids substituted on the carboxyl moiety, with or without further changes within the alkenyl portion of the molecule. We shall certainly see novel specific agonists and antagonists, novel blockers of endocannabinoid transport and metabolism, and novel stable derivatives and new types of drugs. If more cannabinoid receptors are found, the possibility that AEA and 2-AG derivatives, or altogether different molecules, bind to these proteins will have to be tested.

Pharmacology: We shall certainly learn much more on the interaction between the endocannabinoids and neurotransmitters, hormones and cytokines. We may expect to see more publications on the endocannabinoids in the skin or in reproductive, the gastrointestinal and the respiratory systems, whose interactions with AEA and 2-AG have been investigated only to a limited extent. Novel observations on actions in the CNS are certainly to be published. Our knowledge on the endocannabinoids in the immune system is still fragmentary and advances should be expected.

Therapeutical implications: In view of the ubiquitousness of endocannabinoid action, the development of specific drugs may be complicated. Nevertheless one can expect to see efforts to develop specific cannabinoid analgesics, neuroprotective agents, antiglaucoma drugs, antiobesity drugs, drugs acting in the immune system, in movement disorders etc. There are indications that the endocannabinoids may play a role in schizophrenia and possibly in Tourette's syndrome Huntington's chorea, and multiple sclerosis (see Consroe, 1998, for review, and Baker *et al.*, 2000 for a recent example). Furthermore, the possibility that endocannabinoid-derived drugs can be used as anti-cancer agents has also been supported by recent data (De Petrocellis *et al.*, 1998; Galve-Roperh *et al.*, 2000). These initial observations should certainly be clarified.

The above random thoughts are more or less obvious. Should we also expect advances in the chemistry of sleep, memory and, ultimately, emotions?

REFERENCES

Acosta-Urquidi, J. and Chase, R. (1975) "The effects of delta9-tetrahydrocannabinol on action potentials in the mollusc Aplysia," *Canadian Journal of Physiology and Pharmacology* **53**: 793–798.

Allen, A. C., Gammon, C. M., Ousley, A. H., McCarthy, K. D. and Morell, P. (1992) "Bradykinin stimulates arachidonic acid release through the sequential actions of an sn-1 diacylglycerol lipase and a monoacylglycerol lipase," *Journal of Neurochemistry* **58**: 1130–1139.

Arai, Y., Fukushima, T., Shirao, M., Yang, X. and Imai, K. (2000) "Sensitive determination of anandamide in rat brain utilizing a coupled-column HPLC with fluorimetric detection," *Biomedical Chromatography* **14**: 118–124.

Arnone, M., Maruani, J., Chaperon, F., Thiebot, M. H., Poncelet, M., Soubrie, P. and Le Fur, G. (1997) "Selective inhibition of sucrose and ethanol intake by SR 141716, an antagonist of central cannabinoid (CB_1) receptors," *Psychopharmacology* (*Berliner*) **132**: 104–106.

Arreaza, G., Devane, W. A., Omeir, R. L., Sajnani, G., Kunz, J., Cravatt, B. F. and Deutsch, D. G. (1997) "The cloned rat hydrolytic enzyme responsible for the breakdown of anandamide also catalyzes its formation via the condensation of arachidonic acid and ethanolamine," *Neuroscience Letters* **234**: 59–62.

Bachur, N. R. and Udenfriend, S. (1966) "Microsomal synthesis of fatty acid amides," *Journal of Biological Chemistry* **241**: 1308–1313.

Baker, D., Pryce, G., Croxford, J. L., Brown, P., Pertwee, R. G., Huffman, J. W. and Layward, L. (2000) "Cannabinoids control spasticity and tremor in a multiple sclerosis model," *Nature* **404**: 84–87.

Basavarajappa, B. S. and Hungund, B. L. (1999) "Chronic ethanol increases the cannabinoid receptor agonist anandamide and its precursor N-arachidonoylphosphatidylethanolamine in SK-N-SH cells," *Journal of Neurochemistry* **72**: 522–528.

Beaulieu, P., Bisogno, T., Punwar, S., Farquhar-Smith, W. P., Ambrosino, G., Di Marzo, V. and Rice, A. S. C. (2000) "Role of the endogenous cannabinoid system in the formalin test of persistent pain in the rat," *European Journal of Pharmacology* **396**: 85–92.

Beltramo, M., Stella, N., Calignano, A., Lin, S. Y., Makriyannis, A. and Piomelli, D. (1997) "Functional role of high-affinity anandamide transport, as revealed by selective inhibition," *Science* **277**: 1094–1097.

Ben-Shabat, S., Fride, E., Sheskin, T., Tamiri, T., Rhee, M. H., Vogel, Z., Bisogno, T., De Petrocellis, L., Di Marzo, V. and Mechoulam, R. (1998) "An entourage effect: inactive endogenous fatty acid glycerol esters enhance 2-arachidonoyl-glycerol cannabinoid activity," *European Journal of Pharmacology* **353**: 23–31.

Berdyshev, E. V. (1999) "Inhibition of sea urchin fertilization by fatty acid ethanolamides and cannabinoids," *Comparative Biochemistry and Physiology Part C, Pharmacology Toxicology and Endocrinology* **122**: 327–330.

Berdyshev, E. V., Boichot, E., Germain, N., Allain, N., Anger, J. P. and Lagente, V. (1997) "Influence of fatty acid ethanolamides and delta9-tetrahydrocannabinol on cytokine and arachidonate release by mononuclear cells," *European Journal of Pharmacology* **330**: 231–240.

Berdyshev, E. V., Schmid, P. C., Dong, Z. and Schmid, H. H. (2000) "Stress-induced generation of *N*-acylethanolamines in mouse epidermal JB6 P+ cells," *The Biochemical Journal* **346**: 369–374.

Berrendero, F., Sepe, N., Ramos, J. A., Di Marzo, V. and Fernandez-Ruiz, J. J. (1999) "Analysis of cannabinoid receptor binding and mRNA expression and endogenous cannabinoid contents in the developing rat brain during late gestation and early postnatal period," *Synapse* **33**: 181–191.

Bisogno, T., Berrendero, F., Ambrosino, G., Cebeira, M., Ramos, J. A., Fernandez-Ruiz, J. J. and Di Marzo, V. (1999a) "Brain regional distribution of endocannabinoids: implications for their biosynthesis and biological function," *Biochemical and Biophysical Research Communication* **256**: 377–380.

Bisogno, T., Delton-Vandenbroucke, I., Milone, A., Lagarde, M. and Di Marzo, V. (1999b) "Biosynthesis and inactivation of *N*-arachidonoylethanolamine (anandamide) and *N*-docosahexaenoylethanolamine in bovine retina," *Archives Biochemistry and Biophysics* **370**: 300–307.

Bisogno, T., Katayama, K., Melck, D., Ueda, N., De Petrocellis, L., Yamamoto, S. and Di Marzo, V. (1998) "Biosynthesis and degradation of bioactive fatty acid amides in human breast cancer and rat pheochromocytoma cells – implications for cell proliferation and differentiation," *European Journal of Biochemistry* **254**: 634–642.

Bisogno, T., Maurelli, S., Melck, D., De Petrocellis, L. and Di Marzo, V. (1997a) "Biosynthesis, uptake, and degradation of anandamide and palmitoylethanolamide in leukocytes," *Journal of Biological Chemistry* **272**: 3315–3323.

Bisogno, T., Melck, D., De Petrocellis, L. and Di Marzo, V. (1999c) "Phosphatidic acid as the biosynthetic precursor of the endocannabinoid 2-arachidonoylglycerol in intact mouse neuroblastoma cells stimulated with ionomycin," *Journal of Neurochemistry* **72**: 2113–2119.

Bisogno, T., Sepe, N., De Petrocellis, L., Mechoulam, R. and Di Marzo, V. (1997b) "The sleep inducing factor oleamide is produced by mouse neuroblastoma cells," *Biochemical and Biophysical Research Communications* **239**: 473–479.

Bisogno, T., Sepe, N., Melck, D., Maurelli, S., De Petrocellis, L. and Di Marzo, V. (1997c) "Biosynthesis, release and degradation of the novel endogenous cannabimimetic metabolite 2-arachidonoylglycerol in mouse neuroblastoma cells," *The Biochemical Journal* **322**: 671–677.

Bisogno, T., Ventriglia, M., Milone, A., Mosca, M., Cimino, G. and Di Marzo, V. (1997d) "Occurrence and metabolism of anandamide and related acyl-ethanolamides in ovaries of the sea urchin *Paracentrotus lividus*," *Biochimica et Biophysica Acta* **1345**: 338–348.

Boger, D. L., Henriksen, S. J. and Cravatt, B. F. (1998) "Oleamide: an endogenous sleep-inducing lipid and prototypical member of a new class of biological signaling molecules," *Current Pharmaceutical Design* **4**: 303–314.

Bohme, G. A., Laville, M., Ledent, C., Parmentier, M. and Imperato, A. (2000) "Enhanced long-term potentiation in mice lacking cannabinoid CB_1 receptors," *Neuroscience* **95**: 5–7.

Burstein, S. H. (1999) "The cannabinoid acids: nonpsychoactive derivatives with therapeutic potential," *Pharmacology and Therapeutics* **82**: 87–96.

Burstein, S. H., Rossetti, R. G., Yagen, B. and Zurier, R. B. (2000) "Oxidative metabolism of anandamide," *Prostaglandins and Other Lipid Mediators* **61**: 29–41.

Cadas, H., di Tomaso, E. and Piomelli, D. (1997) "Occurrence and biosynthesis of endogenous cannabinoid precursor, *N*-arachidonoyl phosphatidylethanolamine, in rat brain," *Journal of Neuroscience* **17**: 1226–1242.

Calignano, A., La Rana, G., Giuffrida, A. and Piomelli, D. (1998) "Control of pain initiation by endogenous cannabinoids," *Nature* **394**: 277–281.

Carrie, I., Clement, M., de Javel, D., Frances, H. and Bourre, J. M. (2000) "Phospholipid supplementation reverses behavioral and biochemical alterations induced by n-3 polyunsaturated fatty acid deficiency in mice," *Journal of Lipid Research* **41**: 473–480.

Castellano, C., Cabib, S., Palmisano, A., Di Marzo, V. and Puglisi-Allegra, S. (1997) "The effects of anandamide on memory consolidation in mice involve both D1 and D2 dopamine receptors," *Behavioural Pharmacology* **8**: 707–712.

Chapman, K. D. and Moore, T. S. Jr. (1993) "*N*-acylphosphatidylethanolamine synthesis in plants: occurrence, molecular composition, and phospholipid origin," *Archives of Biochemistry and Biophysics* **301**: 21–33.

Chapman, K. D. and Moore, T. S. Jr. (1994) "Isozymes of cottonseed microsomal *N*-acylphosphatidylethanolamine synthase: detergent solubilization and electrophoretic separation of active enzymes with different properties," *Biochimica et Biophysica Acta* **1211**: 29–36.

Chapman, K. D., Lin, I. and De Souza, A. D. (1995) "Metabolism of cottonseed microsomal *N*-acylphosphatidylethanolamine," *Archives of Biochemistry and Biophysics* **318**: 401–407.

Chapman, K. D., Tripathy, S., Venables, B. and Desouza, A. D. (1998) "*N*-Acylethanolamines: formation and molecular composition of a new class of plant lipids," *Plant Physiology* **116**: 1163–1168.

Chapman, K. D., Venables, B., Markovic, R., Blair, R. W. Jr. and Bettinger, C. (1999) "*N*-Acylethanolamines in seeds. Quantification of molecular species and their degradation upon imbibition," *Plant Physiology* **120**: 1157–1164.

Cheer, J. F., Cadogan, A. K. Marsden, C. A., Fone, K. C. and Kendall, D. A. (1999) "Modification of 5-HT2 receptor mediated behaviour in the rat by oleamide and the role of cannabinoid receptors," *Neuropharmacology* **38**: 533–541.

Coburn, A. and Rich, H. (1960) "The concept of egg yolk as a dietary inhibitor of rheumatic susceptibility," *Lancet* **1**: 870–871.

Collin, C., Devane, W. A., Dahl, D., Lee, C. J., Axelrod, J. and Alkon, D. L. (1995) "Long-term synaptic transformation of hippocampal CA1 gamma-aminobutyric acid synapses and the effect of anandamide," *Proceedings of the National Academy of Sciences of the USA* **92**: 10167–10171.

Colombo, G., Agabio, R., Diaz, G., Lobina, C., Reali, R. and Gessa, G. L. (1998) "Appetite suppression and weight loss after the cannabinoid antagonist SR 141716," *Life Sciences* **63**: PL113–117.

Consroe, P. (1998) "Brain cannabinoid systems as targets for the therapy of neurological disorders," *Neurobiology of Disease* **5**: 534–551.

Cravatt, B. F., Giang, D. K., Mayfield, S. P., Boger, D. L., Lerner, R. A. and Gilula, N. B. (1996) "Molecular characterization of an enzyme that degrades neuromodulatory fatty-acid amides," *Nature* **384**: 83–87.

Cravatt, B. F., Prospero-Garcia, O., Siuzdak, G., Gilula, N. B., Henriksen, S. J., Boger, D. L. and Lerner, R. A. (1995) "Chemical characterization of a family of brain lipids that induce sleep," *Science* **268**: 1506–1509.

De Petrocellis, L., Melck, D., Bisogno, T., Milone, A. and Di Marzo, V. (1999) "Finding of the endocannabinoid signalling system in *Hydra*, a very primitive organism: possible role in the feeding response," *Neuroscience* **92**: 377–387.

De Petrocellis, L., Melck, D., Palmisano, A., Bisogno, T., Laezza, C., Bifulco, M. and Di Marzo, V. (1998) "The endogenous cannabinoid anandamide inhibits human breast cancer cell proliferation," *Proceedings of the National Academy of Sciences of the USA* **95**: 8375–8380.

Désarnaud, F., Cadas, H. and Piomelli, D. (1995) "Anandamide amidohydrolase activity in rat brain microsomes. Identification and partial characterization," *Journal of Biological Chemistry* **270**: 6030–6035.

Deutsch, D. G. and Chin, S. A. (1993) "Enzymatic synthesis and degradation of anandamide, a cannabinoid receptor agonist," *Biochemical Pharmacology* **46**: 791–796.

Deutsch, D. G., Goligorsky, M. S., Schmid, P. C., Krebsbach, R. J., Schmid, H. H., Das, S. K., Dey, S. K., Arreaza, G., Thorup, C., Stefano, G. and Moore, L. C. (1997) "Production and physiological actions of anandamide in the vasculature of the rat kidney," *Journal of Clinical Investigation* **100**: 1538–1546.

Devane, W. A. and Axelrod, J. (1994) "Enzymatic synthesis of anandamide, an endogenous ligand for the cannabinoid receptor, by brain membranes," *Proceedings of the National Academy of Sciences of the USA* **91**: 6698–6701.

Devane, W. A., Breuer, A., Sheskin, T., Jarbe, T. U., Eisen, M. S. and Mechoulam, R. (1992a) "A novel probe for the cannabinoid receptor," *Journal of Medicinal Chemistry* **35**: 2065–2069.

Devane, W. A., Dysarz, F. A., Johnson, M. R., Melvin, L. S. and Howlett, A. C. (1988) "Determination and characterization of a cannabinoid receptor in rat brain," *Molecular Pharmacology* **34**: 605–613.

Devane, W. A., Hanus, L., Breuer, A., Pertwee, R. G., Stevenson, L. A., Griffin, G., Gibson, D., Mandelbaum, A., Etinger, A. and Mechoulam, R. (1992b) "Isolation and structure of a brain constituent that binds to the cannabinoid receptor," *Science* **258**: 1946–1949.

Di Marzo, V. (1998) "'Endocannabinoids' and other fatty acid derivatives with cannabimimetic properties: biochemistry and possible physiopathological relevance," *Biochimica et Biophysica Acta* **1392**: 153–175.

Di Marzo, V. and Deutsch, D. G. (1998) "Biochemistry of the endogenous ligands of cannabinoid receptors," *Neurobiology of Disease* **5**: 386–404.

Di Marzo, V. and Fontana, A. (1995) "Anandamide, an endogenous cannabinomimetic eicosanoid: 'killing two birds with one stone'," *Prostaglandins, Leukotrienes and Essential Fatty Acids* **53**: 1–11.

Di Marzo, V., Berrendero, F., Bisogno, T., Gonzalez, S., Cavaliere, P., Romero, J., Cebeira, M., Ramos, J. A. and Fernandez-Ruiz, J. J. (2000b) "Enhancement of anandamide formation in the limbic forebrain and reduction of endocannabinoid contents in the striatum of delta9-tetrahydrocannabinol-tolerant rats," *Journal of Neurochemistry* **74**: 1627–1635.

Di Marzo, V., Bisogno, T., De Petrocellis, L., Melck, D. and Martin, B. R. (1999a) "Cannabimimetic fatty acid derivatives: the anandamide family and other endocannabinoids," *Current Medicinal Chemistry* **6**: 721–744.

Di Marzo, V., Bisogno, T., De Petrocellis, L., Melck, D., Orlando, P., Wagner, J. A. and Kunos, G. (1999b) "Biosynthesis and inactivation of the endocannabinoid 2-arachidonoylglycerol in circulating and tumoral macrophages," *European Journal of Biochemistry* **264**: 258–267.

Di Marzo, V., Bisogno, T., Melck, D. and De Petrocellis, L. (2000c) "Endocannabinoids: new targets for drug development," *Current Pharmceutical Design* **6**: 1361–1380.

Di Marzo, V., Bisogno, T., Melck, D., Ross, R., Brockie, H., Stevenson, L., Pertwee, R. G. and De Petrocellis, L. (1998a) "Interactions between synthetic vanilloids and the endogenous cannabinoid system," *FEBS Letters* **436**: 449–454.

Di Marzo, V., Bisogno, T., Sugiura, T., Melck, D. and De Petrocellis, L. (1998b) "The novel endogenous cannabinoid 2-arachidonoylglycerol is inactivated by neuronal- and basophil-like cells: connections with anandamide," *The Biochemical Journal* **331**: 15–19.

Di Marzo, V., Breivogel, C. S., Tao, Q., Bridgen, D. T., Razdan, R. K., Zimmer, A. M., Zimmer, A. and Martin, B. R. (2000d) "Levels, metabolism, and pharmacological activity of anandamide in CB_1 cannabinoid receptor knockout mice. Evidence for non-CB_1, non-CB_2 receptor-mediated actions of anandamide in mouse brain," *Journal of Neurochemistry* **75**: 2434–2444.

Di Marzo, V., De Petrocellis, L., Bisogno, T. and Melck, D. (1999c) "Metabolism of anandamide and 2-arachidonoylglycerol: an historical overview and some recent developments," *Lipids* **34**: S319–325.

Di Marzo, V., De Petrocellis, L., Bisogno, T., Melck, D. and Sepe, N. (1998c) "Cannabimimetic fatty acid derivatives: biosynthesis and catabolism," in *Proceedings of 4th International Congress on Essential Fatty Acids and Eicosanoids*, edited by R. A. Riemersma, R. Armstrona, R. W. Kelly and R. Wilson, pp. 358–362, AOCS Press.

Di Marzo, V., De Petrocellis, L., Sepe, N. and Buono, A. (1996a) "Biosynthesis of anandamide and related acylethanolamides in mouse J774 macrophages and N18 neuroblastoma cells," *The Biochemical Journal* **316**: 977–984.

Di Marzo, V., De Petrocellis, L., Sugiura, T. and Waku, K. (1996b) "Potential biosynthetic connections between the two cannabimimetic eicosanoids, anandamide and 2-arachidonoylglycerol, in mouse neuroblastoma cells," *Biochemical and Biophysical Research Communications* **227**: 281–288.

Di Marzo, V., Fontana, A., Cadas, H., Schinelli, S., Cimino, G., Schwartz, J. C. and Piomelli, D. (1994) "Formation and inactivation of endogenous cannabinoid anandamide in central neurons," *Nature* **372**: 686–691.

Di Marzo, V., Hill, M. P., Bisogno, T., Crossman, A. R. and Brotchie, J. M. (2000e) "Enhanced levels of endogenous cannabinoids in the globus pallidus are associated with a reduction in movement in an animal model of Parkinson's disease," *The FASEB Journal* **14**: 1432–1438.

Di Marzo, V., Melck, D., Bisogno, T. and De Petrocellis, L. (1998d) "Endocannabinoids: endogenous cannabinoid receptor ligands with neuromodulatory action," *Trends in Neurosciences* **21**: 521–528.

Di Marzo V., Melck, D., De Petrocellis, L. and Bisogno, T. (2000a) "Cannabimimetic fatty acid derivatives in cancer and inflammation," *Prostaglandins and other Lipid Mediators* **61**: 43–61.

Di Marzo, V., Sepe, N., De Petrocellis, L., Berger, A., Crozier, G., Fride, E. and Mechoulam, R. (1998e) "Trick or treat from food endocannabinoids?" *Nature* **396**: 636–637.

di Tomaso, E., Beltramo, M. and Piomelli, D. (1996) "Brain cannabinoids in chocolate," *Nature* **382**: 677–678.

Dye, L. and Blundell, J. E. (1997) "Menstrual cycle and appetite control: implications for weight regulation," *Human Reproduction* **12**: 1142–1151.

Egertova, M., Giang, D. K., Cravatt, B. F. and Elphick, M.R. (1998) "A new perspective on cannabinoid signalling: complementary localization of fatty acid amide hydrolase and the CB_1 receptor in rat brain," *Proceedings of the Royal Society of London Series B. Biological Sciences* **265**: 2081–2085.

Elphick, M. R. (1998) "An invertebrate G-protein coupled receptor is a chimeric cannabinoid/melanocortin receptor," *Brain Research* **780**: 168–171.

Epps, D. E., Grupp, I. L., Grupp, G. and Schwartz, A. (1983) "Protective effects of N-acylethanolamines, an endogenous class of lipid amides, on hypoxic guinea pig heart," *IRCS Medical Science* **11**: 899–900.

Epps, D. E., Mandel, F. and Schwartz, A. (1982) "The alteration of rabbit skeletal sarcoplasmic reticulum function by N-acylethanolamine, a lipid associated with myocardial infarction," *Cell Calcium* **3**: 531–543.

Epps, D. E., Natarajan, V., Schmid, P. C. and Schmid, H. H. O. (1980) "Accumulation of N-acylethanolamine glycerophospholipids in infarcted myocardium," *Biochimica et Biophysica Acta* **618**: 420–430.

Facci, L., Dal Toso, R., Romanello, S., Buriani, A., Skaper, S. D. and Leon, A. (1995) "Mast cells express a peripheral cannabinoid receptor with differential sensitivity to anandamide and palmitoylethanolamide," *Proceedings of the National Academy of Sciences of USA* **92**: 3376–3380.

Felder, C. C., Nielsen, A., Briley, E. M., Palkovits, M., Priller, J., Axelrod, J., Nguyen, D.N., Richardson, J. M., Riggin, R. M., Koppel, G. A., Paul, S. M. and Becker, G. W. (1996) "Isolation and measurement of the endogenous cannabinoid receptor agonist, anandamide, in brain and peripheral tissues of human and rat," *FEBS Letters* **393**: 231–235.

Fernandez-Ruiz, J., Berrendero, F., Hernandez, M. L. and Ramos, J. A. (2000) "The endogenous cannabinoid system and brain development," *Trends in Neurosciences* **23**: 14–20.

Fernandez-Ruiz, J. J., Munoz, R. M., Romero, J., Villanua, M. A., Makriyannis, A. and Ramos, J. A. (1997) "Time course of the effects of different cannabimimetics on prolactin and gonadotrophin secretion: evidence for the presence of CB_1 receptors in hypothalamic structures and their involvement in the effects of cannabimimetics," *Biochemical Pharmacology* **53**: 1919–1927.

Ferrari, F., Ottani, A. and Giuliani, D. (1999) "Cannabimimetic activity in rats and pigeons of HU 210, a potent antiemetic drug," *Pharmacology Biochemistry and Behavior* **62**: 75–80.

Fontana, A., Di Marzo, V., Cadas, H. and Piomelli, D. (1995) "Analysis of anandamide, an endogenous cannabinoid substance, and of other natural *N*-acylethanolamines," *Prostaglandins Leukotrienes and Essential Fatty Acids* **53**: 301–308.

Fride, E. and Mechoulam, R. (1993) "Pharmacological activity of the cannabinoid receptor agonist, anandamide, a brain constituent," *European Journal of Pharmacology* **231**: 313–314.

Galve-Roperh, I., Sanchez, C., Cortes, M. L., del Pulgar, T. G., Izquierdo, M. and Guzman, M. (2000) "Anti-tumoral action of cannabinoids: involvement of sustained ceramide accumulation and extracellular signal-regulated kinase activation," *Nature Medicine* **6**: 313–319.

Ganley, O. H. and Robinson, H. J. (1959) "Antianaphylactic and antiserotonin activity of a compound obtained from egg yolk, peanut oil, and soybean lecithin," *Journal of Allergy* **30**: 415–419.

Ganley, O. H., Graessle, O. E. and Robinson, H. J. (1958) "Anti-inflammatory activity of compounds obtained from egg yolk, peanut oil, and soybean lecithin," *The Journal of Laboratory and Clinical Medicine* **51**: 709–714.

Gaoni, Y. and Mechoulam, R. (1964) "Isolation, structure, and partial synthesis of an active constituent of hashish," *Journal of American Chemical Society* **86**: 1646–1647.

Gardner, E. L. and Vorel, S. R. (1998) "Cannabinoid transmission and reward-related events," *Neurobiology of Disease* **5**: 502–533.

Giang, D. K. and Cravatt, B. F. (1997) "Molecular characterization of human and mouse fatty acid amide hydrolases," *Procoeedings of the National Academy of Sciences of the USA* **94**: 2238–2242.

Giuffrida, A. and Piomelli, D. (1998) "Isotope dilution GC/MS determination of anandamide and other fatty acylethanolamides in rat blood plasma," *FEBS Letters* **422**: 373–376.

Giuffrida, A., Parsons, L. H., Kerr, T. M., Rodriguez de Fonseca, F., Navarro, M. and Piomelli, D. (1999) "Dopamine activation of endogenous cannabinoid signaling in dorsal striatum," *Nature Neuroscience* **2**: 358–363.

Glass, M., Brotchie, J. M. and Maneuf, Y. P. (1997) "Modulation of neurotransmission by cannabinoids in the basal ganglia," *European Journal of Neuroscience* **9**: 199–203.

Goldstein, A. (1976) "Opioid peptides endorphins in pituitary and brain," *Science* **193**: 1081–1086.

Gonzalez, S., Bisogno, T., Wenger, T., Manzanares, J., Milone, A., Berrendero, F., Di Marzo, V., Ramos, J. A. and Fernandez-Ruiz, J. J. (2000) "Sex steroid influence on cannabinoid CB(1) receptor mRNA and endocannabinoid levels in the anterior pituitary gland," *Biochemical and Biophysical Research Communications* **270**: 260–266.

Gonzalez, S., Manzanares, J., Berrendero, F., Wenger, T., Corchero, J., Bisogno, T., Romero, J., Fuentes, J. A., Di Marzo, V., Ramos, J. A. and Fernandez-Ruiz, J. J. (1999) "Identification of endocannabinoids and cannabinoid CB(1) receptor mRNA in the pituitary gland," *Neuroendocrinology* **70**: 137–145.

Gonzalez, S. C., Matsudo, V. K. and Carlini, E. A. (1971) "Effects of marihuana compounds on the fighting behavior of Siamese fighting fish (*Betta splendens*)," *Pharmacology* **6**: 186–199.

Goparaju, S. K., Kurahashi, Y., Suzuki, H., Ueda, N. and Yamamoto, S. (1999a) "Anandamide amidohydrolase of porcine brain: cDNA cloning, functional expression and site-directed mutagenesis(1)," *Biochimica et Biophysica Acta* **1441**: 77–84.

Goparaju, S. K., Ueda, N., Taniguchi, K. and Yamamoto, S. (1999b) "Enzymes of porcine brain hydrolyzing 2-arachidonoylglycerol, an endogenous ligand of cannabinoid receptors," *Biochemical Pharmacology* **57**: 417–423.

Goth, A. and Knoohuizen, M. (1962) "Tissue thromboplastins and histamine release from mast cells," *Life Sciences* **1**: 459–465.

Griffin, G., Tao, Q. and Abood, M. E. (2000) "Cloning and pharmacological characterization of the rat CB(2) cannabinoid receptor," *The Journal of Pharmacology and Experimental Therapeutics* **292**: 886–894.

Gulaya, N. M., Melnik, A. A., Balkov, D. L., Volkov, G. L., Vysotskiy, M. V. and Vaskovskyk, V. E. (1993) "The effect of long-chain *N*-acylethanolamines on some membrane-associated functions of neuroblastoma C1300 N18 cells," *Biochimica et Biophysica Acta* **1152**: 280–288.

Hampson, A. J., Bornheim, L. M., Scanziani, M., Yost, C. S., Gray, A. T., Hansen, B. M., Leonoudakis, D. J. and Bickler, P. E. (1998a) "Dual effects of anandamide on NMDA receptor-mediated responses and neurotransmission," *Journal of Neurochemistry* **70**: 671–676.

Hampson, A. J., Grimaldi, M., Axelrod, J. and Wink, D. (1998b) "Cannabidiol and (−)Delta9-tetrahydrocannabinol are neuroprotective antioxidants," *Proceedings of the National Academy of Sciences of the USA* **95**: 8268–8273.

Hansen, H. S., Lauritzen, L., Moesgaard, B., Strand, A. M. and Hansen, H. H. (1998) "Formation of *N*-acyl-phosphatidylethanolamines and *N*-acetylethanolamines: proposed role in neurotoxicity," *Biochemical Pharmacology* **55**: 719–725.

Hansen, H. S., Lauritzen, L., Strand, A. M., Moesgaard, B. and Frandsen, A. (1995) "Glutamate stimulates the formation of *N*-acylphosphatidylethanolamine and *N*-acylethanolamine in cortical neurons in culture," *Biochimica et Biophysica Acta* **1258**: 303–308.

Hansen, H. S., Lauritzen, L., Strand, A. M., Vinggaard, A. M., Frandsen, A. and Schousboe, A. (1997) "Characterization of glutamate-induced formation of *N*-acylphosphatidylethanolamine and *N*-acylethanolamine in cultured neocortical neurons," *Journal of Neurochemistry* **69**: 753–761.

Hanus, L., Breuer, A., Tchilibon, S., Shiloah, S., Goldenberg, D., Horowitz, M. *et al.* (1999a) "HU-308: a specific agonist for CB$_2$, a peripheral cannabinoid receptor," *Proceedings of the National Academy of Sciences of the USA* **96**: 14228–14233.

Hanus, L., Gopher, A., Almog, S. and Mechoulam, R. (1993) "Two new unsaturated fatty acid ethanolamides in brain that bind to the cannabinoid receptor," *Journal of Medicinal Chemistry* **61**: 3032–3034.

Hanus, L. O., Fales, H. M., Spande, T. F. and Basile, A. S. (1999b) "A gas chromatographic-mass spectral assay for the quantitative determination of oleamide in biological fluids," *Analytical Biochemistry* **270**: 159–166.

Hao, S., Avraham, Y., Mechoulam, R. and Berry, E. M. (2000) "Low dose anandamide affects food intake, cognitive function, neurotransmitter and corticosterone levels in diet-restricted mice," *European Journal of Pharmacology* **392**: 147–156.

Hillard, C. J. and Jarrahian, A. (2000) "The movement of *N*-arachidonoylethanolamine (anandamide) across cellular membranes," *Chemistry and Physics of Lipids* **108**: 123–134.

Hillard, C. J., Edgemond, W. S., Jarrahian, A. and Campbell, W. B. (1997) "Accumulation of N-arachidonoylethanolamine (anandamide) into cerebellar granule cells occurs via facilitated diffusion," *Journal of Neurochemistry* **69**: 631–638.

Hillard, C. J., Wilkison, D. M., Edgemond, W. S. and Campbell, W. B. (1995) "Characterization of the kinetics and distribution of N-arachidonylethanolamine (anandamide) hydrolysis by rat brain," *Biochimica et Biophysica Acta* **1257**: 249–256.

Howlett, A. C. and Fleming, R. M. (1984) "Cannabinoid inhibition of adenylate cyclase. Pharmacology of the response in neuroblastoma cell membrane," *Molecular Pharmacology* **26**: 532–538.

Huang, S. M., Bisogno, T., Petros, T. J., Chang, S. Y., Zavitsonos, P. A. *et al.* (2001) "Identification of a new class of molecules, the arachidonyl aminoacids, and characterization of one member that inhibits pain," *Journal of Biological Chemistry* (in press).

Huidobro-Toro, J. P. and Harris, R. A. (1996) "Brain lipids that induce sleep are novel modulators of 5-hydroxytrypamine receptors," *Proceedings of the National Academy of Sciences of the USA* **93**: 8078–8082.

Izzo, A. A., Mascolo, N., Capasso, R., Germano, M. P., De Pasquale, R. and Capasso, F. (1999) "Inhibitory effect of cannabinoid agonists on gastric emptying in the rat," *Naunyn Schmiedebergs Archives Pharmacology* **360**: 221–223.

Jarái, Z., Wagner, J. A., Goparaju, S. K., Wang, I. and Razdan, R. K. *et al.* (2000) "Cardiovascular effects of Z-arachidonoylglycenol in anesthesized mice," *Hypertension* **35**: 679–684.

Jarái, Z., Wagner, J. A., Varga, K., Lake, K. D., Compton, D. R., Martin, B. R., Zimmer, A. M., Bonner, T. I., Buckley, N. E., Mezey, E., Razdan, R. K., Zimmer, A. and Kunos, G. (1999) "Cannabinoid-induced mesenteric vasodilation through an endothelial site distinct from CB_1 or CB_2 receptors," *Proceedings of the National Academy of Sciences of the USA*, **96**: 14136–14141.

Jarbe, T. U., Hiltunen, A. J. and Mechoulam, R. (1989) "Stereospecificity of the discriminative stimulus functions of the dimethylheptyl homologs of 11-hydroxy-delta8-tetrahydrocannabinol in rats and pigeons," *The Journal of Pharmacology and Experimental Therapeutics* **250**: 1000–1005.

Jarbe, T. U., Hiltunen, A. J., Mathis, D. A., Hanus, L., Breuer, A. and Mechoulam, R. (1993) "Discriminative stimulus effects and receptor binding of enantiomeric pairs of cannabinoids in rats and pigeons; a comparison," *The Journal of Pharmacology and Experimental Therapeutics* **264**: 561–569.

Jarrahian, A., Manna, S., Edgemond, W. S., Campbell, W. B. and Hillard, C. J. (2000) "Structure-activity relationships among anandamide head group analogs for the anandamide transporter," *Journal of Neurochemistry* **74**: 2597–2606.

Kahlich, R., Klima, J., Cihla, F., Frankova, V., Masek, K., Rosicky, M., Matousek, F. and Bruthans, J. (1979) "Studies on prophylactic efficacy of N-2-hydroxyethyl palmitamide (Impulsin) in acute respiratory infections. Serologically controlled field trials," *Journal of Hygiene, Epidemiology, Microbiology and Immunology* **23**: 11–24.

Katayama, K., Ueda, N., Katoh, I. and Yamamoto, S. (1999) "Equilibrium in the hydrolysis and synthesis of cannabimimetic anandamide demonstrated by a purified enzyme," *Biochimica et Biophysica Acta* **1440**: 205–214.

Keefer, E. W., Chapman, K. D. and Gross, G. W. (1999) "Rapid *in vitro* screening and pharmacological evaluation of novel plant-derived cannabimimetic drug candidates," *Society Neuroscience Abstract* **25**: 1197, 482.6.

Kempe, K., Hsu, F. F., Bohrer, A. and Turk, J. (1996) "Isotope dilution mass spectrometric measurements indicate that arachidonylethanolamide, the proposed endogenous ligand of the cannabinoid receptor, accumulates in rat brain tissue post mortem but is contained at low levels in or is absent from fresh tissue," *Journal of Biological Chemistry* **271**: 17287–17295.

Koga, D., Santa, T., Fukushima, T., Homma, H. and Imai, K. (1997) "Liquid chromatographic-atmospheric pressure chemical ionization mass spectrometric determination of anandamide and its analogs in rat brain and peripheral tissues," *Journal of Chromatography B, Biomedical Sciences and Applications* **690**: 7–13.

Koga, D., Santa, T., Hagiwara, K., Imai, K., Takizawa, H., Nagano, T., Hirobe, M., Ogawa, M., Sato, T., Inoue, K. *et al.* (1995) "High-performance liquid chromatography and fluorometric detection of arachidonylethanolamide (anandamide) and its analogues, derivatized with 4-(*N*-chloroformylmethyl-*N*-methyl)amino-7-N,*N*-dimethylaminosulphonyl- 2,1,3-benzoxadiazole (DBD-COCl)," *Biomedical Chromatography* **9**: 56–57.

Kondo, S., Kondo, H., Nakane, S., Kodaka, T., Tokumura, A., Waku, K. and Sugiura, T. (1998) "2-Arachidonoylglycerol, an endogenous cannabinoid receptor agonist: identification as one of the major species of monoacylglycerols in various rat tissues, and evidence for its generation through Ca^{2+}-dependent and -independent mechanisms," *FEBS Letters* **429**: 152–156.

Krsiak, M., Sechserova, M., Perlik, F. and Elis, J. (1972) "Effect of palmitoylethanolamide on alcohol intoxication in man," *Act. Nerv. Super. (Praha)* **14**: 187.

Kruszka, K. K. and Gross, R. W. (1994) "The ATP- and CoA-independent synthesis of arachidonoylethanolamide. A novel mechanism underlying the synthesis of the endogenous ligand of the cannabinoid receptor," *Journal of Biological Chemistry* **269**: 14345–14348.

Kuehl, F. A. Jr., Jacob, T. A., Ganley, O. H., Ormond, R. E. and Meisinger, M. A. P. (1957) "The identification of *N*-(2-hydroxyethyl)-palmitamide as a naturally occurring anti-inflammatory agent," *Journal of American Chemical Society* **79**: 5577–5578.

Kunos, G., Jarái, Z., Varga, K., Liu, J., Wang, L. and Wagner, J. (2000) "Cardiovascular effects of endocannabinoids – The plot thickens," *Prostaglandins other Lipid Mediators* **61**: 71–84.

Kurahashi, Y., Ueda, N., Suzuki, H., Suzuki, M. and Yamamoto, S. (1997) "Reversible hydrolysis and synthesis of anandamide demonstrated by recombinant rat fatty-acid amide hydrolase," *Biochemical and Biophysical Research Communications* **237**: 512–515.

Kuwae, T., Schmid, P. C. and Schmid, H. H. (1997) "Alterations of fatty acyl turnover in macrophage glycerolipids induced by stimulation. Evidence for enhanced recycling of arachidonic acid," *Biochimica et Biophysica Acta* **1344**: 74–86.

Kuwae, T., Shiota, Y., Schmid, P. C., Krebsbach, R. and Schmid, H. H. (1999) "Biosynthesis and turnover of anandamide and other *N*-acylethanolamines in peritoneal macrophages," *FEBS Letters* **459**: 123–127.

Lambert, D. M. and Di Marzo, V. (1999) "The palmitoylethanolamide and oleamide enigmas: are these two fatty acid amides cannabimimetic?" *Current Medicinal Chemistry* **6**: 757–773.

Lambert, D. M., Di Paolo, F. G., Sonveaux, P., Kanyonyo, M., Govaerts, S. J., Hermans, E., Bueb, J., Delzenne, N. M. and Tschirhart, E. J. (1999) "Analogues and homologues of *N*-palmitoylethanolamide, a putative endogenous CB(2) cannabinoid, as potential ligands for the cannabinoid receptors," *Biochimica et Biophysica Acta* **1440**: 266–274.

Landsman, R. S., Burkey, T. H., Consroe, P., Roeske, W. R. and Yamamura, H. I. (1997) "SR141716A is an inverse agonist at the human cannabinoid CB_1 receptor," *European Journal of Pharmacology* **334**: R1-R2.

Ledent, C., Valverde, O., Cossu, G., Petitet, F., Aubert, J. F., Beslot, F., Bohme, G. A., Imperato, A., Pedrazzini, T., Roques, B. P., Vassart, G., Fratta, W. and Parmentier, M. (1999) "Unresponsiveness to cannabinoids and reduced addictive effects of opiates in CB_1 receptor knockout mice," *Science* **283**: 401–404.

Lerner, R. A., Siuzdak, G., Prospero-Garcia, O., Henriksen, S. J., Boger, D. L. and Cravatt, B. F. (1994) "Cerebrodiene: A brain lipid isolated from sleep-deprived cats," *Proceedings of the National Academy of Sciences of the USA* **91**: 9505–9508.

Leweke, F. M., Giuffrida, A., Wurster, U., Emrich, H. M. and Piomelli, D. (1999) "Elevated endogenous cannabinoids in schizophrenia," *Neuroreport* **10**:1665–1669.

Maccarone, M., Valensise, H., Bari, M., Lazzarin, N., Romanici, C. and Finazzi-Agro', A. (2000a) "Relation between decreased anandamide hydrolase concentration in human lymphocytes and miscarriage," *Lancet* **355**: 1326–1329.

Maccarone, M., Bari, M., Lorenzon, T., Bisogno, T., Di Marzo, V. and Finazzi-Agro', A. (2000b) "Anandamide uptake by human endothelial cells and its regulation by nitric oxide,"*Journal of Biological Chemistry* **275**: 13484–13492.

Maccarone, M., Bari, M., Menichelli, A., Del Principe, D. and Finazzi-Agro', A. (1999) "Anandamide activates human platelets through a pathway independent of the arachidonate cascade," *FEBS Letters* **447**: 277–282.

Maccarone, M., Fiorucci, L., Erba, F., Bari, M., Finazzi-Agro', A. and Ascoli, F. (2000c) "Human mast cells take up and hydrolyze anandamide under the control of 5-lipoxygenase and do not express cannabinoid receptors," *FEBS Letters* **468**: 176–180.

Maccarone, M., van der Stelt, M., Rossi, A., Veldink, G. A., Vliegenthart, J. F. and Finazzi-Agro', A. (1998) "Anandamide hydrolysis by human cells in culture and brain," *Journal of Biological Chemistry* **273**: 32332–32339.

Mackie, K., Devane, W. A. and Hille, B. (1993) "Anandamide, an endogenous cannabinoid, inhibits calcium currents as a partial agonist in N18 neuroblastoma cells," *Molecular Pharmacology* **44**: 498–503.

Makriyannis, A. (1995) "The role of cell membranes in cannabinoid activity," in *Cannabinoid Receptors*, edited by R. G. Pertwee, pp. 87–116, London: Academic Press.

Mallet, P. E. and Beninger, R. J. (1998) "The cannabinoid CB$_1$ receptor antagonist SR141716A attenuates the memory impairment produced by delta9-tetrahydrocannabinol or anandamide," *Psychopharmacology (Berliner)* **140**: 11–19.

Mansbach, R. S., Rovetti, C. C., Winston, E. N. and Lowe III, J. A. (1996) "Effects of the cannabinoid CB$_1$ receptor antagonist SR141716A on the behavior of pigeons and rats," *Psychopharmacology (Berliner)* **124**: 315–322.

Martin, B. R. and Lichtman, A. H. (1998) "Cannabinoid transmission and pain perception," *Neurobiology of Disease* **5**: 447–461.

Martin, B. R., Mechoulam, R. and Razdan, R. K. (1999) "Discovery and characterization of endogenous cannabinoids," *Life Sciences* **65**: 573–595.

Mascia, M. S., Obinu, M. C., Ledent, C., Parmentier, M., Bohme, G. A., Imperato, A. and Fratta, W. (1999) "Lack of morphine-induced dopamine release in the nucleus accumbens of cannabinoid CB(1) receptor knockout mice," *European Journal of Pharmacology* **383**: R1–R2.

Matsuda, L. A., Lolait, S. J., Brownstein, M. J., Young, A. C. and Bonner, T. I. (1990) "Structure of a cannabinoid receptor and functional expression of the cloned cDNA," *Nature* **346**: 561–564.

Maurelli, S., Bisogno, T., De Petrocellis, L., Di Luccia, A., Marino, G. and Di Marzo, V. (1995) "Two novel classes of neuroactive fatty acid amides are substrates for mouse neuroblastoma 'anandamide amidohydrolase,'" *FEBS Letters* **377**: 82–86.

McAndrew, R. S., Leonard, B. P. and Chapman, K. D. (1995) "Photoaffinity labeling of cottonseed microsomal *N*-acylphosphatidylethanolamine synthase protein with a substrate analogue, 12-[(4-azidosalicyl)amino] dodecanoic acid," *Biochimica et Biophysica Acta* **1256**: 310–318.

Mechoulam, R. (1986) "The pharmacohistory of *Cannabis sativa*," in: *Cannabinoids as Therapeutic Agents*, edited by R. Mechoulam, pp. 1–19. Boca Raton: CRS Press.

Mechoulam, R., Fride, E., Ben-Shabat, S., Meiri, U. and Horowitz, M. (1998a) "Carbachol, an acetylcholine receptor agonist, enhances production in rat aorta of 2-arachidonoyl glycerol, a hypotensive endocannabinoid," *European Journal of Pharmacology* **362**: R1–R3.

Mechoulam, R., Ben-Shabat, S., Hanus, L., Ligumsky, M., Kaminski, N. E., Schatz, A. R., Gopher, A., Almog, S., Martin, B. R., Compton, D. R., Pertwee, R. G., Griffin, G., Bayewitch, M., Barg, J. and Vogel, Z. (1995) "Identification of an endogenous 2-monoglyceride, present in canine gut, that binds to cannabinoid receptors," *Biochemical Pharmacology* **50**: 83–90.

Mechoulam, R., Devane, W. A. and Glaser, R. (1992) "Cannabinoid geometry and biological activity," in: *Marijuana/Cannabinoids Neurobiology and Neurophysiology*, edited by L. Murphy and A. Bartke, pp. 1–33. Boca Raton: CRS Press.

Mechoulam, R., Feigenbaum, J. J., Lander, N., Segal, M., Jarbe, T. U. C., Hiltunen, A. J. and Consroe, P. (1988) "Enantiomeric cannabinoids: Stereospecificity of psychotropic activity," *Experientia* **44**: 762–764.

Mechoulam, R., Fride, E. and Di Marzo, V. (1998b) "Endocannabinoids," *European Journal of Pharmacology* **359**: 1–18.

Mechoulam, R., Fride, E., Hanus, L., Sheskin, T., Bisogno, T., Di Marzo, V., Bayewitch, M. and Vogel, Z. (1997) "Anandamide may mediate sleep induction," *Nature* **389**: 25–26.

Melck, D., Bisogno, T., De Petrocellis, L., Chuang, Hh., Julius, D., Bifulco, M. and Di Marzo, V. (1999) "Unsaturated long-chain N-acyl-vanillyl-amides (N-AVAMs): vanilloid receptor ligands that inhibit anandamide-facilitated transport and bind to CB_1 cannabinoid receptors," *Biochemical and Biophysical Research Communications* **262**: 275–284.

Mendelson, W. B. and Basile, A. S. (1999) "The hypnotic actions of oleamide are blocked by a cannabinoid receptor antagonist," *Neuroreport* **10**: 3237–3239.

Meng, I. D., Manning, B. H., Martin, W. J. and Fields, H. L. (1998) "An analgesia circuit activated by cannabinoids," *Nature* **395**: 381–383.

Merkler, D. J., Merkler, K. A., Stern, W. and Fleming, F. F. (1996) "Fatty acid amide biosynthesis: a possible new role for peptidylglycine alpha-amidating enzyme and acyl-coenzyme A: glycine N-acyltransferase," *Archives of Biochemistry and Biophysics* **330**: 430–434.

Munro, S., Thomas, K. L. and Abu-Shaar, M. (1993) "Molecular characterization of a peripheral receptor for cannabinoids," *Nature* **365**: 61–65.

Natarajan, V., Schmid, P. C. and Schmid, H. H. (1986) "N-acylethanolamine phospholipid metabolism in normal and ischemic rat brain," *Biochimica et Biophysica Acta* **878**: 32–41.

Paria, B. C., Ma, W., Andrenyak, D. M., Schmid, P. C., Schmid, H. H., Moody, D. E., Deng, H., Makriyannis, A. and Dey, S. K. (1998) "Effects of cannabinoids on preimplantation mouse embryo development and implantation are mediated by brain-type cannabinoid receptors," *Biology of Reproduction* **58**: 1490–1495.

Paria, B. C., Zhao, X., Wang, J., Das, S. K. and Dey, S. K. (1999) "Fatty-acid amide hydrolase is expressed in the mouse uterus and embryo during the periimplantation period," *Biology of Reproduction* **60**: 1151–1157.

Parolaro, D. (1999) "Presence and functional regulation of cannabinoid receptors in immune cells," *Life Sciences* **65**: 637–644.

Paton, W. D. (1975) "Pharmacology of marijuana," *Annuual Review of Pharmacology* **15**: 191–220.

Perlik, F., Raskova, H. and Elis, J. (1971) "Anti-inflammatory properties of N(2-hydroxy-ethyl) palmitamide," *Acta Physiologica Academic Science Hungarica* **39**: 395–400.

Pertwee, R. G., Joe-Adigwe, G. and Hawksworth, G. M. (1996) "Further evidence for the presence of cannabinoid CB_1 receptors in mouse vas deferens," *European Journal of Pharmacology* **296**: 169–172.

Pestonjamasp, V. K. and Burstein, S. H. (1998) "Anandamide synthesis is induced by arachidonate mobilizing agonists in cells of the immune system," *Biochimica et Biophysica Acta* **1394**: 249–260.

Piomelli, D., Beltramo, M., Glasnapp, S., Lin, S. Y., Goutopoulos, A., Xie, X. Q. and Makriyannis, A. (1999) "Structural determinants for recognition and translocation by the anandamide transporter," *Proceedings of the National Academy of Sciences of the USA* **96**: 5802–5807.

Portier, M., Rinaldi-Carmona, M., Pecceu, F., Combes, T., Poinot-Chazel, C., Calandra, B., Barth, F., Le Fur, G. and Casellas, P. (1999) "SR 144528, an antagonist for the peripheral cannabinoid receptor that behaves as an inverse agonist," *The Journal of Pharmacology and Experimental Therapeutics* **288**: 582–589.

Prescott, S. M. and Majerus, P. W. (1983) "Characterization of 1,2-diacylglycerol hydrolysis in human platelets. Demonstration of an arachidonoyl-monoacylglycerol intermediate," *Journal of Biological Chemistry* **258**: 764–769.

Rakhshan, F., Henn, T. A. and Barker, E. L. (1999) "Regulation of uptake of the endogenous cannabinoid anandamide by protein kinase C," *Society of Neuroscience Abstract* **25**: 1196, 482.2.

Rakhshan, F., Day, T. A., Blakely, R. D. and Barker, E. L. (2000) "Carrier-mediated uptake of the endogenous cannabinoid anandamide in RBL-2H3 cells," *The Journal of Pharmacology and Experimental Therapeutics* **292**: 960–967.

Raskova, H., Masek, K. and Linet, O. (1972) "Non-specific resistance induced by palmitoyl-ethanolamide," *Toxicon.* **10**: 485–490.

Reibaud, M., Obinu, M. C., Ledent, C., Parmentier, M., Bohme, G. A. and Imperato, A. (1999) "Enhancement of memory in cannabinoid CB_1 receptor knock-out," *European Journal of Pharmacology* **379**: R1–2.

Richardson, J. D., Aanonsen, L. and Hargreaves, K. M. (1997) "Hypoactivity of the spinal cannabinoid system results in NMDA-dependent hyperalgesia," *Journal of Neuroscience* **18**: 451–457.

Ritenour-Rodgers, K. J., Driscoll, W. J., Merkler, K. A., Merkler, D. J. and Mueller, G. P. (2000) "Induction of peptidylglycine alpha-amidating monooxygenase in N(18)TG(2) cells: a model for studying oleamide biosynthesis," *Biochemical and Biophysical Research Communications* **267**: 521–526.

Rodríguez de Fonseca, F., Del Arco, I., Martin-Calderón, J. L., Gorriti, M. A. and Navarro, M. (1998) "Role of the endogenous cannabinoid system in the regulation of motor activity," *Neurobiology of Disease* **5**: 483–501.

Sagan, S., Venance, L., Torrens, Y., Cordier, J., Glowinski, J. and Giaume, C. (1999) "Anandamide and WIN 55212–2 inhibit cyclic AMP formation through G-protein-coupled receptors distinct from CB_1 cannabinoid receptors in cultured astrocytes," *European Journal of Neuroscience* **11**: 691–699.

Santucci, V., Storme, J. J., Soubrie, P. and Le Fur, G. (1996) "Arousal-enhancing properties of the CB_1 cannabinoid receptor antagonist SR 141716A in rats as assessed by electroencephalographic spectral and sleep-waking cycle analysis," *Life Sciences* **58**: PL103–110.

Schmid, H. H., Schmid, P. C. and Natarajan, V. (1996) "The *N*-acylation-phosphodiesterase pathway and cell signalling," *Chemistry and Physics of Lipids* **80**: 133–142.

Schmid, H. H. O., Schmid, P. C. and Natarajan, V. (1990) "*N*-Acylated glycerophospholipids and their derivatives," *Progress in Lipid Research* **29**: 1–43.

Schmid, P. C., Krebsbach, R. J., Perry, S. R., Dettmer, T. M., Maasson, J. L. and Schmid, H. H. (1995) "Occurrence and postmortem generation of anandamide and other long-chain *N*-acylethanolamines in mammalian brain," *FEBS Letters* **375**: 117–120 plus (1996), **385**: 125–126 (correction).

Schmid, P. C., Paria, B. C., Krebsbach, R. J., Schmid, H. H. and Dey, S. K. (1997) "Changes in anandamide levels in mouse uterus are associated with uterine receptivity for embryo implantation," *Proceedings of the National Academy of Sciences of the USA* **94**: 4188–4192.

Schmid, P. C., Schwartz, K. D., Smith, C. N., Krebsbach, R. J., Berdyshev, E. V. and Schmid, H. H. (2000) "A sensitive endocannabinoid assay. The simultaneous analysis of *N*-acylethanolamines and 2-monoacylglycerols," *Chemistry and Physics of Lipids* **104**: 185–191.

Schuel, H., Burkman, L. J., Picone, R. P. and Makriyannis, A. (1998) "Functional cannabinoid receptors in human sperm," *ICRS Symposium* **59**.

Schuel, H., Chang, M. C., Berkery, D., Schuel, R., Zimmerman, A. M. and Zimmerman, S. (1991) "Cannabinoids inhibit fertilization in sea urchins by reducing the fertilizing capacity of sperm," *Pharmacology, Biochemistry and Behavior* **40**: 609–615.

Schuel, H., Goldstein, E., Mechoulam, R., Zimmerman, A. M. and Zimmerman, S. (1994) "Anandamide (arachidonylethanolamide), a brain cannabinoid receptor agonist, reduces sperm fertilizing capacity in sea urchins by inhibiting the acrosome reaction," *Proceedings of the National Academy of Sciences of the USA* **91**: 7678–7682.

Sepe, N., De Petrocellis, L., Montanaro, F., Cimino, G. and Di Marzo, V. (1998) "Bioactive long chain *N*-acylethanolamines in five species of edible bivalve molluscs. Possible implications for mollusc physiology and sea food industry," *Biochimica et Biophysica Acta* **1389**: 101–111.

Sheskin, T., Hanus, L., Slager, J., Vogel, Z. and Mechoulam, R. (1997) "Structural requirements for binding of anandamide-type compounds to the brain cannabinoid receptor," *Journal of Medicinal Chemistry* **40**: 659–667.

Simiand, J., Keane, M., Keane, P. E. and Soubrie, P. (1998) "SR 141716, a CB_1 cannabinoid receptor antagonist, selectively reduces sweet food intake in marmoset," *Behavioural Pharmacology* **9**: 179–181.

Sinor, A. D., Irvin, S. M. and Greenberg, D. A. (2000) "Endocannabinoids protect cerebral cortical neurons from in vitro ischemia in rats," *Neuroscience Letters* **278**: 157–160.

Soderstrom, K. and Johnson, F. (2000) "CB_1 cannabinoid receptor expression in brain regions associated with zebra finch song control," *Brain Research* **857**: 151–157.

Stefano, G. B., Liu, Y. and Goligorsky, M. S. (1996) "Cannabinoid receptors are coupled to nitric oxide release in invertebrate immunocytes, microglia, and human monocytes," *The Journal of Biological Chemistry* **271**: 19238–19242.

Stefano, G. B., Rialas, C. M., Deutsch, D. G. and Salzet, M. (1998) "Anandamide amidase inhibition enhances anandamide-stimulated nitric oxide release in invertebrate neural tissues," *Brain Research* **793**: 341–345.

Stefano, G. B., Salzet, B. and Salzet, M. (1997a) "Identification and characterization of the leech CNS cannabinoid receptor: coupling to nitric oxide release," *Brain Research* **753**: 219–224.

Stefano, G. B., Salzet, B., Rialas, C. M., Pope, M., Kustka, A., Neenan, K., Pryor, S. and Salzet, M. (1997b) "Morphine- and anandamide-stimulated nitric oxide production inhibits presynaptic dopamine release," *Brain Research* **763**: 63–68.

Stella, N., Schweitzer, P. and Piomelli, D. (1997) "A second endogenous cannabinoid that modulates long-term potentiation," *Nature* **388**: 773–778.

Straiker, A., Stella, N., Piomelli, D., Mackie, K., Karten, K. J. and Maguire, G. (1999) "Cannabinoid CB_1 receptors and ligands in vertebrate retina: localization and function of an endogenous signaling system," *Proceedings of the National Academy of Sciences, USA* **96**: 14565–14570.

Sugiura, T., Kodaka, T., Nakane, S., Kishimoto, S., Kondo, S. and Waku, K. (1998) "Detection of an endogenous cannabimimetic molecule, 2-arachidonoylglycerol, and cannabinoid CB_1 receptor mRNA in human vascular cells: is 2-arachidonoylglycerol a possible vasomodulator?" *Biochemical and Biophysical Research Communications* **243**: 838–843.

Sugiura, T., Kodaka, T., Nakane, S., Miyashita, T., Kondo, S., Suhara, Y., Takayama, H., Waku, K., Seki, C., Baba, N. and Ishima, Y. (1999) "Evidence that the cannabinoid CB_1 receptor is a 2-arachidonoylglycerol receptor. Structure–activity relationship of 2-arachidonoylglycerol, ether-linked analogues, and related compounds," *Journal of Biological Chemistry* **274**: 2794–2801.

Sugiura, T., Kondo, S., Kishimoto, S., Miyashita, T., Nakane, S., Kodaka, T., Suhara, Y., Takayama, H. and Waku, K. (2000) "Evidence that 2-arachidonoylglycerol but not *N*-palmitoylethanolamine or anandamide is the physiological ligand for the cannabinoid CB_2 receptor. Comparison of the agonistic activities of various cannabinoid receptor ligands in HL-60 cells," *Journal of Biological Chemistry* **275**: 605–612.

Sugiura, T., Kondo, S., Kodaka, T., Tonegawa, T., Nakane, S., Yamashita, A., Ishima, Y. and Waku, K. (1996a) "Enzymatic synthesis of oleamide (cis-9,10-octadecenoamide), an endogenous sleep-inducing lipid, by rat brain microsomes," *Biochemical and Molecular Biology International* **40**: 931–938.

Sugiura, T., Kondo, S., Sukagawa, A., Nakane, S., Shinoda, A., Itoh, K., Yamashita, A. and Waku, K. (1995) "2-Arachidonoylglycerol: a possible endogenous cannabinoid receptor ligand in brain," *Biochemical and Biophysical Research Communications* **215**: 89–97.

Sugiura, T., Kondo, S., Sukagawa, A., Tonegawa, T., Nakane, S., Yamashita, A. and Waku, K. (1996b) "Enzymatic synthesis of anandamide, an endogenous cannabinoid receptor ligand, through N-acylphosphatidylethanolamine pathway in testis: involvement of Ca(2+)-dependent transacylase and phosphodiesterase activities," *Biochemical and Biophysical Research Communications* **218**: 113–117.

Sugiura, T., Kondo, S., Sukagawa, A., Tonegawa, T., Nakane, S., Yamashita, A., Ishima, Y. and Waku, K. (1996c) "Transacylase-mediated and phosphodiesterase-mediated synthesis of N-arachidonoyl-ethanolamine, an endogenous cannabinoid-receptor ligand, in rat brain microsomes. Comparison with synthesis from free arachidonic acid and ethanolamine," *European Journal of Biochemistry* **240**: 53–62.

Sulcova, E., Mechoulam, R. and Fride, E. (1998) "Biphasic effects of anandamide," *Pharmacology, Biochemistry and Behavior* **59**: 347–352.

Tanda, G., Pontieri, F. E. and Di Chiara, G. (1997) "Cannabinoid and heroin activation of mesolimbic dopamine transmission by a common mu1 opioid receptor mechanism," *Science* **276**: 2048–2050.

Terranova, J. P., Storme, J. J., Lafon, N., Perio, A., Rinaldi-Carmona, M., Le Fur, G. and Soubrie, P. (1996) "Improvement of memory in rodents by the selective CB$_1$ cannabinoid receptor antagonist SR141716," *Psychopharmacology* (*Berliner*) **126**: 165–172.

Thomas, E. A., Carson, M. J., Neal, M. J. and Sutcliffe, J. G. (1997) "Unique allosteric regulation of 5-hydroxytryptamine receptor-mediated signal transduction by oleamide. *Proceedings of the National Academy of Sciences of the United States of America* **94**: 14115–14119.

Tiger, G., Stenstrom, A. and Fowler, C. J. (2000) "Pharmacological properties of rat brain fatty acid amidohydrolase in different subcellular fractions using palmitoylethanolamide as a substrate," *Biochemical Pharmacology* **59**: 647–653.

Tripathy, S., Venables, B. J. and Chapman, K. D. (1999) "N-Acylethanolamines in signal transduction of elicitor perception. Attenuation of alkalinization response and activation of defense gene expression," *Plant Physiology* **121**: 1299–1308.

Turkanis, S. A. and Karler, R. (1975) "Influence of anticonvulsant cannabinoids on posttetanic potentiation at isolated bullfrog ganglia," *Life Sciences* **17**: 569–578.

Turkanis, S. A. and Karler, R. (1986) "Effects of delta-9-tetrahydrocannabinol, 11-hydroxy-delta-9-tetrahydrocannabinol and cannabidiol on neuromuscular transmission in the frog," *Neuropharmacology* **25**: 1273–1278.

Ueda, N. and Yamamoto, S. (2000) "Anandamide amidohydrolase (fatty acid amide hydrolase) (2000)," *Prostaglandins and other Lipid Mediators* **61**: 19–28.

Ueda, N., Kurahashi, Y., Yamamoto, S. and Tokunaga, T. (1995) "Partial purification and characterization of the porcine brain enzyme hydrolyzing and synthesizing anandamide," *Journal of Biological Chemistry* **270**: 23823–23827.

Ueda, N., Yamanaka, K., Terasawa, Y. and Yamamoto, S. (1999) "An acid amidase hydrolyzing anandamide as an endogenous ligand for cannabinoid receptors," *FEBS Letters* **454**: 267–270.

Van der Kloot, W. (1994) "Anandamide, a naturally-occurring agonist of the cannabinoid receptor, blocks adenylate cyclase at the frog neuromuscular junction," *Brain Research* **649**: 181–184.

Varga, K., Wagner, J. A., Bridgen, D. T. and Kunos, G. (1998) "Platelet- and macrophage-derived endogenous cannabinoids are involved in endotoxin-induced hypotension," *The FASEB Journal* **12**: 1035–1044.

Wagner, J. A., Varga, K. and Kunos, G. (1998) "Cardiovascular actions of cannabinoids and their generation during shock," *Journal of Molecular Medicine* **76**: 824–836.

Wagner, J. A., Varga, K., Ellis, E. F., Rzigalinski, B. A., Martin, B. R. and Kunos, G. (1997) "Activation of peripheral CB_1 cannabinoid receptors in haemorrhagic shock," *Nature* **390**: 518–521.

Walker, J. M., Hohmann, A. G., Martin, W. J., Strangman, N. M., Huang, S. M. and Tsou, K. (1999a) "The neurobiology of cannabinoid analgesia," *Life Sciences* **65**: 665–673.

Walker, J. M., Huang, S. M., Strangman, N. M., Tsou, K. and Sanudo-Pena, M. C. (1999b) "Pain modulation by release of the endogenous cannabinoid anandamide," *Proceedings of the National Academy of Sciences of the USA* **96**: 12198–12203.

Watanabe, K., Matsunaga, T., Nakamura, S., Kimura, T., Ho, I. K., Yoshimura, H. and Yamamoto, I. (1999) "Pharmacological effects in mice of anandamide and its related fatty acid ethanolamides, and enhancement of cataleptogenic effect of anandamide by phenyl-methylsulfonyl fluoride," *Biological and Pharmceutical Bulletin* **22**: 366–370.

Wenger, T., Toth, B. E., Juaneda, C., Leonardelli, J. and Tramu, G. (1999) "The effects of cannabinoids on the regulation of reproduction," *Life Sciences* **65**: 695–701.

Wickens, A. P. and Pertwee, R. G. (1993) "delta 9-Tetrahydrocannabinol and anandamide enhance the ability of muscimol to induce catalepsy in the globus pallidus of rats," *European Journal of Pharmacology* **250**: 205–208.

Wilcox, B. J., Ritenour-Rodgers, K. J., Asser, A. S., Baumgart, L. E., Baumgart, M. A., Boger, D. L., DeBlassio, J. L., deLong, M. A., Glufke, U., Henz, M. E., King III, L., Merkler, K. A., Patterson, J. E., Robleski, J. J., Vederas, J. C. and Merkler, D. J. (1999) "N-acylglycine amidation: implications for the biosynthesis of fatty acid primary amides," *Biochemistry* **38**: 3235–3245.

Williams, C. M. and Kirkham, T. C. (1999) "Anandamide induces overeating: mediation by central cannabinoid (CB_1) receptors," *Psychopharmacology* (*Berliner*) **143**: 315–317.

Yagen, B. and Burstein, S. (2000) "Novel and sensitive method for the detection of anandamide by the use of its dansyl derivative," *Journal of Chromatography B* **740**: 93–99.

Yamaguchi, F., Macrae, A. D. and Brenner, S. (1996) "Molecular cloning of two cannabinoid type 1-like receptor genes from the puffer fish *Fugu rubripes*," *Genomics* **35**: 603–605.

Yang, H. Y., Karoum, F., Felder, C., Badger, H., Wang, T. C. and Markey, S. P. (1999) "GC/MS analysis of anandamide and quantification of N-arachidonoylphosphatidyletha-nolamides in various brain regions, spinal cord, testis, and spleen of the rat," *Journal of Neurochemistry* **72**: 1959–1968.

Zimmer, A., Zimmer, A. M., Hohmann, A. G., Herkenham, M. and Bonner, T. I. (1999) "Increased mortality, hypoactivity, and hypoalgesia in cannabinoid CB_1 receptor knock-out mice," *Proceedings of the National Academy of Sciences of the USA* **96**: 5780–5785.

Zygmunt, P. M., Petersson, J., Andersson, D. A., Chuang, H., Sorgard, M., Di Marzo, V., Julius, D. and Hogestatt, E. D. (1999) "Vanilloid receptors on sensory nerves mediate the vasodilator action of anandamide," *Nature* **400**: 452–457.

Chapter 6

Cannabinoids and endocannabinoids: behavioral and developmental aspects

Ester Fride and M. Clara Sañudo-Peña

ABSTRACT

Marijuana and its major psychoactive component Δ^9-tetrahydrocannabinol (Δ^9-THC), profoundly affect brain function and behavior. Endogenous cannabinoids (endocannabinoids), such as the prototypical anandamide have effects on behavior, similar, but not identical to those of plant-derived or synthetic cannabinoids. Although there is no single, unique effect of cannabinoids on behavior, a battery of behavioral and physiological tests is commonly used to reflect central cannabinoid effects in man. Cannabinoids affect a number of functions including nociception, motor activity, stress and anxiety, feeding and appetite and sleep. Hence, endocannabinoids presumably play a physiological role in these areas. Although marijuana has been considered in the past a "soft drug", there is now widespread agreement that addiction to cannabis develops by activating the same neural substrate as other drugs including cocaine, heroin and alcohol.

Cannabinoid receptors and endocannabinoids have been detected in uterus, fetal tissues and in newborns, implying on one hand, possible irreversible adverse effects to the developing organism when exposed to cannabinoids *in utero* and during lactation. On the other hand, these findings imply possible physiological roles of endocannabinoids during development. Data supporting both conjectures are reviewed.

Key Words: marijuana, cannabis, endocannabinoids, behavior, development, prenatal delayed effects

Three major breakthroughs in cannabis research have paved the way toward the dramatic increase in research activity in this area: first, the isolation of Δ^9-tetrahydrocannabinol (Δ^9-THC), the major psychoactive component of the *Cannabis sativa* plant (Gaoni and Mechoulam, 1964); second, the demonstration that specific receptors for cannabinoids exist in the central nervous system and peripheral organs (Devane *et al.*, 1988; Gerard *et al.*, 1991; Matsuda *et al.*, 1990; Munro *et al.*, 1993) and third, the discovery of the endogenous cannabinoids ("endocannabinoids"), in brain and other organs (Devane *et al.*, 1992; Hanuš *et al.*, 1993; Mechoulam *et al.*, 1995). These milestones will hopefully culminate eventually in a fuller understanding of the three aspects of cannabis which are of theoretical and clinical interest: its medicinal potential, the potential harmful consequences of abuse and the multitude of physiological roles of the endogenous cannabinoid-receptor systems.

Humans consume marijuana in order to achieve euphorogenic effects ("high") or medicinal benefits. Experimentally, a wide variety of effects induced by cannabinoid substances whether derived from the *Cannabis sativa* plant, extracted from brain or other organs, or their synthetic derivatives, have been reported (for reviews see Compton *et al.*, 1996; Howlett, 1995; Ameri, 1999; Mechoulam *et al.*, 1998a,b). These include effects on motor activity, pain perception, cognition, mood and consciousness, as well as appetite-enhancing, cataleptic, antiemetic, hypothermic, hypotensive and immunosuppressant effects (Abood and Martin, 1992; Ameri, 1999; Axelrod and Felder, 1998; Compton *et al.*, 1996; Hall and Solowij, 1998). The pleasurable and relaxing experience of marijuana use (Court, 1998; Hall and Solowij, 1998), together with recent biological evidence for the addictive potential of cannabis (Gardner and Vorel, 1998; Iversen, 2000; Tanda *et al.*, 1997), help explain the continued use of this illicit drug.

The observed consequences of a single, or a limited number of exposures of cannabinoids to the mature organism, whether measured at the behavioral or molecular-physiological level, are usually transient (Court, 1998; Heishman *et al.*, 1997; Fride, 1995; Hall and Solowij, 1998). However, adverse effects of chronic marijuana smoking have been reported, although these are complex (Court, 1998; Hall and Solowij, 1998; Solowij, 1995). For example, there is widespread agreement that chronic cannabis smoking has damaging effects on the respiratory system (Hall and Solowij, 1998; Ashton, 1999). Further, subtle but definitive cognitive impairment has been shown to result from long-term cannabis use. However, adverse effects on reproductive and immune functions have not been proven in humans (Hall and Solowij, 1998; Pope and Yurgelun-Todd, 1996; Ashton, 1999). Thus, in the face of rising marijuana use amongst teenagers (Mathias, 1997), potential long-term adverse effects of marijuana use are still the subject of debate. As for all drugs of abuse, the accumulating information distilling out of the combined research efforts will help educate the teenage and adult public about the consequences of abusing the drug.

One may argue, however, that it is even more critical to gather information on the potential harmful sequellae of marijuana consumption during pregnancy and nursing, for two reasons, one biological, and one philosophical. Philosophically, these offspring have been exposed beyond their control, so that many adverse effects of maternal marijuana smoking cannot be ascribed to the person's own decision to ingest the drug. Biologically, insults during critical periods of early (prenatal and perinatal) development are more likely to have *permanent* consequences compared to adverse effects of exposure in the mature organism. Indeed, alterations in development of brain, behavior and health parameters have been reported as the result of exposure to cannabinoids during the fetal (prenatal) or early postnatal period in humans (Fried, 1996; Fried *et al.*, 1998; Fried and Watkinson, 2000) and in animals (del Arco *et al.*, 2000; Fride and Mechoulam, 1996a; Navarro *et al.*, 1995).

The first part of this chapter will focus on psychoactive effects of cannabinoids in adults. In the second part of this chapter, developmental aspects of the cannabinoid system will be explored.

PHARMACOLOGICAL PROFILE OF CANNABINOIDS

In the absence of a behavioral or physiological response which is unique for cannabinoids, Martin and colleagues (1991) have developed a multiple *in vivo* assay

176   E. Fride and M. C. Sañudo-Peña

for the evaluation of cannabimimetic effects. These procedures have been shown to be predictive of psychoactive cannabinoid activity and to highly correlate with affinity for the – predominantly central- cannabinoid CB_1 receptor (Abood and Martin, 1992; Compton et al., 1993; Martin et al., 1991; Razdan, 1986). The full battery includes a fourfold evaluation in mice (the "tetrad"), a drug discrimination test and catalepsy test in rats, an evaluation of static ataxia in dogs and operant suppression in monkeys (Martin et al., 1991). However, often only parts of the battery are used, such as the mouse tetrad and the two tests in rats (Martin et al., 1991) or mouse tetrad and the dog ataxia test (Little et al., 1989), or the mouse tetrad, the rat discrimination learning and in addition, assessment of psychotomimetic activity in man (Compton et al., 1993), or the mouse tetrad alone (Fride and Mechoulam, 1993; Adams et al., 1995, 1998). Sometimes, only two of the components of the mouse tetrad are used, for example for SAR studies (Adams et al., 1995). Whether employed in full or in part, the test battery produces a characteristic pharmacological profile for exogenous (Martin et al., 1991; Compton et al., 1993; Little et al., 1989) as well as endogenous cannabinoids (endocannabinoids) (Fride and Mechoulam, 1993).

The mouse tetrad

The mouse tetrad consists of 4 simple evaluations, which may be measured in sequence in the same animal: Motor activity in an open field is measured for various lengths of time, but typically for 10 min, by digitized or manual observation. Next, the amount of time in which the mouse is immobile after it is placed on a metal ring of 5.5 cm diameter, held at about 16 cm above a table top, is recorded for 4–5 min. This method was developed by Pertwee (1972) and is taken as a measure of catalepsy. Although an automated version of the ring-catalepsy test has been developed (Martin et al., 1992), this assay is usually performed manually. Third, rectal temperature is measured by a telethermometer. Analgesic (pain-reducing) effects of cannabinoids are measured by the tail flick (e.g. Little et al., 1989) or hotplate method (e.g. Fride and Mechoulam, 1993; Fride et al., 1995; see also Segal, 1986). In the tail flick test, radiant heat is focused on the tail, and the latency which is measured until the animal flicks its tail away, is taken as a measure of nociceptive sensitivity (Tjolsen and Hole, 1997). In the hotplate test, the mouse is placed on a hotplate with its temperature usually fixed at 54 or 55°C. The latency to responses such as jumping or licking a hindpaw, is taken as the nociceptive response (Ankier, 1974). The tail flick test is considered a reflex response at the spinal level, while the hotplate tests pain perception at the higher (supraspinal) levels (Tjolsen and Hole, 1997). In general, both tests are sensitive to cannabinoids. Compton et al., 1993 have shown a high degree of correlation between performance in the mouse tetrad and cannabinoid receptor (CB_1) binding in rat brain membranes. These authors also described a high degree of correlation between CB_1 receptor binding and cannabinoid potencies in the rat drug discrimination test, which is taken to be predictive of cannabimimetic effects in man (Balster and Prescott, 1992). Since they also found a significant correlation between CB_1 receptor binding and psychoactive effect in humans, they suggested that the mouse model may be used to investigate the abuse potential of cannabinoids (Compton et al., 1993).

Drug discrimination

In the drug discrimination paradigm (see review by Balster and Prescott, 1992), laboratory animals learn to recognize the presence of a certain drug (such as nicotine, morphine, LSD, Δ^9-THC) and express the discrimination between the drug under investigation and a control substance ("placebo") in a two choice situation, where the "correct" choice is rewarded. Results from numerous studies indicate that animals will learn to discriminate cannabinoids from drugs from different classes and they will substitute other cannabinoids for Δ^9-THC (same class drug). These observations are compatible with the assessments (see above) that cannabinoids, although they are not associated with one unique type of behavior, produce a characteristic pattern of effects on the central nervous system. Moreover, both discriminative stimulus effects of various cannabinoids and marijuana-intoxication symptoms in humans were found to highly correlate with CB_1 receptor binding. Consequently, it was suggested that the rat model of drug discrimination may be used to predict cannabinoid intoxication in humans (Balster and Prescott, 1992).

PHARMACOLOGICAL PROFILE OF ENDOCANNABINOIDS

The endogenous ligands for the cannabinoid receptors discovered thus far (the "endocannabinoids"), include the "anandamides" (Devane et al., 1992; Hanuš et al., 1993), 2-arachidonoyl glycerol (2-AG, Mechoulam et al., 1995) and noladin ether (Hanuš et al., 2001). Thus far, the prototypical anandamide, arachidonoyl ethanol amide (Devane et al., 1992) is the most thoroughly studied endocannabinoid. Although the overall pharmacological activity is similar to the psychoactive plant constituent Δ^8-THC (Fride and Mechoulam, 1993; Mechoulam and Fride, 1995), it is clear that differences between anadamide and plant-derived and synthetic cannabinoids are present too (Fride et al., 1995; Mechoulam and Fride, 1995; Mackie et al., 1993; Smith et al., 1994; Welch et al., 1995). Behaviorally, it was clear from the initial description of anandamide's effects in the tetrad, that it has partial effects for some of its components (hypothermia and analgesia, see Figure 6.1) (Fride and Mechoulam, 1993; Mechoulam and Fride, 1995). Moreover, when different routes of administering anandamide were compared, a complex pattern of full and partial activities was observed (Smith et al., 1994). Further, Δ^9-THC but not anandamide produced conditioned place avoidance (Mallet and Beninger, 1998a).

Additional behavioral differences include the effects of very low doses of anandamides (0.0001–0.01 mg/kg).

Biphasic effects

Very low doses of anandamide and the synthetic endocannabinoid-like docosahexaenylethanolamide, but not of Δ^9-THC, inhibited pharmacological effects of conventional doses of Δ^9-THC (Fride et al., 1995). Low doses (0.01 mg/kg) of anandamide by itself, showed effects opposite to the pharmacological effects of moderate or high doses (Mechoulam and Fride, 1995; Sulcova et al., 1998). In rats, moderately

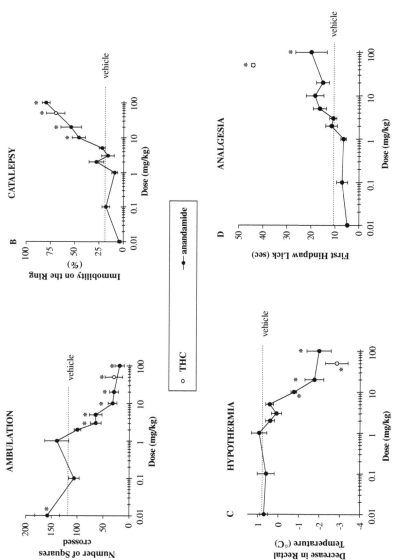

Figure 6.1 Effects of anandamide injected *i.p.* to Sabra mice on ambulation (horizontal motor activity) in an open field (A), catalepsy on a ring (B), change in rectal temperature (C) and nociception on a hot plate (latency to first hindpaw lick in sec) (D). In separate mice, a single high dose (50 mg/kg) of Δ^8-THC (which has, under the present conditions, potencies and maximal effects similar to Δ^9-THC, Fride unpublished) was administered for comparison. The results indicate that anandamide has similar maximal effects for motor inhibition (A) and catalepsy (B), somewhat less for hypothermia and acts as a partial agonist compared to THC in the hot plate test (D) (see Mechoulam and Fride, 1995).

low doses of Δ^9-THC (1–2 mg/kg) were shown to stimulate movement (Sañudo-Peña *et al.*, 2000).

The biphasic effects of anandamide suggest the possibility that the physiological functions of the endocannabinoids may be opposite to many of the experimental, pharmacological observations performed with high doses of cannabinoids. A speculative explanation was offered as a linkage of the CB_1 receptor to G_s type proteins: when agonist concentrations are high (which is usually the case in pharmacological experiments, or after intake of high amounts of cannabis), only G_i protein activation is observable, resulting overall in behavioral depression. By contrast, when agonist concentrations are low, activation of G_s proteins become apparent (Fride *et al.*, 1995), in an analogous fashion to what has been found for opiate receptors (Cruciani *et al.*, 1993). Interestingly, exactly such G_s linkage to the CB_1 receptor has been demonstrated, at least in neurons from the corpus striatum and in CB_1-transfected cells (Glass and Felder, 1997).

Entourage effect

Another phenomenon which can be observed with endocannabinoids (but not with plant-derived or synthetic cannabinoids) is the "entourage effect" (Ben-Shabat *et al.*, 1998). Thus, 2-AG when isolated from brain as well as spleen and gut, is accompanied by several 2-acyl-glycerol esters, two major ones being 2-palmitoyl-glycerol and 2-linoleoyl-glycerol. These two esters do not bind to the cannabinoid receptors. However, they potentiate the apparent binding of 2-AG to its receptor while 2-linoleoyl-glycerol also inhibits the inactivation of 2-AG in neuronal cells. *In vivo*, both esters potentiate 2-AG-induced effects on the tetrad (Ben-Shabat *et al.*, 1998).

Drug discrimination

As noted above ("Pharmacological profile of Cannabinoids"), the drug discrimination paradigm in animals is a good predictor of cannabinoid intoxication in humans (Balster and Prescott, 1992). In order to further characterize the cannabinoid-like properties of the endocannabinoids, anandamide was tested for its ability to substitute for Δ^9-THC or other cannabinoids in the rat drug discrimination test. Initially, it was reported that anandamide substituted for Δ^9-THC, but only at high doses (30–45 mg/kg) which also produced severe immobility (Wiley *et al.*, 1995a). However, this lack of generalization to anandamide could be ascribed to the facile breakdown of anandamide (Di Marzo *et al.*, 1998a), since dose-dependent generalization to Δ^9-THC was observed upon administration of the metabolically stable analogs of anandamide, (R)-methanandamide (Burkey and Nation, 1997; Jarbe *et al.*, 1998) or 2-methyl-arachidonyl-2'-fluoroethylamide (Wiley *et al.*, 1997).

PHARMACOLOGICAL PROFILE OF CANNABINOID RECEPTOR ANTAGONISTS

The synthesis of specific antagonists for both the CB_1 ("SR141716A", Rinaldi-Carmona *et al.*, 1994) and CB_2 ("SR144528", Rinaldi-Carmona *et al.*, 1998) receptors, has greatly contributed to further our understanding of these receptors, and to tease out differences between the CB_1 and CB_2 receptors. For example, application of the CB_2 antagonist in several pharmacological paradigms used to test for cannabinoid effects, helped characterize a new specific CB_2 agonist ("HU-308") and the discovery that not only CB_1, but also CB_2 receptors are involved in blood pressure regulation (Hanuš et al., 1999). Effective SR141716A-induced blockade of the CB_1 receptor-induced activities has been shown for a number of centrally mediated functions (Rinaldi-Carmona et al., 1994; Compton et al., 1996). For example, the CB_1 receptor antagonist blocked Δ^9-THC-induced effects on the tetrad (Adams et al., 1998; Compton et al., 1996; Fride et al., 1998a), Δ^9-THC – or anandamide-induced learning and memory impairment (Brodkin and Moerschbaecher, 1997; Mallet and Beninger, 1998a), and the discriminative properties of Δ^9-THC in rats and monkeys (Wiley et al., 1995b). Intriguingly, anandamide-induced effects on the mouse tetrad were *not* blocked by SR141716A (Adams et al., 1998; Fride et al., 1998a). Furthermore, SR141716A, when administered alone, was also found to display *agonist* effects in the tetrad at high doses (Compton et al., 1996; Fride, unpublished) and *inverse agonist* properties *in vitro* (Shire et al., 1999; Landsman et al., 1997).

Additional intrinsic effects of SR141716A when administered alone, include enhanced arousal (at the expense of REM sleep and slow-wave-sleep, Santucci et al., 1996), increased pain sensitivity (hyperalgesia, Richardson et al., 1997), improved memory (Terranova et al., 1996) and rewarding properties in the conditioned place preference test for incentive motivational effects (Sañudo-Peña et al., 1997). When antagonist-induced effects *in vivo* are opposite to those seen after agonist administration, they can be interpreted as an inverse agonist effect, or alternatively, as an inhibition of an endogenous tone of endocannabinoid release.

CB₁ KNOCKOUT MICE

Modern techniques including the development of animals with directed gene deletion ("receptor knockout mice"), were recently applied to the cannabinoid area. The purpose of such manipulation is to shed additional light on the functions of the cannabinoid receptors, although observations on knockout phenotypes should be interpreted with caution (Gingrich and Hen, 2000).

Two laboratories have independently developed strains of mice where the CB_1 receptor was deleted (Steiner et al., 1999; Zimmer et al., 1999; Ledent et al., 1999). Each laboratory started with a different parent strain. Zimmer and colleagues (1999), developed the knockouts at the National Institutes of Health in the USA from C57BL/6J mice, while Ledent et al. (1999) deleted the CB_1 receptor from CD_1 mice in Belgium. Overall, pharmacological evaluations from both laboratories yielded outcomes as expected. Thus, cannabinoid (THC-, HU210-)induced responses in the tetrad (motor activity, catalepsy, hypothermia and hypoalgesia on the hot-plate, see "Pharmacological profile of cannabinoids") were absent in the CB_1 receptor-deleted (–/–) knockout mice (Zimmer et al., 1999; Ledent et al., 1999). Interestingly, Δ^9-THC-induced hypoalgesia in the tail flick test was present, despite the gene deletion, in both knockout strains (Zimmer et al., 1999; Ledent et al.,

1999). Since the tail flick test is presumably measuring spinal pain perception, while the hotplate tests supraspinal mechanisms of pain (Tjolsen and Hole, 1997, see "Pharmacological profile of cannabinoids"), this observation suggests that mainly higher level pain mechanisms are affected by the CB_1 receptor deletion. This conclusion will have to be reconciled with the spinal component of cannabinoid receptor-mediated pain (Martin and Lichtman, 1998), perhaps by suggesting a non-CB_1 receptor mechanism at the spinal level.

Evaluation of spontaneous behavior and physiological functions in the Belgium knockouts did not uncover differences from "wild-type" (non-knockout parent strain mice) in pain threshold, locomotor activity and body temperature (Ledent et al., 1999), thus suggesting that endocannabinoids do not play a critical role in these functions. By contrast, when spontaneous functions in the NIH knockouts were assessed, inhibition of motor activity, significant catalepsy on the ring, hypoalgesia on the hotplate and in the formalin test for nociception were uncovered, whereas, no differences in body temperature and in the tail flick test for pain perception were detected (Zimmer et al., 1999).

These seemingly paradoxical observations in the NIH knockouts (hypolocomotion, hypoalgesia and catalepsy, are commonly seen when cannabinoids are administered to normal animals), were tentatively ascribed to neuronal reorganization (Steiner et al., 1999; Zimmer et al., 1999). One may also conjecture however, that this observation is compatible with a low dose endocannabinoid-induced tonic *stimulation* of activity (see above, "biphasic effects") which is disrupted in the knockouts.

Furthermore, possible reasons for the discrepancies between the two types of knockout strain, such as different parent strains and hence possibly different compensatory mechanisms should be further investigated.

Additional observations on the CB_1 receptor knockout mice include reduced reinforcing properties and withdrawal effects of cannabinoids (Ledent et al., 1999), absence of morphine-induced dopamine release in the limbic brain (nucleus accumbens) (Mascia et al., 1999) and enhancement of memory (Reibaud et al., 1999).

Overall, the observations on CB_1 receptor-deleted mice have supported previous assessments of a physiological role for the cannabinoid system in motor control, pain perception, temperature regulation, memory and motivational processes.

ADDICTION; TOLERANCE; CRAVING AND REINFORCEMENT; WITHDRAWAL

Tolerance

Tolerance to cannabinoids developed in all species studied, with varying duration and onset, depending for example, on the parameter studied (Compton et al., 1996; Jones et al., 1981; Adams et al., 1976; Fitton and Pertwee, 1982; Jarbe, 1978; Webster et al., 1973). In humans, development of tolerance to the psychoactive effects of marijuana is clearly seen with "heavy" (daily) use, but usually not with casual or moderate use (Compton et al., 1996; Iversen, 2000). Tolerance to anandamide has been shown in animal studies (Fride, 1995; Welch, 1997).

The behavioral tolerance is accompanied, analogous to other classes of drugs, by a decrease in CB_1 receptors in all brain areas which are relevant for the CB_1 receptor-tolerant behaviors (Breivogel *et al.*, 1999; Romero *et al.*, 1997).

Craving and reinforcement

Addictive potential of marijuana was long thought to be very weak or absent (Compton *et al.*, 1996). However, although addictive behaviors such as compulsive drug seeking (due to "craving"), is rarely induced by marijuana use, preparations containing higher Δ^9-THC concentrations such as hashish, have been shown to induce addictive behaviors, especially in populations at risk (Crowley *et al.*, 1998; Gardner and Vorel, 1998). Hence one may speculate that marijuana, as obtained at the turn of the millenium, may be addictive as well, since it often contains much higher concentrations of Δ^9-THC than in the 1960s and 1970s (Ashton, 1999; Iversen, 2000).

From animal studies it has gradually become clear that cannabinoids interact with the same neural substrates which are thought to be responsible for the euphoriant and rewarding effects of other drugs of abuse such as cocaine, opiates and alcohol (Gardner and Lowinson, 1991; Gardner and Vorel, 1998, and references therein). These neural substrates of addiction include the medial forebrain bundle, containing the dopamine pathways, leading from the mesencephalic ventral tegmentum to the nucleus accumbens and the prefrontal cortex. It appears that cannabinoids, like other drugs of abuse, increase dopamine activity in these neural circuits (Chen *et al.*, 1990; Diana *et al.*, 1998; French, 1997; Gardner and Lowinson, 1991; Gessa *et al.*, 1998; Jentsch *et al.*, 1997; Tanda *et al.*, 1997). In behavioral tests of addiction, Δ^9-THC significantly lowered brain reward thresholds in the median forebrain bundle (Gardner and Vorel, 1998; Gardner and Lowinson, 1991; Lepore *et al.*, 1996).

Furthermore, Δ^9-THC was shown to be appetitive in the "conditioned place preference" test, but only after the appropriate timing and dosing (Lepore *et al.*, 1995). Thus aversive effects of cannabinoids have been repeatedly shown too (Chaperon *et al.*, 1998; Gardner and Vorel, 1998; Parker and Gillies, 1995; McGregor *et al.*, 1996; Mallet and Beninger, 1998b; Sañudo Peña *et al.*, 1997). These biphasic effects are well known from human experience; low doses of Δ^9-THC produce a "high" feeling, while high doses may be aversive (Ashton, 1999; Gardner and Vorel, 1998).

Similarly, self administration of cannabinoids has been hard to show in animal studies (Gardner and Vorel, 1998; Compton *et al.*, 1996), possibly due to masking anxiogenic effects of cannabinoids (Chakrabarti *et al.*, 1998; Onaivi *et al.*, 1995; Rodriguez de Fonseca *et al.*, 1996). Confirming this suspicion in a recent study using the synthetic CB_1 receptor antagonist WIN 55,212-2, a robust, but biphasic effect on self-administration in mice was demonstrated, suggesting rewarding effects at lower, and aversive effect at high doses of WIN 55,212-2 (Martellotta *et al.*, 1998).

Thus overall, despite earlier doubts, recent studies have produced convincing evidence for the mesolimbic-mesocortical dopamine system as a substrate for cannabinoid abuse potential. Moreover, a common opioid receptor mechanism appears to mediate both cannabinoid and heroin-induced activation of the mesolimbic dopamine activation (Tanda *et al.*, 1997). It has also recently been shown in an alcohol craving paradigm that SR141716A can block the "craving" for alcohol in rats (Gallate *et al.*, 1999), again suggesting a common abuse mechanism for various types of substances.

In summary, it has become clear that cannabis has addictive properties similar to other drugs of abuse. This realization lends biological support for the controversial "gateway" theory, which states that cannabis often introduces novel users to more destructive and addictive drugs. One should not overlook however, possible genetic variation in cannabis abuse. Thus, studies indicating genetic variation in the reward system (Gardner and Vorel, 1998) and in the emotional effects of cannabinoids (Chakrabarti *et al.*, 1998; Onaivi *et al.*, 1990, 1995) support a genetic predisposition to cannabis abuse. Whether a certain individual will eventually succumb to the addictive potential of cannabis will obviously be the outcome of a combination of various factors.

Withdrawal and dependence

Dependence was conclusively shown when administration of the CB_1 antagonist SR141716A to rats (Aceto *et al.*, 1995, 1996; Diana *et al.*, 1998; Tsou *et al.*, 1995) or mice (Cook *et al.*, 1998) receiving a chronic regimen of cannabinoids, produced obvious behavioral withdrawal symptoms (including "wet dog shakes", facial rubbing, scratching and licking). Interestingly, the withdrawal syndrome in rats was accompanied by a decrease in mesolimbic dopamine activity (Diana *et al.*, 1998; Tanda *et al.*, 1999).

In humans, early uncontrolled and more recently, controlled studies, have also demonstrated dependence and withdrawal symptoms (Compton *et al.*, 1996; Crowley *et al.*, 1998; Haney *et al.*, 1999; Kouri *et al.*, 1999). The fact that a clear abstinence syndrome has only been shown recently, may be related to the generally much higher concentrations of THC found in marijuana cigarettes (joints) (Ashton, 1999; Iversen, 2000).

It has been suggested that tolerance and dependence (Cook *et al.*, 1998) on one hand, and craving and dependence (Gardner and Vorel, 1998) on the other, are related phenomena. Addiction to drugs including cannabis is not only explained by their positive reinforcing qualities (Gardner and Lowinson, 1991; Gardner and Vorel, 1998), but also by the attempt toward off the stress-like symptoms (elevation of corticotropin-releasing factor in the limbic system) experienced during withdrawal (Rodriguez de Fonseca *et al.*, 1997).

STRESS AND ANXIETY

In addition to the euphoriant effects of marijuana smoking, dysphoric effects including anxiety and panic reactions are also commonly observed phenomena induced by cannabis consumption (Ashton,1999). In animal studies, Δ^9-THC (Oinaivi *et al.*, 1990, 1995), HU-210 (a very potent synthetic cannabinoid, Giuliani *et al.*, 2000a) and anandamide (Chakrabarti *et al.*, 1998) induced anxiety in the "plus maze" (fewer entries onto the "open", anxiety-provoking arms). Moreover, both Δ^9-THC and anandamide had an activational effect on the neuroendocrine (hypothalamus-pituitary-adrenal, "HPA") axis, which plays a central role in the stress response (Weidenfield *et al.*, 1994). Thus in that study, depletion of corticotropin-releasing factor (CRF), together with increased serum ACTH and corticosterone were observed. Further support for a direct hypothalamic effect of cannabinoids on the activation of the pituitary-adrenal axis comes from an interesting study by Rodriguez de Fonseca and colleagues (1996). These authors reported that administration of the very potent cannabinoid receptor agonist HU-210, produced a behavioral stress response in the defensive-withdrawal test in rats, which was accompanied by a rise in plasma corticosterone. A CRF-antagonist counteracted the behavioral stress response.

It should be noted that no anxiogenic effects of anandamide were found in the plus maze when rats were investigated (Crawley *et al.*, 1993). Since strain-dependent effects of anandamide in mice in this assay have been reported (Chakrabati *et al.*, 1998), species variation and/or strain differences may be invoked to explain the negative findings in Crawley *et al.* (1993) study.

Concluding from these animal studies, it is clear that cannabinoids have the potential to be potent anxiogenic agents. How does this relate to the human cannabis consumer? When fear or anxiety responses are seen in naïve human consumers, discontinuation of cannabis intake is common (Hall and Solowij, 1998). However, for experienced users, it has also been surmised that part of the continued use of cannabis, is in fact an attempt to ward off the anxiety experienced during withdrawal from the drug (Gardner and Vorel, 1998; Rodriguez de Fonseca *et al.*, 1997).

The prefrontal cortex (PFC)

The prefrontal cortex (PFC) is thought to be a major site for the regulation of anxiety and the response to stress. The PFC is also thought to be the site of action of anti-schizophrenic ("major tranquilizing") drugs (dopamine receptor antagonists). The selective elevation of dopamine turnover in the PFC induced by stress (Herve *et al.*, 1979), further implicates the PFC-dopamine system in the stress response. Interestingly, Δ^9-THC produces schizophrenia-like symptoms in humans (Emrich *et al.*, 1997) and increases dopamine turnover (Jentsch *et al.*, 1997) and presynaptic dopamine efflux in the PFC (Chen *et al.*, 1990) in rodents, which is compatible with the rich distribution of CB_1 receptors in the frontal cortex (see Sañudo-Peña and Fride, this book). Taken together, this information possibly implicates endocannabinoids in the PFC in stress and anxiety. Initial strides have been made to test the hypothesis that endocannabinoids serve as intermediaries between the stress stimulus and the resulting dopamine activation (Fride *et al.*, 1998b). First, the endocannabinoids anandamide and 2-AG were detected in the PFC of mice. Moreover, preliminary observations indicated that the PFC of acutely stressed mice (30 min of noise stress) contained 4 times as much anandamide as those of unstressed mice. Such increase was not seen in the hippocampi of these mice. No differences in levels of 2-AG between stressed and control PFCs were detected (Fride *et al.*, 1998b, see Figure 6.2). Thus, the hypothesis that anandamide in the PFC may function as a "stress mediator" deserves further investigation.

FEEDING AND APPETITE

Marijuana or its major psychotropic constituent Δ^9-THC is used clinically to enhance appetite in acquired immunodeficiency syndrome (AIDS) (Beal *et al.*, 1997; Mattes *et al.*, 1994; Plasse *et al.*, 1991; Struwe *et al.*, 1993; Mechoulam *et al.*, 1998a). The rationale for such use is based both on anecdotal and scientific evidence. Thus cannabis, Δ^9-THC as well as endocannabinoids enhance appetite (Mechoulam *et al.*, 1998b; Gallate *et al.*, 1999; Williams *et al.*, 1998; Williams and Kirkham, 1999), although no effect (Graceffo and Robinson, 1998) or reductions in food intake (Miczek and Dixit, 1980; Sofia and Knobloch, 1976; Compton *et al.*, 1996) have also been reported, mainly in earlier studies (see Compton *et al.*, 1996). These opposite effects on food intake may be explained by a masking anxiogenic effect of the cannabinoids. Indeed, in a recent report (Giuliani *et al.*, 2000b), HU-210 (a very potent CB receptor agonist)-induced reductions in food intake were noted, which may be ascribed to the stress induced by HU-210 (Rodriguez de Fonseca *et al.*, 1996; Giuliani *et al.*, 2000a, see "Stress and Anxiety").

Chronic administration of low doses of anandamide (presumably without stress effects), enhanced food intake in food deprived mice (Hao *et al.*, 2000). Conversely, further support for an appetite-enhancing effect of cannabinoids comes from studies where appetite suppression and weight loss were reported using the CB_1 receptor antagonist SR141716A (Arnone *et al.*, 1997; Colombo *et al.*, 1998). Thus the evidence is clearly tilted – unless masked by an anxiety-induced loss of appetite – toward an appetite-enhancing effect of cannabinoids and resulting weight gain. It has been suggested that the increase in appetite is either socially induced and/or

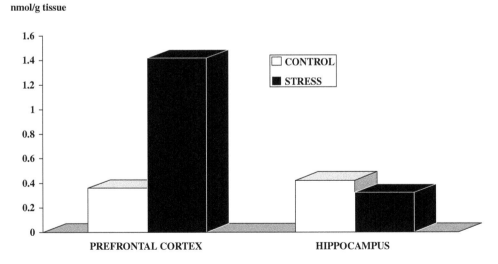

Figure 6.2 Anandamide levels in the prefrontal cortex and hippocampus of acutely stressed (30 min of bell noise) female Sabra mice were measured by GC-MS using SIM (selective ion monitoring) mode with deuterated anandamide as internal standard (Fride *et al.*, 1998). The – preliminary – data presented here, suggest a selective, almost 4-fold increase in anandamide levels in the PFC as a result of stress.

related to an increased craving for sweets and carbohydrate-rich drinks (Compton *et al.*, 1996). Thus, in an experimental situation, human subjects were prone to eat more snack foods when smoking marijuana in social circles, while subjects smoking marijuana in isolation, did not alter food intake (Foltin *et al.*, 1986).

In accordance with a stimulatory effect of cannabinoids on feeding, a hypothalamo-collicular pathway involved in feeding-related behaviors and disinhibited upon activation of CB_1 receptors has been suggested (see Sañudo-Peña and Fride, this book). Be that as it may, the potential for weight gain in conditions such as AIDS and cancer deserves further exploration.

Endocannabinoids and other *N*-acylethanolamines (NAPE) have been detected in several foods including chocolate (di Tomaso *et al.*, 1996), milk, oatmeal, hazelnuts, millet and soy beans (Di Marzo *et al.*, 1998b). The presence of endocannabinoids in chocolate is especially interesting and has been investigated further. Although the endocannabinoid concentrations in chocolate are far too low to induce visible marijuana-like euphoriant effects when taken orally (Di Marzo *et al.*, 1998b), two alternative biological explanations may explain the phenomenon of "chocolate craving". First, there may be enough endocannabinoids present in chocolate to activate the mesolimbic reward system (see "Addiction") required for craving. Intriguingly, the smell of chocolate reduced theta brain waves, presumably reducing attention and promoting a sense of relaxation (Martin, 1998). Second, non-cannabinoid NAPE's found in chocolate, with "entourage" properties (see "Pharmacological profile of endocannabinoids"), i.e., which do not bind the cannabinoid receptor but inhibit the degradation of endocannabinoids and/or potentiate endocannabinoid binding to the CB_1 receptors (Ben-Shabbat *et al.*, 1998), may amplify the activities of the endocannabinoids present in chocolate (Fride *et al.*, 1997). Indeed, oleoyl ethanol amide and lineoyl ethanol amide, both present in chocolate (di Tomaso *et al.*, 1996; Di Marzo *et al.*, 1998b) exhibit psychotropic effects in the "tetrad", presumably by potentiating endocannabinoid activity (Fride *et al.*, 1997).

In summary, the well known appetite enhancing effect of cannabis may be clinically exploited to enhance appetite in AIDS, cancer patients or other patients suffering from wasting diseases. However, the importance of the presence of endocannabinoids in milk may transcend the subtle effects on appetite, mentioned above. Rather, being present in milk, their physiologic importance may lie in their potential role during the neonatal period, when endocannabinoids may play a crucial role in the well being of the newborn. Indeed, initial evidence has accumulated in support of such hypothesis, which will be discussed below (see "Function of the endocannabinoid system in the neonate").

SLEEP

Drowsiness and sleepiness are commonly observed in the later stages of marijuana intoxication. Increased sleeping time and slow wave sleep have been recorded upon acute administration of Δ^9-THC, the major psychoactive ingredient of marijuana (Paton and Pertwee, 1973; Pivik *et al.*, 1972). Anandamide was shown later to increase slow-wave and REM sleep in rats at the expense of wakefulness (Murillo-

Rodriguez *et al.*, 1998), while, conversely, the CB_1 receptor antagonist SR141716A increased wakefulness at the expense of slow-wave and REM sleep (Santucci *et al.*, 1996). These findings support a role for the endocannabinoids in sleep regulation.

Oleamide is a lipid which was first identified in cerebrospinal fluid of sleep-deprived cats (Cravatt *et al.*, 1995), and was later shown to be produced in a mouse neuronal cell line (neuroblastoma $N_{18}TG_2$, Bisogno *et al.*, 1997). Oleamide was shown to induce behavioral characteristics of normal sleep and to increase slow-wave sleep when injected into rats (Cravatt *et al.*, 1995; Basile *et al.*, 1999; Yang *et al.*, 1999). In view of the close structural relationship between oleamide and anandamide, both being fatty acid amides, and both having sleep-promoting properties, the interaction between them was explored (Mechoulam *et al.*, 1997). Although, unlike anandamide, oleamide does not bind to CB_1 receptors (Boring *et al.*, 1996), a cannabinoid-like syndrome in the "tetrad" was observed after injection of oleamide into mice, similarly to anandamide-induced symptoms (Mechoulam *et al.*, 1997). Since oleamide inhibited fatty acid amide hydrolase (FAAH), an enzyme which deactivates anandamide, we proposed that elevation of endogenous anandamide and subsequent enhanced activation of the CB_1 receptor, may explain the cannabinoid-like effects induced by oleamide and perhaps also its sleep-promoting effects (Mechoulam *et al.*, 1997).

Lambert and Di Marzo in a very elucidating review (1999), suggested that anandamide and oleamide may have common as well as distinct pathways of action. Thus when mechanisms of oleamide and anandamide were compared, it appeared that oleamide and anandamide inhibited each other's degradation by binding to the same inactivating enzyme (FAAH). This interaction could form the basis of a cross-regulation between these two compounds (Lambert and Di Marzo, 1999; Boger *et al.*, 1998).

Recently, anandamide has been found to bind to $5\text{-}HT_2$ receptors (Kimura *et al.*, 1998). More specific studies on the interaction of oleamide with serotonin, indicated that this compound, but not anandamide, potentiated serotonin-induced activation of $5\text{-}HT_{1A}$, $5\text{-}HT_{2A}$ and $5\text{-}HT_{2C}$ receptors (Huidobro-Toro and Harris, 1996; Boger *et al.*, 1998). Recently Cheer *et al.* (1999) found supporting evidence for a role of oleamide in $5\text{-}HT_2$ receptor potentiation *in vitro*, and *in vivo* at the behavioral level. Their data further suggest that cannabinoid and $5\text{-}HT_2$ receptors interact at the membrane level, or alternatively, that cannabinoids and oleamide act by allosteric modulation of a cannabinoid binding site on $5\text{-}HT_2$ receptors.

Somewhat paradoxically, $5\text{-}HT_2$ receptor antagonism (with ketanserin) resulted in a cannabinoid-like profile on the "tetrad" (see "Pharmacological profile of cannabinoids"), similar to that of cannabinoids such as Δ^9-THC and anandamide, and also similar to that obtained with oleamide (Fride, 1999).

Taken together, these observations suggest that oleamide and possibly also anandamide may induce at least some of their effects by direct or indirect interaction with $5\text{-}HT_2$ receptors, the exact nature of which will have to be determined in future studies.

Alternative or additional mechanisms by which oleamide and anandamide exert their effects should also be considered. Previously suggested modes of action, including inhibition of gap junctions (Guan *et al.*, 1997; Venance *et al.*, 1995), GABA receptor modulation (Yost *et al.*, 1998) and nonspecific membrane perturbation

(Lerner, 1997; see also Mechoulam *et al.*, 1998b), should be studied further as mechanisms by which endocannabinoids and oleamide are involved in sleep.

DEVELOPMENTAL ASPECTS OF CANNABINOIDS

The cannabinoid system in development

Initial reports studying the first weeks of postnatal life in the rat described a gradual increase in brain CB_1 receptor mRNA (McLaughlin and Abood, 1993) and in the density of CB_1 receptors (Belue *et al.*, 1995; Rodriguez de Fonseca *et al.*, 1993). In later studies, investigating the gestational period, CB_1 receptor mRNA was detected from gestational day 11 in the rat (Buckley *et al.*, 1998). Additional studies have uncovered more complex developmental patterns. Thus, whereas the highest levels of mRNA expression of the CB_1 receptor are seen at adulthood in regions such as the caudate-putamen and cerebellum, other areas such as cerebral cortex and hippocampus display the highest mRNA CB_1 receptor levels between gestational day 21 and postnatal day 5 (i.e., during the last trimester of pregnancy and the first week of life), with the first postnatal day expressing peak levels (Berrendero *et al.*, 1998a; 1999).

Moreover, atypical patterns (i.e., different from those in adult) of CB_1 receptor densities were observed: a transient presence of CB_1 receptors was detected in white matter regions including the corpus callosum and anterior commisure (connecting neuronal pathways between the left and right hemispheres) between gestational day 21 and postnatal day 5, suggesting a role for endocannabinoids in brain development (Romero *et al.*, 1997; Fernandez-Ruiz *et al.*, 2000).

On the other hand, cannabinoid receptor binding in the areas with the densest CB_1 receptor presence in adults (caudate-putamen, cerebral cortex, hippocampus and cerebellum) appears to follow a more classical developmental course, increasing progressively from gestational day 14 throughout the postnatal period until adult levels are reached (Berrendero *et al.*, 1998a; 1999; Fernandez-Ruiz *et al.*, 2000).

These latter data are compatible with the observation in mice that motor depression in an open field and hypoalgesia in response to administration of anandamide or Δ^9-THC are not fully developed until adulthood (Fride and Mechoulam, 1996a). Interestingly, Δ^8-THC at relatively high doses $(18\,mg/m^2)$ prevented vomiting caused by anti-cancer chemotherapy in young children, without producing undesirable cannabimimetic CNS effects (Abrahamov *et al.*, 1995). At such doses one would normally expect very significant cannabimimetic effects, as seen in adults. A tentative explanation based on the data from animals studies (Fride and Mechoulam, 1996a), was offered: on one hand, in the developing organism, the cannabinoid receptor system is not fully developed (hence the lack of cannabimimetic effects). On the other hand, the antiemetic effects are not transmitted through the cannabinoid receptors. The existence of nonspecific effects caused by cannabinoids has been shown previously (Felder *et al.*, 1992; Martin, 1986) and non cannabimimetic cannabinoids with antiemetic properties are indeed known (Feigenbaum *et al.*, 1989). As the vomiting center in the brain, including the chemoreceptor trigger zone in the area postrema, is relatively poor in cannabinoid receptors (Herkenham *et al.*, 1990; Herkenham, 1995; Mailleux and Vanderhaeghen, 1992b), it seems plausible that the antiemetic effects are not receptor mediated, or, at least, are not mediated through the cannabinoid receptor.

Although this clinical success can be explained based on the animal data described above (Fride and Mechoulam, 1996a; Fernandez-Ruiz *et al.*, 2000), the absence of psychoactive cannabinoid effects in juvenile animals and humans, will have to be reconciled with CB_1 receptor density assessments made *post mortem* from human fetal (last trimester) and neonatal brains. These data, although derived from a very limited sample, demonstrated densities in the fetal/neonatal brains, generally higher than (Glass *et al.*, 1997) or similar to (Mailleux and Vanderhaeghen, 1992b) those in adult brains.

Aging

At the other end of the life cycle, a loss of cannabinoid receptors and receptor function were found in the basal ganglia of aging rats (>two years old) (Mailleux and Vanderhaeghen, 1992a; Romero *et al.*, 1998). Since these structures play a pivotal role in motor function (see Sañudo-Peña and Fride, this book), this loss of cannabinoid receptors may explain some of the motor impairments frequently seen in advancing age (Mailleux and Vanderhaeghen, 1992a; Romero *et al.*, 1998).

When additional CB_1 receptor-rich brain regions were investigated in senescent rats, decreases in receptor binding were found in the cerebellum, cerebral cortex and hypothalamus. No changes in CB_1 receptor binding in limbic areas and in the brainstem were detected. On the other hand, increases in CB_1 receptor mRNA were observed in the brainstem (Berrendero *et al.*, 1998b).

EFFECTS OF PERINATAL EXPOSURE TO CANNABINOIDS

The major psychoactive component of marijuana, Δ^9-THC, has been shown to cross the placenta in humans (Blackard and Tennes, 1984) and in rodents (Vardaris *et al.*, 1976). Moreover, during pregnancy, CB_1 receptors have been detected in uterus (Das *et al.*, 1995), blastocytes (Paria *et al.*, 1995) and in fetal tissue from the earliest day studied (day 11, Buckley *et al.*, 1998), thus exposing the developing organism to potential teratogenic effects of cannabinoids throughout gestation. Hence, marijuana consumption by pregnant women which, in North America, has a reported incidence of about 15% (Briggs *et al.*, 1990; Fried and O' Connel, 1987; Wenger *et al.*, 1991), is of major concern vis-a-vis the well being of the offspring of these mothers. However, possible consequences of marijuana smoking during pregnancy for development of the offspring have been found inconsistent as discussed previously (Fried *et al.*, 1999; Fride and Mechoulam, 1996b). For example, transient and permanent changes in prenatally exposed offspring have been observed, including increased incidence of cleft palate (Bloch *et al.*, 1986), changes in somatic growth (Fried, 1976; Hutchings *et al.*, 1987; Wenger *et al.*, 1991), developmental alterations in the pituitary-adrenal axis (Wenger *et al.*, 1991), monoamine neurotransmitters (Walters and Carr, 1988; Dalterio, 1986) and sexual development (Dalterio, 1986;

Dalterio *et al.*, 1986). It has been argued however, that at least some of these effects may be attributed to secondary effects of impaired feeding patterns of the cannabinoid-exposed dams (Hutchings *et al.*, 1989, 1991).

A recent extensive study on gestational drug abuse, including marijuana abuse, did not report increased mortality during the first two years of the offspring's life (Ostrea *et al.*, 1997). However in an ongoing prospective study, Fried and colleagues have found decreased head circumference which only attained statistical significance in early adolescence (Fried *et al.*, 1999).

Likewise, Fried and colleagues, investigating the consequences of prenatal exposure to marijuana in the same well defined human population in Canada, have pointed at subtle but definite cognitive deficiencies in these offspring, becoming apparent only from the age of four years. (Fried, 1996; Fried *et al.*, 1998; Fried and Watkinson, 2000). Thus it seems that prenatal exposure to cannabinoids does result in adverse consequences for the offspring. However, these defects are subtle and are not apparent immediately after birth.

Animal studies have produced evidence that some of the sequelae of prenatal exposure to cannabinoids appear to specifically affect the endocannabinoid-CB_1 receptor system in the offspring. Thus after prenatal exposure to anandamide (Fride and Mechoulam, 1996b) or Δ^9-THC, the density of CB_1 receptors was higher in the brains of offspring of Δ^9-THC-treated mothers (Fride *et al.*, 1996). This was in accordance with observations on the "tetrad", where the performance of the experimental offspring (without any challenge) was similar to that of regular ("naïve") mice, acutely injected with Δ^9-THC (Table 6.1). These observations suggest that prenatal exposure to cannabinoids (Δ^9-THC) causes a specific sensitization of the CB_1 receptor system as apparent in the adult offspring.

In addition, close inspection of a number of studies on the adult offspring of Δ^9-THC-treated mothers, reveals a pattern of alterations which seems to share a number of characteristics with the consequences of prenatal stress (see Fride and Mechoulam, 1996b). These include:

Table 6.1 Effects of prenatal treatment with Δ^9-tetrahydrocannabinol on the adult offspring

Prenatal treatment	(N)	Catalepsy (% immobility)	Locomotion (Number of squares crosssed)	Hypothermia ($\Delta°C$)	Analgesia (hot plate, % MPE[#])
Vehicle	7	15±4	127±11	37.4±0.10	0±08
Δ^9-THC	6	48±5*	78±06*	36.3+0.40*	11±05
Comparative values in normal mice					
Δ^9-THC 5–10 mg/kg	6	57±5	70±16	36.1±0.30	24±09

Source: Pregnant mice (Sabra strain) were injected s. c. daily during the last trimester (week) of pregnancy. The offspring at adulthood (2–3 months old), were subjected to a series of 4 tests (the "tetrad") used to assess cannabinoid activity (Martin *et al.*, 1991). The results were compared to naïve mice, acutely injected with a moderate dose (5–10 mg/kg) of Δ^9-THC. The results indicate that prenatally Δ^9-THC-exposed mice perform "as if" they have been acutely injected with Δ^9-THC; * significantly different from vehicle treated control offspring (p<0.05); [#] reaction times (latency to jump from hot plate or to lick hind paw) was normalized to % MPE=% maximal possible response (see for example Fride and Mechoulam, 1993).

- "Demasculinization" of prenatally stressed males (Dahlof *et al.*, 1978; Ward, 1972) or in males which were prenatally exposed to cannabinoids (Dalterio and Bartke, 1981; Dalterio *et al.*, 1986; Fride and Mechoulam, 1996b).

- Asymmetries (differences between the left and right sides of the brain or body) after prenatal stress (Fride and Weinstock, 1987, 1988, 1989; Sciulli *et al.*, 1979) and after prenatal Δ^9-THC (increase in fluctuating dental asymmetries, Siegel *et al.*, 1977).

- Changes in nigrostriatal dopamine activity after prenatal Δ^9-THC (Fernandez-Ruiz *et al.*, 1992; Navarro *et al.*, 1994; Garcia-Gil *et al.*, 1998) are also reminiscent of the changes in dopamine metabolism caused by prenatal noise and light stress (Fride and Weinstock, 1989).

- Impaired functioning of the HPA axis and increased emotionality after pre- or neonatal cannabinoids (Mokler *et al.*, 1987; Navarro *et al.*, 1995; del Arco *et al.*, 2000) are very similar to the increased corticosterone release, behavioral response to stressful stimuli and increased emotionality observed in prenatally stressed rats (Fride *et al.*, 1985, 1986; Fride and Weinstock, 1988).

- Finally, impairment of the prefrontal cortical dopamine system after prenatal stress (Fride and Weinstock, 1988) may have a common basis with the increased activity and responsivity of the dopamine neurons in the limbic forebrain after perinatal exposure to marijuana or Δ^9-THC (Garcia *et al.*, 1996; Rodriguez de Fonseca *et al.*, 1991). Interestingly, based on their studies on behavioral and cognitive functions in children of mothers who consumed cannabis during pregnancy, Fried and colleagues (Fried, 1996; Fried *et al.*, 1998; Fried and Watkinson, 2000) suggest that perinatal exposure to marijuana may affect prefrontal lobe cognitive ("executive") functioning.

Thus, there is ample evidence to support the hypothesis that prenatal exposure to Δ^9-THC or anandamide induces permanent effects on the offspring which are common to the consequences of stress during pregnancy and are presumably due to changes in the maternal environment. These changes would include alterations in placental blood flow, hormonal and neurotransmitter release, nutritional factors, maternal behavior (Dalterio, 1986) or increases in maternal heart rate and blood pressure and consequently reduced blood supply to the fetus (Zuckerman *et al.*, 1989; see also Fride and Mechoulam, 1996a).

It has also been suggested that perinatal exposure to Δ^9-THC may have effects on the offspring which are common to prenatal influences of various psychotropic drugs including amphetamine, caffeine, cocaine and benzodiazepines (Dow-Edwards, 1989; Fernandez-Ruiz *et al.*, 1992; Fride and Weinstock, 1989; Navarro *et al.*, 1995; Spear *et al.*, 1989). Thus it has been hypothesized that such drugs target development of the same systems, notably the dopaminergic pathways and hypothalamo-pituitary-adrenal (HPA) axis (Fernandez-Ruiz *et al.*, 1992; Wenger *et al.*, 1991; Garcia-Gil *et al.*, 1997).

Taken together one may put forward the hypothesis that prenatal insult, whether due to maternal stress, drugs or other factors, permanently impairs the offsprings' mesolimbic/mesocortical dopamine system. As a result, the mature offspring is less able to cope with stress in adult life, display subtle cognitive deficiencies, and may be more vulnerable to drug addiction. Indeed, as pointed out above (see "Addiction"), the mesolimbic-mesocortical dopamine systems seem to be the substrate of reinforcement and hence of potential drug abuse (Tanda *et al.*, 1997; Gardner and Vorel, 1998).

In accordance with this theory, it has been shown that maternal (gestational and postpartum) exposure to Δ^9-THC caused enhancement of the reinforcing effects of morphine in a place-conditioning test (Navarro *et al.*, 1995) and in morphine self-administration behavior and altered limbic μ opioid binding (Vela *et al.*, 1998). Although the reason why this latter observation was only seen in the female offspring is not clear, a number of sexually dimorphic consequences of perinatal cannabinoid exposure have been reported (Ambrosio *et al.*, 1999; Fride and Mechoulam, unpublished; Navarro *et al.*, 1994).

If these animal data can be generalized to the human population, maternal marijuana smoking during pregnancy may, tragically, increase the likelihood that the offspring will become drug abusers, when reaching adulthood. Of course, genetic predisposition (Oinaivi *et al.*, 1990; Chakrabarti *et al.*, 1998) and postnatal environment (Wakshlak and Weinstock, 1990) are also major factors in determining the eventual phenotype of the offspring. Hence the myriad of relevant factors and their interactions should be the subject of further research into the consequences of maternal drug abuse for the offspring.

FUNCTION OF THE ENDOCANNABINOID SYSTEM IN THE NEONATE

The endocannabinoids appear to play a major role in fertility and reproductive functions, which is described elsewhere in this book (S. K. Dey, "Cannabinoids and reproduction"). Recent work on the medicinal aspects of marijuana have indicated that the plant may be used beneficially to combat weight loss in AIDS and cancer patients by enhancing appetite (Mechoulam *et al.*, 1998b). More recently we have uncovered evidence that the above mentioned observations may just be the "tip of the iceberg" of the critical involvement of the cannabinoid system in growth and in the feeding response of the neonate. Thus, when endocannabinoid function is blocked in newborn mouse pups by the administration of the CB_1 antagonist SR141716A, the pups fail to ingest maternal milk and die within the first week of life. This is not a general toxic effect of SR141716A because co-administration of Δ^9-THC almost completely prevented pup mortality (Fride *et al.*, 1999), indicating that the growth-arresting effect of SR141716A is CB_1 receptor-mediated. Interestingly, the endocannabinoid 2-arachidonoyl glycerol (2-AG) (but not anandamide) reaches peak levels within the first 24 h after birth in the rat (Berrendero *et al.*, 1999; Fernandez-Ruiz *et al.*, 2000). Thus a picture is emerging, where endocannabinoids (at least 2-AG), during a critical period (24 h after birth in the rodent) do not just play a regulatory role, but are an *absolute* requirement for survival and growth of the newborn.

ACKNOWLEDGEMENTS

EF was supported in part, by a grant from the Israeli Ministry of Health. M.C.S.P. was supported by NIH grant #DA 12999.

REFERENCES

Abood, M. E. and Martin, B. R. (1992) "Neurobiology of marijuana abuse," *Trends in Pharmacological Science* **13**: 201–206.

Abrahamov, A., Abrahamov, A. and Mechoulam, R. (1995) "An efficient new cannabinoid antiemetic in pediatric oncology," *Life Sciences* **56**: 2097–2102.

Aceto, M. D., Scates, S. M., Lowe, J. A. and Martin, B. R. (1995) "Cannabinoid precipitated withdrawal by the selective cannabinoid receptor antagonist, SR 141716A," *European Journal of Pharmacology* **282**: R1–R2.

Aceto, M. D., Scates, S. M., Lowe, J. A. and Martin, B. R. (1996) "Dependence on delta 9-tetrahydrocannabinol: studies on precipitated and abrupt withdrawal," *The Journal of Pharmacology and Experimental Therapeutics* **278**: 1290–1295.

Adams, I. B., Compton, D. R. and Martin, B. R. (1998) "Assessment of anandamide interaction with the cannabinoid brain receptor: SR 141716A antagonism studies in mice and autoradiographic analysis of receptor binding in rat brain," *The Journal of Pharmacology and Experimental Therapeutics* **284**: 1209–1217.

Adams, I. B., Ryan, W., Singer, M., Thomas, B. F., Compton, D. R., Razdan, R. K. and Martin, B. R. (1995) "Evaluation of cannabinoid receptor binding and in vivo activities for anandamide analogs," *The Journal of Pharmacology and Experimental Therapeutics* **273**: 1172–1181.

Adams, M. D., Chait, L. D. and Earnhardt, J. T. (1976) "Tolerance to the cardiovascular effects of delta9-tetrahydrocannabinol in the rat," *British Journal of Pharmacology* **56**: 43–48.

Ambrosio, E., Martin, S., Garcia-Lecumberri, C. and Crespo, J. A. (1999) "The neurobiology of cannabinoid dependence: sex differences and potential interactions between cannabinoid and opioid systems," *Life Sciences* **65**: 687–694.

Ameri, A. (1999) "The effects of cannabinoids on the brain," *Progress in Neurobiology* **58**: 315–348.

Ankier, S. I. (1974) "New hot plate tests to quantify antinociceptive and narcotic antagonist activities," *European Journal of Pharmacology* **27**: 1–4.

Arnone, M., Maruani, J., Chaperon, F., Thiebot, M. H., Poncelet, M., Soubrie, P. and Le-Fur, G. (1997) "Selective inhibition of sucrose and ethanol intake by SR 141716, an antagonist of central cannabinoid (CB$_1$) receptors," *Psychopharmacology Berliner* **132**: 104–106.

Ashton, C. H. (1999) "Adverse effects of cannabis and cannabinoids," *British Journal of Anaesthesia* **83**: 637–649.

Axelrod, J. and Felder, C. C. (1998) "Cannabinoid receptors and their endogenous agonist, anandamide," *Neurochemical Research* **23**: 575–581.

Balster, R. L. and Prescott, W. R. (1992) "Delta 9-tetrahydrocannabinol discrimination in rats as a model for cannabis intoxication," *Neuroscience and Biobehavioral Reviews* **16**: 55–62.

Basile, A. S., Hanuš, L. and Mendelson, W. B. (1999) "Characterization of the hypnotic properties of oleamide," *Neuroreport* **10**: 947–951.

Beal, J. E., Olson, R., Lefkowitz, L., Laubenstein, L., Bellman, P., Yangco, B., Morales, J. O., Murphy, R., Powderly, W., Plasse, T .F., Mosdell, K. W. and Shepard, K. V. (1997) "Long-term efficacy and safety of dronabinol for acquired immunodeficiency syndrome-associated anorexia," *Journal of Pain and Symptom Management* **14**: 7–14.

Belue, R. C., Howlett, A. C., Westlake, T. M. and Hutchings, D. E. (1995) "The ontogeny of cannabinoid receptors in the brain of postnatal and aging rats," *Neurotoxicol Teratology* **17**: 25–30.

Ben-Shabat, S., Fride, E., Sheskin, T., Tamiri, T., Rhee, M. H., Vogel, Z., Bisogno, T., De-Petrocellis, L., Di-Marzo, V. and Mechoulam, R. (1998) "An entourage effect: inactive endogenous fatty acid glycerol esters enhance 2-arachidonoyl-glycerol cannabinoid activity," *European Journal of Pharmacology* **353**: 23–31.

Berrendero, F., Garcia-Gil, L., Hernandez, M. L., Romero, J., Cebeira, M., de-Miguel, R., Ramos, J. A. and Fernandez-Ruiz, J. J. (1998a) "Localization of mRNA expression and activation of signal transduction mechanisms for cannabinoid receptor in rat brain during fetal development," *Development* **125**: 3179–3188.

Berrendero, F., Romero, J., Garcia-Gil, L., Suarez, I., De-la-Cruz, P., Ramos, J. A. and Fernandez-Ruiz, J. J. (1998b) "Changes in cannabinoid receptor binding and mRNA levels in several brain regions of aged rats," *Biochimica et Biophysica Acta* **1407**: 205–214.

Berrendero, F., Sepe, N., Ramos, J. A., Di-Marzo, V. and Fernandez-Ruiz, J. J. (1999) "Analysis of cannabinoid receptor binding and mRNA expression and endogenous cannabinoid contents in the developing rat brain during late gestation and early postnatal period," *Synapse* **33**: 181–191.

Bisogno, T., Sepe, N., De Petrocellis, L., Mechoulam, R. and Di Marzo, V. (1997) "The sleep inducing factor oleamide is produced by mouse neuroblastoma cells," *Biochemical and Biophysical Research Communications* **239**: 473–479.

Blackard, C. and Tennes, K. (1984) "Human placental transfer of cannabinoids," *The New England Journal of Medicine* **311**: 797.

Bloch, E., Fishman, R. H., Morrill, G. A. and Fujimoto, G. I. (1986) "The effect of intragastric administration of delta 9-tetrahydrocannabinol on the growth and development of fetal mice of the A/J strain," *Toxicology and Applied Pharmacology* **82**: 378–382.

Boger, D. L., Henriksen, S. J. and Cravatt, B. F. (1998) "Oleamide: an endogenous sleep-inducing lipid and prototypical member of a new class of biological signaling molecules," *Current Pharmaceutical Design* **4**: 303–314.

Boring, D. L., Berglund, B. A. and Howlett, A. C. (1996) "Cerebrodiene, arachidonyl-ethanolamide, and hybrid structures: potential for interaction with brain cannabinoid receptors," *Prostaglandins Leukotrienes and Essential Fatty Acids* **55**: 207–210.

Breivogel, C. S., Childers, S. R., Deadwyler, S. A., Hampson, R. E., Vogt, L. J. and Sim-Selley, L. J. (1999) "Chronic delta9-tetrahydrocannabinol treatment produces a time-dependent loss of cannabinoid receptors and cannabinoid receptor-activated G proteins in rat brain," *The Journal of Neurochemistry* **73**: 2447–2459.

Briggs, G. G., Freeman, R. K. and Yaffe, S. J. (1990) *Drugs in pregnancy and lactation*, 3rd edn, Baltimore: Williams and Williams.

Brodkin, J. and Moerschbaecher, J. M. (1997) "SR141716A antagonizes the disruptive effects of cannabinoid ligands on learning in rats," *The Journal of Pharmacology and Experimental Therapeutics* **282**: 1526–1532.

Buckley, N. E., Hansson, G., Harta, G. and Mezey, E. (1998) "Expression of the CB_1 and CB_2 receptor receptor messenger RNAs during embryonic development in the rat," *Neuro-Science* **82**: 1131–1149.

Burkey, R. T. and Nation, J. R. (1997) "(R)-methanandamide, but not anandamide, substitutes for delta 9-THC in a drug-discrimination procedure," *Experimental and Clinical Psychopharmacology* **5**: 195–202.

Chakrabarti, A., Ekuta, J. E. and Onaivi, E. S. (1998) "Neurobehavioral effects of anandamide and cannabinoid receptor gene expression in mice," *Brain Research Bull* **45**: 67–74.

Chaperon, F., Soubrie, P., Puech, A. J. and Thiebot, M. H. (1998) "Involvement of central cannabinoid (CB_1) receptors in the establishment of place conditioning in rats," *Psychopharmacology Berliner* **135**: 324–332.

Cheer, J. F., Cadogan, A. K., Marsden, C. A., Fone, K. C. and Kendall, D. A. (1999) "Modification of 5-HT2 receptor mediated behaviour in the rat by oleamide and the role of cannabinoid receptors," *Neuropharmacology* **38**: 533–541.

Chen, J. P., Paredes, W., Li, J., Smith, D., Lowinson, J. and Gardner, E. L. (1990) "Delta 9-tetrahydrocannabinol produces naloxone-blockable enhancement of presynaptic basal dopamine efflux in nucleus accumbens of conscious, freely-moving rats as measured by intracerebral microdialysis," *Psychopharmacology Berliner* **102**: 156–162.

Colombo, G., Agabio, R., Diaz, G., Lobina, C., Reali, R. and Gessa, G. L. (1998) "Appetite suppression and weight loss after the cannabinoid antagonist SR 141716," *Life Sciences* **63**: PL113–PL117.

Compton, D. R., Harris, L. S., Lichtman, A. H. and Martin, B. R. (1996) "Marihuana," in *Pharmacological aspects of drug dependence*, edited by C. R. Schuster and M. Kuhar, pp. 83–158. New York: Springer.

Compton, D. R., Rice, K. C., De-Costa, B. R., Razdan, R. K., Melvin, L. S., Johnson, M. R. and Martin, B. R. (1993) "Cannabinoid structure–activity relationships: correlation of receptor binding and in vivo activities," *The Journal of Pharmacology and Experimental Therapeutics* **265**: 218–226.

Cook, S. A., Lowe, J. A. and Martin, B. R. (1998) "CB$_1$ receptor antagonist precipitates withdrawal in mice exposed to Delta 9-tetrahydrocannabinol," *The Journal of Pharmacology and Experimental Therapeutics* **285**: 1150–1156.

Court, J. M. (1998) "Cannabis and brain function," *Journal of Paediatric Child Health* **34**: 1–5.

Cravatt, B. F., Prospero-Garcia, O., Siuzdak, G., Gilula, N. B., Henriksen, S. J., Boger, D. L. and Lerner, R. A. (1995) "Chemical characterization of a family of brain lipids that induce sleep," *Science* **268**: 1506–1509.

Crawley, J. N., Corwin, R. L., Robinson, J. K., Felder, C. C., Devane, W. A. and Axelrod, J. (1993) "Anandamide, an endogenous ligand of the cannabinoid receptor, induces hypomotility and hypothermia in vivo in rodents," *Pharmacology, Biochemistry and Behavior* **46**: 967–972.

Crowley, T. J., Macdonald, M. J., Whitmore, E. A. and Mikulich, S. K. (1998) "Cannabis dependence, withdrawal, and reinforcing effects among adolescents with conduct symptoms and substance use disorders," *Drug and Alcohol Dependence* **50**: 27–37.

Cruciani, R. A., Dvorkin, B., Moris, S. A., Crain, S. M. and Makman, M. H. (1993) "Direct coupling of opiod receptors to both stimulatory and inhibitory guanine nucleotide-binding proteins in F-11 neuroblastoma-sensory neuron hybrid cells," *Proceedings of the National Academy of Sciences of the USA* **90**: 3019–3023.

Dahlof, L. G., Hard, E. and Larsson, K. (1978) "Influence of maternal stress on the development of the fetal genital system," *Physiology and Behavior* **20**: 193–195.

Dalterio, S. L. (1986) "Cannabinoid exposure: effects on development," *Neurobehavioural Toxicology and Teratology* **8**: 345–352.

Dalterio, S. and Bartke, A. (1981) "Fetal testosterone in mice: effect of gestational age and cannabinoid exposure," *The Journal of Endocrinology* **91**: 509–514.

Dalterio, S. L., Michael, S. D. and Thomford, P. J. (1986) "Perinatal cannabinoid exposure: demasculinization in male mice," *Neurobehavioral Toxicology and Teratology* **8**: 391–397.

Das, S. K., Paria, B. C., Chakraborty, I. and Dey, S. K. (1995) "Cannabinoid ligand-receptor signaling in the mouse uterus," *Proceedings of the National Academy of Sciences of the USA* **92**: 4332–4336.

Del-Arco, I., Munoz, R., Rodriguez-De-Fonseca, F., Escudero, L., Martin-Calderon, J. L., Navarro, M. and Villanua, M. A. (2000) "Maternal exposure to the synthetic cannabinoid HU-210: effects on the endocrine and immune systems of the adult male offspring," *Neuroimmunomodulation* **7**: 16–26.

Devane, W. A., Dysarz III, F. A., Johnson, M. R., Melvin, L. S. and Howlett, A. C. (1988) "Determination and characterization of a cannabinoid receptor in rat brain," *Molecular Pharmacology* **34**: 605–613.

Devane, W. A., Hanuš, L., Breuer, A., Pertwee, R. G., Stevenson, L. A., Griffin, G., Gibson, D., Mandelbaum, A., Etinger, A. and Mechoulam, R. (1992) "Isolation and structure of a brain constituent that binds to the cannabinoid receptor," *Science* **258**: 1946–1949.

Diana, M., Melis, M. and Gessa, G. L. (1998) "Increase in meso-prefrontal dopaminergic activity after stimulation of CB$_1$ receptors by cannabinoids," *European Journal of Neurosciences* **10**: 2825–2830.

Diana, M., Melis, M., Muntoni, A. L. and Gessa, G. L. (1998) "Mesolimbic dopaminergic decline after cannabinoid withdrawal," *Proceedings of the National Academy of Sciences of the USA* **95**: 10269–10273.

Di-Marzo, V., Melck, D., Bisogno, T. and De-Petrocellis, L. (1998a) "Endocannabinoids: endogenous cannabinoid receptor ligands with neuromodulatory action," *Trends in Neuroscience* **21**: 521–528.

Di-Marzo, V., Sepe, N., De-Petrocellis, L., Berger, A., Crozier, G., Fride, E. and Mechoulam, R. (1998b) "Trick or treat from food endocannabinoids?" *Nature* **396**: 636–637.

Di Tomaso, E., Beltramo, M. and Piomelli, D. (1996) "Brain cannabinoids in chocolate," *Nature* **382**: 677–678.

Dow-Edwards, D. (1989) "Long-term neurochemical and neurobehavioral consequences of cocaine use during pregnancy," *Annals of the New York Academy of Sciences* **562**: 280–289.

Emrich, H. M., Leweke, F. M. and Schneider, U. (1997) "Towards a cannabinoid hypothesis of schizophrenia: cognitive impairments due to dysregulation of the endogenous cannabinoid system," *Pharmacology, Biochemistry and Behavior* **56**: 803–807.

Feigenbaum, J. J., Richmond, S. A., Weissman, Y. and Mechoulam, R. (1989) "Inhibition of cisplatin-induced emesis in the pigeon by a non-psychotropic synthetic cannabinoid," *European Journal of Pharmacology* **169**: 159–165.

Felder, C. C., Veluz, J. S., Williams, H. L., Briley, E. M. and Matsuda, L. A. (1992) "Cannabinoid agonists stimulate both receptor- and non-receptor-mediated signal transduction pathways in cells transfected with and expressing cannabinoid receptor clones," *Molecular Pharmacology* **42**: 838–845.

Fernandez-Ruiz, J., Berrendero, F., Hernandez, M. L. and Ramos, J. A. (2000) "The endogenous cannabinoid system and brain development," *Trends in Neurosciences* **23**: 14–20.

Fernandez-Ruiz, J. J., Bonnin, A., Cebeira, M. and Ramos, J. A. (1992) "Maternal cannabinoid exposure and brain development: changes in the ontogeny of dopaminergic neurons," in *Neurobiology and neurophysiology of cannabinoids, Biochemistry and physiology of substance abuse*, edited by A. Bartke and L. L. Murphy, pp. 119–164. Boca Raton: CRC press.

Fitton, A. G. and Pertwee, R. G. (1982) "Changes in body temperature and oxygen consumption rate of conscious mice produced by intrahypothalamic and intracerebroventricular injections of delta 9-tetrahydrocannabinol," *The British Journal of Pharmacology* **75**: 409–414.

Foltin, R. W., Brady, J. V. and Fischman, M. W. (1986) "Behavioral analysis of marijuana effects on food intake in humans," *Pharmacology, Biochemistry and Behavior* **25**: 577–582.

French, E. D. (1997) "delta9-Tetrahydrocannabinol excites rat VTA dopamine neurons through activation of cannabinoid CB_1 but not opioid receptors," *Neuroscience Letters* **226**: 159–162.

Fride, E. (1995) "Anandamides: tolerance and cross-tolerance to delta 9-tetrahydrocannabinol," *Brain Research* **697**: 83–90.

Fride, E. (1999) "Anandamide and oleamide: no pot, no sleep," *1999 Symposium on the Cannabinoids, Burlington, Vermont, International Cannabinoid Research Society*, 23.

Fride, E., Barg, J., Levy, R., Saya, D., Heldman, E., Mechoulam, R. and Vogel, Z. (1995) "Low doses of anandamides inhibit pharmacological effects of delta 9-tetrahydrocannabinol," *The Journal of Pharmacology and Experimental Therapeutics* **272**: 699–707.

Fride, E., Ben-Shabat, S. and Mechoulam, R. (1996) "Effects of prenatal Δ^9-THC on mother-pup interactions and CB_1 receptors," *Society of Neurosciences Abstract* **22**: 1683.

Fride, E., Ben-Shabat, S. and Mechoulam, R. (1998a) "Pharmacology of anandamide: Interaction with serotonin systems?" *1998 Symposium on the Cannabinoids, Burlington, Vermont, International Cannabinoid Research Society*, 76.

Fride, E., Ben-Shabat, S. and Mechoulam, R. (1998b) "The anandamide-cannabinoid receptor system and the prefrontal cortex in mice," *1998 Symposium on the Cannabinoids, Burlington, Vermont, International Cannabinoid Research Society*, 44.

Fride, E., Bisogno, T., Di Marzo, V., Bayewitch, M., Vogel, Z. and Mechoulam, R. (1997) "Pharmacology of anandamide: the chocolate-sleep connection," *Society of Neurosciences Abstract* **23**: 1230.

Fride, E., Dan, Y., Feldon, J., Halevy, G. and Weinstock, M. (1986) "Effects of prenatal stress on vulnerability to stress in prepubertal and adult rats," *Physiology and Behavior* **37**: 681–687.

Fride, E., Dan, Y., Gavish, M. and Weinstock, M. (1985) "Prenatal stress impairs maternal behavior in a conflict situation and reduces hippocampal benzodiazepine receptors," *Life Sciences* **36**: 2103–2109.

Fride, E., Ginsburg, Y., Breuer, A., Bisogno, T., Di Marzo, V. and Mechoulam, R. (2001) "Critical role of the endogenous cannabinoid system in mouse pup suckling and growth," *European Journal of Pharmacology* **419**: 207–214.

Fride, E. and Mechoulam, R. (1993) "Pharmacological activity of the cannabinoid receptor agonist, anandamide, a brain constituent," *European Journal of Pharmacology* **231**: 313–314.

Fride, E. and Mechoulam, R. (1996a) "Ontogenetic development of the response to anandamide and delta 9-tetrahydrocannabinol in mice," *Brain Research Developmental Brain Research* **95**: 131–134.

Fride, E. and Mechoulam, R. (1996b) "Developmental aspects of anandamide: ontogeny of response and prenatal exposure," *Psychoneuroendocrinology* **21**: 157–172.

Fride, E. and Weinstock, M. (1987) "Increased interhemispheric coupling of the dopamine systems induced by prenatal stress," *Brain Research Bulletin* **18**: 457–461.

Fride, E. and Weinstock, M. (1988) "Prenatal stress increases anxiety related behavior and alters cerebral lateralization of dopamine activity," *Life Sciences* **42**: 1059–1065.

Fride, E. and Weinstock, M. (1989) "Alterations in behavioral and striatal dopamine asymmetries induced by prenatal stress," *Pharmacology, Biochemistry and Behavior* **32**: 425–430.

Fried, P. A. (1976) "Short and long-term effects of pre-natal cannabis inhalation upon rat offspring," *Psychopharmacology Berliner* **50**: 285–291.

Fried, P. A. (1996) "Behavioral outcomes in preschool and school-age children exposed prenatally to marijuana: a review and speculative interpretation," *NIDA Research Monograph* **164**: 242–260.

Fried, P. A. and O'Connell, C. M. (1987) "A comparison of the effects of prenatal exposure to tobacco, alcohol, cannabis and caffeine on birth size and subsequent growth," *Neurotoxicology and Teratology* **9**: 79–85.

Fried, P. A. and Watkinson, B. (2000) "Visuoperceptual functioning differs in 9- to 12-year olds prenatally exposed to cigarettes and marihuana," *Neurotoxicology and Teratology* **22**: 11–20.

Fried, P. A., Watkinson, B. and Gray, R. (1998) "Differential effects on cognitive functioning in 9- to 12-year olds prenatally exposed to cigarettes and marihuana," *Neurotoxicology and Teratology* **20**: 293–306.

Fried, P. A., Watkinson, B. and Gray, R. (1999) "Growth from birth to early adolescence in offspring prenatally exposed to cigarettes and marijuana," *Neurotoxicology and Teratology* **21**: 513–525.

Gallate, J. E., Saharov, T., Mallet, P. E. and McGregor, I. S. (1999) "Increased motivation for beer in rats following administration of a cannabinoid CB_1 receptor agonist," *European Journal of Pharmacology* **370**: 233–240.

Gaoni, Y. and Mechoulam, R. (1964) "Isolation, structure and partial synthesis of an active constituent of hashish," *Journal of American Chemical Society* **86**: 1646–1647.

Garcia, L., de-Miguel, R., Ramos, J. A. and Fernandez-Ruiz, J. J. (1996) "Perinatal delta 9-tetrahydrocannabinol exposure in rats modifies the responsiveness of midbrain dopaminergic neurons in adulthood to a variety of challenges with dopaminergic drugs," *Drug and Alcohol Dependence* **42**: 155–166.

Garcia-Gil, L., De Miguel, R., Cebeira, M., Villanua, M. A., Ramos, J. A. and Fernandez-Ruiz, J. J. (1997) "Perinatal delta(9)-tetrahydrocannabinol exposure alters the responsiveness of hypothalamic dopaminergic neurons to dopamine-acting drugs in adult rats," *Neurotoxicology and Teratology* **19**: 6477–6487.

Garcia-Gil, L., Ramos, J. A., Rubino, T., Parolaro, D. and Fernandez-Ruiz, J. J. (1998) "Perinatal delta9-tetrahydrocannabinol exposure did not alter dopamine transporter and tyrosine hydroxylase mRNA levels in midbrain dopaminergic neurons of adult male and female rats," *Neurotoxicology and Teratology* **20**: 549–553.

Gardner, E. L. and Lowinson, J. H. (1991) "Marijuana's interaction with brain reward systems: update 1991," *Pharmacology, Biochemistry and Behavior* **40**: 571–580.

Gardner, E. L. and Vorel, S. R. (1998) "Cannabinoid transmission and reward-related events," *Neurobiology of Disease* **5**: 502–533.

Gerard, C. M., Mollereau, C., Vassart, G. and Parmentier, M. (1991) "Molecular cloning of a human cannabinoid receptor which is also expressed in testis," *Journal of Biochemistry* **279**: 129–134.

Gessa, G. L., Melis, M., Muntoni, A. L. and Diana, M. (1998) "Cannabinoids activate mesolimbic dopamine neurons by an action on cannabinoid CB_1 receptors," *European Journal of Pharmacology* **341**: 39–44.

Gingrich, J. A. and Hen, R. (2000) "The broken mouse: the role of development, plasticity and environment in the interpretation of phenotypic changes in knockout mice," *Current Opinion in Neurobiology* **10**: 146–152.

Giuliani, D., Ferrari, F. and Ottani, A. (2000a) "The cannabinoid agonist HU 210 modifies rat behavioural responses to novelty and stress," *Pharmacological Research* **41**: 47–53.

Giuliani, D., Ottani, A. and Ferrari, F. (2000b) "Effects of the cannabinoid receptor agonist, HU 210, on ingestive behaviour and body weight of rats," *European Journal of Pharmacology* **391**: 275–279.

Glass, M., Dragunow, M. and Faull, R. L. (1997) "Cannabinoid receptors in the human brain: a detailed anatomical and quantitative autoradiographic study in the fetal, neonatal and adult human brain," *NeuroScience* **77**: 299–318.

Glass, M. and Felder, C. C. (1997) "Concurrent stimulation of cannabinoid CB_1 and dopamine D2 receptors augments cAMP accumulation in striatal neurons: evidence for a Gs linkage to the CB_1 receptor," *Journal of Neuroscience* **17**: 5327–5333.

Graceffo, T. J. and Robinson, J. K. (1998) "Delta-9-tetrahydrocannabinol (THC) fails to stimulate consumption of a highly palatable food in the rat," *Life Sciences* **62**: PL85–PL88.

Guan, X., Cravatt, B. F., Ehring, G. R., Hall, J. E., Boger, D. L., Lerner, R. A. and Gilula, N. B. (1997) "The sleep-inducing lipid oleamide deconvolutes gap junction communication and calcium wave transmission in glial cells," *The Journal of Cell Biology* **139**: 1785–1792.

Hall, W. and Solowij, N. (1998) "Adverse effects of cannabis," *Lancet* **352**: 1611–1616.

Haney, M., Ward, A. S., Comer, S. D., Foltin, R. W. and Fischman, M. W. (1999) "Abstinence symptoms following smoked marijuana in humans," *Psychopharmacology Berliner* **141**: 395–404.

Hanuš, L., Abu-Lati, S., Fride, E., Breuer, A., Vogel, Z., Shalev, D.E., Kustanovich, I. and Mechoulam, R. (2001) "2-Arachidonyl glyceryl ether, an endogenous agonist of the cannabinoid CB_1 receptor," *Proceedings of the National Academy of Sciences of the USA* **98**: 3662–3665.

Hanuš, L., Breuer, A., Tchilibon S., Shiloah, S., Goldenberg, D., Horowitz, M., Pertwee, R. G., Ross, R. A., Mechoulam, R. and Fride, E. (1999) "HU-308: a specific agonist for CB_2, a peripheral cannabinoid receptor," *Proceedings of the National Academy of Sciences of the USA* **96**: 14228–14233.

Hanuš, L., Gopher, A., Almog, S. and Mechoulam, R. (1993) "Two new unsaturated fatty acid ethanolamides in brain that bind to the cannabinoid receptor," *Journal of Medicinal Chemistry* **36**: 3032–3034.

Hao, S., Avraham, Y. and Berry, E. M. (2000) "Low dose anandamide affects food intake, cognitive function and neurotransmitter levels in diet-restricted mice," *European Journal of Pharmacology* **392**: 147–156.

Heishman, S. J., Arasteh, K. and Stitzer, M. L. (1997) "Comparative effects of alcohol and marijuana on mood, memory, and performance," *Pharmacology, Biochemistry and Behavior* **58**: 93–101.

Herkenham, M. (1995) "Localization of cannabinoid receptors in brain and periphery," in *Cannabinoid receptors*, edited by R. G. Pertwee, pp. 145–166. London: Academic Press.

Herkenham, M., Lynn, A. B., Little, M. D., Johnson, M. R., Melvin, L. S., de-Costa, B. R. and Rice, K. C. (1990) "Cannabinoid receptor localization in brain," *Proceedings of the National Academy of Sciences of the USA* **87**: 1932–1936.

Herve, D., Tassin, J. P., Barthelemy, C., Blanc, G., Lavielle, S. and Glowinski, J. (1979) "Difference in the reactivity of the mesocortical dopaminergic neurons to stress in the BALB/C and C57 BL/6 mice," *Life Sciences* **25**: 1659–1664.

Howlett, A. C. (1995) "Pharmacology of cannabinoid receptors," *Annual Reviews of Pharmacology and Toxicology* **35**: 607–634.

Huidobro-Toro, J. P. and Harris, R. A. (1996) "Brain lipids that induce sleep are novel modulators of 5-hydroxytrypamine receptors," *Proceedings of the National Academy of Sciences of the USA* **93**: 8078–8082.

Hutchings, D. E., Brake, S. C. and Morgan, B. (1989) "Animal studies of prenatal delta-9-tetrahydrocannabinol: female embryolethality and effects on somatic and brain growth," *Annals of the New York Academy of Sciences* **562**: 133–144.

Hutchings, D. E., Fico, T. A., Banks, A. N., Dick, L. S. and Brake, S. C. (1991) "Prenatal delta-9-tetrahydrocannabinol in the rat: effects on postweaning growth," *Neurotoxicology and Teratology* **13**: 245–248.

Hutchings, D. E., Morgan, B., Brake, S. C., Shi, T. and Lasalle, E. (1987) "Delta-9-tetrahydrocannabinol during pregnancy in the rat: I. Differential effects on maternal nutrition, embryotoxicity, and growth in the offspring," *Neurotoxicology and Teratology* **9**: 39–43.

Iversen, L. L. (2000) *The science of marijuana*. New York: Oxford University Press.

Jarbe, T. U. (1978) "Delta 9-tetrahydrocannabinol: tolerance after noncontingent exposure in rats," *Archives of International Pharmacodynamics and Therapeutics* **231**: 49–56.

Jarbe, T. U., Lamb, R. J., Makriyannis, A., Lin, S. and Goutopoulos, A. (1998) "Delta9-THC training dose as a determinant for (R)-methanandamide generalization in rats," *Psychopharmacology Berliner* **140**: 519–522.

Jentsch, J. D., Andrusiak, E., Tran, A., Bowers, M. B. Jr. and Roth, R. H. (1997) "Delta 9-tetrahydrocannabinol increases prefrontal cortical catecholaminergic utilization and impairs spatial working memory in the rat: blockade of dopaminergic effects with HA966," *Neuropsychopharmacology* **16**: 426–432.

Jones, R. T., Benowitz, N. L. and Herning, R. I. (1981) "Clinical relevance of cannabis tolerance and dependence," *Journal of Clinical Pharmacology* **21**: 143S–152S.

Kimura, T., Ohta, T., Watanabe, K., Yoshimura, H. and Yamamoto, I. (1998) "Anandamide, an endogenous cannabinoid receptor ligand, also interacts with 5-hydroxytryptamine (5-HT) receptor," *Biological and Pharmaceutical Bulletin* **21**: 224–226.

Kouri, E. M., Pope, H. G. Jr. and Lukas, S. E. (1999) "Changes in aggressive behavior during withdrawal from long-term marijuana use," *Psychopharmacology Berliner* **143**: 302–308.

Lambert, D. M. and Di-Marzo, V. (1999) "The palmitoylethanolamide and oleamide enigmas: are these two fatty acid amides cannabimimetic?" *Current Medicinal Chemistry* **6**: 757–773.

Landsman, R. S., Burkey, T. H., Consroe, P., Roeske, W. R. and Yamamura, H. I. (1997) "SR141716A is an inverse agonist at the human cannabinoid CB₁ receptor," *European Journal of Pharmacology* **334**: R1–R2.

Ledent, C., Valverde, O., Cossu, G., Petitet, F., Aubert, J. F., Beslot, F., Bohme, G. A., Imperato, A., Pedrazzini, T., Roques, B. P., Vassart, G., Fratta, W. and Parmentier, M. (1999) "Unresponsiveness to cannabinoids and reduced addictive effects of opiates in CB₁ receptor knockout mice," *Science* **283**: 401–404.

Lepore, M., Liu, X., Savage, V., Matalon, D. and Gardner, E. L. (1996) "Genetic differences in delta 9-tetrahydrocannabinol-induced facilitation of brain stimulation reward as measured by a rate-frequency curve-shift electrical brain stimulation paradigm in three different rat strains," *Life Sciences* **58**: PL365–PL372.

Lepore, M., Vorel, S. R., Lowinson, J. and Gardner, E. L. (1995) "Conditioned place preference induced by delta 9-tetrahydrocannabinol: comparison with cocaine, morphine, and food reward," *Life Sciences* **56**: 2073–2080.

Lerner, R. A. (1997) "A hypotheis about the endogenous analogue of general anesthesia," *Proceedings of the National Academy of Sciences of the USA* **94**: 13375–13377.

Little, P. J., Compton, D. R., Mechoulam, R. and Martin, B. R. (1989) "Stereochemical effects of 11-OH-delta 8-THC-dimethylheptyl in mice and dogs," *Pharmacology, Biochemistry and Behavior* **32**: 661–666.

Mackie, K., Devane, W. A. and Hille, B. (1993) "Anandamide, an endogenous cannabinoid, inhibits calcium currents as a partial agonist in N18 neuroblastoma cells," *Molecular Pharmacology* **44**: 498–503.

Mailleux, P. and Vanderhaeghen, J. J. (1992a) "Age-related loss of cannabinoid receptor binding sites and mRNA in the rat striatum," *Neuroscience Letters* **147**: 179–181.

Mailleux, P. and Vanderhaeghen, J. J. (1992b) "Localization of cannabinoid receptor in the human developing and adult basal ganglia. Higher levels in the striatonigral neurons," *Neuroscience Letters* **148**: 173–176.

Mallet, P. E. and Beninger, R. J. (1998a) "Delta9-tetrahydrocannabinol, but not the endogenous cannabinoid receptor ligand anandamide, produces conditioned place avoidance," *Life Sciences* **62**: 2431–2439.

Mallet, P. E. and Beninger, R. J. (1998b) "The cannabinoid CB$_1$ receptor antagonist SR141716A attenuates the memory impairment produced by delta9-tetrahydrocannabinol or anandamide," *Psychopharmacology Berliner* **140**: 11–19.

Martellotta, M. C., Cossu, G., Fattore, L., Gessa, G. L. and Fratta, W. (1998) "Self-administration of the cannabinoid receptor agonist WIN 55,212-2 in drug-naive mice," *Neuro-Science* **85**: 327–330.

Martin, B. R. (1986) "Cellular effects of cannabinoids," *Pharmacological Reviews* **38**: 45–74.

Martin, G. N. (1998) "Human electroencephalographic (EEG) response to olfactory stimulation: two experiments using the aroma of food," *International Journal of PsychoPhysiology* **30**: 287–302.

Martin, B. R., Compton, D. R., Thomas, B. F., Prescott, W. R., Little, P. J., Razdan, R. K., Johnson, M. R., Melvin, L. S., Mechoulam, R. and Ward, S. J. (1991) "Behavioral, biochemical, and molecular modeling evaluations of cannabinoid analogs," *Pharmacology, Biochemistry and Behavior* **40**: 471–478.

Martin, B. R. and Lichtman, A. H. (1998) "Cannabinoid transmission and pain perception," *Neurobiology of Disease* **5**: 447–461.

Martin, B. R., Prescott, W. R. and Zhu, M. (1992) "Quantitation of rodent catalepsy by a computer-imaging technique," *Pharmacology, Biochemistry and Behavior* **43**: 381–386.

Mascia, M. S., Obinu, M. C., Ledent, C., Parmentier, M., Bohme, G. A., Imperato, A. and Fratta, W. (1999) "Lack of morphine-induced dopamine release in the nucleus accumbens of cannabinoid CB(1) receptor knockout mice," *European Journal of Pharmacology* **383**: R1–R2.

Mathias, R. (1997) "Marijuana and tabacco use up again among 8th and 10th graders," *Nida Notes* **12**: 12–13.

Matsuda, L. A., Lolait, S. J., Brownstein, M. J., Young, A. C. and Bonner, T. I. (1990) "Structure of a cannabinoid receptor and functional expression of the cloned cDNA," *Nature* **346**: 561–564.

Mattes, R. D., Engelman, K., Shaw, L. M. and Elsohly, M. A. (1994) "Cannabinoids and appetite stimulation," *Pharmacology, Biochemistry and Behavior* **49**: 187–195.

McGregor, I. S., Issakidis, C. N. and Prior, G. (1996) "Aversive effects of the synthetic cannabinoid CP 55,940 in rats," *Pharmacology, Biochemistry and Behavior* **53**: 657–664.

McLaughlin, C. R. and Abood, M. E. (1993) "Developmental expression of cannabinoid receptor mRNA," *Brain Research Development Brain Research* **76**: 75–78.

Mechoulam, R., Ben-Shabat, S., Hanuš, L., Ligumsky, M., Kaminski, N. E., Schatz, A. R., Gopher, A., Almog, S., Martin, B. R., Compton, D. R. *et al*. (1995) "Identification of an endogenous 2-monoglyceride, present in canine gut, that binds to cannabinoid receptors," *Biochemical Pharmacology* **50**: 83–90.

Mechoulam, R. and Fride, E. (1995) "The unpaved road to the endogenous brain cannabinoid ligands, the anandamides," in *Cannabinoid receptors*, edited by R. Pertwee, pp. 233–258. London: Academic Press.

Mechoulam, R., Fride, E., Hanuš, L., Sheskin, T., Bisogno, T., Di Marzo, V., Bayewitch, M. and Vogel, Z. (1997) "Anandamide may mediate sleep induction," *Nature* **389**: 25–26.

Mechoulam, R., Fride, E. and Di-Marzo, V. (1998a) "Endocannabinoids," *European Journal of Pharmacology* **359**: 1–18.

Mechoulam, R., Hanuš, L. and Fride, E. (1998b) "Towards cannabinoid drugs – revisited," in *Progress in Medicinal Chemistry*, edited by G. P. Ellis, D. K. Luscombe and A. W. Oxford, pp. 199–243. Amsterdam: Elsevier.

Miczek, K. A. and Dixit, B. N. (1980) "Behavioral and biochemical effects of chronic delta 9-tetrahydrocannabinol in rats," *Psychopharmacology Berliner* **67**: 195–202.

Mokler, D. J., Robinson, S. E., Johnson, J. H., Hong, J. S. and Rosecrans, J. A. (1987) "Neonatal administration of delta-9-tetrahydrocannabinol (THC) alters the neurochemical response to stress in the adult Fischer-344 rat," *Neurotoxicology and Teratology* **9**: 321–327.

Munro, S., Thomas, K. L. and Abu-Shaar, M. (1993) "Molecular characterization of a peripheral receptor for cannabinoids," *Nature* **365**: 61–65.

Murillo-Rodriguez, E., Sanchez-Alavez, M., Navarro, L., Martinez-Gonzalez, D., Drucker-Colin, R. and Prospero-Garcia, O. (1998) "Anandamide modulates sleep and memory in rats," *Brain Research* **812**: 270–274.

Navarro, M., Rodriguez-de-Fonseca, F., Hernandez, M. L., Ramos, J. A. and Fernandez-Ruiz, J. J. (1994) "Motor behavior and nigrostriatal dopaminergic activity in adult rats perinatally exposed to cannabinoids," *Pharmacology, Biochemistry and Behavior* **47**: 47–58.

Navarro, M., Rubio, P. and de-Fonseca, F. R. (1995) "Behavioural consequences of maternal exposure to natural cannabinoids in rats," *Psychopharmacology Berliner* **122**: 1–14.

Navarro, M., Rubio, P. and Rodriguez-de-Fonseca, F. (1994) "Sex-dimorphic psychomotor activation after perinatal exposure to (−)-delta 9-tetrahydrocannabinol. An ontogenic study in Wistar rats," *Psychopharmacology Berliner* **116**: 414–422.

Onaivi, E. S., Chakrabarti, A., Gwebu, E. T. and Chaudhuri, G. (1995) "Neurobehavioral effects of delta 9-THC and cannabinoid (CB$_1$) receptor gene expression in mice," *Behavioural Brain Research* **72**: 115–125.

Onaivi, E .S., Green, M. R. and Martin, B. R. (1990) "Pharmacological characterization of cannabinoids in the elevated plus maze," *The Journal of Pharmacology and Experimental Therapeutics* **253**: 1002–1009.

Ostrea, E M. Jr., Ostrea, A. R. and Simpson, P. M. (1997) "Mortality within the first 2 years in infants exposed to cocaine, opiate, or cannabinoid during gestation," *Pediatrics* **100**: 79–83.

Paria, B .C., Das, S. K. and Dey, S. K. (1995) "The preimplantation mouse embryo is a target for cannabinoid ligand-receptor signaling," *Proceedings of the National Academy of Sciences of the USA* **92**: 9460–9464.

Parker, L. A. and Gillies, T. (1995) "THC-induced place and taste aversion in Lewis and Sprague-Dawley rats," *Behavioral Neuroscience* **109**: 71–78.

Paton, W. D. M. and Pertwee, R. G. (1973) "The pharmacology of cannabis in man," in *Marijuana: Chemistry, Pharmacology, Metabolism and Clinical effects*, edited by R. Mechoulam, New York: Academic Press.

Pertwee, R. G. (1972) "The ring test: a quantitative method for assessing the 'cataleptic' effect of cannabis in mice," *The British Journal of Pharmacology* **46**: 753–763.

Pivik, R. T., Zarcone, V., Dement, W. C. and Hollister, L. E. (1972) "Delta-9-tetrahydrocannabinol and synhexl: effects on human sleep patterns," *Clinical Pharmacology and Therapeutics* **13**: 426–435.

Plasse, T. F., Gorter, R. W., Krasnow, S. H., Lane, M., Shepard, K. V. and Wadleigh, R. G. (1991) "Recent clinical experience with dronabinol," *Pharmacology, Biochemistry and Behavior* **40**: 695–700.

Pope, H. G. Jr. and Yurgelun-Todd, D. (1996) "The residual cognitive effects of heavy marijuana use in college students," *Journal of The American Medical Association* **275**: 521–527.

Razdan, R. K. (1986) "Structure–activity relationships in cannabinoids," *Pharmacology Review* **38**: 75–149.

Reibaud, M., Obinu, M. C., Ledent, C., Parmentier, M., Bohme, G. A. and Imperato, A. (1999) "Enhancement of memory in cannabinoid CB_1 receptor knock-out mice," *European Journal of Pharmacology* **379**; R1–R2.

Richardson, J. D., Aanonsen, L. and Hargreaves, K. M. (1997) "SR 141716A, a cannabinoid receptor antagonist, produces hyperalgesia in untreated mice," *European Journal of Pharmacology* **319**, R3–R4.

Rinaldi-Carmona, M., Barth, F., Heaulme, M., Shire, D., Calandra, B., Congy, C., Martinez, S., Maruani, J., Neliat, G., Caput, D. *et al.* (1994) "SR141716A, a potent and selective antagonist of the brain cannabinoid receptor," *FEBS Letters* **350**: 240–244.

Rinaldi-Carmona, M., Barth, F., Millan, J., Derocq, J. M., Casellas, P., Congy, C., Oustric, D., Sarran, M., Bouaboula, M., Calandra, B., Portier, M., Shire, D., Breliere, J. C. and Le-Fur, G. L. (1998) "SR 144528, the first potent and selective antagonist of the CB_2 cannabinoid receptor," *The Journal of Pharmacology and Experimental Therapeutics* **284**: 644–650.

Rodriguez-de-Fonseca, F., Carrera, M. R. A., Navarro, M., Koob, G. F. and Weiss, F. (1997) "Activation of corticotropin-releasing factor in the limbic system during cannabinoid withdrawal," *Science* **276**: 2050–2054.

Rodriguez-de-Fonseca, F., Cebeira, M., Fernandez-Ruiz, J. J., Navarro, M. and Ramos, J. A. (1991) "Effects of pre- and perinatal exposure to hashish extracts on the ontogeny of brain dopaminergic neurons," *NeuroScience* **43**: 713–723.

Rodriguez-de-Fonseca, F., Ramos, J. A., Bonnin, A. and Fernandez-Ruiz, J. J. (1993) "Presence of cannabinoid binding sites in the brain from early postnatal ages," *Neuroreport* **4**: 135–138.

Rodriguez-de-Fonseca, F., Rubio, P., Menzaghi, F., Merlo-Pich, E., Rivier, J., Koob, G. F. and Navarro, M. (1996) "Corticotropin-releasing factor (CRF) antagonist [D-Phe12,Nle21,38,C alpha MeLeu37]CRF attenuates the acute actions of the highly potent cannabinoid receptor agonist HU-210 on defensive-withdrawal behavior in rats," *The Journal of Pharmacology and Experimental Therapeutics* **276**: 56–64.

Romero, J., Berrendero, F., Garcia-Gil, L., de-la-Cruz, P., Ramos, J. A. and Fernandez-Ruiz, J. J. (1998) "Loss of cannabinoid receptor binding and messenger RNA levels and cannabinoid agonist-stimulated [35S]guanylyl-5′O-(thio)-triphosphate binding in the basal ganglia of aged rats," *Neuroscience* **84**: 1075–1083.

Romero, J., Garcia-Palomero, E., Berrendero, F., Garcia-Gil, L., Hernandez, M. L., Ramos, J. A. and Fernandez-Ruiz, J. J. (1997) "Atypical location of cannabinoid receptors in white matter areas during rat brain development," *Synapse* **26**: 317–323.

Romero, J., Garcia-Palomero, E., Castro, J. G., Garcia-Gil, L., Ramos, J. A. and Fernandez-Ruiz, J. J. (1997) "Effects of chronic exposure to delta9-tetrahydrocannabinol on cannabinoid receptor binding and mRNA levels in several rat brain regions," *Brain Research Molecular Brain Research* **46**: 100–108.

Santucci, V., Storme, J. J. and Soubrie, P. L. F. G (1996) "Arousal-enhancing properties of the CB_1 cannabinoid receptor antagonist SR 141716A in rats as assessed by electroencepohalographic spectral and sleep-waking cycle analysis," *Life Sciences* **58**: 103–110.

Sañudo-Peña, M. C. and Fride, E. (2001) "Marijuana and movement disorders," in *Biology of Marijuana*, edited by E. Onaivi, Harwood academic publishers.

Sañudo-Peña, M. C., Romero, J., Seale, G. E., Fernandez-Ruiz, J. J. and Walker, J. M. (2000) "Activational role of cannabinoids on movement," *European Journal of Pharmacology* **391**: 269–274.

Sañudo-Peña, M. C., Tsou, K., Delay, E. R., Hohman, A. G., Force, M. and Walker, J. M. (1997) "Endogenous cannabinoids as an aversive or counter-rewarding system in the rat," *Neuroscience Letters* **223**: 125–128.

Sciulli, P. W., Doyle, W. J., Kelley, C., Siegel, P. and Siegel, M. I. (1979) "The interaction of stressors in the induction of increased levels of fluctuating asymmetry in the laboratory rat," *American Journal of Physical Anthropology* **50**: 279–284.

Segal, M. (1986) "Cannabinoids and analgesia," in *Cannabinoids as therapeutic agents*, edited by R. Mechoulam, pp. 106–120. Boca Raton: CRC Press.

Shire, D., Calandra, B., Bouaboula, M., Barth, F., Rinaldi-Carmona, M., Casellas, P. and Ferrara, P. (1999) "Cannabinoid receptor interactions with the antagonists SR 141716A and SR 144528," *Life Sciences* **65**: 627–635.

Siegel, P., Siegel, M. I., Krimmer, E. C., Doyle, W. J. and Barry III, H. (1977) "Fluctuating dental asymmetry as an indicator of the stressful prenatal effects of delta9-tetrahydrocannabinol in the laboratory rat," *Toxicology and Applied Pharmacology* **42**: 339–344.

Smith, P. B., Compton, D. R., Welch, S. P., Razdan, R. K., Mechoulam, R. and Martin, B. R. (1994) "The pharmacological activity of anandamide, a putative endogenous cannabinoid, in mice," *The Journal of Pharmacology and Experimental Therapeutics* **270**: 219–227.

Sofia, R. D. and Knobloch, L. C. (1976) "Comparative effects of various naturally occurring cannabinoids on food, sucrose and water consumption by rats," *Pharmacology, Biochemistry and Behavior* **4**: 591–599.

Solowij, N. (1995) "Do cognitive impairments recover following cessation of cannabis use?" *Life Sciences* **56**: 2119–2126.

Spear, L. P., Kirstein, C. L. and Frambes, N. A. (1989) "Cocaine effects on the developing central nervous system: behavioral, psychopharmacological, and neurochemical studies," *Annals of the New York Academy of Sciences* **562**: 290–307.

Steiner, H., Bonner, T. I., Zimmer, A. M., Kitai, S. and Zimmer, A. (1999) "Altered gene expression in striatal projection neurons in CB_1 cannabinoid receptor knockout mice," *Proceedings of the National Academy of Sciences of the USA* **96**: 5786–5790.

Struwe, M., Kaempfer, S. H., Geiger, C. J., Pavia, A. T., Plasse, T. F., Shepard, K. V., Ries, K. and Evans, T. G. (1993) "Effect of dronabinol on nutritional status in HIV infection," *The Annals of Pharmacotherapy* **27**: 827–831.

Sulcova, E., Mechoulam, R. and Fride, E. (1998) "Biphasic effects of anandamide," *Pharmacology, Biochemistry and Behavior* **59**: 347–352.

Tanda, G., Loddo, P. and Di-Chiara, G. (1999) "Dependence of mesolimbic dopamine transmission on delta-9-tetrahydrocannabinol," *European Journal of Pharmacology* **376**: 23–26.

Tanda, G., Pontieri, F. E. and Di-Chiara, G. (1997) "Cannabinoid and heroin activation of mesolimbic dopamine transmission by a common mu1 opioid receptor mechanism," *Science* **276**: 2048–2050.

Terranova, J. P., Storme, J. J., Lafon, N. L. F. G., Perio, A. S. P., Rinaldi-Carmona, M., Le-Fur, G. and Soubrie, P. (1996) "Improvement of memory in rodents by the selective CB_1 receptor antagonist, SR141716A," *Psychopharmacology Berliner* **126**: 165–172.

Tjolsen, A. and Hole, K. (1997) "Animal models of analgesia," in *The pharmacology of pain (Handbook of Experimental Pharmacology)*, edited by L. A. Dickenson and J.-M. Besson, pp. 1–19. Heidelberg: Springer.

Tsou, K., Patrick, S. L. and Walker, J. M. (1995) "Physical withdrawal in rats tolerant to delta 9-tetrahydrocannabinol precipitated by a cannabinoid receptor antagonist," *European Journal of Pharmacology* **280**: R13–R15.

Vardaris, R. M., Weisz, D. J., Fazel, A. and Rawitch, A. B. (1976) "Chronic administration of delta-9-tetrahydrocannabinol to pregnant rats: studies of pup behavior and placental transfer," *Pharmacology, Biochemistry and Behavior* **4**: 249–254.

Vela, G., Martin, S., Garcia-Gil, L., Crespo, J. A., Ruiz-Gayo, M., Javier-Fernandez-Ruiz, J., Garcia-Lecumberri, C., Pelaprat, D., Fuentes, J. A., Ramos, J. A. and Ambrosio, E. (1998)

"Maternal exposure to delta9-tetrahydrocannabinol facilitates morphine self-administration behavior and changes regional binding to central mu opioid receptors in adult offspring female rats," *Brain Research* **807**: 101–109.

Venance, L., Piomelli, D., Glowinski, J. and Giaume, C. (1995) "Inhibition by anandamide of gap junctions and intercellular calcium signalling in striatal astrocytes," *Nature* **376**: 590–594.

Wakshlak, A. and Weinstock, M. (1990) "Neonatal handling reverses behavioral abnormalities induced in rats by prenatal stress," *Physiology and Behavior* **48**: 289–292.

Walters, D. E. and Carr, L. A. (1988) "Perinatal exposure to cannabinoids alters neurochemical development in rat brain," *Pharmacology, Biochemistry and Behavior* **29**: 213–216.

Ward, I. L. (1972) "Perinatal stress feminizes and demasculizes the behaviour of males," *Science* **175**: 82–84.

Webster, C. D., LeBlanc, A. E., Marshman, J. A. and Beaton, J. M. (1973) "Acquisitions and loss of tolerance to 1-D9-trans-Tetrahydrocannabinol in rats on an avoidance schedule," *Psychopharmacological Bulletin* **30**: 217–226.

Weidenfeld, J., Feldman, S. and Mechoulam, R. (1994) "Effect of the brain constituent anandamide, a cannabinoid receptor agonist, on the hypothalamo-pituitary-adrenal axis in the rat," *Neuroendocrinology* **59**: 110–112.

Welch, S. P., Dunlow, L. D., Patrick, G. S. and Razdan, R. K. (1995) "Characterization of anandamide- and fluoroanandamide-induced antinociception and cross-tolerance to delta 9-THC after intrathecal administration to mice: blockade of delta 9-THC-induced antinociception," *The Journal of Pharmacology and Experimental Therapeutics* **273**: 1235–1244.

Welch, S. P. (1997) "Characterization of anandamide-induced tolerance: comparison to delta 9-THC-induced interactions with dynorphinergic systems," *Drug Alcohol Dependence* **45**: 39–45.

Wenger, T., Croix, D., Tramu, G. and Leonardelli, J. (1991) "Prenatally administered delta-9-tetrahydrocannabinol temporarily inhibits the developing hypothalamo-pituitary system in rats," *Pharmacology, Biochemistry and Behavior* **40**: 599–602.

Wiley, J. L., Balster, R. and Martin, B. (1995a) "Discriminative stimulus effects of anandamide in rats," *European Journal of Pharmacology* **276**: 49–54.

Wiley, J. L., Lowe, J. A., Balster, R. L. and Martin, B. R. (1995b) "Antagonism of the discriminative stimulus effects of delta 9-tetrahydrocannabinol in rats and rhesus monkeys," *The Journal of Pharmacology and Experimental Therapeutics* **275**: 1–6.

Wiley, J. L., Golden, K. M., Ryan, W. J., Balster, R. L., Razdan, R. K. and Martin, B. R. (1997) "Evaluation of cannabimimetic discriminative stimulus effects of anandamide and methylated fluoroanandamide in rhesus monkeys," *Pharmacology, Biochemistry and Behavior* **58**: 1139–1143.

Williams, C. M. and Kirkham, T. C. (1999) "Anandamide induces overeating: mediation by central cannabinoid (CB$_1$) receptors," *Psychopharmacology Berliner* **143**: 315–317.

Williams, C. M., Rogers, P. J. and Kirkham, T. C. (1998) "Hyperphagia in pre-fed rats following oral D9-THC," *Physiology and Behavior* **65**: 343–346.

Yang, J. Y., Wu, C. F. and Song, H. R. (1999) "Studies on the sedative and hypnotic effects of oleamide in mice," *Arzneimittel-Forschung* **49**: 663–667.

Yost, C. S., Hampson, A. J., Leonoudakis, D., Koblin, D. D., Bornheim, L. M. and Gray, A. T. (1998) "Oleamide potentiates benzodiazepine-sensitive gamma-aminobutyric acid receptor activity but does not alter minimum alveolar anesthetic concentration," *Anesthesia and Analgesia* **86**: 1294–1300.

Zimmer, A., Zimmer, A. M., Hohmann, A. G., Herkenham, M. and Bonner, T. I. (1999) "Increased mortality, hypoactivity, and hypoalgesia in cannabinoid CB$_1$ receptor knock-out mice," *Proceedings of the National Academy of Sciences of the USA* **96**: 5780–5785.

Zuckerman, B., Frank, D. A., Hingson, R., Amaro, H., Levenson, S. M., Kayne, H., Parker, S., Vinci, R., Aboagye, K., Fried, L. E. *et al.* (1989) "Effects of maternal marijuana and cocaine use on fetal growth [see comments]," *The New England Journal of Medicine* **320**: 762–768.

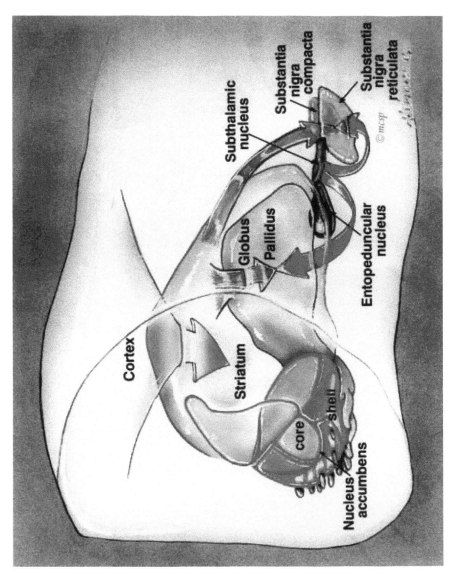

Plate 1 Model of cannabinoid action in the output nuclei of the basal ganglia. Cannabinoids act on both inhibitory (striatal) and excitatory (subthalamic) inputs to the output nuclei (arrows). The noticeable effect on movement will depend on the current level of activity of each input. (*See page 217*)

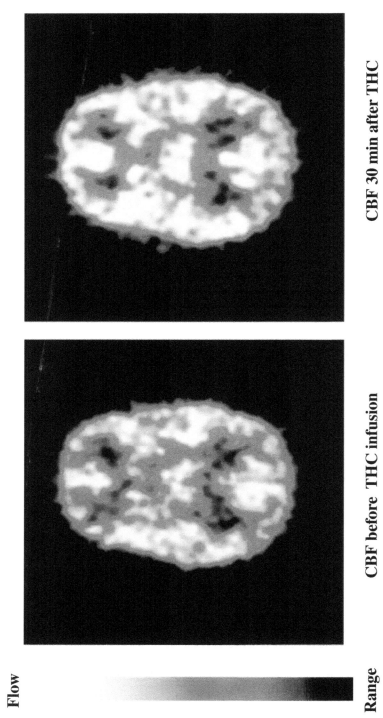

Flow

Range

CBF before THC infusion   **CBF 30 min after THC**

Plate 2 Regional CBF with PET. *(See page 255)*

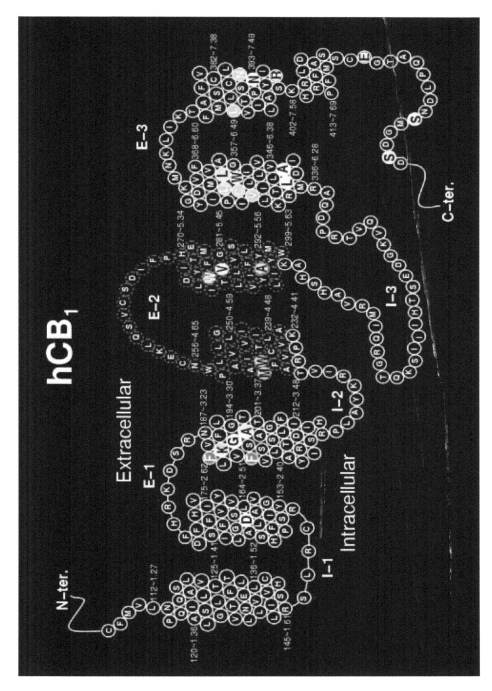

Plate 3 Helix net representation of the human CB$_1$ receptor sequence. (*See page 465*)

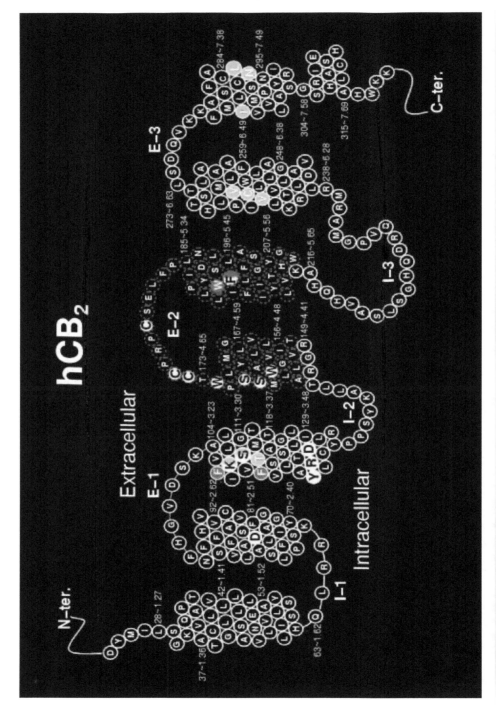

Plate 4 Helix net representation of the human CB$_2$ receptor sequence. (*See page 466*)

Chapter 7

Marijuana and movement disorders

M. Clara Sañudo-Peña and Ester Fride

ABSTRACT

The receptor for cannabinoids is highly expressed in areas of the brain that control movement. Basic research has unveiled a major modulatory role for this new neurochemical system in the brain. Cannabinoids can exert opposite actions at the cellular level and at the level of the circuitry within a nucleus or in between nuclei. They inhibit or excite neurons and oppose excitatory and inhibitory input transmission within the same nucleus. In both cases, their actions depend on the level of ongoing activation and tend to return the system to basal levels of activity. Similar complexity in actions is observed after systemic administration of cannabinoids where relatively low doses enhance motor output while higher doses inhibit movement and furthermore induce catalepsy. Therefore, low doses of cannabinoids would be desirable to treat hypokinesias while higher doses may be relevant in hyperkinetic conditions. Even more desirable would be the use of cannabinoids coadjunctive with lower doses of other drugs already in use in the clinic. This approach might eliminate negative secondary short and long term effects of some of the current treatments. The use of precursors (or inhibitors of degradation or uptake) of the endogenous cannabinoid system would be preferable to the direct administration of agonists. The well established low toxicity, anti-inflammatory, and neuroprotective properties of cannabinoids, together with their neuromodulatory actions in brain motor areas, present this system as an exciting new target for novel pharmacotherapies in movement disorders.

Key Words: basal ganglia, parkinson's disease, huntington's disease, tourette's syndrome, dystonia, spasticity

INTRODUCTION

Marijuana and movement

The recreational and medicinal properties of the marijuana plant have been known by humankind since ancient times. Today, the plant is primarily used for its euphorogenic properties, despite its continuing illegal status. But even for the common user, one of the many physiological effects known to be induced by the active ingredients in the plant is especially striking, that is, the effects on movement. The major recognized effect of marijuana on movement is the induction of hypoactivity. Nevertheless, as will be described below, the hypoactive states induced by these compounds are complex. Even when inducing hypoactivity this is characterized by a state of hyperreflexia uncommon to other drugs inducing depression of the central nervous system. Other than the effects on general activity in humans, cannabinoids are known to induce small impairments in motor coordination (Dewey, 1986; Hall *et al.*, 1994; Hollister, 1986; Martin *et al.*, 1994; Wilson *et al.*, 1994).

Cannabinoids

The marijuana or hemp plant is botanically classified as a member of the family Cannabaceae and the genus *Cannabis*. Accordingly, the compounds with a pharmacological profile similar to the active ingredients in the plant are denominated cannabinoids. The main active principle in the marijuana plant was identified as Δ^9-tetrahydrocannabinol (Mechoulam *et al.*, 1970). Currently, many synthetic compounds are available (Howlett, 1995). Preparations from the plant contain dozens of cannabinergic compounds. Some of these chemicals behave as functional antagonists at the recently cloned cannabinoid receptor while many others are non-cannabinergic compounds (Compton *et al.*, 1993; Dewey, 1986; Feeney, 1979; Grinspoon and Bakalar, 1993; Karniol *et al.*, 1975). This makes it extremely difficult to infer which compound or combination of compounds in the marijuana plant (or receptor/s in the brain) is implicated in producing a physiological effect. Therefore, the experimental data presented in this chapter was obtained employing either Δ^9-tetrahydrocannabinol or one of the synthetic cannabinergic drugs. Nevertheless, several human reports included did employ smoked marijuana.

Cannabinoid receptors and cellular actions of cannabinoids

The first cannabinoid receptor was cloned a decade ago. It was called CB_1. This was soon followed by the cloning of a second subtype of cannabinoid receptor that was accordingly called CB_2. Both receptor subtypes belong to the family of seven transmembrane domain G-protein coupled receptors (Matsuda *et al.*, 1990; Munro *et al.*, 1993). The CB_2 cannabinoid receptor is mainly associated with the immune system. The CB_1 cannabinoid receptor is the receptor expressed by neurons and therefore will be the focus of interest in this chapter. In accordance, the motor effects of cannabinoids seem to be mediated by the neural CB_1 cannabinoid receptor (Rinaldi-Carmona *et al.*, 1994; Compton *et al.*, 1996; Souilhac *et al.*, 1995). In general, the basal effect of activating CB_1 cannabinoid receptors is inhibition of neurotransmission (Howlett *et al.*, 1986; Deadwyler *et al.*, 1993; Mackie and Hille, 1992; Mackie *et al.*, 1995). However, a secondary opposite effect increasing the excitability of cells has also been reported (Axelrod and Felder, 1998; Fride *et al.*, 1995; Glass and Felder, 1997; Netzeband *et al.*, 1999). These opposite effects of the activation of CB_1 receptors have been suggested to depend on the level of neuronal activity. When the cell is activated, the basal action of cannabinoids inhibiting neurotransmission will be obtained. However, when there is already inhibition, the secondary activational action of cannabinoids will be noticeable. Similar opposite effects of cannabinoid action depending on the state of the system are observed within neural circuits and at the systemic level in the control of movement (as discussed below). These complex opposite actions characterize cannabinoids as neuromodulators.

Endogenous cannabinoid system

Although several endogenous compounds that bind to the CB_1 cannabinoid receptor have been isolated from the brain and proposed to be endogenous neurotransmitters (Devane *et al.*, 1992, 1994; Mechoulam *et al.*, 1995; Hanus *et al.*, 2001) there is still no knowledge on which neurons are cannabinergic as in GABAergic or glutamatergic. An enzyme degrading the known endogenous cannabinoid agonists (as well as other amides and esters of arachidonic acid) has been shown to have complementary distribution to some extent to that of the receptor (Egertova *et al.*, 1998; Tsou *et al.*, 1998b). However, the distribution of this enzyme in the central nervous system, unlike that of the receptor, is very widespread. Also, this enzyme is not specific to endogenous cannabinoids. It also breaks down other compounds like oleamide which is involved in sleep and has no direct cannabinergic activity (Cravatt *et al.*, 1995, 1996; Mechoulam *et al.*, 1997). Therefore, its anatomical distribution has provided little insight into the anatomy of the endogenous system. At this time, we are still lacking a complete description of the endogenous system that would greatly facilitate the study of its physiology.

Locomotion

Movement is a fundamental property of animal life. Unlike plants, animals all along the evolutionary scale exhibit movement capabilities to different extents. The highest phylogenetically evolved mammals possess a well-developed locomotor system. This system consists of a complex neuromuscular network. Hundreds of muscles innervated by a similar number of nerves are ultimately under the direct control of the central nervous system. Even the simplest movement requires the coordination of commands all along this network. Thus, the action of any drug on movement would depend on the location site of its receptors in the system. This chapter will mainly focus on the classical motor systems implicated in the control of locomotion disregarding neural control for more discrete types of movements (i.e. movements of the eyes, mouth, etc . . .).

LOCALIZATION OF CB_1 CANNABINOID RECEPTORS IN RELATION TO MOVEMENT

Anatomical techniques

The studies of localization of CB_1 receptors in the brain mentioned below employ three different techniques which produce complementary information. The first one is receptor autoradiography where a radiolabelled agonist to the receptor is used to mark binding sites in the brain. This technique lacks cellular resolution. *In situ* hybridization uses a radiolabelled probe complementary to the mRNA of the receptor under study and allows the identification of neurons that can produce the receptor. This technique gives no information about the localization of the expressed receptor itself. Finally, immunohistochemistry directly labels the localization of the receptor with an antibody raised against it and possesses sub-cellular resolution.

The localization of cannabinoid receptors in the central nervous system is highly conserved among species. We are including, together with the human studies, data from rats, since there has been a lot of research in the cannabinoid field in the laboratory with the latter species. Unless otherwise indicated, the descriptions apply to both rat and human.

Overview of motor systems

The most prominent feature of CB_1 cannabinoid receptor distribution in the brain is its high level of expression in areas involved in the control of movement, which is consistent with the effects of cannabinoids on movement (Glass *et al.*, 1997; Fride *et al.*, 1995; Herkenham *et al.*, 1991a,b,c; Mailleux and Vanderhaeghen, 1992; Pettit *et al.*, 1998; Romero *et al.*, 1995,1996a,b; Sañudo-Peña *et al.*, 1999a; Sulcova *et al.*, 1998; Tsou *et al.*, 1998a). No cannabinoid receptors are found in striate muscle. The brain areas with the highest levels of cannabinoid receptors include the basal ganglia and the cerebellum, classically referred to as the extrapyramidal motor system. The basal ganglia together with the vestibulo-cerebellar system are implicated in the maintenance of muscle tone and equilibrium where they basically exert opposite actions. Together, they provide an adequate basal state for movement to occur. Also, moderate levels of CB_1 binding exist in what was originally called the pyramidal motor system – the one with its origin in the motor cortices. Other motor areas involved in the above mentioned circuits or their outputs are the red nucleus, superior colliculus, and reticulospinal systems. The red nucleus is part of both the pyramidal and cerebellar systems and together they are implicated in the fine tuning of movement. The superior colliculus is a center of sensory integration and provides movement output from both the pyramidal and extrapyramidal systems. The final brain links for the control of movement are the reticulospinal systems of the brainstem. Nevertheless, movement in the brain can be elicited from many other areas not classically associated with motor control such as the amygdala, hippocampus, or hypothalamus (all containing cannabinoid receptors and their encoding mRNA) among others. They add an emotional, motivational or cognitive component to the motor output which is beyond the scope of this review. The final stage for the production of movement is the ventral horn of the spinal cord, the site of origin of the motorneurons ultimately innervating the muscles, which also contains CB_1 cannabinoid receptors. Below is a more accurate description of CB_1 cannabinoid receptors along these circuits.

Basal ganglia

The basal ganglia (masses of grey matter located deep in the cerebral hemispheres) comprise a group of brain nuclei involved in many human movement disorders such as Huntington's disease, Parkinson's disease, Hemiballism, or Tourette's syndrome. It will be discussed extensively in this chapter. The main input structures of the basal ganglia are the striatum and the subthalamic nucleus. Both receive extensive innervation from the cortex and thalamus, project to the output nuclei of the basal ganglia, and are topographically organized. Otherwise, they are very different structures. The striatum (grooved) is a big mass of grey matter while the subthalamic nucleus is small. The former is an inhibitory source to the output nuclei, its main neurotransmitter being GABA. The latter is an excitatory source to the output nuclei, its main transmitter being glutamate. Finally, the striatal output system is mainly silent and only gets activated in a phasic fashion. On the contrary, the excitatory input from the subthalamic nucleus to the output nuclei is tonically active. The output nuclei of the basal ganglia comprise the substantia nigra reticulata, the globus pallidus (external segment of the globus pallidus in primates) and the entopeduncular nucleus (internal segment of the globus pallidus in primates). These nuclei are considered output structures of the basal ganglia towards the production of movement.

Figure 7.1 is a very simplified diagram of the basal ganglia circuits. The output system from the striatum is anatomically segregated and classified as direct and indirect output pathways (Kawaguchi et al., 1990). The direct pathway projects to the substantia nigra reticulata and endopeduncular nucleus (omitted from the diagram in the interest of clarity), both of which project to the motor thalamus and brain stem. The indirect pathway projects to the globus pallidus that in turn sends a massive inhibitory projection to the subthalamic nucleus. The last structure, as previously mentioned, is an excitatory source to the substantia nigra reticulata. Therefore, through excitation of either the striatal direct or indirect output pathway the final result will be inhibition of the substantia nigra reticulata which in turn produces movement. Motor output can be obtained very readily from the substantia nigra reticulata, that has a direct output to the superior colliculus.

Both direct and indirect striatal output pathways utilize the inhibitory neurotransmitter GABA, but these pathways contain markedly different levels of neuropeptides and dopamine receptor subtypes. Striatopallidal neurons contain mainly enkephalin and D_1 dopamine receptors, whereas striatonigral and striato-entopeduncular neurons contain mainly dynorphin, substance P, and D_2 dopamine receptors (Gerfen and Young, 1988; LeMoine et al., 1991; LeMoine and Bloch, 1995). Several lines of evidence suggest that dopamine can induce movement from the striatum by inhibition of the striatopallidal pathway through D_2 receptors or activation of the striatonigral pathway through D_1 receptors (Cooper et al., 1995; Gerfen, 1995). Either action produces movement consistent with the different molecular actions of D_2 and D_1 dopamine receptor types on their respective output pathways (Costall et al., 1972; Cooper et al., 1995; Gerfen et al., 1990, 1991; Graybiel, 1990; Herrera-Marschitz et al., 1985a,b; Herrera-Marschitz and Ungerstedt, 1987; Keefe and Gerfen, 1995; Nisenbaum et al., 1994; Robertson et al., 1989, 1990). Dopamine also has direct actions at the output nuclei. For instance, dopamine acting via D_1 receptors releases GABA from striatal terminals at the substantia nigra reticulata (Graybiel, 1990; You et al., 1994). Both dopamine actions through D_1 dopamine receptors, activation of the striatonigral pathway or at the substantia nigra reticulata itself, stimulate movement by inhibiting this output nucleus. The activation of dopamine D_2 receptors in the striatum will inhibit

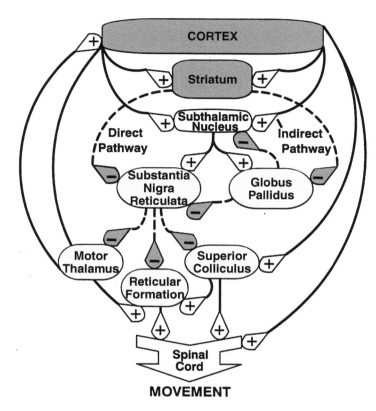

Figure 7.1 Simplified schematic of the basal ganglia circuitry. The gray nuclei and axonal endings illustrate inhibitory actions (−), while the white nuclei and axonal endings illustrate excitatory actions (+). See text for details.

the striatopallidal pathway which in turn will increase the inhibitory action of pallidal neurons both at the subthalamic nucleus and at the substantia nigra reticulata. Both actions would end up inhibiting the substantia nigra reticulata and thus producing movement.

Cannabinoid receptor distribution

Cortex

In general, the cortex shows two bands of neurons that contain mRNA for CB_1 cannabinoid receptors. A superficial band of labeling corresponding to layers II and III, and a deeper band corresponding to layers V and VI. This hybridization signal follows a rostrocaudal decreasing gradient in the rat. Many neurons including pyramidal are lightly labelled in these two bands while less numerous non pyramidal neurons express high levels of labeling. In addition, scarce to highly labeled neurons exist in lamina I and scarce positive neurons for CB_1

mRNA are present in the subcortical white matter (Mailleux and Vanderhaeghen, 1992). The binding sites for cannabinoids are all over the cortex with a bilaminar higher density in layers I and VI following a similar rostrocaudal gradient (Mailleux and Vanderhaeghen, 1992; Jansen *et al.*, 1992; Herkenham *et al.*, 1991c). The cells and fibers expressing the receptor follow a similar bilaminar pattern of distribution to that outlined for the mRNA (Tsou *et al.*, 1998a). The gradient of receptors in the cortex differs between rats and humans. Similar to rats, the levels of receptors in humans are highest in frontal areas. In contrast, the levels are lower in human primary sensory and motor cortex than secondary sensory and motor regions. Additionally, the left (dominant) hemisphere is enriched in receptors in areas associated with verbal language functions (i.e., Wernickes's area) (Glass *et al.*, 1997).

Basal ganglia

The levels of CB_1 cannabinoid receptors in the basal ganglia output nuclei are the highest in the brain. Both the striatum and subthalamic nucleus contain mRNA and express the CB_1 cannabinoid receptor. Also, they are the source of CB_1 cannabinoid receptors to the globus pallidus, entopeduncular nucleus and substantia nigra reticulata (Herkenham *et al.*, 1991b,c; Mailleux and Vanderhaeghen, 1992; Sañudo-Peña and Walker, 1997). The output nuclei show a dense network of afferent immunoreactive fibers but lack intrinsic CB_1 cannabinoid receptors (Figure 7.2). No somas labeled with the antibody against CB_1 cannabinoid receptors or mRNA for the receptor were observed in these nuclei. In general, there is a gradient of CB_1 cannabinoid receptors in the basal ganglia where they are mainly associated with motor versus limbic areas. In this sense, the striatum shows higher density of these receptors in the lateral part and both the dorsal striatum and globus pallidus show higher levels of receptors than the ventral striatum (nucleus accumbens) and ventral pallidum (Glass *et al.*, 1997; Herkenham *et al.*, 1991c; Mailleux and Vanderhaeghen, 1992; Pettit *et al.*, 1998; Tsou *et al.*, 1998a).

Superior colliculus

The antibody against CB_1 cannabinoid receptors labeled cells in the intermediate layers of the superior colliculus. Similarly, higher levels of binding for cannabinoids were observed in the intermediate grey layer of the superior colliculus than in the rest of the layers in this structure. CB_1 cannabinoid receptors are also observed in fibers that form the predorsal bundle and the collicular commisure as well as in numerous transverse fibers that are preferentially concentrated in the superficial grey layer (Herkenham *et al.*, 1991c; Sañudo-Peña *et al.*, 2000a).

Red nucleus

The red nucleus has very sparse binding and labeling for CB_1 cannabinoid receptors (Herkenham *et al.*, 1991c). Labeled fibers could also be seen in the ventral tegmental decussation.

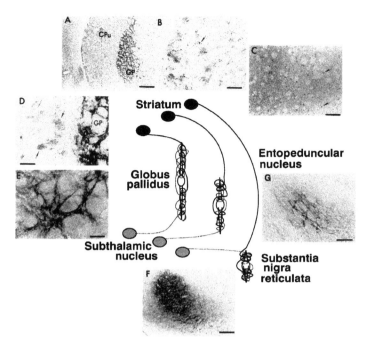

Figure 7.2 The schematic in the center of the figure illustrates the pattern of arborization of striatal and subthalamic fibers within the basal ganglia output nuclei. The terminal portion of striatal and subthalamic axons closely entwine virtually the entire extent of the dendrites of nigral, entopeduncular, and pallidal neurons. Thus the terminal arborizations most closely reflect the patterns of orientation of the dendritic fields of recipient neurons. Most of the neurons located in the pallidum and entopeduncular nucleus have dendritic fields whose main axes is oriented dorsoventrally while most of the substantia nigra neurons have their main axis oriented principally along the rostrocaudal plane (Parent and Hazrati, 1995a,b). (A) Coronal section of the rat brain at the septo-fimbrial level with a very densely stained globus pallidus for CB_1 cannabinoid receptors (GP), while the striatum (CPU) and the cortex are moderately stained. In the CPU and the GP, there is a lateral-to-medial density gradient, the lateral region being the more densely labeled. (B) In the CPU, there are numerous elliptical moderately stained CB_1 immunoreactive neurons, 10–15 mm in long axis, with scant punctuated cytoplasm, and unindented unstained nuclei. They are in the size range and shape typical of medium-sized spiny neurons. The number of these neurons is higher in the rostral and lateral part of the CPU. (C) In the medial part of the rostral CPU, there are numerous intensely stained immunoreactive fiber bundles coursing medially and caudally into the GP. (D) As they approach the GP, the bundles group together and become larger. (E) In the GP, a dense fine unbeaded CB_1-like immunoreactive nerve fiber meshwork is traversed by the large immunonegative fascicles. Similar to the GP a meshwork of fibers inmunolabeled for CB_1 receptors is observed in the entopeduncular nucleus. (G) In the substantia nigra reticulata (SNr) the CB_1-like immunoreactivity is shown as fine dots due to cross section of the projection fibers. (F) In some parasagital sections (not shown), an almost continuous band of CB_1 immunoreactivity can be seen from the CPU through the GP, entopeduncular nucleus to the SNr. The immunoreactivity in these three target areas occurred in unbeaded fine axons; no immunoreactive neurons were found in these areas (Tsou *et al.*, 1998). Scale bars = 500 μm, (A); 200 μm, (C); 50 μm (B–G).

Cerebellum

The molecular layer of the cerebellum exhibits very high levels of binding for cannabinoids. In contrast, the deep cerebellar nuclei show the lowest levels of binding in the brain and do not contain the mRNA for the receptor. The granular layer has sparse levels of receptors but expresses the mRNA for it and the axons of granule cells are a source of receptors to the molecular layer. Also intrinsic neurons in the molecular layer contain mRNA for the cannabinoid receptor. Basket cells express these receptors at their terminals surrounding Purkinje neurons. In contrast, Purkinje neurons do not express the receptor (Glass *et al.*, 1997; Herkenham *et al.*, 1991a,c; Mailleux and Vanderhaeghen, 1992; Pacheco *et al.*, 1993; Pettit *et al.*, 1998; Tsou *et al.*, 1998a).

Thalamus

The thalamus in general has scarce binding for cannabinoids in the rat and low levels in humans where the motor thalamic nuclei (ventral anterior and ventral lateral) show very low densities. Also, it is almost devoid of mRNA for the cannabinoid receptor with the exception of slightly labeled neurons in the medial portion of the lateral habenula (Glass *et al.*, 1997; Herkenham *et al.*, 1991c; Mailleux and Vanderhaeghen, 1992; Tsou *et al.*, 1998a).

Brain stem

Very sparse binding for cannabinoids is observed in the reticular formation. The brain stem overall has slight levels of mRNA for the cannabinoid receptor. The very low levels of these receptors in the brainstem may explain the low toxicity of these compounds (Glass *et al.*, 1997; Herkenham *et al.*, 1991c; Mailleux and Vanderhaeghen, 1992; Tsou *et al.*, 1998a).

Spinal cord and dorsal root ganglion

Numerous fibers labeled for CB_1 receptors are found in the spinal cord and are especially numerous in the dorsal horn. At least half of the binding in the dorsal horn has a presynaptic origin on dorsal root ganglion input terminals. Fibers extending from the white matter into the grey matter are observed under the central canal. Cells with a very light sheet of immunoreactivity are observed throughout the grey matter of the spinal cord. Very lightly labeled neurons and their processes are observed in the ventral horn. The amount of immunoreactivity for CB_1 cannabinoid receptors is much higher in the dorsal root ganglion than in the spinal cord. Many neurochemically different cells in the dorsal root ganglion express the receptor with varying intensities. Also, both dorsal and ventral roots show labeled fibers as does the peripheral nerve which agrees with the reported axonal flow of CB_1 cannabinoid receptors in peripheral nerves (Glass *et al.*, 1997; Herkenham *et al.*, 1991c; Hohmann and Herkenham, 1999a,b; Hohmann *et al.*, 1999; Mailleux and Vanderhaeghen, 1992; Pettit *et al.*, 1998; Sañudo-Peña *et al.*, 1999a; Tsou *et al.*, 1998a).

MOTOR EFFECTS INDUCED BY ACTIVATION OF CB₁ CANNABINOID RECEPTORS

Behavioral measures of movement; turning

When a treatment that affects movement is applied bilaterally in the brain, the result obtained is increase or decrease in locomotor activity depending on the effect of the treatment. However, when the treatment is applied unilaterally (only in one side of the brain) the resulting increase in movement will be expressed as contralateral turning. The imbalance created between the two sides of the brain, higher motor output in the treated versus the untreated side, makes the animal turn towards the opposite side (contralateral) to the manipulation. Conversely, when the treatment administered unilaterally decreases movement, this will be expressed as ipsilateral turning. That is, the animal will turn towards the same (ipsi) side on which the manipulation has taken place. Turning correlates well with the cellular activation or inhibition of basal ganglia nuclei and for that reason has been extensively used in the study of basal ganglia physiology. As mentioned before, the cells in the substantia nigra reticulata are tonically active and serve to inhibit movement. Following the reasoning outlined above, cellular activation in the substantia nigra reticulata leads to inhibition of movement when it occurs bilaterally and ipsilateral turning when it occurs unilaterally (Figure 7.3). Conversely, inhibition of the substantia nigra reticulata increases movement when the treatment is bilateral and produces contralateral turning when the treatment is unilateral. Opposite effects on movement are obtained in the globus pallidus external and internal (entopeduncular nucleus in rodents) segments (Sañudo-Peña *et al.*, 1996, 1998a,b; Burbaud *et al.*, 1998).

Basal ganglia

Within a nucleus

Administration of a cannabinoid agonist into the substantia nigra pars reticulata stimulates movement that is expressed as contralateral rotation (Sañudo-Peña *et al.*, 1996). This effect is possibly due to the inhibition by the cannabinoid agonist of the release of glutamate from subthalamic terminals in the substantia nigra reticulata. Since cannabinoids block the excitatory effect that the stimulation of the subthalamic nucleus has on the activity of the neurons in the substantia nigra reticulata, thus returning the system to basal levels of activity (Sañudo-Peña and Walker, 1997). This action indirectly inhibits the neurons in this nucleus which leads to movement. Also in accordance, cannabinoids administered into the substantia nigra reticulata increase the turning induced by GABA agonists (Wickens and Pertwee, 1995). That is, cannabinoids further inhibit the neurons in this nucleus. The presynaptic inhibition by cannabinoid agonists of neurotransmitter release was inferred from studies reporting inhibition by cannabinoids of calcium channels (Mackie and Hille, 1992; Mackie *et al.*, 1995) and has recently been optically visualized (Kim and Thayer, 2000). The fact that the effect of cannabinoids at subthalamic terminals is the basal notice-

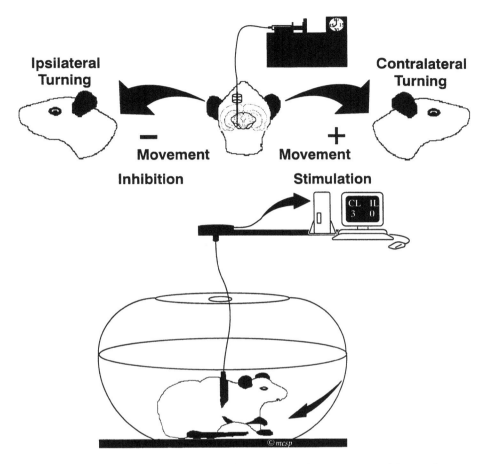

Figure 7.3 Illustration of the turning model employed to study the activation or inhibition of movement after unilateral application of a treatment in a brain site. The animal receives a microinjection of a compound unilaterally and is immediately placed in a rotometer. The ipsilateral or contralateral number of turns is registered by a computer.

able action in the substantia nigra reticulata may be due to the tonic nature of this input (Robledo and Feger, 1991).

By contrast, the inhibitory striatal input to the substantia nigra reticulata is mainly silent (Wilson, 1993), a fact that will mask any action of cannabinoids at this site under basal conditions. However, when the striatal input to the substantia nigra becomes activated, a secondary inhibitory action of cannabinoids at this site is observed. The inhibitory effect of cannabinoids on the striatal input to the substantia nigra reticulata has also been directly observed with electrophysiological techniques (Miller and Walker, 1995; Chan and Yung, 1998; Chan *et al.*, 1998). Again, cannabinoids return the system to basal levels of activity by blocking the inhibition of the neurons in the substantia nigra reticulata induced by the stimulation of the

striatum. Accordingly, the stimulatory effects on movement of intranigral administration of a D$_1$ dopamine agonist, that releases GABA from striatal terminals, can
be reversed by a cannabinoid agonist. Nevertheless, the basal effect of cannabinoids at their primary site of action, the tonically active subthalamic input, is
always noticeable (Sañudo-Peña et al., 1996).

Similar experiments and results were obtained in the globus pallidus which
shares a similar array of inputs with the substantia nigra reticulata. As was the case
for the substantia nigra reticulata, it appears that the output neurons of the globus
pallidus are inhibited by local administration of a cannabinoid which in this structure leads to ipsilateral rotation (Sañudo-Peña and Walker, 1998a). The inhibition
of pallidal neurons by the cannabinoid and in turn the induction of ipsilateral
turning is consistent with the finding that cannabinoids enhance the catalepsy
produced by pallidal microinjections of GABA agonists (Pertwee and Wickens,
1991). A similar mechanism, of inhibition of the tonic excitatory input from the
subthalamic nucleus, as for the substantia nigra reticulata might account for the
turning behavior produced by microinjections of cannabinoids in the globus
pallidus. In support of this possibility, glutamate antagonists that enhanced the
catalepsy induced by cannabinoids (Kinoshita et al., 1994), induced ipsilateral rotation when injected into the globus pallidus (Yamaguchi et al., 1986). The secondary
action of cannabinoids at the inhibitory striatal input is also observable in the globus
pallidus. Cannabinoids block the inhibitory action that the stimulation of the striatum has on the activity of the neurons in the globus pallidus (Miller et al., 1996).

In summary (see Figure 7.4), as previously mentioned for the cellular actions,
cannabinoids can exert opposite effects within the subtantia nigra reticulata or the
globus pallidus where they oppose the actions of both major excitatory and inhibitory sources. The noticeable action will thus depend on the current state of the
system and tend to return the system toward its basal levels of activity. The third
output nucleus, the endopeduncular nucleus, has a similar input arrangement to
the globus pallidus and substantia nigra reticulata which would indicate similar
cannabinoid actions in this nucleus. Both cellular actions (see "Cannabinoid receptors and cellular actions of cannabinoids" this chapter) and the actions at the
circuitry level within a nucleus suggest that cannabinoids play a major modulatory
role in the basal ganglia output system.

Between nuclei

A similar complex pattern of opposite effects on movement to that observed when
cannabinoids are administered within a nucleus of the basal ganglia is observed
when considering cannabinoid actions in the basal ganglia as a whole. In this
sense, cannabinoids activate movement when administered into the substantia
nigra reticulata (Sañudo-Peña et al., 1996) or when administered into the striatum
(Souilhac et al., 1995; Sañudo-Peña et al., 1998a). The last action is apparently
mediated by inhibition of GABA release from recurrent axons of the medium
spiny neurons themselves or from those of striatal interneurons (Szabo et al.,
1998). The opposite action, inhibition of motor output is obtained when they are
microinjected into the globus pallidus (Sañudo-Peña and Walker, 1998) or the
subthalamic nucleus (Miller et al., 1998). Reproducing at the level of the basal

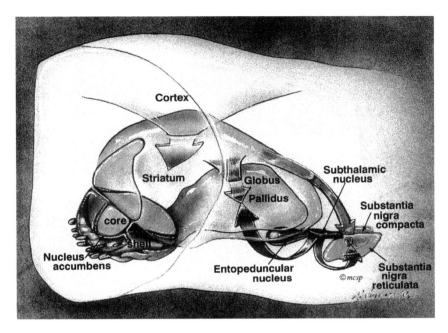

Figure 7.4 Model of cannabinoid action in the output nuclei of the basal ganglia. Cannabinoids act on both inhibitory (striatal) and excitatory (subthalamic) inputs to the output nuclei (arrows). The noticeable effect on movement will depend on the current level of activity of each input. (*See Color plate 1*)

ganglia as a whole the opposite effects observed at the cellular and within single nuclei levels.

Superior colliculus

The output of the substantia nigra reticulata to the superior colliculus is directed towards the site of origin of the crossed descending output system (Williams and Faull, 1988). This output system has its origin in the lateral aspect of the intermediate layers (Redgrave *et al.*, 1986), a site where most, if not all, motor systems in the brain converge. Movement induced from this site acquires biologically relevant significance since this pathway mediates approach/pursuit responses that have been interpreted as predatory behavior (Dean *et al.*, 1989). Unilateral electrical or chemical stimulation of the lateral intermediate layers of the superior colliculus induces contralateral turning (Dean *et al.*, 1986; Speller and Wetsby, 1996). The substantia nigra reticulata is tonically inhibiting this area and removal of this inhibition produces movement (Dean *et al.*, 1989; Williams and Faull, 1988).

Cannabinoids strongly activate movement when administered into the lateral intermediate layers of the superior colliculus (Sañudo-Peña *et al.*, 1998a,b). Since activation of cannabinoid receptors inhibits the excitability of neurons, the direct action of the cannabinoid agonist on the cells of origin of this collicular output system cannot account for the behavioral effect observed. An indirect disinhibitory action is suggested similar to the inhibition of the release of GABA from striatal terminals observed in the substantia nigra reticulata and in the globus pallidus. In this sense, the superior colliculi of both hemispheres are connected by a commisure. The cells of origin of this commisure mostly reside in the intermediate layers and in turn project to almost mirror-symmetrical areas of the contralateral colliculus. At least some of this tecto-tectal cells are GABAergic and some terminate directly on large efferent neurons whose axons originate the crossed collicular output pathway (Behan and Kime, 1996; Magalhaes-Castro et al., 1978). Therefore, an inhibitory action of the cannabinoid agonist on the release of the inhibitory transmitter from these terminals could indirectly excite the cells in the intermediate layers including those originating the predorsal bundle therefore inducing movement ipsilaterally, and those giving rise to the contralateral inhibitory commisural pathway inhibiting movement contralaterally (Sañudo-Peña and Walker, 1998b). Another alternative would be an inhibitory action of the cannabinoid agonist on a different inhibitory collicular input. There is an extensive projection from the hypothalamus to the superior colliculus. Interestingly, the ventromedial nucleus of the hypothalamus innervates the intermediate and deep layers of the superior colliculus and has been referred to as a satiety center suppressing the stereotypic movements that comprise exploration and feeding behaviors (Canteras et al., 1994; Hetherington and Ranson, 1942; Rieck et al., 1986; Stellar and Stellar, 1985). Since stimulation of the crossed output pathway of the superior colliculus stimulates movement resembling predatory behavior including locomotion biting and gnawing movements it could be an output hypothalamic area mediating this behavior. The ventromedial hypothalamic nucleus expresses mRNA for CB_1 receptors (Mailleux and Vanderhaeghen, 1992) suggesting the receptor is expressed on its terminals in target areas as is the case for striatal and subthalamic neurons. The inhibitory action of cannabinoids on the release of neurotransmitters from ventromedial hypothalamic terminals may be releasing the crossed output collicular pathway (Sañudo-Peña et al., 1998c). This would be in accordance with the well known stimulatory action of cannabinoids in feeding (Williams and Rogers, 1998; see Chapter 6, this book). In summary, the crossed output system of the superior colliculus is another brain site, together with the striatum or substantia nigra reticulata, where cannabinoids act to stimulate movement.

Systemic

Cannabinoids administered systemically decrease movement at very low doses in an autoreceptor-like effect followed by a dose dependent stimulating effect on activity that is interrupted by the appearance of rigidity and catalepsy (Figure 7.5, Sañudo-Peña et al., 2000c). The initial inhibition of movement observed with a very low dose of the compound may result from an autoreceptor mechanism because autoreceptors normally show a much higher affinity for ligands and act to inhibit the endogenous system (Disko et al., 1998; Fride et al., 2001; Mao et al., 1996). In accordance with this, the same low dose of Δ^9-tetrahydrocannabinol increased 2-deoxyglucose uptake in a general manner all over the brain (Margulies and Hammer, 1991). Since cerebral metabolism as measured by 2-deoxyglucose uptake reflects activation of terminals as opposite to cell bodies (Schwartz *et al.*, 1979; Kadekaro *et al.*, 1987) and cannabinoid receptor agonists inhibit neurotransmitter release which is the opposite effect observed in the 2-deoxyglucose study, it supports an autoreceptor effect at this dose. These findings suggest that the endogenous cannabinoid system has an activational role in movement.

The dose dependency of the increase in movement by higher doses confirms the stimulatory role of the cannabinoid receptor agonist in movement which is interrupted by the appearance of catalepsy. The dose that induced the higher levels of activity in this study reduced 2-deoxyglucose uptake in a general manner (Margulies and Hammer, 1991) this time, indicating the inhibitory action of the cannabinoid receptor agonists on neurotransmitter release. Accordingly, a decrease in locomotor activity was observed in a study of knockout animals for the neural

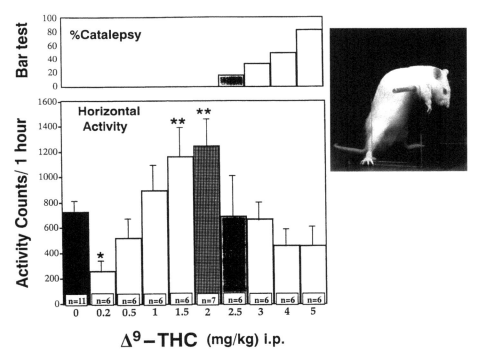

Figure 7.5 Upper part: Percentage of animals within each dose group of Δ^9-tetrahydrocannabinol that exhibit catalepsy. Catalepsy is measured as descent latencies of one minute and over from a bar (see insert at the upper right). Lower part: Dose-curve of systemic administration of Δ^9-tetrahydrocannabinol effects on horizontal activity in rats measured as the mean number of beam breaks ±SEM in an activity monitor that an animal made along an horizontal plane during the hour long observation period. There is an increase in activity with relatively low doses (1–2 mg/kg) of the cannabinoid receptor agonist; * significantly different from the rest of the groups except the ones receiving 4 or 5 mg/kg of Δ^9-tetrahydrocannabinol, $p < 0.05$; ** significantly different from the rest of the groups except the one receiving 1 mg/kg of Δ^9-tetrahydrocannabinol, $p < 0.05$.

cannabinoid receptor which supports the activational role of the endogenous system on movement (Zimmer *et al.*, 1999). The CB_1 receptor knockout displays alterations in the basal ganglia (Steiner *et al.*, 1999). The stimulatory effect of cannabinoids on movement can also be observed immediately after administration of higher doses of cannabinoid receptor agonists while later after administration, high doses of cannabinoid receptor agonists inhibit movement and produce catalepsy (Dewey, 1986; Hollister, 1986). Opposite dose-dependent effects on movement have also been reported for the endogenous cannabinoid anandamide (Sulcova *et al.*, 1998).

As mentioned before, cannabinoids inhibit GABA release from striatal terminals at the output nuclei of the basal ganglia. However, they also inhibit glutamate release from subthalamic terminals at the output nuclei of the basal ganglia, and this action would resemble an effect opposite (GABA-like) to the former one. The increase in movement could be related to the inhibitory effect of cannabinoids on the glutamatergic transmission in the basal ganglia. The subthalamic input is tonically active and thus would be the primary determinant of the action of a cannabinoid. As the dose of the cannabinoid is increased the secondary action of cannabinoids blocking the GABAergic transmission (striatal) in the basal ganglia would produce an opposite inhibitory effect on movement. The simultaneous increase and decrease in motor output may result in rigidity and catalepsy. In summary, cannabinoids have an activational role in movement that is overridden at higher doses by the major modulatory actions of these compounds counteracting opposite systems (Sañudo-Peña *et al.*, 2000c).

THERAPEUTIC USE OF CANNABINOIDS ON MOVEMENT DISORDERS

The low toxicity of cannabinoids, the high levels of their receptors in motor areas and their modulatory actions in the control of movement in the basal ganglia, together with the existence of endogenous ligands, suggest that this new neurochemical system may be important in the normal control of movement and provides a novel aim for pharmacotherapy in movement disorders. The beneficial effects of cannabinoids as antiinflamatories or neuroprotectants (Dewey, 1986; Sinor *et al.*, 2000), though not directly related to the neural mechanisms underlying motor impairments might add extra benefits in counteracting cause/effects in movement disorders. On the other hand, the euphorogenic, sedative or anxiolytic action of cannabinoids (Dewey, 1986; Hall *et al.*, 1994; Hollister, 1986) interfere sometimes with the interpretation of the real improvement in the motor disease, since anxiety might be an important cause/effect of the disorder itself, while euphoria or cognitive impairments are undesirable side effects. The literature regarding clinical use of cannabinoids in the treatment of movement disorders is not very extensive. It has mainly been prompted by popular claims on the beneficial effects of marijuana in different clinical conditions. The recent study by the Institute of Medicine of the National Academy of Sciences on Marijuana and Medicine (Joy *et al.*, 1999) has prompted intense efforts in basic research to reveal the clinical potential of these compounds. The potential of cannabinoids in neurological disorders has recently been extensively reviewed (Consroe, 1998).

Movement disorders

Movement disorders are defined as neurological syndromes where there is either an excess of movement (hyperkinesia, dyskinesia) or a paucity of voluntary and automatic movement unrelated to weakness or spasticity (hypokinesia (decrease amplitude of movement), bradykinesia (slowness of movement), akinesia (loss of movement)). The parkinsonian syndromes are the most common cause of such paucity of movement. Most movement disorders are associated with pathological alterations in the basal ganglia. Bradykinesia, rigidity and rest tremor have been associated with the substantia nigra reticulata. Ballism with the subthalamic nucleus. Chorea and dystonia with the striatum. Other disorders, like intention tremor, ataxia and impair coordination have been associated with the cerebellum, while other tremors have been related to the cerebral cortex, brain stem and spinal cord (Fahn *et al.*, 1997). We will focus only on those movement disorders where cannabinoids might have potential therapeutic uses.

Parkinson's disease

The term parkinsonism is applied to neurologic syndromes in which patients exhibit some combination of tremor at rest, rigidity, bradykinesia, sudden transient inability to move and loss of postural reflexes. It is the most common of all movement disorders. The major pathologic abnormality in Parkinson's disease is the degeneration of the dopaminergic neurons in the substantia nigra compacta. Dopamine replacement is the current most effective treatment for parkinsonism but it becomes less efficient with time and induces undesirable secondary effects such as the development of dyskinesias (Albin *et al.*, 1989; DeLong, 1990; Fahn *et al.*, 1997).

Dopamine denervation, as in Parkinson's disease, chronically shuts down the striatal output system and by all the previously mentioned mechanisms disinhibits both the subthalamic nucleus and the substantia nigra reticulata neurons ultimately inducing rigidity and preventing movement (Obeso *et al.*, 1997). Lesions or inactivation of the subthalamic nucleus are effective in reducing parkinsonian symptoms (Bergman *et al.*, 1990; Benazzouz *et al.*, 1993).

To follow up the data presented above of cannabinoid actions in the basal ganglia, similar studies were performed in a rat model of Parkinson's disease (Sañudo-Peña *et al.*, 1998b). The administration of a cannabinoid agonist into the globus pallidus or striatum of lesioned animals induced the same relative amount of turning and in the same direction as in intact animals. However, in the substantia nigra reticulata, the contralateral turning induced by the cannabinoid agonist increased over nine-fold in the lesioned compared to the intact animals. This result is in accordance with a cannabinoid action at the subthalamonigral pathway, because in this animal model of Parkinson's disease, stimulation of the subthalamic nucleus leads to a greatly exaggerated excitatory response in the substantia nigra reticulata, but not in other areas innervated by the subthalamic nucleus, such as the globus pallidus (Robledo and Feger, 1991). Therefore, the cannabinoid action specifically aimed the hyperactivity of the subthalamic nucleus at the substantia nigra reticulata. Furthermore, cannabinoids and dopamine oppose each other's effects acting through D_2 dopamine receptors in the striatum, globus pallidus and substantia nigra reticulata (Giuffrida et al., 1999; Sañudo-Peña et al., 1996, 1998a; Sañudo-Peña and Walker, 1998) and acting through D_1 receptors in the striatum and substantia nigra reticulata of intact animals (Giuffrida et al., 1999; Sañudo-Peña et al., 1998), but neither effect occurs in the substantia nigra reticulata of animals with dopamine lesions (Sañudo-Peña et al., 1999b). Therefore, the specific inhibitory action of cannabinoids on the hyperactive subthalamonigral terminals may be significant for Parkinson's disease in conjunctive treatment with dopamine agonists.

As mentioned before, the systemic effects of cannabinoids parallel their complex opposite actions at the cellular, circuitry and system levels. Cannabinoid agonists have been reported to have no effect or to further exacerbate parkinsonian symptoms in animals (Sakurai et al., 1985) and to have no effects in humans (Frankel et al., 1990). In the animal study, Δ^9-tetrahydrocannabinol induced ipsilateral turning at lower doses and catalepsy at higher doses. The authors interpreted the result as a stimulatory action of the cannabinoid on movement at the intact side. Alternatively, it might also be interpreted as an exacerbation of the parkinsonian symptoms at the lesioned side. In this respect, it should be pointed out that systemic anticholinergics and NMDA antagonists, both used in the clinic to ameliorate parkinsonian symptoms, induce ipsilateral turning in the same rat model of Parkinson's disease, while the intranigral administration of NMDA antagonists induced contralateral turning (Sakurai et al., 1985; St-Pierre and Bedard, 1994). Therefore, cannabinoids in this animal model are not dissimilar in their actions to other agents already in use in the clinic in this movement disorder.

The study in humans employed smoked marijuana (Frankel et al., 1990). The use of a cannabinoid agonist and in the dose-range where they stimulate movement (see Muller-Vahl et al., 1999a) when administered systemically should be used instead. Alternatively, agents aim at the endogenous system, such as precursors on the biosynthetic pathways or inhibitors of breakdown or reuptake could be used. The major modulatory action of cannabinoids might be able to block the development of the undesirable effects of dopaminergic agents, the most powerful antiparkinsonian drugs, and/or lower the amounts of these drugs needed to obtain improvement, again increasing the therapeutic extent in time during dopamine replacement. It should also be noted that cannabinoids have been proven to be effective as muscle relaxant agents, and that effect could be mediated centrally (basal ganglia) but also at the peripheral level (see cannabinoid effects on spasticity). Finally, the known effect of cannabinoids inhibiting the release of acetylcholine (Domino, 1981; Gessa et al., 1998) might contribute to their beneficial effects in Parkinson's disease.

Dystonia

This is the most common movement disorder after the parkinsonian syndromes. Dystonia is defined as a syndrome of sustained muscle contractions, frequently causing twisting and repetitive movements or abnormal postures. Dystonic contractions are usually aggravated during voluntary movement (action dystonia). Hemidystonias are associated with abnormalities in the contralateral striatum (Marsden *et al.*, 1985). In dystonia, not only the striatopallidal pathway degenerates (as in Huntington's chorea) but also the striatonigral pathway (Reiner *et al.*, 1988). Dystonia can be observed early in the onset of parkinsonism and at the late stages of Huntington's disease and is treated with dopaminergic, anticholinergic, GABA-B or antihistaminergic agents and more rarely with antidopaminergic compounds (Cardoso, 1997). The beneficial effects of cannabinoids in dystonic movement disorders in humans (Consroe *et al.*, 1986; Sandyk *et al.*, 1986) are in accordance with animal data where a full cannabinoid agonist (WIN55,212-2) produced antidystonic effects on its own and lowered the effective levels of other antidystonic drugs in mutant dystonic hamsters (Richter and Loscher, 1994). The lack of inhibitory striatal control over the output nuclei could be compensated with the primary inhibitory action of cannabinoids acting at subthalamic terminals. Also, the inhibitory action of cannabinoids on the release of acetylcholine and their similar actions to GABA-B agonists could contribute to their effectiveness in reducing dystonia (Hampson and Deadwyler, 1999; Romero *et al.*, 1996a; Sañudo-Peña *et al.*, 2000b).

Tourette's syndrome

Tics are involuntary, rapid, brief, purposeless, repetitive, and stereotyped movements and vocalizations that compose the core symptoms of Tourette's syndrome. Tics are usually increased by emotional states. The neuroanatomical substrate involve cortico-limbic-basal ganglia and brain stem sites. The former areas mediating the cognitive and motor symptoms while the last site being important in the vocalizations (Eidelberg *et al.*, 1997; Singer, 1997). Hyperactivity of the dopaminergic system has been proposed as cause of the syndrome since it can be treated with antidopaminergic compounds. Benzodiazepines are also effective. Several reports claimed beneficial effects of smoked marijuana in attenuating Tourette's syndrome (Sandyk and Awerbuch, 1988; Heeming and Yellowlees, 1993). Recently, the first clinical trial employing Δ^9-tetrahydrocannabinol confirmed the reduction in all the symptoms of the syndrome (Muller-Vahl *et al.*, 1999b). The opposite effects of dopamine and cannabinoid agonists in the basal ganglia agree with the counteraction by cannabinoids of an excessive dopaminergic function, as does cannabinoid enhancement of benzodiazepine actions in motor systems (Pertwee and Greentree, 1988; Pertwee and Wickens, 1991).

Huntington's disease

Huntington's disease is an autosomal dominant condition and the leading cause of adult onset chorea. Chorea (dance) refers to an irregular, nonrhythmic, rapid, unsustained involuntary movement that flows from one body part to another. The timing, direction, and distribution are not patterned but random and changing. Degeneration of the striatopallidal pathway is the major feature in Huntington's disease which is consistent with the loss of cannabinoid receptors observed in the globus pallidus of Huntington's disease patients in early stages of the disease (Richfield and Herkenham, 1994). The loss of inhibition of the pallidal output neurons is associated with the chorea observed in this early stages (Albin *et al.*, 1992; Reiner *et al.*, 1988). As mentioned above, the primary action of cannabinoids in the globus pallidus is inhibition of pallidal neurons by their actions at subthalamic terminals. Therefore, providing an alternative mechanism to inhibit pallidal output neurons and in turn inhibit the excessive production of movement seen in chorea. However, no effect (Consroe *et al.*, 1991) or enhancement of choreatic movements was observed in Huntington's disease after the administration of cannabinoids (Muller-Vahl *et al.*, 1999a). The former study employed an atypical cannabinoid analog with extremely low affinity for the cannabinoid receptor (Compton *et al.*, 1993; Dewey, 1986; Feeney, 1979; Karniol *et al.*, 1975) and the latter study showed the activational role of cannabinoids in movement when administered systemically. This dose-range property could be exploited in hypokinetic disorders such as parkinsonism. A higher dose of a cannabinoid agonist, more in the dose range that was active reducing the symptoms of Tourette's syndrome (Muller-Vahl *et al.*, 1999a), would be appropriate in hyperkinetic disorders such as chorea.

Spasticity

Spasticity is characterized by an increase in muscle tone with a velocity-dependent increase when the muscle is passively stretched. It develops after supraspinal or spinal lesions of descending motor systems. The increased muscle tone of spastic muscles is caused by reflex activity at the level of the spinal cord due to the removal of supraspinal inhibitory influences on these circuits (North, 1991). Therefore, the site of action of antispastic drugs is primarily at the spinal cord, dorsal root ganglion and/or neuromuscular junction. Alternatively, the output of the basal ganglia is a central site controlling spasticity which is relevant in conditions such as multiple sclerosis where no complete removal of supraspinal control exists (Turski *et al.*, 1990). So far, the most consistently reported beneficial effect of cannabinoids in movement disorders is as a muscle relaxant. Marijuana improved the spasticity that resulted from paraplegia, quadriplegia, or multiple sclerosis (Clifford, 1983; Consroe *et al.*, 1997; Grinspoon and Bakalar, 1993; Meinck *et al.*, 1989; Petro and Ellenberger, 1981). Recently, the data obtained in human studies on the antispastic effects of cannabinoids has been reproduced in an animal model of multiple sclerosis. In this study, the cannabinoid agonist WIN55,212-2 ameliorated both tremor and spasticity while cannabinoid antagonism exacerbated both symptoms (Baker *et al.*, 2000). Furthermore, cannabinoids inhibit the development of chronic relapsing experimental allergic encephalomyelitis in rodents which is a model of multiple sclerosis (Baker *et al.*, 1990). The muscle relaxant effect of cannabinoids at the spinal level could be due to inhibition of primary afferent input to spinal cord reflex circuits as well as direct inhibition of motorneurons at the spinal or peripheral sites. At the supraspinal level, cannabinoids' primary action at the substantia nigra reticulata is inhibitory as mentioned above, and inhibition of this site decreases muscle tone (Turski *et al.*, 1990).

SUMMARY

The major modulatory actions of cannabinoids in brain areas involved in many human movement disorders is consistent with their putative therapeutic use. Cannabinoids have been recently proven beneficial in the treatment of dystonia. Also, they improve the motor and cognitive impairments in Tourette's syndrome. Other movement disorders where these compounds might have therapeutic uses include Parkinson's and Huntington's disease. Finally, the well known muscle relaxant properties of cannabinoids are relevant in the treatment of spasticity that results from paraplegia, quadriplegia, or multiple sclerosis. The low toxicity, anxiolytic, analgesic, antiinflamatory and neuroprotectant properties of cannabinoids favor the design of novel therapies targeting this novel neurochemical system.

ACKNOWLEDGMENTS

M.C.S.P. dedicates this chapter to the late Prof. Kang Tsou with gratitude for his generosity sharing his immense knowledge and for his continuous encouragement. E.F. was supported by a grant from the Israeli Ministry of Health and M.C.S.P. was supported by a grant from the National Institute of Health (DA12999) M.C.S.P. would also like to thank N.I.H. for the gift of Δ^9-THC, Prof. K. Mackie (Washington University) for providing the antibody against CB_1 receptors, Sanofi Recherche (Montpellier, France) for the gift of SR141716A, and Pfizer (Groton, CT) for the gift of CP55,940.

REFERENCES

Albin, R. L., Reiner, A., Anderson, K. D., Dure IV, L. S., Handelin, B., Balfour, R., Whetsell, W. O. Jr., Penney, J. B. and Young, A. B. (1992) "Preferential loss of striato-external pallidal projection neurons in presymptomatic Huntington's disease," *Annals of Neurology* **31**: 425–430.

Albin, R. L., Reiner, A. B. and Penney, J. B. (1989) "The functional anatomy of basal ganglia disorders," *Trends in Neuroscience* **12**: 366–375.

Axelrod, J. and Felder, C. C. (1998) "Cannabinoid receptors and their endogenous agonist, anandamide," *Neurochemical Research* **23**: 575–581.

Baker, D., Pryce, G., Croxford, J. L., Brown, P., Pertwee, R. G., Huffman, J. W. and Layward, L. (1990) "Induction of chronic relapsing experimental allergic encephalomyelitis in Biozzi mice," *Journal of Neuroimmunology* **28**: 261–270.

Baker, D., Pryce, G., Croxford, J. L., Brown, P., Pertwee, R. G., Huffman, J. W. and Layward, L. (2000) "Cannabinoids control spasticity and tremor in a multiple sclerosis model," *Nature* **404**: 84–87.

Behan, M. and Kime, N. M. (1996) "Spatial distribution of tectotectal connections in the cat," *Progress in Brain Research* **112**: 131–142.

Benazzouz, A., Gross, C., Feger, J., Boraud, T. and Bioulac, B. (1993) "Reversal of rigidity and improvement of motor performance by subthalamic high-frequency stimulation in MPTP-treated monkeys," *European Journal of Neuroscience* **5**: 382–389.

Bergman, H., Wichmann, T. and DeLong, M. R. (1990) "Reversal of experimental parkinsonism by lesions of the subthalamic nucleus," *Science* **249**: 1436–1438.

Burbaud, P., Bonnet, B., Guehl, D., Lagueny, A. and Bioulac, B. (1998) "Movement disorders induced by gamma-aminobutyric agonist and antagonist injections into the internal globus pallidus and substantia nigra pars reticulata of the monkey," *Brain Research* **780**: 102–107.

Canteras, N. S., Simerly, R. B. and Swanson, L. W. (1994) "Organization of projections from the ventromedial nucleus of the hypothalamus: a phaseolus vulgaris-leucoagglutinin study in the rat," *Journal of Comparative Neurology* **348**: 41–79.

Cardoso, F. (1997) "Dystonia and dyskinesia," *The Psychiatric Clinics of North America* **20**: 821–838.

Chan, P. K. Y. and Yung, W. H. (1998) "Occlusion of the presynaptic action of cannabinoids in rat substantia nigra pars reticulata by cadmiun," *Neuroscience Letters* **249**: 57–60.

Chan, P. K. Y., Chan, S. C. Y. and Yung, W.-H. (1998) "Presynaptic inhibition of GABAergic inputs to rat substantia nigra pars reticulata neurons by a cannabinoid agonist," *Neuroreport* **9**: 671–675.

Clifford, D. B. (1983) "Tetrahydrocannabinol for treatment of multiple sclerosis," *American Neurology* **13**: 669–671.

Compton, D. R., Rice, K. C., De Costa, B. R., Razdan, R., Melvin, L. S., Johnson, M. R. and Martin, B. R. (1993) "Cannabinoid structure–activity relationships: correlation of receptor binding and in vivo activities," *Journal of Pharmacology and Experimental Therapeutics* **265**: 218–226.

Compton, D. R., Aceto, M. D., Lowe, J. and Martin, B. R. (1996) "In vivo characterization of a specific cannabinoid receptor antagonist (SR141716A): inhibition of D9-tetrahydrocannabinol-induced responses and apparent agonist activity," *Journal of Pharmacology and Experimenetal Therapeutics* **277**: 586–594.

Consroe, P. (1998) "Brain cannabinoid systems as targets for the therapy of neurological disorders," *Neurobiology of Disease* **5**: 534–551.

Consroe, P., Laguna, J., Allender, J., Snider, S., Stern, L., Sandyk, R., Kennedy, K. and Schram, K. (1991) "Controlled clinical trial of cannabidiol in Huntington's disease," *Pharmacology, Biochemistry and Behavior* **40**: 701–708.

Consroe, P., Musty, R., Rein, J., Tillery, W. and Pertwee, R. (1997) "The Perceived effects of smoked cannabis on patients with multiple sclerosis," *European Neurology* **38**: 44–48.

Consroe, P., Sandyk, R. and Snider, S. R. (1986) "Open label evaluation of cannabidiol in dystonic movement disorders," *International Journal of Neuroscience* **30**: 277–282.

Cooper, A. J., Moser, B. and Mitchell, I. J. (1995) "A subset of striatopallidal neurons are Fosimmunopositive following acute monoamine depletion in the rat," *Neuroscience Letters* **187**: 189–192.

Costall, B., Naylor, R. J. and Olley, J. E. (1972) "Catalepsy and circling behaviour after intracerebral injections of neuroleptic, cholinergic and anticholinergic agents into the caudate-putamen, globus pallidus and substantia nigra of rat brain," *Neuropharmacology* **11**: 645–663.

Cravatt, B. F., Giang, D. K., Mayfield, S. P., Bolger, D. L., Lerner, R. A. and Guila, N. B., (1996) "Molecular characterization of an enzyme that degrades neuromodulatory fatty acid amides," *Nature* **384**: 84–87.

Cravatt, B. F., Prospero-Garcia, O., Siuzdak, G., Gilul, N., Henriksen, S., Bolger, D. L. and Lerner, R. A. (1995) "Chemical characterization of a family of brain lipids that induce sleep," *Science* **268**: 1506–1509.

Deadwyler, S. A., Hampson, R. E., Bennett, B. A., Edwards, T. A., Mu, J., Pacheco, M. A., Ward, S. J. and Childers, S. R. (1993) "Cannabinoids modulate potassium current in culture hippocampal neurons," *Receptors Channels* **1**: 121–134.

Dean, P., Redgrave, P. and Westby, G. W. M. (1989) "Event or emergency? two response systems in the mammalian superior colliculus," *Trends in Neuroscience* **12**: 137–147.

Dean, P., Redgrave, P., Sahibzaba, N. and Tsuji, K. (1986) "Head and body movements produced by electrical stimulation of the superior colliculus in rats: effects of interruption of crossed tectoreticulospinal pathway," *Neuroscience* **19**: 367–380.

DeLong, M. R. (1990) "Primate models of movement disorders of basal ganglia origin," *Trends in Neuroscience* **13**: 281–285.

Devane, W. A. and Axelrod, J. (1994) "Enzymatic synthesis of anandamide, an endogenous ligand for the cannabinoid receptor, by brain membranes," *Proceedings of the National Academy of Sciences of USA* **91**: 6698–6701.

Devane, W. A., Hanus, L., Breuer, A., Pertwee, R. G., Stevenson, L. A., Griffin, G., Gibson, D., Mandelbaum, A., Etinger, A. and Mechoulam, R. (1992) "Isolation and structure of a brain constituent that binds to the cannabinoid receptor," *Nature* **258**: 1946–1949.

Dewey, W. L. (1986) "Cannabinoid pharmacology," *Pharmacological Reviews* **38**: 151–178.

Disko, U., Haaf, A., Heimrich, B. and Jackisch, R. (1998) "Postnatal development of muscarinic autoreceptors modulating acetylcholine release in the septohippocampal cholinergic system II. Cell body region: septum," *Brain Research. Developmental Brain Research* **108**: 31–37.

Domino, E. F. (1981) "Cannabinoids and the cholinergic system," *Journal of Clinical Pharmacology* **21**: 249S–255S.

Eidelberg, D., Moeller, J. R., Antonini, A., Kazumata, K., Dhawan, V., Budman, C. and Feigin, A. (1997) "The metabolic anatomy of Tourette's syndrome," *Neurology* **48**: 927–934.

Egertova, M., Giang, D. K., Cravatt, B. F. and Elphick, M. R. (1998) "A new perspective on cannabinoid signalling: complementary localization of fatty acid amide hydrolase and the CB_1 receptor in rat brain," *Proceedings of the Royal Society of London Series B* **265**: 2081–2085.

Frankel, J. P., Hughes, A., Lees, A. J. and Stern, G. M. (1990) "Marijuana for Parkinsonian Tremor," *Psychiatry* **53**: 436.

Fride, E., Barg, J., Levy, R., Saya, D., Heldman, E., Mechoulam, R. and Vogel, Z. (1995) "Low doses of anandamides inhibit pharmacological effects of delta 9-tetrahydrocannabinol," *The Journal of Pharmacology and Experimental Therapeutics* **272**: 699–707.

Fride, E., Ginzburg, Y., Breuer, A., Bisogno, T., Di Marzo, V., Mechoulam, R. (2001) "Critical role of the endogenous cannabinoid system in mouse pup suckling and growth," *European Journal of Pharmacology* **419**: 207–214.

Fahn, S., Greene, P. E., Ford, B. and Bressman, S. B. (eds) (1997) *Handbook of movement disorders*. Blackwell Science, Current Medicine, Inc., Philadelphia, PA, USA.

Feeney, D. M. (1979) "Marijuana and epilepsy: paradoxical anticonvulsant and convulsant effects," in *Marihuana: biological effects. Analysis, metabolism, cellular responses, reproduction and brain*, edited by G. G. Nahas and W. D. M. Paton, pp. 643–657. Oxford: Pergamon Press.

Gerfen, C. R. (1995) "Dopamine receptor function in the basal ganglia," *Clinical Neuropharmacology* **18**: S162–S177.

Gerfen, C. R., Engber, T. M. and Mahan, L. C. (1990) "D1 and D2 dopamine receptor-regulated gene expression of striatonigral and striatopallidal neurons," *Science* **250**: 1429–1432.

Gerfen, C. R., McGinty, J. F. and Young, W. S. (1991) "Dopamine differentially regulates dynorphin, substance P, and enkephalin expression in striatal neurons: in situ hybridization histochemical analysis," *Journal of Neuroscience* **11**: 1016–1031.

Gerfen, C. R. and Young, W. S. (1988) "Distribution of striatonigral and striatopallidal peptidergic neurons in both patch and matrix compartments: an in situ hybridization histochemistry and fluorescent retrograde tracing study," *Brain Research* **460**: 161–167.

Gessa, G. L., Casu, M. A., Carta, G. and Mascia, M. S. (1998) "Cannabinoids decrease acetylcholine release in the medial-prefrontal cortex and hippocampus, reversal by SR141716A," *European Journal of Pharmacology* **355**: 119–124.

Giuffrida, A., Parsons, L. H., Kerr, T. M., Rodriguez de Fonseca, F., Navarro, M. and Piomelli, D. (1999) "Dopamine activation of endogenous cannabinoid signalling in dorsal striatum," *Nature Neuroscience* **2**: 358–363.

Glass, M., Dragunow, M. and Faull, R. L. M. (1997) "Cannabinoid receptors in the human brain: a detailed anatomical and quantitative autoradiographic study in the fetal, neonatal and adult human brain," *Neuroscience* **77**: 299–318.

Glass, M. and Felder, C. C. (1997) "Concurrent stimulation of cannabinoid CB_1 and dopamine D2 receptors augments cAMP accumulation in striatal neurons evidence for a Gs linkage to the CB_1 receptor," *Journal of Neuroscience* **17**: 5327–5333.

Graybiel, A. M. (1990) "Neurotransmitters and neuromodulators in the basal ganglia," *Trends in Neuroscience* **13**: 244–253.

Grinspoon, L. and Bakalar, J. B. (1993) *Marijuana, the forbidden medicine*. New Heaven: Yale University Press.

Hall, W., Salowij, N. and Lemon, J. (ed.) (1994) "Cannabis the Drug," in *The health and psychological consequences of cannabis use*. National Drug Strategy Monograph Series **25**: 29–40. Canberra, Australia, Australian Government Publishing Service.

Hampson, R. E. and Deadwyler, S. A. (1999) "Cannabinoids, hippocampal function and memory," *Life Science* **65**: 715–723.

Hanus, L., Abu-Lafi, S., Fride, E., Breuer, A., Vogel, Z., Shalev, D. E., Kustanovich, I., Mechoulam, R. (2001) "Z-Arachidonyl glyceryl ether an endogenous agonist of the cannabinoid CB_1 receptor," *Proceedings of the National Academy of Sciences (USA)* **98**: 3662–3665.

Hemming, M. and Yellowlees, P. M. (1993) "Effective treatment of Tourette's syndrome with marijuana," *Journal of Clinical Psychopharmacology* **7**: 389–391.

Herkenham, M., Groen, B. G. S., Lynn, A. B., de Costa, B. R. and Richfield, E. K. (1991a) "Neuronal localization of cannabinoid receptors and second messengers in mutant mouse cerebellum," *Brain Research* **552**: 301–310.

Herkenham, M., Lynn, A. B., de Costa, B. R. and Richfield, E. K. (1991b) "Neuronal localization of cannabinoid receptors in the basal ganglia of the rat," *Brain Research* **547**: 267–274.

Herkenham, M., Lynn, A. B., Johnson, M. R., Melvin, L. S., de Costa B. R. and Rice, K. C. (1991c) "Characterization and localization of a cannabinoid receptor in rat brain: a quantitative in vitro autoradiographic study," *Journal of Neuroscience* **11**: 563–583.

Herrera-Marschitz, M., Forster, C. and Ungerstedt, U. (1985a) "Rotational behavior elicited by intracerebral injections of apomorphine and pergolide in 6-hydroxy-dopamine-lesioned rats. I. Comparison between systemic and intrastriatal injections," *Acta Physiologica Scandinava* **125**: 519–527.

Herrera-Marschitz, M., Forster, C. and Ungerstedt, U. (1985b) "Rotational behavior elicited by intracerebral injections of apomorphine and pergolide in 6-hydroxy-dopamine-lesioned rats. I. The striatum of the rat is heterogeneously organized for rotational behavior," *Acta Physiologica Scandinava* **125**: 529–535.

Herrera-Marschitz, M. and Ungerstedt, U. (1987) "The dopamine-g aminobutyric acid interaction in the striatum of the rat is differently regulated by dopamine D-1 and D-2 types of receptor: evidence obtained with rotational behavioural experiments," *Acta Physiologica Scandinava* **129**: 371–380.

Hetherington, A. W. and Ranson, S. W. (1942) "The spontaneous activity and food intake of rats with hypothalamic lesions," *American Journal of Physiology* **136**: 609–617.

Hohmann, A. G. and Herkenham, M. (1999a) "Cannabinoid receptors undergo axonal flow in sensory nerves," *Neuroscience* **92**: 1171–1175.

Hohmann, A. G. and Herkenham, M. (1999b) "Localization of central cannabinoid CB_1 receptor messenger RNA in neuronal subpopulations of rat dorsal root ganglia: a double-label in situ hybridization study," *Neuroscience* **90**: 923–931.

Hohmann, A. G., Briley, E. M. and Herkenham, M. (1999c) "Pre- and postsynaptic distribution of cannabinoid and mu opioid receptors in rat spinal cord," *Brain Research* **822**: 17–25.

Hollister, L. E. (1986) "Health aspects of cannabis," *Pharmacological Reviews* **38**: 1–20.

Howlett, A. C., Qualy, J. M. and Khachatrian, L. L. (1986) "Involvement of Gi in the inhibition of adenylate by cannabimimetic drugs," *Molecular Pharmacology* **29**: 307–313.

Howlett, A. C. (1995) "Pharmacology of cannabinoid receptors," *Annual Review of Pharmacology and Toxicology* **35**: 607–634.

Jansen, E. M., Haycock, D. A., Ward, S. J. and Seybold, V. S. (1992) "Distribution of cannabinoid receptors in rat brain determined with aminoalkylindoles," *Brain Research* **575**: 93–102.

Joy, J. E., Watson, J. Jr. and Benson, J. A. Jr. (eds) (1999) *Marijuana and Medicine.* Washington, DC: Natl. Acad Press.

Kadekaro, M., Vance, W. H., Terrell, M. L., Gary, H. Jr., Eisenberg, H. M. and Sokoloff, L. (1987) "Effects of antidromic stimulation of the ventral root on glucose utilization in the ventral horn of the spinal cord of the rat," *Proceedings of the National Academy of Sciences of USA* **84**: 5492–5495.

Karniol, I. G., Shirakawa, I., Kasinski, N., Pfeferman, A. and Carlini, E. A. (1975) "Cannabidiol interfieres with the effects of delta-9-tetrahydrocannabinol in man," *European Journal of Pharmacology* **28**: 172–177.

Kawaguchi, Y., Wilson, C. J. and Emson, P. C. (1990) "Projection subtypes of rat neostriatal matrix cells revealed by intracellular injection of biocytin," *Journal of Neuroscience* **10**: 3421–3438.

Keefe, K. A. and Gerfen, C. R. (1995) "D1-D2 dopamine receptor synergy in striatum: effects of intrastriatal infusions of dopamine agonists and antagonists on inmediate early gene expression," *Neuroscience* **66**: 903–913.

Kinoshita, H., Hasegawa, T., Kameyama, T., Yamamoto, I. and Nabeshima, T. (1994) "Competitive NMDA antagonists enhance the catalepsy induced by delta 9-tetrahydrocannabinol in mice," *Neuroscience Letters* **174**: 101–104.

Kim, D. J. and Thayer, S. A. (2000) "Activation of CB_1 cannabinoid receptors inhibits neurotransmitter release from identified synaptic sites in rat hippocampal cultures," *Brain Research* **852**: 398–405.

Le Moine, C. and Bloch, B. (1995) "D1 and D2 dopamine receptor gene expression in the rat striatum: sensitive cRNA probes demonstrate prominent segregation of D1 and D2 mRNAs in distinct neuronal populations of the dorsal and ventral striatum," *Journal of Comparative Neurology* **35**: 418–426.

Le Moine, C., Norman, E. and Bloch, B. (1991) "Phenotypical characterization of the rat striatal neurons expressing the D1 dopamine receptor gene," *Proceedings of the National Academy of Sciences of USA* **88**: 4205–4209.

Mackie, K. and Hille, B. (1992) "Cannabinoid inhibit N-type calcium channels in neuroblastoma-glioma cells," *Proceedings of the National Academy of Sciences of USA* **89**: 3825–3829.

Mackie, K., Lai, Y., Westenbroek, R. and Mitchell, R. (1995) "Cannabinoids activate an inwardly rectifying potassium conductance and inhibit Q-type calcium currents in AtT20 cells transfected with rat brain cannabinoid receptor," *Neuroscience* **15**: 6552–6561.

Magalhaes-Castro, H. H., de Lima, A. D., Saraiva, P. E. S. and Magalhaes-Castro, B. (1978) "Horseradish peroxidase labeling of cat tectotectal cells," *Brain Research* **148**: 1–13.

Mailleux, P. and Vanderhaeghen, J. J. (1992) "Distribution of neuronal cannabinoid receptor in the adult rat brain: A comparative receptor binding radioautography and in situ hybridization histochemistry," *Neuroscience* **48**: 655–688.

Marsden, C. D., Obeso, J. A., Zarranz, J. J. and Lang, A. E. (1985) "The anatomical basis of symptomatic hemidystonia," *Brain* **108**: 463–483.

Mao, A., Freeman, K. A. and Tallarida, R. J. (1996) "Transient loss of dopamine autoreceptor control in the presence of highly potent dopamine agonists," *Life Sciences* **59**: PL317–324.

Margulies, J. E. and Hammer, R. P. (1991) "D^9-Tetrahydrocannabinol alter cerebral metabolism in a biphasic, dose-dependent manner in rat brain," *European Journal of Pharmacology* **202**: 373–378.

Martin, B. R., Welch, S. P. and Abood, M. (1994) "Progress toward understanding the cannabinoid receptor and its second messenger systems," *Advances in pharmacology* **25**: 341–397.

Matsuda, L. A., Lolait, S. J., Brownstein, M., Young, A. and Boner, T. I. (1990) "Structure of a cannabinoid receptor and functional expression of the cloned cDNA," *Nature* **346**: 561–564.

Mechoulam, R., Shani, A., Edery, H. and Grunfeld, Y. (1970) "Chemical basis of hashsish activity," *Science* **169**: 611–612.

Mechoulam, R., Fride, E., Hanus, L., Sheskin, T., Bisogno, T., Di Marzo, V., Bayewitch, M. and Vogel, Z. (1997) "Anandamide may mediate sleep induction," *Nature* **389**: 25–26.

Mechoulam, R., Ben-Shabat, S., Hanus, L., Ligumsky, M., Kaminski, N. E., Schatz, A. R., Gopher, A., Almog, S., Martin, B. R., Compton, D. R., Pertwee, R. G., Griffin, G., Bayewitch, M., Barg, J. and Vogel, Z. (1995) "Identification of an endogenous Z-monoglyceride, present in canine gut, that binds to cannabinoid receptors," *Biochemical Pharmacology* **50**: 83–90.

Meinck, H. M., Schonle, P. W. and Conrad, B. (1989) "Effects of cannabinoids on spasticity and ataxia in multiple sclerosis," *Journal of Neurology* **236**: 120–122.

Miller, A., Sañudo-Peña, M. C. and Walker, J. M. (1998) "Ipsilateral turning behavior induced by unilateral microinjections of a cannabinoid into the rat subthalamic nucleus," *Brain Research* **793**: 7–11.

Miller, A. and Walker, J. M. (1995) "Effects of a cannabinoid on spontaneous and evoked neuronal activity in the substantia nigra pars reticulata," *European Journal of Pharmacology* **279**: 179–185.

Miller, A. and Walker, J. M. (1996) "Electrophysiological effects of a cannabinoid on neural activity in the globus pallidus," *European Journal of Pharmacology* **304**: 29–35.

Muller-Vahl, K. R., Schneider, U. and Emrich, H. M. (1999a) "Nabilone increases choreatic movements in Huntington's disease," *Movement Disorders* **14**: 1038–1040.

Muller-Vahl, K. R., Schneider, U., Kolbe, H. and Emrich, H. M. (1999b) "Treatment of tourette's syndrome with delta-9-tetrahydrocannabinol," *American Journal of Psychiatry* **156**: 495.

Munro, S., Thomas, K. L. and Abu-Shaar, M. (1993) "Molecular characterization of a peripheral receptor for cannabinoids," *Nature* **365**: 61–65.

North, J. (1991) "Trends in the pathophysiology and pharmacotherapy of spasticity," *Journal of Neurology*, **238**, 131–139.

Netzeband, J. G., Conroy, S. M., Parsons, K. L. and Gruol, D. L. (1999) "Cannabinoids enhance NMDA-elicited Ca^{2+} signals in cerebellar granule neurons in culture," *Journal of Neuroscience* **19**: 8765–8777.

Nisembaum, L. K., Kitai, S. T. and Gerfen, C. R. (1994) "Dopaminergic and muscarinic regulation of striatal enkephalin and substance P messenger RNAs following striatal dopamine denervation: effects of systemic and central administration of quinpirole and scopolamine," *Neuroscience* **63**: 435–449.

Obeso, J. A., Rodriguez, M. C. and DeLong, M. R. (1997) "Basal ganglia pathophysiology: A critical review," *Advances in Neurology* **74**: 3–18.

Pacheco, M. A., Ward, S. J. and Childers, S. R. (1993) "Identification of cannabinoid receptors in cultures of rat cerebellar granule cells," *Brain Research* **603**: 102–110.

Parent, A. and Hazrati, L.-N. (1995a) "Functional anatomy of the basal ganglia. I. The cortico-basal ganglia-thalamo-cortical loop," *Brain Research Reviews* **20**: 91–127.

Parent, A. and Hazrati, L.-N. (1995b) "Functional anatomy of the basal ganglia. II. The place of subthalamic nucleus and external pallidum in basal ganglia circuitry," *Brain Research Reviews* **20**: 128–154.

Pettit, D. A. D., Harrison, M. P., Olson, J. M., Spencer, R. F. and Cabral, G. A. (1998) "Immunohistochemical localization of the neural cannabinoid receptor in the rat brain," *Journal of Neuroscience Research* **51**: 391–402.

Petro, D. J. and Ellenberger, C. (1981) "Treatment of human spasticity with D⁹-Tetrahydro-cannabinol," *Journal of Clinical Pharmacology* **21**: 413S–416S.

Pertwee, R. G. and Greentree, S. G. (1988) "Delta-9-tetrahydrocannabinol-induced catalepsy in mice is enhanced by pretreatment with flurazepam or chlordiazepoxide," *Neuropharmacology* **27**: 485–491.

Pertwee, R. G. and Wickens, A. P. (1991) "Enhancement by chlordiazepoxide of catalepsy induced in rats by intravenous or intrapallidal injections of enantiomeric cannabinoids," *Neuropharmacology* **30**: 237–244.

Redgrave, P., Odekunle, A. and Dean, P. (1986) "Tectal cells of origin of predorsal bundle in the rat: location and segregation from ipsilateral descending pathways," *Experimental Brain Research* **63**: 279–293.

Reiner, A., Albin, R. and Anderson, K. D. (1988) "Differential loss of striatal projection neurons in Huntington's disease," *Proceedings of the National Academy of Sciences USA* **85**: 5733–5737.

Richfield, E. K. and Herkenham, M. (1994) "Selective vulnerability in Huntington's disease: preferential loss of cannabinoid receptors in lateral globus pallidus," *Annals of Neurology* **36**: 577–584.

Richter, A. and Loscher, W. (1994) "(+)-WIN55,212-2 a novel cannabinoid receptor agonist, exerts antidystonic effects in mutant dystonic hamsters," *European Journal of Pharmacology* **264**: 371–377.

Rieck, R. W., Huerta, M. F., Harting, J. K. and Weber, J. T. (1986) "Hypothalamic and ventral thalamic projections to the superior colliculus in the cat," *Journal of Comparative Neurology* **243**: 249–265.

Rinaldi-Carmona, M., Barth, F., Heaulme, M., Shire, D., Calandra, B., Congy, C., Martinez, S., Maruani, J., Neliat, G., Caput, D., Ferrara, P., Soubrie, P., Breliere, J. C. and Le Fur, G. (1994) "SR141716A, a potent and selective antagonist of the brain cannabinoid receptor," *FEBS Letters* **350**: 240–244.

Robertson, H. A., Peterson, M. R., Murphy, K. and Robertson, G. S. (1989) "D1-dopamine receptor agonists selectively activate striatal c-fos independent of rotational behavior," *Brain Research* **503**: 346–349.

Robertson, G. S., Vincent, S. R. and Fibiger, H. C. (1990) "Striatonigral projection neurons contain D1 dopamine receptor-activated c-fos," *Brain Research* **523**: 288–290.

Robledo, P. and Feger, J. (1991) "Acute monoaminergic depletion in the rat potentiates the excitatory effect of the subthalamic nucleus in the substantia nigra pars reticulata but not in the pallidal complex," *Journal of Neural Transmission* **86**: 115–126.

Romero, J., de Miguel, R., García-Palomero, E., Fernández-Ruiz, J. J. and Ramos, J. A. (1995) "Time-course of the effects of anandamide, the putative endogenous cannabinoid receptor ligand, on extrapyramidal function," *Brain Research* **694**: 223–232.

Romero, J., García-Palomero, E., Fernández-Ruiz, J. J. and Ramos, J. A. (1996a) "Involvement of GABA_B receptors in the motor inhibition produced by agonists of brain cannabinoid receptors," *Behavioral Pharmacology* **7**: 299–302.

Romero, J., García-Palomero, E., Lin, S. Y., Ramos, J. A., Makriyannis, A. and Fernández-Ruiz, J. J. (1996b) "Extrapyramidal effects of methanadamide, an analog of anandamide, the endogenous CB₁ receptor ligand," *Life Sciences* **15**: 1249–1257.

Sakurai, Y., Ohta, H., Shimazoe, T., Kataoka, Y., Fujiwara, M. and Ueki, S. (1985) "D⁹-Tetrahydrocannabinol elicited ipsilateral circling behavior in rats with unilateral nigral lesion," *Life Sciences* **37**: 2181–2185.

Sandyk, R., Snider, S. R., Consroe, P. and Elias, S. M. (1986) "Cannabidiol in dystonic movement disorders," *Psychiatry Research* **18**: 291.

Sandyk, R. and Awerbuch, G. (1988) "Marijuana and Tourette's syndrome," *Journal of Clinical Psychopharmacology* **8**: 444–445.

Sañudo-Peña, M. C., Patrick, S. L., Patrick, R. L. and Walker, J. M. (1996) "Effects of intranigral cannabinoids on rotational behavior in rats: interactions with the dopaminergic system," *Neuroscience Letters* **206**: 21–24.

Sañudo-Peña, M. C. and Walker, J. M. (1997) "Role of the subthalamic nucleus on cannabinoid actions in the substantia nigra of the rat," *Journal of Neurophysiology* **77**: 1635–1638.

Sañudo-Peña, M. C. and Walker, J. M. (1998a) "Effects of intrapallidal cannabinoids on rotational behavior in rats: interactions with the dopaminergic system," *Synapse* **28**: 27–32.

Sañudo-Peña, M. C. and Walker, J. M. (1998b) "A novel neurotransmitter system involved in the control of motor behavior by the basal ganglia," *Annals of the New York Academy of Sciences* **860**: 475–479.

Sañudo-Peña, M. C., Force, M., Tsou, K., Miller, A. S. and Walker, J. M. (1998a) "Effects of intrastriatal cannabinoids on rotational behavior in rats: interactions with the dopaminergic system," *Synapse* **30**: 221–226.

Sañudo-Peña, M. C., Patrick, S. L., Patrick, R. L. and Walker, J. M. (1998b) "Cannabinoid control of movement in the basal ganglia in an animal model of parkinson disease," *Neuroscience Letters* **248**: 171–174.

Sañudo-Peña, M. C., Tsou, K. and Walker, J. M. (1999a) "Motor Actions of cannabinoids in the basal ganglia output nuclei," *Life Sciences* **65**: 703–713.

Sañudo-Peña, M. C., de Miguel, R., Romero, J., Huang, S. M., Fernandez-Ruiz, J. J., Ramos, J. A., Tsou, K. and Walker, J. M. (1999b) "Cannabinoids oppose dopamine actions on movement in the substantia nigra reticulata of intact animals but they do not in a rat model of Parkinson's disease," *Society for Neuroscience Abstracts* **25**: 377.

Sañudo-Peña, M. C., Tsou, K., Romero, J., Mackie, K. and Walker, J. M. (2000a) "Role of the superior colliculus in the motor effects of cannabinoids and dopamine," *Brain Research* **853**: 207–214.

Sañudo-Peña, M. C., Romero, J., Fernandez Ruiz, J. J. and Walker, J. M. (2000b) "Activational role of cannabinoids on movement," *European Journal of Pharmacology* **391**: 269–274.

Sañudo-Peña, M. C., Walker, J. M. and McLemore, G. L. (2000c) "Blockade of cannabinoid withdrawal syndrome expression by a GABA-B agonist," *Abstracts of the 29th Annual Meeting of New England Pharmacologists*.

Schwartz, W. J., Smith, C. B., Davidsen, L., Savaki, H., Sokoloff, L., Mata, M., Fink, D. J. and Gainer, H. (1979) "Metabolic mapping of functional activity in the hypothalamo-neurohypophysial system of the rat," *Science* **205**: 723–725.

Singer, H. S. (1997) "Neurobiology of Tourette's syndrome," *Neurologic Clinics* **15**: 357–379.

Sinor, A. D., Irvin, S. M. and Greenberg, D. A. (2000) "Endocannabinoids protect cerebral cortical neurons from in vitro ischemia in rats," *Neuroscience Letters* **278**: 157–160.

Souilhac, J., Poncelet, M., Rinaldi-Carmona, M., Le-Fur, G. and Soubrie, P. (1995) "Intrastriatal injection of cannabinoid receptor agonists induced turning behavior in mice," *Pharmacology, Biochemistry and Behavior* **51**: 3–7.

St-Pierre, J. A. and Bedard, P. J. (1994) "Intranigral but not intrastriatal microinjections of the NMDA antagonist MK-801 induces contralateral circling in the 6-OHDA rat model," *Brain Research* **660**: 255–260.

Speller, J. M. and Westby, G. W. M. (1996) "Bicuculline-induced circling from the rat superior colliculus is blocked by GABA microinjection into the deep cerebellar nuclei," *Experimental Brain Research* **110**: 425–434.

Steiner, H., Bonner, T. I., Zimmer, A. M., Kitai, S. T. and Zimmer, A. (1999) "Altered gene expression in striatal projection neurons in CB_1 cannabinoid receptor knockout mice," *Proceedings of the National Academy of Science of USA* **96**: 5786–5790.

Stellar, J. R. and Stellar, E. (1985) *The neurobiology of motivation and reward.* New York: Springer.

Sulcova, E., Mechoulam, R. and Fride, E. (1998) "Biphasic effects of anandamide," *Pharmacology, Biochemistry and Behavior* **59**: 347–352.

Szabo, B., Dorner, L., Pfreundtner, C., Norenberg, W. and Starke, K. (1998) "Inhibition of GABAergic inhibitory postsynaptic currents by cannabinoids in rat corpus striatum," *Neuroscience* **85**: 395–403.

Tsou, K., Brown, S., Mackie, K., Sañudo-Peña, M. C. and Walker, J. M. (1998a) "Immunohistochemical distribution of cannabinoid CB_1 receptors in the rat central nervous system," *Neuroscience* **83**: 393–411.

Tsou, K., Nogueron, M. I., Muthian, S., Sañudo-Peña, M. C., Hillard, C. J., Deutsch, D. G. and Walker, J. M. (1998b) "Fatty acid amide hydrolase that degrades anandamide is located preferentially in large neurons in the rat central nervous system as revealed by immunohistochemistry," *Neuroscience Letters* **254**: 137–140.

Turski, L., Klockgether, T., Schwarz, M., Turski, W. A. and Sontag, K. -H. (1990) "Substantia nigra: a site of action of muxcle relaxant drugs," *Annals of Neurology* **28**: 341–348.

Wickens, A. P. and Pertwee, R. G. (1995) "Effect of D^9-Tetrahydrocannabinol on circling in rats induced by intranigral muscimol administration," *European Journal of Pharmacology* **282**: 251–254.

Williams, M. N. and Faull, R. L. M. (1988) "The nigrotectal projection and tectospinal neurons in the rat. A light and electron microscopic study demonstrating a monosynaptic nigral input to identified tectospinal neurons," *Neuroscience* **25**: 533–562.

Williams, C. M., Rogers, P. J. and Kirkham, T. C. (1998) "Hyperphagia in pre-fed rats following oral delta9-THC," *Physiology and Behavior* **65**: 343–346.

Wilson, C. J. (1993) "The generation of natural firing patterns in neostriatal neurons," *Progress in Brain Research* **99**: 277–298.

Wilson, W. H., Ellinwood, E. H., Mathew, R. J. and Johnson, K. (1994) "Effects of marijuana on performance of a computerized cognitive-neuromotor test battery," *Psychiatry Research* **51**: 115–125.

Yamaguchi, K., Nabeshima, T. and Kameyama, T. (1986) "Role of dopaminergic and GABAergic mechanisms in discrete brain areas in phencyclidine-induced locomotor stimulation and turning behavior," *Journal of Pharmacobiodynamics* **9**: 975–998.

You, Z. B., Herrera-Marschitz, M., Nylander, I., Goiny, M., O'Connor, W. T., Ungerstedt, U. and Terenius, L. (1994) "The striatonigral dynorphin pathway of the rat studied with in vivo microdialysis-II. Effects of dopamine D_1 and D_2 receptor agonists," *Neuroscience* **63**: 427–434.

Zimmer, A., Zimmer, A. M., Hohmann, A. G., Herkenham, M. and Bonner, T. I. (1999) "Increased mortality, hypoactivity, and hypoalgesia in cannabinoid CB_1 receptor knockout mice," *Proceedings of the National Academy of Sciences of USA* **96**: 5780–5785.

Chapter 8

Effects of marijuana on brain: function and structure

William H. Wilson and Roy J. Mathew

ABSTRACT

According to the White House, in 1995, 77% of all current illicit drug users were marijuana smokers. Approximately 57% of current illicit drug users limited consumption exclusively to marijuana (United States Congress, The National Drug Control Strategy, February, 1997). There is national debate over whether to legalize the use of marijuana; some speaking for (Kassirer, 1997), while others are opposed (Nahas *et al.*, 1997). In the midst of this ongoing debate a better understanding of the potential health effects of marijuana use and the neurobiologic mechanisms that underlie these effects is of considerable interest.

This chapter presents the evidence and discusses the question of whether brain development and differentiation may be affected by marijuana use before and during adolescence. The possibility that marijuana may alter normal developmental changes does not preclude the possibility that it may also have neurotoxic effects, including the possibility that early adolescence may be a period when the brain may be more vulnerable to such insults. The human and animal data presented previously support possible neurotoxic effects, perhaps permanent changes in brain neurochemistry, morphology and funtion following chronic exposure to cannabinoids.

Key Words: marijuana, PET, CBF depersonalization, MRI, [133]Xenon

ASSESSING THE EFFECTS OF MARIJUANA ON BRAIN

Like most drugs of abuse, marijuana has wide ranging effects on human behavior. It is known to induce euphoria, anxiety, lethargy, drowsiness, confusion, memory defects, altered time sense, depersonalization, impaired neuro-motor performance and psychotic symptoms (Brill and Nahas, 1984; Institute of Medicine, 1982; Mendelson *et al.*, 1976; Maykut, 1985; Hollister, 1986; Mathew and Wilson, 1991). These behavioral effects tend to vary considerably from subject to subject and seem to be influenced by a variety of non-specific factors including the personality of the subject, pre-existing mood state, expectancy, social setting in which the drug is consumed, previous exposure to the drug, sex, mental health, etc.

Over the past two decades we have been actively involved in studying the effects of substances of abuse on behavior. Our modus operandi in this work has been to measure changes in behavior and correlate these to measures of brain function. In the normal brain, cerebral blood flow (CBF) and metabolism (CMR) are closely coupled with brain function. Research conducted in animals and human subjects has confirmed this hypothesis (Mathew *et al.*, 1985; Risberg, 1980; Phelps *et al.*, 1982; Mazziotta, 1985). This relationship is fundamental to the study of regional brain function by looking at CBF or CMR. With the development of non-invasive techniques with improved spatial resolution, study of the effects of central nervous system acting drugs on CBF and CMR became easier and more meaningful. Acute and chronic effects of a variety of psychopharmacological agents have been studied extensively. In spite of the high prevalence of marijuana smoking, until recently, very little information was available on its effects on CBF.

Multiple effects of marijuana on blood flow

CBF measurement is important to marijuana research as it prompts the opportunity to identify brain regions associated with the behavioral changes, thus providing further insight into these effects. Unfortunately, marijuana (and several other CNS acting drugs) induce CBF changes through mechanisms other than alterations in brain function (Mathew and Wilson, 1991).

Effects on vascular smooth muscle

A number of CNS acting drugs are vasoactive, i.e. they act upon the smooth muscle in the blood vessel wall. Caffeine-induced cerebral vasoconstriction and increase in CBF induced by amyl nitrite are examples of this (Mathew and Wilson, 1985; Mathew *et al.*, 1989). Although marijuana has not been demonstrated to have a vascular effect in the brain, it does influence blood vessels elsewhere in the body. For example, its dilatory effect on the conjunctival and muscle blood vessels of the eye are well known (Hollister *et al.*, 1981; Ohlsson *et al.*, 1980; Weiss *et al.*, 1972). The "red" seen after marijuana smoking is known to last for several hours (Hollister *et al.*, 1981; Ohlsson *et al.*, 1980). It is conceivable that marijuana might have a similar dilatory effect on cerebral blood vessels.

Changes in general circulation and respiration

A number of physiological changes associated with marijuana are relevant to CBF. Marijuana may increase respiratory rate and alter blood carbon dioxide levels. CBF is acutely sensitive to alterations in carbon dioxide. Even modest changes in carbondioxide are associated with large changes in CBF, i.e. an increase in arterial carbondioxide levels by 1 mm of mercury partial pressure will be accompanied by a 3–4% increase in CBF (Maximilian *et al.*, 1980; Mathew and Wilson, 1988).

Increase in pulse rate is one of the most consistent physiological changes seen after marijuana consumption. Fortunately, changes in pulse and blood pressure are unlikely to influence CBF substantially. In the normal brain, CBF is insulated from modest changes in blood pressure through autoregulatory mechanisms (Strandgaard and Paulson, 1984).

Autonomic changes

The drug clearly has autonomic effects. Cerebral blood vessels receive vasoconstrictive sympathetic fibers from the superior cervical ganglion (Edvinsson, 1982; Busija and Heistead, 1984). Sympathetic stimulation can produce CBF reduction. However, cerebral blood vessels in human subjects are believed to be less sensitive to autonomic activation as compared to several animal species. It also needs to be pointed out that the argument that marijuana induces tachycardia via vagal blockade (and not sympathetic stimulation) has been put forward (Mathew *et al.*, 1992).

Carbon monoxide production

Marijuana smoking has been reported to increase blood levels of carbon monoxide (Aranow and Cassidy, 1979; Mathew *et al.*, 1992). This is of importance to CBF research. Carbon monoxide production and formation of carboxy hemoglobin interfere with transport and delivery of oxygen to the brain. The brain compensates for cerebral anoxia through a variety of mechanisms including increased oxygen extraction from the blood and cerebral vasodilation. Thus, increased carbon monoxide production during marijuana smoking can lead to a CBF increase.

Variability of effects

Further complicating the study of the acute effects of marijuana is the variability in those effects. Volkow *et al.* (1991a) studied the effects of THC on cerebral metabolism of glucose (CMR). In one study, CMR was measured during baseline and 30–40 min after 2 mg of THC, given intravenously (Volkow *et al.*, 1991a). Post THC changes in global CMR were variable, some showing increases and some decreases. All subjects however, were reported to have showed an increase in normalized cerebellar metabolism which correlated with intoxication and plasma THC levels. In another report, CMR was described under resting conditions, 24 h after an intravenous injection of 2 mg of THC (Volkow and Fowler, 1993). Cerebellar CMR showed consistent increase which correlated with plasma THC and intoxication. There were some apparent differences between marijuana users and controls on CMR responses to THC with the marijuana users showing less changes in the cerebellum and more changes in the prefrontal cortex after THC.

In a previous study, we measured CBF (Mathew *et al.*, 1989) in experienced (EXP) (minimum of 10 "joints" per week for past 3 years) and inexperienced (INX) marijuana smokers (no exposure to marijuana for a minimum of 3 years). CBF was measured with the [133]Xenon inhalation technique before and after smoking a marijuana cigarette (THC=2.2%) and a placebo (marijuana after THC extraction) cigarette. The INX after smoking marijuana reported dysphoria including severe anxiety. Anxiety as measured by the Profile of Mood States (POMS) (McNair *et al.*, 1971) increased from 6.5 (SD 2.6) before to 18.2 (SD 10.2) after marijuana. EXP, on the other hand, did not have any adverse reactions. Their POMS anxiety score dropped from 9.8 (SD 3.3) to 6.9 (SD 2.7). In INX, post-marijuana CBF (global) decreased, but in EXP it increased significantly. Anxiety and CBF had an inverse relationship, in keeping with the results of other experiments (Mathew and Wilson, 1990). The effect of marijuana on CBF in the absence of severe anxiety seemed to be one of increase.

Measurement techniques in our studies of brain function

The ^{133}Xenon inhalation technique

The first isotope clearance technique developed by Ingvar and Lassen (1961) represented significant advancement. For the first time it was possible to measure regional flow values. The technique involved intracarotid injection of ^{133}Xenon dissolved in saline followed by recording of progressive decline in radioactivity over the ipse-lateral cerebral cortex by a system of scintillation detectors or gamma camera applied over the scalp. The clearance curves were analyzed with a bicompartmental model which permitted separation of flow to the gray matter from the slower clearing white matter. The intracarotid isotope injection resulted in high isotope concentration in the ipse-lateral hemisphere which improved the spatial resolution of the technique to a considerable extent.

Obrist and associates modified the intracarotid injection technique to make it totally non-invasive (Obrist *et al.*, 1975). ^{133}Xenon gas was administered through a face mask instead of intracarotid injection. The CBF measurement consisted of a one minute inhalation of ^{133}Xenon and air mixture followed by monitoring of the removal of the isotope from different parts of the brain by 32 scintillation detectors mounted on a helmet and applied to the scalp (Figure 8.1). The scalp clearance curves are analyzed by a bicompartmental model. The technique was associated with low radiation exposure and therefore, it could be repeated back to back on the same subject several times (similar to the intracarotid injection technique). Test re-test reliability was good. It measured CBF to both hemispheres

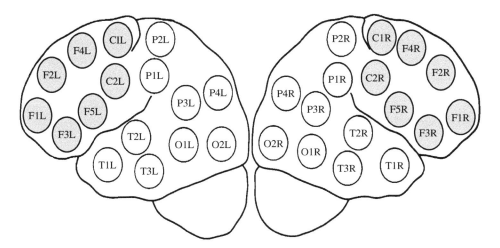

Figure 8.1 Detector configuration – shaded circles indicate anterior brain regions.

and in both carotid and vertibro-basilar systems. The primary disadvantages of the technique were limited spatial resolution, reduced sensitivity to changes in gray matter flow (inferior to the intracarotid injection technique in both respects) inability to measure subcortical CBF and regional CMR and the "look-through" effect (areas of low perfusion overlapped by higher perfusion areas).

Positron emission tomography

With the advent of tomographic techniques, measurement of subcortical flow and metabolism became possible. Positron emission tomography (PET) involves administration of a radiotracer followed by reconstruction of tomographic slices of the brain depicting isotope concentrations in various cortical and subcortical brain regions. Several radionuclides, such as ^{11}C, ^{13}N, ^{15}O and ^{18}F are available at most PET centers with cyclotrons. Other positron emitting nuclides have been produced and used on a more limited basis.

PET techniques which utilized ^{15}O labeled water as tracer provided the possibility of absolute quantitation of CBF. The autoradiographic version of the technique was associated with low levels of radiation exposure and can be repeated in the same subject several times, back to back. However, absolute quantitation requires arterial puncture with multiple blood draws to determine the input function (arterial blood time–activity curves) which are then used to compute absolute CBF. Many questions and study designs do not require absolute CBF, and thus, many CBF studies only use counts. PET is a sophisticated CBF measurement technique, but also suffers from shortcomings (Phelps et al., 1982; Mazziotta, 1985). In spite of improved spatial resolution, the resolution is still insufficient to image small brain regions satisfactorily. The cerebral cortex is less than 2 mm thick and current PET scanners do not have the spatial resolution necessary to image cerebral cortex satisfactorily.

PET scans have been used extensively to quantify cerebral metabolic rate of glucose (CMR) with the ^{18}F deoxy-glucose method (Sokoloff and Smith, 1983; Alavi et al., 1986). The technique is well validated and found useful in a wide range of research projects. The major disadvantages of the technique are that it quantifies glucose metabolism over a 20 min period during the uptake of the ^{18}F deoxy-glucose (and therefore is unsuitable to examine rapid changes in CMR after the administration of a drug) and that it cannot be repeated back to back because of the half-life of ^{18}F. Long intervals between 2 CMR glucose measurements is likely to lower stability of the CMR values across measurements and increase the influence of non-specific factors.

Factors affecting PET scanning

Image registration

Fully utilized PET techniques require a standardized, accurate and reproducible method for anatomical localization of CBF information. Various approaches have been used to address the extremely complex range of problems associated with the alignment and anatomic localization of PET data (Mazziotta and Koslow, 1987;

Mazziotta *et al.*, 1991). We register the PET image using each subject's own MRI, following methods developed by Chen and Pelizzari (Chen *et al.*, 1989; Pelizzari *et al.*, 1989). The software developed by these investigators uses a surface matching technique to register cross-sectional images using the brain surface contours from the MRI and PET scans. Algorithms are used to find the coordinate transformation (including 3-dimensional translation and rotation) that optimally matches the two models, using the known relative pixel sizes. Accuracy of this technique is in the order of 1–2 mm when performed with head phantom and human studies (Pelizzari *et al.*, 1989; Turkington *et al.*, 1995). This allows one to define regions of interest on the anatomically accurate MRI scan and then transform these regions into the coordinate system of the PET scan.

Positioning and alignment

Because of the importance of these factors between MRI and PET imaging studies, specific attention must be devoted to these issues. The need for critical positioning and repositioning is largely a result of the non-uniform sampling of data sets with many older PET systems. New generation scanners with approximately symmetrical 3-dimensional resolution and significantly larger field of view (FOV of 15 cm), reduce the need for exact positioning since reformatting can be performed with 3-dimensional interpolation.

Attenuation correction

Many photons scatter before leaving the body. If photons scatter (or are absorbed) this leads to a loss in detected events that otherwise would have been recorded. This attenuation has several detrimental effects: overall loss of counts, leading to higher imaging noise, image non-uniformity, due to higher attenuation of photons from some regions than others, and distortions, due to the differential attenuation of photons from a particular source. An attenuation correction is necessary both to obtain uniformity within images and to achieve absolute accuracy in the radioactivity measurement. A transmission scan is done on each subject to provide this correction information. This transmission scan provides the necessary correction factors. This correction is valid for all scans on a subject as long as there is no movement.

Partial volume corrections

Brain structures whose size is small relative to the system resolution will appear to have lower activity (fewer counts) and therefore CBF. This effect is made less severe by the relatively high resolution of newer PET scanners. A correction for the decreasing volume of these structures may be useful in providing more accurate estimates of CBF values. This correction can be applied to isolated gray matter regions by using a dilated ROI to measure activity in the region and that which has blurred into neighboring pixels, and then dividing by the known size of the region, as determined from the MR image. Currently available PET scanners are able to record data from 35 slices simultaneously over a 15 cm field of view with a reconstruction resolution of about 5 mm axial and 5 mm transverse (FWHM) (DeGrado *et al.*, 1994).

Autoradiographic ^{15}O-water cerebral blood flow determination

Several methods have been described for the measurement of regional cerebral blood flow (rCBF) with PET (e.g. Frackowiak *et al.*, 1980; Huang *et al.*, 1983). The majority of techniques have used ^{15}O-water as the tracer either by administration intravenously or by inhalation of ^{15}O-carbon dioxide. The validity and limitation of ^{15}O-water as a flow tracer is well documented (Herscovitch *et al.*, 1983; Ledi *et al.*, 1986; Raichle *et al.*, 1983). The advantages of using ^{15}O-water are its simplicity of production, short half-life, tissue:blood partition coefficient being less dependent on pathology, and ease of mathematical modeling. A disadvantage of using ^{15}O-water is that it is not strictly freely diffusible. Of the four methods (steady state, autoradiographic, integrated projection, build-up) that have been described to measure CBF with PET, the autoradiographic technique is most widely used. We have implemented the ^{15}O-water autoradiographic CBF technique in our laboratory as described by Meyer (1989) and Hoffman and Coleman (1992).

PET data acquisition

PET scanners have multiple rings of detectors that are usually separated by shielding (septa) that minimizes the amount of radiation which originated in a different plane hitting a detector. Radiation coming from a different plane than the one a particular detector results in extra counts called "scattered events". There are two modes of data acquisition with PET referred to as 2D and 3D. Volume imaging and 3D imaging have the same meaning when applied to PET data acquisition. Three-dimensional (3D) acquisition in PET permits a large increase in sensitivity (measured counts per unit injected dose) compared to the 2D mode. This increase in sensitivity can be used to reduce injected dose, decrease scan time, improve data quality, or some combination of the three. For ^{15}O-water studies the dose is reduced (compared to 2D acquisition), leading to a significant reduction in radiation exposure to the subject. The injected dose in 3D mode (compared to 2D) would drop from about 50 mCi to about 12 mCi per scan. This means that four scans can be done for about the same level of radiation exposure. An increase in detected scattered events and increased processing time are the main disadvantages. Newer scanners have retractable septa, as well as the necessary acquisition and reconstruction features to make imaging in this 3D mode possible. A 3D scatter correction scheme has to be implemented (Stearns, 1995). Some quantitative error (less than 5%) is expected to remain in the images after scatter correction (Turkington *et al.*, 1996). However, the errors for studies such as ^{15}O-water CBF studies should be systematic, and thus, consistent within image sets, i.e. from one scan to the next on an individual, and to some degree, between subjects. The 3D mode can not be used to improve image quality by increasing the count rates, because with ^{15}O-water imaging (where the scan time is fixed by blood-flow kinetics), detector count rate limitations require that the dose be lowered, due to scanner count rate limitations. Instead, a smaller dose must be administered ($\sim$12 mCi) (Sadato *et al.*, 1995), resulting in comparable image noise to much higher dose 2D studies. The reduction in radiation exposure to the subjects is usually considered worth these disadvantages. For our current CBF studies we acquire emission data in 3D mode.

Three-dimensional imaging also has become routine for (^{18}F) FDG studies, both quantitative and non-quantitative. Unlike CBF studies, FDG studies use the same dose (10 mCi) in 3D as in 2D, allowing shorter scans and/or improved images. The shorter scan time is a particular advantage for some subject groups who are unable or unwilling to remain still for the longer scan time in 2D mode.

Segmentation

Unlike the xenon procedure above, it is necessary to define the gray and white matter on the MRI to get gray matter blood flow with PET. The MR (proton density) image can be utilized to establish segmentation criteria to determine CSF, white and gray matter based on signal intensity. Use of segmentation criteria to identify tissue types has been shown to be a reliable method (DeCarli *et al.*, 1992; Byrum *et al.*, 1996). This process essentially consists of identifying signal levels on the MRI for predominantly gray matter, white and CSF, then using these activity levels to establish criteria. A pixel is included as gray matter if its signal strength is at or above that cut point. CBF for each pixel is determined and thereby blood flow to gray matter.

CHANGES IN MENTAL STATE ASSOCIATED WITH MARIJUANA

Our work on the acute changes in mental state associated with marijuana intoxication has focused on three effects commonly reported with marijuana. The objective has been to better understand the patterns of regional changes in brain function associated with the effects of marijuana. It should be noted that none of the changes take place independently of other changes. Thus, we have taken the approach of measuring multiple dependent variables and through appropriate statistical techniques, sorting out the relationships among these multiple dependent variables and regional changes in brain function.

An important basis for these studies has been the identification and mapping of cannabinoid receptor sites. Pertwee (1997) noted that the distribution of cannabinoid receptor binding sites is densest in those areas where behavioral studies have shown effects, including cognition and memory (cerebral cortex and hippocampus) and motor function, time sense and movement (cerebellum and basal ganglia). In the rat brain, binding sites have been found to be distributed highest in the substantia nigra pars raticulata, globus pallidus and the molecular layer of the cerebellum. They are distributed next highest in areas including the hippocampus. In the cerebral cortex the cingulate gyrus, frontal and parietal lobes have the highest densities. Binding sites are sparse in the amygdala, thalamus and hypothalamus. Similar distribution patterns have been reported in humans (Pertwee, 1997). It has been reported that in many areas the binding site density exceeds that of neuropeptides and is similar to the density of benzodiazepines, striatal dopamine and whole-brain glutamate (Herkenham *et al.*, 1990). These findings about regional variation in density of receptor sites, while suggestive, should be viewed with the understanding that there are recognized species differences. It has been reported, for example, that humans have a substantially reduced receptor expression in the cerebellum compared to rats (Matsuda *et al.*, 1993). Matsuda (1997) has suggested that this difference is consistent with the findings that humans have a reduced motor disturbance after exposure to marijuana.

Intoxication related euphoria or high

Marijuana intoxication is accompanied by a wide variety of behavioral changes including the euphoric mood state known as a "High" (Gold, 1989; Mathew *et al.*, 1993; Mathew and Wilson, 1993). Since acute, marijuana-induced behavioral changes are reversible, it would seem safe to assume that such changes may be brought about via alterations in global and/or regional brain function. The physiological basis for euphoria is believed to be increased levels of brain arousal (Edvinsson, 1982; Thomas, 1982; Paulson *et al.*, 1990). Arousal refers to generalized diffuse activation of the brain. In a non-drug induced state this may be mediated by the brain stem reticular activating system. Conditions characterized by increased brain arousal such as epileptic seizures and moderate degrees of anxiety are associated with CBF increase while low arousal states such as slow wave sleep and coma are accompanied by decreased CBF (Paulson *et al.*, 1990; Schacter and Singer, 1962). Arousal related CBF changes are most marked in the frontal regions (Mathew and Wilson, 1990; Ingvar and Lassen, 1976; Mathew, 1989) which is consistent with the heavy concentration of cannabinoid receptor sites.

Time sense

The ability to judge the passage of time and separate past, present and future are fundamental requirements for normal mental operations. There are two varieties of time sense, physical and personal. The first is determined by physical events such as the sunrise and sunset while the second is a subjective judgement of the passage of time. Temporal integration is a subjective framework through which an individual views and orients himself/herself. It is intimately related to memory formation, thought process and goal directed behavior. Temporal disintegration is a general term which refers to alterations in the rate, sequential ordering and goal-directedness of thinking processes.

In acute organic states, disorder of time perception is clearly shown in temporal disorientation (disorientation to time). Disorders of time perception are also seen in patients with temporal lobe lesions. A variety of drugs, especially hallucinogens, also are known to interfere with time sense (Tart, 1969; Kenna and Sedman, 1964; Huxley, 1963).

Marijuana seems to induce temporal disintegration more consistently than any other drug (Clark *et al.*, 1970; Tinklenberg *et al.*, 1976; Vachon *et al.*, 1974; Tinklenberg *et al.*, 1972). Melges and associates did a number of studies on temporal disintegration associated with marijuana smoking (Melges *et al.*, 1971; Melges *et al.*, 1970a; Melges *et al.*, 1974; Melges *et al.*, 1970b). The major effect was increased concentration on the present and the experience of slowing of the passage of time (Melges *et al.*, 1971; Melges *et al.*, 1974).

Depersonalization

Melges *et al.* (1970a) found an association between the fragmentation of temporal experience and drug induced depersonalization. Depersonalization is characterized by a feeling of detachment or estrangement from one's self, of being an outside observer, or as if living in a dream or a movie. Various types of sensory anesthesia, lack of affective response and a sensation of lacking control of one's actions, including speech, are often present. Depersonalization experience is an associated feature of such mental conditions as fatigue (Mayer-Gross, 1935), sleep deprivation (Bliss *et al.*, 1959), sensory deprivation (Reed and Sedman, 1964), anxiety (Noyes and Kletti, 1977; Roth and Argyle, 1988; Cassano *et al.*, 1989), depression (Fish, 1967), schizophrenia (Ackner, 1954), drug abuse (Good, 1989), and meditation (Castillo, 1990). It has also been reported in temporal-lobe epilepsy (Penfield and Kristiensen, 1951) and temporal-lobe migraine (Simpson, 1969). Depersonalization disorder is persistent or recurrent episodes of depersonalization, sufficiently severe to cause marked distress, in the absence of other psychiatric disorders (American Psychiatric Association, DSM 4th ed, 1994; Chee and Wong, 1990). In spite of its ubiquitousness, little is known about the brain mechanisms which underlie depersonalization, although a number of hypotheses have been advanced (Chee and Wong, 1990; Mellor, 1988; Sedman, 1970). Induction of depersonalization in laboratory settings is not easy and, therefore, the number of physiological studies of this phenomenon are few. Available studies suggest an association with increased levels of brain arousal (Lader, 1975; Kelly and Walter, 1968). Although difficult to induce using behavioral techniques, certain pharmacological agents are known to induce this phenomenon consistently (Good, 1989). Marijuana and its active ingredient tetrahydrocannabinol (THC) are known to induce an altered state of awareness associated with depersonalization (Johnson, 1990; Moran, 1986; Melges *et al.*, 1970a; Mathew *et al.*, 1993).

ACUTE CHANGES AFTER MARIJUANA

Each of the studies presented below were reviewed and approved by the Institutional Review Board (IRB) at Duke University Medical Center. Only volunteers with previous exposure to marijuana were recruited. All volunteers signed an approved written informed consent form prior to their participation in a study.

Studies with ^{133}Xenon and smoking marijuana

Studies of marijuana's effects on blood flow in animals has shown mixed results with some showing increased blood flow and others showing decrease or no change. Similar conflicting findings have been reported in humans with glucose metabolism (e.g. Volkow *et al.*, 1995). Part of the conflicting results may be due to the different effects of marijuana on psychological state which in turn drives regional variation in CBF. The present study was undertaken to examine the roles played by different behavioral and physiologic factors on CBF after smoking marijuana (Mathew and Wilson, 1993).

Subjects

Thirty-five normal male subjects (Table 8.1) participated in the study. Individuals with a history of recent marijuana exposure (and therefore unlikely to experience severe anxiety) were recruited through local advertising. Physical and psychiatric disorders were excluded with physical examination and psychiatric interview. Subjects with

Table 8.1 Subject description for marijuana CBF study

Age	27.1 ± 8.0 yr.
Trait anxiety score	32.1 ± 8.0
Beck depression score	1.3 ± 2.7
Began THC use	17.8 ± 3.5 yr.
No. THC CIGS smoked	84 ± 108.6/yr.
Tobacco	29.4% Yes
Alcohol use	90.6% Yes
Other drug use	58.8% Yes

current substance abuse (other than marijuana) and alcoholism were also not accepted for the study. Participants were instructed to avoid marijuana and medications (prescription and over the counter) for 2 weeks before the first visit to the laboratory until the end of the project. Urine drug screens were performed during each visit to the laboratory to confirm this. They were also told not to consume any alcohol for 24 h and nicotine and caffeine for 2 h before each laboratory visit.

Marijuana administration

Subjects had three visits to the laboratory, during which they smoked one of three marijuana cigarette: high potency (THC = 3.55%), low potency (THC = 1.75%) and a placebo cigarette (marijuana after THC extraction). The order of administration of marijuana and placebo was double blind basis and randomized. The visits were separated by a minimum of one week so that residual effects from one visit would not affect the measurements taken during the next one. The elimination half life for THC in experienced marijuana smokers is about 56 h. Thus, 7 days represent about three half lifes. Marijuana smoking routine was not standardized, but the subjects smoked all of the cigarette over a 10 min period. During each visit to the laboratory, the following measurements were taken.

CBF Measurements

On each day CBF was measured four times with the ^{133}Xenon inhalation technique before and 30 min, 60 min and 120 min after smoking the cigarettes.

Quantification of physiological and behavioral changes

Venous blood samples were drawn for plasma THC assays (radioimmunoassay) before and 5, 10, 20, 40, 130 and 170 min after smoking. End tidal carbon dioxide levels ($PECO_2$) were monitored during each CBF measurement. End tidal Carbon monoxide levels, pulse and respiratory rate were assessed before each CBF measurement.

The following rating scales were also administered prior to each CBF measurement: Profile of Mood States (POMS) (McNair *et al.*, 1971), Depersonalization Inventory (DPI) (Dixon, 1963) and Temporal Disintegration Inventory (TDI) (Melges *et al.*, 1970a). A 10 cm analog scale was used to assess degrees of "high" before each CBF measurement.

Statistical techniques

Procedures of analysis included multivariate analysis of variance (MANOVA) using Statistical Analysis System (SAS) procedure General Linear Models for all repeated measures variables and repeated measures ANOVA. Post hoc analyses between means made use of Tukey's procedure (Kirk, 1982). We also employed multiple regression and canonical correlation analyses to examine relationships between and among variables.

CBF changes

Analysis of the hemispheric mean CBF using a Drug by Hemisphere by Time model indicated a significant Drug by Time interaction (F=2.53, p≤.05) with significant global increases in CBF following low and high dose marijuana (Figure 8.2). Post-hoc tests for each drug based on global mean flow found no differences across time for placebo; for low dose baseline differed from 30 min and 30 min differed from 120 min, and for high dose baseline differed from 30 and 60 min but not from 120 min and 30 min differed from 120 min. All post hoc differences were <.01 level. CO_2 is a potent cerebral vasodilator (Maximilian *et al.*, 1980). Adjusting CBF for $PECO_2$ values with a predetermined correction factor (3% CBF per mm $PECO_2$), did not change the results.

The [133]Xenon inhalation technique measures gray matter perfusion to the cerebral cortex underlying the scalp scintillation detectors in each hemisphere. Although the technique does not have fine spatial resolution, it is possible to separate blood flow to large brain regions. We calculated flow to the anterior (frontal and central) and posterior (temporal, parietal and occipital) regions for each hemisphere and submitted these to a Drug by Time by Hemisphere by Anterior-Posterior (A-P) model (Figure 8.3). In addition to the main effect of A-P (F=372.8, p ≤ .0001), this analysis indicated several significant interactions including Drug by

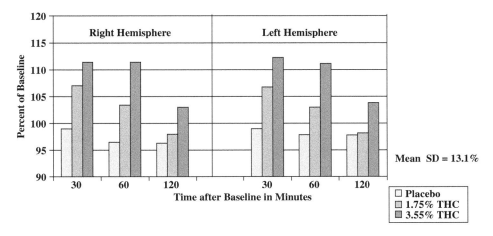

Figure 8.2 Percent change in hemispheric mean CBF after smoking marijuana.

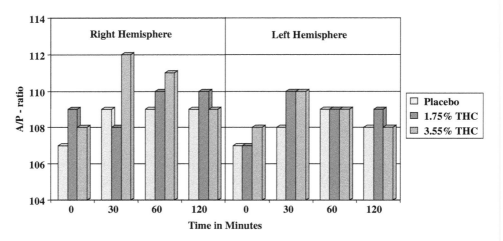

Figure 8.3 Anterio-posterior gradient of CBF after smoking marijuana or placebo.

A-P (F = 3.26, p ≤ .05), A-P by Time (F = 7.52, p ≤ .001) and Drug by Time (F = 2.50, p ≤ .05). CBF increases were stronger in the anterior part of each hemisphere, consistent with the relative density of cannabinoid receptor sites.

Behavioral changes

All ratings scales were analyzed with Drug by Time models. Intoxication ratings ("high") indicated a significant Drug by Time interaction (F = 10.95, p ≤ .001). The greatest increase (Figure 8.4) in ratings was at 30 min post 3.55% THC.

POMS factors anxiety and confusion showed significant changes. Anxiety appeared to show the most change, at 30 min. following 3.55% THC, (Drug by Time F = 3.51, p ≤ .01). Although statistically significant, the actual increase in anxiety was modest (baseline mean = 4.9 SD 5.1, 30 min mean = 6.7 SD 6.8). Similar changes were seen after low dose marijuana but not after placebo. The Confusion factor did not indicate a significant interaction, but there was a significant main effect of drug (F = 5.16, p ≤ .02) with scores after high dose being elevated most. DPI and TDI findings are presented below.

Physiological and pharmacological changes

Analysis of THC plasma levels (ng/ml) indicated a highly significant Dose by Time interaction (F = 17.71, p < .001) (Table 8.2). Peak plasma levels were at 5 min for both doses of THC. Analyses conducted for the systolic blood pressures indicated slight but statistically significant Drug by Time effects (F = 3.08, p < .02). Consistent with many previous reports (Brill and Nahas, 1984) pulse rate (Table 8.2) increased in a fashion consistent with dose, leading to a highly significant Drug by Time interaction (F = 14.79, p < .001). Analysis of respiration rate indicated a significant time effect for the three drug conditions, with a slight but statistically

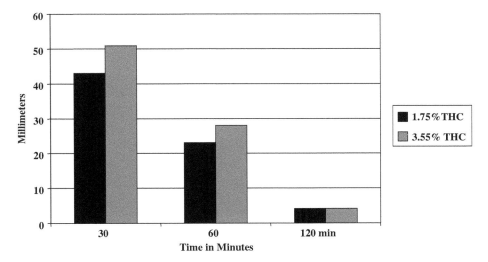

Figure 8.4 Intoxication ratings after smoking marijuana measured with analog rating scale.

significant increase from baseline with the greatest increase being at 60 min ($F = 3.06$, $p < .05$). $PECO_2$ did not change across time and there was no Drug by Time effect. This variable was not entered into any of the regression models, but was used to correct the CBF measures across time. Analysis of carbon monoxide levels indicated a significant time effect, but no differences between drugs and placebo, which indicates that there was no differential effects of CO.

Changes in time sense and its correlates after marijuana smoking

In this group of 35 subjects, disturbance of time sense was measured with the Temporal Disintegration Inventory (TDI), a rating scale developed by Melges *et al.* (1970a). The TDI has a scoring range from −35 to +35 with the lower values

Table 8.2 Physiological measures following smoking marijuana

Times		Plasma levels THC				Pulse rate			
		B(*)	30	60	120	B(*)	30	60	120
Placebo	M	3.2	3.8	3.4	3.1	73.9	71.6	67.2	65.4
	SD	2.3	3.1	2.3	2.2	12.5	11.5	12.4	12.4
1.75%	M	3.5	36.0	13.9	10.0	75.2	92.9	73.3	67.2
	SD	2.2	21.4	8.7	7.9	12.8	19.6	13.7	11.7
3.55%	M	3.9	52.2	17.7	12.2	71.4	98.3	79.5	69.9
	SD	3.9	45.0	12.3	8.1	11.5	17.4	16.5	12.5

Drug × Time: $F = 17.71$, $p \le .001$ Drug × Time: $F = 14.79$, $p \le .001$
Plasma Levels in ng/ml   Pulse rate in beats/min
(*) B = baseline, 30, 60 and 120 min.   (*) B = baseline, 30, 60 and 120 min.

(−35) indicating better temporal integration. One sub-scale consists of 14 statements which can be used to compute a total score and two sub-scales. Eight statements relate to temporal distinction – that is orderly indexing of memories as past perceptions, as present experiences and as future expectations without confusion of these temporal categories. The remaining 6 items relate to goal directedness – that is adjusting plans of action to reach goals. Plans of action are hierarchies of sequential acts; goals are desired outcomes of these acts. As described above, measurements were made before and 30, 60 and 120 min after smoking marijuana or placebo.

Results

Temporal Disintegration Inventory's (TDI) total score and sub-scales were analyzed using a Drug by Time multivariate model. Analysis of this scale indicated a statistically significant Drug by Time interaction (F=3.65, p < .05) with most marked changes in the high dose condition at 30 min after smoking (Table 8.3). Both sub-scales also showed significant Drug by Time interactions on repeated measures ANOVA (Table 8.3). It should be noted that there was essentially no change across the placebo condition. The changes after marijuana were in the direction of increased disturbance of time sense.

Relationships among temporal disintegration and other variables were explored with a multiple regression analysis, performed with the data from all time points for all subjects. This model predicted scores on the TDI using a hierarchical model which initially removed subject variability, and then rating scale and CBF regions were allowed to enter in a stepwise additive model. Separate analyses were

Table 8.3 Temporal disintegration scale component scores

Total score (a) Time (b)	Placebo		1.75% THC		3.55% THC	
	Mean	SD	Mean	SD	Mean	SD
B	−19.6	9.2	−20.0	9.2	−20.2	8.3
30	−19.0	9.9	−15.7	10.60	−13.5	12.0*
Repeated measures anova F = 3.65, p < .01						
Component 1: Temporal distinction Time (a)						
B	−11.6	4.6	−11.6	5.3	−11.9	5.1
30	−11.4	5.0	−8.4	6.6*	−8.0	7.3**
Repeated measures anova F = 4.05, p < .001						
Component 2: Goal directedness Time (b)						
B	−8.0	5.9	−8.4	4.9	−8.3	4.5
30	−7.6	5.8	−7.3	5.1	−5.5	5.4*
Repeated measures anova F = 3.08, p < .007						

(a) Scores range from −35 (best) to +35 (worst); (b) Time B = baseline 30 = 30 min post smoking. Differences compared to placebo change at same time point; * −p ≤ .05; ** −p ≤ .01.

Table 8.4 Multiple regression to predict temporal
disintegration

	Partial r*	p ≤
Brain Region CBF		
Temporal right	0.12	.02
Parietal left	0.16	.002
Frontal right	0.10	.05
Rating Scales		
Depersonalization	0.52	.001
Confusion**	0.19	.001
Intoxication	0.14	.007

* Partial correlation with all variables entered; ** Factor
from the Profile of Mood States.

conducted with the rating scales and CBF. Three regions (temporal and frontal
right and parietal left) entered for CBF (Table 8.4). Three rating scale variables
also co-varied significantly with changes in TDI: depersonalization, Confusion
(a factor of the Profile of Mood States) and self ratings of intoxication.

Depersonalization and its correlates after marijuana smoking

Depersonalization involves an alteration in the perception or experience of the self
in which the usual sense of one's own reality is temporarily lost or changed. Marijuana
is known to induce an altered state of awareness associated with depersonalization
(Johnson, 1990; Moran, 1986). In this study we used a rating scale developed by
Dixon(1963) to assess depersonalization. Melges and associates found an association
between temporal disintegration (altered time sense) and depersonalization after
oral administration of marijuana extract containing 20, 40 and 60 mg of tetrahydrocannabinol (THC) in 8 normal volunteers (Melges *et al.*, 1970a).

Data analysis

To examine the relationships between depersonalization and CBF and among
these two and the other rating scales and physiological measures, multiple regression analyses were employed to determine partial correlations with the effects of
all other variables in the model removed. We report partial correlations (pr)
because this is the unique correlation of one independent variable (e.g. ratings)
with the dependent variable (i.e., CBF) when other variables have been removed.
In these regression models, analyses were performed with the data from all time
points for all subjects. Subject differences were regressed first, so that covariation of
rating scales and other measures across time could be analyzed. These models
predicted DPI using a hierarchical approach, initially removing subject variability
and then rating scale and physiological variables were allowed to enter on a stepwise additive model.

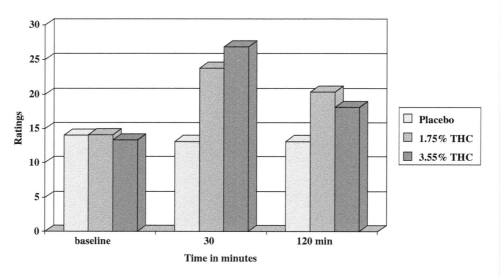

Figure 8.5 Depersonalization ratings after smoking marijuana or placebo.

Results

Analysis of the DPI total score, using a Drug by Time multivariate model (Figure 8.5) indicated a statistically significant Drug by Time interaction ($F = .249$, $p \leq .05$) with the greatest change (relative to the corresponding baseline) occurring at 30 min following 3.55% THC cigarettes. This pattern was also the case for the 1.75% THC cigarettes. There was no change from baseline following placebo.

Regression analyses

Multiple regression analysis was performed with the data from all time points for all subjects. This model predicted DPI using a hierarchical model which initially removed subject variability and then, rating scale and CBF variables were allowed to enter on a stepwise additive model. Again we conducted separate analyses for CBF and rating scales. Presented in Table 8.5 are six variables that had significant partial correlations (Pr) with DPI in their respective models. The most significant predictor (largest pr) was TDI with $pr = 0.52$, $p \leq .001$.

A separate multiple regression analysis was conducted to examine the relationship between regional changes in CBF and DPI to determine whether there was any significant co-variation with regional CBF. Again a hierarchical model was employed to force the removal of subject variability before the CBF variables were entered. For DPI, two regions entered significantly (Frontal Right-partial $r = 0.23$, $p \leq 0.001$ and Parietal Left-partial $r = -0.16$, $p < .003$) with the parietal change being negatively related to change in DPI.

Table 8.5 Multiple regression to predict depersonalization

	Partial r*	p ≤
Brain Regions		
Frontal right	0.23	.001
Parietal left	−0.16	.003
Rating Scales		
Temporal disintegration	0.52	.001
Tension**	0.14	.006
Intoxication	0.24	.001
Somatic sensation	0.14	.006

* Partial correlation with all variables entered; ** Factor from the Profile of Mood States.

Canonical correlational analysis

Canonical correlation analysis is a procedure that looks for the best weighted linear combination of predictor variables on one side of the equation and another weighted linear combination of criterion variables on the other side of the equation. These weighted combinations are called canonical variables. In this canonical correlation analysis we employed the data (both the 10 regional CBF values and rating scales) from all subjects for all time points on all three drug conditions. This permitted us to look at the relationship between change in regional CBF across time and rating scales in a simultaneous equation. We removed subject means from the data so that the relationship would be based on the pattern of change (Table 8.6). The first canonical correlation (0.405) was significant (p ≤ .001). On the CBF side of the equation, the structure coefficients (correlations of the

Table 8.6 Structure coefficients between variables and their canonical variable and the associated canonical variable

	CBF (a)	Scale (b)		Scale	CBF
Fr(c)	0.71	0.29	Dpi	0.69	0.28
Cr	0.50	0.20	State anxiety	0.46	0.18
Tr	0.66	0.27	Somatic sensation	0.47	0.19
Pr	0.47	0.19	TDI	0.55	0.22
Or	0.55	0.22	Intoxication	0.89	0.36
Fl	0.64	0.26	Tension	0.37	0.15
Cl	0.39	0.16	Confusion	0.59	0.24
Tl	0.44	0.18			
Pl	0.37	0.15			
Ol	0.50	0.20			

Canonical Correlation 0.405 p < .001; (a) CBF canonical variable of regional CBF; (b) SCALE canonical variable of scales; (c) FR to OL – Frontal right to occipital left.

individual variables with their respective canonical variable which indicates which variables or regions define the canonical variable) were highest (rank order) for frontal right, temporal right and frontal left and smallest for parietal left. On the rating scale side, the self rating of "high" was greatest followed by depersonalization, confusion and temporal disintegration. Thus, changes in CBF in the frontal and temporal regions particularly on the right side were most heavily correlated with a pattern of psychological changes assessed by these four rating scales.

Studies with positron emission tomography and THC infusion

Depersonalization and intoxication

We examined the relationship between depersonalization induced by marijuana smoking and associated changes in regional cerebral blood flow (CBF) measured with the ^{133}Xenon inhalation technique (Obrist *et al.*, 1975). While marijuana increased CBF especially in the non-dominant hemisphere and frontal lobes, no clear association between depersonalization and regional CBF was identified. This may be because the ^{133}Xenon inhalation technique has poor spatial resolution compared to positron emission tomography (PET) and does not provide information on subcortical flow. In the study described below, we utilized PET and THC infusion (Mathew *et al.*, 1999).

Subjects

Study participants were screened by the first author, who excluded significant physical or psychiatric disorders, abuse or addiction to any drug other than marijuana during the previous 6 months, current use of any prescribed or unprescribed medication, current vascular disorders including migraine and heavy alcohol use (more than two drinks per day for males and one drink per day for females). Substance abuse was evaluated according to the structured clinical interview for DSM-III-R. While all were reported to have used marijuana, none met the criteria for abuse or dependence for alcohol or other drugs. All subjects completed the Beck Depression Inventory (Beck and Beck, 1972), State-Trait Anxiety Inventory (Spielberger *et al.*, 1970), Drug Abuse Screening Test (DAST) (Skinner, 1982) and the Brief Michigan Alcoholism Screening Test (MAST) (Selzer, 1971). All participants were predominately right-handed, verified with the Harris Test of Lateral Dominance (Harris, 1974). None of the participants scored positive on MAST and DAST. Medical history, physical examination, electrocardiogram and clinical laboratory tests were obtained on all participants and yielded normal results. Pregnancy was excluded in women of child-bearing potential with plasma HCG tests.

The 59 subjects had a mean age of 31.8 years and a median of 32. There were 33 males (mean age of 33.5, SD 8.3 years), and 26 females (mean age of 29.8, SD 6.6 years). Of the 59 subjects, 17 either currently smoked cigarettes (N = 14) or had in the past but did not at present (N = 3), and all reported use of less than a pack per day. Most subjects were social drinkers (49 out of 59). On average, they began using marijuana in their teen years (16.8, SD 3.6 years). ANOVA indicated no difference in age of onset of marijuana use by dose-group nor were there differences by gender. The mean Beck Depression Score was 1.4, SD 2 and the Trait Anxiety Scores of this STAI was 30.1, SD 6.1. The mean MAST was 3, SD 4.8 and the mean DAST 2.9, SD 2.5. There were no differences among the three groups on Beck, Trait Anxiety, MAST or DAST scores by ANOVA.

THC administration

Using a double-blind, randomized between groups design, the subjects were assigned (males and females independently) to one of three groups: placebo infusion (11 m/10 f), low-dose (0.15 mg/min) THC infusion (13 m/6 f), or high-dose (0.25 mg/min) THC infusion (9 m/10 f). Subjects received an IV infusion for 20 min of either THC suspended in human albumin or human albumin as described by Perez-Reyes *et al.* (1972). Chi-square test and analysis of variants (ANOVA) indicated no differences in age and sex distributions of the subjects by dose-group. They were required to abstain from marijuana for 2 weeks, alcohol for 24 h and nicotine and caffeine for four hours before visits. A urine drug screen was conducted before PET scans which indicated no recent use for any participant.

Magnetic resonance imaging (MRI)

MRI with both T_2-weighted and T_1-weighted images was performed and used for region identification in PET scans. Transaxial MRI scans were performed with a General Electric 1.5 Tesla Sigma System. Computer algorithms (Chen *et al.*, 1989; Turkington *et al.*, 1995) were used to anatomically align and register PET images with the MRI. Regions of interest (ROI) were defined on each slice of the MRI and then transferred to the PET images. Accuracy of the computer algorithms used to anatomically align and register PET and MRI is in the order of 1–2 mm (Turkington *et al.*, 1995). One person, blind to group assignment, did all region identification for all subjects, using a set of rules for identifying the regions of interest (Talairach and Tournoux, 1988). The MRI also was utilized to establish segmentation criteria to separate white from gray matter. Use of segmentation criteria to identify tissue types has been shown to be a reliable method (DeCarli *et al.*, 1992; Byrum *et al.*, 1996).

PET scan

Measurements of CBF utilized ^{15}O-water and the autoradiographic method (Eichling *et al.*, 1974; Frackowiak *et al.*, 1980; Huang *et al.*, 1983). The first PET scan was obtained under resting conditions. Following this baseline scan, the infusions were given under double-blind conditions. PET scans were repeated at 30, 60, 90 and 120 min after the start of the THC infusion. CBF measurements were performed in a semi-dark room with eyes and ears unoccluded. Subjects were instructed not to move. Arterial blood samples were drawn during each scan for the determination of the ^{15}O-water input function.

A volume weighted mean CBF for gray matter was determined for each ROI by computing the CBF for each pixel, summing all pixels across all slices containing the ROI and dividing by the number of pixels in the ROI volume. Pixels from the PET images were then included in an area weighted mean gray matter CBF only if their signals on the MRI were in the gray matter range.

Rating scales

The rating scales, administered before the first PET scan and repeated after other PET scans, included the Analog Intoxication Scale, Depersonalization Inventory (DPI) (Dixon, 1963), Somatic Sensation Scale (Tyrer, 1976) and the anxiety-tension sub-scale of Profile of Mood States (POMS) (McNair et al., 1971). A 10 ml venous blood sample was drawn following each PET scan to determine plasma levels of Δ^9-tetrahydrocannabinol.

Data analysis

The analysis had as its objective the evaluation of regional differences in CBF change over time in response to THC or albumin-placebo infusion, and examination of the relationship between regional CBF change to ratings of depersonalization. Analysis of variance (ANOVA) with a repeated measures model was employed to compare regional CBF over time between groups in the following model: Dose Group (placebo, 0.15 mg/min and 0.25 mg/min) by Hemisphere by ROI by Time (baseline, 30, 60, 90 and 120 min). In this model hemisphere, region and time were treated as multiple (or repeated) measures. The Greenhouse-Geiser (GG) correction of the degrees of freedom was employed to provide a conservative F-test of the repeated measures factors. Post hoc contrasts between means were carried out with the Least Significant Difference test (LSD-test), which keeps the alpha level constant for all pairwise contrasts (Kirk, 1982). Regression analysis was employed to relate change in CBF over time to change in ratings. The model employed forced subject variables and plasma levels (as covariates) into the model first and then allowed rating scales data to enter stepwise.

Results

Presented in Figure 8.6 are two PET images for a subject at baseline and 30 min post THC. This image shows the significant increase in CBF in most frontal cortical regions.

Analysis of cortical regions using a dose-group by hemisphere by region, by time model indicated a dose-group by time, by region interaction (F = 2.81; df = 48,1056; p ≤ .001). We further analyzed this interaction for regional differences by conducting a simple main effects analysis on the individual region using a Group by Hemisphere by Time model. These analyses indicated significant dose-group by time interactions for frontal (p ≤ .001), parietal (p ≤ .011), occipital (p ≤ .001), insula (p ≤ .006) and anterior cingulate (p < .027). We carried out post-hoc

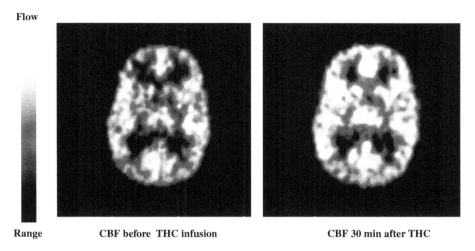

Figure 8.6 Regional CBF with PET. (*See Color plate 2*)

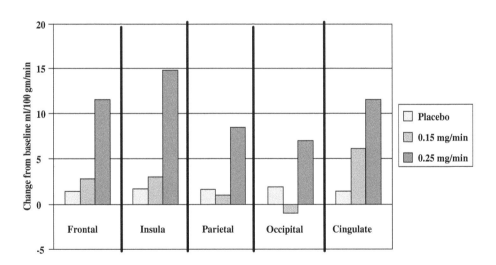

Figure 8.7 Regional differences in CBF response 30 min after THC or placebo infusion.

analyses on the change from baseline and then compared placebo to high and low-dose THC groups and low to high-dose groups. These analyses indicated THC produced increased CBF in most brain regions at 30 min for both dose-groups compared to the placebo group, with somewhat greater differences after high-dose THC. There was a gradient of cortical response as seen in Figure 8.7. The largest increase was in the anterior cingulate and insula. Comparisons between low and high-dose THC found significantly greater increases in the right frontal and right insula.

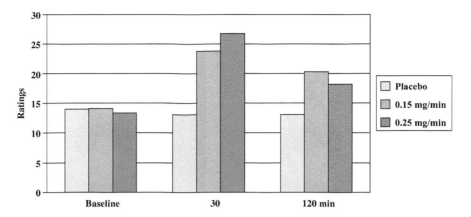

Figure 8.8 Depersonalization ratings after THC or placebo infusions.

The sub-cortical regions included basal ganglia, thalamus, amygdala and hippo-campus. All sub-cortical regions showed decrease from baseline once the global effect was removed. Post-hoc LSD analyses indicated a significant reduction of relative CBF compared to placebo.

Analysis of variance of the rating scales using a dose-group by time model showed significant effects for depersonalization (Figure 8.8) (F = 11.15, df = 4,112, p ≤ .001), Somatic Sensation (F = 4.89, df = 4,112, p ≤ .001), anxiety (F = 3.44, df = 4,112, p ≤ .001) and intoxication (F = 24.86, df = 4,112, p ≤ .001).

In order to determine the relative relationship between depersonalization ratings and regional CBF, after accounting for other variables, we computed hierarchical regression analyses of each ROI, utilizing the absolute CBF. In these analyses, we were modeling CBF over time after first removing global effects, subject specific variation and plasma level of THC. Then we allowed rating scale data to enter in a stepwise fashion. Subject specific variability was removed by including subject as a dummy variable (Cohen, 1988).

DPI had a strong first order correlation with ratings of intoxication across time (r = .59); therefore, we report partial correlations (pr) because this is the unique cor-relation of one independent variable (e.g. ratings) with the dependent variable (i.e., CBF) when other variables have been removed. This permits an examination of which ratings show significant covariation over time with CBF. Depersonalization showed significant positive partial correlations with two ROI (right frontal: pr = 0.20, p ≤ .05 and right cingulate: pr = 0.32, p ≤ .002). Intoxication ratings also had significant cor-relation with right frontal (pr = .41, p ≤ .001), but not with cingulate blood flow.

Plasma level (Figure 8.9) was also analyzed and as expected, showed a signific-ant dose-group by time interaction (F = 56.4, df = 8, 196, p < .001).

Time sense after tetrahydrocannabinol administration

Marijuana intoxication is known to cause changes in the perception of time (Clark *et al.*, 1970; Mathew and Wilson, 1996; McNair *et al.*, 1971; Melges *et al.*, 1970a; Melges

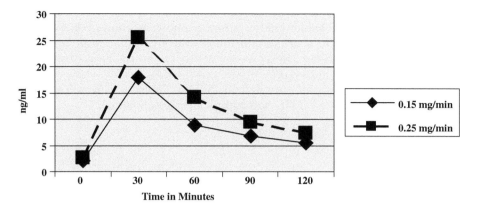

Figure 8.9 Plasma levels after THC infusion.

et al., 1970b; Melges *et al.*, 1971; Spielberger *et al.*, 1970; Tinklenberg *et al.*, 1972; Tinklenberg *et al.*, 1976). An association between time perception and cerebellar function has been reported (Keele and Ivry, 1991). Studies with normal subjects and those with previous marijuana use have reported increased cerebellar metabolism following THC administration (Vachon *et al.*, 1974; Volkow *et al.*, 1991b). In addition, marijuana users were found to have decreased cerebellar metabolism at baseline. Although marijuana intoxication was found to correlate with CMR (Vachon *et al.*, 1974; Volkow *et al.*, 1991a; Volkow *et al.*, 1991b), no information is available on the relationship between marijuana induced alterations in time sense and cerebellar activity. In our previous study, utilizing marijuana smoking and the [133]Xenon inhalation procedure we found that there was a change in time perception (Mathew *et al.*, 1993), but no association with regional CBF, which may be due to the lack of spatial resolution of this technique. We were also not able to measure cerebellar blood flow with this technique. We utilized PET to study the effects of THC infusion on CBF and its effects on the perception of time in 46 volunteers (Mathew *et al.*, 1998).

Subjects

Subject recruitment and screening were the same as previously described for the study above. Data from 46 subjects (22 male and 24 female) with a mean age of 29.9 + 6.5 years are reported below. All had previously used marijuana with a mean estimated use of 147 ± 165.2 "joints" per year. All had a high school degree or better. Analyses indicated no differences in any of these variables among the three study conditions at baseline.

Procedures

Subjects came to the PET Laboratory after abstaining from marijuana for two weeks, alcohol for 24 h, nicotine and caffeine for 4 hours before the visit. A urine drug screen was conducted before PET scans which were negative on all participants. The subjects were randomly assigned (males and females were randomized separately) to receive double-blind either an intravenous infusion of Δ^9-tetrahydrocannabinol (0.15 mg/min or 0.25 mg/min for 20 min) or human albumin (placebo condition) as described (Perez-Reyes *et al.*, 1972). Venous blood samples were obtained for plasma THC measurement after each CBF. Δ^9-THC was assayed with radioimmunoassay using a kit supplied by NIDA. Subjects were allowed to go home only after the effects of THC had completely disappeared.

Rating

Scales were administered before the first measurement and repeated after other scans. The scales administered included an analog intoxication scale, Depersonalization Inventory (Dixon, 1963), Temporal Disintegration Inventory (Melges *et al.*, 1970a) and the Tension-Anxiety subscale of the Profile of Mood states (POMS) (McNair *et al.*, 1971).

PET scans

CBF measurements using ^{15}O-water were performed as previously described. The first PET scan was obtained under resting conditions, then scans were repeated at 30, 60, 90 and 120 min.

Magnetic resonance imaging

MRI was performed for region identification and segmentation criteria. Analyses were based on area weighted mean CBF from the ROI. Mean blood flow also was determined for each posterior lobe of the cerebellum.

Data analysis

MANOVA was employed to compare groups in a Group by Hemisphere by Time model for the cerebellar flow; and a Group by Hemisphere by Region (frontal, temporal, parietal, occipital, insula, anterior cingulate) by Time model for the cortical and sub-cortical flow data. The hemisphere, region and time factors were treated as multiple (or repeated) dependent measures. This analysis was followed by ANOVA analyses when appropriate. These analyses included baseline, 30 and 60 min data only. The approximate F based on the Wilk's lambda statistic is reported for the multivariate data. Multiple regression analysis over time, which initially removed global flow changes and effects of plasma level, was employed and partial correlations determined between cerebellar blood flow and TDI.

Results

Mean cerebellar and regional cerebral blood flows are presented in Table 8.7. Table values are change-from-baseline at 30 min post infusion. The MANOVA analysis of cerebellar flow indicated a significant Group by Time interaction

Table 8.7 Change in cerebellar and cerebral blood flow 30 min after Delta-9 THC

	Placebo		0.15 mg/min		0.25 mg/min	
	Mean	Std dev	Mean	Std dev	Mean	Std dev
Left hemisphere						
Frontal	2.19	3.28	2.92	10.01	11.66	13.31
Temporal	2.21	4.12	1.68	9.88	10.19	11.65
Parietal	2.76	4.28	1.30	9.30	9.51	12.78
Occipital	3.04	5.76	−.80	9.58	8.38	14.17
Insula	2.30	7.30	4.13	14.37	16.81	15.18
A-cingulate	5.35	5.73	6.25	16.26	11.42	15.54
Sub-cortical						
Basal ganglia	0.41	6.32	1.73	12.71	11.24	13.00
Thalamus	1.19	7.48	−.96	12.38	10.55	18.25
Amygdala	4.51	7.31	3.16	11.00	8.27	11.75
Hippocampus	2.63	7.31	−1.20	10.67	10.66	14.29
Global mean	1.47	3.77	1.75	8.74	10.44	13.05
Cerebellar	2.78	4.99	1.78	10.30	13.73	17.39
Right hemisphere						
Frontal	2.71	4.30	2.34	10.00	13.46	13.96
Temporal	3.18	4.42	1.58	10.69	11.07	12.04
Parietal	2.72	5.79	.97	8.81	9.74	13.26
Occipital	3.36	5.71	−.98	10.76	7.72	13.89
Insula	5.08	6.50	2.72	14.82	17.35	17.91
A-cingulate	0.97	7.21	5.54	14.56	16.13	14.84
Sub-cortical						
Basal ganglia	2.48	5.25	1.50	15.20	11.24	12.84
Thalamus	3.34	6.58	−1.61	12.32	13.11	18.57
Amygdala	4.60	6.08	0.87	11.15	7.13	9.68
Hippocampus	1.89	5.39	−.89	11.00	9.14	10.83
Global mean	1.96	4.55	1.47	8.89	11.5	13.89
Cerebellar	2.36	5.01	1.99	10.70	13.56	17.71

(Wilk's lambda $F = 3.02$, df = 4,84, $p \leq .022$). There was no difference between left and right lobes by group over time. Cortical CBF for six ROI (frontal, temporal, parietal, occipital, insula and anterior cingulate) in each hemisphere, were analyzed in a Dose-group (placebo, low and high dose THC) by Hemisphere by ROI by Time (baseline, 30 and 60 min) MANOVA model which indicated a Group by Region by Time interaction (Wilk's lambda $F = 2.47$, df = 20,68, $p \leq .003$). Post hoc analyses of these regional values indicated significant change in frontal, cingulate and insula.

Analysis of four sub-cortical regions (basal ganglia, thalamus, amygdala and hippocampus) was also carried out. This analysis indicated a significant Dose-group by Region by Time interaction (Wilk's lambda $F = 1.95$, df = 12,78, $p < .040$). These data indicate a significant increase in blood flow following THC compared to placebo (Table 8.7).

To examine further the effects of THC on cerebellar flow, we computed standardized change from baseline (absolute change divided by standard error of the mean of the baseline CBF) in CBF for those 31 subjects who received THC. We then grouped subjects as to whether their standardized change fell within or outside one SEM unit. These data indicated approximately 60% of the subjects had an increase greater than 1 SEM in blood flow, while about 24% had a decrease greater than 1 SEM. When we examined standardized change on the basis of dose, greater numbers of subjects showed positive change in CBF following higher than low dose THC. Thus, consistent with previous reports (Volkow *et al.*, 1991a; Volkow *et al.*, 1996), there was variability in response to THC, with a number of subjects showing a decrease in CBF and cerebellar blood flow after THC.

Possible differences between those who did or did not have an increase in CBF after THC were analyzed for the 31 THC-subjects by grouping them according to change for the cerebellar flow: increased ($\geq +1$ SEM) CBF (n = 20) versus decreased or no change ($< +1$ SEM) (n = 11). We then examined differences between these two groups for age, sex, plasma levels and rating scale changes from baseline to the 30 min time point. The decrease-CBF group included 6 male and 5 female and the increase-CBF group included 11 male and 9 female subjects. Change in only one variable (Temporal Disintegration Inventory) separated these two groups significantly (Decrease CBF: TDI mean = 12.1818 ± 11.7; increase CBF: TDI mean = 1.1000 ± 10.1). Those who had a decrease in CBF had a greater increase in TDI ($\dagger = 2.68$, df = 29, p $\leq$ .012). Analysis of TDI total score by the three study dose conditions (placebo, low, high THC) failed to show a significant change in TDI, probably because of the above finding.

We also examined possible differences in the time course of regional cortical and cerebellar blood flow in a MANOVA analysis of those who received THC, using a Region (frontal, temporal, parietal, occipital, anterior cingulate, insula and cerebellar) by Hemisphere by Time (30 and 60 min). This analysis also included as a grouping factor whether subjects had an increase or decrease of cerebellar flow (as in the above analysis). There was a significant Group by Region by Time interaction (Wilk's lambda F = 2.54, df = 12,18, p $\leq$ 0.036). Presented in Figure 8.10 are the global and cerebellar CBF values at 30 and 60 min, expressed as percent of baseline to make differences from baseline easier to compare. Computation of global CBF does not include blood flow to the cerebellum. The maximum increase was seen for at 30 min post infusion, with CBF moving towards baseline at 60 min.

Regression analysis was utilized to examine the relationship of CBF change over time in the cerebellum to change in other brain regions for the 31 who received THC. We computed partial correlations after removing the effects of plasma level and global CBF. The ROI to show the greatest positive partial correlation (pr) was the pons (Right pr = 0.417, p $\leq$ .001; Left pr = .441, p < .001) followed by the temporal lobes (Right pr = .267, p < .001; Left pr = .242, p < .002).

Partial correlations (pr) of change of CBF over time of each ROI with the change in TDI found none of the correlations were significant; however, two regions had correlations that approached significance (i.e. p < 0.10) left cerebellar pr = −.217 and right cerebellar pr = −.235. When we added the measure of anxiety as a covariate, the right cerebellar lobe became significant (pr = −.261, p $\leq$ .044) and the left cerebellum also approached significance (pr = −.245, p $\leq$ .059).

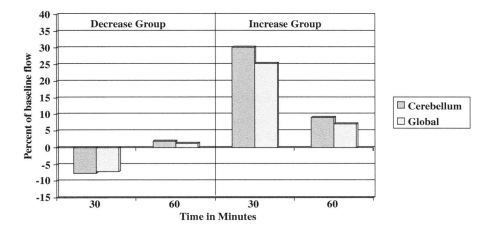

Figure 8.10 Global and cerebellum change in blood flow for all subjects after THC infusion.

Summary and implications of acute effects

Following marijuana consumption there are consistent effects on brain function including a feeling of intoxication, alteration in the perception of the passage of time and distortion of the sense of self. Associated with these are increases in anxiety and confusion. We have shown that following marijuana intoxication (both after smoking and after infusion of THC) there are significant increases in CBF over most of the brain. However, there are clear regional differences, and in particular in the frontal cortex, cingulate and cerebellum. This pattern of CBF change is consistent with the reported distribution of cannabinoid receptor sites, but the changes are not a simple pharmacokinetic effect. These CBF changes are more highly correlated with changes in mental state than plasma levels. Changes in the perception of intoxication are correlated with changes in frontal function, depersonalization is correlated with CBF increase in right frontal and right cingulate and alteration of time sense was correlated with temporal and cerebellar flow. Interestingly, the subjects with the greatest increase of temporal disintegration had a reduction in cerebellar flow, while those who had an increase in cerebellar blood flow did not have a significant increase in the scores on the Temporal Disintegration Inventory.

These studies and those of numerous other investigators show that marijuana and its active ingredient THC have pronounced effects on brain function. It has been shown to affect memory function (Wilson *et al.*, 1994) and other cognitive and motor processes (Solowij, 1999). The findings on the effects of alteration of time sense and depersonalization and their associated effects on brain raise the possibility that these effects may play a role in other known effects of marijuana such as memory impairment. This possibility is suggested by the findings recently demonstrating that patients with depersonalization disorder performed significantly worse on measures of short-term memory and attention (Guralnik *et al.*, 2000). It also suggests other potential areas of effects. For example, while marijuana is known to have effects on driving related behaviors (Wilson *et al.*, 1994) these effects have been shown to be worse when marijuana is combined with alcohol (Robbe, 1998).

Chronic effects related to marijuana use

Drug induced phenomena may be subsumed under acute effects, withdrawal, chronic/residual (possibly reversible) effects and permanent effects that may be associated with morphological changes. Demarcation among these categories is clearer with certain drugs but less so with others. For example, in the case of ethyl alcohol, phenomena related to intoxication (acute effect), withdrawal (delirium tremens in extreme cases), reversible functional changes (cognitive changes including memory defects) and gross morphological changes (diffuse brain atrophy with ventricular and sulcal dilation) are well described. However, in the case of marijuana, such distinctions are less clear. Many studies are available on marijuana's acute effects. While greater attention has been focused recently on determining potential chronic effects of marijuana, much less is known about potential chronic morphological brain changes in humans.

Animal studies

Animal data seem to support chronic effects, perhaps permanent changes in brain neurochemistry, morphology and function following exposure to cannabinoids during critical periods of development. Data show that early exposure may have chronic effects on brain development. The interface between morphology and function involves neurochemistry. Walters and Carr (1988) reported that perinatal exposure to marijuana and other cannabinoids leads to a reduction of tyrosine hydroxylase activity in the striatum in rats. Tyrosine hydroxylase determines the rate of dopamine synthesis and is a rate limiting factor in the production. The regulation of transmitter synthesis is controlled by factors that influence the activity of tyrosine hydroxylase (Lovenberg *et al.*, 1978). It also has been shown that early postnatal exposure to cannabinoids alters beta-endorphin levels in adult rats (Kumar *et al.*, 1990).

Fernández-Ruiz *et al.*(1992) reported a series of studies with rats exposed to Δ^9-THC during pre-, peri- and postnatal periods of development. They studied the effects of maternal consumption of cannabinoids during gestation and lactation on the development of nigrostriatal, mesolimbic and tuberoinfundibular dopamine systems. In these studies, pregnant rats were fed with hashish extracts daily from day five of gestation to postnatal day 24. Pups were then studied until day 40. Changes were reported in the ontogeny of all three dopamine systems, and there was evidence of sex differences in these effects, with greater effects in males. Changes included differences in the number of D_1 and D_2 receptors (high in males and low in females) and decrease in tyrosine hydroxylase activity in males. Changes seen at day 40 (16 days after exposure ended) were most likely not dependent on the presence of drug, because it has been reported that the half-life of THC and metabolites in rat is approximately 17 h (Klausner and Dingell, 1971). An important point from these data is that the investigators concluded that exposure to cannabinoids during critical periods of development led to marked changes in the functional expression of dopamine neurons in adulthood, and that there were sex differences in these effects.

Chronic administration of relatively large doses of marijuana has been reported to lead to changes in behaviors believed related to hippocampal function (e.g. Stiglick and Kalant, 1985) and structure (Scallet et al., 1987; Landfield et al., 1988). Other studies (reviewed by Scallet, 1991) which exposed rats to marijuana reported changes in the CA3 region neurons of the hippocampus. The CA3 area is a high density cannabinoid receptor binding site. Reported changes included short broken axon-dedritic connections, small neuron size and not being separated from extracellular space by intact membranes. Importantly, the degree of histological change was greater in peripubertal animals than in young adults. Compton and Martin (1995) noted that these changes could have come about by the binding of cannabinoids to myelin basic protein, to which they bind with high affinity (Nye et al., 1988).

For rhesus monkeys, the results of chronic exposure to marijuana smoke is less clear. It has been reported that chronic exposure to marijuana smoke produced neurophysiological and ultrastructural changes in several brain areas including the hippocampus (Harper et al., 1977; Heath et al., 1980). However, other investigators have reported that a 1-year exposure to inhalation of marijuana smoke in late adolescent rhesus monkey did not lead to long-term neuro-histological and electronmicroscopic changes of the brain to CA3 region neurons. The animals were examined 7 months after stopping marijuana exposure (Slikker et al., 1992). Compton and Martin (1995) noted that these monkeys were only exposed for one year, and to compare to the rat data would need to have been exposed for three years.

Human functional studies

In chronic marijuana users, a clinical condition, loosely described as "amotivational syndrome" associated with apathy, reduced drive and ambition, impaired ability to carry out complex tasks, loss of initiative and effectiveness, failure to pursue long-term plans, difficulty in concentrating and decline in school and work performance has been reported (McGlothlin and West, 1968; Gold, 1989; Nahas, 1984). Although firm objective data in support of this syndrome are lacking, several data based neuropsychological studies are available on the effects of chronic cannabis use. While several earlier studies showed impaired performance (Agarwal et al., 1975; Mendhiratta et al., 1988; Wig and Varma, 1977), others failed to find significant changes (Rubin and Comitas, 1975; Satz et al., 1976; Block, 1996; Bowman and Pihl, 1973; Ray et al., 1978; Schaeffer et al., 1981). Pope et al. (1995) have reviewed the data of residual neurocognitive effects.

In a more recent study, Block and Ghoneim (1993) compared adult marijuana users (144 subjects) and non-users (72 subjects) matched on intellectual functioning before the onset of drug use (based on standardized tests administered during the fourth grade of grammar school) on their performance on the twelfth grade versions of these tests and other computerized cognitive tests. Heavy marijuana use (seven or more times weekly) was associated with deficits in mathematical skills and verbal expression in the Iowa Tests of Educational Development and selective impairments in memory retrieval processes. Impairments depended on the frequency of marijuana use.

Leon-Carrion (1990) compared 23 male chronic cannabis users (18–27 years) with 24 control subjects who had never smoked or used cannabis on the Wechsler Adult Intelligence Scale (Wechsler and Wechsler, 1981) and found heavy marijuana smokers showed significant impairment on 6 out of the 11 sub-scales of the WAIS. Schwartz (1991) studied six cannabis-dependent adolescents 48 h after hospitalization and after six weeks of monitored abstinence and compared them with two control groups. There was a significant impairment of test of visual and auditory memory, and the short-term memory impairment remained even after six weeks of abstinence. Deahl (1991) determined that available research provided sufficient evidence to conclude that people who smoke cannabis experience short-term memory deficits that persist after several weeks of abstinence. Mendhiratta et al. (1988) reported on a re-evaluation of heavy, long-term marijuana and hashish smokers and a control group after ten years. They reported that users had continued use during this time. Approximately, half of the subjects previously studied were re-tested. They concluded that consistent with their earlier report, users performed more poorly than controls, and there was evidence for continued deterioration of performance which was greater for users.

In a carefully designed study (Pope and Yargelun-Todd, 1996) which included monitoring of marijuana use via overnight stay in the research center, 65 heavy users and 64 light users of marijuana (defined on the basis of frequency of use over a 30 day period) were compared on a battery of neuropsychological measures. Differences were present between heavy and light users in performance on the Wisconsin Card Sorting Task (Heaton et al., 1993) even after a day of abstinence. The investigators were careful to note that it could not be determined if these differences were due in part to drug residue effects, withdrawal effects or to chronic neurotoxicity.

Fletcher et al. (1996) studied two groups of long-term marijuana users in Costa Rica and compared them to age matched control groups. The older group had used marijuana on average for 34 years, while the younger group had consumed for an average of 8 years. They reported that the older long-term users performed worse than non-users on memory and attention tasks.

According to Solowij et al. (1991), long-term cannabis users (nine subjects) performed significantly worse on a complex auditory selective attention task as compared to non-cannabis users. In a second study, Solowij (1995) studied event related potentials (ERP) in a group of 32 marijuana users who had used marijuana for a mean of 6.69 years. Users were categorized as heavy or light users. Users, who were required to abstain from marijuana for 24 h were compared with a group of 16 controls. The analysis indicated that reaction time to targets was significantly longer and the proportion of correctly detected targets was significantly lower in marijuana users than controls. The largest differences were between heavy users and controls. Analysis of the ERP wave form indicated a significant problem for selective attention which was related to years of marijuana use. They concluded that both frequency and duration of use were important in this measure of selective attention. Solowij (1995) examined ERP in a group of 28 ex-cannabis users who had previously used cannabis for at least 5 years, but had ceased for at least 6 weeks (mean 2 years) prior to testing. These data failed to show evidence of improvement with increasing length of abstinence, thus suggesting significant long-term changes. An excellent review of Solowij's work and other findings related to marijuana are found in her book on cannabis(1999).

Studies utilizing quantitative EEG have shown persistent changes in chronic marijuana users compared to controls. Struve *et al.* (1994) in a replication of their earlier work studied a group of chronic smokers (who were withdrawn from drug during the time of testing) and compared them to a control group. There were significant changes in the alpha activity (relative power) particularly over the frontal lobes. It should be noted that long-term changes in function are also found following alcohol withdrawal, where abnormal EEG patterns persists in some patients for over one year (e.g. Bennett *et al.*, 1960).

However, other studies failed to find impairment following moderate use of marijuana (Chait and Perry, 1994), and insignificant differences in behavioral ratings and cognitive function the morning following smoking a measured number of puffs of either placebo or marijuana cigarettes (Chait, 1990).

Human morphological studies

Few studies of morphological function have been conducted. The first report of anatomical changes consistent with brain atrophy in marijuana smokers (Campbell *et al.*, 1971) was conducted on patients who had otherwise been identified for evaluation for neurological disorder or cognitive impairment, making the interpretation of these data difficult. However, there have been three other studies to use computed tomography (CT) scans (Co *et al.*, 1977; Hannerz and Hindmarsh, 1982; Kuehnle *et al.*, 1977) in heavy marijuana users compared to normal controls. All three of these studies examined ventricle measurements, but none could confirm differences from controls. None of these studies made measurements of gray or white matter, thus making comparison with our findings impossible, except for lateral ventricle volume, which we also found was not affected. In our study (Wilson *et al.*, 2000), the ventricle as a percentage of WBV was 2.2% for the group, which is consistent with other reports (DeCarli *et al.*, 1992). It is also important to note that CT scanning is of significantly lower quality for differentiation of soft tissue compared to MRI and presents no means of quantitative analysis of gray and white matter.

Potential effects of marijuana use in early adolescence

We have recently published findings suggestive of brain morphological changes in marijuana smokers (Wilson *et al.*, 2000). MRI was utilized to determine brain volumes for gray, white and lateral ventricles, and PET was used to determine

CBF in 57 subjects with a history of marijuana use. The relationships of three variables – age at first use, duration of use (defined as current age minus age at first use) and current level of use – to brain volume measures (whole brain, gray matter, white matter and lateral ventricle volumes expressed as a percentage of whole brain volume), global CBF and body size were evaluated.

Subjects

Volunteers were recruited through local advertising, and all volunteers signed a written informed consent prior to their participation in the study. They were screened by a psychiatrist (RJM) who excluded significant physical or psychiatric disorders, abuse or addiction to any drug other than marijuana during the previous 6 months, current use of any prescribed or unprescribed medication and current vascular disorders including migraine. Males with alcohol use of more than two drinks per day and females with more than one drink per day were also excluded. Substance abuse was evaluated according to the Structured Clinical Interview for DSM-III-R (Spitzer et al., 1980). Volunteers were required to have used marijuana either in the past or at present. None met criteria for abuse or dependence for other substances. All subjects completed the Drug Abuse Screening Test-DAST (Skinner, 1982), Brief Michigan Alcoholism Screening Test MAST (Pokorny et al., 1972), Beck Depression Inventory (Beck and Beck, 1972), and the Trait Anxiety scores of the State-Trait anxiety scale (Spielberger et al., 1970). Participants were predominantly right handed. This was verified by using the Harris Test for lateral Dominance (Harris, 1974). Medical history, physical examination, electrocardiogram and clinical laboratory tests were obtained on all. None gave history or findings suggestive of neurological disorder, brain damage, head trauma, epilepsy, recurrent vascular headache or cognitive deficit. Pregnancy was excluded with plasma HCG tests. Subjects completed an interview including questions about age at first use of marijuana, current use and a checklist of the number of different types of drugs tried including alcohol and nicotine.

A total of 57 were included in the study, ranging in age from 19 to 48 (mean 31.3 ± 7) years. There were 32 males (mean age of 33.5 ± 8.3) and 25 females (mean age of 29.8 ± 6.6). A total of 17 subjects either currently smoked cigarettes (n = 14) or had but now did not (n = 3). All reported use of less than a pack per day. There was no sex difference in nicotine use. Most used alcohol (48/57) or had but stopped (2/57); but there was no age-of-onset or sex difference for alcohol use. Most began using marijuana in their teens (16.8 ± 3.6 years with a median of 16 years). ANOVA indicated no difference in age of first marijuana use by sex, or in duration of use (defined as current age minus age at first use). All completed high school with 54.4% completing college. Most were employed (77.8%) or in school (16.1%). There were no differences (chi-square) for education or employment by sex. The mean Beck Depression score was 1.4 ± 2.0, and the Trait Anxiety scores of the State-Trait anxiety scale was 30.1 ± 6.1. The mean MAST was 3.0 ± 4.8 and the mean DAST was 2.9 ± 2.5. The most common other drugs to have been tried were LSD (39/57), cocaine (34/57), amphetamine (21/57) and psilocybin (23/57).

Magnetic resonance imaging

MRI utilized both T_2-weighted and T_1-weighted images. Transaxial MRI scans were performed with a 1.5 Tesla system. The MR (proton density) image was utilized to establish segmentation criteria to determine whole brain (cerebrum) volume (WBV) (which excluded CSF and cerebellum volumes) and gray matter. Determination of the lateral ventricle volumes utilized T_2. Use of segmentation criteria to identify tissue types has been shown to be a reliable method (DeCarli *et al.*, 1992; Byrum *et al.*, 1996). White matter volume was determined by subtracting gray plus ventricle from whole brain volume. Segmentation criteria were established on an individual basis. This process consisted of identifying signal levels on the MRI for predominantly gray matter and predominantly white, then using the midpoint of these activity levels as the cut criteria. A pixel was included as gray matter if its signal strength was at or above that cut point. ROI volumes were determined for each MRI slice (taking into consideration the slice thickness) and summed across all slices. Lateral ventricular volumes were determined by outlining the ventricles on each appropriate T_2-weighted MRI slice, and using segmentation criteria to compute the volume.

PET scans

CBF measurements were performed via PET scans, utilizing the autoradiographic method using procedures described previously. Determination of the ^{15}O-water input function was made, thus permitting absolute quantification of CBF.

Data analysis

For analyses on age of first use, the 57 subjects were stratified on the basis of their sex and age of starting (FIRST) marijuana use: early (before age 17) or late (age 17 or later). Mean ages at first use were as follows: early males = 14.5 ± 1.3, n = 13; late males = 19.3 ± 2.7, n = 19; early females = 14.5 ± 1.3, n = 16 and late females = 19.1 ± 2.0, n = 9. There was no difference in age of first use by sex. We included current age and current level of marijuana use in all analyses as covariates.

Duration of use

In each of these analyses, we examined the possible relationship of duration of use (defined as current age minus age at first use) by establishing two groups based on the median length of use. The same analyses were carried out with this variable as for age of first use. None of these analyses found significant main effects of duration, nor were there significant interactions with it. Thus, we are only presenting the FIRST USE variable.

Brain volumes

As expected whole brain (cerebrum) volume (WBV) showed a sex difference (F = 35.42, df = 1,51, p < = .001), although there were no main effect for FIRST

Table 8.8 Height, weight and MRI volume measurements

Age first use		Subjects grouped by sex and age at first use of marijuana			
		Males		Females	
		<17	≥17	<17	≥17
Height*	M	69.0	72.3	64.8	64.9
	SD	2.7	3.0	2.8	2.8
Weight*	M	171.1	191.7	138.6	145.2
	SD	19.1	26.6	22.1	28.0
Cerebrum (A)	M	1295.8	1316.1	1149.9	1097.9
	SD	130.0	90.6	110.9	86.5
Gray	M	688.0 (53%)	729.7 (55%)	631.4 (55%)	631.5 (58%)
	SD	70.3	68.9	58.0	82.3
White	M	583.0 (45%)	554.5 (43%)	495.5 (43%)	440.3 (40%)
	SD	76.6	68.0	82.9	53.3
Ventricles	M	24.8 (2%)	31.8 (2%)	23.0 (2%)	26.1 (2%)
	SD	6.5	15.2	9.9	11.2

Definitions of variables: cerebrum volume, gray, white and ventricle volumes in cubic centimeters. Cerebrum is whole brain volume. * Height is in inches and weight is in pounds. Numbers in ( ) are per cent of cerebrum volume.

USE, nor an interaction. Therefore, to adjust for this sex difference, we expressed gray matter, white matter and ventricular volumes as a percentage of WBV, and all subsequent analyses were based on the percent volumes. The whole brain, gray matter, white matter and ventricular volumes expressed as cubic millimeters and as percents of WBV are presented in Table 8.8.

ANCOVAs indicated significant FIRST USE main effects for percent whole brain gray matter ($F = 5.940$, df $= 1,51$, $p <= .018$), but no sex differences or interactions. Early users had smaller volumes. Percent whole brain white matter also showed a FIRST USE main effect ($F = 7.334$, df $= 1,51$, $p <= .009$), those starting early having greater volume.

The analysis of percent lateral ventricular volumes did not show significant differences. It should be noted that in our study the lateral ventricle as a percentage of WBV was 2.2% for the whole group, which is consistent with other reports (DeCarli *et al.*, 1992).

We determined cortical gray matter volumes for six ROI (frontal, temporal, parietal and occipital lobes, cingulate gyrus and insula) in each hemisphere and expressed these volumes as a percentage of WBV. These data were analyzed in a Sex by FIRST USE multivariate analysis of covariance (MANCOVA) model with ROI and hemisphere treated as repeated measures, and age and current marijuana use as covariates. This analysis indicated a significant FIRST main effect ($F = 4.53$, df $= 1,51$, $p \le .038$), a FIRST USE by Hemisphere interaction ($F = 6.08$, df $= 1,53$, $p \le .017$) and a FIRST USE by Hemisphere by ROI interaction (Pillais approximate $F = 2.48$, df $= 5,49$, $p \le .044$). The multivariate test of the four-way interaction of Sex by FIRST USE by Hemisphere by Region approached significance ($p \le .071$), and there was a significant Sex by Region interaction (Pillais approximate $F = 3.23$, df $= 5,49$, $p \le .013$). Females tended to have slightly greater

Table 8.9 Regional volumes of gray matter that show group differences subjects grouped by sex and age of first use of marijuana

Age Groups	Males				Females			
	<17		≥17		<17		≥17	
	Mean	sd	Mean	sd	Mean	sd	Mean	sd
Left Hemisphere								
FRNT(V)	112.2	18.8	119.3	20.9	111.2	15.0	103.2	17.5
FRNT(%)	8.6	1.0	9.1	1.5	9.7	1.3	9.4	1.2
PARI(V)	66.9	15.6	74.4	16.9	71.5	11.7	74.3	9.4
PARI(%)	5.1	1.0	5.7	1.2	6.2	0.9	6.8	0.9
Right Hemisphere								
FRNT(V)	120.7	19.3	131.4	22.8	116.0	18.5	117.4	17.1
FRNT(%)	9.3	1.0	10.0	1.7	10.1	1.4	10.6	0.9
PARI(V)	60.3	15.3	68.0	17.8	65.4	10.3	68.9	10.1
PARI(%)	4.7	1.1	5.2	1.3	5.7	0.9	6.3	0.9

FRNT(V) – frontal volume; FRNT(%) – frontal % whole brain volume; PARI(V) – parietal volume; PARI(%) – parietal % whole brain volume. Volumes are expressed as cubic millimeters.

percentages for regional gray matter volumes. To analyze the above three-way interaction, we computed separate repeated measures ANCOVAs for each ROI using a Sex by FIRST by Hemisphere model. These analyses indicated that the frontal gray matter ROI showed a significant FIRST by Hemisphere interaction (F = 11.99, df = 1,53, p ≤ .001), and the parietal gray matter region there was a significant main effects of FIRST (F = 4.89, df = 1,51, p ≤ .032) Table 8.9.

CBF

ANCOVA of whole brain gray matter CBF (ml/100 gm/min) indicated that males had lower flow (F = 5.940, df = 1,51, p ≤ .018), and there was a significant interaction of FIRST by Sex (F = 4.35, df = 1,51, p ≤ .042). Post hoc analyses on global CBF showed that blood flows of males who started early were significantly higher than later starting males, but there was no difference between the two groups of females, nor was there a difference between early males and early females (Table 8.9). It has been suggested that sex differences in cerebral metabolism may be accounted for by total brain volume, and if one corrects for this the sex differences go away (Yoshii et al., 1988). With this in mind, we repeated the ANCOVA on global CBF adding WBV as a covariate, and this analysis indicated a stronger statistical interaction between First and Sex (F = 5.03, df = 1,50, p ≤ .029), and the sex main effect remained essentially unchanged (F = 5.56, df = 1,50, p ≤ .013). There was no significant relationship between CBF and duration of use.

Height and weight

Presented in Table 8.8 are the means and sd for height and weight, grouped by sex and age of first use. ANOVA of current age indicated no group differences.

Analysis of height indicated significant sex (F=56.50, p≤.001) and FIRST (F = 5.07, p≤ .028) main effects, and a significant Sex by FIRST interaction (F = 3.98, p≤ .050). Post hoc analyses indicated Early Male were shorter than Late Male, but there was no difference between Early and Late Female. There were also Sex (F=38.82, p≤ .001) and FIRST (F=5.44, p≤ .023) main effects for weight, but no significant interaction.

Summary of chronic effects

Analyses indicated a significant relationship to age of first use of marijuana for height, weight, global CBF and percentages of gray matter volume. Morphological measures indicated that subjects who had started smoking earlier (EM and FM) had lower body weight, reduced height, reduced percent cortical gray matter, and increased whole brain white matter, but whole brain and ventricular volumes were not different. In addition, the mean CBF for males who started smoking before age 17 was significantly higher than males who started after 17, and their mean CBF was not different from females. Males who started later had significantly lower mean CBF than females consistent with many reports (e.g. Mathew *et al.*, 1986; Esposito *et al.*, 1996)

Possible mechanisms for effects on brain morphology

The data we present show group differences related to age of first use of marijuana. Early users were defined as those who started before the age of 17. The effects manifest themselves in 3-dimensions. (1) Both males and females who are early users have reduced cortical gray matter (as a percentage of whole brain volume) when compared to those who started use later. Differences are greater over the frontal cortex; (2) The CBF in males who are early users is significantly higher compared to other males, and is not different from females; (3) There are differences in body size with early users (particularly males) being smaller (height and weight). All of these differences were significantly related to use of marijuana during early adolescence, a time when there are substantial changes in body growth, sexual maturation and brain development (as discussed below).

It is important to note that the effects on brain morphology, body size and CBF were not statistically related to duration of marijuana use or to current level of use. In fact, the Early vs. Late marijuana users did not differ on duration of use or current level of use. Thus, while these variables may have played some role, we do not believe that the results can be explained on this basis.

Potential effects on brain morphological development

It is generally agreed that adolescence is a time of significant increase in the release of gonadal and pituitary hormones. One of the well established effects of THC is a suppression of release of pituitary hormones, including prolactin, growth hormone, leutinizing hormone and gonadotropin (e.g. Rettori *et al.*, 1988; Fernández-Ruiz *et al.*, 1992; Wenger *et al.*, 1992). THC also has an inhibitory effect on serum testosterone levels, and it has been shown to reduce testicular weight (Wenger *et al.*, 1992). Smoking marijuana has been shown to significantly reduce luteinizing hormone (LH) in human males (Cone *et al.*, 1986) and females (Mendelson *et al.*, 1986) by as much as 30%. In males LH stimulates the interstitial endocrinocytes in the testes for production and secretion of testosterone. It has been shown that THC can reduce the pubertal body weight growth in male rats (Gupta and Elbracht, 1983), and can delay sexual maturation in rats (Field and Tyrey, 1990). It should be noted that other drugs may have effects on hormones and may alter development (Ward, 1992).

During adolescence there are also significant developmental changes ongoing in the human brain (Dekaban and Sadowsky, 1978; Dobbing and Sands, 1973; Jernigan *et al.*, 1991; Reiss *et al.*, 1996). The study by Reiss *et al.* (1996, page 1763) has shown "Prominent age related changes in gray matter, white matter and CSF . . . ", which " . . . appear to reflect ongoing maturation and remodeling." The changes include increases in myelinization and decreases in gray matter. Huttenlocher (1979) demonstrated that there is a significant increase in frontal cortex synaptic density in humans between birth and age 2 and that it then declines to adult levels between ages 2 and 16. Purves (1988) has suggested that due to the rapid somatic growth during adolescence, there may be a need for neural plasticity to persist to accommodate these changes. Thus, data show that during adolescence there is an active process of brain maturation involving both gray and white matter.

To these observations must be added findings concerning the effect of pituitary and gonadal hormones on brain development and function. While it has been known that gonadal hormones play a role in prenatal brain development (e.g. Hoffman and Coleman, 1992; Witelson, 1991), it has also been shown that they may play a significant role, postnatally, in sex differences in function and possibly structure. Clark and Goldman-Rakic (1989) demonstrated that postnatal treatment of normal female rhesus monkeys with testosterone had a significant "masculinizing" influence on the maturation of their orbital prefrontal cortex. Those females treated with testosterone performed a task (on which male monkeys normally do better) at a level equal to males, but females not so treated did more poorly than males on the task. Clark and Goldman-Rakic (1989) concluded that gonadal hormones may play a significant role in modifying the brain at times when there is "dynamic change" ongoing in the cortex. It is clear from the work of Jernigan *et al.* (1990, 1991) and Reiss *et al.* (1996) that adolescence is a time of active brain development with one component of this activity being a significant change in gray matter. This is important because it suggests a possible process where a reduction of pituitary and gonadal hormones (possibly resulting from marijuana use) may affect change in development and subsequently function. Clark and Goldman-Rakic (1989) suggested the possibility that androgens may ". . . exert some protective influence on cortical cells in a specific locale, in a manner similar to the influence of androgens on cell death in other systems. . . ." Nordeen *et al.* (1985) have shown that androgens prevent normally occurring cell death in spinal nerves. It may be that using marijuana during early adolescence has the potential to alter the course of brain development, due to effects on gonadal and pituitary hormones, that are not present at later stages of development.

Of interest here also would be the possible effects of marijuana on sexual differentiation of the brain. It has been demonstrated that there are sexually dimorphic structural differences in normal human brain (e.g. Schlaepfer *et al.*, 1995). It is generally recognized that normal brain development follows a different course depending on sex (e.g. Hoffman and Swabb, 1991; Witelson, 1991). Sexual differences in brain organization are dependent on hormone levels occurring at critical periods. Prenatal exposure to marijuana administered to female rats has been shown to alter gonadal function (Nahas, 1984). Studies show the testes actively produce hormones beginning in utero (Huhtaniemi, 1989), and marijuana has been reported to decrease plasma levels of testosterone (Nahas, 1984). Marijuana has been shown to have significant effects on weight (smaller), height (shorter) and head circumference (smaller) of newborns to mothers' who used it during pregnancy and while breast feeding (Tuchmann-Duplessis, 1993). It is interesting to note that males may be more at risk of the effects of drugs because of the effects on hormones. As Kimura (1992) has noted, "... the default form of the organism is female." It is the application of male hormones that masculinize the brain. It is in this context that we wish to note that males who start early have a more female like global CBF. It is generally accepted that females have a higher resting CBF than males (Gur *et al.*, 1982; Mathew *et al.*, 1986; Esposito *et al.*, 1996), and in the present study this expected main effect of sex was confirmed. However, CBF of males who started using marijuana early was not different from females, but was significantly higher than males who started smoking after age 17. It has not been clearly established why females have a higher CBF.

These data present the question of whether brain development and differentiation may be affected by marijuana use before and during adolescence. The possibility that marijuana may alter normal developmental changes does not preclude the possibility that it may also have neurotoxic effects, including the possibility that early adolescence may be a period when the brain may be more vulnerable to such insults. The human and animal data presented previously support possible neurotoxic effects, perhaps permanent changes in brain neurochemistry, morphology and function following chronic exposure to cannabinoids.

ACKNOWLEDGMENTS

The project was supported by grants from NIDA, DA 04985, DA 10215 and DA11775. We wish to thank the staff in the Cerebral Blood Flow Laboratory and the PET and MRI facilities at Duke for their considerable efforts in this research. We also wish to thank Dr. Timothy Turkington for his review of the PET scanning procedures.

REFERENCES

Ackner, B. (1954) "Depersonalization: 1. Etiology and phenomenology; 2. Clinical syndromes," *Journal of Mental Science* **100**: 836–872.

Agarwal, A. K., Sethi, B. B. and Gupta, S. C. (1975) "Physical and cognitive effects of chronic bhang (cannabis) intake," *Indian Journal of Psychiatry* **17**: 1–7.

Alavi, A., Dann, R., Chawluk, J., Alavi, J., Kushner, N. and Reivich, M. (1986) "Positron emission tomography of regional cerebral glucose metabolism," *Seminars in Nuclear Medicine* **16**: 2.

American Psychiatric Association. (1994) *Diagnostic and Statistical Manual of Mental Disorders*, 4th edn. Washington, D. C.: American Psychiatric Association.

Aranow, W. S. and Cassidy, J. (1979) "Effect of marijuana and placebo marijuana smoking on angina pectoris," *New England Journal of Medicine* **291**: 65.

Beck, A. T. and Beck, R. W. (1972) "Screening depressed patients in family practice," *Postgraduate Medicine* **52**: 81–85.

Bennett, A. E., Mowery, G. L. and Fort, J. T. (1960) "Brain damage from chronic alcoholism: the diagnosis of intermediate stage of alcohol brain disease," *American Journal Psychiatry* **116**: 705–711.

Bliss, E. L., Clark, L. D. and West, C. D. (1959) "Studies of sleep deprivation – relationship to schizophrenia," *Archives of Neurological Psychiatry* **81**: 348–359.

Block, R. I. (1996) "Does heavy marijuana smoking impair human cognition and brain function?" *Journal of American Medical Association* **275**: 560–561.

Block, R. I. and Ghoneim, M. M. (1993) "Effects of marijuana use on human cognition," *Psychopharmacology* **110**: 219–228.

Bowman, M. and Pihl, R. O. (1973) "Cannabis: psychological effects of chronic heavy use. A controlled study of intellectual functioning in chronic users of high potency cannabis," *Psychopharmacologia* **29**: 159–170.

Brill, H. and Nahas, G. G. (1984) "Cannabis intoxication and mental illness," in *Marijuana in Science and Medicine*, edited by G. G. Nahas, p. 263. New York: Raven Press.

Busija, D. W. and Heistead, D. D. (1984) "Factors involved in the physiological regulation of the cerebral circulation," *Physiological Biochemistry Pharmacology* **101**: 161.

Byrum, C. E., MacFall, J. R., Charles, H. C., Chitilla, V. R., Boyko, O. B., Upchurch, L., Smith, J. S., Rajagopalan, P., Passe, T., Kim, D., Xanthakos, S. and Krishnan, K. R. K. (1996) "Accuracy and reproducibility of brain and tissue volumes using a magnetic resonance segmentation method," *Psychiatry Research: Neuroimaging* **67**: 215–234.

Campbell, A. M. G., Evans, M., Thomsom, J. and Williams, M. (1971) "Cerebral atrophy in young cannabis smokers," *Lancet* **2**: 1219–1225.

Cassano, G. B., Petracca, A. and Perugi, G. *et al.* (1989) "Derealization and panic attacks: A clinical evaluation in 150 patients with panic disorder – agoraphobia," *Comprehensive Psychiatry* **30**: 5–12.

Castillo, R. J. (1990) "Depersonalization and meditation," *Psychiatry* **53**: 158–168.

Chait, L. D. (1990) "Subjective and behavioral effects of marijuana the morning after smoking," *Psychopharmacology* **100**: 328–333.

Chait, L. D. and Perry, J. L. (1994) "Acute and residual effects of alcohol and marijuana, alone and in combination, on mood and performance," *Psychopharmacology* **115**: 340–349.

Chee, K. T. and Wong, K. E. (1990) "Depersonalization syndrome – a report of nine cases," *Singapore Medical Journal* **31**: 331–334.

Chen, C. T., Pelizzari, C. A., Chen, T. Y., Hu, X., Lebin, D. N. and Cooper, M. D. (1989) "Integration of brain images from multiple modalities," *Journal of Cerebral Blood Flow Metabolism* **9**(1): S402.

Clark, A. S. and Goldman-Rakic, P. S. (1989) "Gonadal hormones influence the emergence of cortical function in nonhuman primates," *Behavioral Neuroscience* **103**(6): 1287–1295.

Clark, L. D., Hughes, R. and Nakashima, E. N. (1970) "Behavioral effects of marijuana: experimental studies," *Archives of General Psychiatry* **23**: 193–198.

Co, B. T., Goodwin, D., Gado, M., Mikhael, M. and Hill, S. (1977) "Absence of cerebral atrophy in chronic cannabis users: evaluation by computerized transaxial tomography," *Journal of the American Medical Association* **237**(12): 1229–1230.

Cohen, J. (1988) *Statistical Power Analysis for the Behavioral Sciences*, 2nd edn. Hillsdale, New Jersey: Lawrence Erlbaum Associates.

Compton, D. R. and Martin, B. R. (1995) "Marijuana Neurotoxicology," in *Handbook of Neurotoxicology*, edited by L. W. Chang and R. S. Dyer. New York: Marcel Dekker, Inc.

Cone, E. J., Johnson, R. E., Moore, J. D. and Roache, J. D. (1986) "Acute effects of smoking marijuana on hormones, subjective effects and performance in male human subjects," *Pharmacology, Biochemistry and Behavior* **24**: 1749–1754.

Deahl, M. (1991) "Cannabis and memory loss," *British Journal of Addiction* **86**: 249–252.

DeCarli, C., Maisog, J., Murphy, D. G. M., Teichberg, D., Rapoport, S. I. and Horwits, B. (1992) "Method for quantification of brain, ventricular, and subarachnoid CSF volumes from MR images," *Journal Computer Assisted Tomography* **16**(2): 274–284.

DeGrado, T. R., Turkington, T. G., Williams, J. J., Stearns, C. W., Hoffman, J. M. and Coleman, R. E. (1994) "Performance characteristics of whole-body PET scanners," *Journal Nuclear Medicine* **35**: 1398–1406.

Dekaban, A. S. and Sadowsky, D. (1978) "Changes in brain weights during the span of human life: relation of brain weights to body heights and weights," *Annals of Neurology* **4**: 345–356.

Dixon, J. C. (1963) "Depersonalization phenomenon in a sample population of college students," *British Journal of Psychiatry* **109**: 371–375.

Dobbing, J. and Sands, J. (1973) "Quantitative growth and development of human brain," *Archives of Disease in Childhood* **48**: 757–767.

Edvinsson, L. (1982) "Sympathetic control of cerebral circulation," *Trends in Neuroscience* **5**: 425.

Eichling, J. O., Raichle, M. E. and Grubb, R. L. *et al.* (1974) "Evidence of the limitations of water as a freely diffusible tracer in brain of monkey," *Circulation Research* **35**: 358–364.

Esposito, G., Van Horn, J. D., Weinberger, D. R. and Berman, K. F. (1996) "Sex differences in cerebral blood flow as a function of cognitive state with PET," *Journal of Nuclear Medicine* **37**: 559–564.

Ethelberg, S. (1950) "Symptomatic 'cataplexy' or chalastic fits in cortical lesion of the frontal lobe," *Brain* **73**: 499–512.

Fernández-Ruiz, J. J., Rodriguez de Fonseca, F., Navarro, M. and Ramos, J. A. (1992) "Maternal cannabinoid exposure and brain development: changes in the ontogeny of dopamine neurons," in *Marijuana/Cannabinoids: Neurobiology and Neurophysiology*, edited by L. Murphy and A. Bartke. Boca Raton: CRC Press, Inc.

Field, E. and Tyrey, L. (1990) "Delayed sexual maturation during prepubertal cannabinoid treatment: importance of the timing of treatment," *Journal of Pharmacological Experimental Therapy* **254**(1):171–175.

Fish, F. (1967) *Clinical psychopathology, Signs and Symptoms in Psychiatry*. Bristol: John Wright and Sons.

Fletcher, J., Page, J., Francis, D., Copeland, K. and Naus, M. *et al.* (1996) "Cognitive correlates of long-term cannabis use in Costa Rican men," *Archives of General Psychiatry* **53**: 1051–1057.

Frackowiak, R. S., Lenzi, G. L., Jones, G. and Heather, J. D. (1980) "Quantitative measurement of regional cerebral blood flow and oxygen metabolism in man using ^{15}O and positron emission tomography: Theory, procedure and normal values," *Journal of Computer Assisted Tomography* **4**: 727–736.

Gold, M. S. (1989) *Marijuana*, New York: Plenum Medical Book Company.

Good, M. I. (1989) "Substance-induced dissociative disorders and psychiatric nosology," *Journal of Clinical Psychopharmacology* **9**: 88–93.

Gupta, D. and Elbracht, C. (1983) "Effects of tetrahydrocannabinols on pubertal body weight spurt and sex hormones in developing male rats," *Research Experimental Medicine* **182**: 95–104.

Gur, R. C., Gur, R. E. and Obrist, W. D. *et al*. (1982) "Sex and handedness differences in cerebral blood flow during rest and cognitive activity," *Science* **217**: 659–661.

Guralnik, O., Schmeidler, J. and Simeon, D. (2000) "Feeling unreal: Cognitive processed in Depersonalization," *American Journal of Psychiatry* **157**: 103–109.

Hannerz, J. and Hindmarsh, T. (1982) "Neurological and neuroradiological examination of chronic cannabis smokers," *Annals of Neurology* **13**(2): 207–210.

Harper, J. W., Heath, R. G. and Meyers, W. A. (1977) "Effects of cannabis sativa on ultra-structure of the synapse in monkey brain," *Journal on Neuroscience Research* **3**: 87.

Harris, A. J. (1974) *Harris Test of Lateral Dominance*, New York: Psychological Corp.

Heath, R. G., Fitzjarrell, A. T., Fontana, C. J. and Garey, R. E. (1980) "*Cannabis sativa*: effects on brain function and ultrastructure in Rhesus monkeys," *Biological Psychiatry* **15**: 657.

Heaton, R. K., Chelune, G. J., Talley, J. L., Kay, G. G. and Curtiss, G. (1993) *Wisconsin Card Sorting Test Manual*, Odessa, FLA: Psychological Assessment Resources.

Herkenham, M., Lynn, A. B., Little, M. D., Johnson, M. R., Melvin, L. S., de Costa, R. B. and Rice, K. C. (1990) "Cannabinoid receptor localization in brain," *Proceeding of the National Academy of Sciences of the USA* **87**: 1932–1936.

Herscovitch, P., Markham, J. and Raichle, M. E. (1983) "Brain blood flow measured with intravenous $H^2{}^{15}O$: Theory and error analysis," *Journal Nuclear Medicine* **24**: 782–789.

Hoffman, J. M. and Coleman, R. E . (1992) "Perfusion quantification using positron emission tomography," *Investigative Radiology* **27**: 522–526.

Hoffman, M. A. and Swabb, D. F. (1991) "Sexual dimorphism of the human brain: myth and reality," *Experimental Clinical Endocrinology* **98**: 161–170.

Hollister, L. E. (1986) "Health aspects of cannabis," *Pharmacological Review* **38**: 1.

Hollister, L. E., Gillespie, H. K., Ohlsson, A., Lindgren, J. E., Whalen, A. and Agurell, S. (1981) "Do plasma concentrates of delta 9 tetrahydrocannabinol reflect the degree of intoxication?," *Journal of Clinical Pharmacology* **21**: 1715–1755.

Huang, S. C., Carson, R. E., Hoffman, E. J., Carson, J., MacDonald, N., Barrio, J. R. and Phelps, M. E. (1983) "Quantitative measurement of local cerebral blood flow in humans by positron emission tomography and ^{15}O-water," *Journal Cerebral Blood Flow Metabolism* **3**: 141–153.

Huhtaniemi, I. (1989) "Endocrine function and regulation of the fetal and neonatal testis," *The International Journal of Developmental Biology* **33**: 117–123.

Huttenlocher, P. R. (1979) "Synaptic density in human frontal cortex – developmental changes and effects of aging," *Brain Research* **163**: 195–205.

Huxley, A. (1963) *The Doors of Perception*, New York: Harper and Row Publishers, Inc.

Ingvar, D. H. and Lassen, N. A. (1961) "Quantitative determination of cerebral blood flow in man," *Lancet* **2**: 806.

Ingvar, D. H. and Lassen, N. A. (1976) *Brain Metabolism and Cerebral Disorders*, edited by H. E. Himwich, pp. 181–206. New York: Spectrum Publishing.

Institute of Medicine. (1982) "*Marijuana and Health*," Report of a study by a committee of the Institute of Medicine, Division of Health Sciences Policy, pp. 112. Washington, DC: National Academy Press.

Jernigan, T. L. and Tallal, P. (1990) "Late childhood changes in brain morphology observable with MRI," *Developmental Medicine Child Neurology* **32**: 379–385.

Jernigan, T. L., Trauner, D. A., Hasselink, J. R. and Tallal, P. A. (1991) "Maturation of human cerebrum observed in vivo during adolescence," *Brain* **114**: 2037–2049.

Johnson, B. A. (1990) "Psychopharmacological effects of cannabis," *British Journal of Hospital Medicine* **43**: 114–122.

Kassirer, J. P. (1997) "Federal foolishness and marijuana," *New England Journal of Medicine* **336**(5): 366–367.

Keele, S. W. and Ivry, R. B. (1991) "Does the cerebellum provide a common computation for diverse tasks: a timing hypothesis," *Annals of New York Academy of Science* **608**: 179–211.

Kelly, D. H. W. and Walter, C. J. S. (1968) "The relationship between clinical diagnoses and anxiety assessed by forearm in blood flow and other measurements," *British Journal of Psychiatry* **114**: 611–626.

Kenna, J. C. and Sedman, G. (1964) "The subjective experience of time during lysergic acid diethylamide (LSD-25) intoxication," *Psychopharmacologia* **5**: 280–288.

Kimura, D. (1992) "Sex differences in the brain," *Scientific American* **267**(3): 118–125.

Kirk, R. E. (1982) *Experimental Design: Procedures for the Behavioral Sciences*, 2nd edn. Belmont, California: Brooks/Cole Publishing Co.

Klausner, H. A. and Dingle, J. V. (1971) "The metabolism and excretion of delta-9-tetra-hydrocannabinol in the rat," *Life Sciences* **10**: 49.

Kuehnle, J., Mendelson, J., Davis, K. and New, P. (1977) "Computed tomographic examination of heavy marijuana smokers," *Journal of American Medical Association* **237**: 1231–1232.

Kumar, A. M., Haney, M., Becker, T., Thompson, M., Kream, R. and Miczek, K. (1990) "Effect of early exposure to delta-9-tetrahydrocannabinol on the levels of opiate peptides, gonadotropin-releasing hormone and substance P in the adult male rat brain," *Brain Research* **525**: 78.

Lader, M. (1975) *The Psychophysiology of Mental Illness*. London: Routledge and Kegan-Paul.

Landfield, P. W., Cadwallader, L. B. and Vinsant, S. (1988) "Quantitative changes in hippo-campal structure following long-term exposure to delta-9-tetrahydrocannabinol: Possible mediation by glucocorticoid systems," *Brain Research* **443**: 47–62.

Ledi, H., Kanno, I., Miura, S., Murakami, M., Takahashi, T. K. and Uemera, K . (1986) "Error analysis of quantitative cerebral blood flow measurement using $H^{215}O$ autoradiography and positron emission tomography with respect to the dispersion of the input function," *Journal of Cerebral Blood Flow Metabolism* **6**: 536–545.

Leon-Carrion, J. (1990) "Mental performance in long-term heavy cannabis use: A preliminary report," *Psychology Reports* **67**: 947–952.

Lovenberg, W., Ames, M. and Lerner, P. (1978) "Mechanism of short-term regulation of tyrosine hydroxylase," in *Psychopharmacology: A Generation of Progress*, edited by M. Lipton, A. DiMascio and K. F. Killam, pp. 247–260. New York: Raven Press.

Mathew, R. J. (1989) "Hyperfrontality of regional cerebral blood flow distribution in normals during resting wakefulness: Fact or artifact?" *Biological Psychiatry* **26**: 717–724.

Mathew, R. J. and Wilson, W. H. (1985) "Caffeine-induced changes in cerebral circulation," *Stroke* **16**: 814.

Mathew, R. J. and Wilson, W. H. (1988) "Cerebral blood flow changes induced by CO_2 in anxiety," *Psychiatry Research* **23**: 285.

Mathew, R. J. and Wilson, W. H. (1990) "Anxiety and cerebral blood flow," *American Journal of Psychiatry* **147**: 838–847.

Mathew, R. J. and Wilson, W. H. (1991) "Substance abuse and cerebral blood flow," *American Journal of Psychiatry* **148**: 292.

Mathew, R. J. and Wilson, W. H. (1993) "Acute changes in cerebral blood flow after smoking marijuana," *Life Sciences* **52**: 757–767.

Mathew, R. J. and Wilson, W. H. (1996) "Depersonalization and cerebral blood flow," Presented at the Royal College of Psychiatrists Regional Meeting, Hyderabad, India, November 28 to December 1.

Mathew, R. J., Margolin, R. A. and Kessler, R. M. (1985) "Cerebral function, blood flow and metabolism: A new vista in psychiatric research," *Integrative Psychiatry* **3**: 214.

Mathew, R. J., Wilson, W. H. and Tant, S. (1986) "Determinants of resting regional cerebral blood flow in normal subjects," *Biological Psychiatry* **21**: 907–914.

Mathew, R. J., Wilson, W. H. and Tant, S. R. (1989) "Acute changes in cerebral blood flow associated with marijuana smoking," *Acta Psychiatrica Scandinavica* **79**: 118–128.

Mathew, R. J., Wilson, W. H., Humphreys, D. F., Lowe, J. V. and Weithe, K. E. (1992) "Regional cerebral blood flow after marijuana smoking," *Biological Psychiatry* **32**: 164–169.

Mathew, R. J., Wilson, W. H., Humphreys, D., Lowe, J. V. and Weithe, K. E. (1993) "Depersonalization after marijuana smoking," *Biological Psychiatry* **33**: 431–441.

Mathew, R. J., Wilson, W. H., Turkington, T. G. and Coleman, R. E. (1998) "Cerebellar activity and disturbed time sense after THC," *Brian Research* **797**: 183–189.

Mathew, R. J., Wilson, W. H., Chiu, N. Y., Turkington, T. G., DeGrado, T. R. and Coleman, R. E. (1999) "Regional cerebral blood flow and depersonalization after tetrahydrocannabinol administration," *Acta Psychiatrica Scandinavica* **100**: 67–75.

Matsuda, L. A. (1997) "Molecular aspects of cannabinoid receptors," *Critical Reviews in Neurobiology* **11**(2&3): 143–166.

Matsuda, L. A., Lolait, S. J., Brownstein, M. J. and Bonner, T. I. (1993) "The THC receptor and its implications," in *Biological Basis of Substance Abuse*, edited by S. G. Korenman and J. D. Barchas, p. 95. New York: Oxford University Press, Inc.

Maximilian, V. A., Prohovnik, I. and Risberg, J. (1980) "Cerebral hemodynamic response to mental activation during normo- and hyper-carbia," *Stroke* **11**: 342.

Mayer-Gross, W. (1935) "On depersonalization," *British Journal of Medical Psychology* **15**: 98–122.

Maykut, M. O. (1985) "Health consequences of acute and chronic marijuana use," *Progress in Neuro-Psychopharmacology and Biological Psychiatry* **9**: 209.

Mazziotta, J. C. (1985) "PET scanning principles and applications," *Disc Neuro Science* **11**: 9.

Mazziotta, J. C., Pelizzari, C. C., Chen, G. T., Bookstein, F. L. and Balentino, D. (1991) "Region of interest issues: The relationship between structure and function in the brain," *Journal of Cerebral Blood Flow Metabolism* **11**: A51–A56.

Mazziotta, J. C. and Koslow, S. H. (1987) "Assessment of goals and obstacles in data acquisition and analysis from emission tomography: Report of a series of international workshops," *Journal Cerebral Blood Flow Metabolism* **7**: S1–S31.

McGlothlin, W. H. and West, L. J. (1968) "The marijuana problem: An over-view," *American Journal of Psychiatry* **125**: 370–378.

McNair, D. M., Lorr, M. and Doppleman, L. F. (1971) *Manual for the Profile of Mood States*. San Diego, California: Educational and Industrial Testing Service.

Melges, F. T., Tinklenberg, J. R., Hollister, L. E. and Gillespie, H. K. (1970a) "Temporal disintegration and depersonalization during marijuana intoxication," *Archives of General Psychiatry* **23**: 204–210.

Melges, F. T., Tinklenberg, J. R., Hollister, L. E. and Gillespie, H. K. (1970b) "Marijuana and temporal disintegration," *Science* **168**: 118–120.

Melges, F. T., Tinklenberg, J. R., Holister, L. E. and Gillespie, H. K. (1971) "Marijuana and the temporal span of awareness," *Archives of General Psychiatry* **24**: 564–567.

Melges, F. T., Tinklenberg, J. R., Deardorff, M., Davis, N. H., Anderson, R. E. and Owen, C. A. (1974) "Temporal disorganization and delusional-like ideation," *Archives of General Psychiatry* **30**: 855–861.

Mellor, C. S. (1988) "Depersonalization and self-perception," *British Journal of Psychiatry* **153**(supplement 2): 15–19.

Mendelson, J. H., Mello, N. K., Ellingboe, J., Skupny, A. S. T., Lex, B. W. and Griffin, M. (1986) "Marihuana smoking suppresses luteinizing hormone in women," *The Journal of Pharmacology and Experimental Therapeutics* **237**(3): 862–866.

Mendelson, J. H., Babor, T. F., Kuehnle, J. C., Rossi, A. M., Bernstein, J. G., Mello, N. K. and Greenberg, I. (1976) "Behavioral and biologic aspects of marijuana use," *Annals of the New York Academy of Sciences* **282**: 186.

Mendhiratta, S. S., Varma, S. K., Dang, R., Malhotra, A. K. and Nehra, R. (1988) "Cannabis and cognitive function: A re-evaluation study," *British Journal Addiction* **83**(7): 749–753.

Meyer, E. (1989) "Simultaneous correction for tracer arrival delay and dispersion in CBF measurements by the $H_2^{15}O$ autoradiographic method and dynamic PET," *Journal Nuclear Medicine* **30**: 1069–1078.

Moran, C. (1986) "Depersonalization and agoraphobia associated with marijuana use," *British Journal Medical Psychology* **59**: 187–196.

Nahas, G. G. (1984) "Toxicology and Pharmacology," in *Marijuana in Science and Medicine*, New York: Raven Press.

Nahas, G. G., Sutin, K., Manger, W. and Hyman, G. (1997) "Marijuana is the wrong Medicine," *Wall Street Journal*, March 11, 1997, in *Marijuana and Medicine*, edited by G. G. Nahas, K. M. Sutin, D. Harvey, S. Agurell, pp. 61–62. Totowa, New Jersey: Humana Press, 1999.

Nordeen, E. J., Nordeen, K. W., Senegelaub, D. R. and Arnold, A. P. (1985) "Androgens prevent normally occurring cell death in a sexually dimorphic spinal nucleus," *Science* **229**(4714): 671–673.

Noyes, R. J. R. and Kletti, R. (1977) "Depersonalization response to life-threatening danger," *Comprehensive Psychiatry* **18**: 375–384.

Nye, J. S., Voglmaier, S., Martenson, R. E. and Snyder, S. H. (1988) "Myelin basic protein is an endogenous inhibitor of the high-affinity cannabinoid binding site in brain," *Journal Neurochemistry* **50**: 1170–1178.

Obrist, W. D., Thompson, H. K. Jr., Wang, H. S. and Wilkinson, W. E. (1975) "Regional cerebral blood flow estimated by [133]Xenon inhalation," *Stroke* **6**: 245.

Ohlsson, A., Lindgren, J. E., Wahlen, A., Agurell, S., Hollister, L. E. and Gillespie, H. K. (1980) "Plasma delta-9-tetrahydrocannabinol concentrations and clinical effects after oral and intravenous administration and smoking," *Clinical Pharmacology and Therapeutics* **28**(3): 409–416.

Paulson, O. B., Strandgaard, S. and Edvinsson, L. (1990) "Cerebral autoregulation," *Cerebrovascular and Brain Metabolism Reviews* **2**(2): 161–192.

Pelizzari, C. A., Chen, G. T. Y., Spelbring, D. R., Weichsebaum, R. R. and Chen, C. T. (1989) "Accurate three-dimensional registration of CT, PET and/or MR images of the brain," *Journal Computer Assisted Tomography* **13**: 20–26.

Penfield, W. and Kristiensen, K. (1951) *Epileptic Seizure Patterns,* Springfield, Illinois: Charles C. Thomas.

Perez-Reyes, M., Timmons, M. C. and Lipton, M. A. *et al.* (1972) "Intravenous injection in man of delta 9-tetrahydrocannabinol and 11-OH-Delta 9-tetrahydrocannabinol," *Science* **177**: 633–635.

Pertwee, R. G. (1997) "Pharmacology of cannabinoid CB_1 and CB_2 Receptors," *Pharmacological Therapy* **74**(2): 129–180.

Phelps, M. E., Mazziotta, J. C. and Huang, S. C. (1982) "Study of cerebral function with positron emission tomography," *Journal Cerebral Blood Flow Metabolism* **2**: 13.

Pokorny, A. D., Miller, B. A. and Kaplan, H. B. (1972) "The brief MAST: A shortened version of the Michigan Alcoholism Screening Test," *American Journal Psychiatry* **129**(3): 342–345.

Pope, H. G., Gruber, A. J. and Yargelun-Todd, D. (1995) "The residual neuropsychological effects of cannabis: the current status of research," *Drug and Alcohol Dependence* **38**: 25–34.

Pope, H. G. and Yargelun-Todd, D. (1996) "The residual effects of heavy marijuana use in college students," *Journal of the American Medical Association* **275**(7): 521–527.

Purves, D. (1988) *Body and Brain: A Tropic Theory of Neural Connections,* Cambridge, MA and London: Harvard University Press.

Raichle, M. E., Martin, W. R., Herscovitch, P., Mintun, M. A. and Markham, J. (1983) "Brain blood flow measured with intravenous $H_2^{15}O$: Implementation and Validation," *Journal Nuclear Medicine* **24**: 790–798.

Ray, R., Prabhu, G. G. and Mohan, D. (1978) "The association between chronic cannabis use and cognitive functions," *Drug Alcohol Dependance* **3**: 365–368.

Reed, G. F. and Sedman, G. (1964) "Personality and depersonalization under sensory deprivation condition," *Perceptual and Motor Skills* **18**: 659–660.

Reiss, A. L., Abrams, M. T., Singer, H. S., Ross, J. L. and Denckla, M. B. (1996) "Brain development, sex and IQ in children. A volumetric imaging study," *Brain* **119**: 1763–1774.

Rettori, V., Wenger, T., Snyder, G., Dalterio, S. and McCann, S. M. (1988) "Hypothalamic action of delta-9-tetrahydrocannabinol to inhibit the release of prolactin and growth hormone in rat," *Neuroendocrinology* **47**: 498–510.

Risberg, J. (1980) "Regional cerebral blood flow measurements by ^{133}Xenon inhalations: Methodology and applications in neuropsychology and psychiatry," *Brain Language* **9**: 9.

Robbe, H. (1998) "Marijuana's impairing effects on driving are moderate when taken alone but severe when combined with alcohol," *Human Psychopharmacol. Clinical Exp.* **13**: S70-S78.

Roth, M. and Argyle, E. N. (1988) "Anxiety, panic and phobic disorders: An overview," *Journal Psychiatry Research* **22**: 33–54.

Rubin, V. and Comitas, L. (1975) *Ganja in Jamaica: A medical anthropological Study of Chronic Marijuana Use*, The Hague: Mouton and Compnay.

Sadato, N., Carsons, R. E., Daube-Witherspoon, M. E., Campbell, G., Hallett, M. and Herscovitch, P. (1995) "Optimization of non-invasive activation studies with O-15 water and 3D PET," *Journal of Nuclear Medicine* **36**: 82P.

Satz, P., Fletcher, J. and Sutker, L. (1976) "Neuropsychologic, intellectual and personality correlates of chronic marijuana use in native Costa Ricans in Chronic Cannabis Use," *Annals of New York Academy of Science* **282**: 266–306.

Scallet, A. C. (1991) "Neurotoxicology of cannabis and THC: A review of chronic exposure studies in animals," *Pharmacology Biochemistry and Behavior* **40**: 671–676.

Scallet, A. C., Uemura, E., Andrews, A., Ali, S. F., McMillan, D. E., Paule, M. G., Brown, R. M. and Slikker, W. J. R. (1987) "Morphometric studies of the rat hippocampus following chronic delta-9-tetrahydrocannabinol (THC)," *Brain Research* **436**: 193–198.

Schacter, S. and Singer, J. E. (1962) "Cognitive, social and physiological determinates of emotional states," *Psychological Review* **69**: 379–399.

Schaeffer, J., Andrysiak, T. and Ungerleides, J. T. (1981) "Cognition and long-term use of ganja (cannabis)," *Science* **213**: 465–466.

Schlaepfer, T., Harris, G., Tien, A., Peng, L., Lee, S. and Pearlson, G. (1995) "Structural differences in the cerebral cortex of healthy female and male subjects: a magnetic resonance imaging study," *Psychiatry Research: Neuroimaging* **61**: 129–135.

Schwartz, R. H. (1991) "Heavy marijuana use and recent memory impairment," *Psychiatric Annal* **21**: 80–82.

Sedman, G. (1970) "Theories of depersonalization: A re-appraisal," *British Journal of Psychiatry* **117**: 1–14.

Selzer, M. L. (1971) "The Michigan Alcoholism Screening Test: The quest for a new diagnostic instrument," *American Journal Psychiatry* **127**: 1653–1658.

Simpson, J. A. (1969) "The clinical neurology of temporal lobe disorders," in *Current Problems in Neuropsychiatry: Schizophrenia, epilepsy, the temporal lobe*, edited by R. N. Herrington. *British Journal of Psychiatry*, Special Publication No. **4**. Ashford: Royal Medical Psychological Association, Headly Bros.

Skinner, H. A. (1982) "Drug Abuse Screening Test," *Addiction Behaviors* **7**: 363.

Slikker, W., Paule, M. G., Ali, S. F., Scallett, A. C. and Bailey, J. R. (1992) "Behavioral, neurochemical and neurohistological effects of chronic marijuana smoke exposure in the nonhuman primate," In *Marijuana/Cannabinoids: Neurobiology and Neurophysiology*, edited by L. Murphy and A. Bartke, Chapter 7. Boca Raton: CRC Press, Inc.

Sokoloff, L. and Smith, C. (1983) "Biochemical Principles for the Measurement of Metabolic Rate *In Vivo*," in *Positron Emission Tomography of the Brain*, edited by W.-D. Heiss, M. E. Phelps. New York: Springer-Verlag.

Solowij, N. (1995) "Do cognitive impairments recover following cessation of cannabis use?" *Life Sciences* **56**(23/24): 2119–2126.

Solowij, N. (1999) "Cannabis and Cognitive Function," Cambridge: Cambridge University Press.

Solowij, N., Michie, P. T. and Fox, A. M. (1991) "Effects of long-term cannabis use on selective attention: An event-related potential study. Special Issue: Pharmacological, Chemical, Biochemical and Behavioral Research on Cannabis and the Cannabinoids," *Pharmacology, Biochemistry and Behavior* **40**: 683–688.

Spielberger, C. D., Gorsuch, R. L. and Lushene, R. D. (1970) "*STAI Manual*," Palo Alto, California: Consulting Psychologist Press.

Spitzer, R. L., Williams, J. B. and Skodol, A. E. (1980) "DSM-III: the major achievements and an overview," *American Journal of Psychiatry* **137**: 151–164.

Stearns, C. W. (1995) "Scatter correction method for 3D PET using 2D fitted Gaussian functions," *Journal of Nuclear Medicine* **36**: 105P.

Stiglick, A. and Kalant, H. (1985) "Residual effects of chronic cannabis treatment on behavior in mature rats," *Psychopharmacology* **85**: 436.

Strandgaard, S. and Paulson, O. B. (1984) "Cerebral autoregulation," *Stroke* **15**: 413.

Struve, F., Straumanis, J. and Patrick, G. (1994) "Persistent topographic quantitative EEG sequelae of chronic marijuana use: A replication and initial discriminant function analysis," *Clinical Electroencephalography* **25**(2): 63–75.

Talairach, J. and Tournoux, P. (1988) *A stereotactic coplanar atlas of the human brain*. Stuttgart: Thieme.

Tart, C. T. (1969) *Altered states of consciousness: A book of readings*. New York: John Wiley and Sons, Inc.

Thomas, D. J. (1982) "Whole blood viscosity and cerebral blood flow," *Stroke* **13**(3): 285–287.

Tinklenberg, J. R., Roth, W. T. and Kopell, B. S. (1976) "Marijuana and ethanol: Differential effects on time perception, heart rate and subject response," *Psychopharmacology* **49**: 275–279.

Tinklenberg, J. R., Kopell, B. S., Melges, F. T. and Hollister, L. E. (1972) "Marijuana and alcohol: Time production and memory functions," *Archives of General Psychiatry* **27**: 812–815.

Tuchmann-Duplessis, H. (1993) "Effects of cannabis on reproduction," in *Cannabis; Physiopathology, Epidemiology, Detection*, edited by G. Nahas and C. Latour. Boca Raton: CRC Press.

Turkington, T. G., Hoffman, J. M., Jaszczak, R. J., MacFall, J. R., Harris, C. G., Kilts, C. D., Pelizzari, C. A. and Coleman, R. E. (1995) "Accuracy of surface fit registration for PET and MR images using full and incomplete brain surfaces," *Journal of Computer Assisted Tomography* **19**(1): 117–124.

Turkington, T. G., Hamblen, S. M., Hawk, T. C., DeGrado, T. R. and Coleman, R. E. (1996) "A quantitative evaluation of 3D PET in FDG brain imaging," *Journal of Nuclear Medicine*, **37**(5), No. 993, 219P.

Tyrer, P. (1976) *The Role of Bodily Feelings in Anxiety*, London: Oxford University Press.

United States Congress. Committee on Government Reform. Subcommittee on Criminal Justice, Drug Policy and Human Resources. (1997) "The National Drug Control Strategy," February, **197**: Washington: U.S. G.P.O.

Vachon, L., Sulkowski, A. and Rich, E. (1974) "Marijuana effects on learning, attention and time estimation," *Psychopharmacologia* **39**: 1–11.

Volkow, N. D. and Fowler, J. S. (1993) *Cannabis; Physiopathology, Epidemiology, Detection*, edited by G. G. Nahas and C. Latour. Boca Raton, FLA: CRC Press.

Volkow, N. D., Gillespie, H., Mullani, N., Tancredi, L., Grant, C., Ivanovic, M. and Hollister, L. (1991a) "Cerebellar metabolic activation by delta-9-tetrahydrocannabinol in human brains: A study with positron emission tomography and ^{18}F-2-fluoro-2-deoxyglucose," *Psychiatry Research: Neuroimaging* **40**: 69–78.

Volkow, N. D., Gillespie, H., Mullani, N., Tancredi, L., Hollister, L., Ivanovic, M. and Grant, C. (1991b) "Use of positron emission tomography to investigate the action of marijuana in the human brain," in *Orthophysiology of Illicit Drugs, Cannabis, Cocaine, Opiates*, edited by G. G. Nahas and C. Latouri, pp. 3–12. Oxford: Pergamon.

Volkow, N. D., Gillespie, H., Tancredi, L. and Hollister, L. (1995) "The effects of marijuana in the human brain measured with regional brain glucose metabolism," *In Sites of Drug Action in the Human brain*, edited by A. Begon and N. D. Volkow, pp. 75–86. Boca Raton, FL: CRC Press.

Volkow, N. D., Gillespie, H., Mullani, N., Tancredi, L., Grant, C., Valentine, A. and Hollister, L. (1996) "Brain glucose metabolism in chronic marijuana users at baseline and during marijuana intoxication," *Psychiatry Research: Neuroimaging* **67**: 29–38.

Walters, D. E. and Carr, L. A. (1988) "Perinatal exposure to cannabinoids alters neurochemical development of the brain," *Pharmacological Biochemistry of Behavior* **29**: 213.

Ward, O. B. (1992) "Fetal drug exposure and sexual differentiation of males," in *Handbook of Neurobiology: Vol. 11, Sexual Differentiation*, edited by Gerall, Moltz and Ward, Chapter 6. New York: Plenum Press.

Wechsler, D. and Wechsler, A. (1981) *Adult Intelligence Scale-Revised*. Cleveland, OH: Psychological Corp.

Weiss, J. L., Watanabe, A. M., Lemberger, L., Tamarkin, N. R. and Cardon, P. V. (1972) "Cardiovascular effects of delta-9-tetrahydrocannabinol in man," *Pharmacological Therapy* **13**: 671.

Wenger, T., Croix, D., Tramu, G. and Leonardelli, J. (1992) "Effects of delta-9-tetrahydocannabinol on pregnancy, puberty and the neuroendocrine system," in *Marijuana/Cannabinoids: Neurobiology and Neurophysiology*, edited by L. Murphy and A. Bartke. Boca Raton, FLA: CRC Press, Inc.

Wig, N. N. and Varma, V. K. (1977) "Patterns of long-term heavy cannabis use in North India and its effects on cognitive functions: A preliminary report," *Drug, Alcohol Dependency* **2**: 211–219.

Wilson, W. H., Ellinwood, E. H., Mathew, R. J. and Johnson, K. (1994) "Effects of marijuana on performance of a computerized cognitive-neuromotor test battery," *Psychiatry Research* **51**: 115–125.

Wilson, W. H., Mathew, R. J., Turkington, T., Hawk, T. C., Coleman, R. E. and Provenzale, J. (2000) "Brain morphological changes and early marijuana use: A Magnetic Resonance and Positron Emission Tomography study," *Journal of Addictive Diseases* **19**(1): 1–22.

Witelson, S. F. (1991) "Neural sexual mosaicism: sexual differentiation of the human temporoparietal region for functional asymmetry," *Psychoneuroendocrinology* **16**(1–3): 131–153.

Yoshii, F., Barker, W. W., Chang, J. Y., Loewenstein, D., Apicella, A., Smith, D., Ginsberg, M. D., Pascal, S. and Duara, R. (1988) "Sensitivity of cerebral glucose metabolism to age, gender, brain volume, brain atrophy, and cerebrovascular risk factors," *J Cereb Blood Flow Metab* **8**: 654–661.

Chapter 9

Marijuana and cannabinoid effects on immunity and AIDS

Guy A. Cabral

ABSTRACT

Marijuana and its major psychoactive component, Δ^9-tetrahydrocannabinol (THC), have been shown to alter resistance to bacterial, protozoan, and viral infections in experimental animals and *in vitro* systems. These alterations have occurred correlative to modifications in the functional activities of a diverse array of cellular and humoral factors of the immune system. In addition, marijuana and THC, as well as other cannabinoids, have been reported to directly target the functional activities of lymphocytes, macrophages, natural killer cells, and other immunocytes from rodents and humans. It has been proposed that these activities are operative through both receptor and non-receptor mediated modes. Reports that marijuana and THC alter anti-microbial activity *in vivo* and *in vitro*, indicate that marijuana use presents a potential risk of decreased resistance to infections in humans. However, few controlled longitudinal epidemiological and immunological studies have been undertaken to correlate the immunosuppressive effects of marijuana smoke or cannabinoids on the incidence of infections or disease in humans.

Key Words: AIDS, cannabinoids, Δ^9-tetrahydrocannabinol, immunity, infections, THC

INTRODUCTION

Marijuana, *Cannabis sativa*, is a highly complex substance which contains in excess of 400 chemical entities including Δ^9-tetrahydrocannabinol (THC), its major psychoactive component. THC has been reported to have therapeutic potential in terms of its anti-nociceptive properties, ability to reduce intraocular pressure and bronchial constriction, and action as an anti-convulsant and antiemetic agent (Munson and Fehr, 1983; Dewey, 1986). Accumulating experimental evidence indicates, also, that marijuana and cannabinoids can alter functional activities of the immune system. Studies extending back to the 1970s, in which *in vitro* and *in vivo* experimental models have been used, have indicated that marijuana or THC impairs cell-mediated immunity, humoral immunity, and cellular defenses against a variety of infectious agents (reviewed in: Cabral and Dove Pettit, 1998; Friedman and Klein, 1999). Compromised resistance in mice, rats, and guinea pigs to infection with amebae, herpes simplex virus, Friend Leukemia virus, *Listeria monocytogenes*, *Staphylococcus albus*, *Treponema pallidum*, and *Legionella pneumophila* has been reported. These observations suggest that marijuana and THC exert a broad spectrum of effects and that they target multiple elements of the immune system since host responsiveness to such a wide array of infectious agents involves the interplay of diverse cellular and humoral elements. To date, however, there is a paucity of experimental evidence which links directly the use of marijuana in a recreational or therapeutic mode to compromised host resistance in humans. Few longitudinal epidemiological studies using biological or clinical approaches have been undertaken or completed. Furthermore, epidemiological data which have been obtained have been difficult to interpret as a result of confounding multiple drug use in participants and have yielded contradictory results.

EFFECTS ON HOST RESISTANCE USING ANIMAL MODELS OF INFECTION

Decreased resistance to infection with viruses

Guinea pigs and mice have been used extensively to document the effects of cannabinoids on resistance to infectious agents and to define elements of the immune system targeted by these compounds. Morahan *et al*. (1979) demonstrated that THC administered intraperitoneally to BALB/c mice decreased resistance to *Listeria monocytogenes* or herpes simplex virus type 2 (HSV-2), pathogens infectious also in humans. Animals inoculated intravenously with *Listeria* exhibited a significant dose-dependent increase in mortalities following administration of THC at doses as low as 38 mg/kg. However, doses of 150 to 200 mg/kg were required to effect maximal suppression of host resistance to *Listeria*. The decreased resistance produced by THC was similar to that produced by flumethazone, a known immunosuppressive steroid. Furthermore, sodium pentobarbitol, which causes CNS depression and anesthesia, had only a slight effect indicating that the decrease in host resistance was apparently due to the immunosuppressive properties of THC rather than to effects on the central nervous system (CNS). Treatment of mice with 1-methyl-Δ^8-THC or with cannabidiol also decreased host resistance, although not to the same extent as THC. A similar dose-dependent decrease in host resistance was exerted by THC following intravenous administration of HSV-2. The doses and treatment regimen with cannabinoids that effected decreases in resistance were similar to those which caused suppression of the delayed-type hypersensitivity response to sheep erythrocytes, a barometer of cell-mediated immunity. However, in these studies, the doses that exerted decreases in resistance were relatively high and THC was administered parenterally rather than through inhalation of marijuana smoke to mimic the natural route of exposure. Nevertheless, Morahan *et al*. (1979) demonstrated that cannabinoids had the capacity to suppress host resistance to *Listeria* and to HSV-2 and that this decrease in resistance correlated with drug-induced dysfunction in immune responsiveness in which macrophages and T-lymphocytes played a prominent role.

The studies of Morahan *et al*. (1979) were confirmed, and extended, by those of Mishkin and Cabral (1985) and Cabral *et al*. (1986) who employed guinea pig and murine models of human genital herpes virus infection. Guinea pigs treated with THC (2–25 mg/kg) and infected intravaginally with HSV-2 experienced rapid onset of primary disease marked by greater severity and number of herpetic lesions as compared with similarly infected, non drug-treated controls (Cabral *et al*., 1986).

Drug-treated animals also shed greater amounts of virus from the vagina and experienced a greater frequency and severity of recurrent herpetic infection. In addition, Mishkin and Cabral (1985) used a vaginal HSV-2 mouse infectivity model to demonstrate that THC delayed the onset and decreased the magnitude of the delayed hypersensitivity response (DHR) to intravaginally administered HSV-2. A reduction of T lymphocyte-dependent cell-mediated immunity in association with THC administration to experimental animals has been reported also by Klykken *et al.* (1977) and Smith *et al.* (1978).

In addition to effecting a decrease in T lymphocyte-dependent cell-mediated immunity, THC has been reported to inhibit antibody responses directed against HSV-2 (Mishkin and Cabral, 1985). Significantly, lower titers of anti-HSV-2 complement-fixing antibody were recorded in animals given THC at 15 mg/kg or 100 mg/kg or administered 200 mg/kg cyclophosphamide, a potent inhibitor of antibody production. Similarly, THC effected a significant decrease in the production of anti-HSV-2 neutralizing antibody. Furthermore, THC (5–100 mg/kg) decreased significantly the production of interferon in response to HSV-2 (Cabral, Lockmuller and Mishkin, 1986). The diminished interferon titers to HSV-2 persisted through 24 h following infection (Figure 9.1).

Cannabinoids also have been reported to alter resistance and immune responsiveness to retroviruses (Specter *et al.*, 1991). Spleen cells obtained from mice infected with Friend Leukemia Virus (FLV) and pretreated with THC (7.5–10 µg/ml) exhibited decreased lymphocyte blastogenic response and natural killer (NK) cytotoxicity as compared with those from animals receiving only THC or virus.

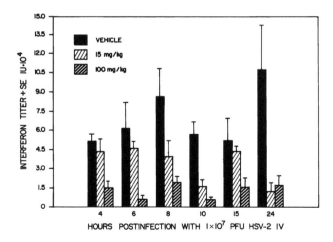

Figure 9.1 Alpha/beta interferon titers recorded at 4–24 h post-HSV-2 inoculation. Animals (n = 8/group) were administered vehicle (emulphor:ethanol:saline, 1:1:18) or THC (15 or 100 mg/kg) intraperitoneally and were injected intravenously with 1×10^7 plaque-forming units of HSV-2. Significant suppression (P < 0.05) of HSV-2-induced interferon over a 24 h period was noted in serum of animals dosed with 100 mg/kg THC at each recorded time period. Significant suppression (P < 0.05) of alpha/beta interferon was noted for animals dosed with 15 mg/kg THC at 10, 15, and 24 h post-HSV-2 injection.

Furthermore, when FLV and THC were co-administered to mice concurrently infected with herpes simplex virus, mortality attributed to the retroviral infection occurred significantly more rapidly than in the absence of the drug. Based on these results, Specter *et al.* (1991) speculated that marijuana could serve as a co-factor, possibly in conjunction with opportunistic pathogens, in the progression of infection with retroviruses. However, extrapolation of these data as applicable to infection with the human immunodeficiency virus (HIV) warrants further investigation.

Collectively, the data suggest that THC and marijuana alter host resistance to virus infections in animal models by targeting antiviral defenses which come into play early in the infection process. Furthermore, reports of inhibitory effects on antiviral antibody responses imply that elements of the immune system which play a role in limiting virus spread from initial sites of infection may be affected.

Decreased resistance to infection with bacteria

Klein *et al.* (1994) reported that THC induces significantly increased mortality in mice infected with *Legionella pneumophila*, the causative agent of Legionnaires' disease. In their experiments, the investigators tested whether the cannabinoid altered secondary immunity to *Legionella* since such immune responsiveness is critical for host survival upon challenge with this pathogen and formulates the basis for vaccine prophylaxis. *Legionella*-primed mice challenged with a secondary lethal dose of *Legionella* survived the secondary challenge infection. In contrast, significantly increased mortalities were obtained for animals subjected to the same *Legionella* infection and challenge regimen but receiving THC three weeks prior to the *Legionella* challenge. Klein *et al.* (1994) concluded that THC administration at the time of primary infection suppressed development of secondary immunity to *Legionella*. Furthermore, in a separate study, Klein *et al.* (1993) reported that THC induces cytokine-mediated mortality of mice infected with *Legionella*. Mice receiving two injections of THC (8 mg/kg), one 24 h before and the other 24 h after a sublethal injection of *Legionella*, experienced acute collapse and death. However, neither one nor two injections of THC given to animals before infection with *Legionella* resulted in death. The THC-induced mortality resembled cytokine-mediated shock. Moreover, acute phase sera from these animals contained significantly elevated levels of tumor necrosis factor (TNF) and interleukin 6 (IL6) implicating these cytokines as causative, at least in part, of the enhanced mortality. Mice receiving a normally sublethal injection of *Legionella* and administered anti-TNFα, anti-IL6, or a mixture of anti-IL1α and anti-IL1β antibodies 1 h before the second THC injection, were protected from THC-induced mortalities. Of the antibodies introduced into animals, those for IL6 were shown to be the most effective in rendering protection. Similar experiments performed on cultured splenocytes obtained from mice infected with *Legionella* and administered THC demonstrated alterations in levels of cytokines which are attributable to T lymphocyte subsets (Newton *et al.*, 1994). Splenocytes from THC-treated infected animals stimulated in culture with mitogen were deficient in interferon gamma (IFNγ) production. In addition, increased production of antibody to *Legionella* of the IgG_1 isotype, as compared to that for the IgG_{2a} isotype, was observed in sera of infected mice treated with THC. Furthermore, THC treatment of cultured, normal splenocytes with mitogen resulted in production of relatively higher levels of interleukin 4 (IL4) as compared with those for IFNγ. Collectively, these results indicated that THC modifies significantly the course of primary and secondary infection by *Legionella* and alters selectively the profile of cytokine production. Furthermore, the data suggest that THC induces a shift from Th1 to Th2 lymphocyte subtype activity, with resultant disruption of the network of cytokines which play a central and pivotal role in ablating infection with *Legionella*.

Recently, Massi *et al.* (1998) studied the effect of acute (1 h) and chronic (7 and 14 days) exposure to THC on immune parameters in male Swiss mice. It was reported that acute exposure to THC (10 mg/kg) resulting from subcutaneous administration had no effect on the splenocyte proliferative response to Concanavalin A or natural killer cell activity. However, a significant decrease in interleukin 2 (IL2) production was noted. On the other hand, chronic (7 days) administration, for which mice were shown to be tolerant to THC-induced analgesia, yielded a profile of immune responsiveness in which the splenocyte proliferative response was inhibited, nitric oxide (NO) activity was diminished, and IL2 and IFNγ levels were reduced. Thus THC administration in an acute versus a chronic mode, at least by a subcutaneous route of administration, apparently effects a distinctive profile of altered immune functional and biochemical parameters.

Protection against infection

In contrast to the body of data indicating that cannabinoids exacerbate host resistance to infection, it has been reported that cannabinoids have the potential to exert protective effects for select pathological conditions. Gallily *et al.* (1997) indicated that Dexanabinol (HU-211), a synthetic non-psychotropic cannabinoid, improved neurological outcomes in rodent models of brain trauma, ischemia, and meningitis. HU-211 was shown to suppress TNFα production and to rescue mice and rats from endotoxic shock after lipopolysaccharide (LPS, *Escherichia coli* 055:B5) injection. Bass *et al.* (1996) tested the efficacy of HU-211 in combination with antimicrobial therapy in reducing brain damage in a rat model of pneumococcal meningitis. *Streptococcus pneumonia*-infected rats were treated with saline, ceftriaxone (100 mg/kg), or with a combination of ceftriaxone (100 mg/kg) and HU-211 (5 mg/kg intravenously). Brain edema and blood-brain barrier impairment were significantly ($p < 0.05$) reduced for infected animals receiving combination therapy as compared with control animal groups. In addition, HU-211 has been reported to reduce the inflammatory response in the brain and spinal cord in rats used as experimental models of autoimmune encephalomyelitis (Achiron *et al.*, 2000). It was suggested from these latter studies that Dexanabinol could prove useful as an alternative mode of treatment of acute relapses of multiple sclerosis.

Relevance of doses used in animal studies

In the various studies involving animal models, the doses of cannabinoids which have been used have ranged from 0.2 mg/kg to 100 mg/kg. Rosenkrantz (1976) determined the relevance of THC doses and routes of administration used in rats and mice as compared with those absorbed by humans. Using cannabinoid levels and body surface area as markers, it was estimated that a dose 10–12 times greater was required in mice than in humans to elicit similar effects. Using such extrapolation, 100 mg/kg in a mouse corresponds approximately to 8–10 mg/kg in humans. Hembree *et al.* (1976) have reported that such a high dose in humans is achievable by heavy marijuana smokers.

EFFECTS ON HOST RESISTANCE USING IN VITRO MODELS OF INFECTION

Decreased resistance to infection with viruses

Studies which have been performed with *in vitro* models of infection have yielded results consistent with those obtained with *in vivo* models. *In vitro* studies have included the use of fully constituted primary cell populations such as splenocytes as well as purified macrophages and lymphocytes from rodents and humans. Immortalized continuous cell types which exhibit macrophage-like or T lymphocyte-like properties also have been employed. Cabral and Vásquez (1991, 1992) reported that THC affects the capacity of macrophages to respond to herpes simplex viruses in culture. THC (10^{-5} M–10^{-7} M) exerted a dose-dependent effect on macrophage extrinsic antiviral activity to HSV-2 (Figure 9.2), an activity which is characterized by the ability of macrophages to restrict the replication of viruses in adjacent uninfected permissive xenogeneic cells in an interferon-independent fashion (Morahan *et al.*, 1977). The inhibitory effect was exerted on a variety of murine macrophage-like cells including RAW264.7, J774A.1, and P388D1. The macrophage-like cells regained their extrinsic antiviral activity in a time-related fashion following removal of THC. In contrast, THC had no effect on intrinsic antiviral activity, an activity which is characterized by the capacity of macrophages to ingest and degrade virus and thereby maintain a nonpermissive state for productive virus infection (Stevens and Cook, 1971; Selgrade and Osborne, 1974). THC has been demonstrated also to decrease cytotoxic T lymphocyte (CTL) activity against herpes simplex virus type 1 (HSV-1) (Fischer-Stenger, Updegrove and Cabral, 1992). Spleen cells from C3H/HeJ (H-2^k) mice primed with HSV-1 and administered THC (15 mg/kg and 100 mg/kg) were deficient in cytolytic activity against HSV-1-infected murine L929 (H-2^k) target cells *in vitro*. In their experiments, THC *in vivo* exposure had little effect on the number of T lymphocytes expressing the Lyt-2 or L3T4 antigens. Furthermore, Nomarski optics microscopy revealed that the CTLs from the drug-treated, virus-primed mice were able to bind specifically to HSV-1-infected target cells. THC affected CTL cytoplasmic polarization and granule reorientation toward the CTL-virus-infected cell interface. It was concluded that THC elicited dysfunction in CTLs by altering effector cell-target cell postconjugation events resulting in failure of delivery of "lethal hit" molecules to the virus-infected cells.

Decreased resistance to infection with protozoa

THC also has been reported to affect macrophage functional activities against *Naegleria fowleri*, a free-living ameba which can cause a fatal disease in humans

A

Extrinsic Anti-HSV2 Activity
Vehicle Control

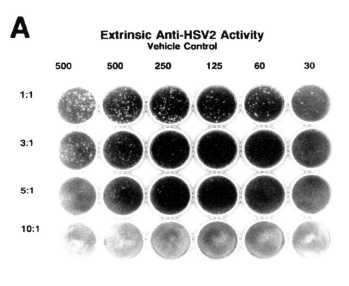

B

Extrinsic Anti-HSV2 Activity
10^{-6} M THC

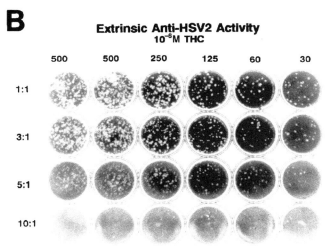

Figure 9.2 Effect of THC on macrophage extrinsic antiviral activity. P388D1 macrophage-like cells were treated with (A) vehicle or with (B) 10^{-6} M THC for 48 h. Macrophages, then, were added to HSV-2-infected Green monkey kidney (Vero) cell monolayers to yield effector cell:target cell (E:T) ratios of 5:1, 3:1, and 1:1. The number over each column designates the calculated number of plaque-forming units (pfu) of HSV-2 added to each Vero cell monolayer in that column. The number assigned to each row designates the macrophage:Vero cell E:T ratio. (A) Co-cultures containing P388D1 macrophages treated with vehicle (0.1% ethanol). There is a decrease in the number of virus plaques in direct correlation with increasing E:T ratios. At an E:T ratio of 5:1 a minimal number of plaques was elicited in all of the co-cultures regardless of the input number of infectious HSV-2. (B) Co-cultures containing P388D1 macrophages pretreated (48 h) with 10^{-6} M THC. Note the increase in the number of plaques in all co-cultures at all E:T ratios indicative of a decrease in extrinsic antiviral activity.

known as Primary Amebic Meningoencephalitis (PAME) (Marciano-Cabral, 1988). Peritoneal macrophages from mice receiving the macrophage activator *Bacillus Calmétte-Guérin* (BCG) in concert with THC (25–100 mg/kg) exhibited a drug dose-related reduction *in vitro* in their capacity to lyse the amebae (Burnette-Curley *et al.*, 1993). Consistent with results obtained from studies to assess effects of THC on CTL antiviral activity, Nomarski optics microscopy, scanning electron microscopy, and radiolabeling binding studies demonstrated that macrophages retained the capacity to attach to their target amebae but failed to lyse them.

Decreased resistance to infection with bacteria

Arata *et al.* (1992) have reported that THC affects macrophage functional activities *in vitro* against *Legionella pneumophila*. Treatment of macrophages from A/J mice with THC (8–10 µM) resulted in enhanced growth of *Legionella* within these cells. Furthermore, THC treatment overcame macrophage restriction of the growth of *Legionella* which is normally induced by macrophage activation with bacterial lipopolysaccharide.

EFFECTS ON IMMUNE CELL FUNCTIONAL ACTIVITIES IN VITRO

Early studies

A considerably larger body of data is available concerning the *in vitro* effects of cannabinoids and marijuana on immune cell parameters elicited in response to various mitogens and/or immune modulators. Studies conducted since the early 1970s have reported that cannabinoids and marijuana affect the functional activities of a variety of immune cells from rodents and humans including B lymphocytes (Zimmerman *et al.*, 1977; Smith *et al.*, 1978; Baczynsky and Zimmerman, 1983; Klein and Friedman, 1990; Nahas and Osserman, 1991; Kaminski *et al.*, 1992), T lymphocytes (Nahas *et al.*, 1974; Gupta *et al.*, 1974; Peterson, Graham and Lemberger, 1976; Nahas, Morishima and Desoize, 1977; Klein *et al.*, 1985; Cabral *et al.*, 1987; Klein *et al.*, 1991; Lee *et al.*, 1995), macrophages (Mann *et al.*, 1971; Drath *et al.*, 1979; Lopez-Cepero *et al.*, 1986; Cabral and Mishkin, 1989; Burstein *et al.*, 1994), and natural killer cells (Specter *et al.*, 1986; Patel *et al.*, 1985; Klein *et al.*, 1987; Kawakami *et al.*, 1988).

Effects on macrophages

Recently, Baldwin *et al.* (1997) evaluated the function of alveolar macrophages recovered from the lungs of nonsmokers and habitual smokers of tobacco, marijuana, or crack cocaine. Macrophages recovered from marijuana smokers were deficient in the ability to phagocytose *Staphylococcus aureus* and were severely limited in the capacity to kill bacteria and tumor cells. Experiments in which NG-monomethyl-L-arginine monoacetate, an inhibitor of nitric oxide synthase, was used suggested that macrophages from marijuana smokers were not able to use nitric oxide (NO) as an antibacterial effector molecule. Furthermore, macrophages from marijuana smokers, but not from smokers of tobacco or cocaine, produced lower levels of TNFα, GMC-SF, and IL6 when stimulated with lipopolysaccharide in culture when compared with alveolar macrophages obtained from control subjects. Based on these observations, it was concluded that habitual exposure of the lung to marijuana impaired select functions of alveolar macrophages including their capacity to produce cytokines.

McCoy *et al.* (1995) conducted a series of studies using THC and other cannabinoids to define the site of action within macrophages at which these compounds exerted their effects. The ability of macrophages exposed to THC to process and present soluble protein antigens was investigated by the stimulation of antigen-specific murine helper T cell hybridomas to secrete interleukin-2 (IL2). The T cell response to hen egg lysozyme (HEL) was dramatically reduced after a 24 h pretreatment of murine peritoneal macrophages or of a murine macrophage hybridoma with THC (Figure 9.3). In contrast, THC exposure did not alter the capacity of peritoneal macrophages or the macrophage hybridoma to process chicken ovalbumin and augmented their presenting cell function of a pigeon cytochrome *c* response. The level of T cell activation with peptides of lysozyme and cytochrome *c*, which do not require processing, was inhibited only at the highest concentrations of THC, suggesting that THC mainly affected antigen processing and not antigen presentation. It was concluded that THC exerted a differential effect on the capacity of macrophages to process antigens that are necessary for CD4+ T lymphocytes and that the nature of these effects was dependent on the intrinsic conformation of the antigen itself. Matveyeva *et al.* (2000) extended these studies to demonstrate that the THC induced impairment of HEL processing by macrophages was due, at least in part, to a selective increase in aspartyl cathepsin D proteolytic activity. It was suggested that upregulation of cathepsin D activity resulted in "over-processing" of HEL yielding peptides below the critical octapeptide to undecapeptide range requisite for antigen presentation. In addition, Clements, Cabral and McCoy (1996) demonstrated that THC suppresses a fixation-resistant co-stimulatory signal to helper T cells and does so, in part, by diminishing expression of macrophage heat-stable antigen.

Effects on lymphocytes, other immunocytes, and cytokines

Kusher *et al.* (1994) assessed the effect of THC (0.005–5.0 μg/ml) on the synthesis of TNFα by human large granular lymphocytes (LGL) in culture. These investigators reported that THC at physiological levels down-regulated TNFα production and diminished LGL cytolytic activity against K562 tumor cells. Furthermore, it was suggested that since the NK/polymorphonuclear neutrophil axis represents an important early defense against the opportunistic fungus *Candida albicans*, repression of this system by THC could contribute to susceptibility to infections with opportunistic pathogens.

Srivastava *et al.* (1998) examined the effects of cannabinoids *in vitro* on cytokine production. It was found that cannabinoids exerted a multiplicity of alterations in levels of cytokines from various immune cells. Using human T, B, eosinophilic, and CD8+ NK cell lines as *in vitro* models, exposure for 24 h to THC or the

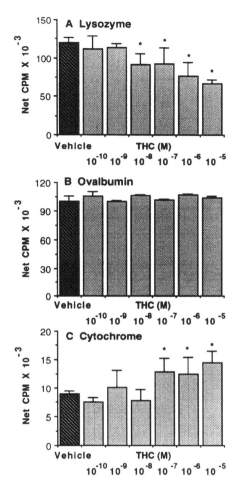

Figure 9.3 Antigen-dependent effects of THC on T-cell activation after a 24 h pretreatment of presenting cells (McCoy, K. L., Gainey, D. and Cabral, G. A., 1995). Murine Clone 63 (MHC $A_\beta^d : A_\alpha^d$ plus $E_\beta^k : E_\alpha^k$) cells as the antigen-presenting cells were preincubated with 0.1% ethanol as vehicle or with various concentrations of THC for 24 h. T-cells and antigen were added to the cultures, and the secretion of IL2 by the T cells was measured after 24 h. Culture supernatants were collected and assayed for IL2 by incubating the IL2-dependent cell line CTLL-2 with 25% culture supernatants at 37 C for 18 h. The wells were pulsed with 1 μCi of ^{3}[H]thymidine, harvested after another 6 h, and radiolabel incorporation was measured by liquid scintillation. Values are the mean cpm $\times 10^{-3}$ in experimental cultures minus the mean cpm in medium control $\pm$ S. D.* denotes significantly different (p < 0.05) from vehicle control. (A) Response of hen egg lysozyme-specific T-cell hybridoma 9.30.B2 ($A_\beta^d : A_\alpha^d$) to 200 μM hen egg lysozyme. (B) Response of chicken ovalbumin-specific T-cell hybridoma 3DO.54.8 ($A_\beta^d : A_\alpha^d$) to 50 μM chicken ovalbumin. (C) Response of pigeon cytochrome c-specific $E_\beta^k : E_\alpha^k$-restricted T-cell hybridoma 2B4.11 to 500 μM cytochrome c. THC inhibited processing of HEL, had no effect on that of ovalbumin, and augmented processing of cytochrome c.

relatively non-psychotropic cannabinoid cannabidiol (CBD) (2.5–10 μg/ml) exerted a variety of effects on immune cells exhibiting distinctive phenotypic markers. THC decreased constitutive production of the CXC chemokine interleukin 8 (IL8), of the CC chemokines MIP-1α, MIP-1β, and RANTES, and of phorbol ester-stimulated production of TNFα, granulocyte-macrophage colony-stimulating factor (GM-CSF), and IFNγ by NK cells. In addition, THC inhibited expression of MIP-1β in human T-lymphotropic virus 1 (HTLV-1)-positive B lymphocytes. In contrast, THC augmented levels of IL8, MIP-1α, and MIP-1β in B lymphocytes and IL8 and MIP-1β in eosinophils. Both CBD and THC inhibited the production of interleukin 10 (IL10) in HUT-78 T cells. Thus, cannabinoids modulated the production of cytokines, and of constitutively-expressed as well as inducibly-expressed chemokines, in purified human immune cell populations. However, the effects of THC and CBD were neither uniform in action nor consistent across cell lineages.

THC, and possibly other cannabinoids, can exert biphasic effects on immune cells. Pross et al. (1992) assessed the effect of THC on T lymphocyte stimulation with anti-CD3 antibody and revealed that lower drug concentrations increased proliferation while higher concentrations inhibited the response. Augmentation effects of cannabinoids have been observed also by Derocq et al. (1995) who reported that human tonsillar B cells exposed to nanomolar concentrations of cannabinoid exhibited enhanced growth. The cannabinoid-enhancing B cell proliferation was inhibited by pertussis toxin suggesting that a G protein-coupled receptor process was involved. Furthermore, the absence of an antagonist effect by SR141716A, an antagonist specific for the CB_1 cannabinoid receptor, together with the observation that human B cells displayed large amounts of CB_2 receptor mRNA, suggested that the growth enhancing activity was mediated through the CB_2 receptor. These observations of augmentation of cellular activities are consistent with those reported by McCoy et al. (1995) for THC-induced enhancement of macrophage processing of cytochrome c, and indicate that cannabinoids have the potential to exert both inhibitory and augmentative effects on immune cell functions, the nature of which may be predicated on the concentration of drug to which cells are exposed and the fundamental character of the cellular process examined. In addition, biphasic effects of cannabinoids with respect to immune cell lineages have been observed by Klein et al. (1985). These investigators demonstrated that THC concentrations in the micromolar range suppressed mouse splenocyte proliferation to T cell mitogens and to the B cell mitogen LPS. However, B cells appeared to be more sensitive than T cells to the effects of THC.

MECHANISMS BY WHICH MARIJUANA AND CANNABINOIDS ALTER IMMUNE CELL FUNCTION AND HOST RESISTANCE TO INFECTION

Multiple modes of action

Marijuana and cannabinoids have been reported to exert a wide range of in vivo and in vitro effects on a diverse array of immune cell types. Cannabinoids have been shown to augment (McCoy *et al.*, 1995; Derocq *et al.*, 1995; Srivastava *et al.*, 1998) or inhibit immune cell functions (reviewed in: Munson *et al.*, 1976; Cabral and Dove Pettit, 1998). These disparate effects may result from either receptor or non-receptor mediated modes of action. Felder *et al.* (1992) demonstrated that cannabinoid agonists stimulated receptor and non-receptor-mediated signal transduction pathways in fibroblast cell lines which had been transfected with a recombinant cannabinoid receptor expression vector and which expressed cannabinoid receptors. Experiments using the synthetic cannabinoid receptor agonist CP55940 indicated that the cloned receptors coupled to the inhibition of cAMP accumulation as anticipated for the involvement of a cannabinoid receptor-linked event. However, CP55940 also stimulated the increase of free arachidonic acid in a non-stereoselective fashion indicative of the absence of a functional linkage to a cannabinoid receptor for this cellular activity. Thus, cannabinoids can stimulate signaling pathways through both receptor- and non-receptor-mediated modes and do so within the same cell.

Changes effected through alterations in membranes

At high concentrations (i.e. 10^{-5} M or greater), THC and other cannabinoids can cause membrane perturbation and disrupt cell membranes (Figure 9.4). Physical disruption of cellular membranes may affect translational events and post-translational modifications such as glycosylation, phosphorylation, and proteolytic cleavage of precursor molecules. In addition, intracellular and extracellular communication through signaling molecules may be affected. Furthermore, since THC is a highly lipophilic molecule, its interaction with cellular membranes may alter membrane fluidity with consequent alterations in selective permeability (Wing *et al.*, 1985). Changes in surface-membrane selective permeability, with attendant increases in intracellular sodium, have been proposed as a mode by which viruses effect a shut-down of host cell macromolecular synthesis (Carrasco and Smith, 1976; Garry *et al.*, 1979). THC and other cannabinoids may have a similar effect on cells. Such alterations in membranes may account for the reported inhibition of protein synthesis (Cabral and Mishkin, 1989; Cabral and Fischer-Stenger, 1994) and of molecular precursor transport by THC (Desoize *et al.*, 1979). The relatively high concentrations of THC which would account for these effects are achievable in humans in the context of immune cells which populate and circulate through the lung, an organ which would be exposed directly to marijuana smoke.

Changes effected through cannabinoid receptors

THC at lower concentrations, and at sites distal to the lung, may affect immune cell functions by signaling through cannabinoid receptors. Cannabinoid receptors have been identified both within the brain and cells of the immune system (Figure 9.5). Two cannabinoid receptors, CB_1 and CB_2, have been identified. The CB_1 has been localized to neuronal tissues (Matsuda *et al.*, 1990) and testis (Galiègue *et al.*, 1995), and to a lesser extent to immune cells. In contrast, the CB_2 has been observed only in cells of the immune system. Both receptors are coupled to a pertussis toxin-sensitive G_i/G_o protein (Howlett *et al.*, 1986; Matsuda *et al.*, 1990) to

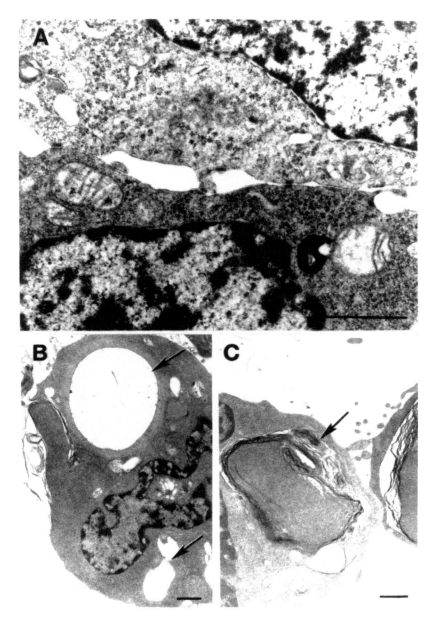

Figure 9.4 Transmission electron micrograph demonstrating that exposure to high concentrations of Δ^9-tetrahydrocannabinol (THC) results in membrane perturbation. Murine peritoneal macrophages were exposed to vehicle (0.1% ethanol) or THC (10^{-5} M) for 24 h. (A) Cells treated with vehicle. (B) Cells treated with THC exhibiting large intracytoplasmic vacuoles (arrows). (C) Cells treated with THC exhibiting large intracytoplasmic membranous whorls (arrow) indicative of membrane damage. The bars represent 1 μm.

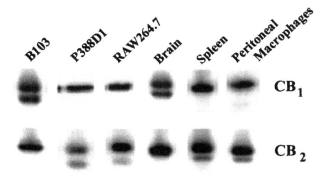

Figure 9.5 Identification of CB_1 and CB_2 mRNA by Mutagenic Reverse Transcription-Polymerase Chain Reaction (MRT-PCR) (Dove Pettit *et al.*, 1996). For detection of CB_1 mRNA, total RNA was subjected to reverse transcription using an oligonucleotide primer containing a single base mismatch generating a unique *Msp*I restriction site. The reverse transcription products then were amplified by PCR using a pair of highly conserved oligonucleotide primers specific for human, rat, and mouse CB_1 sequences. The PCR amplification products were digested with *Msp*I and subjected to electrophoretic separation through a 3% agarose gel and the DNA was transferred to nylon membrane. A similar approach was applied for detection of CB_2 mRNA, except that a primer conserved for mouse and rat CB_2 sequence and containing an unique *Hind*III restriction site was used for reverse transcription. PCR was performed using conserved primers for mouse and rat CB_2, and PCR products were digested with *Hind*III. The blots were hybridized with a ^{32}P-labeled random-primed rat CB_1 (for mouse or rat amplicons) or a mouse or rat CB_2 fragment. The CB_1 cDNA fragment was generated by amplification from a pCD-SKR6 template (Matsuda *et al.*, 1990) using the CB_1 primers. The murine and rat CB_2 cDNA fragments were generated by amplification from a mouse and a rat CB_2 template (gift from T. I. Bonner, NIMH Bethesda MD) using the CB_2 primers. (Top panel) Southern blot analysis of MRT-PCR products amplified from total RNA from rat brain, rat B103 neuroblastoma cells, murine P388D1 and RAW264.7 macrophage-like cells, rat spleen, and thioglycolate-elicited murine peritoneal macrophages. CB_1 mRNA was detected in rat brain and rat B103 neuroblastoma cells as demonstrated by the presence of two products following digestion with *Msp*I. The upper band (i.e. larger digestion product) represents product derived from genomic DNA templates, while the lower band (i.e. smaller digestion product) represents that derived from mRNA templates. (Bottom panel) CB_2 mRNA was detected in total RNA from murine P388D1 and RAW264.7 macrophage-like cells, rat spleen, and murine peritoneal macrophages as demonstrated by the presence of two products following digestion with *Hind*III. No CB_2 mRNA was detected in total RNA from rat brain or rat B103 neuroblastoma cells.

inhibit adenylate cyclase activity resulting in decreases in levels of cAMP (Howlett *et al.*, 1990; Felder *et al.*, 1992; Felder *et al.*, 1995) and initiate mitogen-activated protein kinase (MAPK) and immediate early gene signaling pathways (Bouaboula *et al.*, 1993, 1995, 1996). However, in contrast to the CB_1, no modulation of N-type calcium channels (Mackie and Hille, 1992) and G-protein regulated inwardly rectifying K+ channels (Childers and Deadwilder, 1996) has been observed for the CB_2 (Felder *et al.*, 1995).

The presence of CB_2 receptors exclusively within immune cells suggests a role for these receptors in their activities. Transcripts (i.e. mRNAs) for the CB_2 have been found in spleen and tonsils (Galiègue *et al.*, 1995; Munro *et al.*, 1993) and other immune tissues (Munro *et al.*, 1993; Bouaboula *et al.*, 1993). However, in all cases reported to date, levels of message for the CB_2 exceed those for the CB_1. The distribution pattern of levels of CB_2 mRNA displays major variation in human blood cell populations with a rank order of B lymphocytes > natural killer (NK) cells >> monocytes > polymorphonuclear neutrophils > T8 lymphocytes > T4 lymphocytes (Galiègue *et al.*, 1995). A rank order for levels of CB_2 transcripts similar to that for primary human cell types has been recorded for human cell lines belonging to the myeloid, monocytic, and lymphoid lineages (Galiègue *et al.*, 1995). In addition, the presence of cognate protein has been demonstrated in rat lymph nodes, Peyer's Patches, and spleen (Lynn and Herkenham, 1994). Table 9.1 summarizes the reported distribution of cannabinoid receptors in immune cells and tissues.

Kaminski *et al.* (1992, 1994) provided evidence which implicated a functional linkage between cannabinoid receptors and cannabinoid-mediated alterations in the activities of immune cells. It was noted that suppression of the humoral immune response by cannabinoids was mediated partially by inhibition of adenylate cyclase through a pertussis toxin sensitive Guanine nucleotide binding protein (G protein) coupled mechanism. THC and the synthetic bicyclic cannabinoid CP55940 inhibited the lymphocyte proliferative response and the sheep erythrocyte IgM antibody-forming cell response of murine splenocytes to phorbol-12-myristate-13-acetate (PMA) plus the calcium ionophore ionomycin. Also, Jeon *et al.* (1996) indicated that LPS-inducible NO release by the murine macrophage-like cell

Table 9.1 Distribution of cannabinoid receptors in immune cells and tissues

Cell Type/Tissue	Species	Receptor	Reference
B lymphocytes	Human	CB_2	Galiègue *et al.* (1995)
T4 lymphocytes	Human	CB_2	Galiègue *et al.* (1995)
T8 lymphocytes	Human	CB_2	Galiègue *et al.* (1995)
Leukocytes	Human	CB_2	Bouaboula *et al.* (1993)
Macrophages	Human	CB_2	Galiègue *et al.* (1995)
Microglia	Rat	CB_1	Waksman *et al.* (1999)
Mononuclear Cells	Human, rat	CB_2	Galiègue *et al.* (1995)
			Facci *et al.* (1995)
Mast cells	Rat	CB_2	Facci *et al.* (1995)
Natural Killer Cells	Human	CB_2	Galiègue *et al.* (1995)
Peyer's Patches	Rat	CB^*	Lynn and Herkenham (1994)
Spleen	Human,	CB_1, CB_2	Kaminski *et al.* (1992)
	Mouse, rat	CB^*	Munro *et al.* (1993)
			Galiègue *et al.* (1995)
			Facci *et al.* (1995)
			Lynn and Herkenham (1994)
Thymus	Human	CB_2	Galiègue *et al.* (1995)
Tonsils	Human	CB_2	Galiègue *et al.* (1995)
Lymph Nodes	Rat	CB^*	Lynn and Herkenham (1994)

* Cannabinoid receptor subtype not discriminated.

line RAW264.7 was suppressed by THC and other agonists by mechanisms which implicated the involvement of cannabinoid receptors. It was indicated that the attenuation of inducible NO gene expression by THC was mediated through the inhibition of nuclear factor-κB/Rel activation. In addition, Burstein et al. (1994) have presented evidence that indicates that THC-induced arachidonic acid release from mouse peritoneal cells occurs through a series of events that are consistent with a receptor-mediated process involving the stimulation of one or more phospholipases.

Recently, Waksman et al. (1999) provided evidence that cannabinoids can affect immune cell function through the CB_1. The synthetic cannabinoid CP55940 inhibited the production of inducible NO by neonatal rat microglial cells (Figure 9.6), cells which constitute the resident macrophages of the brain. The inhibitory effect was stereoselective in that the dose-dependent inhibition of NO release was exerted by the cannabinoid receptor high affinity binding enantiomer CP55940 while a minimal effect was exerted by the low affinity binding paired enantiomer CP56667. Furthermore, reversal in CP55940-mediated inhibition of NO release was effected when microglial cells were pretreated with the CB_1-selective antagonist SR141716A. In contrast, Stefano et al. (1996) demonstrated that cannabinoid receptor agonists exerted an opposite effect on the production and/or release of constitutively expressed NO. Cannabinoid receptor agonists increased constitutive NO levels in cultures of human monocytes. As in the case of inducible NO, this effect was inhibited by the CB_1 antagonist SR141716A suggesting that the CB_1 was involved in the augmentation process. On the other hand, McCoy et al. (1999) implied that a functional linkage existed between cannabinoid mediated inhibition of the processing of hen egg lysozyme(HEL) by macrophages and the CB_2. In their studies, processing of HEL was inhibited by THC and other cannabinoid agonists. Stereoselective cannabinoid enantiomers showed a differential inhibitory effect for the bioactive enantiomer CP55940 versus that of its less bioactive enantiomeric pair CP56667. Furthermore, the CB_1-selective antagonist SR141716A did not block the inhibitory effect of the cannabinoid agonist while the CB_2-selective antagonist SR144528 did. More recently, Massi et al. (2000) reported that both types of cannabinoid receptors are involved in mediating NK cell cytolytic activity. Inhibition of NK cell activity by THC was partially reversed by both the CB_1 and the CB_2 antagonists, although the CB_1 antagonist was more effective. In addition, the CB_1 and the CB_2 antagonists completely reversed THC-mediated inhibition of IFNγ production.

Few studies have addressed the issue of the effects of cannabinoids on infectious processes in the context of a functional linkage to a cannabinoid receptor. Noe et al. (1998), using syncytial formation as a barometer of infection, reported that cannabinoid receptor agonists enhanced syncytia formation in MT-2 cells infected with cell free Human Immunodeficiency virus MN strain (HIV-1MN). Recently, Gross et al. (2000) implicated the CB_1 receptor as linked functionally to cannabinoid effects on Brucella suis growth within macrophages. The CB_1-selective antagonist SR141716A effected a dose-dependent inhibition of the intracellular multiplication of this Gram-negative bacterium. However, the nonselective CB_1/CB_2 cannabinoid receptor agonists CP55940 or WIN55212-2 reversed the SR141716A-mediated effect. These results suggested a beneficial application of a CB_1 antagonist as an inhibitor of macrophage infection by the intracellular

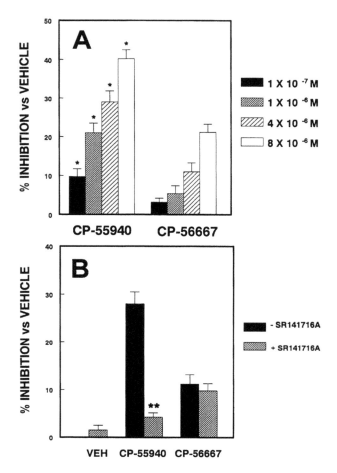

Figure 9.6 Inhibition of neonatal rat cortical microglial inducible nitric oxide release by the synthetic cannabinoid agonist CP55940. (A) Differential inhibition of NO release by the cannabinoid agonist CP55940 versus its paired enantiomer CP56667. Microglial cells were pretreated with drug or vehicle for 8 h, treated with 20 μg/ml LPS plus 10 U/ml IFNγ, and culture supernatants were assayed for nitrite 24 h later. Nitrite release from vehicle-treated cultures was 25.4 ± 3.3 [μM/10^6 cells/ml]. Results are expressed as percent inhibition versus vehicle and are the mean $\pm$ S.E.M. of triplicate wells. The high affinity cannabinoid CP55940 exerted a dose-dependent inhibition of NO release from rat microglial cells. The drug dose-dependent inhibition was significantly greater ($p < 0.05$, Student's t-test) than that exerted by its paired enantiomer CP56667 at each comparable concentration. (B) Reversal of CP55940-mediated inhibition of NO release by the CB_1-specific antagonist SR141716A. Microglial cells were pretreated with 5×10^{-7} M SR141716A prior to exposure to 5×10^{-6} M CP55940 or CP56667 and LPS/IFNγ activation. Results (mean $\pm$ S.E.M. of triplicate wells) are expressed as percent inhibition versus vehicle control (**$p < 0.01$ versus SR141716A). Nitrite accumulation in vehicle-treated cultures was 29.3 ± 3.5 [μM/10^6 cells]. Collectively, these results are consistent with a functional linkage between the CB_1 receptor and cannabinoid-mediated inhibition of inducible NO production by neonatal rat cortical microglial cells.

pathogen *Brucella suis*. Thus, data from several laboratories suggest that both CB_1 and CB_2 receptors may be linked functionally to cannabinoid agonist-mediated effects on immune cells. However, the relative contributions of the two cannabinoid receptors to individual functional activities within specified immune cell types, or within individual cells co-expressing CB_1 and CB_2, remain to be defined. The present availability of CB_1 and CB_2 specific antagonists (Rinaldi-Carmona *et al.*, 1994; Rinaldi-Carmona *et al.*, 1998), as well as of CB_1 and CB_2 knock-out mice which already have provided valuable insight regarding functional activities such as enhancement of memory, hypoalgesia, vasodilation, and embryonic development (Reibaud *et al.*, 1999; Zimmer *et al.*, 1999; Jarai *et al.*, 1999; Buckley *et al.*, 1998), affords the opportunity to systematically identify signal transductional systems within immune cells which serve as specific targets of cannabinoids.

EFFECTS OF CANNABINOIDS AND MARIJUANA ON HUMAN INFECTIONS AND AIDS

Background to human studies

Although many studies have addressed the effects of cannabinoids and marijuana on host resistance and immunity using *in vitro* and *in vivo* models, there have been few which have assessed directly the effects of marijuana usage in humans. The scarcity of data applies particularly to evaluation of effects of marijuana, used either in a recreational or therapeutic mode, among humans who have various immune deficiencies including those associated with infection with the human immunodeficiency virus (HIV). Epidemiological studies similar to those which have been performed to assess effects of tobacco or marijuana have not been carried out in human populations in relation to decreased resistance to infections or other pathology associated with immune dysfunction. The multiple epidemiological studies which have been conducted to assess the incidence of marijuana use among human population groups, especially among adolescents, have not included evaluation of biological and immunological markers associated with barometers of immune competence. The limited number of studies which have been performed to date have yielded limited and often contradictory data as to effects of cannabinoids on human immunity and host resistance to infection.

Studies reporting minimal effects on immunity and host resistance

A number of human clinical trials in which peripheral blood lymphocytes from healthy individuals and from chronic marijuana smokers were obtained have revealed no differences in mitogen-induced proliferation of rosette formation which was used as a measure of cell-mediated immunity functional competence (Kaklamani *et al.*, 1978; Rachelefsky and Opelz, 1977; Lau *et al.*, 1976). Marijuana use was found to be associated with an increase in the percentage of CD4+ T lymphocytes in the peripheral circulation with a mean CD4:CD8 ratio of 1.95 in marijuana smokers versus one of 1.27 in control subjects (Wallace *et al.*, 1988). However, other immune function tests which were performed, such as proliferation, were found to be normal. In addition, Wallace *et al.* (1998) examined risk factors and outcomes associated with identification of *Aspergillus* in respiratory specimens from individuals with HIV disease as part of a study to evaluate pulmonary complications of HIV infection. It was indicated that a substantially greater proportion of patients with *Aspergillus* as compared with control subjects died during the study. However, the use of cigarettes and marijuana was found not to be associated with *Aspergillus* respiratory infection. DiFranco *et al.* (1996), through the San Francisco Men's Health Study (SFMHS), evaluated the association of specific recreational drugs and alcohol with laboratory predictors of AIDS. Participants in the study were evaluated at entry into the program in 1984 and in the context of the development of AIDS during six years of follow-up. No substantial association could be obtained between the use of marijuana and the development of AIDS among HIV-infected men. In addition, it has been reported that cannabinoid use in a therapeutic mode exerts few deleterious effects, at least as they relate to immune competence and resistance to infection (Timpone *et al.*, 1997).

Studies reporting deleterious effects on immunity and host resistance

In contrast, other studies have suggested that cannabinoids or marijuana exert deleterious effects as they relate to the immune system and resistance to infection in humans. Gross *et al.* (1991) indicated that marijuana consumption altered responsiveness of human Papillomavirus (HPV) to systemic recombinant interferon alpha 2a treatment. Tindall *et al.* (1988) conducted immunoepidemiological studies using univariant and multivariant analyses and implied an association between marijuana use and progression of HIV infection. Caiaffa *et al.* (1994) indicated that smoking illicit drugs such as marijuana, cocaine, or crack, *Pneumocystis carinii* pneumonia, and immunosuppression increased risk of bacterial pneumonia in HIV-seropositive users. More recently, Whitfield *et al.* (1997) examined the impact of ethanol and Marinol/marijuana usage on HIV+/AIDS patients undergoing azidothymidine, azidothymidine/dideoxycytidine, or dideoxyinosine therapy. In HIV+/AIDS patients with the lowest CD4+ counts (those not on DDI monotherapy), utilization of Marinol/marijuana did not seem to have a deleterious effect. However, Marinol/marijuana usage was associated with depressed CD4+ counts and elevated amylase levels within the DDI subgroup. Furthermore, Marinol/marijuana use was associated with declining health status in both the AZT and AZT/DDC groups.

SUMMARY

The cumulative data obtained through cell culture studies using various immune cell populations extracted from animals or humans, together with those obtained using animal models of infection, are consistent with the proposition that marijuana and cannabinoids alter immune cell function and can exert deleterious effects on resistance to infection in humans. Both receptor-mediated and non-receptor mediated modes of action have been proposed to account for the effects of cannabinoids. However, few controlled longitudinal epidemiological and immunological studies have been undertaken to correlate the immunosuppressive effects of marijuana smoke or cannabinoids on the incidence of infections or viral disease in humans. Clearly, additional investigation to resolve the long-term immunological consequences of cannabinoid and marijuana use as they relate to resistance to infections in humans is warranted.

REFERENCES

Achiron, A., Miron, S., Lavie, V., Margalit, R. and Biegon, A. (2000) "Dexanabinol (HU-211) effect on experimental autoimmune encephalomyelitis: implications for the treatment of acute relapses of multiple sclerosis," *Journal of Neuroimmunology* **102(1)**: 26–31.

Arata, S., Newton, C., Klein, T. and Friedman, H. (1992) "Enhanced growth of *Legionella pneumophila* in tetrahydrocannabinol-treated macrophages," *Proceedings of the Society for Experimental Biology and Medicine* **199**: 65–67.

Baczynsky, W. O. T. and Zimmerman, A. M. (1983) "Effects of delta-9-tetrahydrocannabinol, cannabinol and cannabidiol on the immune system in mice, *In vitro* investigation using cultured mouse splenocytes," *Pharmacology* **26(1)**: 12–19.

Baldwin, G. C., Tashkin, D. P., Buckley, D. M., Park, A. M., Dubinett, S. M. and Roth, M. D. (1997) "Marijuana and cocaine impair alveolar macrophage function and cytokine production," *American Journal of Respiratory and Critical Care Medicine* **156(5)**: 1606–1613.

Bass, R., Engelhard, D., Trembovler, V. and Shohami, E. (1996) "A novel nonpsychotropic cannabinoid, HU-211, in the treatment of experimental pneumococcal meningitis," *The Journal of Infectious Diseases* **173**: 735–738.

Bouaboula, M., Poinot-Chazel, C., Marchand, J., Canat, X., Bourrie, B., Rinaldi-Carmona, M. *et al.* (1996) "Signaling pathway associated with stimulation of CB$_2$ peripheral cannabinoid receptor. Involvement of both mitogen-activated protein kinase and induction of Krox-24 expression," *European Journal of Biochemistry* **237(3)**: 704–711.

Bouaboula, M., Poinot-Chazel, C., Mourichoud, G., Arnaud, S., Roux, P., Cassellas, P. *et al.* (1995) "Activation of mitogen-activated protein kinases by stimulation of the central cannabinoid receptor CB$_1$," *Biochemical Journal* **312**: 637–641.

Bouaboula, M., Rinaldi, M., Carayon, P., Carillon, C., Delpech, B., Shire, D., Le Fur, G. and Casellas, P. (1993) "Cannabinoid receptor expression in human leukocytes," *European Journal of Biochemistry* **214**: 173–180.

Buckley, N. E., Hansson, S., Harta, G. and Mezey, E. (1998) "Expression of the CB$_1$ and CB$_2$ receptor messenger RNAs during embryonic development of the rat," *Neuroscience* **82(4)**: 1131–1149.

Burnette-Curley, D., Marciano, F. M., Fischer, K. and Cabral, G. A. (1993) "Delta-9-tetrahydrocannabinol inhibits cell contact-dependent cytotoxicity of *Bacillus* Calmétte-Guérin-activated macrophages," *International Journal of Immunopharmacology* **15**: 371–382.

Burstein, S., Budrow, J., Debatis, M., Hunter, S. A. and Subramanian, A. (1994) "Phospholipase participation in cannabinoid-induced release of free arachidonic acid," *Biochemical Pharmacology* **48**: 1253–1264.

Cabral, G. A. and Dove Pettit, D. (1998) "Drugs and immunity: Cannabinoids and their role in decreased resistance to infectious disease," *Journal of Neuroimmunology* **83**: 116–123.

Cabral, G. A. and Fischer-Stenger, K. (1994) "Inhibition of macrophage inducible protein expression by delta-9-tetrahydrocannabinol," *Life Sciences* **54**: 1831–1844.

Cabral, G. A. and Mishkin, E. M. (1989) "Delta-9-tetrahydrocannabinol inhibits macrophage protein expression in response to bacterial immunomodulators," *Journal of Toxicology and Environmental Health* **26**: 175–182.

Cabral, G. and Vásquez, R. (1991) "Effects of marijuana on macrophage function," in *Drugs of abuse, immunity, and immunodeficiency*, edited by H. Friedman, S. Specter and T. Klein, pp. 93–105. New York: Plenum Press.

Cabral, G. A. and Vásquez, R. (1992) "Delta-9-tetrahydrocannabinol suppresses macrophage extrinsic antiherpesvirus activity," *Proceedings of the Society for Experimental Biology and Medicine* **199**: 255–263.

Cabral, G. A., Lockmuller, J. C. and Mishkin, E. M. (1986) "Delta-9-tetrahydrocannabinol decreases alpha/beta interferon response to herpes simplex virus type 2 in the B6C3F1 mouse," *Proceedings of the Society for Experimental Biology and Medicine* **181**: 305–311.

Cabral, G. A., McNerney, P. J. and Mishkin, E. M. (1987) "Effect of micromolar concentrations of delta-9-tetrahydrocannabinol on herpes simplex virus type 2 replication in vitro," *Journal of Toxicology and Environmental Health* **21(3)**: 277–293.

Cabral, G., Mishkin, E. M., Marciano, F. M., Coleman, R. E., Harris, L. S. and Munson, A. E. (1986) "Effect of delta-9-tetrahydrocannabinol on herpes simplex virus type 2 vaginal infection in the guinea pig," *Proceedings of the Society for Experimental Biology and Medicine* **182**: 181–186.

Caiaffa, W. T., Vlahov, D., Graham, N. M., Astemborski, J., Solomon, L., Nelson, K. E. and Munoz, A. (1994) "Drug smoking, *Pneumocystis carinii* pneumonia, and immunosuppression increase risk of bacterial pneumonia in human immunodeficiency virus-seropositive injection drug users," *American Journal of Respiratory and Critical Care Medicine* **150**: 1493–1498.

Carrasco, L. and Smith, A. E. (1976) "Sodium ions and the shut-off of host cell protein synthesis by picornaviruses," *Nature* **264**: 807–809.

Childers, S. R. and Deadwyler, S. A. (1996) "Role of cyclic AMP in the actions of cannabinoid receptors," *Biochemical Pharmacology* **52**: 819–827.

Clements, D., Cabral, G. A. and McCoy, K. L. (1996) "Delta-9-tetrahydrocannabinol selectively inhibits macrophage co-stimulatory activity and down-regulates heat-stable antigen expression," *The Journal of Pharmacology and Experimental Therapeutics* **277**: 1315–1321.

Derocq, J. M., Segui, M., Marchand, J., Le Fur, G. and Casellas, P. (1995) "Cannabinoids enhance human B-cell growth at low nanomolar concentrations," *FEBS Letters* **369**: 177–182.

Desoize, B., Leger, C. and Nahas, G. G. (1979) "Plasma membrane inhibition of macromolecular precursor transport by THC," *Biochemical Pharmacology* **28**: 1113–1118.

Dewey, W. L. (1986) "Cannabinoid Pharmacology," *Pharmacology Reviews* **38(2)**: 151–178.

DiFranco, M. J., Sheppard, H. W., Hunter, D. J., Tosteson, T. D. and Ascher, M. S. (1996) "The lack of association of marijuana and other recreational drugs with progression to AIDS in the San Francisco Men's Health Study," *Annals of Epidemiology* **6(4)**: 283–289.

Dove Pettit, D. A., Anders, D. L., Harrison, M. P. and Cabral, G. A. (1996) "Cannabinoid receptor expression in immune cells," *Advances in Experimental Medicine and Biology* **402**: 119–129.

Drath, D. B., Shorey, J. M., Price, L. and Huber, G. L. (1979) "Metabolic and functional characteristics of alveolar macrophages recovered from rats exposed to marijuana smoke," *Infection and Immunity* **25**: 268–272.

Facci, L., Dal Tosso, R., Romanello, S., Buriani, A., Skaper, S. D. and Leon, A. (1995) "Mast cells express a peripheral receptor and differential sensitivity to anandamide and palmitoylethanolamide," *Proceedings of the National Academy of Sciences of the USA* **92**: 3376–3380.

Felder, C. C., Joyce, K. E., Briley, E. M., Mansouri, J., Mackie, K., Blond, O., Lai, Y., Ma, A. L. and Mitchell, R. L. (1995) "Comparison of the pharmacology and signal transduction of the human cannabinoid CB_1 and CB_2 receptors," *Molecular Pharmacology* **48**: 443–450.

Felder, C. C., Veluz, J. S, Williams, H. L., Briley, E. M. and Matsuda, L. A. (1992) "Cannabinoid agonists stimulate both receptor- and non-receptor-mediated signal transduction pathways in cells transfected with and expressing cannabinoid receptor clones," *Molecular Pharmacology* **42**: 838–845.

Fischer-Stenger, K., Updegrove, A. W. and Cabral, G. A. (1992) "Delta-9-tetrahydrocannabinol decreases cytotoxic T lymphocyte activity to herpes simplex virus type 1-infected cells," *Proceedings of the Society for Experimental Biology and Medicine* **200**: 422–430.

Friedman, H. and Klein, T. W. (1999) "Marijuana and immunity," *Science and Medicine* **6(2)**: 12–21.

Galiègue, S., Mary, S., Marchand, J., Dussosoy, D., Carrière, D., Carayon, P., Bouaboula, M., Shire, D., Le Fur, G. and Casellas, P. (1995) "Expression of central and peripheral cannabinoid receptors in human immune tissues and leukocyte subpopulations," *European Journal of Biochemistry* **232**: 54–61.

Gallily, R., Yamin, A., Waksman, Y., Ovadia, H., Weidenfeld, J., Bar-Joseph, A. *et al.* (1997) "Protection against septic shock and suppression of tumor necrosis factor α and nitric oxide production by Dexanabinol (HU-211), a non-psychotropic cannabinoid," *The Journal of Pharmacology and Experimental Therapeutics* **283**: 918–924.

Garry, R. F., Bishop, J. M., Parker, S., Westbrook, K., Lewis, G. and Waite, M. R. F. (1979) "Na^+ and K^+ concentrations and the regulation of protein synthesis in sindbis virus-infected chick cells," *Virology* **96**: 108–120.

Gross, A., Terraza, A., Marchant, J., Bouaboula, M., Ouahrani-Bettache, S., Liautard, J. P. *et al.* (2000) "A beneficial aspect of a CB_1 cannabinoid receptor antagonist: SR141716A is a potent inhibitor of macrophage infection by the intracellular pathogen *Brucella suis*," *Journal of Leukocyte Biology* **67(3)**: 335–344.

Gross, G., Roussaki, A., Ikenberg, H. and Dress, N. (1991) "Genital warts do not respond to systemic recombinant interferon alpha-2a treatment during cannabis consumption," *Dermatology* **183**: 203–207.

Gupta, G., Grieco, M. and Cushman, P. (1974) "Impairment of rosette-forming T-lymphocytes in chronic marihuana smokers," *New England Journal of Medicine* **291**: 874–876.

Howlett, A. C., Champion, T. M., Wilken, G. H. and Mechoulam, R. (1990) "Stereochemical effects of 11-OH-delta 8-tetrahydrocannabinol-dimethylheptyl to inhibit adenyl cyclase and bind to the cannabinoid receptor," *Neuropharmacology* **29**: 161–165.

Howlett, A. C., Qualy, J. M. and Khachchatrian, L. L. (1986) "Involvement of Gi in the inhibition of adenylate cyclase by cannabimimetic drugs," *Molecular Pharmacology* **29**: 307–313.

Jarai, Z., Wagner, J. A., Varga, K., Lake, K. D., Compton, D. R., Martin, B. R. *et al.* (1999) "Cannabinoid-induced mesenteric vasodilation through an endothelial site distinct from CB_1 or CB_2 receptors," *Proceedings of the National Academy of Sciences of the USA* **96(24)**: 14136–14141.

Jeon, Y. J., Yang, K. H., Pulaski, J. T. and Kaminski, N. E. (1996) "Attenuation of inducible nitric oxide gene expression by Δ9-tetrahydrocannabinol is mediated through the inhibition of nuclear factor-κB/rel activation," *Molecular Pharmacology* **50**: 334–341.

Kaklamani, E., Trichopoulos, D., Koutselinis, A., Drouga, M. and Karalis, D. (1978) "Hashish smoking and T-lymphocytes," *Archives of Toxicology* **40**: 97–101.

Kaminski, N. E., Abood, M. E., Kessler, F. K., Martin, B. R. and Schatz, A. R. (1992) "Identification of a functionally relevant cannabinoid receptor on mouse spleen cells that is involved in cannabinoid-mediated immune modulation," *Molecular Pharmacology* **42**: 736–742.

Kaminski, N. E., Koh, W. S., Yang, K. H., Lee, M. and Kessler, F. K. (1994) "Suppression of the humoral immune response by cannabinoids is partially mediated through inhibition of adenylate cyclase by a pertussis toxin-sensitive G-protein coupled mechanism," *Biochemical Pharmacology* **48**: 1899–1908.

Kawakami, Y., Klein, T. W., Newton, C., Djeu, J. Y., Specter, S. and Friedman, H. (1988) "Suppression of delta-9-tetrahydrocannabinol of interleukin 2-induced lymphocyte proliferation and lymphokine-activated killer cell activity," *International Journal of Immunopharmacology* **10**: 485–488.

Klein, T. W. and Friedman, H. (1990) In *Drugs of Abuse and Immune Function*, edited by R. Watson, pp. 87–111. Boca Raton: CRC Press.

Klein, T. W., Kawakami, Y., Newton, C. and Friedman, H. (1991) "Marijuana components suppress induction and cytolytic function of murine cytotoxic T cells *in vitro* and *in vivo*," *Journal of Toxicology and Environmental Health* **32**: 465–477.

Klein, T. W., Newton, C. and Friedman, H. (1994) "Resistance to *Legionella pneumophila* suppressed by the marijuana component, tetrahydrocannabinol," *Journal of Infectious Diseases* **169**: 1177–1179.

Klein, T. W., Newton, C. and Friedman, H. (1987) "Inhibition of natural killer cell function by marijuana components," *Journal of Toxicology and Environmental Health* **20**: 321–332.

Klein, T., Newton, C., Widen, R. and Friedman, H. (1993) "Delta-9-tetrahydrocannabinol injection induces cytokine-mediated mortality of mice infected with *Legionella pneumophila*," *The Journal of Pharmacology and Experimental Therapeutics* **267**: 635–640.

Klein, T. W., Newton, C., Widen, R. and Friedman, H. (1985) "The effect of delta-9-tetrahydrocannabinol and 11-hydroxy-delta 9-tetrahydrocannabinol on T-lymphocyte and B-lymphocyte mitogen responses," *Journal of Immunopharmacology* **7**: 451–466.

Klykken, P. C., Smith, S. H., Levy, J. A., Razdan, R. and Munson, A. E. (1977) "Immunosuppressive effects of 8,9-epoxyhexahydrocannabinol (EHHC)," *The Journal of Pharmacology and Experimental Therapeutics* **201**: 573–579.

Kusher, D. I., Dawson, L. O., Taylor, A. C. and Djeu, J. Y. (1994) "Effect of the psychoactive metabolite of marijuana, delta-9-tetrahydrocannabinol (THC), on the synthesis of tumor necrosis factor by human large granular lymphocytes," *Cellular Immunology* **154(1)**: 99–108.

Lau, R. J., Tubergen, D. G., Barr, M. Jr., Domino, E. F., Benowitz, N. and Jones, R. T. (1976) "Phytohemagglutinin-induced lymphocyte transformation in humans receiving delta-9-tetrahydrocannabinol," *Science* **192**: 805–807.

Lee, M., Yang, K .H. and Kaminski, N. E. (1995) "Effects of putative receptor ligands, anandamide and 2-arachidonyl-glycerol, on immune function in B6C3F1 mouse splenocytes," *The Journal of Pharmacology and Experimental Therapeutics* **275**: 529–536.

Lopez-Cepero, M., Friedman, M., Klein, T. and Friedman, H. (1986) "Tetrahydrocannabinol-induced suppression of macrophage spreading and phagocyte activity *in vitro*," *Journal of Leukocyte Biology* **39**: 679–686.

Lynn, A. B. and Herkenham, M. (1994) "Localization of cannabinoid receptors and nonsaturable high-density cannabinoid binding sites in peripheral tissues of the rat: implications for receptor-mediated immune modulation by cannabinoids," *The Journal of Pharmacology and Experimental Therapeutics* **268**: 1612–1623.

Mackie, K. and Hille, B. (1992) "Cannabinoids inhibit N-type calcium channels in neuroblastoma-glioma cells," *Proceedings of the National Academy of Sciences of the USA* **89**: 3825–3829.

Mann, P. E. G., Cohen, A. B., Finley, T. N. and Ladman, A. J. (1971) "Alveolar macrophages. Structural and functional differences between non-smokers and smokers of marijuana *in vitro*," *Laboratory Investigation* **25**: 111–120.

Marciano-Cabral, F. (1988) "Biology of *Naegleria* spp.," *Microbiological Reviews* **52**: 114–133.

Massi, P., Fuzio, D., Vigano, D., Sacerdote, P. and Parolaro, D. (2000) "Relative involvement of cannabinoid CB(1) and CB(2) receptors in the Delta9-tetrahydrocannabinol-induced inhibition of natural killer activity," *European Journal of Pharmacology* **387(3)**: 343–347.

Massi, P., Sacerdote, P., Ponti, W., Fuzio, D., Manfredi, B., Vigano, D., Rubino, T., Bardotti, M. and Parolaro, D. (1998) "Immune function alterations in mice tolerant to delta-9-

tetrahydrocannabinol: functional and biochemical parameters," *Journal of Neuroimmunology* **92(1–2)**: 60–66.

Matsuda, L. A., Lolait, S. J., Brownstein, M. J., Young, A. C. and Bonner, T. I. (1990) "Structure of a cannabinoid receptor and functional expression of the cloned cDNA," *Nature* **346**: 561–564.

Matveyeva, M., Hartman, C. B., Harrison, M. T., Cabral, G. A. and McCoy, K. L. (2000) "Delta⁹-tetrahydrocannabinol selectively increases aspartyl cathepsin D proteolytic activity and impairs lysozyme processing by macrophages," *International Journal of Immunopharmacology* **22(5)**: 373–381.

McCoy, K. L., Gainey, D. and Cabral, G. A. (1995) "Δ⁹-Tetrahydrocannabinol modulates antigen processing by macrophages," *The Journal of Pharmacology and Experimental Therapeutics* **273**: 1216–1223.

McCoy, K. L., Matveyeva, M., Carlisle, S. J. and Cabral, G. A. (1999) "Cannabinoid inhibition of the processing of intact lysozyme by macrophages: Evidence for CB₂ receptor participation," *The Journal of Pharmacology and Experimental Therapeutics* **289**: 1620–1625.

Mishkin, E. M. and Cabral, G. (1985) "Delta-9-tetrahydrocannabinol decreases host resistance to herpes simplex virus type 2 vaginal infection in the B₆C₃F₁ mouse," *Journal of General Virology* **66**: 2539–2549.

Morahan, P. S., Breinig, M. C. and McGeorge, M. B. (1977) "Immune responses to vaginal or systemic infection of BALBc mice with herpes simplex virus type 2," *Journal of Immunology* **119**: 2030–2036.

Morahan, P. S., Klykken, P. C., Smith, S. H., Harris, L. S. and Munson, A. E. (1979) "Effects of cannabinoids on host resistance to *Listeria monocytogenes* and herpes simplex virus," *Infection and Immunity* **23**: 670–674.

Munro, S., Thomas, K. L. and Abu-Shaar, M. (1993) "Molecular characterization of a peripheral receptor for cannabinoids," *Nature* **365**: 61–65.

Munson, A. E., Levy, J. A., Harris, L. S. and Dewey, W. L. (1976) "Effects of delta 9-tetrahydrocannabinol on the immune system," in *The Pharmacology of Marijuana*, edited by M. C. Braude and S. Szara, pp. 187–197. New York: Raven Press.

Munson, A. E. and Fehr, K. O. (1983) "Immunological Effects of Cannabis," in *Cannabis and Health Hazards*, edited by K. O. Fehr and H. Kalant, pp. 257–353. Toronto: Alcoholism and Drug Addiction Research Foundation.

Nahas, G. G., Morishima, A. and Desoize, B. (1977) "Effects of cannabinoids on macromolecular synthesis and replication of cultured lymphocytes," *Federation Proceedings* **36**: 1748–1752.

Nahas, G. G. and Osserman, E. F. (1991) "Altered serum immunoglobulin concentration in chronic marijuana smokers," *Advances in Experimental Medicine and Biology* **288**: 25–32.

Nahas, G. G., Suciu-Foca, N., Armand, J. P. and Morishima, A. (1974) "Inhibition of cellular mediated immunity in marihuana smokers," *Science* **183**: 419–420.

Newton, C. A., Klein, T. W. and Friedman, H. (1994) "Secondary immunity to *Legionella pneumophila* and Th1 activity are suppressed by delta-9-tetrahydrocannabinol injection," *Infection and Immunity* **62**: 4015–4020.

Noe, S. N., Nyland, S. B., Ugen, K., Friedman, H. and Klein, T. W. (1998) "Cannabinoid receptor agonists enhance syncytia formation in MT-2 cells infected with cell free HIV-1MN," *Advances in Experimental Medicine and Biology* **437**: 223–229.

Patel, V., Borysenko, M., Kumar, M. S. A. and Millard, W. J. (1985) "Effects of acute and subchronic delta-9-tetrahydrocannabinol administration on the plasma catecholamine, beta-endorphin, and corticosterone levels and splenic natural killer cell activity in rats," *Proceedings of the Society for Experimental Biology and Medicine* **180**: 400–404.

Peterson, B. H., Graham, J. and Lemberger, L. (1976) "Marihuana, tetrahydrocannabinol and T-cell function," *Life Sciences* **19**: 395–400.

Pross, S. H., Nakano, Y., Widen, R., McHugh, S., Newton, C. A., Klein, T. W. and Friedman, H. (1992) "Differing effects of delta-9-tetrahydrocannabinol (THC) on murine spleen cell populations dependent upon stimulators," *International Journal of Immunopharmacology* **14**: 1019–1027.

Rachelefsky, G. S. and Opelz, G. (1977) "Normal lymphocytes function in the presence of delta-9-tetrahydrocannabinol," *Clinical Pharmacology and Therapeutics* **21**: 44–46.

Reibaud, M., Obinu, M. C., Ledent, C., Parmentier, M., Bohme, G. A. and Imperato, A. (1999) "Enhancement of memory in cannabinoid CB$_1$ receptor knock-out mice," *European Journal of Pharmacology* **379(1)**: R1–R2.

Rinaldi-Carmona, M., Barth, F., Heaulme, M., Shire, D., Calandra, B., Congy, C. *et al.* (1994) "Sr141716A, a potent and selective antagonist of the brain cannabinoid receptor," *FEBS Letters* **350**: 240–244.

Rinaldi-Carmona, M., Barth, F., Millan, J., Derocq, J. M., Casellas, P., Congy, C. *et al.* (1998) "SR144528, the first potent and selective antagonist of the CB$_2$ cannabinoid receptor," *The Journal of Pharmacology and Experimental Therapeutics* **284**: 644–650.

Rosenkrantz, H. (1976) "The immune response and marihuana," in *Marihuana: Biological effects, analysis, cellular responses, reproduction and brain*, edited by G. G. Nahas, W. D. M. Paton, and J. Idanpaan-Heikkila, pp. 441–456. New York: Pergamon Press.

Selgrade, M. K. and Osborn, J. E. (1974) "Role of macrophages in resistance to murine cytomegalovirus," *Infection and Immunity* **10**: 1383–1390.

Smith, S. H., Harris, L. S., Uwaydah, I. M. and Munson, A. E. (1978) "Structure–activity relationships of natural and synthetic cannabinoids in suppression of humoral and cell-mediated immunity," *The Journal of Pharmacology and Experimental Therapeutics* **207**: 165–170.

Specter, S. C., Klein, T. W., Newton, C., Mondragon, M., Widen, R. and Friedman, H. (1986) "Marijuana effects on immunity: suppression of human natural killer cell activity by delta-9-tetrahydrocannabinol," *International Journal of Immunopharmacology* **8**: 741–745.

Specter, S., Lancz, G., Westrich, G. and Friedman, H. (1991) "Delta-9-tetrahydrocannabinol augments murine retroviral induced immunosuppression and infection," *International Journal of Immunopharmacology* **13**: 411–417.

Srivastava, M. D., Srivastava, B. I. and Brouhard, B. (1998) "Delta-9-tetrahydrocannabinol and cannabidiol alter cytokine production by human immune cells," *Immunopharmacology*, **40(3)**: 179–185.

Stefano, G. B., Liu, Y. and Goligorsky, M. S. (1996) "Cannabinoid receptors are coupled to nitric oxide release in invertebrate immunocytes, microglia, and human monocytes," *Journal of Biological Chemistry* **271**: 19238–19242.

Stevens, J. G. and Cook, M. L. (1971) "Restriction of herpes simplex virus by macrophages. An analysis of the cell-virus integration," *Journal of Experimental Medicine* **133**: 19–38.

Timpone, J. G., Wright, D. J., Li, N., Egorin, M. J., Enama, M. E., Mayers, J. *et al.* (1997) "The safety and pharmokinetics of single-agent and combination therapy with Megestrol acetate and Dronabinol for the treatment of HIV wasting syndrome," *AIDS Research and Human Retroviruses* **13**: 305–315.

Tindall, B., Cooper, D. A., Donovan, B., Barnes, T., Philpot, C. R., Gold, J. and Penny, R. (1988) "The Sydney AIDS Project: development of acquired immunodeficiency syndrome in a group of HIV seropositive homosexual men," *Australian and New Zealand Journal of Medicine* **18(1)**: 8–15.

Waksman, Y., Olson, J. M., Carlisle, S. J. and Cabral, G. A. (1999) "The central cannabinoid receptor (CB$_1$) mediates inhibition of nitric oxide production by rat microglial cells," *The Journal of Pharmacology and Experimental Therapeutics* **288**: 1357–1366.

Wallace, J. M., Lim, R., Browdy, B. L., Hopewell, P. C., Glassroth, J., Rosen, M. J. *et al.* (1998) "Risk factors and outcomes associated with identification of *Aspergillus* in respiratory specimens from persons with HIV disease. Pulmonary complications of HIV infection study group," *Chest* **114(1)**: 131–137.

Wallace, J. M., Tashkin, D. P., Oishi, J. S. and Barbers, R. G. (1988) "Peripheral blood lymphocyte subpopulations and mitogen responsiveness in tobacco and marijuana smokers," *Journal of Psychoactive Drugs* **20(1)**: 9–14.

Whitfield, R. M., Bechtel, L. M. and Starich, G. H. (1997) "The impact of ethanol and Marinol/marijuana usage on HIV+/AIDS patients undergoing azidothymidine, azidothymidine/dideoxycytidine, or dideoxyinosine therapy," *Alcoholism, Clinical and Experimental Research* **21(1)**: 122–127.

Wing, D. R., Leuschner, J. T. A., Brent, G. A., Harvey, D. J. and Paton, W. D. M. (1985) "Quantification of *in vivo* membrane associated delta-9-tetrahydrocannabinol and its effects on membrane fluidity," in *Proceedings of the 9th International Congress of Pharmacology 3rd satellite symposium on cannabis*, edited by D. J. Harvey, pp. 411–418. Oxford: IRL Press.

Zimmer, A., Zimmer, A. M., Hohman, A. G., Herkenham, M. and Bonner, T. I. (1999) "Increased mortality, hypoactivity, and hypoalgesia in cannabinoid CB$_1$ receptor knockout mice," *Proceedings of the National Academy of Sciences of the USA* **96(10)**: 5780–5785.

Zimmerman, S., Zimmerman, A. M., Cameron, I. L. and Lawrence, H. L. (1977) "delta-1-tetrahydrocannabinol, cannabidiol and cannabinol effects on the immune response of mice," *Pharmacology* **15**: 10–23.

Chapter 10

Marijuana and cognitive function

Nadia Solowij

ABSTRACT

There is now a good evidence that heavy and long-term use of marijuana can result in subtle impairments of memory, attention and executive function and that the functioning of prefrontal cortical, hippocampal and cerebellar regions can become compromised. Human research has used increasingly sophisticated and sensitive techniques to examine the cognitive consequences of marijuana use and their neural concomitants, improving upon the methodology that produced equivocal results in past studies. The discovery of the endogenous cannabinoid system, a decade ago spurred a vast amount of animal research on the effects of cannabinoids on receptor and overall brain function. This research has demonstrated alterations in the functioning of the brain in regions and in neuromodulator systems (e.g. dopaminergic and cholinergic) that are crucial for cognitive processes. Further research is required to elucidate the parameters of human use that may lead to clinically significant dysfunction and to investigate individual susceptibilities to impairments. There is evidence that the alterations in cognition and brain function may persist following cessation of use but the extent to which they might recover with prolonged abstinence has yet to be determined. This chapter reviews the literature pertinent to marijuana and cognitive function with a focus on the most recent animal and human research.

Key Words: cannabis, cognitive function, memory, cannabinoid receptor, long-term drug effects

INTRODUCTION

Over the years there has been much debate about whether heavy, frequent or prolonged use of marijuana may lead to a deterioration in cognitive function that persists well beyond any period of acute intoxication. Concerns in the community are well founded given that marijuana is the most popular recreational drug among young people with use often commencing in the mid teen years. The extent to which the use of marijuana may interfere with scholastic achievement and the psychological and emotional maturation of young users is uncertain, despite several decades of use and concomitant research in many parts of the world. It is generally accepted both within the scientific community and among users themselves that cognition is impaired during the acute intoxication after smoking or ingestion of marijuana, and this is well supported by animal and human laboratory research (for reviews regarding acute effects see Beardsley and Kelly, 1999; Chait and Pierri, 1992; Solowij, 1998). Questions remain about possible cumulative effects following years of regular intoxication and whether the functions of the brain may be altered in the long term. There is much anecdotal evidence that the cognitive function of long-term users may be compromised, and users themselves express concerns that their drug use is adversely affecting their memory function and their ability to concentrate. These concerns can become a prime motivator in seeking treatment to assist them to quit using the drug. The scientific evidence from past research has been inconclusive. However, recent research with improved methodology continues to demonstrate definite but subtle impairments of memory, attention and higher cognitive function associated with regular use of marijuana – sometimes with heavy use, sometimes only with long-term use. In recent years there has been an explosion of studies examining the workings of the cannabinoid receptor in the brain and its endogenous and exogenous ligands. This chapter will briefly review the literature pertinent to cognitive function, with a focus on findings from the most recent research. For more detailed reviews of earlier literature the reader is referred to Solowij, 1998. The term marijuana will be used here to encompass the range of preparations of cannabis plant matter in human usage.

EFFECTS ON THE CENTRAL NERVOUS SYSTEM

Marijuana exerts its most prominent effects on the central nervous system (CNS). The discovery of the cannabinoid receptor (CB_1) in the brain and the endogenous cannabinoids that bind to this receptor (Matsuda *et al.*, 1990; Devane *et al.*, 1992; Stella *et al.*, 1997) confirmed the direct activation of the CNS by marijuana, or its primary psychoactive constituent, Δ^9-tetrahydrocannabinol (THC). Cannabinoid receptors are widely distributed throughout the brain with high density in regions known to be involved in cognition such as cerebral cortex, hippocampus and cerebellum (Glass *et al.*, 1997; Herkenham *et al.*, 1990; Tsou *et al.*, 1998). There has been a surge in research to determine the physiological roles of the cannabinoid receptors and their endogenous ligands. A large number of animal studies have now confirmed that anandamide, the first endogenous cannabinoid that was discovered, is involved in functions such as analgesia, sleep, memory and motor control. The second confirmed endogenous ligand, 2-AG, is present in brain in amounts 170 times greater than anandamide, and the actions of both anandamide and 2-AG are very similar to those of THC (Childers and Breivogel, 1998; Martin *et al.*, 1999; see also, Chapter 5 (Di Marzo), Chapter 6 (Fride) and Chapter 17 (Reggio) of this volume). Human studies of the acute effects of marijuana also suggest that the system is involved in regulating mood, emotion, memory, attention and other cognitive functions. It is possible that the prolonged use of exogenous cannabinoids (e.g. marijuana) may alter such functions, and indeed the endogenous cannabinoid system and receptor itself. The evidence to date from both human and animal research suggests that they are not grossly impaired in the long term but that there are definite alterations in their function.

ANIMAL RESEARCH

Animal research into the effects of marijuana on CNS function has typically administered known quantities of cannabinoids to animals for a specified time and then examined performance on various tasks, before using histological and morphometric methods to study the brains of the exposed animals. Animal research enables specificity of attribution to cannabinoid effects by excluding various confounding factors. Extrapolation of the findings directly to humans is, however, difficult because of differences in brain and behavior, patterns of use, routes of administration, dosage and methods of assessment. In general, the results of studies with primates produce results that most closely resemble the likely effects in humans.

Overall, surprisingly few animal studies of neurotoxicity have been published and the results have been equivocal. A recent study showed that large doses of THC applied directly to cultured rat hippocampal neurons resulted in significant toxicity and cell death (Chan *et al.*, 1998). There is some evidence for long-term changes in hippocampal ultrastructure and morphology in rodents and monkeys following chronic administration of THC. Several studies of rhesus monkeys administered doses comparable to those used by human heavy marijuana users have reported permanent alterations in brain function and ultrastructure after only a few months exposure (e.g. Heath *et al.*, 1980), and brain atrophy, particularly in frontal regions, with long-term exposure (e.g. 5 years, McGahan *et al.*, 1984). Other animal studies have found no major abnormalities (or even neuronal protection: e.g. Hampson *et al.*, 1998b; see further below) but there have been large differences in the methodology employed, including the type of cannabinoid administered, the dosage and the duration over which it was administered. While recent research has concentrated on acute studies to elucidate the mechanisms of action and the functional roles of the endogenous cannabinoid system, past studies have provided convincing evidence that chronic administration of large doses of THC leads to neurobehavioral toxicity in animals that is characterised by lasting impairment in learning and memory function, EEG and biochemical alterations, impaired motivation, lethargy, sedation, depression and aggressive irritability (see Adams and Martin, 1996; Solowij, 1998).

Cannabinoids interact with most neuromodulator systems that underlie information processing in the brain and anandamide interacts with the dopaminergic system (see Martin and Cone, 1999; Pertwee, 1992). Cannabinoid receptors inhibit noradrenergic neurons (Schlicker and Gothert, 1998) and modify other neurotransmitter systems via their effects on interneurons since they are located on interneurons in many brain regions (e.g. substantia nigra, globus pallidus and hippocampus; Hampson and Deadwyler, 1998; Katona *et al.*, 1999). There is a paucity of studies investigating alterations in brain chemistry following very long-term treatment with cannabinoids akin to that of human chronic use, but if any few irreversible effects are suggested by some (Slikker *et al.*, 1992). One research team has shown that perinatal exposure to THC alters not only the normal development of dopaminergic neurons in the hypothalamus and midbrain of postnatal and peripubertal rats, but that there is a persistent alteration in the activity of these neurons and in the behavior of adult rats perinatally exposed to THC (Garcia *et al.*, 1996; Garcia-Gil *et al.*, 1997; Navarro *et al.*, 1995, 1996). Indeed it is now believed that endogenous cannabinoids actually contribute to brain development through the activation of second-messenger coupled cannabinoid receptors (Fernandez-Ruiz *et al.*, 2000).

Acute administration of cannabinoids to rats has been shown to dose-dependently increase the firing rate and burst firing of mesoprefrontal dopaminergic neurons arising in the ventral tegmentum and projecting to the prefrontal cortex (Diana *et al.*, 1998). Increased concentration of dopamine (and norepinephrine) in prefrontal cortex following the acute administration of THC has been shown to be directly related to impaired performance on a spatial delayed alternation working memory task, the impairment being prevented by the dopaminergic modulator HA966 (Jentsch *et al.*, 1997). Following repeated exposure to THC, however, prefrontal cortical dopamine metabolism was reduced (Jentsch *et al.*, 1998), and reduced dopaminergic transmission in the limbic system was associated with withdrawal from chronic cannabinoid administration in rats (Diana *et al.*, 1998). Both reduced and excessive dopaminergic activity are detrimental to working memory performance. Jentsch and Taylor (1999) review the evidence that suggests that altered dopaminergic activity, impaired frontal cortical inhibitory response control and cognitive dysfunction resulting from chronic drug use, together with impulsivity and altered incentive motivational processes due to limbic/amygdalar dysfunction, may underlie further drug-seeking behavior.

The cannabinoid receptor plays an important role in regulating the neural activity critical for memory processing (Hampson and Deadwyler, 1999) among other cognitive functions. A recent study found that cannabinoid receptors may sequester G-proteins and thus prevent other G-protein coupled receptors from transducing their biological signals (Vasquez and Lewis, 1999). Anandamide modulates NMDA receptor activity (Hampson *et al.*, 1998a) and increases protein tyrosine phosphorylation in rat hippocampal cultured neurons thus exerting neurotrophic effects and playing a role in synaptic plasticity (Derkinderen *et al.*, 1996). Cannabinoids have been shown to alter GABAergic receptor activation (Hampson *et al.*, 1998c), inhibiting GABA(A) synaptic transmission (Hoffman and Lupica, 2000), glutamatergic transmission (Shen *et al.*, 1996; Shen and Thayer, 1999) and long-term potentiation (LTP) formation in the rat hippocampus (LTP being a neural model for the cellular basis underlying learning and memory processes; Collins *et al.*, 1995; Stella *et al.*, 1997; Terranova *et al.*, 1995). Lévénès *et al.* (1998) demonstrated cannabinoid inhibition of glutamatergic transmission at parallel fiber-Purkinje cell synapses in rat cerebellum, and impaired long-term depression, which may explain the cerebellar dysfunction caused by cannabinoids. Bohme *et al.* (2000) have shown that invalidating the CB_1 receptor gene in mice leads to larger hippocampal LTP and therefore an enhanced capacity to strengthen synaptic connections crucial for memory formation. The same team showed that CB_1 knock-out mice were able to retain memory in an object recognition task for at least 48 h while control mice lost the capacity to retain memory after 24 h (Reibaud *et al.*, 1999). These results confirm the crucial role that cannabinoid receptors play in memory storage and retrieval processes.

Research from the laboratory of Hampson and Deadwyler has elucidated some of the mechanisms of action of exogenous and endogenous cannabinoids on memory processing. They showed that cannabinoid administration to rats dose-dependently impaired performance in short-term spatial memory (delayed nonmatch to sample) tasks in a manner similar to damage to or even complete removal of the hippocampus (see Hampson and Deadwyler, 1999). The mechanism appeared to be the inhibition of neural activity in hippocampal and retrohippocampal areas brought about by combined cannabinoid and GABA receptor activation. The authors propose a hippocampal circuit underlying task performance whereby the distribution of cannabinoid receptors suggests that they would exert their primary influence on the encoding but not retrieval processes of memory, but high dose cannabinoid activation of the widely distributed GABA receptors would impair all aspects of memory processing. They describe further research that showed that activation of hippocampal cannabinoid receptors also disrupted the retention of trial specific information, while activation of retrohippocampal cannabinoid receptors on GABAergic neurons resulted in a shift in behavior to the use of an alternate response strategy. Hampson and Deadwyler (1998, 1999) postulate the role of endogenous cannabinoids as selectively blocking or reducing the encoding of stimuli when appropriate or advantageous, thus regulating the ability to retain information across delays and preventing learned information from being over-written by new stimuli, and enabling the forgetting of stored stimuli following retrieval, but also, endogenous cannabinoids may selectively regulate the engagement of the entire hippocampal formation in memory processing, and thus not only modulate the encoding of information but also the behavioral strategies by which that information is used or the means by which it is retrieved. Exogenous cannabinoids (e.g. THC) override the normal function of the endogenous cannabinoids by disrupting the encoding of information, retention processes, and perhaps the selective switching of the hippocampus, when it is not appropriate nor advantageous to do so. Winsauer et al. (1999) describe the disruptive effects of several cannabinoids on the acquisition of new behavior and performance accuracy of rhesus monkeys learning complex discriminations.

Earlier studies had confirmed that cannabinoids impair working memory through a cannabinoid receptor mechanism, and determined a hippocampal locus of memory impairment. Lichtman, Dimen and Martin (1995) reported that intra-cerebral administration of cannabinoids such as CP55,940 into the cannabinoid receptor rich hippocampus of rats disrupted spatial working memory in the radial-arm maze task but more compelling support for receptor-mediated THC-induced memory impairment was provided by the dose-dependent reversal of such impairment by the cannabinoid antagonist SR141716A (Lichtman and Martin, 1996). Collins et al. (1995) had shown that the inhibitory effects of the potent cannabinoid HU-210 on LTP in rat hippocampus were blocked by SR141716A. Furthermore, Terranova et al. (1996) demonstrated in rats and mice actual improvement of short-term working memory and of memory consolidation, and the abolishment of memory disturbance induced by retroactive inhibition or that associated with aging, by the administration of the above antagonist in a dose-dependent fashion in the absence of any pretreatment with cannabinoids. The authors commented that SR141716A did not enhance retrieval, but facilitated the memory processes involved immediately after acquisition and during consolidation. The results of this research suggest that the endogenous cannabinoid system is involved in forgetting and in the memory deterioration associated with aging (see further below). However, it should be noted that this facilitation of memory was partially antagonized by scopolamine, a muscarinic antagonist that impairs memory, which implies a connection between the blockade of cannabinoid receptors and the facilitation of cholinergic transmission (Terranova *et al.*, 1996).

Gifford and Ashby (1996) suggested that an endogenous substance might inhibit the release of acetylcholine through activation of the cannabinoid receptor. Gessa *et al.* (1998) used microdialysis in freely moving rats to demonstrate lasting cannabinoid receptor-mediated inhibition of acetylcholine release in medial-prefrontal cortex and hippocampus after administration of doses of THC that were relevant to human usage. Tolerance did not develop to the inhibition of hippocampal acetylcholine following repeated exposure to THC over 7 days (Carta *et al.*, 1999). However, Lagalwar *et al.* (1999) showed that while anandamide also inhibits the brain muscarinic acetylcholine receptor, this inhibition was not mediated by the cannabinoid receptor. Lichtman and Martin (1996) found that SR141716A did not alleviate scopolamine-induced impairment on the radial-arm maze in rats suggesting that cannabinoids and cholinergic drugs do not impair spatial memory through a common serial pathway. Presburger and Robinson (1999) demonstrated selective and specific cannabinoid receptor-mediated impairment of visuo-spatial attention in rats, whereby the accuracy of stimulus detection was disrupted by THC in a manner different to the impairment produced by scopolamine or glutamatergic (NMDA receptor) antagonism. Mechoulam *et al.* (1999) report that an acetylcholine receptor agonist enhanced the production of the endocannabinoid 2-AG in rat aorta. Large amounts of 2-AG and high concentration of cannabinoid receptors have been found in the retina of rhesus monkeys and a number of other vertebrate species, suggesting a role for the endogenous cannabinoid system in retinal physiology signaling and perhaps for vision in general (Straiker *et al.*, 1999). Cannabinoid receptors appear to even be involved in bird song learning (Soderstrom and Johnson, 2000).

While the effects of THC and anandamide are similar in many respects, there are a number of qualitative differences in their actions (e.g. McGregor *et al.*, 1998). Hampson *et al.* (1998a) showed that anandamide modulates NMDA receptor activity and neurotransmission in hippocampus in a manner unlike THC and unaffected by a CB_1 antagonist. Carriero *et al.* (1998) confirmed that the rank order of potency of various cannabinoids for suppressing lever pressing in rats was consistent with the rank order of affinities for the CB_1 receptor shown by these drugs (CP55,940 > WIN55,212 (two potent synthetic cannabinoids) > THC > AM 356 (methanandamide, a more stable analog of anandamide)). Early studies had reported that anandamide failed to impair memory function in rats in the manner of THC and other psychoactive cannabinoids (Crawley *et al.*, 1993; Lichtman *et al.*, 1995). However, the lack of demonstrable memory impairment following administration of anandamide was primarily due to its rapid metabolism (Deutsch and Chin, 1993), together with its lower affinity and the nature of the tasks employed in different studies with different routes of administration. When rats were pretreated with a protease inhibitor, anandamide dose-dependently impaired working memory in a delayed non-match to sample task (Mallet and Beninger, 1996) and this anandamide-induced memory impairment was attenuated by

SR141716A (Mallet and Beninger, 1998). Terranova *et al.* (1995) demonstrated inhibition of hippocampal LTP by anandamide and reversal by SR141716A. Murillo-Rodriguez *et al.* (1998) showed that both anandamide and its metabolic precursor arachidonic acid deteriorated memory consolidation when administered intra-cerebroventricularly to rats. Castellano *et al.* (1997) showed that anandamide dose-dependently impaired retention when administered within a short period of time post-training (when the memory trace is susceptible to modulation) and that these effects were antagonized by dopamine receptor agonists. In another study they showed that these effects on memory consolidation were strain specific; anand-amide improved retention in one strain of mice but impaired it in another (Castellano *et al.*, 1999). This study also showed antagonism of these effects by naltrexone in both strains, providing suggestive evidence that endogenous cannabinoids affect memory processes through opioid systems as well as interacting with the dopamin-ergic system. Tanda *et al.* (1997) had previously reported the activation of mesolimbic dopamine transmission by THC through an opioid receptor mechan-ism, and Manzanares *et al.* (1998) showed that subchronic cannabinoid administra-tion (5 days) increased opioid gene expression in various brain regions, suggesting "an interaction between the cannabinoid and enkephalinergic systems that may be part of a molecular integrative response to behavioral and neurochemical alter-ations that occur in cannabinoid drug abuse". A recent study found that very low doses of anandamide improved the maze performance of food deprived rats and reversed the neurotransmitter changes induced by severe diet restriction (Hao *et al.*, 2000). Giuffrida *et al.* (1999) have shown that neural activity stimulates the release of anandamide but not 2-AG in the dorsal striatum of freely moving rats, and that this release is potentiated by the administration of a dopamine D-2-like agonist. Anandamide may mediate sleep induction as it is potentiated by the sleep-inducing lipid oleamide (Mechoulam *et al.*, 1997); oleamide must interact with cannabinoid receptors as its action has been shown to be blocked by the canna-binoid antagonist SR14176A (Mendelson and Basile, 1999). Previous studies had determined a role for the endogenous cannabinoid system in the sleep/wakefulness cycle (e.g. Santucci *et al.*, 1996; Murillo-Rodriguez *et al.*, 1998). This research may explain the drowsiness that often accompanies the latter stages of marijuana intoxi-cation, the use of marijuana to facilitate sleep induction, and the sleep difficulties encountered by marijuana users withdrawing from the drug.

Although most of the recent animal studies reviewed above have involved only acute (or subchronic) administration of cannabinoids, their findings are helping to elucidate the actions of cannabinoids in the brain, which will in turn enable a more thorough understanding of its long-term effects. Rats chronically administered cannabinoids (e.g. for up to three months) have shown deficits in memory and learning that persisted sometimes for up to several months after administration of cannabinoids ceased (see Solowij, 1998), although tolerance to such effects has also been shown to occur fairly rapidly (e.g. after 21–35 days, Deadwyler *et al.*, 1995; 21 days, Sim *et al.*, 1996; or even 11 days of administration, Rubino *et al.*, 1994). Toler-ance to the various pharmacological effects of cannabinoids occurs to varying degrees (for example little occurs to the subjective high in humans); this might in part be explained by the fact that dopamine neurons in different brain regions develop a differential response to chronic cannabinoid treatment (Wu and

French, 2000). Tolerance is often mediated by a down-regulation of receptors and reduced binding, which have been demonstrated following prolonged administration of cannabinoids (Oviedo *et al.*, 1993; De Fonseca *et al.*, 1994), but are apparently not irreversible (Westlake *et al.*, 1991). Other studies have demonstrated tolerance to THC without any alteration of cannabinoid receptor binding (Abood *et al.*, 1993) or with increased binding in hippocampus and cerebellum but decreased binding in striatum following chronic administration (Romero *et al.*, 1995). Receptor internalization has also been shown to occur rapidly following agonist binding and receptor activation (Hsieh *et al.*, 1999). A biochemical basis of cannabinoid tolerance has also been demonstrated with large decreases in G-protein activation throughout the brain (and most dramatically in hippocampus) and profound desensitization of cannabinoid-activated signal transduction mechanisms after treatment with high dose THC for only 21 days, thus showing that effects on receptor function may occur without consistent changes in the number of receptor binding sites (Sim *et al.*, 1996). Another recent study showed that pronounced changes occur in the expression of cannabinoid receptor mRNA as a function of exposure to THC over 21 days (Zhuang *et al.*, 1998). Different brain regions with high receptor densities showed a different time course of change and differential changes with mostly decreased message in striatum, mostly increased message in hippocampus and a biphasic response in cerebellum. Changes in all three regions occurred only after 7 days of THC treatment and a regional divergence was apparent only after 3 days, but all three regions returned to pretreatment mRNA expression levels by 21 days of treatment. Zhuang *et al.* (1998) reported that a major implication of their results was that "cannabinoid systems in the brain appear to continue to operate at reduced levels of efficiency (decreased receptor number desensitization) following prolonged drug exposure sufficient to generate behavioral and/or physiological tolerance" (p. 148). The truly long-term consequences for receptor–ligand binding, receptor number and particularly function that may result from much longer exposure to cannabinoids, and the extent of reversibility of effects following such exposure, have not yet been determined to any degree of accuracy. The evidence is largely consistent with recovery from memory deficits some time after administration of cannabinoids ceases, but the results of animal studies may not reflect the gradual changes that could occur at the cannabinoid receptor and to the endogenous system, and indeed other neuromodulator systems, over a period of much longer exposure to the drug akin to that of human chronic usage.

Animal research has also examined changes in cannabinoid receptors with age. Mailleux and Vanderhaeghen (1992) reported age-related losses in cannabinoid receptor binding sites and mRNA in the rat striatum. Belue *et al.* (1995) demonstrated progressively increased binding capacity in rat striatum, cerebellum, cortex and hippocampus from birth through to adulthood, which they interpreted as reflecting either "an increased differentiation of neurons into cells possessing cannabinoid receptors, or an increase in the number of receptors on cell bodies or projections in regions undergoing developmental changes". Once the adult levels had been reached, binding activity in a whole brain preparation neither increased nor declined with normal aging; it would be interesting to see if the same results would have been obtained, had other specific sites such as prefrontal cortex been examined and if rats had been chronically administered cannabinoids. These studies require replication and pave the way for further exploration of aging phenomena as they may interact with chronic exposure to the drug (e.g. age-related cognitive decline).

HUMAN RESEARCH

There is no evidence from human studies of any structural brain damage following prolonged exposure to cannabinoids. The most recent study using sophisticated technology and measurement techniques showed that frequent but relatively short-term use of marijuana does not produce any structural brain abnormalities or global or regional changes in brain tissue volume or composition assessable by magnetic resonance imaging (MRI) (Block *et al.*, 2000a). Existing methods of brain imaging may not be sensitive enough to detect the long-term subcellular or biochemical alterations that might be produced in the CNS in the absence of any distortion of gross cell architecture. The most convincing evidence on brain alterations would come from post-mortem studies. Recent studies of human brain post-mortem have reported reduced binding and decreased cannabinoid receptor density with disease and normal aging, but no studies have yet examined the brains of healthy long-term users of marijuana.

Westlake *et al.* (1994) showed that binding of the potent cannabinoid agonist CP55,940 was substantially reduced in the caudate and hippocampus of Alzheimer's brains, with lesser reductions in the substantia nigra and globus pallidus. Reduced binding was associated also with increasing age and/or general disease processes resulting in cortical pathology and thus not specific to Alzheimer's. They claimed that receptor losses were not associated with overall decrements in levels of cannabinoid receptor gene expression. Biegon and Kerman (1995) found that increasing age is associated with a decrease in the density of cannabinoid receptors in prefrontal cortex, particularly in cingulate cortex and the superior frontal gyrus. The authors discussed their findings in terms of indirect modulation of dopaminergic activity by cannabinoids, and suggested that the age-related decline in receptor density in prefrontal regions may contribute to decreased drug-seeking behavior with increasing age. These findings have implications for elucidating the role of the endogenous cannabinoid system in the cognitive decline that occurs with age. As this age-related decrease in cannabinoid receptors was found in the normal human brain, it would be interesting for further research to examine receptor density in the prefrontal cortex of chronic marijuana users.

Copolov and colleagues (1999) reported preliminary findings from immunohistochemical studies of post-mortem tissue from a small sample of marijuana using and non-using schizophrenic subjects and non-drug using controls that the density of cannabinoid receptors in frontal cortex or caudate-putamen was not altered by marijuana use or schizophrenia. The concentration of tyrosine hydroxylase was increased in subjects with schizophrenia and marijuana use, and the dopamine transporter was unaltered in these subjects but significantly decreased in schizophrenic subjects with no history of marijuana use. Interestingly, a recent study found elevated levels of anandamide in the CSF of a small schizophrenic sample

(Leweke *et al.*, 1999). This may reflect a compensatory mechanism for the elevated dopamine levels in the brain and might also explain why persons with schizophrenia often smoke marijuana in an attempt to alleviate their symptoms. Alternatively, it was suggested that persons predisposed to schizophrenia may have an imbalance of endogenous cannabinoids which may contribute to the pathogenesis of schizophrenia. This would be of particular concern to marijuana users if it were shown that the long-term ingestion of exogenous cannabinoids also lead to an imbalance in endogenous cannabinoids. To date, there have been no thorough, well-controlled studies investigating alterations of neuromodulator levels in human marijuana users. Musselman *et al.* (1994) found no alteration of monoamine levels in cerebrospinal fluid (CSF) of subjects denying marijuana use but with cannabinoids detected in their urines; this study was riddled with factors that would not enable a proper conclusion to be drawn. Markianos and Stefanis (1982) had previously reported that dopamine levels dropped and norepinephrine rose in long-term users during a three day period of abstinence but the opposite effects were observed after smoking on the fourth day.

Altered brain function and metabolism in humans have been demonstrated following acute and chronic use of marijuana by research utilising cerebral blood flow (CBF), positron emission tomography (PET), and electroencephalographic (EEG) techniques. A number of studies have reported increases in CBF following acute administration of cannabinoids and reduced global levels of CBF in chronic users in the unintoxicated state, while PET studies have suggested altered brain function in prefrontal and cerebellar regions in chronic users (see Chapter 8 (Mathew), this volume; Loeber and Yurgelun-Todd, 1999). In the most recent carefully controlled study using PET and sensitive analytic techniques, Block and colleagues (2000b) found that following more than 26h of supervised abstinence frequent marijuana users showed substantially lower resting levels of brain blood flow (up to 18%) than controls in a large region of posterior cerebellum, with a similar but reduced effect in prefrontal cortex. Users showed elevated blood flow in the right anterior cingulate. The mean level of marijuana use in this sample was 17 times per week for 3.9 years. The authors interpreted the cerebellar hypoactivity as reflecting functional changes due to the frequent use of marijuana and postulated possible direct or indirect effects on cognitive function, particularly since the cerebellum has been linked to an internal timing system and alterations of time sense are common following marijuana smoking. Increased cerebellar brain glucose metabolism has been reported to follow acute administration of THC to human volunteers and correlate with the degree of subjective intoxication (Volkow *et al.*, 1996), but subjects who showed a *decrease* in cerebellar CBF were those who experienced a significant alteration in time sense (Mathew *et al.*, 1998). Increased metabolism and CBF have also been observed in prefrontal cortex during acute intoxication (see Chapter 8 (Mathew) this volume; Loeber and Yurgelun-Todd, 1999; Volkow *et al.*, 1996). In contrast, decreased regional CBF was found in prefrontal cortex and hippocampus of rats administered THC, but no changes were apparent in cerebellum (Bloom *et al.*, 1997), suggesting differences between human effects and animal models.

The findings of Block *et al.*'s study (2000b) were in accord with the only previous PET study of marijuana users in the unintoxicated state (Volkow *et al.*, 1996).

Another recent study of chronic heavy marijuana users diagnosed with attention deficit/hyperactivity disorder reported prefrontal cortical and temporal lobe hypoperfusion from clinical judgements of single photon emission computed tomography (SPECT) images (Amen and Waugh, 1998). Some of these changes were evident in users who had been abstinent for more than 6 months. However, this finding was not supported as being associated with heavy marijuana use according to Block et al.'s (2000b) high resolution measurements. In reviewing the combined literature on neuroimaging and animal receptor and neurochemical models, Loeber and Yurgelun-Todd (1999) concluded that the metabolism of component regions of the frontopontocerebellar network are altered by both acute and chronic exposure to marijuana through modulation of the cannabinoid and dopamine systems. They propose that chronic use results in changes at the receptor level that lead to an alteration in the dopamine system which then leads to a global reduction in brain metabolism, particularly in the frontal lobe and cerebellum. This reduction appears to be reversed on acute exposure to marijuana (Loeber and Yurgelun-Todd, 1999) and the reversal of chronic effects by acute exposure is supported by other neurochemical and electrophysiological research findings (e.g. Markianos and Stefanis, 1982; Solowij et al., 1995).

In contrast to brain imaging in a resting state, Block's team have also used PET techniques to measure cerebral blood flow during the performance of verbal memory recall tasks (Block et al., 1999; in press) and during a selective attention task (Block et al., 2000c). Memory-related blood flow in frequent marijuana users showed decreases relative to controls in prefrontal cortex, increases in memory-relevant regions of cerebellum, and altered lateralization in hippocampus. The greatest differences between users and controls occurred in brain activity related to episodic memory encoding. Buschke's selective reminding procedure was used in learning and relearning lists of words and users required more list presentations to achieve perfect recall. They did not differ in the total number of words recalled during PET but they did show an increased recency effect (far better recall of words at the end of the list than those in the middle) relative to controls. This suggested that users may rely more on short-term memory than episodic memory encoding and retrieval, implying an altered distribution of memory processes, and this could contribute to poor learning over multiple trials. Users showed no hippocampal laterality effect during the memory tests while controls showed greater activity in the left (language dominant) hippocampus than the right hippocampus. Users showed decreases in most of the prefrontal regions measured (or small increases), whereas controls showed increased prefrontal activity, and these group differences were more apparent during recall of new lists than previously learned lists. The authors suggested that the left hemispheric rCBF changes were therefore more related to episodic memory encoding than retrieval, but changes in the right hemisphere might have been related to episodic memory retrieval processes. Marijuana users showed decreased activation in a number of prefrontal areas related to working memory, but increased activity when recalling newly learned lists in regions of the cerebellum that may be involved in memory and attention. Despite this increased cerebellar activation compared to controls, marijuana users still showed an underlying hypoactivity in posterior cerebellar rCBF, as had previously been reported in resting conditions (Block et al., 2000b).

There were a number of smaller differences (increases and decreases) between the activations of users and controls in many other brain regions. In the auditory selective attention task, thus far reported only in abstract form (Block *et al.*, 2000c), chronic use of marijuana was found to alter the normal pattern of attention-related hemispheric asymmetry with greater left hemisphere activation regardless of the ear to which attention was directed. Yurgelun-Todd *et al.* (1999) also reported in abstract form the results of a functional MRI study applying a cognitive challenge paradigm where marijuana users displayed a pattern of decreased dorsal lateral prefrontal cortex and increased anterior cingulate activation compared to controls after 28 days abstinence. These are the first studies to have used functional brain imaging to examine brain activity in marijuana users during the performance of an actual cognitive task and the alterations in cognition and brain activation observed have important implications for further research, particularly to investigate their potential recovery following prolonged abstinence. There has been some evidence to suggest that altered cognition and brain function may persist for two years or more after cessation of chronic use (Solowij, 1998; see below). Marijuana users do, however, report noticeable improvements after cessation of use; they may report feeling that they have come out of a fog and have clearer, less muddy thinking (Gruber *et al.*, 1997; Lundqvist, 1995; Solowij *et al.*, 1995).

Previous research that combined the assessment of cognition and brain function in chronic marijuana users had utilised brain event-related potentials (ERPs) derived from the EEG recorded during a complex cognitive task of selective attention believed to be subserved by the frontal lobes and possibly the cerebellum. Solowij and colleagues (see Solowij, 1998) conducted a series of controlled studies with chronic users in the unintoxicated state which showed that the ability to focus attention and filter out complex irrelevant information, as indexed by alterations in ERPs recorded from frontal regions of the brain, was progressively impaired with the number of years of marijuana use, regardless of frequency of use. This suggested that the impairment was the result of gradual changes occurring in the brain as a result of cumulative exposure to marijuana. This deficit was evident also in a group of ex-marijuana users with a mean duration of abstinence of two years (range 3 months to 6 years), and was related to their past use of marijuana even after controlling for confounding variables (Solowij, 1995). This was the first study to examine the cognitive and brain function of past users of marijuana after such a protracted period of self-reported abstinence and the results suggest an enduring impairment of selective attention. In this paradigm there was no apparent recovery of function with increasing abstinence, and a similar lack of recovery was found in a single case study monitored for 6 weeks following cessation of marijuana use (Solowij *et al.*, 1995). Further research to replicate and extend these findings is in progress. In current marijuana users only, speed of information processing, as indexed by the latency of the P300 component of the ERP, was impaired with increasing frequency of use; this was interpreted as a short term effect due to accumulated cannabinoids. These differential effects of the frequency and duration of marijuana use have not been investigated consistently in other studies. Clearly the parameters of marijuana use that are associated with short- or long-lasting cognitive and brain dysfunction have not been fully elucidated and there has been much debate about the attribution of deficits to lingering acute effects, drug residues, abstinence effects or lasting changes caused by chronic use (Pope *et al.*, 1995; Solowij, 1998).

Marked alterations in EEG have been shown to manifest in association with long-term marijuana use but the cognitive consequences of such alterations are not known. Struve and colleagues used quantitative EEG techniques to detect significant increases in absolute power, relative power and interhemispheric coherence of EEG alpha and theta activity, primarily in fronto-central cortex, in daily marijuana users of up to 30 years duration compared to short-term users and nonusers (e.g. Struve *et al.*, 1994; see also Solowij, 1998). The results suggest that there may be a gradient of quantitative EEG change associated with progressive increases in the total cumulative exposure of daily marijuana use which may indicate organic change. Recent studies from this group have examined more cognitively linked components of the evoked potential. Patrick *et al.* (1997) found no difference between very strictly screened marijuana users and controls in early and middle latency evoked potentials in a simple paradigm but in another study (Patrick *et al.*, 1999) reported a reduced P50 auditory sensory gating response which they interpreted as being due to possible hippocampal dysfunction. Another recent study used ERPs to investigate the acute effects of THC on the recognition of emotionally charged words (Leweke *et al.*, 1998). An overall decrease in recognition rate was found under THC regardless of the emotional charge of a word, but ERPs revealed an enhanced implicit memory response to positively charged words which the authors interpreted as being congruous with the drug-induced mood. Johnson *et al.* (1997) investigated the association between the cannabinoid receptor gene (CNR1) and the P300 amplitude of evoked potentials and found a significant decrease in P300 amplitude that was most marked in the frontal lobes of alcohol and drug addicts who were homozygous for the CNR1 greater than or equal to 5 repeat alleles. They had previously linked this homozygocity to drug dependence (Comings *et al.*, 1997). Other recent studies have found polymorphisms in the human cannabinoid receptor gene which may prove useful for investigating neuropsychiatric disorders that could be related to metabolic disturbances of the endogenous cannabinoid system (e.g. Gadzicki *et al.*, 1999).

In other recent studies of cognitive function, Pope and Yurgelun-Todd (1996) used a select set of neuropsychological tests to demonstrate greater impairment on attentional and executive functions in heavy marijuana users than in light users. This study also controlled for a variety of potentially confounding variables, such as estimated levels of premorbid cognitive functioning, and use of alcohol and other substances. Heavy marijuana users were more susceptible to interference, made more perseverative responses, had poorer recall and showed deficient learning. This study and another by the same team (Pope *et al.*, 1997) found some gender-specific cognitive effects of heavy marijuana use: female heavy users remembered fewer items and made more errors than female light users in a visuospatial memory task, whereas male heavy users were more impaired in attentional/interference tasks and in delayed recall. Insufficient attention has been given to the study of possible sex differences in response to marijuana use and females have been greatly underrepresented in previous research of this kind.

These studies have been important in not only demonstrating cognitive impairments associated with marijuana use, but in showing that they increase with duration and/or frequency of use. The results suggest that marijuana use may compromise some memory functions in humans, but that it may be the attentional/executive system that is mainly impaired with the most pronounced effects on the abilities to shift and/or sustain attention, processes associated with the prefrontal cortex and also the cerebellum, and processes important for efficient memory function. The results of another recent study supported this conclusion, and suggested an interaction with the mental deterioration that occurs with normal aging (Fletcher *et al.*, 1996). In a 17-year follow-up of chronic users, older long-term users were found to perform more poorly than older nonusers on complex short-term memory tests involving learning lists of words and on complex tasks of selective and divided attention associated with working memory. No differences were found between younger users and nonusers. The effects on cognition of marijuana use and aging were also investigated by Elwan *et al.* (1997). The marijuana users had significantly poorer attention, as measured by the Paced Auditory Serial Addition Test and the Trail Making Test, than non-users, and performance on both tests declined with age, but there appeared to be no interactions between age and marijuana use. Since an interaction between the effects of marijuana use and increasing age might be anticipated from the findings of research on the cannabinoid receptor, further human studies are required.

There is nevertheless continued concern that adolescents may be at greater risk of experiencing the adverse effects of marijuana use on psyche and cognition (Hall and Solowij, 1998). Adolescence is a particularly vulnerable period when foundations are laid for important educational, intellectual, emotional and maturational developments, concurrent with maturation of the CNS and reproductive system. There is evidence for impaired educational attainment among adolescents who use marijuana, and early commencement of marijuana use (e.g. prior to the age of 16) contributes to poor psychosocial adjustment (Fergusson and Horwood, 1997) and impairs specific attentional functions (Ehrenreich *et al.*, 1999) in adulthood. Two studies of marijuana-using adolescents have found evidence of short-term memory deficits, and following abstinence of 4–6 weeks still a generalized impairment of memory (Millsaps *et al.*, 1994; Schwartz *et al.*, 1989).

There is also evidence that prenatal exposure to cannabinoids can affect cognitive development in children. Fried and colleagues (1998), reporting on the most recent follow-up of children exposed to marijuana in utero, found that from ages four to twelve, the exposed children performed more poorly than controls on tasks of executive function that required attention, impulse control, concept formation, visual analysis and hypothesis testing. In line with cognitive electrophysiological and neuropsychological studies of adult marijuana users discussed above, these findings implicate the prefrontal cortex as being particularly vulnerable to the effects of marijuana. Another recent study provides evidence that childhood executive cognitive functioning capacity is a salient predictor of subsequent drug use in adolescence, with deficient executive functioning preceding marijuana use (Aytaclar *et al.*, 1999), however this was observed in a high risk group whose fathers had substance use disorders.

The long-term cognitive effects of marijuana are quite specific and subtle, requiring sensitive measures to detect them. A recent epidemiological study with a potentially powerful longitudinal approach failed to detect any cognitive decline associated with marijuana use during three study waves in 1981, 1982 and 1993–96 (Lyketsos *et al.*, 1999). This was most likely due to the lack of sensitivity of the assessment instrument used: the Mini Mental State Examination tests a restricted set of very simplistic cognitive functions used for discriminating patients with moderate to severe deficits and does not adequately measure functions such as ability to attend to relevant input, ability to solve abstract problems, and ability to retain information over prolonged time intervals (Lezak, 1995; Spreen and Strauss, 1998). A further problem with this study was that any effect of marijuana use would have been greatly diffused by the inclusion as heavy users of people who may have had an isolated incident of smoking daily or more often for over 2 weeks during any one of the study wave periods but who had not used previously or since.

Long-term heavy use of marijuana may interfere with motivation, in particular with setting and achieving long term aspirations, or even short-term goals (Kouri *et al.*, 1995; Reilly *et al.*, 1998; Solowij *et al.*, 1995); the idea of an amotivational syndrome has long been touted. Evidence for such a syndrome associated with chronic marijuana use has been equivocal; chronic users may become apathetic, lethargic, withdrawn and unmotivated, but these are probably merely symptoms of depression, cognitive impairment, and chronic intoxication (Hall and Solowij, 1997; Hall *et al.*, 1994). Solowij (1998) argued that possible differences in motivation between users and controls were unlikely to explain the differences in cognitive performance and brain function detected. Withdrawal symptoms following cessation of daily marijuana use may also occur, with nervousness, anxiety, restlessness, sleep disturbances and changes in appetite experienced to varying degrees (Wiesbeck *et al.*, 1996). These have also been proposed to underlie the differences in cognition found in studies where users are required to abstain for a defined period prior to testing – a requirement of *all* studies trying to disentangle the acute or short-term residual effects of marijuana from more enduring impairments. Animal research and studies of cannabinoid receptor alterations support the hypothesis that the cognitive deficits could manifest as a result of long-term use of marijuana, however further research on long-term effects should continue to control for a variety of potential confounds.

Just as there has been a predominance of acute animal studies since the discovery of the endogenous cannabinoid system, human research on acute effects of marijuana on cognition and brain function continues too. This is important to further elucidate our understanding of the actions of cannabinoids in the human brain and consider their implications for society, particularly with the growing calls for medical marijuana. Marijuana impairs driving behavior for at least an hour after smoking, but there is also evidence that users are aware of their impairment and tend to compensate for it by driving more cautiously (Smiley, 1999). A recent study reported impaired perceptual motor speed and accuracy, two important parameters of driving ability, immediately after smoking marijuana (Kurzthaler *et al.*, 1999). Heishman *et al.* (1997) compared varying doses of marijuana and alcohol and found that both drugs produced comparable impairment in digit-symbol substitution and word recall tests, but had no effect in some tests of time perception and reaction time. A number of other recent studies of acute administration have been discussed above.

There is much that remains unknown about the various interrelationships between exogenous ingested cannabinoids and the endogenous cannabinoid system and indeed other neuromodulator systems. Despite the evidence for neurotoxicity from animal studies and the human evidence that long-term or heavy use of marijuana adversely affects cognitive and brain function, cannabinoids such as THC and cannabidiol have also, intriguingly, been shown to have neuroprotective antioxidant properties (e.g. Hampson *et al.*, 1998b) – they are capable of *protecting* neurons from damage. The potential for therapeutic application to a variety of neurological conditions is promising and research clearly suggests other beneficial effects that cannabinoids can have on CNS function, including analgesia, reduction of intra-cranial pressure following head trauma, anticonvulsant and antispasticity effects (see Pertwee, 2000). Cannabidiol or synthetic cannabinoids, such as dexanabinol (HU-211), are better suited for therapeutic purposes as they have no psychoactivity (Leker *et al.*, 1999; Nagayama *et al.*, 1999; Shohami *et al.*, 1995, 1997). If relatively large doses of psychoactive cannabinoids are taken, aside from the precautions inherent during a phase of acute intoxication, there is a risk that some residual effects on cognition could be experienced for hours later and even the following day that would affect performance of complex tasks requiring high levels of cognitive ability. If the nature of the indication for which cannabinoids were prescribed was a chronic condition requiring long-term use of psychoactive can-nabinoids, deterioration of cognitive ability is likely to occur to varying degrees. Further research is essential to determine the extent to which the alterations of brain function that have now been observed in association with marijuana use translate into clinically significant deficits. There has been a paucity of research investigating individual differences in susceptibility to cognitive impairments associated with marijuana use; this information could be combined with the findings of biochem-ical research to possibly prevent cognitive impairments both in recreational and medicinal users of marijuana. At present, the evidence suggests that the effects on cognitive function are relatively subtle and should not impact greatly upon any decision to medicate where the relief brought about by cannabinoids is marked and where other medications have not been successful. In general, however, occa-sional use for only a short period of time would be advisable in order to avoid or minimise adverse effects on cognition, subtle though they may be.

CONCLUSION

It is apparent from many years of research that long-term use of marijuana does not result in gross cognitive deficits. However, recent reviewers agree that there is now sufficient evidence that it leads to a more subtle and selective impairment of higher cognitive functions (Block, 1996; Hall *et al.*, 1994; Pope *et al.*, 1995; Solowij, 1998) which arises from altered functioning of the frontal lobe, hippocampus and cerebellum (Block *et al.*, in press; Hampson and Deadwyler, 1999; Loeber and Yurgelun-Todd, 1999; Solowij, 1998). The findings from recent methodologically rigorous research provide evidence for impaired learning, organization and integration of complex information in tasks involving various mechanisms of attention, memory processes and executive function. It is not clear to what extent the alterations in brain function and cognitive impairments as detected in laboratory testing might impact upon daily life, although users themselves complain of problems with memory, concentration, loss of motivation, paranoia, depression, dependence and lethargy (Reilly *et al.*, 1998; Solowij, 1998). Schwenk (1998) has argued that there is no clear causal relationship between marijuana use and job performance. The nature of the cognitive deficits as assessed by psychological testing suggests that long-term users would perform reasonably well in routine tasks of everyday life, although they may be more distractible and short-term memory may be compromised. Difficulties are likely to be encountered in performing complex tasks that are novel or that cannot be solved by automatic application of previous knowledge, or with tasks that rely heavily on a memory component or require strategic planning and multi-tasking.

There is ongoing debate as to whether the deficits detected are due to the accumulation of cannabinoids (a brief drug residue effect lasting up to 24h), the withdrawal from cannabinoids (since most studies have required users to abstain for a number of hours prior to testing), or whether they may indeed represent a more lasting alteration of brain function. The degree of cognitive impairment is associated either with frequency of marijuana use (being maximal in heavy users), or may increase with the duration of marijuana use. The latter suggests that gradual long-term changes may occur in the brain. Recent findings of altered functioning of the endogenous cannabinoid system, and other neuromodulator systems with which it interacts, support the notion that long-term changes may occur, although changes in receptor function due to chronic drug exposure should potentially normalize over time with abstinence. It is not yet possible to specify levels of marijuana use that are safe, hazardous and harmful in terms of the risk of cognitive impairment or psychological harm, and indeed it may prove difficult to do so due to wide ranging individual susceptibility to such harms, and until the implications of altered brain function are fully understood. In general, prolonged heavy use of marijuana (daily or near daily use) places users at greatest risk of adverse health and psychological consequences (Hall *et al.*, 1994). Limited evidence suggests that some of the cognitive impairments and alterations in brain function that are associated with long-term or heavy use do persist but may become attenuated with abstinence. The extent to which they might recover with prolonged abstinence is unknown, but research in progress will soon help to shed some light on this crucial question.

REFERENCES

Abood, M. E., Sauss, C., Fan, F., Tilton, C. L. and Martin, B. R. (1993) "Development of behavioral tolerance to Δ^9-THC without alteration of cannabinoid receptor binding or mRNA levels in whole brain," *Pharmacology Biochemistry and Behavior* **46**: 575–579.

Adams, I. B. and Martin, B. R. (1996) "Cannabis: Pharmacology and toxicology in animals and humans," *Addiction* **91**: 1585–1614.

Amen, D. G. and Waugh, M. (1998) "High resolution brain SPECT imaging of marijuana smokers with AD/HD," *Journal of Psychoactive Drugs* **30**: 209–214.

Aytaclar, S., Tarter, R. E., Kirisci, L. and Lu, S. (1999) "Association between hyperactivity and executive cognitive functioning in childhood and substance use in early adolescence," *Journal of the American Academy of Child and Adolescent Psychiatry* **38**: 172–178.

Beardsley, P. M. and Kelly, T. H. (1999) "Acute effects of cannabis on human behavior and central nervous system functions," in *The Health Effects of Cannabis*, edited by H. Kalant, W. Corrigall, W. Hall and R. Smart, pp. 129–169. Toronto: Addiction Research Foundation, Centre for Addiction and Mental Health.

Belue, R. C., Howlett, A. C., Westlake, T. M. and Hutchings, D. E. (1995) "The ontogeny of cannabinoid receptors in the brain of postnatal and aging rats," *Neurotoxicology and Teratology* **17**: 25–30.

Biegon, A. and Kerman, I. (1995) "Quantitative autoradiography of cannabinoid receptors in the human brain post-mortem," in *Sites of Drug Action in the Human Brain*, edited by A. Biegon and N. D. Volkow, pp. 65–74. Boca Raton: CRC Press.

Block, R. I. (1996) "Does heavy marijuana use impair human cognition and brain function?" *Journal of the American Medical Association* **275**: 560–561.

Block, R. I., O'Leary, D. S., Ehrhardt, J. C., Augustinack, J. C., Ghoneim, M. M., Arndt, S. and Hall, J. A. (2000a) "Effects of frequent marijuana use on brain tissue volume and composition," *NeuroReport* **11**: 491–496.

Block, R. I., O'Leary, D. S., Hichwa, R. D., Augustinack, J. C., Boles Ponto, L. L., Ghoneim, M. M. *et al.* (1999) "Effects of chronic marijuana use on regional cerebral blood flow during recall," *Society for Neuroscience Abstracts* **25**: 2077.

Block, R. I., O'Leary, D. S., Hichwa, R. D., Augustinack, J. C., Boles Ponto, L. L., Ghoneim, M. M. *et al.* (2000b) "Cerebellar hypoactivity in frequent marijuana users," *NeuroReport* **11**: 749–753.

Block, R. I., O'Leary, D. S., Hichwa, R. D., Augustinack, J. C., Boles Ponto, L. L., Ghoneim, M. M. *et al.* (2000c) "Effects of frequent marijuana use on attention-related regional cerebral blood flow," To be presented at the *30th Annual Meeting of the Society for Neuroscience* New Orleans, LA, November 4–9.

Block, R. I., O'Leary, D. S., Hichwa, R. D., Augustinack, J. C., Boles Ponto, L. L., Ghoneim, M. M. *et al.* (in press) "Effects of frequent marijuana use on memory-related regional cerebral blood flow," *Pharmacology Biochemistry and Behavior*.

Bloom, A. S., Tershner, S., Fuller, S. A. and Stein, E. A. (1997) "Cannabinoid-induced alterations in regional cerebral blood flow in the rat," *Pharmacology Biochemistry and Behavior* **57**: 625–631.

Bohme, G. A., Laville, M., Ledent, C., Parmentier, M. and Imperato, A. (2000) "Enhanced long-term potentiation in mice lacking cannabinoid CB_1 receptors," *Neuroscience* **95**: 5–7.

Carriero, D., Aberman, J., Lin, S. Y., Hill, A., Makriyannis, A. and Salamone, J. D. (1998) "A detailed characterization of the effects of four cannabinoid agonists on operant lever pressing," *Psychopharmacology* **137**: 147–156.

Carta, G., Nava, F. and Gessa, G. L. (1999) "Inhibition of hippocampal acetylcholine release after acute and repeated Delta(9)-tetrahydrocannabinol in rats," *Brain Research* **809**: 1–4.

Castellano, C., Cabib, S., Palmisano, A., Di Marzo, V. and Puglisi-Allegra, S. (1997) "The effects of anandamide on memory consolidation in mice involve both D-1 and D-2 dopamine receptors," *Behavioural Pharmacology* **8**: 707–712.

Castellano, C., Ventura, R., Cabib, S. and Puglisi-Allegra, S. (1999) "Strain-dependent effects of anandamide on memory consolidation in mice are antagonized by naltrexone," *Behavioural Pharmacology* **10**: 453–457.

Chait, L. D. and Pierri, J. (1992) "Effects of smoked marijuana on human performance: A critical review," in *Marijuana/Cannabinoids: Neurobiology and Neurophysiology*, edited by L. Murphy and A. Bartke, pp. 387–423. Boca Raton: CRC Press.

Chan, G.C.-K., Hinds, T. R., Impey, S. and Storm, D. R. (1998) "Hippocampal neurotoxicity of Δ^9-tetrahydrocannabinol," *The Journal of Neuroscience* **18**: 5322–5332.

Childers, S. R. and Breivogel, C. S. (1998) "Cannabis and endogenous cannabinoid systems," *Drug and Alcohol Dependence* **51**: 173–187.

Collins, D. R., Pertwee, R. G. and Davies, S. N. (1995) "Prevention by the cannabinoid antagonist, SR141716A, of cannabinoid-mediated blockade of long-term potentiation in the rat hippocampal slice," *British Journal of Pharmacology* **115**: 869–870.

Comings, D. E., Muhleman, D., Gade, R., Johnson, P., Verde, R., Saucier, G. and MacMurray, J. (1997) "Cannabinoid receptor gene (CNR1): Association with IV drug use," *Molecular Psychiatry* **2**: 161–168.

Copolov, D., Bradbury, R., Dong, P., Dean, B. and Lim, A. (1999) "The interaction between cannabis and dopamine in the brain: Possible mechanisms related to psychosis," Paper presented at the *Inaugural International Cannabis and Psychosis Conference*, Melbourne, Australia: 16–17 February.

Crawley, J., Corwin, R., Robinson, J., Felder, C., Devane, W. and Axelrod, J. (1993) "Anandamide, an endogenous ligand of the cannabinoid receptor, induces hypomotility and hyperthermia *in vivo* in rodents," *Pharmacology Biochemistry and Behavior* **46**: 967–972.

Deadwyler, S. A., Heyser, C. J. and Hampson, R. E. (1995) "Complete adaptation to the memory disruptive effects of delta-9-THC following 35 days of exposure," *Neuroscience Research Communications* **17**: 9–18.

De Fonseca, F. R., Gorriti, M., Fernandez-Ruiz, J. J., Palomo, T. and Ramos, J. A. (1994) "Downregulation of rat brain cannabinoid binding sites after chronic Δ^9-tetrahydrocannabinol treatment," *Pharmacology Biochemistry and Behavior* **47**: 33–40.

Derkinderen, P., Toutant, M., Burgaya, F., LeBert, M., Siciliano, J. C., deFranciscis, V. *et al.* (1996) "Regulation of a neuronal form of focal adhesion kinase by anandamide," *Science* **273**: 1719–1722.

Deutsch, D. G. and Chin, S. A. (1993) "Enzymatic synthesis and degradation of anandamide, a cannabinoid receptor agonist," *Biochemical Pharmacology* **46**: 791–796.

Devane, W. A., Hanus, L., Breuer, A., Pertwee, R. G., Stevenson, L. A., Griffin, G., Gibson, D., Mandelbaum, A., Etinger, A. and Mechoulam, R. (1992) "Isolation and structure of a brain constituent that binds to the cannabinoid receptor," *Science* **258**: 1946–1949.

Diana, M., Melis, M. and Gessa, G. L. (1998) "Increase in meso-prefrontal dopaminergic activity after stimulation of CB_1 receptors by cannabinoids," *European Journal of Neuroscience* **10**: 2825–2830.

Diana, M., Melis, M., Muntoni, A. L. and Gessa, G. L. (1998) "Mesolimbic dopaminergic decline after cannabinoid withdrawal," *Proceedings of the National Academy of Sciences of the USA* **95**: 10269–10273.

Ehrenreich, H., Rinn, T., Kunert, H. J., Moeller, M. R., Poser, W., Schilling, L. *et al.* (1999) "Specific attentional dysfunction in adults following early start of cannabis use," *Psychopharmacology* **142**: 295–301.

Elwan, O., Hassan, A. A. H., Naseer, M. A., Elwan, F., Deif, R., El Serafy, O. *et al.* (1997) "Brain aging in a sample of normal Egyptians: cognition, education, addiction and smoking," *Journal of the Neurological Sciences* **148**: 79–86.

Fergusson, D. and Horwood, J. (1997) "Early onset cannabis use and psychosocial adjustment in young adults," *Addiction* **92**: 279–296.

Fernandez-Ruiz, J., Berrendero, F., Hernandez, M. L. and Ramos, J. A. (2000) "The endogenous cannabinoid system and brain development," *Trends in Neurosciences* **23**: 14–20.

Fletcher, J. M., Page, J. B., Francis, D. J., Copeland, K., Naus, M. J., Davis, C. M. *et al.* (1996). "Cognitive correlates of long-term cannabis use in Costa Rican men," *Archives of General Psychiatry* **53**: 1051–1057.

Fried, P. A., Watkinson, B. and Gray, R. (1998) "Differential effects on cognitive functioning in 9- to 12-year olds prenatally exposed to cigarettes and marihuana," *Neurotoxicology and Teratology* **20**: 293–306.

Gadzicki, D., Muller-Vahl, K. and Stuhrmann, M. (1999) "A frequent polymorphism in the coding exon of the human cannabinoid receptor (CNR1) gene," *Molecular and Cellular Probes* **13**: 321–323.

Garcia, L., de Miguel, R., Ramos, J. A. and Fernandez-Ruiz, J. J. (1996) "Perinatal Delta(9)-tetrahydrocannabinol exposure in rats modifies the responsiveness of midbrain dopaminergic neurons in adulthood to a variety of challenges with dopaminergic drugs," *Drug and Alcohol Dependence* **42**: 155–166.

Garcia-Gil, L., de Miguel, R., Munoz, R. M., Cebeira, M., Villanua, M. A., Ramos, J. A. and Fernandez-Ruiz, J. J. (1997) "Perinatal Delta(9)-tetrahydrocannabinol exposure alters the responsiveness of hypothalamic dopaminergic neurons to dopamine-acting drugs in adult rats," *Neurotoxicology and Teratology* **19**: 477–487.

Gessa, G. L., Casu, M. A., Carta, G. and Mascia, M. S. (1998) "Cannabinoids decrease acetylcholine release in the medial-prefrontal cortex and hippocampus, reversal by SR 141716A," *European Journal of Pharmacology* **355**: 119–124.

Gifford, A. N. and Ashby, C. R. J. (1996) "Inhibition of acetylcholine release from hippocampal slices by the cannabimimetic aminoalkylindole WIN 55212-2, and evidence for the release of an endogenous cannabinoid inhibitor," *Journal of Pharmacology and Experimental Therapeutics* **277**: 1431–1436.

Giuffrida, A., Parsons, L. H., Kerr, T. M., de Fonseca, F. R., Navarro, M. and Piomelli, D. (1999) "Dopamine activation of endogenous cannabinoid signaling in dorsal striatum," *Nature Neuroscience* **2**: 358–363.

Glass, M., Dragunow, M. and Faull, R. L. (1997) "Cannabinoid receptors in the human brain: A detailed anatomical and quantitative autoradiographic study in the fetal, neonatal and adult human brain," *Neuroscience* **77**: 299–318.

Gruber, A. J., Pope, H. G., Jr. and Oliva, P. (1997) "Very long-term users of marijuana in the United States: a pilot study," *Substance Use & Misuse* **32**: 249–264.

Hall, W. and Solowij, N. (1998) "Adverse effects of cannabis," *The Lancet*, **352**: 1611–1616.

Hall, W. and Solowij, N. (1997) "Long-term cannabis use and mental health," *British Journal of Psychiatry* **171**: 107–108.

Hall, W., Solowij, N. and Lemon, J. (1994) *The Health and Psychological Consequences of Cannabis Use*. National Drug Strategy Monograph Series No. 25, Canberra: Australian Government Publishing Service.

Hampson, A. J., Bornheim, L. M., Scanziani, M., Yost, C. S., Gray, A. T., Hansen, B. M. *et al.* (1998a) "Dual effects of anandamide on NMDA receptor-mediated responses and neurotransmission," *Journal of Neurochemistry* **70**: 671–676.

Hampson, A. J., Grimaldi, M., Axelrod, J. and Wink, D. (1998b) "Cannabidiol and (−)Δ^9-tetrahydrocannabinol are neuroprotective antioxidants," *Proceedings of the National Academy of Sciences of the USA* **95**: 8268–8273.

Hampson, R. E. and Deadwyler, S. A. (1998) "Role of cannabinoid receptors in memory storage," *Neurobiology of Disease* **5**: 474–482.

Hampson, R. E. and Deadwyler, S. A. (1999) "Cannabinoids, hippocampal function and memory," *Life Sciences* **65**: 715–723.

Hampson, R. E., Mu, J., Kirby, M. T., Zhuang, S.-Y. and Deadwyler, S. A. (1998c) "Role of cannabinoid receptors on GABAergic neurons in the rat hippocampus," *Society of Neuroscience Abstracts* **24**: 1245.

Hao, S. Z., Avraham, Y., Mechoulam, R. and Berry, E. M. (2000) "Low dose anandamide affects food intake, cognitive function, neurotransmitter and corticosterone levels in diet-restricted mice," *European Journal of Pharmacology* **392**: 147–156.

Heath, R. G., Fitzjarrell, A. T., Fontana, C. J. and Garey, R. E. (1980) "*Cannabis sativa*: effects on brain function and ultrastructure in rhesus monkeys," *Biological Psychiatry* **15**: 657–690.

Heishman, S. J., Arasteh, K. and Stitzer, M. L. (1997) "Comparative effects of alcohol and marijuana on mood, memory and performance," *Pharmacology Biochemistry and Behavior* **58**: 93–101.

Herkenham, M., Lynn, A. B., Little, M. D., Johnson, M. R., Melvin, L. S., De Costa, B. R. and Rice, K. C. (1990) "Cannabinoid receptor localization in brain," *Proceedings of the National Academy of Sciences of the USA* **87**: 1932–1936.

Hoffman, A. F. and Lupica, C. R. (2000) "Mechanisms of cannabinoid inhibition of GABA(A) synaptic transmission in the hippocampus," *Journal of Neuroscience* **20**: 2470–2479.

Hsieh, C., Brown, S., Derleth, C. and Mackie, K. (1999) "Internalization and recycling of the CB_1 cannabinoid receptor," *Journal of Neurochemistry* **73**: 493–501.

Jentsch, J. D., Andrusiak, E., Tran, A., Bowers, M. B. and Roth, R. H. (1997) "Δ^9-Tetrahydrocannabinol increases prefrontal cortical catecholaminergic utilization and impairs spatial working memory in the rat: blockade of dopaminergic effects with HA966," *Neuropsychopharmacology* **16**: 426–432.

Jentsch, J. D. and Taylor, J. R. (1999) "Impulsivity resulting from frontostriatal dysfunction in drug abuse: implications for the control of behavior by reward-related stimuli," *Psychopharmacology* **146**: 373–390.

Jentsch, J. D., Verrico, C. D., Le, D. and Roth, R. H. (1998) "Repeated exposure to delta(9)-tetrahydrocannabinol reduces prefrontal cortical dopamine metabolism in the rat," *Neuroscience Letters* **246**: 169–172.

Johnson, J. P., Muhleman, D., MacMurray, J., Gade, R., Verde, R., Ask, M. *et al.* (1997) "Association between the cannabinoid receptor gene (CNR1) and the P300 event-related potential," *Molecular Psychiatry* **2**: 169–171.

Katona, I., Sperlagh, B., Sik, A., Kafalvi, A., Vizi, E. S., Mackie, K. and Freund, T. F. (1999) "Presynaptically located CB_1 cannabinoid receptors regulate GABA release from axon terminals of specific hippocampal interneurons," *Journal of Neuroscience* **19**: 4544–4558.

Kouri, E., Pope, H. G., Yurgelun-Todd, D. and Gruber, S. (1995) "Attributes of heavy vs. occasional marijuana smokers in a college population," *Biological Psychiatry* **38**: 475–481.

Kurzthaler, I., Hummer, M., Miller, C., Sperner-Unterweger, B., Gunther, V., Wechdorn, H. *et al.* (1999) "Effect of cannabis use on cognitive functions and driving ability," *Journal of Clinical Psychiatry* **60**: 395–399.

Lagalwar, S., Bordayo, E. Z., Hoffman, K. L., Fawcett, J. R. and Frey, W. H. (1999) "Anandamides inhibit binding to the muscarinic acetylcholine receptor," *Journal of Molecular Neuroscience* **13**: 55–61.

Leker, R. R., Shohami, E., Abramsky, O. and Ovadia, H. (1999) "Dexanabinol; a novel neuroprotective drug in experimental focal cerebral ischemia," *Journal of the Neurological Sciences* **162**: 114–119.

Lévénès, C., Daniel, H., Soubrié, P. and Crépel, F. (1998) "Cannabinoids decrease excitatory synaptic transmission and impair long-term depression in rat cerebellar Purkinje cells," *Journal of Physiology* **510**: 867–879.

Leweke, F. M., Giuffrida, A., Wurster, U., Emrich, H. M. and Piomelli, D. (1999) "Elevated endogenous cannabinoids in schizophrenia," *NeuroReport* **10**: 1665–1669.

Leweke, M., Kampmann, C., Radwan, M., Dietrich, D. E., Johannes, S., Emrich, H. M. and Munte, T. F. (1998) "The effects of tetrahydrocannabinol on the recognition of emotionally charged words: An analysis using event-related brain potentials," *Neuropsychobiology* **37**: 104–111.

Lezak, M. D. (1995) *Neuropsychological Assessment*, 3rd edn. New York: Oxford University Press.

Lichtman, A. H. and Martin, B. R. (1996) "Delta-9-tetrahydrocannabinol impairs spatial memory through a cannabinoid receptor mechanism," *Psychopharmacology* **126**: 125–131.

Lichtman, A. H., Dimen, K. R. and Martin, B. R. (1995) "Systemic or intrahippocampal cannabinoid administration impairs spatial memory in rats," *Psychopharmacology* **119**: 282–290.

Loeber, R. T. and Yurgelun-Todd, D. A. (1999) "Human neuroimaging of acute and chronic marijuana use: Implications for frontocerebellar dysfunction," *Human Psychopharmacology-Clinical and Experimental* **14**: 291–301.

Lundqvist, T. (1995) *Cognitive Dysfunctions in Chronic Cannabis Users Observed During Treatment: An Integrative Approach*, Stockholm: Almqvist and Wiksell International.

Lyketsos, C. G., Garrett, E., Liang, K.-Y. and Anthony, J. C. (1999) "Cannabis use and cognitive decline in persons under 65 years of age," *American Journal of Epidemiology* **149**: 794–800.

Mailleux, P. and Vanderhaeghen, J. J. (1992) "Age-related loss of cannabinoid receptor binding sites and mRNA in the rat striatum," *Neuroscience Letters* **147**: 179–181.

Mallet, P. E. and Beninger, R. J. (1996) "The endogenous cannabinoid receptor agonist anandamide impairs memory in rats," *Behavioral Pharmacology* **7**: 276–284.

Mallet, P. E. and Beninger, R. J. (1998) "The cannabinoid CB_1 receptor antagonist SR141716A attenuates the memory impairment produced by Δ^9-tetrahydrocannabinol or anandamide," *Psychopharmacology* **140**: 11–19.

Manzanares, J., Corchero, J., Romero, J., Fernandez-Ruiz, J. J., Ramos, J. A. and Fuentes, J. A. (1998) "Chronic administration of cannabinoids regulates proenkephalin mRNA levels in selected regions of the rat brain," *Molecular Brain Research* **55**: 126–132.

Markianos, M. and Stefanis, C. (1982) "Effects of acute cannabis use and short-term deprivation on plasma prolactin and dopamine-beta-hydroxylase in long-term users," *Drug and Alcohol Dependence* **9**: 251–255.

Martin, B. R. and Cone, E. J. (1999) "Chemistry and pharmacology of cannabis," in *The Health Effects of Cannabis*, edited by H. Kalant, W. Corrigall, W. Hall and R. Smart, pp. 19–68. Toronto: Addiction Research Foundation, Centre for Addiction and Mental Health.

Martin, B. R., Mechoulam, R. and Razdan, R. K. (1999) "Discovery and characterization of endogenous cannabinoids," *Life Sciences* **65**: 573–595.

Mathew, R. J., Wilson, W. H., Turkington, T. G. and Coleman, R. E. (1998) "Cerebellar activity and disturbed time sense after THC," *Brain Research* **797**: 183–189.

Matsuda, L. A., Lolait, S. J., Brownstein, M., Young, A. and Bonner, T. I. (1990) "Structure of a cannabinoid receptor and functional expression of the cloned cDNA," *Nature* **346**: 561–564.

McGahan, J. P., Dublin, A. B. and Sassenrath, E. (1984) "Long-term delta-9-tetrahydrocannabinol treatment: Computed tomography of the brains of rhesus monkeys," *American Journal of Diseases of Children* **138**: 1109–1112.

McGregor, I. S., Arnold, J. C., Weber, M. F., Topple, A. N. and Hunt, G. E. (1998) "A comparison of Delta(9)-THC and anandamide induced c-fos expression in the rat forebrain," *Brain Research* **802**: 19–26.

Mechoulam, R., Fride, E., Ben-Shabat, S., Meiri, U. and Horowitz, M. (1999) "Carbachol, an acetylcholine receptor agonist, enhances production in rat aorta of 2-arachidonoyl glycerol, a hypotensive endocannabinoid," *European Journal of Pharmacology* **362**: R1–R3.

Mechoulam, R., Fride, E., Hanus, L., Sheskin, T., Bisogno, T., Di Marzo, V. *et al.* (1997) "Anandamide may mediate sleep induction," *Nature* **389**: 25–26.

Mendelson, W. B. and Basile, A. S. (1999) "The hypnotic actions of oleamide are blocked by a cannabinoid receptor antagonist," *Neuroreport* **10**: 3237–3239.

Millsaps, C. L., Azrin, R. L. and Mittenberg, W. (1994) "Neuropsychological effects of chronic cannabis use on the memory and intelligence of adolescents," *Journal of Child and Adolescent Substance Abuse* **3**: 47–55.

Murillo-Rodriguez, E., Sanchez-Alavez, M., Navarro, L., Martinez-Gonzalez, D., Drucker-Colin, R. and Prospero-Garcia, O. (1998) "Anandamide modulates sleep and memory in rats," *Brain Research* **812**: 270–274.

Musselman, D. L., Haden, C., Caudle, J. and Kalin, N. H. (1994) "Cerebrospinal fluid study of cannabinoid users and normal controls," *Psychiatry Research* **52**: 103–105.

Nagayama, T., Sinor, A. D., Simon, R. P., Chen. J., Graham, S. H., Jin, K. L. and Greenberg, D. A. (1999) "Cannabinoids and neuroprotection in global and focal cerebral ischemia and in neuronal cultures," *Journal of Neuroscience* **19**: 2987–2995.

Navarro, M., de Miguel, R., de Fonseca, F. R., Ramos, J. A. and Fernandez-Ruiz, J. J. (1996) "Perinatal cannabinoid exposure modifies the sociosexual approach behavior and the mesolimbic dopaminergic activity of adult male rats," *Behavioural Brain Research* **75**: 91–98.

Navarro, M., Rubio, P. and de Fonseca, F. R. (1995) "Behavioural consequences of maternal exposure to natural cannabinoids in rats," *Psychopharmacology* **122**: 1–14.

Oviedo, A., Glowa, J. and Herkenham, M. (1993) "Chronic cannabinoid administration alters cannabinoid receptor binding in rat brain: a quantitative autoradiographic study," *Brain Research* **616**: 293–302.

Patrick, G., Straumanis, J. J., Struve, F. A., Fitz-Gerald, M. J., Leavitt, J. and Manno, J. E. (1999) "Reduced p50 auditory gating response in psychiatrically normal chronic marihuana users: A pilot study," *Biological Psychiatry* **45**: 1307–1312.

Patrick, G., Straumanis, J. J., Struve, F. A., Fitz-Gerald, M. J. and Manno, J. E. (1997) "Early and middle latency evoked potentials in medically and psychiatrically normal daily marihuana users: A paucity of significant findings," *Clinical Electroencephalography* **28**: 26–31.

Pertwee, R. G. (1992) "*In vivo* interactions between psychotropic cannabinoids and other drugs involving central and peripheral neurochemical mechanisms," in *Marijuana / Cannabinoids: Neurobiology and Neurophysiology*, edited by L. Murphy and A. Bartke, pp. 165–218. Boca Raton: CRC Press.

Pertwee, R. G. (2000) "Neuropharmacology and therapeutic potential of cannabinoids," *Addiction Biology* **5**: 37–46.

Pope, H. G., Gruber, A. J. and Yurgelun-Todd, D. (1995) "The residual neuropsychological effects of cannabis: the current status of research," *Drug and Alcohol Dependence* **38**: 25–34.

Pope, H. G. Jr., Jacobs, A., Mialet, J. P., Yurgelun-Todd, D. and Gruber, S. (1997) "Evidence for a sex-specific residual effect of cannabis on visuospatial memory," *Psychotherapy and Psychosomatics* **66**: 179–184.

Pope, H. G. and Yurgelun-Todd, D. (1996) "The residual cognitive effects of heavy marijuana use in college students," *Journal of the American Medical Association* **275**: 521–527.

Presburger, G. and Robinson, J. K. (1999) "Spatial signal detection in rats is differentially disrupted by Delta-9-tetrahydrocannabinol, scopolamine, and MK-801," *Behavioral Brain Research* **99**: 27–34.

Reibaud, M., Obinu, M. C., Ledent, C., Parmentier, M., Bohme, G. A. and Imperato, A. (1999) "Enhancement of memory in cannabinoid CB_1 receptor knock-out mice," *European Journal of Pharmacology* **379**: R1–R2.

Reilly, D., Didcott, P., Swift, W. and Hall, W. (1998) "Long-term cannabis use: characteristics of users in an Australian rural area," *Addiction* **93**: 837–846.

Romero, J., Garcia, L., Fernandez-Ruiz, J., Cebeira, M. and Ramos, J. (1995) "Changes in rat brain cannabinoid binding sites after acute or chronic exposure to their endogenous agonist, anandamide, or to Δ^9-tetrahydrocannabinol," *Pharmacology Biochemistry and Behavior* **51**: 731–737.

Rubino, T., Massi, P., Patrini, G., Venier, I., Giagnoni, G. and Parolaro, D. (1994) "Chronic CP 55, 940 alters cannabinoid receptor mRNA in the rat brain: an in situ hybridization study," *Neuroreport* **5**: 2493–2496.

Santucci, V., Storme, J. J., Soubrié, P. and Le Fur, G. (1996) "Arousal-enhancing properties of the CB₁ cannabinoid receptor antagonist SR 141716A in rats as assessed by electroencephalographic spectral and sleep-waking cycle analysis," *Life Sciences* **58**: 103–110.

Schlicker, E. and Gothert, M. (1998) "Interactions between the presynaptic alpha(2)-autoreceptor and presynaptic inhibitory heteroreceptors on noradrenergic neurones," *Brain Research Bulletin* **47**: 129–132.

Schwartz, R. H., Gruenewald, P. J., Klitzner, M. and Fedio, P. (1989) "Short-term memory impairment in cannabis-dependent adolescents," *American Journal of Diseases of Children* **143**: 1214–1219.

Schwenk, C. R. (1998) "Marijuana and job performance: comparing the major streams of research," *Journal of Drug Issues* **28**: 941–970.

Shen, M., Piser, T. M., Seybold, V. S. and Thayer, S. A. (1996) "Cannabinoid receptor agonists inhibit glutamatergic synaptic transmission in rat hippocampal cultures," *Journal of Neuroscience* **16**: 4322–4334.

Shen, M. X. and Thayer, S. A. (1999) "Delta(9)-tetrahydrocannabinol acts as a partial agonist to modulate glutamatergic synaptic transmission between rat hippocampal neurons in culture," *Molecular Pharmacology* **55**: 8–13.

Shohami, E., Gallily, R., Mechoulam, R., Bass, R. and Ben-Hur, T. (1997) "Cytokine production in the brain following closed head injury: Dexanabinol (HU-211) is a novel TNF-alpha inhibitor and an effective neuroprotectant," *Journal of Neuroimmunology* **72**: 169–177.

Shohami, E., Novikov, M. and Bass, R. (1995) "Long-term effect of HU-211, a novel noncompetitive NMDA antagonist, on motor and memory functions after closed-head injury in the rat," *Brain Research* **674**: 55–62.

Sim, L. J., Hampson, R. E., Deadwyler, S. A. and Childers, S. R. (1996) "Effects of chronic treatment with delta(9)-tetrahydrocannabinol on cannabinoid-stimulated [S-35] GTP-gamma-S autoradiography in rat brain," *Journal of Neuroscience* **16**: 8057–8066.

Slikker, W. Jr., Paule, M. G., Ali, S. F., Scallet, A. C. and Bailey, J. R. (1992) "Behavioral, neurochemical, and neurohistological effects of chronic marijuana smoke exposure in the nonhuman primate," in *Marijuana / Cannabinoids: Neurobiology and Neurophysiology*, edited by L. Murphy and A. Bartke, pp. 219–273. Boca Raton: CRC Press.

Smiley, A. (1999) "Marijuana: On-road and driving-simulator studies," in *The Health Effects of Cannabis*, edited by H. Kalant, W. Corrigall, W. Hall and R. Smart, pp. 171–191. Toronto: Addiction Research Foundation, Centre for Addiction and Mental Health.

Spreen, O. and Strauss, E. (1998) *A Compendium of Neuropsychological Tests. Administration, Norms and Commentary*, 2nd edn. New York: Oxford University Press.

Soderstrom, K. and Johnson, F. (2000) "CB₁ cannabinoid receptor expression in brain regions associated with zebra finch song control," *Brain Research* **857**: 151–157.

Solowij, N. (1995) "Do cognitive impairments recover following cessation of cannabis use?" *Life Sciences* **56**: 2119–2126.

Solowij, N. (1998) *Cannabis and Cognitive Functioning*. Cambridge: Cambridge University Press.

Solowij, N., Grenyer, B. F. S., Chesher, G. and Lewis, J. (1995) "Biopsychosocial changes associated with cessation of cannabis use: A single case study of acute and chronic cognitive effects, withdrawal and treatment," *Life Sciences* **56**: 2127–2134.

Stella, N., Schweitzer, P. and Piomelli, D. (1997) "A second endogenous cannabinoid that modulates long-term potentiation," *Nature* **388**: 773–778.

Straiker, A., Stella, N., Piomelli, D., Mackie, K., Karten, H. J. and Maguire, C. (1999) "Cannabinoid CB₁ receptors and ligands in vertebrate retina: Localization and function of an endogenous signaling system," *Proceedings of the National Academy of Sciences of the USA* **96**: 14565–14570.

Struve, F. A., Straumanis, J. J. and Patrick, G. (1994) "Persistent topographic quantitative EEG sequelae of chronic marihuana use: A replication study and initial discriminant function analysis," *Clinical Electroencephalography* **25**: 63–75.

Tanda, G., Pontieri, F. E. and di Chiara, G. (1997) "Cannabinoid and heroin activation of mesolimbic dopamine transmission by a common mu 1 opioid receptor mechanism," *Science* **276**: 2048–2050.

Terranova, J.-P., Michaud, J.-C., Le Fur, G. and Soubrié, P. (1995) "Inhibition of long-term potentiation in rat hippocampal slices by anandamide and WIN55212-2: reversal by SR141716 A, selective antagonist of CB_1 cannabinoid receptors," *Naunyn-Schmiedeberg's Archives of Pharmacology* **352**: 576–579.

Terranova, J. P., Storme, J. J., Lafon, N., Pério, A., Rinaldi-Carmona, M., Le Fur, G. and Soubrié, P. (1996) "Improvement of memory in rodents by the selective CB_1 cannabinoid receptor antagonist, SR 141716," *Psychopharmacology* **126**: 165–172.

Tsou, K., Brown, S., Sanudopena, M. C., Mackie, K. and Walker, J. M. (1998) "Immuno-histochemical distribution of cannabinoid CB_1 receptors in the rat central nervous system," *Neuroscience* **83**: 393–411.

Vasquez, C. and Lewis, D. L. (1999) "The CB_1 cannabinoid receptor can sequester G-proteins, making them unavailable to couple to other receptors," *Journal of Neuroscience* **19**: 9271–9280.

Volkow, N. D., Gillespie, H., Mullani, N., Tancredi, L., Grant, C., Valentine, A. and Hollister, L. (1996) "Brain glucose-metabolism in chronic marijuana users at baseline and during marijuana intoxication," *Psychiatry Research: Neuroimaging* **67**: 29–38.

Westlake, T. M., Howlett, A. C., Ali, S. F., Paule, M. G. and Scallett, W., Jr. (1991) "Chronic exposure to delta-9-tetrahydrocannabinol fails to irreversibly alter brain cannabinoid receptors," *Brain Research* **544**: 145–149.

Westlake, T. M., Howlett, A. C., Bonner, T. I., Matsuda, L. A. and Herkenham, M. (1994) "Cannabinoid receptor binding and messenger RNA expression in human brain: An in vitro receptor autoradiography and in situ hybridization histochemistry study of normal aged and Alzheimer's brains," *Neuroscience* **63**: 637–652.

Wiesbeck, G. A., Schuckit, M. A., Kalmijn, J. A., Tipp, J. E., Bucholz, K. K. and Smith, T. L. (1996) "An evaluation of the history of a marijuana withdrawal syndrome in a large population," *Addiction* **91**: 1469–1478.

Winsauer, P. J., Lambert, P. and Moerschbaecher, J. M. (1999) "Cannabinoid ligands and their effects on learning and performance in rhesus monkeys," *Behavioral Pharmacology* **10**: 497–511.

Wu, X. F. and French, E. D. (2000) "Effects of chronic Delta(9)-tetrahydrocannabinol on rat midbrain dopamine neurons: an electrophysiological assessment," *Neuropharmacology* **39**: 391–398.

Yurgelun-Todd, D. A., Gruber, S. A., Hanson, R. A., Baird, A. A., Renshaw, P. F. and Pope, H. G. (1999) "Residual effects of marijuana use: An fMRI study," in *Problems of Drug Dependence 1998: Proceedings of the 60th Annual Scientific Meeting of the College on Problems of Drug Dependence*, NIDA Research Monograph 179, edited by L. S. Harris, p. 78. Bethesda, MD: National Institute on Drug Abuse.

Zhuang, S.-Y., Kittler, J., Grigorenko, E. V., Kirby, M. T., Sim, L. J., Hampson, R. E. *et al.* (1998) "Effects of long-term exposure to Δ^9-THC on expression of cannabinoid receptor (CB_1) mRNA in different rat brain regions," *Molecular Brain Research* **62**: 141–149.

Chapter 11

Marijuana and endocrine function

Laura L. Murphy

ABSTRACT

Marijuana and its cannabinoids have profound effects on hormone secretion in humans and experimental animals. The most studied aspect of the cannabinoid effects on endocrine function is how marijuana and cannabinoids affect the hormones of reproduction. Cannabinoids inhibit the hormones of the hypothalamic-pituitary-gonadal axis, i.e. gonadotropin-releasing hormone (GnRH), luteinizing hormone (LH)/follicle-stimulating hormone (FSH), and the sex steroids estrogen, progesterone and testosterone. Inhibition of this axis may be responsible for the ability of marijuana to cause anovulation, oligospermia, and changes in sexual behavior. Marijuana and cannabinoids activate the hypothalamic-pituitary-adrenal axis and promote the release of corticotropin-releasing hormone (CRH), adrenocorticotropin hormone (ACTH), and adrenal corticosteroids. There is less information regarding the ability of marijuana and cannabinoids to inhibit the secretion of prolactin, growth hormone, thyroid hormones, and insulin. Cannabinoid receptors mediate the effects of cannabinoids on endocrine function. The brain is the most likely site of cannabinoid action, however the pituitary gland and gonads may also be directly affected. Much remains to be learned regarding how marijuana affects endocrine function and, importantly, the role endogenous cannabinoid compounds may have in endocrine regulation.

Key Words: marijuana, cannabinoids, endocrine, hormones, reproduction, stress

INTRODUCTION

Within the last decade it has become increasingly evident that the long-reported effects of marijuana smoking on endocrine function (Mechoulam, 1986) are specific effects mediated by cannabinoid receptors. Activation of these receptors by exogenous intake of cannabinoids through marijuana smoking or prescription Δ^9-tetrahydrocannabinol (THC), or activation by endogenous cannabinoid compounds, affects the neuroendocrine regulation of hormone secretion, ultimately altering endocrine function (Table 11.1). This chapter will discuss what is currently known regarding marijuana and cannabinoids and their effects on hormone secretion, sites and mechanisms of their action, and the consequent effects of altering hormone levels.

Table 11.1 Effect of marijuana and cannabinoids on endocrine function

Effect	Site of Action	Result
↓ LH, ↓ FSH	↓ GnRH release	Suppression of estradiol, progesterone and testosterone
↓ prolactin	↑ dopamine release ↓ prolactin release	Decrease lactation (?)
↓ growth hormone	↑ somatostatin release	Decrease growth/metabolism (?)
↓ TSH	↓ TSH release (?)	Decrease T_3/T_4 release
↑ ACTH	↑ CRH release	Increase corticosteroid release

SEX HORMONES

Male

In the male, the gonadotropins luteinizing hormone (LH) and follicle-stimulating hormone (FSH) are synthesized and released from the anterior pituitary gland in response to the hypothalamic peptide releasing hormone GnRH or gonadotropin-releasing hormone. Once released, LH and FSH bind to receptors within the testes where these hormones maintain spermatogenesis and testosterone secretion. Exposure to cannabinoids can result in altered levels of LH, FSH and testosterone in the male. In human males, cannabis smoking has been shown to decrease serum LH, FSH, and testosterone levels when compared to pre-smoking baseline hormone levels (Cone et al., 1986) or when compared to nonsmoking individuals (Kolodny et al., 1974, 1976). An increased incidence of low sperm count, or oligospermia, has been reported in men who were heavy marijuana smokers (Kolodny et al., 1974; Hembree et al., 1976).

In laboratory animals, cannabinoid exposure produces a predominantly inhibitory effect on gonadotropin and testicular steroid release in the male. Acute treatment with Δ^9-tetrahydrocannabinol (THC) produces a consistent and significant dose- and time-related decrease in circulating LH and testosterone levels in male rodents (Steger et al., 1990). As shown in Figure 11.1, an acute intravenous injection of THC inhibits the pulsatile pattern of LH release in castrate, testosterone-treated male rats, thus reducing basal levels of LH. In the male rhesus monkey, an acute dose of THC produced a 65% reduction in plasma testosterone levels by 60 min of treatment that lasted for approximately 24 h (Smith et al., 1976). Long-term cannabinoid exposure in male mice disrupts spermatogenesis and can induce aberrations in sperm morphology (Zimmerman et al., 1999). Both long-term and acute THC exposure inhibits male sexual behavior (i.e. libido and copulatory performance) and gonadotropin responses in the male rodent to sexually-receptive female rats (Murphy et al., 1994). Together, all of these studies suggest that THC inhibits LH and FSH secretion, consequently decreasing testosterone production, altering spermatogenesis, and influencing male sexual behavior.

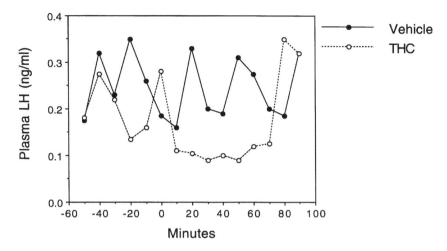

Figure 11.1 Pulsatile LH secretion in 2 representative castrate male rats treated with testosterone. At time 0, rats received either intravenous vehicle (solid line/circle) or 0.5 mg/kg body weight of THC (dashed line/open circle) and sequential blood samples were taken through jugular cannulae every 10 min for 50 min pre- and 90 min post-treatment. Note the complete suppression of LH pulsatility in the THC-treated rat.

Females

The gonadotropins LH and FSH in the female, when released from the anterior pituitary gland, act on the ovaries to stimulate the synthesis and release of the sex steroids estrogen and progesterone, and maintain folliculogenesis. The gonadal steroids complete the feedback loop by exerting a primarily negative feedback action on the hypothalamus and pituitary gland, to inhibit GnRH and LH/FSH secretion, respectively. Humans and monkeys exhibit an average 28-day menstrual cycle characterized by a follicular phase, during which ovarian follicles develop and recruitment of the preovulatory follicle occurs, the preovulatory phase, where there is maturation and ovulation of the dominant follicle, and the luteal phase, which is dominated by the progesterone-secreting corpus luteum. The effects of THC exposure during the different phases of the menstrual cycle have been studied in both humans and monkeys. When a marijuana cigarette was smoked by women during the luteal phase, there was a 30% suppression of plasma LH levels within an hour of smoking when compared to placebo-smoking control subjects (Mendelson *et al.*, 1986). However, there were no changes in plasma LH levels following marijuana smoking by women in their follicular phase of the cycle (Mendelson *et al.*, 1986), or in post-menopausal women (Mendelson *et al.*, 1985a). Interestingly, there was a significant increase in plasma LH levels when women smoked the marijuana cigarette during the periovulatory phase of her cycle (Mendelson and Mello, 1984). In the rhesus monkey, an acute injection of THC during the luteal phase of the menstrual cycle significantly reduced circulating progesterone levels (Smith *et al.*, 1983). When THC was administered for

18 days during the follicular and preovulatory phases of the monkey's menstrual cycle, the preovulatory estrogen and LH surges were blocked and ovulation did not occur during the treatment cycle or in the following post-treatment cycle (Asch et al., 1981). The long-term exposure of female monkeys to thrice weekly injections of THC resulted in a disruption of menstrual cycles that lasted for several months and was characterized by decreased LH and sex steroid hormone levels and anovulation (Smith et al., 1983).

Female rats typically exhibit a 4- or 5-day estrous cycle, with the preovulatory gonadotropin surge occurring during the afternoon/evening of proestrus followed by ovulation during the early morning of estrus. In female rats, acute THC exposure decreased basal LH levels (Tyrey, 1980) and blocked the preovulatory LH and FSH surges when administered during the afternoon of proestrus (Ayalon et al., 1977). Together, the results in the female indicate that cannabinoids inhibit the gonadotropin release necessary for maintaining ovarian function, consequently suppressing estrogen release, blocking ovulation, and decreasing progesterone levels. It is interesting that the levels of circulating estrogen and/or progesterone in the cycling female may dictate the ability of cannabinoid exposure to alter gonadotropin release. Estrogen and progesterone may differentially modulate the expression and density of cannabinoid receptors in hypothalamus and pituitary (Rodríguez de Fonseca et al., 1994; González et al., 2000), consequently altering cannabinoid responsiveness.

Prolactin is an anterior pituitary hormone that serves many physiological functions in the female, including the stimulation of milk production and the maintenance of lactation after pregnancy in mammals. In laboratory animals, THC has a predominantly inhibitory effect on prolactin secretion. Acute cannabinoid exposure inhibits basal prolactin release in monkeys (Asch et al., 1979) and rats (Hughes et al., 1981), and blocks the prolactin surge that occurs on the afternoon of proestrus (Ayalon et al., 1977) or in response to suckling in rats (Tyrey and Hughes, 1984). The inhibitory prolactin response to cannabinoids can be preceded by a transient, yet significant, stimulation of prolactin release in the female rat (Bonnin et al., 1993). In human females, plasma prolactin levels were significantly reduced when THC was given during the luteal phase but not the follicular phase of the menstrual cycle (Mendelson et al., 1985b), again suggesting a role for sex steroids in modulating responsiveness to cannabinoids.

STRESS HORMONES

The hypothalamic-pituitary-adrenal axis is the major endogenous hormonal system responsible for maintaining homeostatic balance in response to stress. Beginning in the central nervous system, corticotropin-releasing hormone (CRH), synthesized in the hypothalamic parvocellular paraventricular nuclei of the hypothalamus, regulates the release of adrenocorticotropin hormone (ACTH) from the anterior pituitary gland. ACTH release stimulates the synthesis and secretion of the adrenal glucocorticoids (cortisol in humans and corticosterone in rats), which, in turn, inhibit CRH and ACTH release, thus completing the negative feedback loop necessary for regulation of the stress axis. Smoking marijuana or the administration

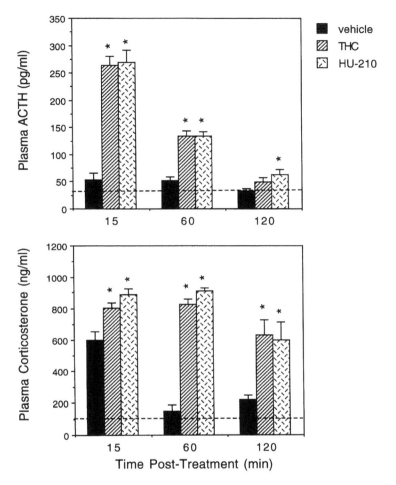

Figure 11.2 The effect of cannabinoids on ACTH and corticosterone release in ovariectomized female rats. Rats received intravenous vehicle (solid bar), 0.5 mg/kg body weight of THC (striped bar), or 20 μg/kg of HU-210 at time 0 and sequential blood samples were taken at 15, 60, and 120 min post-treatment. The dashed line near base of bars represents basal hormone levels in these animals. *p < 0.05 vs vehicle hormone levels at same time point. Note the significant increases in ACTH and corticosterone immediately after cannabinoid administration.

of an acute dose of cannabinoid stimulates the release of CRH (Weidenfeld *et al.*, 1994), ACTH (Rodríguez de Fonseca *et al.*, 1992a; Jackson and Murphy, 1997) and glucocorticoids (Cone *et al.*, 1986; Rodríguez de Fonseca *et al.*, 1992a) in experimental animals and humans. An acute dose of THC or the specific cannabinoid receptor ligand HU210, produces a significant increase in ACTH and corticosterone levels in ovariectomized female rats (Figure 11.2). In male rhesus monkeys, acute exposure to marijuana smoke produced a 178% increase in urinary cortisol and a 31% increase in plasma cortisol levels over pre-exposure hormone levels (Bailey *et al.*, 1990). Although some tolerance developed to the marijuana smoke exposure after 1 month, cortisol levels remained elevated throughout the year-long treatment period. Human users of marijuana exposed to an acute dose of oral THC (Dax *et al.*, 1989) or marijuana cigarettes (Cone *et al.*, 1986), exhibited either no change or an increase in circulating cortisol levels, respectively. However, in men and women who were chronic marijuana smokers, basal cortisol levels were not different when compared to a group of non-marijuana smokers (Block *et al.*, 1991). The reason for the differences in corticosteroid response to cannabinoids in humans is unclear. However, a rapid development of tolerance to cannabinoid has been shown to occur in animal studies (Rodríguez de Fonseca *et al.*, 1992a; Bailey *et al.*, 1990) which may help explain the differences in stress hormone responses to cannabinoids in the human. In the drug-naïve individual, however, it is clear that cannabinoids stimulate the hypothalamic-pituitary-adrenal axis initiating a stress hormone response.

OTHER HORMONES

Less is known regarding cannabinoid effects on other hormonal systems. Several hormones involved either directly or indirectly in the regulation of metabolism, i.e., growth hormone (GH), thyroid-stimulating hormone (TSH), and insulin, are affected as a result of cannabinoid exposure. The secretion of GH by the anterior pituitary gland is regulated by two hormones from the hypothalamus, GH-releasing hormone (GHRH), which stimulates GH secretion, and somatostatin, which inhibits the release of GH. Systemic or intracerebroventricular administration of cannabinoid decreases plasma GH levels in experimental animals (Dalterio *et al.*, 1981; Rettori *et al.*, 1988; Martín-Calderón *et al.*, 1998). The THC-induced decrease in GH levels may be preceded by a transient but significant increase in GH release (Rettori *et al.*, 1988). *In vitro* studies have demonstrated that THC causes significant stimulation of somatostatin release from hypothalamic fragments (Rettori *et al.*, 1988) suggesting that cannabinoid-induced stimuation of somatostatin is at least partly responsible for the inhibition of GH by cannabinoids.

The synthesis and release of thyroid hormones from the thyroid gland are under the control of the pituitary hormone TSH. The secretion of TSH is regulated by direct negative feedback of circulating thyroid hormones on the pituitary as well as by the stimulatory effect of thyrotropin-releasing hormone (TRH) from the hypothalamus. Acute administration of THC in rats reduced serum TSH levels by 90% within 60 min of treatment and produced a maximal suppression of thyroid hormone levels by 6 h (Hillard *et al.*, 1984). THC treatment did not alter hypothalamic levels of TRH suggesting that cannabinoids directly influence TSH or thyroid hormone release (Hillard *et al.*, 1984).

Glucose tolerance may also be impaired following acute exposure to THC in humans (Hollister, 1986). The decreased glucose tolerance was accompanied by an increase in GH, which implies that GH may be interfering with insulin action. In rabbits, acute THC administration produced hyperglycemia via an epinephrine-dependent pathway (Hollister, 1986). Furthermore, in a study using isolated pancreatic islets from rats, basal insulin secretion and the insulin response to a submaximal dose of glucose were significantly increased following THC treatment (Laychock *et al.*, 1986).

MECHANISM AND SITE OF CANNABINOID ACTION ON HORMONE RELEASE

Brain

The brain is the most likely site at which cannabinoids alter hormone secretion (Murphy *et al.*, 1998). Cannabinoids such as THC bind to dense populations of cannabinoid CB_1 receptors in the brain and, notably, the hypothalamus (Herkenham *et al.*, 1990). Autoradiography studies have demonstrated the presence of CB_1 receptors in several hypothalamic nuclei, including the paraventricular nuclei, a predominant site of CRH neuronal cell bodies (Fernández-Ruiz *et al.*, 1997). The ability of specific cannabinoid receptor ligands, i.e. CP55940, WIN55212, HU210, and anandamide, to inhibit LH and prolactin and stimulate ACTH release in laboratory animals indicates that cannabinoid receptors most likely mediate the neuroendocrine actions of THC/cannabinoids (Murphy *et al.*, 1998). Treatment with HU210 produced a dose-related decrease in plasma GH levels indicating that cannabinoid receptors most likely mediate effects on GH release as well (Martín-Calderón *et al.*, 1998). Cannabinoid receptors may be involved in the modulation of neurotransmitter and/or neuropeptide release (Breivogel and Childers, 1998). There is good evidence that the neurotransmitter/neuropeptide systems involved in regulation of GnRH and CRH release, i.e. norepinephrine, dopamine, serotonin, GABA, opioids, are significantly altered following acute cannabinoid exposure (Miguel *et al.*, 1998; Martín-Calderón *et al.*, 1998; Murphy *et al.*, 1998). Treating rats with either anandamide, CP55940 or THC produced a rapid increase in expression of the immediate-early gene *c-fos* in stress-responsive nuclei of the rat brain, i.e. the paraventricular hypothalamus and central nucleus of the amygdala (Herkenham and Brady, 1994; Wenger *et al.*, 1997). These studies suggest that cannabinoids, either directly or indirectly via neuromodulators, cause neuronal activation within brain nuclei leading to CRH activation and release. That hypothalamic releasing factors are altered as a result of cannabinoid exposure is further supported by findings that hypothalamic levels of CRH and GnRH are decreased and increased, respectively, indicating increased release of CRH and suppression of GnRH release following cannabinoid treatment (Wenger *et al.*, 1987; Weidenfeld *et al.*, 1994).

Activation of the tuberoinfundibular dopaminergic neurons (TIDA), the primary hypothalamic input regulating pituitary prolactin secretion, is considered to be responsible for the decrease in plasma prolactin levels observed after acute cannabinoid exposure. Autoradiographic analysis has demonstrated the presence of cannabinoid receptors in the arcuate nucleus, a hypothalamic region where TIDA neurons are located (Fernández-Ruiz *et al.*, 1997). Acute cannabinoid exposure significantly increases the activity of the TIDA neuronal system and increases dopamine release, resulting in decreased prolactin secretion from the pituitary (Rodríguez de Fonseca *et al.*, 1992b; Martín-Calderón *et al.*, 1998).

Pituitary gland

There is some evidence that cannabinoids may regulate pituitary hormone secretion by a direct action on the pituitary itself. A diffuse population of CB_1 cannabinoid receptors have been identified in the anterior pituitary (González et al., 1999, 2000) and the endogenous cannabinoid 2-arachidonoyl-glycerol can be quantified from rat anterior pituitary extracts (González et al., 1999). Pituitary levels of cannabinoid receptor and endocannabinoid can be regulated by the sex steroid hormones estrogen and testosterone (González et al., 2000). Cannabinoid receptors may be localized on prolactin-secreting (lactotropes) and gonadotropin-secreting (gonadotropes) pituitary cells (Wenger et al., 1999). Although there is little evidence of a direct pituitary effect of cannabinoids in the stimulation or suppression of most pituitary hormones, the endocannabinoid anandamide can directly modulate basal prolactin secretion in anterior pituitary cell cultures (González et al., 2000).

Gonads

Cannabinoids may have a direct gonadal action that, particularly with long-term cannabinoid exposure, may play a significant role in the occurrence of lowered sex steroid hormone levels in males and females and also abnormal sperm number and morphology in the male. This is supported by findings that high-dose marijuana smoking in humans has been reported to decrease sperm number in the absence of a corresponding decrease in LH, FSH or testosterone levels (Hembree et al., 1976). Cannabinoid receptor mRNA is found within human testes (Gerard et al., 1991) and ovaries (Galiègue et al., 1995). That cannabinoids directly affect gonadal function can be shown in in vitro studies where treatment with THC directly inhibited the Sertoli cell response to FSH in male rats (Newton et al., 1993), inhibited testosterone production from isolated mouse Leydig cells (Burstein et al., 1978; Dalterio et al., 1985), and directly suppressed female rat granulosa cell function (Adashi et al., 1983). Germ cells may be one site where cannabinoid receptors are localized. Sperm cells obtained from the sea urchin exhibited a dose-related inhibition of the acrosome reaction when exposed to cannabinoids (Schuel et al., 1994).

REFERENCES

Adashi, E. Y., Jones, P. B. C. and Hsueh, A. J. W. (1983) "Direct antigonadal activity of cannabinoids: Suppression of rat granulosa cell functions," *American Journal of Physiology* **244**: E177–E185.

Asch, R. H., Smith, C. G., Siler-Khodr, T. M. and Pauerstein, C. (1979) "Acute decreases in serum prolactin concentrations caused by delta-9-tetrahydrocannabinol in nonhuman primates," *Fertility and Sterility* **32**: 571–574.

Asch, R. H., Smith, C. G., Siler-Khodr, T. M. and Pauerstein, C. J. (1981) "Effects of Δ^9-tetrahydrocannabinol during the follicular phase of the rhesus monkey (Macaca mulatta)," *Journal of Clinical Endocrinology and Metabolism* **52**: 50–55.

Ayalon, D., Nir, I., Cordova, T., Bauminger, S., Puder, M., Naor, Z. et al. (1977) "Acute effects of Δ^1-tetrahydrocannabinol on the hypothalamo-pituitary-ovarian axis in the rat," *Neuroendocrinology* **23**: 31–42.

Bailey, J. R., Paule, M. G., Ali, S. F., Scallet, A. C. and Slikker, W. (1990) "Chronic marijuana smoke exposure in the rhesus monkey II: Effects on urinary cortisol, plasma cortisol and plasma testosterone concentrations," *International Cannabis Research Society Annual Meeting Abstract* **5**: 12.

Block, R. I., Farinpour, R. and Schlechte, J. A. (1991) "Effects of chronic marijuana use on testosterone, luteinizing hormone, follicle stimulating hormone, prolactin and cortisol in men and women," *Drug and Alcohol Dependence* **28**: 121–128.

Bonnin, A., Ramos, J. A., Rodríguez de Fonseca, F., Cebeira, M. and Fernández-Ruiz, J. J. (1993) "Acute effects of Δ^9-tetrahydrocannabinol on tuberoinfundibular dopaminergic activity, anterior pituitary sensitivity to dopamine and prolactin release vary as a function of estrous cycle," *Neuroendocrinology* **58**: 280–286.

Breivogel, C. S. and Childers, S. R. (1998) "The functional neuroanatomy of brain cannabinoid receptors," *Neurobiology and Disease* **5**: 417–431.

Burstein, S., Hunter, S. A., Shoupe, T. S. and Taylor, P. (1978) "Cannabinoid inhibition of testosterone synthesis by mouse Leydig cells," *Research Communication in Chemical Pathology and Pharmacology* **19**: 557–560.

Cone, E. J., Johnson, R. E., Moore, J. D. and Roache, J. D. (1986) "Acute effects of smoking marijuana on hormones, subjective effects and performance in male human subjects," *Pharmacology Biochemistry and Behavior* **24**: 1749–1754.

Dalterio, S. L., Michael, S. D., Macmillan, B. T. and Bartke, A. (1981) "Differential effects of cannabinoid exposure and stress on plasma prolactin, growth hormone and corticosterone levels in male mice," *Life Sciences* **28**: 761–766.

Dalterio, S., Bartke, A. and Mayfield, D. (1985) "Effects of Δ^9-tetrahydrocannabinol on testosterone production in vitro: Influence of Ca^{++}, Mg^{++}, or glucose," *Life Sciences* **37**: 605–612.

Dax, E. M., Pilotte, N. S., Adler, W. H., Nagel, J. E. and Lange, W. R. (1989) "The effects of 9-ene-tetrahydrocannabinol on hormone release and immune function," *Journal of Steroid Biochemistry* **34**: 263–270.

Fernández-Ruiz, J. J., Muñoz, R. M., Romero, J., Villanúa, M. A., Makriyannis, A. and Ramos, J. A. (1997) "Time course of the effects of different cannabimimetics on prolactin and gonadotrophin secretion: Evidence for the presence of CB_1 receptors in hypothalamic structures and their involvement in the effects of cannabimimetics," *Biochemical Pharmacology* **53**: 1919–1927.

Galiègue, S., Mary, S., Marchand, J., Dussossoy, D., Carrière, D., Carayon, P. *et al.* (1995) "Expression of central and peripheral cannabinoid receptors in human immune tissues and leukocyte subpopulations," *European Journal of Biochemistry* **232**: 54–61.

Gerard, C. M., Mollereau, C., Vassart, G. and Parmentier, M. (1991) "Molecular cloning of a human cannabinoid receptor which is also expressed in testis," *Biochemistry Journal* **279**: 129–134.

González, S., Manzanares, J., Berrendero, F., Wenger, T., Corchero, J., Bisogno, T. *et al.* (1999) "Identification of endocannabinoids and cannabinoid CB_1 receptor mRNA in the pituitary gland," *Neuroendocrinology* **70**: 137–145.

González, S., Bisogno, T., Wenger, T., Manzanares, J., Milone, A., Berrendero, F. *et al.* (2000) "Sex steroid influence on cannabinoid CB_1 receptor mRNA and endocannabinoid levels in the anterior pituitary gland," *Biochemistry and Biophysics Research Communication* **270**: 260–266.

Hembree, W. C., Zeidenberg, P. and Nahas, G. G. (1976) "Marihuana's effects on human gonadal function," in *Marihuana, Chemistry, Biochemistry and Cellular Effects*, edited by G. G. Nahas, pp. 521–532. New York: Springer-Verlag.

Herkenham, M., Lynn, A. B., Johnson, M. R., Melvin, L. S., deCosta, B. R. and Rice, K. C. (1990) "Characterization and localization of cannabinoid receptors in rat brain: A quantitative in vitro autoradiographic study," *Journal of Neuroscience* **11**: 563–583.

Herkenham, M. and Brady, L. S. (1994) "Δ^9-tetrahydrocannabinol and the synthetic cannabinoid CP55940 induce expression of c-fos mRNA in stress-responsive nuclei of rat brain," *Society for Neuroscience Abstract* **20**: 1676.

Hillard, C. J., Farber, N. E., Hagen, T. and Bloom, A. S. (1984) "The effects of THC on serum thyrotropin levels in the rat," *Pharmacology Biochemistry and Behavior* **20**: 547–550.

Hollister, L. E. (1986) "Health aspects of cannabis," *Pharmacological Reviews* **38**: 1–20.

Hughes, C. L., Everett, J. W. and Tyrey, L. (1981) "Δ^9-tetrahydrocannabinol suppression of prolactin secretion in the rat: Lack of direct pituitary effect," *Endocrinology* **109**: 876–880.

Jackson, A. L. and Murphy, L. L. (1997) "Role of the hypothalamic-pituitary-adrenal axis in the suppression of luteinizing hormone release by delta-9-tetrahydrocannabinol," *Neuroendocrinology* **65**: 446–452.

Kolodny, R. C., Masters, W. H., Kolodny, R. M. and Toro, G. (1974) "Depression of plasma testosterone levels after chronic intensive marihuana use," *New England Journal of Medicine* **290**: 872–874.

Kolodny, R. C., Lessin, P., Toro, G., Masters, W. H. and Cohen, S. (1976) "Depression of plasma testosterone with acute marijuana administration," in *The Pharmacology of Marijuana*, edited by M. C. Braude and S. Szara, pp. 217–225. New York: Raven Press.

Laychock, S. G., Hoffman, J. M., Meisel, E. and Bilgin, S. (1986) "Pancreatic islet arachidonic acid turnover and metabolism and insulin release in response to delta-9-tetrahydrocannabinol," *Biochemical Pharmacology* **35**: 2003–2008.

Martín-Calderón, J. L., Muñoz, R. M., Villanúa, M. A., del Arco, I., Moreno, J. L., Rodríguez de Fonseca, F. and Navarro, M. (1998) "Characterication of the acute endocrine actions of (−)-11-hydroxy-delta-8-tetrahydrocannabinol-dimethylheptyl (HU-210), a potent synthetic cannabinoid in rats," *European Journal of Pharmacology* **344**: 77–86.

Mechoulam, R., (1986) "The pharmacohistory of Cannabis sativa," in *Cannabinoids as Therapeutics Agents*, edited by R. Mechoulam, pp. 1–19, CRC Press Inc.: Boca Raton, FL.

Mendelson, J. H. and Mello, N. K. (1984) "Effects of marijuana on neuroendocrine hormones in human males and females," in *Marijuana Effects on the Endocrine and Reproductive Systems*, edited by M. C. Braude and J. P. Ludford, pp. 97–109. Washington, DC: US Government Printing Office.

Mendelson, J. H., Cristofaro, P., Ellingboe, J., Benedikt, R. and Mello, N. K. (1985a) "Acute effects of marihuana on luteinizing hormone in menopausal women," *Pharmacology Biochemistry and Behavior* **23**: 765–768.

Mendelson, J. H., Mello, N. K. and Ellingboe, J. (1985b) "Acute effects of marihuana smoking on prolactin levels in human females," *Journal of Pharmacology and Experimental Therapeutics* **232**: 220–222.

Mendelson, J. H., Mello, N. K., Ellingboe, J., Skupny, A. S. T., Lex, B. W. and Griffin, M. (1986) "Marihuana smoking suppresses luteinizing hormone in women," *Journal of Pharmacology and Experimental Therapeutics* **237**: 862–866.

Miguel, R., Romero, J., Muñoz, R. M., Garcia-Gil, L., Gonzalez, S., Villanúa, M. A. *et al.* (1998) "Effects of cannabinoids on prolactin and gonadotrophin secretion: Involvement of changes in hypothalamic gamma-aminobutyric acid (GABA) inputs," *Biochemical Pharmacology* **56**: 1331–1338.

Murphy, L. L., Gher, J., Steger, R. W. and Bartke, A. (1994) "Effects of Δ^9-tetrahydrocannabinol on copulatory behavior and neuroendocrine responses of male rats to female conspecifics," *Pharmacology Biochemistry and Behavior* **48**: 1011–1017.

Murphy, L. L., Muñoz, R. M., Adrian, B. A. and Villanúa, M. A. (1998) "Function of cannabinoid receptors in the neuroendocrine regulation of hormone secretion," *Neurobiology of Disease* **5**: 432–446.

Rettori, V., Wenger, T., Snyder, G., Dalterio, S. and McCann, S. M. (1988) "Hypothalamic action of delta-9-tetrahydrocannabinol to inhibit the release of prolactin and growth hormone in rat," *Neuroendocrinology* **47**: 498–503.

Rodríguez de Fonseca, F., Murphy, L. L., Bonnin, A., Eldridge, J. C., Bartke, A. and Fernández-Ruiz, J. J. (1992a) "Δ^9-Tetrahydrocannabinol administration affects anterior pituitary, k corticoadrenal and adrenomedullary functions in male rats," *Neuroendocrinology (Life Science Advances)*, **11**: 147–156.

Rodríguez de Fonseca, F., Fernández-Ruiz, J. J., Murphy, L. L., Cebeira, M., Steger, R. W., Bartke, A. *et al.* (1992b) "Acute effects of Δ-9-tetrahydrocannabinol on dopaminergic activity in several rat brain areas," *Pharmacology Biochemistry and Behavior* **42**: 269–275.

Rodríguez de Fonseca, F., Cebeira, M., Ramos, J. A., Martin, M. and Fernández-Ruiz, J. J. (1994) "Cannabinoid receptors in rat brain areas: Sexual differences, fluctuations during estrous cycle and changes after gonadectomy and sex steroid replacement," *Life Sciences* **54**: 159–170.

Schuel, H., Goldstein, E., Mechoulam, R., Zimmerman, A. M. and Zimmerman, S. (1994) "Anandamide (arachidonylethanolamide), a brain cannabinoid receptor agonist, reduces fertilizing capacity in sea urchins by inhibiting the acrosome reaction," *Proceedings of the National Academy of Sciences of the USA* **91**: 7678–7682.

Smith, C. G., Moore, C. E., Besch, N. F. and Besch, P. K. (1976) "The effect of marihuana (delta-9-tetrahydrocannabinol) on the secretion of sex hormones in the mature male rhesus monkey," *Clinical Chemistry* **22**: 1184.

Smith, C. G., Almirez, R. G., Berenberg, J. and Asch, R. H. (1983) "Tolerance develops to the disruptive effects of Δ^9-tetrahydrocannabinol on primate menstrual cycle," *Science* **219**: 1453–1455.

Steger, R. W., Murphy, L. L., Bartke, A. and Smith, M. S. (1990) "Effects of psychoactive and nonpsychoactive cannabinoids on the hypothalamic-pituitary axis of the adult male rat," *Pharmacology Biochemistry and Behavior* **37**: 299–302.

Tyrey, L. (1980) "Δ^9-Tetrahydrocannabinol: A potent inhibitor of episodic luteinizing hormone secretion," *Journal of Pharmacology and Experimental Therapeutics* **213**: 306–308.

Tyrey, L. and Hughes, C. L. (1984) "Inhibition of suckling-induced prolactin secretion by delta-9-tetrahydrocannabinol," in *The Cannabinoids: Chemical, Pharmacologic and Therapeutic Aspects*, edited by S. Agurell, W. L. Dewey and R. E. Wilette, pp. 487–495. San Diego: Academic Press.

Weidenfeld, J., Feldman, S. and Mechoulam, R. (1994) "Effect of the brain constituent anandamide, a cannabinoid receptor agonist, on the hypothalamo-pituitary-adrenal axis in the rat," *Neuroendocrinology* **59**: 110–112.

Wenger, T., Rettori, V., Snyder, G. D., Dalterio, S. and McCann, S. M. (1987) "Effects of delta-9-tetrahydrocannabinol on the hypothalamic-pituitary control of luteinizing hormone and follicle-stimulating hormone secretion in adult male rats," *Neuroendocrinology* **46**: 488–493.

Wenger, T., Jamali, K. A., Juaneda, C., Leonardelli, J. and Tramu, G. (1997) "Arachidonyl ethanolamide (anandamide) activates the parvocellular part of hypothalamic paraventricular nucleus," *Biochemistry and Biophysics Research Communication* **237**: 724–728.

Wenger, T., Fernández-Ruiz, J. J. and Ramos, J. A. (1999) "Immunocytochemical demonstration of CB_1 cannabinoid receptors in the anterior lobe of the pituitary gland," *Journal of Neuroendocrinology* **11**: 873–878.

Zimmerman, A. M., Zimmerman, S. and Raj, A. Y. (1999) "Effects of cannabinoids on spermatogenesis in mice," in *Marihuana and Medicine*, edited by G. G. Nahas, K. M. Sutin, D. J. Harvey and S. Agurell, pp. 347–357. New Jersey: Humana Press.

Chapter 12

Embryonic cannabinoid receptors are targets for natural and endocannabinoids during early pregnancy

B. C. Paria and S. K. Dey

ABSTRACT

Cannabinoid exerts its effects via interactions with two types of cannabinoid receptors, brain-type receptor (CB_1) and spleen-type receptor (CB_2). While both receptors are expressed in pre-implantation mouse embryos, our findings suggest that the effects of cannabinoids on embryo development and implantation are primarily mediated by CB_1. The levels of CB_1 in mouse blastocyst are much higher than in the brain. Furthermore, the mouse uterus has anandamide synthesizing and hydrolyzing capacities that are differentially regulated during the peri-implantation period. The uterus also contains high levels of endogenous cannabinoid, anandamide. Natural, synthetic and endogenous cannabinoid ligands interfere with preimplantation embryo development and blastocyst implantation, and these effects are completely reversed by specific CB_1 antagonists. These results suggest that preimplantation mouse embryos are possible targets for cannabinoid ligand-receptor signaling during early pregnancy in the mouse.

Key Words: cannabinoid, anandamide, embryo, mouse, implantation

Marijuana and its cannabinoid derivatives induce a wide spectrum of central and peripheral effects (Dewey, 1986; Martin *et al.*, 1995). Among others, significant concerns about habitual marijuana smoking are the adverse effects on reproductive and developmental functions including pregnancy failure, retarded embryonic development, fetal loss, and reduced fertilizing capacity of sperm (Bloch *et al.*, 1978; Smith and Asch, 1987; Chang *et al.*, 1993; Schuel *et al.*, 1994). Many of the central and peripheral effects of cannabinoids are mediated by recently identified G-protein-coupled brain-type (CB_1) and spleen-type (CB_2) cannabinoid receptors (Howlett, 1995; Matsuda *et al.*, 1990; Munro *et al.*, 1993). Furthermore, two endogenous cannabinomimetic lipid derivatives, anandamide (*N*-arachidonoylethanolamide) and 2-AG (sn-2-arachidonoylglycerol) have been isolated from brain and other tissues (Devane *et al.*, 1992; Felder *et al.*, 1992; Schmid *et al.*, 1997; Mechoulam *et al.*, 1995; Sugiura *et al.*, 1995; Stella *et al.*, 1997). These compounds bind with high affinity to brain-type and spleen-type cannabinoid receptors and mimic most of the effects of $(-)\Delta^9$-tetrahydrocannabinol (–THC), a psychoactive derivative of marijuana.

The CB_1 and CB_2 genes were previously shown to be primarily expressed in the brain and spleen, respectively (Devane *et al.*, 1992; Munro *et al.*, 1993). However, there is now evidence that other tissues also express these receptors. The tissues that also express these receptors are testis, spleen and peripheral blood leukocytes (reviewed in Paria *et al.*, 1999). The expression of cannabinoid receptors in the spleen and leukocyte has been associated with the anti-inflammatory and immuno-suppressive roles of cannabinoids (Kaminski *et al.*, 1992; Bouaboula *et al.*, 1993). The observation of reduced fertilizing capacity of sperm exposed to cannabinoid ligands is consistent with the detection of CB_1 mRNA in the testis and cannabinoid binding sites in the sperm (Chang *et al.*, 1993; Schuel *et al.*, 1994). However, the effects and mode of action of cannabinoids in embryo and uterus remained largely undefined and controversial, in spite of the numerous reports published in this field during the last two decades (Bloch *et al.*, 1978; Smith and Asch, 1987). This review article describes ligand-receptor signaling in embryo–uterine interactions during early pregnancy in the mouse.

EXPRESSION OF CANNABINOID RECEPTORS IN THE PREIMPLANTATION MOUSE UTERUS AND EMBRYO

To examine CB_1 or CB_2 mRNA expression in the mouse uterus and embryo, reverse transcription-coupled PCR was employed. In the preimplantation mouse embryo, CB_1 mRNA was primarily detected from the 4-cell through the blastocyst stages, whereas CB_2 mRNA was present from the 1-cell through the blastocyst stages (Figure 12.1) (Paria *et al.*, 1995). RT-PCR also detected CB_1 mRNA in the pregnant uterus. In contrast, RT-PCR could not detect CB_2 mRNA in the uterus, although this mRNA was detected in the rat and mouse spleen (Das *et al.*, 1995). The presence of cannabinoid receptors both in the uterus and preimplantation embryo was also confirmed by radioligand binding studies. Scatchard

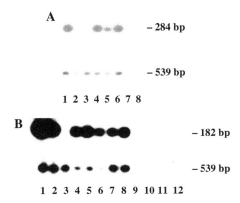

Figure 12.1 Analysis of CB_1 and CB_2 transcripts in the preimplantation mouse embryo. (A) Southern blot analysis of RT-PCR-amplified products of CB_1 (284 bp) or β-actin (539 bp). Lanes: 1, mouse brain; 2–6, embryos at one-cell, two-cell, four-cell, eight-cell/morula, and blastocyst stages, respectively; 7, mouse brain RNA without RT reaction; 8, primer control. (B) Southern blot analysis of RT-PCR-amplified products of CB_2 (182 bp) or β-actin (539 bp). Lanes: 1, rat spleen; 2, mouse spleen; 3, day 1 pregnant uterus; 4–8, embryos at one-cell, two-cell, four-cell, eight-cell/morula, and blastocyst stages, respectively; 9–11, rat spleen, mouse spleen and mouse blastocyst RNA without RT reaction; 12, primer control. Reprinted with the permission from ref. Paria *et al.* 1995.

analysis suggested that both the blastocyst and uterus contain a single class of high affinity binding sites (Das *et al.*, 1995; Yang *et al.*, 1996). However, the levels of cannabinoid receptors are much higher in the mouse embryo than in the uterus or brain. This is consistent with high levels of autoradiographic binding sites for ^{3}H-anandamide in the preimplantation embryo (Paria *et al.*, 1995). At the blastocyst stage, the binding sites were primarily localized in trophectoderm cells, but not in inner cell mass cells (Paria *et al.*, 1995). To biochemically characterize the CB_1 in the embryo, we developed a rabbit antipeptide antibody against the N-terminal region of CB_1 and examined the receptor protein by Western blotting and immunohistochemistry. Western blotting in mouse blastocyst samples detected a major 54 kDa band which is consistent with the predicted size of the CB_1 (Matsuda *et al.*, 1990). The 59 kDa band in the rat brain or mouse blastocyst suggests possible glycosylation of the receptor protein (Figure 12.2) (Song and Howlett, 1995). Immunohistochemistry was also employed to detect the distribution of CB_1 protein in preimplantation mouse embryos. Little or no immunoreactive CB_1 was detected in 1-cell embryos, while distinct signals were evident in embryos from 2-cell through blastocyst stages. At the morula stage, immunoreactive CB_1 was detected primarily in outside cells, while it was predominantly detected in trophectoderm cells of blastocysts with little or no reactivity in the inner cell mass (Yang *et al.*, 1996). The pattern of immunostaining is consistent with our previous observation of autoradiographic distribution of anandamide binding sites in preimplantation mouse embryos (Paria *et al.*, 1995). These results suggest that preimplantation mouse embryo expresses high levels of CB_1 receptors. However, the abundance of CB_2 in the blastocyst is not known.

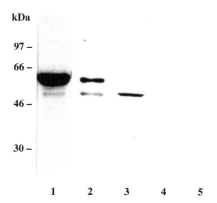

Figure 12.2 Western blot analysis of CB_1 of mouse blastocysts or brain membranes. Day 4 blastocyst preparation and brain membranes were immunoblotted using rabbit antipeptide antibodies to CB_1. Lane 1, Day 4 mouse blastocysts (59 and 54 kDa bands); lane 2, rat brain membranes (59 and 54 kDa bands); lane 3, mouse brain membranes (54 kDa band); Lanes 4 and 5, rat or mouse brain membranes immunoblotted with antibodies preneutralized with excess of antigenic peptide, respectively. Two hundred Day 4 blastocysts were used in this experiment. Molecular weight standards ($\times 10^{-3}$) are indicated. Reprinted with permission from ref. Yang *et al.*, 1996.

ANANDAMIDE IN THE PERIIMPLANTATION MOUSE UTERUS

The identification of the endogenous cannabinoid, anandamide, in the brain led us to examine its synthesis and hydrolysis, as well as its levels in the pregnant mouse uterus. The mechanism of anandamide synthesis remains an open question. Anandamide can be synthesized by either enzymatic condensation of free arachidonic acid and ethanolamine (Deutsch and Chin, 1993; Kruszka and Gross, 1994; Devane and Axelrod, 1994; Ueda *et al.*, 1995), or by the transacylation-phosphodiesterase pathway (Di Marzo *et al.*, 1994; Sugiura *et al.*, 1996). Our findings of relatively low substrate requirement (Km of $3.8\,\mu$M and $1.2\,$mM for arachidonic acid and ethanolamine, respectively) for uterine anandamide synthase activity suggests direct N-acylation of ethanolamine as a possible pathway for anandamide synthesis in the mouse uterus (Paria *et al.*, 1996). The levels of synthase activity remained virtually unchanged, while those of the amidase activity exhibited modest fluctuations on days 1–5 of pregnancy. In contrast, significant increases in uterine synthase activity with concomitant decreases in amidase activity were observed on day 5 (nonreceptive phase) of pseudopregnancy as compared to that observed on day 4 (receptive phase) of pregnancy or pseudopregnancy (Paria *et al.*, 1996). In separated implantation and interimplantation sites on days 5–7, an inverse relationship was noted between the two enzyme activities. While levels of synthase activity were lower at the implantation sites, the levels were higher at the interimplantation sites. In contrast, the reverse was true for the amidohydrolase activity (Figure 12.3). Interestingly, the levels of total anandamide in the uterus (Figure 12.3) also showed fluctuation during early pregnancy as observed for anandamide synthase and amidohydrolase activities (Schmid *et al.*, 1997). Moreover, mouse uterus contains the highest levels of anandamide ever found in any mammalian tissues. The levels were 1345 pmol/μmol lipid P (20 nmol/g tissue) in day 7 interimplantation sites, whereas the brain samples of the same mice contained only 10–14 pmol/g tissue. Moreover, the levels of anandamide are lower in implantation sites, but higher at interimplantation sites. Thus, the findings of CB_1 in the embryo and anandamide in the uterus suggest that a ligand-receptor signaling with an endocannabinoid is operative between the uterus and blastocyst.

EFFECTS OF CANNABINOID AGONISTS ON PREIMPLANTATION EMBRYO DEVELOPMENT

To examine whether cannabinoid ligands influence preimplantation embryo development, 2-cell embryos were cultured in the presence or absence of various synthetic, natural or endogenous cannabinoid agonists (Paria *et al.*, 1995; Yang *et al.*, 1996). All of the agonists [(−)THC, CP 55,940, Win 55212-2, 2-AG and anandamide] exhibited inhibition of embryonic development to blastocysts (Paria *et al.*, 1995, 1998). The developmental arrest primarily occurred between the 4-cell and 8-cell stages. The failure of (+)THC and arachidonic acid to interfere with embryonic development suggests that the effects of cannabinoid agonists on embryo development was not due to non-specific toxic effects. In contrast, a CB_2 agonist,

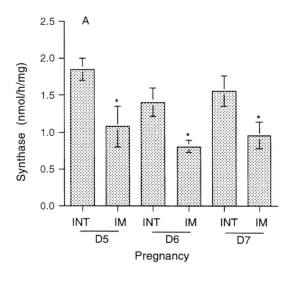

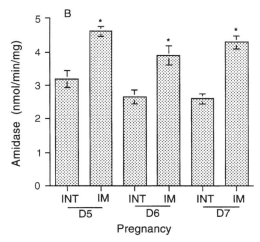

Figure 12.3 The anandamide synthase and amidase activities in uterine implantation and interimplantation sites on days 5–7 (D5–D7) of pregnancy. Microsomal proteins
(100 µg/point) from surgically separated implantation (IM) and interimplantation
(INT) sites were incubated under standard optimal conditions as based on initial
enzyme kinetics to measure the (A) synthase and (B) amidase activity. The synthase
activities in the interimplantation sites were always higher (*$p < 0.05$) than those in
the implantation sites. In contrast, the amidase activity in the implantation sites were
higher (*$p < 0.05$) than those in the interimplantation sites. Values are mean SD and
analyzed by student t-test. Reprinted with permission from ref. Paria et al., 1996.

AM 663, had no inhibitory effects on preimplantation embryo development.
Adverse effects of cannabinoids on embryo development were reversed when
2-cell embryos were cultured in the presence of a CB_1 selective antagonist
SR141716A or AM 251 plus the same concentrations of cannabinoid agonists

(Paria *et al.*, 1998). SR141716A or AM 251 alone had no deleterious effects on embryonic development. The adverse effects of cannabinoid agonists were also noted in trophectoderm cell numbers of blastocysts that escaped the developmental arrest (Yang *et al.*, 1996). The CB_1 antagonist, SR141716A, was also effective in reversing this adverse effect of cannabinoids.

To study effects of anandamide on blastocyst growth and hatching *in vitro*, eight cell embryos recovered on day 3 (1000–1030) were cultured in groups (10–12 embryos/group) for 84 h in 25 μl of Whitten's medium in the absence or presence of anandamide and/or SR141716A (Schmid *et al.*, 1997). Anandamide inhibited zona-hatching of blastocysts *in vitro*, and these detrimental effects of anandamide were reversed by SR141716A (Table 12.1). In contrast, blastocysts exposed to the same low levels of cannabinoid agonists exhibited accelerated trophoblast differentiation with respect to fibronectin-binding activity and trophoblast outgrowth. Again, these effects resulted from activation of embryonic CB_1. However, there was a differential concentration-dependent effect of cannabinoids on the trophoblast with inhibition of differentiation at higher doses of cannabinoids (Wang *et al.*, 1999). Collectively, these results suggest that cannabinoid effects on preimplantation embryo are differentially executed depending on the developmental stages and are mediated primarily via embryonic CB_1.

EFFECTS OF THC ON IMPLANTATION

We observed that single or repeated injections of (–)THC failed to affect implantation in the mouse (Paria *et al.*, 1992). Similarly, continuous infusion of (–)THC (20 μg/h) alone via miniosmotic pumps to pregnant mice failed to prevent implantation (Table 12.2). We suspected that (–)THC was rapidly metabolized to inactive forms *in vivo* and did not reach a critical level in the uterus to affect embryo development and implantation. Indeed, (–)THC plus combined treatment of metyrapone and clotrimazol (50 mg/kg), known inhibitors of cytochrome P450 enzyme system

Table 12.1 Effects of anandamide on blastocyst hatching *in vitro*

Treatment	No. of embryos Cultured	No. of embryos developed to blastocysts	Hatched blastocysts	
			No.	%
Anandamide, nM				
0	42	41	24	58.5
10	42	41	22	53.7
20	45	43	14	32.68*
Anandamide (20 nM) Plus SR (10 nM)	43	43	29	67.4
SR, 10 nM	31	31	21	67.7

Eight-cell embryos (10–20 embryos per group) were cultured in the presence or absence of anandamide and/or SR141716A (SR) for 48 h. Each experiment was repeated 3–4 times. *p < 0.05 (χ^2 test) compared with other groups. Reprinted with permission from ref. Schmid *et al.*, 1997.

Table 12.2 Effects of infusion of (−)THC or (+)THC on implantation

Treatments	No. of Mice	No. of mice with IS	No. of mice without IS	No. of IS	No. of blastocysts recovered
(−)THC	4	4	0	11.5 ± 2.0	0
(−)THC + Met + Clot	13	1	12	3	91 (62)
(+)THC + Met + Clot	6	5	1	10.3 ± 1.7	5
(−)THC + Met + Clot + SR	10	10	0	13.2 ± 1.3	0
Met + Clot	5	4	1	8.3 ± 2.0	8
Met + Clot + SR	4	4	0	9.5 ± 3.3	0
(−)THC + Clot	7	1	6	10	43 (26)
(−)THC + Met	7	7	0	11.0 ± 1.4	0
(+)THC + Clot	4	4	0	11.0 ± 2.1	0
Clot	4	4	0	11.3 ± 1.6	0

Miniosmotic pumps containing (−)THC (active), (+)THC (less active), (−)THC + SR141716A (SR) or (+)THC + SR were placed subcutaneously under the back skin from days 2–5 of pregnancy. The release rate of (−)THC or (+)THC was 20 μg/h, while that of SR was 5 μg/h. The cytochrome P-450 inhibitors metyrapone (Met) plus clotrimazole (Clot) (50 mg/kg each), or Met and Clot each (100 mg/kg) alone were injected twice daily intraperitoneally from days 2 to 4 of pregnancy. Implantation sites (IS) were examined by blue dye method on day 5. (−)THC, but not (+)THC, inhibited implantation in the presence of cytochrome P-450 inhibitors and this inhibition was reversed by a CB_1-R antagonist, SR141716A. Numbers within parantheses indicate the number of zona-encased blastocysts. Reprinted with permission from ref. Paria et al., 1998.

(Bonin *et al.*, 1994), inhibited implantation in 12 of 13 (92%) mice. Co-administration of SR141716A (5 μg/h) with (−)THC completely reversed the implantation-inhibitory effects of (−)THC plus the cytochrome P450 inhibitors. In contrast, infusion of (+)THC with clotrimazole and metyrapone was not effective in inhibiting implantation (Paria *et al.*, 1998). A large number of blastocysts (68%) recovered from mice treated with (−)THC plus metyrapone and clotrimazole was zona-encased.

DISCUSSION

The highlights of our studies are that (1) the mouse uterus during early pregnancy has the capacity to synthesize and degrade an endogenous cannabinoid, anandamide; and (2) the preimplantation mouse embryo expresses functional CB_1 receptors and respond to natural, synthetic or endogenous cannabinoid ligands *in vitro*. However, CB_1 and CB_2 mRNAs are differentially expressed in the preimplantation mouse embryo in a temporal fashion, the significance of which is not yet clear (Paria *et al.*, 1995). The CB_2 mRNA could be of maternal origin that persists through the blastocyst stage, while the accumulation of CB_1 mRNA appears to be associated with the activation of the embryonic genome. Identification of anandamide binding sites by autoradiography and Scatchard analysis of ^{3}H-anandamide suggests a high affinity single class of cannabinoid binding sites in the blastocyst. It is apparent from our studies that mouse blastocysts have many more high affinity cannabinoid receptors than in the mouse brain (Paria *et al.*, 1995; Yang *et al.*, 1996). The inhibitory effects of cannabinoid agonists on preimplantation embryo development at low nanomolar concentrations are consistent with the presence of a large population of high affinity receptors in the embryo. The results of Western blotting and immunocytochemistry show that the blastocyst has a higher abundance of CB_1 than the brain.

The reversal of the cannabinoid-induced inhibition of embryonic development by a CB_1 antagonist (SR141716A) strongly suggests that the brain-type receptors are primarily responsible for the observed effects of cannabinoids and they appear to affect the outer cells of embryos that constitute the trophectoderm. This is consistent with the expression of CB_1 in trophectoderm cells of the blastocysts. Although the CB_2 mRNA is expressed, it is not yet known whether this receptor mRNA is efficiently translated in the preimplantation mouse embryo, or whether this receptor subtype has any functions in the preimplantation embryo. The availability of specific antibodies, or a specific antagonist or agonist to CB_2 are required to explore the interactions of cannabinoid agonists with this receptor subtype and its roles in pre-implantation embryos. In this respect, our recent studies showed that a selective CB_2-R agonist, AM 663, even at 20 nM showed no deleterious effects on embryo development (Paria *et al.*, 1998). Furthermore, the effects of endocannabinoids on embryo development were not reversed by SR144528, a CB_2 antagonist.

Embryonic cannabinoid receptors are coupled to G_i proteins (Paria *et al.*, 1995). However, the physiological significance of cannabinoid receptors in the preimplantation embryo is not yet clearly understood. In this respect, it should be noted that G_i-like proteins are present in the preimplantation mouse embryo (Jones and Schultz, 1990). If cannabinoid ligands are available to a embryo during its normal development, it may modulate the intracellular concentration of cAMP and/or Ca^{++} in the embryo. These two second messengers, involved in important signal transduction pathways, are implicated in cell proliferation, differentiation and gene expression. In this respect, cAMP has been implicated in zygotic gene activation and blastocyst expansion (Manejwala *et al.*, 1989; Poueymirou and Schultz, 1989), while intracellular Ca^{++} plays an important role in cell polarity and embryonic compaction necessary for morula to blastocyst transformation (Pakrashi and Dey, 1984; Ducibella and Anderson, 1975). The failure of embryos to proceed beyond the 8-cell stage after exposure to cannabinoid ligands in culture could be due to the inhibition of Ca^{++} channels resulting from the activation of the cannabinoid receptors. Therefore, tight regulation of the levels of cAMP and Ca^{++} is likely to be critical for normal embryonic development and growth. Although the embryonic arrest with exposure to cannabinoids *in vitro* is consistent with *in vivo* observation of retarded embryonic development and pregnancy failure after chronic exposure to exogenous cannabinoids (Nahas and Latour, 1992; Rosenkrantz, 1979; Dalterio and Bartke, 1981), it is still unknown whether *in vivo* effects of cannabinoids are mediated directly via these embryonic receptors or by some other mechanisms.

Mouse uterus contains high levels of anandamide (Schmid *et al.*, 1997) and has the capacity to synthesize anandamide (Paria *et al.*, 1996). We have also preliminary evidence that this tissue contains 2-AG. We observed that lower levels of anandamide are associated with uterine receptivity for implantation and higher levels are correlated with uterine refractoriness to implantation. Thus, we suspect that increased levels of a cannabinoid agonist in the target tissue would interfere with blastocyst implantation. Indeed, infusion of CP55,940 (a synthetic cannabinoid) via miniosmotic pumps during the preimplantation period, prevented implantation and this inhibition was reversed by co-administration of CP55,940 with SR141716A (Paria *et al.*, 1998; Schmid *et al.*, 1997). However, single or multiple injections of (–)THC (Paria *et al.*, 1992) or continuous infusion of (–)THC during the preimplantation period failed to affect the implantation process. These results suggest that either the process of implantation is unresponsive to (–)THC or this cannabinoid is rapidly metabolized to its inactive forms and/or cleared from the system. The rapid metabolism and/or clearance appear to be the most reasonable explanation, because infusion of (–)THC interfered with implantation only in the presence of cytochrome P450 inhibitors. These observations suggest that under normal conditions females can protect against the adverse effect of cannabinoid during early pregnancy by the P450-linked enzyme system present in the uterus and possibly in the embryo. Since CB_1 is present both in the preimplantation embryo and uterus, it is possible that the adverse effects of (–)THC on implantation are mediated via the uterus and/or embryo. The observed inhibition of implantation by (–)THC could be due to failure of the uterus to achieve the receptive state. On the other hand, the recovery of large number of zona-encased blastocysts suggests that (–)THC interferes with blastocyst growth and attachment with the uterus. Thus, cannabinoids may affect both the embryo and the uterus, and disturb the synchronous development of the blastocyst and uterus for successful implantation. Finally, the reversal of (–)THC-induced inhibition of implantation by SR141716A strongly suggests that these effects are mediated via CB_1. In conclusion, these results place the uterus and embryo as important and physiologically relevant targets for cannabinoid ligand-receptor signaling.

FUTURE DIRECTIONS

Synchronized development of the uterus to the receptive state and embryos to the blastocyst stage is necessary for successful embryo implantation. In this respect, the levels of uterine endocannabinoids and CB_1 in the embryo may play an important role in synchronizing these two critical events necessary for embryo development and implantation. Further investigation is also required to define whether local reduction in the level of anandamide at the implantation sites is necessary for successful implantation and survival of the embryo. Since growth factors, cytokines and prostaglandins can mediate steroid actions in the uterus, it is important to examine whether these factors influence ligand-receptor signaling with endocannabinoids during implantation. Investigation using knock-out mice for the CB_1 and/or CB_2 genes will be valuable in defining many of the effects of cannabinoids during early pregnancy.

ACKNOWLEDGEMENTS

This work was supported by grants from National Institute of Drug Abuse (DA06668). The center grants (HD 0252 and HD 33994) provided access to core facilities.

REFERENCES

Bloch, E., Thysen, B., Morrill, G. A., Gardner, E. and Fujimoto, G. (1978) "Effects of canna-binoids on reproduction and development," *Vitamins and Hormones* **36**: 203–258.

Bonin, A., Miguel, R. D., Fernandez-Ruiz, J. J., Cebeira, M. and Ramos, J. A. (1994) "Possible role of the cytochrome P-450-linked monooxygenase system in preventing delta-tetrahydrocannabinol-induced stimulation of tuberoinfundibular dopaminergic activity in female rats," *Biochemical Pharmacology* **48**: 1387–1392.

Bouaboula, M., Rinaldi, M., Carayon, P., Carillon, C., Delpech, B., Shire, D., LeFur, G. and Gasellas, P. (1993) "Cannabinoid-receptor expression in human leukocytes," *European Journal of Biochemistry* **214**: 173–180.

Chang, M. C., Berkery, D., Schuel, R., Laychock, S. G., Zimmerman, A. M., Zimmerman, S. and Schuel, H. (1993) "Evidence for a cannabinoid receptor in sea urchin sperm and its role in blockade of the acrosome reaction," *Molecular Reproduction and Development* **36**: 507–516.

Dalterio, S. and Bartke, A. (1981) "Fetal testosterone in mice: effect of gestational age and cannabinoid exposure," *Journal of Endocrinology* **91**: 509–514.

Das, S. K., Paria, B. C., Chakraborty, I. and Dey, S. K. (1995) "Cannabinoid ligand-receptor signaling in the mouse uterus," *Proceedings of the National Academy of sciences of the USA* **92**: 4332–4336.

Deutsch, D. D. and Chin, S. A. (1993) "Enzymatic synthesis and degradation of anandamide, a cannabinoid receptor agonist," *Biochemical Pharmacology* **46**: 791–796.

Devane, W. A., Hanus, L., Breuer, A., Pertwee, R. G., Stevenson, L. A., Griffin, G., Gibson, D., Mandelbaum, A., Etinger, A. and Mechoulam, M. (1992) "Isolation and structure of a brain constituent that binds to the cannabinoid receptor," *Science* **258**: 1946–1949.

Devane, W. A. and Axelrod, J. (1994) "Enzymatic synthesis of anandamide, the endogenous ligand for the cannabinoid receptor, by brain membrane," *Procceedings of the National Academy of sciences of the USA* **91**: 6698–6701.

Dewey, W. L. (1986) "Cannabinoid pharmacology," *Pharmacological Reviews* **38**: 151–178.

Di Marzo, V., Fontana, A., Cadas, H., Schinelli, S., Cimino, G., Schwartz, J. C. and Piomelli, D. (1994) "Formation and inactivation of endogenous cannabinoid anandamide in central neurons," *Nature* **372**: 686–691.

Ducibella, T. and Anderson, E. (1975) "Cell shape and membrane change in the eight-cell mouse embryo: prerequisites for morphogenesis of the blastocyst," *Developmental Biology* **47**: 45–48.

Felder, C. C., Veluz, J. S., Williams, H. L., Briley, E. M. and Matsuda, L. A. (1992) "Canna-binoid agonists stimulate both receptor- and non-receptor-mediated signal transduction pathways in cells transfected with and expressing cannabinoid receptor clones," *Molecular Pharmacology* **42**: 838–845.

Howlett, A. C. (1995) "Pharmacology of cannabinoid receptors," *Annual Review of Pharmacology and Toxicology* **35**: 607–634.

Jones, J. and Schultz, R. M. (1990) "Pertussis toxin-catalyzed ADP-ribosylation of a G-protein in mouse oocytes, eggs and preimplantation embryos: developmental changes and possible functional roles," *Developmental Biology* **139**: 250–262.

Kaminski, N. E., Abood, M. E., Kessler, F. K., Martin, B. R. and Schatz, A. R. (1992) "Iden-tification of a functionally relevant cannabinoid receptor on mouse spleen cells that is involved in cannabinoid-mediated immune modulation," *Molecular Pharmacology* **42**: 736–742.

Kruszka, K. K. and Gross, R. W. (1994) "The ATP- and CoA-independent synthesis of arachidonoylethanolamide," *Journal of Biological Chemistry* **269**: 14345–14348.

Manejwala, F., Kaji, E. and Schultz, R. M. (1989) "Development of activatable adenylate cyclase in the preimplantation mouse embryo and role for cyclic AMP in blastocoel formation," *Cell* **46**: 95–103.

Martin, B. R., Compton, D. R., Prescott, W. R., Barrett, R. L. and Razdan, R. K. (1995) "Pharmacological evaluation of dimethylheptyl analogs of delta 9-THC: reassessment of the putative three-point cannabinoid-receptor interaction," *Drug Alcohol Dependance* **37**: 231–240.

Matsuda, L. A., Lolait, S. J., Brownstein, M. J., Young, A. C. and Bonner, T. I. (1990) "Structure of a cannabinoid receptor and functional expression of the cloned cDNA," *Nature* **346**: 561–564.

Mechoulam, R., Ben Shabat, S., Hanus, L., Ligumsky, M., Kaminski, N. E., Schatz, A. R., Gopher, A., Almog, S., Martin, B. R., Compton, D. R., Pertwee, R. G., Griffin, G., Bayewitch, M., Barg, J. and Vogel, Z. (1995) "Identification of an endogenous 2-monoglyceride, present in canine gut, that binds to cannabinoid receptors," *Biochemical Pharmacology* **50**: 83–90.

Munro, S., Thomas, K. L. and Abu-Shaar, M. (1993) "Molecular characterization of a peripheral receptor for cannabinoids," *Nature* **365**: 61–65.

Nahas, G. and Latour, C. (1992) "The human toxicity of marijuana," *The Medical Journal of Australia* **156**: 495–497.

Poueymirou, W. T. and Schultz, R. M. (1989) "Regulation of mouse preimplantation embryo development: inhibition of synthesis of proteins in the two-cell embryos that require transcription by inhibitors of cAMP-dependent protein kinase," *Developmental Biology* **133**: 588–599.

Pakrasi, P. L. and Dey, S. K. (1984) "Role of calmodulin in blastocyst formation in the mouse," *Journal of Reproduction Fertilty* **71**: 513–517.

Paria, B. C., Kapur, S. and Dey, S. K. (1992) "Effects of 9-ene-tetrahydrocannabinol on uterine estrogenicity in the mouse," *The Journal of Steroid Biochemistry and Molecular biology* **42**: 713–719.

Paria, B. C., Das, S. K. and Dey, S. K. (1995) "The preimplantation mouse embryo is a target for cannabinoid ligand-receptor signaling," *Proceedings of the National Academy of Sciences of the USA* **92**: 9460–9464.

Paria, B. C., Deutsch, D. D. and Dey, S. K. (1996) "The uterus is a potential site for anandamide synthesis and hydrolysis: differential profiles of anandamide synthase and hydrolase activities in the mouse uterus during the periimplantation period," *Molecular Reproduction and Development* **45**: 183–192.

Paria, B. C., Ma, W., Andrenyak, D. M., Schmid, P. C., Schmid, H. H. O., Moody, D. E., Deng, H., Makriyannis, A. and Dey, S. K. (1998) "Effects on preimplantation mouse embryo development and implantation are mediated by brain-type cannabinoid receptors," *Biology of Reproduction* **58**: 1490–1495.

Paria, B. C., Das, S. K. and Dey, S. K. (1999) "Cannabinoid ligand-receptor signaling during early pregnancy in the mouse," in *Marihuana and Medicine*, edited by G. G. Nahas, K. M. Sutin, D. J. Harvey and S. Agurell, pp. 393–409. Humana Press: New Jersey.

Rosenkrantz, H. (1979) In *Marihuana: Biological effects*, edited by G. G. Nahas and W. D. Paton, pp. 479–499. Oxford: Pergamon.

Schmid, P. C., Paria, B. C., Krebsbach, R. J., Schmid, H. H. and Dey, S. K. (1997) "Changes in anandamide levels in mouse uterus are associated with uterine receptivity for embryo implantation," *Proceedings of the National Acadamy of Sciences of the USA* **94**: 4188–4192.

Schuel, H., Goldstein, E., Mechoulam, R., Zimmerman, A. M. and Zimmerman, S. (1994) "Anandamide (arachidonoylethanolamide), a brain cannabinoid receptor agonist, reduces sperm fertilizing capacity in sea urchins by inhibiting the acrosome reaction," *Proceedings of the National Acadamy of Sciences of the USA* **91**: 7678–7682.

Smith, C. G. and Asch, R. H. (1987) "Drug abuse and reproduction," *Fertility and Sterility* **48**: 355–373.

Song, C. and Howlett, A. C. (1995) "Rat brain cannabinoid receptors are *N*-linked glycosylated proteins," *Life Sciences* **56**: 1983–1989.

Stella, N., Schweitzer, P. and Piomelli, D. (1997) "A second endogenous cannabinoid that modulates long-term potentiation," *Nature* **388**: 773–777.

Sugiura, T., Kondo, S., Sukagawa, A., Nakane, S., Shinoda, A., Itoh, K., Yamashita, A. and Waku, K. (1995) "2-arachidonoylglycerol: a possible endogenous cannabinoid receptor ligand in brain," *Biochemical and Biophysical Research Communications* **215**: 89–97.

Sugiura, T., Konda, S., Sukagawa, A., Tonegawa, A., Nakane, S., Yamashita, A. and Waku, K. (1996). "Enzymatic synthesis of anandamide, an endogenous cannabinoid receptor ligand, through *N*-acylphosphatidylethanolamine pathway in testis: involvement of Ca^{++} dependent transacylase and phosphodiesterase activities," *Biochemical and Biophysical Research Communications* **218**: 113–117.

Ueda, N., Kurahashi, Y., Yamamoto, S. and Tokunaga, T. (1995) "Partial purification and characterization of the porcine brain enzyme hydrolyzing and synthesizing anandamide," *The Journal of Biological Chemistry* **270**: 23823–23827.

Wang, J., Paria, B. C., Dey, S. K. and Armant, D. R. (1999) "Stage-specific excitation of cannabinoid receptor exhibits differential effects on mouse embryonic development," *Biology of Reproduction* **60**: 839–844.

Yang, Z.-M., Paria, B. C. and Dey, S. K. (1996) "Activation of brain-type cannabinoid receptors interfere with preimplantation mouse embryo development," *Biology of Reproduction* **55**: 756–761.

Antiemetic action of Δ^9-tetrahydrocannabinol and synthetic cannabinoids in chemotherapy-induced nausea and vomiting

Nissar A. Darmani

ABSTRACT

In the last two decades, there have been considerable advances in our understanding of emetic circuits and the mechanisms by which chemotherapeutics produce emesis. Cancer chemotherapy is often accompanied by severe nausea and vomiting, which can lead to refusal or delay of treatment by patients. A number of different classes of antiemetics have been employed clinically to control chemotherapy-induced emesis. These include: dopamine D_2 antagonists (butyrophenones, phenothiazines), cannabinoids, corticosteroids, substituted benzamides, serotonin $5-HT_3$- and tachykinin NK_1-antagonists. Of these, serotonin $5-HT_3$ antagonists appear to be most effective for the prevention of the acute phase of chemotherapy-induced emesis. Recent clinical trials suggest that NK_1 antagonists have broadspectrum antiemetic properties controlling both the acute and delayed phases of emesis in cancer patients.

In this chapter, the findings of 40 clinical trials concerning the antiemetic properties of three well established cannabinoids (Δ^9-THC, nabilone and levonantradol) are discussed. Since most of these studies differ in study design and suffer from methodological inadequacies, full comparison and interpretation of the attained results is difficult. In the main, it is clear that these cannabinoids possess significant antiemetic properties in patients receiving chemotherapy. Both Δ^9-THC and its tested synthetic analogs are superior to placebo, and equivalent or superior to prochlorperazine, but may be inferior to high-dose metoclopramide as antiemetic. The antiemetic efficacies of these cannabinoids relative to more selective and potent $5-HT_3$ antagonists or NK_1 antagonists in cancer patients receiving chemotherapy remains to be investigated. As yet, the tested synthetic cannabinoids offer neither a superior therapeutic index nor a better profile of side effects than the naturally occurring cannabinoid, Δ^9-THC. Furthermore, there is no clear evidence that smoked marijuana is a more efficacious antiemetic than the oral form of Δ^9-THC. Side effects of these cannabinoids are of moderate nature and are generally well tolerated, but limit cannabinoid use in the elderly or when high doses are administered.

Until recently, the receptor mechanism by which structurally diverse cannabinoids produce their antiemetic action was not known. Studies from this laboratory have shown that low to moderate doses of the CB_1 receptor antagonist SR141716A (1–5 mg/kg) and not the CB_2 antagonist SR144528, reverses the antiemetic effects of Δ^9-THC (Darmani, 2001c) and WIN55,212-2 (Darmani, 2001e) against cisplatin-induced vomiting in the least shrew (*Cryptotis parva*). In addition, larger doses of SR141716A (≥ 10 mg/kg) were shown to induce emesis in the least shrew in a dose- and route-dependent manner (Darmani, 2001a). These findings suggest an important role of endocannabinoids in vomiting circuits. Indeed, the endocannabinoid 2-arachidonoylglycerol (2-AG) is a potent emetogenic agent and its emetic effects can be blocked by diverse cannabinoid agonists as well as the CB_1 receptor antagonist SR141716A (Darmani, 2001d). The emetic action of the endocannabinoid 2-AG and the antiemeic effects of xenobiotic cannabinoids are mediated via cannabinoid CB_1 receptors. As revealed in this chapter cannabinoids appear to be broad-spectrum antiemetics in animal models of vomiting and may therefore be useful in other clinical emesis settings besides prevention of chemotherapy-induced nausea and vomiting.

Key Words: marijuana, Δ^9-tetrahydrocannabinol, CP55,940, WIN 55, 212-2, SR141716A, emesis, CB_1 receptor, clinical trial, nabilone, levonantradol

INTRODUCTION

Nausea and vomiting are the common side effects associated with cancer chemotherapy that profoundly affects the patient's quality of life and may lead to refusal of further chemotherapy treatment (Coats *et al.*, 1983). In the interest of enhancing patient compliance and for humanitarian reasons, it is essential to develop new drugs which can prevent the discomfort of chemotherapy without decreasing its effectiveness. In the past two decades, there have been significant advances in the understanding of the mechanisms by which chemotherapy produces emesis. This new knowledge has led to the introduction of clinically useful antiemetics such as serotonin 5-HT_3 receptor antagonists in early 1990s, which has revolutionized the treatment of the acute phase of chemotherapy-induced nausea and vomiting. However, 5-HT_3 antagonists seem to be ineffective in the treatment of the delayed phase of emesis (Cubeddu, 1996; Hesketh, 1996). More broadspectrum antiemetics, such as tachykinin NK_1 receptor antagonists currently under clinical trials, seem to be effective for prevention of both phases of emesis produced by chemotherapeutics (Navari *et al.*, 1999). Other classes of useful antiemetics for patients receiving chemotherapy are: dopamine D_2 antagonists (such as butyrophenones and phenothiazines), corticosteroids, substituted benzamides and cannabinoids (Mitchelson, 1992).

At least six different cannabinoids have been evaluated for their antiemetic properties in clinical trials. These include: smoked marijuana, the most active constituent of marijuana plant, Δ^9-tetrahydrocannabinol (Δ^9-THC), the delta-8 isomer of Δ^9-THC (Δ^8-THC), and synthetic cannabinoids such as nabilone, levonantradol and nonabine. Most of the studies evaluating the antiemetic effects Δ^9-THC have used the oral form of Δ^9-THC rather than smoked marijuana. The use of smoked marijuana as a therapeutic agent is presently a matter of extensive debate in the United States. Efforts to legalize smoked marijuana as medicine have raised many medical, ethical, legal and political issues. Marijuana's potential as medicine *per se* is seriously undermined by the fact that people smoke it, thereby increasing their chance of cancer and respiratory diseases. In addition, there is no scientific evidence that the antiemetic efficacy of smoked marijuana is superior to its oral form. Despite the side effects of Δ^9-THC and other cannabinoids delivered in any form, significant evidence supports the selective use of these cannabinoids for the treatment of nausea and vomiting in some patients treated with chemotherapy. As discussed in this chapter, advances in cannabinoid biology and pharmacology in the last decade provide opportunities for the development of potent synthetic cannabinoids that lack significant psychoactivity but may possess more potent antiemetic as well as other therapeutic benefits. Furthermore, recent animal emesis studies in this laboratory provide new insights in the mechanism of antiemetic action of Δ^9-THC and related synthetic agents which may lead to the introduction of more useful cannabinoid antiemetics.

MECHANISMS OF CHEMOTHERAPY-INDUCED NAUSEA AND VOMITING AND THE CURRENT STATUS OF ANTIEMETICS

In order to understand how antiemetics prevent nausea and vomiting, it is essential to briefly review the mechanisms by which emesis occurs. Emesis is a complex reflex that is developed to different degrees in diverse species. It requires coordination by the vomiting center (VC) which is located in the reticular formation of the medulla (Brunton, 1996). The VC is a collection of recipient and effector nuclei that includes part of the nucleus tractus solitarius (NTS), the dorsal motor nucleus of the vagus nerve, and the area of postrema which contains the chemoreceptor trigger zone (CTZ) in the floor of the fourth ventricle. The VC receives input from the CTZ, from the vestibular apparatus, from higher brain and cortical structures, and from visceral afferents that originate from areas such as the heart, testis and the gastrointestional tract. The CTZ is accessible to emetic substances (including humoral factors) in the circulation as it lacks a complete blood–brain barrier. The initial manifestation of the emetic response often involves nausea, in which both gastric tone and gastric peristalsis are reduced or absent; and the tone of the duodenum and upper jejunum is increased such that their contents reflux. Ultimately the upper portion of the stomach relaxes while the pylorus constricts, and the coordinated contraction of the diaphragm and abdominal muscles leads to expulsion of gastric contents.

Role of neurotransmitters involved in chemotherapy-induced emesis

Several well-established neurotransmitters (acetylcholine, dopamine, histamine and serotonin) are known to act via their specific receptors to induce emesis. Indeed, selective activation of serotonergic 5-HT_3-, dopaminergic D_2- or muscarinic M_1 receptors in both CTZ and NTS (and also histamine H_1 receptors in the NTS), can induce vomiting. More recently, it has been suggested that the neuropeptide, substance P, also plays an important role in the emetic reflex (Bountra *et al.*, 1996). Several classes of drugs, acting as antagonists of the cited receptors, alleviate the symptoms of nausea or vomiting. However, none of these antagonists is completely effective in all cases of vomiting. Thus, the choice of an antiemetic agent depends upon the etiology of nausea and emesis. Chemotherapy-induced nausea and vomiting are commonly classified as anticipatory, acute or delayed. Anticipatory nausea and vomiting (ANV) are learned responses to chemotherapy that develop in up to 25% of patients. Chemotherapeutic agents such as cisplatin cause an intense acute phase of emesis in patients which is complete within the first 12 h of administration. The term delayed emesis describes vomiting which occurs over 2–5 days postchemotherapy treatment (Cubeddu, 1996). Furthermore, the receptor mechanisms which cause these types of nausea and vomiting are thought to be different.

Since antihistamine and anticholinergic agents are ineffective in the treatment of chemotherapy-induced emesis, only the serotonergic, dopaminergic and tachykinin emetic mechanisms will be discussed further. The majority of the body's serotonin (5-hydroxytryptamine = 5-HT) is contained in the enterochromaffin (EC) or mast cells of the gastrointestinal tract (Gaginella, 1995). As these cells are in close proximity to $5\text{-}HT_3$ receptors on the vagal afferent terminals, it has been postulated that chemotherapeutics induce emesis in both animals and man by releasing 5-HT from EC cells, which then stimulates vagal afferents to initiate the emetic reflex (Reviews: Naylor and Rudd, 1996; Cubeddu, 1996). Furthermore, drugs that reduce the presynaptic stores of 5-HT, decrease the emetic response in both animals and man (Andrews et al., 1992; Barns et al., 1988; Cubeddu, 1996). Although there are more than 14 different serotonin receptor subtypes, only $5\text{-}HT_3$- and possibly $5\text{-}HT_4$-receptors are involved in emesis induced by chemotherapeutic agents (Barns and Sharp, 1999). There is strong clinical and experimental evidence that $5\text{-}HT_3$ receptor antagonists are effective antiemetics for the acute phase of chemotherapy-induced vomiting (Cubeddu, 1996).

The cited evidence in conjunction with the findings that abdominal vagotomy and splanchnectomy abolish cytotoxic-induced emesis, suggest that chemotherapeutic drugs have a peripheral site of action (Andrews et al., 1990a,b). However, several other studies point to a central site of action since direct microinjections of $5\text{-}HT_3$ receptor antagonists into the CNS can prevent cisplatin-induced emesis in ferrets (Higgins et al., 1989; Yoshida et al., 1992), dogs (Gidda et al., 1995) or cats (Smith et al., 1988). In addition, complete inhibition of cisplatin-induced emesis was observed following ablation of the area of postrema in cats (McCarthy and Borison, 1984) and dogs (Bhandari et al., 1989). Furthermore, a quaternary $5\text{-}HT_3$ receptor antagonist zatosteron-quat, with limited access to the CNS, failed to prevent cisplatin-induced emesis in the dog (Gidda et al., 1995). Moreover, zatosteron-quat appears to potently block the induced vomiting following its intracerebroventricular injection suggesting that the antiemetic action of $5\text{-}HT_3$ receptor antagonists in the dog lies within the CNS. Interestingly, peripheral administration of quaternary forms of either zatosteron or tropisetron can prevent cisplatin-induced emesis in the ferret (Gidda et al., 1991). This indicates a peripheral site of action is more important in this species. Furthermore, the peripherally acting quaternary form of 5-HT (5-HTQ) can induce emesis in the shrew more potently and more rapidly relative to the selective $5\text{-}HT_3$ agonist mCPBG which should penetrate the CNS more rapidly (Darmani, 1998). In this context, central injection of the $5\text{-}HT_3$ receptor agonist 2-methyl-5-HT, does not induce emesis in the ferret (Higgins et al., 1989). Thus, it seems that 5-HT probably acts in a synergistic manner on both central and peripheral $5\text{-}HT_3$ sites to induce emesis, whereas antagonism of one or both of these sites may prevent the induced emesis.

A role for the dopaminergic D_2 receptors for the induction of emesis in patients receiving chemotherapy rests largely on the antiemetic effectiveness of various dopamine D_2 antagonists. Their antiemetic action seems to be due to antagonism of D_2 receptors in the CRTZ (Reviews: Mitchelson, 1992; Allan, 1992). However, they are not completely effective clinically, particularly against cisplatin-induced emesis. Recent basic evidence indicates that the D_2 site is not the only dopamine receptor involved in the production of emesis and dopamine D_3 sites may also have an important role (Darmani *et al.*, 1999; Yoshida *et al.*, 1992; Yoshikawa *et al.*, 1996). However, the contribution of D_3 receptors in the emetic action of chemotherapeutic agents remains unknown. Although currently 5-HT_3 receptor antagonists are the most useful agents for the prevention of the acute phase of chemotherapy-induced vomiting, they are not particularly effective for the delayed phase of the induced emesis (Cubeddu, 1996; Hesketh, 1996; Verweij *et al.*, 1996; Ossi *et al.*, 1996). On the other hand, addition of dopamine D_2 antagonists seem to potentiate the clinical antiemetic effectiveness of 5-HT_3 antagonists in both acute and delayed emesis (Herrstedt *et al.*, 1993; Herrstedt and Dombernowsky, 1994). Addition of corticosteroids have also been shown to potentiate the antiemetic efficacy of 5-HT_3 antagonists (Munstedt *et al.*, 1999; Perez, 1998).

The tachykinins constitute a family of neuropeptides comprising substance P, neurokinin A and neurokinin B (Betancur *et al.*, 1997). Tachykinins interact with at least three receptor subtypes termed NK_1, NK_2 and NK_3. Substance P binds preferentially to NK_1 receptor, whereas neurokinin A and neurokinin B are preferred endogenous ligands for NK_2 and NK_3 receptors respectively. Since the mid 1990s, it has been shown that several selective NK_1 antagonists possess broadspectrum antiemetic action against a wide variety of emetogens including chemotherapeutics in several animal models of emesis (Bountra *et al.*, 1996). They are also effective in preventing both the acute and delayed forms of emesis in patients receiving chemotherapy (Kris *et al.*, 1997; Navari *et al.*, 1999), or recovering from anesthesia and surgery (Diemunsch *et al.*, 1999). Thus, NK_1 antagonists seem to have the advantage of broadspectrum antiemetic action as well as preventing chemotherapy-induced delayed emesis which the 5-HT_3 antagonists poorly control.

ANTIEMETIC PROPERTIES OF CLINICALLY-USEFUL CANNABINOIDS

During 1970s and 1980s, dopamine D_2 antagonists were the mainstay of antinauseant-antiemetic agents in cancer patients receiving chemotherapy. However, they had limited clinical success. Because of anecdotal reports by young cancer patients that smoking marijuana alleviated the nausea and vomiting caused by chemotherapeutic agents, both government and industry sponsored clinical trials were initiated to test the antiemetic potential of cannabinoids. These trials began in 1975 and have continued up to 1995 (Tables 13.1–13.3; Abrahamov *et al.*, 1995; Archer *et al.*, 1983; Staquet *et al.*, 1981). A recent MEDLINE search yielded 194 titles on the antiemetic properties of marijuana and cannabinoids (Voth and Schwartz, 1997). This literature suggests that Δ^9-THC is a useful antiemetic for nausea and vomiting associated with cancer chemotherapy. Thus far, the following cannabinoids have been tested for their antiemetic efficacy against chemotherapy-induced nausea and vomiting in clinical trials: smoked marijuana, Δ^9-THC, Δ^8-THC, nabilone, levonantradol and nonabine (Tables 13.1–13.3; Abrahamov *et al.*, 1995; Archer *et al.*, 1983). Except for Δ^9-THC (dronabinol), none of these agents is currently clinically available in the United States. The clinical success of 5-HT_3 receptor antagonists in 1990s and the current advances in the prevention of chemotherapy-induced nausea and emesis with NK_1 antagonists have virtually frozen further antiemetic clinical research with cannabinoids. Unlike dopamine D_2-, serotonin 5-HT_3- and tachykinin NK_1-receptor antagonists, the cited cannabinoids act as agonists of cannabinoid CB_1 and CB_2 receptors. Moreover, the mechanism(s) by which cannabinoids produce their clinical antiemetic effect(s) is currently unknown. Cannabinoids with their recently discovered potent analogs and endogenous ligands, provide a unique potential for the further development of new antiemetic agents in our present armamentarium to combat chemotherapy-induced emesis. The following will initially summarize the important findings of clinical trials regarding the antiemetic efficacy and the potential side effects of most often clinically studied cannabinoids such as Δ^9-THC, nabilone and levonantradol. Secondly, the possible mechanisms by which cannabinoids prevent emesis as revealed by animal models of vomiting will be discussed.

Clinical trials with Δ^9-THC

Δ^9-THC is the most extensively studied antiemetic cannabinoid. Table 13.1 summarizes the antiemetic efficacy of orally administered Δ^9-THC against chemotherapy-induced nausea and vomiting from 15 clinical trials. In these studies, patients were undergoing treatment with a variety of chemotherapeutic agents for a plethora of tumors and haematologic malignancies. The administered doses of Δ^9-THC were between 5–15 mg/m^2 and its first dose was administered 1–24 h prior to the start of chemotherapy. Subsequent doses of Δ^9-THC were given at fixed intervals before, during and after chemotherapy. Approximately 80% of the studies were randomized in a double-blind crossover design. The total number of patients who completed the cited clinical trials is 749 and most were adult males and females. However, the median age tended to vary in different trials. For example, Ekert *et al.* (1979) tested children and young adults (5–19 years), whereas Frytak and colleagues' (1979) patient population were mainly elderly (median age 61 years). Although these clinical trials clearly differ in study design and techniques for emesis evaluation, they do provide insight into the antiemetic potential of Δ^9-THC.

Sallan and coworkers (1975), using a randomized, double-blind, placebo-controlled trial, were the first group to show that oral Δ^9-THC has significant antiemetic efficacy relative to placebo (p < 0.001). Of the 15 treatment cycles, no vomiting occurred during 5 courses of chemotherapy and at least 50% reduction occurred during 7 other courses. In the remaining 3 courses, less than 50% protection was observed. Its antiemetic activity correlated with a subjective "psychological high" since no patient vomited while experiencing elation, easy laughing and heightened awareness. The association between "psychological high" and antiemetic responses has also been noted in other studies (Chang *et al.*, 1979; Lucas and Laszlo, 1980; Orr *et al.*, 1980; Sallan *et al.*, 1980). However, attainment of "high" was not associated with antiemetic efficacy in a study by Frytak and colleagues (1979). Moreover, Garb *et al.* (1980) have shown that co-administration of prochlorperazine prevents Δ^9-THC's "high" but not its antiemetic action. Sallan *et al.*'s study (1975) showed that Δ^9-THC can be safely administered orally at 10 mg dosage every 4 h for at least three doses. In a similar study design, Chang and coworkers (1979) confirmed the superiority of oral plus inhaled Δ^9-THC over placebo in patients with osteogenic sarcoma who were receiving high-dose methotrexate (p < 0.001). Furthermore, the antiemetic activity was shown to correlate with both Δ^9-THC's plasma concentration and its

Table 13.1 Clinical trials determining the antiemetic efficacy of orally-administered Δ^9-tetrahydrocannabinol

Investigator	Patients evaluated	Median age and range in years	Chemotherapy	Study design[a]	Antiemetic[b]	Results[b]
Lane et al. (1990)	19	55 (25–64)	Various	R Db	THC oral 10 mg, 4 × for 2–6 days. First dose 24 h before chemotherapy vs. PCP oral 10 mg, 4 × for 2–6 days vs. THC + PCP, 4 × for 2–6 days.	THC + PCP ≥ THC > PCP in reducing both the duration of nausea and vomiting (p < 0.01).
Gralla et al. (1984)	30	58 (45–70)	Mainly cisplatin	R C Db	THC oral 10 mg/m² every 3 h, 5 ×. First dose 1.5 h before cisplatin vs. MCP i.v. 2 mg/kg, every 3 h, 5 ×.	MCP > THC in attenuating vomiting episodes (p < 0.02).
Ungerleider et al. (1982)	214	47 (18–82)	Various	R C Db	THC oral 7.5–12 mg, 4 ×. 1st dose 7h prior chemotherapy and then every 4 h vs. PCP oral 10 mg, drug delivery as above.	THC = PCP.
Chang et al. (1981)	8	41 (17–58)	Adriamycin + Cytoxan	R C Db	THC oral 10 mg/m², every 3 h, 5 ×. Patients whom vomited also received THC cigarettes. vs. placebo.	THC = placebo.
Neidhart et al. (1981)	52	43	Various	Pr R C Db	THC oral 10 mg, 8 ×, every 3–4 h vs. haloperiodol 2 mg, 8 × every 3–4 h.	THC = haloperiodol.
Orr and McKernan (1981)	55	46 (22–71)	Various	R C Db	THC oral 7 mg/m² every 4 h, 4 × vs. PCP oral 7 mg/m² every 4 h, 4 × vs. placebo. First dose given 1h before chemotherapy.	THC > PCP > Placebo (p < 0.005).
Sweet et al. (1981)	25	51 (22–67)	Various	Uncontrolled pilot study	THC oral 5 mg/m² every 8 h, 6 ×, first dose started 24 h prior to chemotherapy.	Less nausea and vomiting.
Garb et al. (1980)	10	Adults	Various	R C Db two pronged study Open label	THC + PCP (or thioethylperazine) oral 10mg, every 6 h, 8 ×. First dose started 24 h before chemotherapy vs. PCP + vehicle. Doses of THC and PCP were gradually increased, starting 24–48 h prior chemotherapy and during first day of chemotherapy.	THC + PCP > PCP + vehicle. Higher doses of THC + phenothiazines gave better protection against vomiting.

Reference	N	Age	Chemotherapy	Study design[a]	THC regimen[b]	Results
Lucas and Laszlo (1980)	53	Adults	Various	Nonrandomized, Hc	THC oral 15 mg/m^2, every 6 h, 4 x, starting 1h prior chemotherapy. Changed to 5 mg/m^2, every 4h, 11 x, starting 12h before to 24h after.	THC effective in 72% of patients but complete blockade in 19%.
Sallan et al. (1980)	84	32.5 (9–70)	Various	R C Db	THC oral 10 or 15 mg/m^2 every 4h, 3 x, starting 1h before chemotherapy vs. PCP oral 10 mg, delivery schedule as above.	THC > PCP. Younger patients responded better (p < 0.004).
Chang et al. (1979)	15	24 (15–49)	Methotrexate with leucovorin	P C R Db	THC oral, 10 mg/m^2, every 3 h, 5 x, first dose 2 h before chemotherapy. THC also given via inhalation.	THC > placebo (p < 0.001). Correlated with plasma THC concentration.
Ekert et al. (1979)	33	5–19, children	Various	R C Db	THC oral 10 mg/m^2, 4 x, first dose 2 h before chemotherapy and every 8 h following chemotherapy vs. MCP oral 5–10 mg, delivery schedule as above vs. PCP oral 5–10 mg, drug delivery as above.	THC superior to both MCP and PCP.
Frytak et al. (1979)	116	61 (21–70)	5-fluorouracil + semustine	R Db	THC oral 15 mg, on day 1, 2 h before chemotherapy, then at 2 and 8 h after chemotherapy. For 3 additional days 3 x daily vs. PCP oral 10 mg, delivery schedule as above vs. placebo.	THC = PCP > placebo.
Kluin-Neleman et al. (1979)	11	36.4	Various	R C Db	THC oral 10 mg/m^2, 3 x, first dose 2 h before chemotherapy and then repeated 4 and 8 h later vs. placebo.	THC > placebo.
Sallan et al. (1975)	20	29.5 (18–76)	Various	R C Db	THC oral 10 mg/m^2, 3 x, every 4 h, first dose 2 h THC > placebo (p < 0.001).	THC > placebo (p < 0.001).

^{a} For study design: C = crossover, Db = double-blind, Hc = historical control, Pr = prospective, R = randomized; ^{b} For drugs: DXM = dexamethazone, MCP = metoclopramide, PCP = prochlorperazine, THC = Δ^9-tetrahydrocannabinol.

induced "high". However, the study of Frytak *et al.* (1979) failed to show a correlation between antiemetic activity and plasma concentration of Δ^9-THC. Another important finding of Chang *et al.*'s study (1979) is that repeated administration of Δ^9-THC may cause development of tolerance to its antiemetic properties. Similar trends toward development of antiemetic tolerance have been anecdotally reported (Sallan *et al.*, 1980; Sweet *et al.*, 1981). A further Δ^9-THC/placebo trial has also confirmed the antiemetic potential of oral Δ^9-THC (Kluin-Neleman *et al.*, 1979). However, in a randomized, double-blind, placebo-controlled trial in cancer patients receiving adriamycin and cytoxan chemotherapy, Δ^9-THC failed to produce significant antiemetic activity (Chang *et al.*, 1981). Although this finding suggests that Δ^9-THC is probably an effective antiemetic against specific chemotherapeutics, a recent survey of clinical trials has found no antiemetic pattern for Δ^9-THC for any particular type of tumor or chemotherapy (Voth and Schwartz, 1997).

Several studies have evaluated the antiemetic efficacy of Δ^9-THC relative to a phenothiazine such as prochlorperazine. At this period of time, prochlorperazine was considered to be one of the more effective antiemetics. However, lack of satisfactory efficacy of prochlorperazine led to the clinical search for better antiemetics. In a randomized double-blind study, Frytak *et al.* (1979) showed that although oral Δ^9-THC was significantly more effective than placebo, its antiemetic efficacy only equaled prochlorperazine. A similar conclusion was reached by the study of Ungerleider and coworkers (1982). However, three other studies have clearly demonstrated the antiemetic superiority of Δ^9-THC over prochlorperazine (Ekert *et al.*, 1979; Orr *et al.*, 1980; Sallan *et al.*, 1980). This inconsistency among the discussed studies is suggested to be due to age differences among the patients in the cited trials. Indeed, patients in the latter studies were either children (5–19 years) or mainly young adults with a median age of less than 33 years, whereas the former studies involved older patients. Two other trials have investigated the combined antiemetic efficacy of Δ^9-THC and prochlorperazine versus each agent administered with a placebo (Garb *et al.*, 1980; Lane *et al.*, 1990). Addition of prochlorperazine not only attenuated the side effects of Δ^9-THC, but also potentiated its antiemetic efficacy ($p < 0.01$). Indeed, several parameters of nausea and vomiting (duration, frequency and intensity) were reported to be significantly less in the combination antiemetic therapy.

In early 1980s, metoclopramide was used in conventional doses as an antiemetic agent for a wide range of indications (Raybold *et al.*, 1995). However, at low to moderate doses, metoclopramide had limited antiemetic efficacy against chemotherapy regimens. Its mechanism of action was ascribed to blockade of dopamine D_2 receptors in the brainstem. Ekert *et al.* (1979) compared the antiemetic efficacy of oral Δ^9-THC ($10\,mg/m^2$) to oral low-dose metoclopramide (5–10 mg) and oral prochlorperazine (5–10 mg) in cancer patients receiving chemotherapy. They found that the antiemetic property of Δ^9-THC was greater than metoclopramide and prochlorperazine. Metoclopramide is also a moderate antagonist of serotonin 5-HT$_3$ receptors, and at high-dose regimens (2–3 mg, i.v.) is an effective antiemetic against cisplatin-induced vomiting when compared with either placebo or prochlorperazine (Gralla, 1983). A comparative, randomized and double-blind study of high-dose metoclopramide versus Δ^9-THC revealed significant differences favoring the metoclopramide-treated group (Gralla *et al.*, 1984). Indeed, complete control of emesis occurred in 47% of metoclopramide-treated patients, whereas only 13% of

Δ^9-THC-exposed patients did not experience vomiting (p < 0.02). Furthermore, the mean number of emetic episodes were 2 and 8 respectively (p < 0.01). As yet, there has been no comparative study to evaluate the clinical benefits of Δ^9-THC relative to selective and more potent 5-HT$_3$ antagonists. As discussed earlier, although 5-HT$_3$ receptor antagonists are the most useful agents for prevention of the acute phase of chemotherapy-induced emesis, the delayed phase of this emesis is not responsive to such drugs. In this context, one interesting advantage of cannabinoids is that many of the patients who are protected during the acute phase of emesis also respond well during the delayed phase (Abrahamov et al., 1995; Chan et al., 1987; Dalzell et al., 1986). Thus, it will be of interest to determine the combined antiemetic effects of Δ^9-THC and selective serotonin 5-HT$_3$ antagonists.

Although oral Δ^9-THC clinical trials stemmed from anecdotal reports of antiemetic effectiveness of inhaled marijuana, only two studies have investigated the antiemetic potential of smoked marijuana. In an uncontrolled trial, Vinciguera and colleagues (1988) found that smoked marijuana was effective in patients in whom available conventional antiemetics had failed. However, smokers were required to completely smoke four cigarettes per day and to inhale each puff deeply and then hold the smoke for 10 sec. More than 20% of patients dropped out of the smoking group before the end of the study and 22% of the remaining patients reported no benefit in this self-rating study. In an earlier, randomized, double-blind, and crossover study, comparing oral Δ^9-THC with smoked marijuana, Levitt et al. (1984) found that oral Δ^9-THC was more effective for nausea and vomiting than smoked marijuana in 35% versus 20% of patients. However, a larger percentage of patients (45%) expressed no preference for either agent.

A more recent open label pediatric clinical trial has taken a different approach by using Δ^8-THC which is a double bond isomer of Δ^9-THC (Abrahamov et al., 1995). Δ^8-THC is less psychotropic and more stable than Δ^9-THC as it does not oxidize to cannabinol and thus has a very long shelf life. Δ^8-THC was given orally (18 mg/m^2) two hours before the start of chemotherapy to 8 children (3–13 years) with various haematologic malignancies. The dose was repeated every 6 h for 24 h. Prevention of nausea and vomiting was complete, regardless of the antineoplastic therapy. Furthermore, no delayed vomiting occurred and few side effects were observed. Although the number of pediatric cancer patients was small in this study, the total number of treatments is considerable (480 times) as most patients underwent several treatment cycles.

Clinical trials with nabilone

Nabilone is a synthetic cannabinoid which was developed by the Lilly laboratories. The general structure of nabilone is similar to Δ^9-THC as they are both dibenzopyrans, with a dimethyl at the position 6, and a hydroxyl at the position 1. However, nabilone differs from Δ^9-THC as it contains a dimethylheptyl side chain (instead of pentyl) at the position 3, and a ketone (instead of methyl) in the position 9. In both animals (Razdan, 1986) and humans (Archer et al., 1986), the psychoactivity of nabilone is approximately 10 times greater than Δ^9-THC, but it possesses a similar spectrum of activity. Nabilone is marketed under the trade name Cesamet in Austria, Canada and the U.K. for the control of nausea and emesis associated with chemotherapy. Although once marketed in the U.S., Cesamet is no longer available there.

Table 13.2 summarizes both the clinical study design and the antiemetic efficacy of nabilone in 14 trials. Nabilone was administered orally at 0.5–2 mg dosage. Although in a few studies, the first nabilone dose was administered 30 min prior chemotherapy, in most studies patients received their initial dose 12 h before chemotherapy. Over 93% of the cited studies are randomized, double-blind, and of crossover design. The total number of patients who completed the clinical trials is 548 and most were over 50 years old. However, 3 studies involved either children (1–17 years) (Dalzell et al., 1986), or younger adults with median age of less than 33 years (Einhorn et al., 1981; Herman et al., 1979). Over 90% of the studies were carried out in 1980s following the initial finding that Δ^9-THC was an effective antiemetic agent.

Several placebo-controlled, randomized, double-blind and crossover trials have confirmed that nabilone possesses significant antiemetic properties in cancer patients receiving chemotherapy. For example, in the study of Jones et al. (1982), nabilone reduced the mean frequency of vomiting from 18 in placebo-exposed patients to 7.2 episodes ($p < 0.001$). It also reduced the severity of nausea by a significant degree ($p < 0.001$). In addition, 67% of patients preferred nabilone over placebo. Using a similar study design, Levitt (1982) reported similar findings in that over 80% of nabilone-treated patients experienced less vomiting and 72% less nausea. Moreover, 78% of patients preferred nabilone.

Eight of the cited studies (Table 13.2) evaluated the efficacy of orally administered nabilone against prochlorperazine or other dopamine D_2 antagonists. In a randomized controlled study involving 113 patients, Herman and coworkers (1979) showed that nabilone is significantly more efficacious than prochlorperazine in reducing both the frequency of vomiting and the intensity of nausea ($p < 0.01$). Indeed, nabilone caused 72% reduction in the frequency of vomiting versus 32% in patients receiving prochlorperazine. In addition, 75% of patients preferred nabilone for emesis protection ($p < 0.001$). Several other studies have confirmed these initial findings (Einhorn et al., 1981; Johansson et al., 1982; Ahmedzai et al., 1983; Niiranen and Matson, 1985; Chan et al., 1987). However, in the study of Steele and coworkers (1980), nabilone was found to possess a similar antiemetic efficacy to prochlorperazine in patients who had received high-dose cisplatin regimens. Although in two of the cited trials (Steele et al., 1980; Ahmedzai et al., 1983), patients failed to show a significant preference for these antiemetics, in the remaining studies 66–75% of patients significantly preferred nabilone as their available antiemetic of choice ($p < 0.05$–0.001). Unlike the discussed trend towards development of tolerance for the antiemetic action of Δ^9-THC upon its repeated administration, the antiemetic efficacy of nabilone seems to significantly increase ($p < 0.05$–0.01) following its daily administration (Ahmedzai et al., 1983; Einhorn et al., 1981). Nabilone also appears to be a more efficacious antiemetic relative to the peripherally acting dopamine D_2 antagonist domperidone (Dalzell et al., 1986).

Other investigators have tried to enhance the antiemetic properties of nabilone by co-administering it with other available useful antiemetics. In a randomized, double-blind, crossover study, Cunningham et al. (1985) showed that although addition of prochloperazine to the nabilone regimen did not enhance its antiemetic efficacy, it did reduce nabilone's CNS side effects. In another controlled study involving 70 patients who had received cisplatin or carboplatin chemotherapy, a combination of nabilone and prochlorperazine had essentially similar antiemetic

Table 13.2 Clinical trials determining the antiemetic efficacy of nabilone

Investigator	Patients evaluated	Median age and range in years	Chemotherapy	Study design[a]	Antiemetic[b]	Results[b]
Cunningham et al. (1988)	70	42 (18–68)	Cisplatin or Carboplatin	R C open	Nabilone oral 2 mg + PCP 5 mg, 3–4 ×, first dose 6 h prior chemotherapy, then every 12 h vs. MCP i.v. 2 mg and then infusion of 3 mg MCP + dexamethasone i.v. 20 mg over 8 h.	MCP + Dexamethasone ≥ Nabilone + PCP in preventing nausea and vomiting.
Chan et al. (1987)	30	11.8 (3.5–17.8), children	Various	R C Db	Nabilone, oral, 0.5–2 mg according to weight, 1–3 × daily. First dose 12 h prior chemotherapy vs. PCP oral, 2.5–10 mg according to weight, 2–3 × daily. First dose 12 h prior chemotherapy.	Nabilone > PCP (p < 0.003) in reducing retching and vomiting episodes.
Niiranen and Mattson (1987)	32	63 (44–79)	Various	R C Db	Nabilone, oral, 2 mg, 2 × daily. First dose 12 h prior, 2nd dose 0.5 h before and 3 rd dose 12 h after chemotherapy vs. DXM, oral, 8 mg or placebo with first dose of nabilone. Then either 10 mg DXM or saline i.v. 0.5 h before and 2 and 6 h after chemotherapy.	Nabilone + DXM > Nabilone or DXM alone in reducing vomiting episodes.
Crawford and Buckman (1986)	7	Adults	Cisplatin	R C Db	Nabilone + placebo, oral 1 mg, 4 ×, 1st and 2nd dose 4 and 2 h prior chemotherapy, then every 8 h vs. MCP + placebo, i.v. 1 mg/kg every 3 h.	Nabilone = MCP.
Dalzell et al. (1986)	18	8 (1–17)	60% Vincristine and others	R C Db	Nabilone, oral, 0.5–1 mg according to weight, 2–3 × daily. First dose 12 h prior and last dose 24 h after chemotherapy vs. domperidone, oral, 5–15 mg according to weight 3 × daily. Delivery schedule as above.	Nabilone > domperidone (P < 0.01) in reducing vomiting frequency and severity of nausea.
Cunningham et al. (1985)	34	55 (39–76)	Various	R C Db	Nabilone (2 mg) + PCP (5 mg), oral 4 ×, first dose 12 h prior chemotherapy and then at 12 h intervals vs. nabilone (2 mg) + placebo, delivery schedule as above.	Nabilone + PCP = Nabilone. PCP reduced CNS side effects of nabilone.
Nirranen and Mattson (1985)	24	61 (48–78)	Various	R C Db	Nabilone, oral, 1 mg, 3–4 ×, first and second dose 12 and 1 h prior chemotherapy, and at 12 h intervals vs. PCP oral 7.5mg, delivery as above.	Nabilone > PCP in decreasing vomiting frequency (p < 0.05).

Table 13.2 (Continued)

Investigator	Patients evaluated	Median age and range in years	Chemotherapy	Study design[a]	Antiemetic[b]	Results[b]
Ahmedzai et al. (1983)	28	58 (27–72)	Various	R C Db	Nabilone oral 2 mg, 2 × daily, first dose 12 h prior chemotherapy and then at 12 h intervals vs. PCP oral 10 mg, 3 × daily, at 8 h intervals. First dose 12 h prior to chemotherapy.	Nabilone > PCP in reducing nausea, vomiting and retching (p < 0.001–0.05).
Johansson et al. (1982)	18	Adults (18–70)	Various	R C Db	Nabilone oral 2 mg, 4 × at 12 h intervals, first dose 12 h prior to chemotherapy vs. PCP oral 10 mg, 4 ×, drug delivery as above.	Nabilone > PCP in reducing nausea and vomiting (p < 0.001).
Jones et al. (1982)	24	Adults	Various	Pr R C Db	Nabilone oral 2 mg, 4 × at 12 h intervals, first dose 12 h prior chemotherapy vs. placebo.	Nabilone > placebo in reducing nausea and vomiting (p < 0.001).
Levitt (1982)	36	Older adults (17–78)	Various	R C Db	Nabilone oral 2 mg, 4 × at 12 h intervals, first dose 12 h prior chemotherapy vs. placebo.	Nabilone > placebo in reducing nausea and vomiting (p < 0.001).
Einhorn et al. (1981)	77	28 (15–74)	Various	Pr R C Db	Nabilone oral 2 mg every 6 h as needed, first dose 30 min prior chemotherapy vs. PCP oral 10 mg, drug delivery as above.	Nabilone > PCP in reducing nausea and vomiting (p < 0.001–0.05).
Steele et al. (1980)	37	50 (19–65)	Various	R C Db	Nabilone oral 2 mg, every 12 h for 3–5 ×, first dose 12 h prior chemotherapy vs. PCP oral 10 mg, drug delivery as above.	Nabilone ≥ PCP depends upon the chemotherapeutic agent used.
Herman et al. (1979)	113	33 (15–74)	Various, most received cisplatin	R C Db	Nabilone oral 2 mg, 4 ×, started either 16 h or 0.5 h prior chemotherapy, every 8 h for various days for different chemotherapies vs. PCP oral 10 mg, drug delivery as above.	Nabilone > PCP in reducing intensity of nausea and frequency of vomiting (p < 0.01–0.001).

[a] For study design: C = crossover, Db = double-blind, Hc = historical control, Pr = prospective, R = randomized; [b] For drugs: DXM = dexamethazone, MCP = metoclopramide, PCP = prochlorperazine, THC = Δ^9-tetrahydrocannabinol.

efficacy to high-dose metoclopramide plus dexamethasone therapy (Cunningham *et al.*, 1988). Indeed, the mean frequency of vomitings with metoclopramide and dexamethasone was 3.45 ± 0.78 compared to 3.92 ± 0.54 with nabilone and prochlorperazine ($p = 0.051$). In addition, nausea was prevented in 36% and 23% of patients respectively ($p > 0.05$). On the other hand, the scores for emesis on the linear analog scale were significantly better ($p < 0.018$) for metoclopramide and dexamethasone combination (2.5 ± 0.32) compared to nabilone plus prochlorperazine (3.51 ± 0.37). However, none of these parameters were different in the two antiemetic combination therapies in patients who had received carboplatin. Furthermore, there was no significant patient preference for metoclopramide and dexamethasone for the entire study. Moreover, a significant portion of the patients receiving carboplatin preferred the nabilone and prochlorperazine combination ($p < 0.013$). By itself, orally administered nabilone appears to be as effective as intravenously administered metoclopramide in terminally ill cancer patients receiving cisplatin with other chemotherapeutics (Crawford and Buckman, 1986). However, only 7 patients completed this study because of progression of the disease. Niiranen and Matson (1987) showed that addition of dexamethasone potentiates the antiemetic efficacy of nabilone. Indeed, the mean episodes of vomiting (3.3) was significantly reduced when dexamethasone was added to the nabilone regimen (mean = 1.8 vomits) ($p < 0.001$). However, the intensity of nausea as assessed by the patients was the same for both treatments. Two thirds of the patients preferred the combined antiemetic therapy. This study concluded that addition of dexamethasone to nabilone enhanced the therapeutic yield of nabilone.

Clinical trials with levonantradol

Levonantradol is another synthetic cannabinoid. It was synthesized by Pfizer researchers, and belongs to the "nonclassical cannabinoids group". Levonantradol is a tricyclic analog of Δ^9-THC where the oxygen in its pyran ring is replaced by a nitrogen atom. Levonantradol is one of the four stereoisomers of nantradol. Levonantradol possesses analgesic (Johnson and Melvin, 1986), cannabimimetic (Levitt, 1986; Compton *et al.*, 1991) and antiemetic action (Table 13.3; Johnson and Melvin, 1986) in both animals and humans. This compound is stereoselective in its cannabinoid effects and is 30 times more potent than Δ^9-THC (Little *et al.*, 1988). Currently, it is not clinically available as an antiemetic. Table 13.3 summarizes the clinical antiemetic efficacy potential of levonantradol from 10 clinical trials. In these studies, levonantradol (0.5–2 mg) was administered either intramuscularly or orally. In nearly all studies, the initial dose of levonantradol was administered 1–2 h prior to chemotherapy, and thereafter was administered every 4 h. Only 40% of the cited studies are randomized, double-blind and of crossover design. The total number of patients who completed the clinical trials is 236 and the majority of patients were over 40 years old.

Six of the cited studies were mainly open, dose-ranging trials, and generally concluded that levonantradol is an effective antiemetic that can be safely administered between 0.5–2 mg doses either orally or intramuscularly (Table 13.3). Complete antiemetic effects were observed in 10–27% of patients, whereas partial protection from chemotherapy-induced nausea and vomiting was apparent in 36–90% of patients. Moreover, in a randomized double-blind and placebo-controlled

Table 13.3 Clinical trials determining the antiemetic efficacy of levonantradol

Investigator	Patients evaluated	Median age and range in years	Chemotherapy	Study design[a]	Antiemetic[b]	Results[b]
Citron et al. (1985)	26	57 (28–67)	Various	R C Db	Levonantradol i.m. 1 mg, 7 x, every 4 h, first dose 2 h prior chemotherapy **vs.** Δ^9-THC oral 15 mg, drug delivery as above.	Levonantradol = Δ^9-THC, side effects similar.
Heim et al. (1984)	45	49 (18–73)	Various	R C Db	Levonantradol i.m. 0.5 mg, 3 x, 1 h before and 2 and 6 h after chemotherapy **vs.** MCP 10 mg, drug delivery as above.	Levonantradol > metocolopromide.
Stambaugh et al. (1984)	12	Adults	Various	R Db dose ranging study	Levonantradol i.m. 0.5, 1, 1.5 or 2 mg. First study dose 2 h prior chemotherapy and then 3 x every 4 h **vs.** placebo.	Levonantradol > placebo (p < 0.01).
Sheidler et al. (1984)	16	Adults (18–70)	Various	R C Db	Levonantradol i.m. 1 mg, 4 x, first dose 2 h prior chemotherapy and then every 4 h **vs.** PCP oral 10 mg, drug delivery as above.	Levonantradol = PCP.
Stuart-Harris et al. (1983)	22	49 (20–70)	Various	Open, dose-escalating	Levonantradol i.m. 0.5 mg, 1 h before chemotherapy and then at 4 h intervals if necessary.	Effective antiemetic in 50% of patients resistant to conventional antiemetics.

Reference	n	Age	Study design[a]	Treatment[b]	Comment
Welsh et al. (1983)	20	Adults	Open, dose-ranging	Levonantradol oral 0.25, 0.5, 0.75 or 1 mg up to 6 x at 4 h intervals. First dose 30 min prior chemotherapy.	Effective antiemetic at 0.7 and 1 mg.
Cronin et al. (1981)	28	33 (11–68)	Open, dose finding study	Levonantradol i.m. 0.5 mg, first dose 2 h prior chemotherapy and then every 4 h. Dose increased in 0.5 mg increments until vomiting stopped or toxicity occurred **vs.** placebo.	89% complete or partial response from 0.5–1.5 mg.
Diasio et al. (1981)	22	47 (22–63)	Open, dose finding study	Levonantradol oral 0.5–1.5 mg, 4–7 x, first dose 2 h prior chemotherapy and then every 4 h.	Orally well tolerated and good antiemetic efficacy at 1 mg dosage.
Heim et al. (1981)	12	Adults	Open	Levonantradol i.m. 1 mg, 3 x, first dose 2 h prior chemotherapy and then every 4 h.	Potent antiemetic in hospitalized patients but has high incidence of side effects.
Laszlo et al. (1981)	33	40 (18–66)	Open, dose finding study	Levonantradol oral 0.5–2 mg, 5 x, first dose 2 h prior chemotherapy and then every 4 h. If patient did not respond to lower doses, then dose increased incrementally by 0.5 mg.	Orally an effective antiemetic.

[a] For study design: C = crossover, Db = double-blind, Hc = historical control, Pr = prospective, R = randomized; [b] For drugs: DXM = dexamethazone, MCP = metoclopramide, PCP = prochlorperazine, THC = Δ^9-tetrahydrocannabinol.

dose-ranging study, Stambaugh *et al.* (1984) confirmed that levonantradol has significant antiemetic action relative to placebo (p < 0.01). However, this study failed to show a significant dosage effect among the different dosages, although all the tested doses had significant antiemetic action versus the placebo (p < 0.01). The optimum antiemetic dose of levonantradol seems to be between 0.5 and 1 mg. There appears to be no correlation between age and antiemetic efficacy of levonantradol (Heim *et al.*, 1981; Cronin *et al.*, 1981; Stuart-Harris *et al.*, 1983).

Citron and colleagues (1985) compared the antiemetic efficacy of 1mg intramuscularly (i.m.) administered levonantradol against oral administration of 15 mg Δ^9-THC in a randomized, double-blind crossover study in cancer patients receiving chemotherapy. The antiemetic efficacy of levonantradol was similar to that of Δ^9-THC. Indeed, the mean number of emetic episodes with levonantradol was 2 versus 3 for Δ^9-THC (p=0.06). No vomiting was seen in 28% of levonantradol-treated subjects, whereas complete blockade of emesis was seen in 20% of patients who had received Δ^9-THC. The degree of nausea prevention was also similar for both agents. Fifty two percent of patients indicated no preference for either drug. Of the 48% of patients expressing a preference, 20% preferred levonantradol and 28% preferred Δ^9-THC. In a similar study design, Sheidler *et al.* (1984) concluded that levonantradol (i.m., 1 mg) is as effective as orally administered prochlorperazine (10 mg). In another controlled study, Heim and coworkers (1984) have shown that the antiemetic efficacy of levonantradol (i.v., 0.5 mg) is significantly (p < 0.05) better (62% less nausea and 58% less vomiting) than the low-dose metoclopramide (10 mg oral) regimen in chemotherapy-exposed patients. However, the antiemetic action of each drug was incomplete in most patients and antiemetic combination was recommended for further trials.

Side effects of cannabinoid antiemetics in clinical trials

Nausea and vomiting are the most debilitating aspects of cancer chemotherapy. Emesis disrupts various domains of health related quality of life. Indeed, patients receiving full protection from emesis have better physical and social function scores as well as experiencing less fatigue and anorexia than those who have had one or more episodes of vomiting (Osoba *et al.*, 1996). In this scenario not only the patient is affected, but also the healthcare professionals who are involved in providing care to the patients since nausea and vomiting are considered as iatrogenic. Antiemetics are supportive-care agents in patients receiving chemotherapy. Such drugs should be free of excessive undesirable side effects which otherwise would further compound patient's suffering as well as complicating chemotherapy.

In addition to its antiemetic properties, Δ^9-THC also possesses a profile of common side effects which includes: experiencing a "psychological high", dysphoric reactions (such as hallucinations, fear, panic attacks and anxiety), sedation, dizziness, orthostatic hypotension and dry mouth. The "psychological high" appears to be one of its most common side effects and can occur in 20–80% of study subjects. However, the relationship between attainment of a high with Δ^9-THC and antiemetic effect seems to be controversial, a notion that some studies confirm (Chang *et al.*, 1979; Lucas and Laszlo, 1980; Orr and McKernan, 1981; Sallan *et al.*, 1980) and others reject (Frytak *et al.*, 1979; Garb *et al.*, 1980; Gralla *et al.*, 1984). Minimal to excessive degrees of sedation can occur in 26–80% of study population which is an important factor to consider when Δ^9-THC is given on an outpatient basis. A number of patients have discontinued the use of Δ^9-THC because of severe dysphoric reactions. Although some of the studies cited in Table 13.1 have reported that 2–6% of patients experience dysphoric effects, other trials suggest a larger (20–32%) dysphoric patient population. Two important autonomic side effects of Δ^9-THC are orthostatic hypotension and dry mouth, which occur in 25–53% and 11–80% of cancer patients respectively, as reported in some but not all of the cited trials.

The antiemetic use of Δ^9-THC is limited because of its erratic oral absorption and euphoric effects. Thus, the discussed synthetic cannabinoids with better pharmacokinetic profiles were introduced. The profile of side effects of nabilone is similar to Δ^9-THC. However, oral administration of nabilone causes a relatively lesser prevalence of "psychological high" (3–14% of subjects) and dysphoric reactions (0–13% of patients) in most of the trials cited in Table 13.2. On the other hand, the prevalence of inducing sedation (7–80% of patients) and dry mouth (0–84% of patients) are similar to that of Δ^9-THC. Vertigo is another problem associated with nabilone administration. The advantage of levonantradol over Δ^9-THC and nabilone is that levonantradol can be administered both orally and intramuscularly. The toxicity profile and the prevalence of its side effects are similar to that of Δ^9-THC.

POSSIBLE MECHANISMS OF ANTIEMETIC ACTION OF CANNABINOIDS

Animal models of emesis and the established cannabinoid antiemetics

In the field of antiemetic research, as has been the case of marijuana in general, clinical research has often preceded animal experiments. From the discussion of the numerous cited clinical trials, it is clear that Δ^9-THC and its tested synthetic analogs have demonstrated efficacy and safety as moderate antiemetics in patients receiving chemotherapy. Scant published animal studies also support the antiemetic properties of these cannabinoids (London *et al.*, 1979; McCarthy and Borison, 1981; McCarthy *et al.*, 1984). However, unlike the case for the dopamine D_2-, serotonin 5-HT_3- and tachykinin NK_1-receptor antagonists, until recently the receptor mechanism(s) responsible for the antiemetic action of Δ^9-THC and its analogues was not known. In addition, it is neither clear if the antiemetic action of Δ^9-THC is a common property of all classes of cannabinoid agonists (including endocannabinoids), nor it is understood whether their antiemetic action is related to cannabinoid psychoactivity. Basic research on the antiemetic action of cannabinoids has been hampered by at least 3 factors: (1) competing and more efficacious antiemetics such as 5-HT_3- and NK_1-antagonists have become clinically available; (2) lack of direct evidence for the existence of cannabinoid receptors; and (3) lack of availability of inexpensive animal models of emesis.

The recent renaissance in cannabinoid receptor biology has led to the identification of at least two distinct types of cannabinoid receptors (CB_1 and CB_2), for which potent and selective antagonists (SR 141716A and SR 144528, respectively) have been developed (Rinaldi-Carmona *et al.*, 1994, 1998). In addition, several hundred naturally occurring as well as synthetic cannabinoid agonists are available. Cannabinoid agonists can be classified according to their chemical structure into four main groups. The first of these is the "classical cannabinoid group" which is made of dibenzopyran derivatives and includes Δ^9-THC. The second is the "nonclassical cannabinoid group", which consists of bicyclic and tricyclic analogs of Δ^9-THC that lack a pyran ring (e.g. CP55,940). The third group of cannabimimetic compounds are aminoalkylindoles, and the prototype of this group is the pravadoline derivative, WIN 55, 212-2. The fourth is the "eicosanoid group", which contains arachidonic acid derivatives such as endogenous cannabinoids, anandamide and 2-arachidonoyl-glycerol (2-AG). Although WIN 55, 212-2 shows a modest degree of selectivity for cannabinoid CB_2 receptors, the others appear to be essentially nonselective in regard to CB_1 and CB_2 receptors. However, newer compounds with better cannabinoid receptor subtype selectivity have been introduced and more are under development (Pertwee, 1997).

Unlike the relatively large body of clinical reports, only a few animal studies on the antiemetic effects of cannabinoids have appeared in the literature prior to the year 2001. Nabilone seems to be ineffective in preventing cisplatin-induced vomiting in the dog (Gylys et al., 1979). However, several cannabinoids (nabilone, Δ^9-THC, 7-hydroxy-Δ^9-THC and N-methyllevonantradol) can block chemotherapy-induced emesis in the cat (McCarthy and Borison, 1981; McCarthy et al., 1984; London et al., 1979). Moreover, both nabilone and Δ^9-THC are effective in preventing cisplatin-induced vomiting in the pigeon (Stark, 1982), and in a similar manner Δ^9-THC is an effective antiemetic in the shrew (Darmani, 2000a). These, as well as other studies, provide evidence that different species can exhibit differential sensitivity to both emetics and antiemetics (King, 1990).

Most animal vomiting studies are confined to large animals such as cats, dogs or ferrets. Utilization of such large animals is not cost effective, and therefore alternative models have been found. Indeed, in the 1980s, Japanese investigators introduced a smaller animal (adult being 50–100 g in weight), the house musk shrew (Suncus murinus), as an experimental vomiting model (Matsuki et al., 1988). Suncus murinus is endogenous to Asia and Africa. Shrews are placed in the order of insectivora and are among the most ancient animals (Churchfield, 1990). Shrews are considered to be closer to man than rodents, lagomorphs and carnivores in the phylogenetic system. Unlike Suncus murinus, the least shrew (Cryptotis parva), is much smaller (adult weighing 4–6 g), and is found in Central and North America (Figure 13.1). It is relatively easy and inexpensive to test vomiting and the antiemetic effects of various drugs in this species (Darmani, 1998; Darmani et al., 1999). Moreover, doses of cisplatin required to induce emesis within a couple of hours of its injection, are often toxic and test animals die within one week (Darmani, 1998, London et al., 1979; Torii et al., 1991). Thus, antiemetic screening studies require an animal model that is not cost prohibitive, and the least shrew satisfies this requirement.

Site of action and the role of cannabinoid CB_1 receptors in emesis

Until recently, an animal model of emesis had not been employed to investigate the receptor mechanism(s) by which Δ^9-THC and other cannabinoids prevent

Figure 13.1 Shows three least shrews (*Cryptotis parva*) in an open-top clear polycarbonate cage (20 × 18 × 21 cm) lined with heated dry loam soil and wood chippings. Each shrew weighs between 3–4 g. One shrew is standing on the top of a wooden nest box (9 × 7 × 6 cm) filled with dry heated grass. The front of the wooden box has an entry hole (1.7 cm diameter) through which shrews can enter and leave the nest box. On the front left hand side of the next box is a food bowl containing a large ball of food. The food mixture consists of two-thirds dry cat food (PMI Nutrition Cat formula) and one-third canned cat food (Kozy Kitten) in sufficient water to give the mixture a paste-like consistency. The shrew nearest to the food bowl has chewed two-thirds of a 2 cm long meal worm (*Tenebrio* sp). On the right side of the nest box is a water bottle with lick tube.

emesis produced by chemotherapeutics. Currently, experiments are in progress in this laboratory which deals with this issue. While the antiemetic effects of Δ^9-THC appear to be receptor-mediated, it is unclear whether CB_1 and/ or CB_2 receptors are involved. Involvement of CB_1 receptor appears most likely since Δ^9-THC produces most of its other effects via this receptor (Pertwee, 1997). If activation of the CB_1 receptor prevents emesis, then CB_1 receptor blockade may induce vomiting. This appears to be the case, as recent findings from this laboratory show that intraperitoneal or subcutaneous administration of the cannabinoid CB_1 receptor antagonist SR 141716A but not the CB_2 receptor antagonist SR 144528, produces emesis in a dose-dependent manner in the least shrew (Figure 13.2). SR 141716A appears to be a more potent emetic agent when it is administrated via the intraperitoneal route ($ED_{50} = 12.3 \pm 1.7$ mg/kg) as its 10 and 20mg/kg doses significantly increased the frequency of vomitings [1.67 ± 0.5 (p < 0.05) and 5.7 ± 0.5 (p < 0.001), respectively] relative to control group. On the other hand, via the subcutaneous

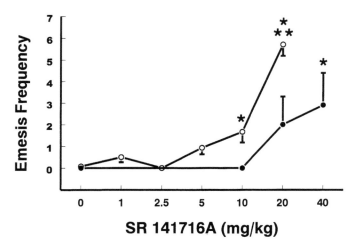

Figure 13.2 Represents the intraperitoneal (○) and the subcutaneous (●) emotogenic dose–response curves of the cited doses of the selective CB₁ receptor antagonist SR 141716A in potentiating the mean frequency of emesis (± SEM) in the least shrew in the 60 min post injection observation period. The Kruskal-Wallis non-parametric ANOVA test showed that significant increases in the mean frequency of emesis occurred as the SR 141716A dosage was increased via the intraperitoneal (p < 0.0001), or subcutaneous (p < 0.003) routes. Dunn's multiple comparisons posthoc test was used to determine the significance for specific doses (*p < 0.05; ***p < 0.001; n = 8–15 shrews per group).

route, SR 141716A is a less efficacious emetic ($ED_{50} = 17.9 \pm 1.1$ mg/kg), and only its 40 mg/kg dose produced a significant number of vomits (2.9 ± 1.5, $p < 0.05$). The percentage of animals vomiting in response to SR 141716A administration has also been shown to increase in a similar route-dependent fashion (ED_{50} 5.5 ± 1.3 and 20 ± 1.02 mg/kg respectively) (Darmani, 2001a). These findings suggest that SR 141716A is either an inverse agonist at CB_1 receptors, or it antagonizes the anti-emetic action of an endogenous ligand(s) acting on CB_1 receptors. Several other studies have already used these two notions to explain the behavioral (locomotor activity, head-twitch response, scratchings) and biochemical effects produced by SR 141716A administration (Compton *et al.*, 1996; Cook *et al.*, 1998; Darmani and Pandya 2000; Landsman *et al.*, 1997). If SR 141716A-induced emesis is a CB_1 receptor-mediated phenomenon, then Δ^9-THC and other cannabinoid agonists should prevent the induced vomiting. Indeed, pretreatment with either CP55,940; WIN 55, 212-2 or Δ^9-THC; reduced the frequency of SR 141716A (20 mg/kg, i.p.)-induced emesis in the least shrew in a dose-dependent manner (Figure 13.3) with the following respective ID_{50} order: $0.28 \pm 2.17 < 3.38 \pm 1.3 < 12.6 \pm 4$ mg/kg. Both Δ^9-THC and WIN 55, 212-2 are able to prevent cisplatin-induced vomiting in the least shrew in a dose-dependent manner (Darmani, 2001c,e). The antiemetic potency of WIN 55, 212-2 was similar to Δ^9-THC against cisplatin-induced emesis. Both cannabinoids reduced the frequency of the induced vomiting at doses lower than those required to significantly affect several locomotor parameters in the least

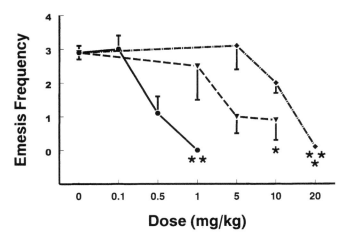

Figure 13.3 Shows the capacity of increasing doses of intraperitoneally administered cannabinoids [CP55,940 = ●; WIN 55, 212–2 = ▼; and Δ^9-THC = ◆] to reduce the frequency (mean & SEM) of vomitings elicited by the intraperitoneal administration of a 20 mg/kg dose of the cannabinoid CB_1 receptor antagonist SR 141716A. The cited doses of each cannabinoid were administered to different groups of shrews 10 min prior to the administration of SR 141716A. The frequency of emesis was recorded for 30 min following the injection of SR 141716A. The Kruskal-Wallis nonparametric ANOVA test showed that the cited cannabinoids significantly attenuated the frequency of the induced vomiting ($p < 0.0003$; $p < 0.014$; $p < 0.0002$; respectively). Dunn's multiple comparisons posthoc test was used to determine significance for the specific doses of cannabinoids (*$p < 0.05$; **$p < 0.01$; ***$p < 0.001$; n = 6–9 per group).

shrew. The antiemetic and sedative actions of cannabinoids are mediated by different loci but both effects are probably produced by cannabinoid CB_1 receptors since these events were reversed by nonemetic doses of SR 141716A. Recent, indirect evidence in nonemetic species also support the role of CB_1 receptors in emesis. For example, different classes of cannabinoid agonists (anandamide, methanandamide, nabilone, Δ^9-THC and WIN 55, 212-2) reduce both the GI motility and intestinal transit in rodents (Calignano *et al.*, 1997; Colombo *et al.*, 1998; Izzo *et al.*, 1999; Krowicki *et al.*, 1999; Shook and Burks, 1989). Since alterations in the gut tone is a factor in the production of emesis (see earlier discussion), cannabinoid-induced reduction in GI tone may help to prevent emesis. Furthermore, the latter effects were potently blocked by SR 141716A at doses which had no effect by itself on the GI function. However, larger doses of SR 141716A promoted GI motility and defication in these studies. More recently, interesting roles have been proposed for endocannabinoids in the modulation of vomiting (Darmani, 2001d). Indeed, the endocannabinoid 2-arachidonoylglycerol (2-AG) was shown to be a potent inducer of vomiting ($ED_{50} = 0.48$ mg/kg, i.p.) in the least shrew. Furthermore, Δ^9-THC, its synthetic analogs and the endocannabinoid anandamide reduced the 2-AG-induced emesis with the rank order potency: CP55,940 > WIN 55, 212-2 > Δ^9-THC > anandamide. The more stable analog of anandamide, methanandamide, also seems to block vomiting produced by 2-AG (Darmani, 2001d) or morphine (Van Sickle, 2001). Thus, 2-AG appears to be an emetic endocannabinoid, whereas anandamide possesses antiemetic properties.

The discussed antiemetic ID_{50} potency order of the tested cannabinoids in preventing SR 141716A-induced vomiting, mirrors their receptor affinity rank order for both CB_1 and CB_2 receptors (CP55,940 < WIN 55, 212-2 < Δ^9-THC) obtained from rodent tissue (Matsuda, 1997; Pertwee, 1997). Furthermore, these rank orders are consistent with the reported ED_{50} potency values of these agents for the tetrad of behaviours in mice (CP55,940 < WIN 55, 212-2 < D^9-THC) which is considered as a measure of the relative psychoactivity of these agents (Abood and Martin, 1992). Thus, these findings suggest that at least in the case of SR 141716A-induced emesis, as expected, the most potent tested cannabinoid, CP55,940, also possesses the greatest degree of antiemetic activity. Obviously, one of the goals of cannabinoid research is to identify those agents which lack significant cannabinoid psychoactivity but possess potent antiemetic as well as other useful medicinal properties. Apart from the present study, very little is known about the antiemetic structure activity relationship (SAR) of different cannabinoids in any species. It seems possible to separate the cannabimimetic psychoactivity of cannabinoids from their antiemetic properties. Indeed, McCarthy et al. (1984) have shown that 7-hydroxy-Δ^9-THC, an active metabolite of Δ^9-THC, is less antiemetic but more cannabimimetic than its parent compound in the cat. More recently, it has been shown that the nonpsychoactive enantiomer (HU 211) of the potent cannabinoid HU 210 can prevent cisplatin-induced emesis in the pigeon (Feigenbaum et al., 1989). However, as HU 211 stereoselectively blocks the NMDA receptor, its antiemetic action is probably related to glutamate NMDA-receptor antagonism. Indeed, such antagonists prevent emesis produced by various emetic stimuli including cisplatin-induced vomiting (Lehmann and Karrberg, 1996; Lucot, 1998). Other studies indicate that dosage titration can differentially separate the various actions of some cannabinoids. For example, Δ^9-THC is almost equipotent in producing the tetrad of behaviors in mice, whereas the newly established cannabinoid analogs exhibit different potencies (Abood and Martin, 1992). Thus, CP55,940 is 10 times more effective in reducing motor activity than producing catalepsy, and WIN 55, 212-2 is four times more potent in producing hypoactivity than antinociception. In addition, several cannabinoids (CP55,940; HU 210; WIN 55, 212-2; Δ^9-THC; and Δ^8-THC) were found to be 3–30 times more potent in blocking the ability of the $5-HT_{2A/C}$ agonist DOI to produce the ear-scratch response than the head-twitch response in mice (Darmani, 2000b).

The sites of antiemetic action of cannabinoids may involve both central and peripheral mechanisms since the CB_1 receptor or its mRNA is found both in the brain structures that control emesis (e.g. nucleus tractus solitarius and lower brain stem, which is the site of origin of GI parasympathetic preganglionic neurons) as well as in the myenteric plexus of the gut (Glass et al., 1997; Lopez-Redondo et al., 1997; Mailleux and Vanderhaegen, 1992; Matsuda et al., 1993; Pertwee et al., 1996; Tsou et al., 1998). Indeed, it has been shown that activation of CB_1 receptors by various cannabinoids (e.g. Δ^9-THC; CP 55, 940) inhibits the electrically-evoked contractions of myenteric plexus muscle preparation via the inhibition of acetylcholine release (Pertwee et al., 1996). The latter effect was abolished by SR 141716A, but opposite effects were produced when it was given alone. Furthermore, both peripheral (i.v.) and central (dorsal surface of medulla) administration of Δ^9-THC decreases gastric motor function in rats and SR 141716A potently blocks the effect (Krowicki et al., 1999).

Possible role of other neurotransmitter systems in the antiemetic properties of cannabinoids

Although the cholinergic neurotransmitter system per se is not directly involved in chemotherapy-induced vomiting (see section 2.1), dopaminergic and serotonergic mechanisms do appear to be important downstream in cannabinoids' antiemetic actions. Indeed, in the feline, nabilone dose-dependently prevents emesis produced by the dopamine D_2 receptor agonist apomorphine (London et al., 1979). In a similar manner, Δ^9-THC prevents vomiting produced by dopamine D_2/D_3 receptor agonists such as apomorphine, quinpirole, quinelorane and 7-OH DPAT in the least shrew (Darmani, unpublished observations). In this context, the least shrew seems to be an excellent dopamine animal model of emesis since the cited selective and nonselective dopamine D_2 receptor agonists can potently induce emesis in this species, whereas D_2 antagonists prevent the induced behavior (Darmani et al., 1999). As discussed earlier, clinical findings further underscore the role of blockade of the dopaminergic system in the antiemetic properties of cannabinoids since a combination of a cannabinoid with a D_2 antagonist appears to be a superior antiemetic regimen than when each drug is given alone to patients receiving chemotherapy (Garb et al., 1980; Lane et al., 1990). Other lines of evidence also indicate interactions between these two neurotransmitter systems. For example, the aminoalkylindole cannabinoid WIN 55, 212-2, can reverse the dopamine D_2 receptor-mediated alleviation of akinesia in the reserpine-treated model of Parkinson's disease (Maneuf et al., 1997). Secondly, the nonclassical cannabinoid CP 55, 940, at doses which do not produce catalepsy, can potentiate the cataleptic effect of the D_2 antagonist raclopride (Anderson et al., 1996).

Involvement of serotonergic mechanisms in the antiemetic properties of cannabinoids is highlighted by several studies. For example, nanomoler concentrations of several classes of cannabinoids (CP55,940; CP56,667; anandamide and WIN 55, 212–2), block the 5-HT_3 receptor-mediated inward currents induced by 5-HT_3 receptor agonists in a dose-dependent but noncompetitive manner in rat nodose ganglion neurons (Fan, 1995). Secondly, cannabinoids reduce 5-HT synthesis, levels and turnover in several regions of rodent brain (Bannergee et al., 1975; Ho et al., 1971; Johnson et al., 1976; Molina-Holgado et al., 1993; Taylor and Fennessy, 1982). Moreover, the CB_1 antagonist SR 141716A, precipitates withdrawal-like behaviors both in cannabinoid tolerant rats (Aceto et al., 1995; Tsou et al., 1998) and drug naive rodents (Cook et al., 1998; Darmani and Pandya, 2000). Many of the induced behaviors have serotonergic origin. Not only the selective serotonin 5-HT_{2A} antagonist SR 46349B, but also Δ^9-THC, prevented some of these SR 141716A-induced behaviors in mice (Darmani and Pandya, 2000). Furthermore, several cannabinoids (CP 55, 940; HU 210; WIN 55, 212-2; Δ^9-THC and Δ^8-THC) potently block the head-twitch and ear-scratch behaviors produced by the 5-HT_{2A} agonist DOI, in mice (Darmani, 2000b). Thus, it seems that cannabinoids generally reduce serotonin function via CB_1 receptors. SR 141716A not only induces vomiting (Darmani, 2001a), but also produces some of these behaviors in the lesser shrew (Darmani, unpublished findings). As in the case of SR 141716A-induced serotonergic behaviors, structurally diverse cannabinoids also block SR 141716A-induced emesis (Darmani, 2001a). Despite the discussed serotonergic findings, as yet there is no basic or clinical study to either compare the antiemetic efficacy of selective 5-HT_3

receptor antagonists against cannabinoids in chemotherapy-induced vomiting, or to determine whether a combination of these antiemetics may have synergistic action. In this context, the least shrew has already been evaluated as an excellent animal model for the induction emesis for 5-HT$_3$ receptor agonists (Darmani, 1998). Furthermore, the latter study shows that 5-HT$_3$ receptor antagonists can prevent emesis produced by the chemotherapeutic agent cisplatin.

Cannabinoid agonists such as WIN 55, 212-2 and methanandamide are also effective antiemetics in preventing morphine-induced emesis in the ferret (Simoneau et al., 2001; Van Sickle 2001). In addition, the antiemetic effects of these cannabinoids were reversed by the CB$_1$-(AM 251) but not by the CB$_2$-(AM 630) receptor antagonist. Although the effect of high doses of AM 251 was not investigated, a 5 mg/kg dose of this CB$_1$ antagonist failed to evoke emesis, but was shown to potentiate the frequency of morphine-induced emesis in the ferret (Van Sickle, 2001). As discussed earlier, the well investigated CB$_1$ receptor antagonist SR 141716A but not the CB$_2$ receptor antagonist SR 144528 produces significant emesis by itself at doses greater than 10 mg/kg (Darmani 2001a). Both AM 251 and SR 141716A also act as inverse agonists and their emetic effects could be caused by inhibition of the binding of an antiemetic endocannabinoid or via a reduction in the activity of a constitutively active cannabinoid CB$_1$ receptor.

Finally, tachykinin NK$_1$ receptors may also play a role in the antiemetic action of cannabinoids. Recent results from this laboratory show that the NK$_1$ antagonist CP 94, 994 attenuated the ability of the CB$_1$ antagonist, SR 141716A, to produce scratching and head-twitching behaviors in mice (Darmani and Pandya, 2000). As already discussed, such antagonists possess broad spectrum antiemetic action and are potent blockers of chemotherapy-induced vomiting both in animals and man (Bountra et al., 1996; Navari et al., 1999). In addition, chronic administration of Δ^9-THC increases mRNA levels of the preferential tachykinin NK$_1$ agonist, substance P (Mailleux and Vanderhaegen, 1994).

SUMMARY AND CONCLUSIONS

At least six naturally occurring and synthetic cannabinoids have been investigated for their antiemetic properties in man. However, only Δ^9-THC, nabilone and levonantradol have been studied extensively in more than 40 clinical trials involving over 1529 patients who had completed these studies. Although a large number of these trials differ in their study design and suffer from methodological inadequacies, it is generally clear that cannabinoids possess significant antiemetic properties in cancer patients receiving chemotherapeutics. Though cannabinoids appear to be a more efficacious class of antiemetics than dopamine D$_2$ receptor antagonists for the prevention of chemotherapy-induced vomiting, the efficacy of tested cannabinoids to date seems not to be as high as the more potent antiemetics such as the selective 5-HT$_3$ receptor antagonists. However, one interesting advantage of cannabinoids is that many of the patients who are protected from the acute phase of emesis, also respond well during the delayed phase of chemotherapy-induced emesis which 5-HT$_3$ receptor antagonists poorly control (Abrahamov et al., 1995; Chan et al., 1987; Dalzell et al., 1986). Currently, clinicians mainly prescribe available cannabinoids only as an adjunct antiemetic for those patients who are refractory to standard antiemetics.

The current lack of interest in the clinical utility of discussed cannabinoids as an antiemetic will probably remain because of their moderate antiemetic efficacy and side effects unless more potent and safer cannabinoids are developed. As yet, no clinical study has demonstrated a better margin of safety for one cannabinoid versus another. Indeed, at effective antiemetic doses, the tested synthetic cannabinoids such as nabilone and levonantradol produce a degree and profile of side effects similar to that of the naturally occurring cannabinoid Δ^9-THC. Even though there is no information regarding the therapeutic index of CP 55, 940 in man; animal emesis studies indicate that this cannabinoid may possess superior antiemetic efficacy. For example, relative to Δ^9-THC, the index of psychoactivity of CP 55, 940 (i.e. its ED_{50} in producing the tetrad of behaviors) only varies from 4 to 25 times in mice; whereas the antiemetic ID_{50} of this cannabinoid is 45 times greater than Δ^9-THC in the least shrew. If the ratio of relative ED_{50} values of these cannabinoids in producing the tetrad of behaviors in the least shrew is similar to mice, then this agent represents a significant improvement in the antiemetic potential of cannabinoids. Furthermore, the author's more recent findings have revealed that different endocannabinoids may possess emetic (e.g. 2-AG) and antiemetic (e.g. anandamide) properties. Manipulation of chemical structure of these agents may lead to the development of new antiemetics.

The concept of developing therapeutically useful analogs of Δ^9-THC has been attractive because of their mild dependency potential as well as their relatively low toxicities in both animals and humans. Thus, significant potential exists for new cannabinoids to provide more effective antiemetic regimens either by themselves or in combination with other agents (serotonin $5\text{-}HT_3$-, dopamine D_2- or tachykinin NK_1-antagonists). Indeed, as discussed in this chapter, there is substantial line of evidence to show that cannabinoids modulate the function of several neurotransmitter systems downstream of cannabinoid receptors. Furthermore, recent emesis studies from this laboratories suggest that the antiemetic action of cannabinoids in the least shrew is most likely to be mediated via CB_1 receptors. A significant number of GI related published studies in rodents support this proposal. Thus, development of selective CB_1 agonists which lack significant psychoactivity may provide a new avenue for the prevention of chemotherapy-induced nausea and vomiting. Moreover, the discussed cannabinoids seem to have a broad-spectrum antiemetic action since they are effective in preventing emesis produced by a variety of stimuli. A further advantage of Δ^9-THC and other cannabinoid antiemetics is that they stimulate appetite in both animals and man (Beal *et al.*, 1997; Gallate *et al.*, 1999; Gorter, 1999; Haney *et al.*, 1999; Rahminiwati and Nishimura, 1999; Williams and Kirkham, 1999; Williams *et al.*, 1998). This property of cannabinoids is important in the light of the fact that anorexia and cachexia are diagnosed in more than two-thirds of all cancer patients with advanced disease, and are independent risk factors for morbidity and mortality (Gorter, 1999). Furthermore, the cited studies show that the appetite-stimulating properties of Δ^9-THC, and its endogenous (anandamide) and synthetic (CP55,940) analogs is mediated via CB_1 receptors.

ACKNOWLEDGMENTS

This work was supported by grants from the National Institute on Drug Abuse (DA 12605) and KCOM's Strategic Fund. The author would like to thank Professor R. Theobald for his helpful suggestions and R. Chronister for typing the manuscript.

REFERENCES

Abood, M. E. and Martin, B. R. (1992) "Neurobiology of marijuana abuse," *Trends in Pharmacological Sciences* **13**: 201–206.

Abrahamov, A., Abrahamov, A. and Mechoulam, R. (1995) "An efficient new cannabinoid antiemetic in pediatric oncology," *Life Sciences* **56**: 2097–2102.

Aceto, M. D., Scates, S. M., Lowe, J. A. and Martin, B. R. (1995) "Cannabinoid precipitated withdrawal by the selective cannabinoid receptor antagonist, SR 141716A," *European Journal of Pharmacology* **282**: R1–R2.

Ahmedzai, S. Claryle, D. L., Calder, I. T. and Moran, F. (1983) "Antiemetic efficacy and toxicity of nabilone, a synthetic cannabinoid, in lung cancer chemotherapy," *British Journal of Cancer* **48**: 657–663.

Allan, S. G. (1992) "Antiemetics," *Gastroenterology Clinics of North American* **21(3)**: 597–611.

Anderson, J. J., Kask, A. M. and Chase, T. N. (1996) "Effects of cannabinoid receptor stimulation and blockade on catalepsy produced by dopamine receptor antagonists," *European Journal of Pharmacology* **295**: 163–168.

Andrews, P. R. L., Bhandari, P. and David, C. J. (1992) "Plasticity and modulation of the emesis reflux," in *Mechanisms and Control of Emesis*, edited by A. I. Bianchhi, L. Grelot, A. D. Miller and G. L. King, pp. 275–284. London: Libbey Eurotext.

Andrews, P. R. L., Bhandari, P., Garland, S., Bingham, S., Davis, C. J., Hawthorn, J. *et al.* (1990a) "Does retching have a function? An experimental study in the ferret," *Pharmacodyn. Ther.* **9**: 135–152.

Andrews, P. R. L., David, C. L., Bingham, S., Davison, H. I. M., Hawthorn, J. and Maskell, L. (1990b) "The abdominal visceral innervation and the emetic reflex: pathways, pharmacology and plasticity," *Canadian Journal of Physiology and Pharmacology* **68**: 325–345.

Archer, C. B., Amlot, P. L. and Trouce, J. R. (1983) "Antiemetic effect of nonabine in cancer chemotherapy: a double blind study comparing nonabine and chlorpromazine," *British Medical Journal* **286**: 350–351.

Archer, R. A., Stark, P. and Lemberger, L. (1986) "Nabilone," in *Cannabinoids and Therapeutic Agents*, edited by R. Mechoulam, pp. 85–98. Boca Raton, Fl: CRC Press.

Bannergee, S. P., Snyder, S. H. and Mechoulam, R. (1975) "Cannabinoids: influence on neurotransmitter uptake in rat brain synaptosomes," *The Journal of Pharmacology and Experimental Therapeutics* **194**: 74–81.

Barns, J. M., Barns, N. M., Costall, B., Naylor, R. J. and Tattersal, F. D. (1988) "Reserpine, para-chlorophenylalamine and fenfluramine antagonize cisplatin-induced emesis in the ferret," *Neuropharmacology* **27**: 783–790.

Barns, J. M. and Sharp, T. (1999) "A review of central 5-HT receptors and their function," *Neuropharmacology* **38**: 1083–1152.

Beal, J. E., Olsen, R., Lefkowitz, L., Laubenstein, L., Bellman, P., Yangco, B. *et al.* (1997) "Long-term efficacy and safety of dronabinol for acquired immunodeficiency syndrome-associated anorexia," *Journal of Pain and Symptom Management* **14**: 7–14.

Betancur, C., Azzi, M. and Rosténe, W. (1997) "Nonpeptide antagonists of neuropeptide receptors: tools for research and therapy," *Trends in Pharmacological Sciences* **18**: 372–386.

Bhandari, P., Gupta, Y. K., Seth, S. D. and Chugh, A. (1989) "Cisplatin-induced emesis: effect of chemoreceptor trigger zone ablation in dogs," *Asia Pacific Journal of Pharmacology* **4**: 209–211.

Bountra, C., Gale, J. D., Gardner, C. J., Jordan, C. C., Kilpatrick, G. J., Twissell, D. J. and Wand, P. (1996) "Towards understanding the aetiology and pathophysiology of the emetic reflex: Novel approaches to antiemetic drugs," *Oncology* **53**(Supp.1): 102–109.

Brunton, L. (1996) "Agents affecting gastrointestinal water flux and motility; emesis and antiemetics; bile acids and pancreatic enzymes," in *Goodman and Gilman's the Pharmacological Basis of Therapeutics*, edited by J. G. Hardman, L. E. Limbird, P. B. Molinoff, R. W. Rudden and A. G. Gilman, pp. 917–936. New York: McGraw-Hill.

Calignano, A., La Rana, G., Makriyannis, A., Lyn, S. Y., Betramo, M. and Piomelli, D. (1997) "Inhibition of intestinal motility by anandamide, an endogenous cannabinoid," *European Journal of Pharmocology* **340**: R7–R8.

Chan, H. S. L., Correia, J. A. and Macleod, S. M. (1987) "Nabilone versus prochlorperazine for control of cancer chemotherapy-induced emesis in children: a double blind, crossover trial," *Pediatrics* **79**: 946–952.

Chang, A. E., Shiling D. J., Stillman, R. C., Goldberg, N. H., Seipp, C. A., Barofsky, I. *et al.* (1981) "A prespedive evaluation of delta-9-tetrahydrocannabinol as an antiemetic in patients receiving adriamycin and cytotaxan chemotherapy," *Cancer* **47**: 1746–1751.

Chang, A. E., Shiling D. J., Stillman, R. C., Goldberg, N. H., Seipp, C. A., Barofsky, I. *et al.* (1979) "Delta-9-tetrahydrocannabinol as an antiemetic in cancer patients receiving high-dose methotrexate: A prospective, randomized evaluation," *Annals of Internal Medicine* **91**: 819–824.

Churchfield, S. (1990) *The Natural History of Shrews*. Cornel University Press, Comstock Publishing Associates, Ithaca, NY.

Citron, M. L., Herman, T. S., Vreeland, F., Krasnow, S. H., Fossieck, B. E., Hardwood, S. *et al.* (1985) "Antiemetic efficacy of levonantradol compared to delta-9-tetrahydrocannabinol for chemotherapy-induced nausea and vomiting," *Cancer Treatment Report* **69**: 109–111.

Coats, A., Abraham, S., Kay, S. B., Sowerbutts, T., Frewin, C., Fox, R. M. *et al.* (1983) "On the receiving end, patient perception of the side-effects of cancer chemotherapy," *European Journal of Clinical Oncology* **19**: 203–208.

Colombo, G., Agabio, R., Lobina, C., Reali, R. and Gessa, G. L. (1998) "Cannabinoid modulation of intestinal propulsion in mice," *European Journal of Pharmacology* **344**: 67–69.

Compton, D. R., Aceto, M. D., Low, J. and Martin, B. R. (1996) "In vivo characterization of a specific cannabinoid receptor antagonist (SR 141716A): inhibition of Δ^9-tetrahydrocannabinol-induced responses and apparent agonist activity," *Journal of Pharmacology and Experimental Therapeutics* **277**: 586–594.

Compton, D. R., Melvin, L. S., Johnson, M. R. and Martin, B. R. (1991) "Nonclassical cannabinoid chemical structures: classification of bicyclic and related nantradol analogs as cannabimimetic agents," *FASEB Journal* **5**: 703–711.

Cook, S. A., Lowe, J. A. and Martin, B. R. (1998) "CB_1 receptor antagonist precipitates withdrawal in mice exposed to Δ^9-tetrahydrocannabinol," *Journal of Pharmacology and Experimental Therapeutics* **285**: 1150–1156.

Crawford, S. M. and Buckman, R. (1986) "Nabilone and metoclopramide in the treatment of nausea and vomiting due to cisplatinum: a double blind study," *Medical Oncology and Tumor Pharmacotherapy* **3**: 39–42.

Cronin, C. M., Sallan, S. E., Gelber, R., Lucas, V. S. and Laszlo, J. (1981) "Antiemetic effect of intramuscular levonantradol in patients receiving anticancer chemotherapy," *Journal of Clinical Pharmacology* **21**: 43S–50S.

Cubeddu, L. X. (1996) "Serotonin mechanisms in chemotherapy-induced emesis in cancer patients," *Oncology* **53**(Suppl.1): 18–25.

Cunningham, D., Bradley, C. J., Forrest, G. J., Hutcheon, A. W., Adams, L., Snedon, M. *et al.* (1988) "A randomized trial of oral nabilone and prochlorperazine compared to intravenous metochlopramide and dexamethazone in the treatment of nausea and vomiting induced by chemotherapy regimens containing cisplatin or cisplatin analogs," *European Journal of Cancer and Clinical Oncology* **24**: 685–689.

Cunningham, D., Forrest, G. J., Soukop, M., Gilchrist, N. L. and Calder, I. T. (1985) "Nabilone and prochlorperazine: a useful combination for emesis induced by cytotoxic drugs," *British Medical Journal* **291**: 864–865.

Dalzell, A. M., Bartlett, H. and Lilleyman, J. S. (1986) "Nabilone: an alternative antiemetic for cancer chemotherapy," *Archives of Disease in Childhood* **161**: 502–505.

Darmani, N. A. (2001a) "Δ^9-Tetrahydrocannabinol and synthetic cannabinoids prevent emesis produced by the cannabinoid CB_1 receptor antagonist/inverse agonist SR 141716A," *Neuropsychopharmacology* **24**: 198–203.

Darmani, N. A. (2001b) "Cannabinoids of diverse structure inhibit two DOI-induced $5\text{-}HT_{2A}$ receptor-mediated behaviors in mice," *Pharmacology, Biochemistry and Behavior* **68**: 311–317.

Darmani, N. A. (2001c) "Δ^9-tetrahydrocannabinol differentially suppresses cisplatin-induced emesis and indeces of motor function via cannabinoid CB_1 receptors in the least shrew," *Pharmacology, Biochemistry and Behavior* **69**: 239–249.

Darmani, N. A. (2001d) "The potent emetogenic effects of the endocannabinoid, 2-AG (2-arachidonoylglycerol) are blocked by Δ^9-THC and other cannabinoids. *Journal of Pharmacology and Experimental Therapeutics* (in press).

Darmani, N. A. (2001e) "The cannabinoid CB_1 receptor antagonist SR 141716A reverse the antiemetic and motor depressant actions of WIN 55,212-2," *European Journal of Pharmacology* (in press).

Darmani, N. A. (1998) "Serotonin $5\text{-}HT_3$ receptor antagonists prevent cisplatin-induced emesis in Cryptotis parva: a new experimental model of emesis," *Journal of Neural Transmission* **105**: 1143–1154.

Darmani, N. A. and Pandya, D. K. (2000) "Involvement of other neurotransmitters in behaviors induced by the cannabinoid CB_1 receptor antagonist SR 141716A in naive mice," *Journal of Neural Transmission* (in press).

Darmani, N. A., Zhao, W. and Ahmad, B. (1999) "The role of D_2 and D_3 dopamine receptors in the mediation of emesis in Cryptotis parva (the least shrew)," *Journal of Neural Transmission* **106**: 1045–1061.

Diasio, R. B., Ettinger, D. S. and Satterwhite, B. E. (1981) "Oral levonantradol in the treatment of chemotherapy-induced emesis: preliminary observations," *Journal of Clinical Pharmacology* **21**: 81S–85S.

Diemunsch, P., Schoeffler, P., Bryssine, B., Cheli-Muller, L. E., Lees, J., McQAuade, B. A. *et al.* (1999) "Antiemetic activity of the NK_1 receptor antagonist GR 205171 in the treatment of established postoperative nausea and vomiting after major gynaecological surgery," *British Journal of Anaesthesia* **82**: 274–276.

Einhorn, L. H., Nagy, C., Fuvnas, B. and Williams, S. D. (1981) "Nabilone: an effective antiemetic in patients receiving cancer chemotherapy," *Journal of Clinical Pharmacology* **21**: 64S–69S.

Ekert, H., Waters, K. D., Jurk, I. H., Mobilia, J. and Loughnan, P. (1979) "Amelioration of cancer chemotherapy-induced nausea and vomiting by delta-9-tetrahydrocannabinol," *The Medical Journal of Australia* **2**: 657–659.

Fan, P. (1995) "Cannabinoid agonists inhibit the activation of $5\text{-}HT_3$ receptors in rat nodose ganglion neurons," *Journal of Neurophysiology* **73**: 907–910.

Feigenbaum, J. J., Richmond, S. A., Weissman, Y. and Mechoulam, R. (1989) "Inhibition of cisplatin-induced emesis in the pigeon by a non-psychotropic synthetic cannabinoid," *European Journal of Pharmacology* **169**: 159–165.

Frytak, S., Moertel, C. G., O'Fallon, J. R., Rubin, J., Creagan, E. T., O'Connell, M. J. *et al.* (1979) "Delta-9-tetrahydrocannabinol as an antiemetic for patients receiving cancer chemotherapy: a comparison with prochlorperazine and a placebo," *Annals of internal medicine* **91**: 825–830.

Garb, S., Beers, A. L., Bogand, M., McMahon, R. T., Mangalik, A., Ashman, A. C. *et al.* (1980) "Two-pronged study of tetrahydroncannabinol (THC) prevention of vomiting from cancer chemotherapy," *IRCS Medical Sciences* **8**: 203–204.

Gaginella, T. S. (1995) "Serotonin in the intestinal tract: A synopsis," in *Serotonin and Gastrointestinal Function*, edited by T. S. Gaginella and J. J. Galligan, pp. 1–5. Boca Raton: CRC Press.

Gallate, J. E., Saharov, T., Mallet, P. E. and McGregor, I. S. (1999) "Increased motivation for beer in rats following administration of a cannabinoid CB_1 receptor agonist," *European Journal of Pharmacology* **370**: 233–240.

Gidda, J. S., Evans, D. C., Cohen, M. L., Wong, D. T., Robertson, D. W. and Parli, C. J. (1995) "Antagonism of serotonin$_3$ (5-HT_3) receptors within the blood–brain barrier prevents cisplatin-induced emesis in dogs," *Journal of Pharmacology and Experimental Therapeutics* **273**: 695–701.

Gidda, J. S., Evans, D. C., Krushink, J. H. and Robertson, D. W. (1991) "Differential effects of quaternized 5-HT_3 receptor antagonists in dogs and ferrets," in *Serotonin: 5-HT CNS Receptors and Brain Function*, edited by P. B. Bradley, pp. 165–172.

Glass, M., Dragunow, M. and Faul, R. L. (1997) "Cannabinoid receptors in the human brain: a detailed anatomical and quantitative autoradiographic study in the fetal, neonatal and adult human brain," *Neuroscience*, **77**: 299–318.

Gorter, R. W. (1999) "Cancer cachexia and cannabinoids," *Forsch Komplem* **3**(Suppl. 6): 21–22.

Gralla, R. J. (1983) "Metoclopramide. A review of antiemetic trials," *Drugs* **25**: 163–173.

Gralla, R. J., Tyson, L. B., Bordin, L. A., Clark, R. A., Kelson, D. P., Kris, M. G. *et al.* (1984) "Antiemetic therapy: a review of recent studies and a report of random assignment trial comparing metochlopramide with delta-9-tetrahydrocannabinol," *Cancer Treatment Reports* **68**: 163–172.

Gylys, J. A., Doran, K. M. and Buyniski, J. P. (1979) "Antagonism of cisplatin induced emesis in the dog," *Research Communications in Chemical Pathology and Pharmacology* **23**: 61–68.

Haney, M., Ward, A. S., Comer, S. D., Foltin, R. W. and Fischmann, M. W. (1999) "Abstinence symptoms following oral THC administration to humans," *Psychopharmacology* (Berl) **141**: 385–394.

Heim, M. E., Queisser, W. and Altenburg, H. P. (1984) "Randomized crossover study of the antiemetic activity of levonantradol and metoclopramide in cancer patients receiving chemotherapy," *Cancer Chemotherapy and Pharmacology* **13**: 123–125.

Heim, M. E., Romer, W. and Queisser, W. (1981) "Clinical experience with levonantradol hydrochloride in the prevention of cancer chemotherapy-induced nausea and vomiting," *Journal of Clinical Pharmacology* **21**: 86S–89S.

Herman, T. S., Einhorn, L. H., Jones, S. E., Nagy, C., Chester, A. B., Dean, J. *et al.* (1979) "Superiority of nabilone over prochlorperazine as an antiemetic in patients receiving cancer chemotherapy," *The New England Journal of Medicine* **300**: 1295–1297.

Herrstedt, J. and Dombernowsky, P. (1994) "Treatment of drug-induced nausea and vomiting," *Ugeskr Laeger* **156**: 453–460.

Herrstedt, J., Sigsgaard, T., Boesgaard, M., Jensen, T. P. and Dombernowsky, P. (1993) "Ondansetron plus metopimazine compared with ondansetron alone in patients receiving moderately emetogenic chemotherapy," *The New England Journal of Medicine* **328**: 1076–1080.

Hesketh, P. (1996) "Management of cisplatin-induced delayed emesis," *Oncology* **53** (Suppl. 1): 73–77.

Higgins, G. A., Kilpatrick, G. J., Bunce, K. T., Jones, B. J. and Tyers, M. B. (1989) "5-HT_3 receptor antagonists injected into the area postrema inhibit cisplatin-induced emesis in the ferret," *British Journal of Pharmacology* **97**: 247–255.

Ho, B. T., Taylor, D., Englert, L. F. and McIsaac, W. M. (1971) "Neurochemical effects of L-Δ^9-tetrahydrocannabinol in rats following repeated inhalation," *Brain Research* **31**: 233–236.

Izzo, A. A., Mascolo, N., Borrelli, F. and Capasso, F. (1999) "Defecation, intestinal fluid accumulation and mobility in rodents: implications of cannabinoid CB_1 receptors," *Naunyn-Schmiedeberg's Archives of Pharmacology* **359**: 65–70.

Johansson, R., Kilkku, P. and Groenroos, M. (1982) "A double-blind, controlled trial of nabilone vs. prochlorperazine for refractory emesis induced by cancer chemotherapy," *Cancer Treatment Reviews* **9**(Suppl. B): 25–33.

Johnson, K. M., Ho, B. T. and Dewy, W. L. (1976) "Effects of Δ^9-tetrahydrocannabinol on neurotransmitter accumulation and release in rat brain forebrain synaptosomes," *Life Sciences* **19**: 347–356.

Johnson, M. R. and Melvin, L. S. (1986) "The discovery of nonclassical cannabinoid analgesics," in *Cannabinoids as Therapeutic Agents*, edited by R. Mechoulam, pp. 121–135. Boca Raton, FL: CRC Press.

Jones, S. E., Durant, J. R., Greco, F. A. and Robertone, A. (1982) "A multi-institutional phase III study of nabilone vs. placebo in chemotherapy-induced nausea and vomiting," *Cancer Treatment* **Rev. 9**(Suppl. B): 45–48.

King, G. L. (1990) "Animal models in the study of vomiting," *Canadian Journal of Pharmacology* **68**: 260–268.

Kluin-Neleman, J. C., Neleman, F. A., Meuwissen, O. J. and Maes, R. A. (1979) "Delta-9-tetrahydrocannabinol (THC) as an antiemetic in patients treated with cancer chemotherapy: a double-blind cross-over trial against placebo," *Veterinary and Human Toxicology* **21**: 338–340.

Kris, M. G., Radford, J. E., Pizzo, B. A., Inabinet, R., Hesketh, A. and Hesketh P. J. (1997) "Use of an NK_1 receptor antagonist to prevent delayed emesis after cisplatin," *Journal of the National Cancer Institute* **89**: 817–818.

Krowicki, Z. K., Moersschbaecher, J. M., Winsauer, P. J., Digvalli, S. V. and Hornby, P. J. (1999) "Δ^9-tetrahydrocannabinol inhibits gastric motility in the rat through cannabinoid CB_1 receptors," *European Journal of Pharmacology* **371**: 187–196.

Landsman, R. S., Burkey, T. H., Consroe, P., Roeske, W. R. and Yammamura, H. I. (1997) "SR 141716A is an inverse agonist at the human cannabinoid CB_1 receptor," *European Journal of Pharmacology* **334**: R1–R2.

Lane, M., Smith, F. E., Sullivan, R. A. and Plasse, T. F. (1990) "Dronabinol and prochlorperazine alone and in combination as antiemetic agents for cancer chemotherapy," *American Journal of Clinical Oncology* **13**: 480–486.

Laszlo, J., Lucas, V. S., Hanson, D. C., Cronin, C. M. and Sallan, S. E. (1981) "Levonantradol for chemotherapy-induced emesis. Phase 1–11 oral administration," *Journal of Clinical Pharmacology* **21**: 51S–56S.

Lehmann, A. and Karrberg, L. (1996) "Effects of N-methyl-D-aspartate receptor antagonists on cisplatin-induced emesis in the ferret," *Neuropharmacology* **35**: 475–481.

Levitt, M. (1982) "Nabilone vs. placebo in the treatment of chemotherapy-induced nausea and vomiting in cancer patients," *Cancer Treatment Reviews* **9**(Suppl. B): 49–53.

Levitt, M., Faiman, C., Hawks, R. and Wilson, A. (1984) "Randomized double blind comparison of delta-9-tetrahydrocannabinol (THC) and marijuana as chemotherapy antiemetics," *Proceedings of the American Society of Clinical Oncology* **3**: 91.

Levitt, M. (1986) "Cannabinoids as antiemetics in cancer chemotherapy," in *Cannabinoids as Therapeutic Agents*, edited by R. Mechoulam, pp. 71–85. Boca Raton, FL: CRC Press.

Little, P. J., Compton, D. R., Johnson, M. R., Melvin, L. S. and Martin, B. R. (1988) "Pharmacology and stereoselectivity of structurally novel cannabinoids in mice," *Journal of Pharmacology and Experimental Therapeutics* **247**: 1046–1051.

London, S. W., McCarthy, L. E. and Borison, H. L. (1979) "Suppression of cancer chemotherapy-induced vomiting in the cat by nabilone, a synthetic cannabinoid," *Proceedings of the Society for Experimental Biology and Medicine* **160**: 437–440.

Lopez-Redondo, F., Lees, G. M. and Pertwee, R. G. (1977) "Effects of cannabinoid receptor ligands on electrophysiological properties of myenteric neurones of the guinea-pigileum," *British Journal of Pharmacology* **122**: 330–334.

Lucas, V. S. and Laszlo, J. (1980) "Δ^9-tetrahydrocannabinol for refractory vomiting induced by cancer chemotherapy," *Journal of the American Medical Association* **243**: 1241–1243.

Lucot, J. B. (1998) "Effects of N-methyl-D-aspartate antagonists on different measures of motion sickness in cats," *Brain Research* **47**: 407–411.

Mailleux, P. and Vanderhaegen, J.-J. (1992) "Distribution of neuronal cannabinoid receptor in the adult rat brain: a comparative receptor binding radioautography and in situ hybridization histochemistry," *Neuroscience* **48**: 655–668.

Mailleux, P. and Vanderhaegen, J.-J. (1994) "Δ^9-tetrahydrocannabinol regulates substance P and enkephalin mRNAs levels in the caudate-putamen," *European Journal of Pharmacology* **267**: R1–R3.

Maneuf, Y. P., Crossman, A. R. and Brotchi, J. M. (1997) "The cannabinoid receptor agonist WIN 55, 212-2 reduces D_2 but not D_1, dopamine receptor-mediated alleviation of akinesia in the reserpine-treated rat model of Parkinsons disease," *Experimental neurology* **148**: 265–270.

Matsuda, L. A. (1997) "Molecular aspects of cannabinoid receptors," *Critical Reviews in Neurobiology* **11**: 143–166.

Matsuda, L. A., Bonner, T. I. and Lolait, S. J. (1993) "Localization of cannabinoid receptor mRNA in rat brain," *The Journal of Comparative Neurology* **327**: 535–550.

Matsuki, N., Ueno, S., Kaji, T., Ishihara, A., Wang, C.-H. and Saito, J. (1988) "Emesis induced by cancer chemotherapeutic agents in the Suncus murinus: a new experimental model," *Japanese Journal of Pharmacology* **48**: 303–306.

McCarthy, L. E. and Borison, H. L. (1981) "Antiemetic activity of N-methylevonantradol and nabilone in cisplatin-treated cats," *Journal of Clinical Pharmacology* **21**: 30S–37S.

McCarthy, L. E. and Borison, H. L. (1984) "Cisplatin-induced vomiting eliminated by ablation of the area postrema in cats," *Cancer Treatment Report* **68**: 401–404.

McCarthy, L. E., Flora, K. P. and Vishnuvajjala, R. (1984) "Antiemetic properties and plasma concentrations of delta-9-tetrahydrocannabinol against cisplatin vomiting in cats," in *The Cannabinoids: Chemical, Pharmacological and Therapeutic Aspects*, edited by S. Agurell, W. L. Dewey and R. E. Willette, pp. 895–902. London: Academic Press.

Mitchelson, F. (1992) "Pharmacological agents affecting emesis: a review (part 1)", *Drugs* **43**: 295–315.

Molina-Holgado, F., Molina-Holgado, E., Leret, M. L., González, M. I. and Reader, T. A. (1993) "Distribution of indolamines and [^{3}H] paraoxetine binding in rat brain regions following acute or perinatal Δ^9-tetrahydrocannabinol treatments," *Neurochemical Research* **18**: 1183–1191.

Munstedt, K., Muller, H., Blauth-Eckmeyer, E., Stenger, K., Zygmunt, M. and Nahrson, H. (1999) "Role of dexamethasone dosage in combination with 5-HT_3 antagonists for prophylaxis of acute chemotherapy-induced nausea and vomiting," *British Journal of Cancer* **79**: 637–639.

Navari, R. M., Reinhardt, R. R., Gralla, R. J., Kris, M. G., Hesketh, P. J., Khojastch, A. *et al.* (1999) "Reduction of cisplatin-induced emesis by a selective neurokinin-1-receptor antagonist," *New England Journal of Medicine* **340**: 190–195.

Naylor, R. J. and Rudd, J. A. (1996) "Mechanisms of chemotherapy/radiotherapy-induced emesis in animal models," *Oncology*, **53**(Suppl. 1): 8–17.

Neidhart, J. A., Gagen, M. M., Wilson, H. E. and Young, D. C. (1981) "Comparative trial of the antiemetic effects of THC and haloperidol," *Journal of Clinical Pharmacology* **21**: 38S–42S.

Niiranen, A. and Mattson, K. (1985) "A crossover comparison of nabilone and prochlorperazine for emesis induced by cancer chemotherapy," *American Journal of Clinical Oncology* **8**: 336–340.

Niiranen, A. and Mattson, K. (1987) "Antiemetic efficacy of nabilone and dexamethasone. A randomized study of patients with lung cancer receiving chemotherapy," *American Journal of Clinical Oncology* **10**: 325–329.

Orr, L. E., McKernan, J. F. and Bloome, B. (1980) "Antiemetic effect of tetrahydrocannabinol compared with placebo and prochloroperazine in chemotheraphy-associated nausea and emesis," *Archives of Internal Medicine* **140**: 1431–1433.

Orr, L. E. and McKernan, J. F. (1981) "Antiemetic effect of Δ^9-tetrahydrocannabinol in chemotherapy-associated nausea and emesis as compared to placebo and compazine," *Journal of Clinical Pharmacology* **21**: 76S–80S.

Osoba, D., Zee, B., Warr, D., Kaizer, L., Latreille, J. and Pater, J. (1996) "Quality of life studies in chemotherapy-induced emesis," *Oncology* **5**(Suppl. 1): 92–95.

Ossi, M., Anderson, E. and Freeman, A. (1996) "5-HT_3 receptor antagonists in the control of cisplatin-induced delayed emesis," *Oncology* **53**(Suppl. 1): 78–85.

Perez, E. A. (1998) "Use of dexamethasone with 5-HT_3-receptor antagonists for chemotherapy-induced nausea and vomiting," *Cancer Journal from Scientific American* **4**: 72–77.

Pertwee, R. G. (1997) "Pharmacology of cannabinoid CB_1 and CB_2 receptors," *Pharmacology and Therapeutics* **74**: 129–180.

Pertwee, R. G., Fernando, S. R., Nash, J. E. and Coutts, A. A. (1996) "Further evidence for the presence of cannabinoid CB_1 receptors in guinea-pig small intestine," *British Journal of Pharmacology* **118**: 2199–2205.

Rahminiwati, M. and Nishimura, M. (1999) "Effects of delta-9-tetrahydrocannabinol and diazepam on feeding behavior in mice," *The Journal of Veterinary Medical Science* **61**: 351–355.

Raybold, H. E., Grundy, D. and Andrews, P. L. R. (1995) "Serotonin and the afferent innervation of the gastrointestinal tract," in *Serotonin and Gastrointestinal Function*, edited by T. S. Gaginella and J. J. Galligan, pp. 127–147. Boca Raton: CRC Press.

Razdan, R. (1986) "Structure activity relationships in cannabinoids," *Pharmacological Reviews* **38**: 75–149.

Rinaldi-Carmona, M., Barth, F., Héalume, M., Shire, D., Calandra, G., Congy, C. *et al.* (1994) "SR 141716A, a potent and selective antagonist of the brain cannabinoid receptor," *FEBBS Letters* **350**: 240–244.

Rinaldi-Carmona, M., Barth, F., Millan, J., Derocq J. M., Casellas, P., Congy, C. *et al.* (1998) "SR 144528, the first potent and selective antagonist of CB_2 cannabinoid receptor," *Journal of Pharmacology and Experimental Therapeutics* **284**: 644–650.

Sallan, S. E., Cronin, C., Zelen, M. and Zinberg, N. E. (1980) "Antiemetics in patients receiving chemotherapy for cancer: a randomized comparison of delta-9-tetrahydrocannabinol and prochlorperazine," *New England Journal of Medicine* **302**: 135–138.

Sallan, S. F., Zinberg, N. E. and Frei, E. (1975) "Antiemetic effect on delta-9-tetrahydrocannabinol in patients receiving cancer chemotherapy," *New England Journal of Medicine* **293**: 795–797.

Sheidler, V. R., Ettinger, D. S., Diasio, R. B., Enterline, J. P. and Brown, M. D. (1984) "Double-blind multiple-dose crossover study of the antiemetic effect of intramuscular levonantradol compared to prochlorperazine," *The Journal of Clinical Pharmacology* **24**: 155–159.

Shook, J. E. and Burks, T. F. (1989) "Psychoactive cannabinoids reduce gastrointestinal propulsion and motility in rodents," *The Journal of Pharmacology and Experimental Therapeutics* **249**: 444–449.

Simoneau, I. I., Hamza, M. S., Mata, H. P., Siegel, E. M., Vanderah, T. W., Porreca, F., Makriyannis, A. and Malan, P. (2001) "The cannabinoid agonist WIN 55, 212-2 suppresses opioid-induced emesis in ferrets," *Anesthesiology* **94**: 882–887.

Smith, W. L., Callahan, E. M. and Alphen, R. S. (1988) "The emetic activity of centrally administered cisplatin in cats and its antagonism by zacopride," *The Journal of Pharmacy and Pharmacology* **40**: 142–143.

Stambaugh, J. E., McAdams, J. and Vreeland, F. (1984) "Dose ranging evaluation of the antiemetic efficacy and toxicity of intramuscular levonantradol in cancer subjects with chemotherapy-induced emesis," *Journal of Clinical Pharmacology* **24**: 480–485.

Staquet, M., Bron, D., Rozencweig, M. and Kenis, Y. (1981) "Clinical studies with a THC analog (BRL-4664) in the prevention of cisplatin-induced vomiting," *Journal of Clinical Pharmacology* **21**: 61S–63S.

Stark, P. (1982) "The pharmacologic profile of nabilone: a new antiemetic agent," *Cancer Treatment Reviews* **9**(Suppl. B): 11–16.

Steele, N., Gralla, R. J., Braun, D. W. and Young, C. W. (1980) "Double-blind comparison of the antiemetic effects of nabilone and prochlorperazine on chemotherapy-induced emesis," *Cancer Treatment Report* **64**: 219–224.

Stuart-Harris, R. C., Mooney, C. A. and Smith, I. E. (1983) "Levonantradol: a synthetic cannabinoid in the treatment of severe chemotherapy-induced nausea and vomiting resistant to conventional anti-emetic therapy," *Clinical Oncology* **9**: 143–146.

Sweet, D. L., Miller, N. J., Weddington, W., Seney, E. and Sushelsky, L. (1981) "Δ^9-tetrahydrocannabinol as an antiemetic for patients receiving cancer chemotherapy. A pilot study," *Journal of Clinical Pharmacology* **21**: 70S–75S.

Taylor, D. A. and Fennessy, M. R. (1982) "Time course of the effects of chronic Δ^9-tetrahydrocannabinol on behavior, body temperature, brain amines and withdrawal-like behavior in the rat," *The Journal of Pharmacy and Pharmacology* **34**: 240–245.

Torii, Y., Saito, H. and Matsuki, N. (1991) "Selective blockade of cytotoxic drug-induced emesis by 5-HT$_3$ receptor antagonists in Sulcus murinus," *Japanese Journal of Pharmacology* **55**: 107–113.

Tsou, K., Brown, S., Sanudo-Pena, M. C., Mackie, K. and Walker, J. M. (1998) "Immunohistochemical distribution of cannabinoid CB$_1$ receptors in the rat central nervous system," *Neuroscience* **83**: 393–411.

Ungerleider, J. T., Andrysiak, T., Fairbanks, L., Goodnight, J., Sarna, G. and Jamison, K. (1982) "Canabis and cancer chemotherapy. A comparison of oral delta-9-THC and prochlorperazine," *Cancer* **50**: 636–645.

Van Sickle, M. D., Oland, L. D., Ho, W., Hillard, C. J., Mackie, K., Davison, J. S. and Sharkey, K. A. (2001) "Cannabinoids inhibit emesis through CB$_1$ receptors in the brain stem," *Gastroenterology* **121**: 767–774.

Verweij, J., de Wit, R. and de Mulder, P. H. M. (1996) "Optimal control of acute cisplatin-induced emesis," *Oncology* **53**(Suppl. 1): 56–64.

Vinciguerra, V., Moore, T. and Brennan, E. (1988) "Inhalation marijuana as an antiemetic for cancer chemotherapy," *The New York State Medical Journal* **88**: 525–527.

Voth, E. A. and Schwartz, R. H. (1997) "Medical applications of delta-9-tetrahydrocannabinol and marijuana," *Annals of Internal Medicine* **126**: 791–798.

Welsh, J., Stuart, F., Sangster, G., Milstead, R., Kaye, S., Cash, H. *et al.* (1983) "Oral levonantradol in the control of cancer chemotherapy-induced emesis," *Cancer Chemotherapy and Pharmacology* **11**: 66–67.

Williams, C. M. and Kirkham, T. C. (1999) "Anandamide induces overeating: mediation by central cannabinoid (CB1) receptors," *Psychopharmacology* (Berl) **143**: 315–317.

Williams, C. M., Rogers, P. J. and Kirkham, T. C. (1998) "Hyperphagia in pre-fed rats following oral delta-9-THC," *Physiology and Behavior* **65**: 343–346.

Yoshida, N., Omoya, H. and Ito, T. (1992) "DAT-582, a novel 5-HT$_3$ receptor antagonist, is a potent and long lasting antiemetic agent in ferret and dog," *Journal of Pharmacology and Experimental Therapeutics* **260**: 1159–1165.

Yoshikawa, T., Yoshida, N. and Hosoki, K. (1996) "Involvement of dopamine D$_3$ receptor in the area postrema in R(+)-7-OH DPAT-induced emesis in the ferret," *European Journal of Pharmacology* **301**: 143–149.

Chapter 14

Cannabis and prostaglandins: an overview

Sumner H. Burstein

ABSTRACT

Several areas where the cannabinoids (CB) and other *Cannabis* components are interrelated to the prostaglandins (PG) are reviewed. These include (1) the actions of cannabinoids as stimulators of eicosanoid synthesis in both *in vitro* and *in vivo* systems; (2) The endocannabinoids which represent a novel class of eicosanoids; (3) The cannabinoid acids that act as inhibitors of eicosanoid synthesis and share some of the actions of the primary cannabinoids but antagonize other actions; and (4) The non-cannabinoid constituents of *Cannabis* such as flavones and the volatile oil fraction as well as pyrolysis products that occur in marijuana smoke. In addition to better characterizing some of the actions of the cannabinoids, these observations on PG–CB relationships provide a possible hypothesis for the mechanisms involved in certain of these actions.

Key Words: cannabinoid, eicosanoid, prostaglandin, tetrahydrocannabinol, endocannabinoid, arachidonic acid

INTRODUCTION

Relationships between cannabinoids (CB) and prostaglandins (PG) were first reported almost three decades ago (Burstein and Raz, 1972; Burstein *et al.*, 1973) when it was shown that CBs had an inhibitory effect on PG synthesis in subcellular preparations with cyclooxygenase (COX) activity (see Figure 14.1 for structures). These studies were modeled after the seminal work of Vane and coworkers (Flower *et al.*, 1972) on the mechanism of action of aspirin in which many of the effects of the NSAIDS could be understood by a reduction in PG synthesis. It is interesting to note that Paton *et al.* (1972) once remarked that *Cannabis* "would have amounted to a modest substitute for aspirin in the days before aspirin was available". Subsequent work has shown that the anti-inflammatory effects of CBs are primarily due to their acid metabolites (reviewed in Burstein, 1999).

Several reviews on the subject of CB–PG relationships have been published in the years since 1972 (Burstein, 1977, 1985, 1987, 1991, 1992, 1999; Martin, 1986; Burstein and Hunter, 1981b; Howes and Osgood, 1976; Burstein *et al.*, 1995a). These have emphasized different aspects of the literature such as the possible role of PGs in the actions of CBs. A wider ranging coverage of the topic will be presented in this chapter. The fact that the known endocannabinoids are derivatives of arachidonic acid (Figure 14.1) has added a new dimension to this subject since it

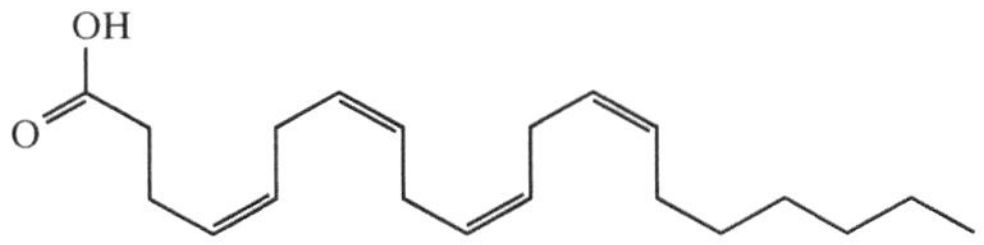

ARACHIDONIC ACID

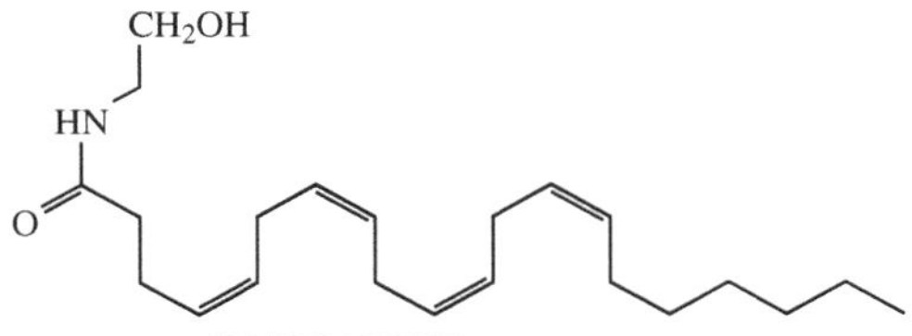

ANANDAMIDE

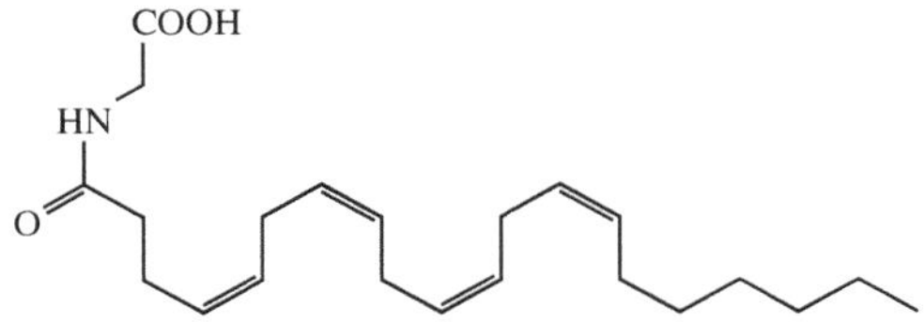

N-ARACHIDONYL GLYCINE (NAGly)

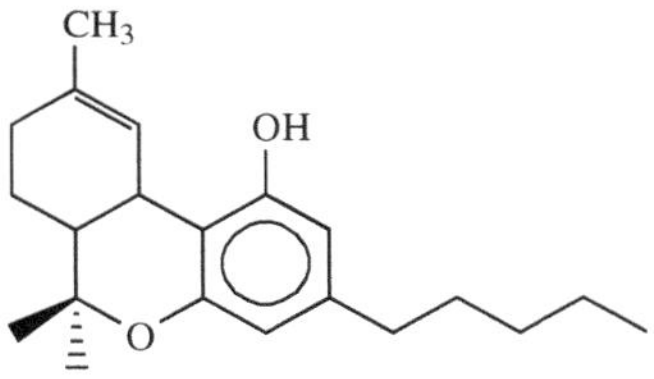

THC

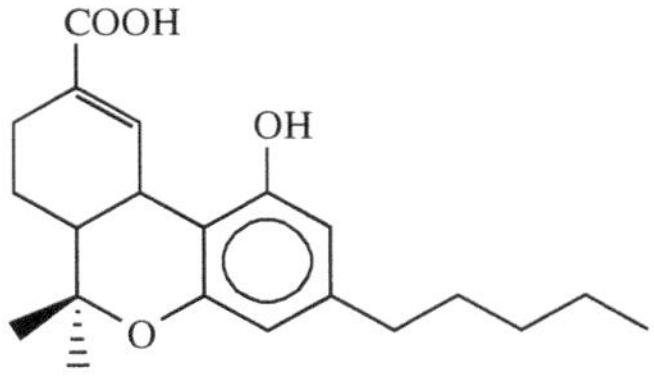

THC-11-OIC ACID

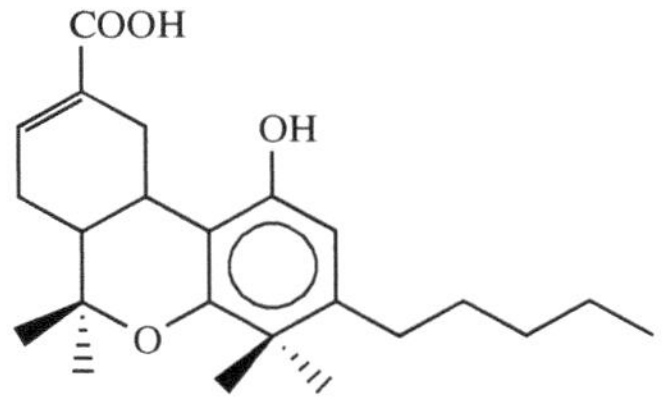

AJULEMIC ACID

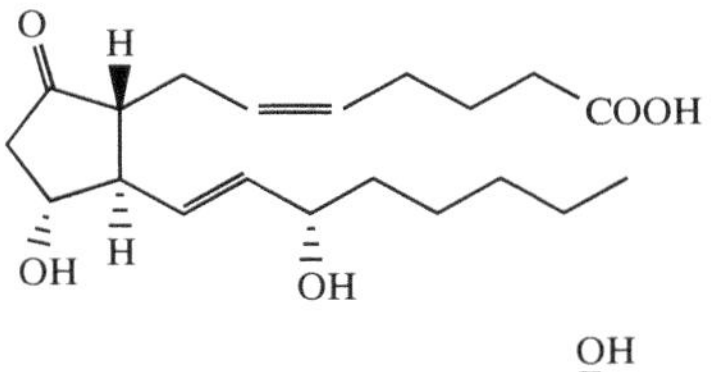

PROSTAGLANDIN E$_2$

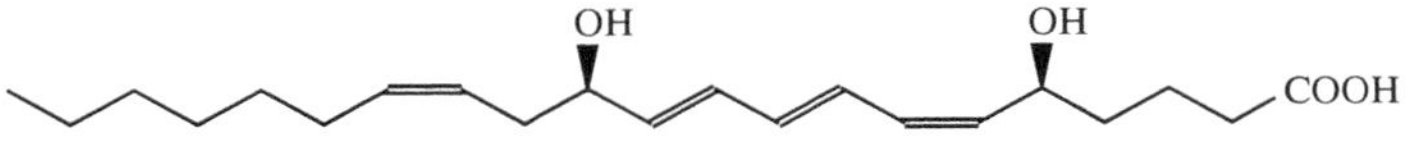

LEUKOTRIENE B$_4$

Figure 14.1 The structures of molecules discussed in this paper.

could be argued that CBs are, in fact, novel members of the eicosanoid family. Several aspects of anandamide literature will also be included in this review.

EXOCANNABINOIDS AS AGONISTS IN PG SYNTHESIS

In vitro preparations

As mentioned above, under certain conditions the primary cannabinoids (THC, CBD etc.) and some of their metabolites inhibit the synthesis of eicosanoids. However, in most instances where systems were exposed to cannabinoids stimulatory effects were observed. The basis of these actions seems to reside in the ability of cannabinoids to mobilize phospholipid bound arachidonic acid, which then enters the cascade of reactions leading to various eicosanoids (Figure 14.2). This suggests either a direct or indirect action of these cannabinoids on one or more of the cellular phospholipases. Several reports using subcellular preparations with phospholipase activity have been published that lend support to this hypothesis (Burstein and Hunter, 1981b; Hunter *et al.*, 1984, 1986; Evans *et al.*, 1987). The activation of phospholipases is generally considered to be a major physiological control point for the regulation of tissue levels of eicosanoids in response to inflammatory and other stimuli. Thus, the cannabinoids would be another example of a class of agonists for this process.

Intact cell models

Cells in culture are useful systems in which to study the effects of cannabinoids on arachidonic acid metabolism and eicosanoid synthesis (Figure 14.2). In the

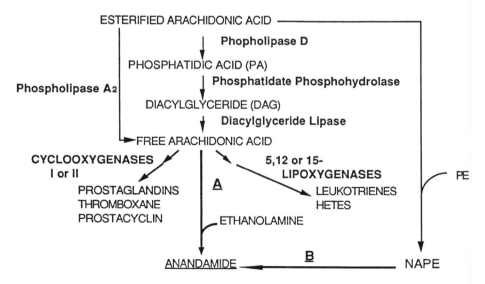

Figure 14.2 Biosynthesis of anandamide and the classical eicosanoids. NAPE = *N*-acyl-phosphatidyl ethanolamine; HETE = hydroxyeicosatetraenoic acid; PE = phosphatidyl ethanolamine (Burstein *et al.*, 2000).

appropriate cell type, one could expect to find the entire signal transduction pathway, that is, ligand–receptor interaction to the cascade responsible for eicosanoid production. To observe an effect, a cell model must be chosen that contains adequate levels of each of the components needed for this process. For example, the expression levels of CB_1 and CB_2 may vary considerably from one cell type to another. Moreover, since these are both G-protein coupled receptors, the specific G-protein heterotrimeric complex needed must also be present at a level adequate to transmit the activated receptor signal. This latter point is especially relevant when cells transfected with either CB_1 or CB_2 are being studied. The overexpressed levels of receptor may not find enough G-protein present to show a true ligand response.

Despite the potential problems cited above, much useful data on CB–PG interactions has been gathered using cell culture models. Studies have been published using HeLa cells (Burstein and Hunter, 1978), WI-38 human lung fibroblasts (Burstein *et al.*, 1982a, 1983, 1986; Hunter *et al.*, 1984), a primary culture of resident mouse peritoneal macrophages (Burstein *et al.*, 1984) and cultured neurons (Chan *et al.*, 1998). In all cases the observed increase in eicosanoid production was paralleled by a stimulation of arachidonic acid release. Structure–activity relationship studies were also examined and have yielded mixed results. When the PG stimulation of THC and a series of its metabolites was compared with published data on "psychoactivity", a close correlation was observed. However, when THC and several of the primary cannabinoids were similarly compared, a poor correlation was seen. These observations might now be explained by the existence of at least two cannabinoid receptor types.

One of the hallmarks of the cannabinoids is the development of tolerance to many of their *in vivo* actions. Interestingly, an analogous effect was found when WI-38 cells were treated repeatedly with THC (Burstein *et al.*, 1985). Daily administration of THC resulted in a successive decrease in PG synthesis along with a decrease in arachidonic acid mobilization.

The involvement of G-proteins in CB stimulated PG synthesis was suggested in models using mouse peritoneal cells and S49 lymphocytes (Audette *et al.*, 1991). The peritoneal cell preparation consists primarily of macrophages in which the CB_2 receptor is highly expressed. This model was also used to demonstrate possible roles for phopholipases A_2 and D in the THC stimulation of arachidonic acid release (Burstein *et al.*, 1994). Direct evidence for the activation of phopholipase A_2 was obtained by Western blot analysis in a gel shift assay.

Effects of cannabinoids on prostaglandin levels in the CNS

In rats, THC at 2 mg/kg caused a specific 33% reduction of PGE_2 when measured in a bioassay procedure (Coupar and Taylor, 1982). This was accompanied by the induction of hypothermia and catatonia followed by sedation. Two points need to be made on these observations. First, is the non-specific nature of the measurements and second, is the possible involvement of THC metabolites in the inhibitory effect on PG levels. By contrast, in a study using radioimmunoassay, similar doses of THC in the rat produced 2–3 fold increases in brain levels of

PGE_2 (Bhattacharya, 1986). Thus, as was the case with *in vitro* studies, the experimental conditions appear to play an important role in the direction of the THC effect on PG levels.

The above findings were supported in a report showing both increases and decreases in the amounts of THC-induced PG synthesis in various brain regions (Reichman *et al.*, 1987). THC-induced brain levels of PGE_2 in the rat could be reduced by the prior administration of i.v. antiserum to PGE_2 (Hunter *et al.*, 1991). This finding suggests the possibility that the brain PGs may be of peripheral origin since it can be expected that the antiserum would not cross the blood–brain barrier.

One study on PG–CB interactions in humans has been published (Perez-Reyes *et al.*, 1991). Blood levels of PGs when measured by immunoassay were seen to increase about twofold shortly after volunteers smoked a standardized marijuana cigarette. Both the increase in PG concentration and several of the effects of the drug were reduced by the prior administration of indomethacin, a potent inhibitor of eicosanoid synthesis. In particular, the THC-induced impairment of time estimation was reduced greatly by indomethacin.

Similar effects were observed with eicosanoid inhibitors in the THC-induced cataleptic response in mice, suggesting that this action may involve mediation by prostaglandins (Fairbairn and Pickens, 1979; Burstein *et al.*, 1987). In addition, the cataleptic effect was restored by the administration of PGE_2. Further evidence for the role of eicosanoids is found in the observation that mice fed a diet deficient in arachidonic acid exhibited a reduced cataleptic response to THC whereas exogenous arachidonic acid restored catalepsy (Fairbairn and Pickens, 1980). Finally, additional support for the involvement of PGs in THC-induced catalepsy is provided by the report that anti-PGE_2 antibodies suppress THC-induced responses in mice (Burstein *et al.*, 1989).

Further evidence suggesting that prostaglandins have a role in the behavioral effects of THC were recently reported by Yamaguchi *et al.* (2001). Using the rat lever-pressing performance model, they found that the effects of both THC and HU-210 were reduced by several diverse COX inhibitors. Moreover, the cannabinoid response was mimicked by the i.c.v. administration of PGE_2. It is further pointed out that both the CB_1 and PGE receptors are highly expressed in the substantia nigra region in the brain.

Prostaglandins and cannabinoids in peripheral systems

Since their discovery, the prostaglandins have been known to have important roles in the reproductive systems. This has prompted investigators looking for mechanisms to explain the effects of cannabinoids on reproduction to study a possible role for the prostaglandins (Jordan and Castracane, 1976; Ayalon *et al.*, 1977; Dalterio *et al.*, 1978, 1981; Rettori *et al.*, 1990). Cannabinoids appear to modulate PG levels in male and female models where they show both stimulatory and inhibitory actions. In another type of endocrine system, the pancreatic islet, where THC exerts an effect, the involvement of eicosanoids was suggested (Laychock *et al.*, 1986).

Another body system where the prostaglandins play a major role is in the cardiovascular system. Both THC-induced hypotension and bradycardia could be inhibited by the prior administration of aspirin in dogs (Burstein *et al.*, 1982b). The inhibition of THC-induced synthesis of vasoactive prostaglandins could explain this type of response. In a related effect, aspirin inhibited the effects of THC on lung perfusion pressure (Kaymacalan and Turker, 1975).

Intraocular pressure is sensitive to levels of prostaglandins in the eye, and the pressure lowering effect of THC is well known. Evidence has been published supporting the idea that the beneficial effect of THC in glaucoma may be explained by a lowering of intraocular PGE_2 (Green and Kim, 1976, 1977; Green and Podos, 1974). Finally, there is some evidence that there are interactions between cannabinoids and prostaglandins that impact on intestinal physiology (Jackson *et al.*, 1976a,b; Coupar and Taylor, 1983).

Cyclic AMP and prostaglandins in cannabinoid action

Prostaglandins have a profound effect on cyclic AMP metabolism through their interaction with adenyl cyclase. Thus, it might be expected that the well-documented effect of cannabinoids on hormone stimulated adenyl cyclase activity may involve mediation by prostaglandins. The evidence reported thus far is inconclusive and suggests that a complex interrelationship exists (Kelly and Butcher, 1973, 1979a,b; Hillard and Bloom, 1983; Howlett, 1984; Howlett and Fleming, 1984).

ENDOCANNABINOIDS: EICOSANOIDS WITH CANNABINOID ACTIVITIES

A large and rapidly growing body of literature exists on the endocannabinoids that would be far beyond the scope of this chapter to review. Instead, only those areas deemed relevant to the possible relationships between cannabinoids and eicosanoids will be mentioned.

Structural and functional comparisons

Prostaglandins are members of a larger group of naturally occurring substances called eicosanoids. The latter are defined by their common biosynthetic origin, namely, eicosatetraenoic acid commonly known as arachidonic acid. The pathways leading to the eicosanoids are shown in Figure 14.2 and involve phospholipases, cyclooxygenases and lipoxygenases as the major mediators. The eicosanoids all originate from arachidonate-containing cellular phospholipid storage pools and are rapidly synthesized following an appropriate physiological or pathological stimulus. Unlike many transmitter molecules, they are not stored to any extent prior to fulfilling their biological function. Generally, they act locally although there may be exceptions where circulating levels are important for expression of their actions.

The most important endogenous cannabinoids discovered thus far are derived from arachidonic acid and, as such, by definition belong to the eicosanoid family.

There may be deeper and more significant relationships since there are data showing that anandamide can stimulate PG synthesis as an *in vitro* (Wartmann *et al.*, 1995) or *in vivo* (Ellis *et al.*, 1995) agonist. Moreover, anandamide can, in principle, serve as a precursor for eicosanoid synthesis either by the direct action of COX and LOX, or by supplying free arachidonic acid for eicosanoid synthesis through the action of FAAH (Pratt *et al.*, 1998).

Stimulation of anandamide synthesis

Much attention has been directed toward the hydrolysis of anandamide as a possible regulatory mechanism, however, the question of agonists for endocannabinoid synthesis has not been studied extensively. The possibility that THC serves as such an agonist is raised by a report showing that in cultured N18TG2 cells $16 \mu M$ THC causes a 100% increase in anandamide synthesis (Burstein and Hunter, 1995b). Interestingly, this was accompanied by a similar increase in the mobilization of free arachidonic acid. Similar findings were reported in a model using rat cortical astrocytes (Shivachar *et al.*, 1996). The effect was antagonized by SR141716a suggesting the involvement of CB_1. A subsequent report using antisense technology gave supporting evidence that this effect was mediated by either CB_1 or CB_2 depending on the cell type that was used (Hunter and Burstein, 1997).

A later study done in the same laboratory gave evidence that non cannabinoid physiological agonists were also effective in stimulating anandamide synthesis in RAW264.7 mouse macrophages (Pestonjamasp and Burstein, 1998). These included platelet activating factor (PAF), bacterial lipopolysaccharide (LPS) and anandamide itself. There appeared to be a positive correlation between the release of arachidonic acid and anandamide synthesis for all three agonists. On the other hand, nitric oxide, which increased free arachidonic acid, had little effect on anandamide synthesis. These data suggest that the stimulation of anandamide production may be part of the inflammatory response sequence of events. The precise role of anandamide in this process is not apparent at this time.

COX-2 substrate

While free arachidonic acid is required for COX-1 mediated conversion to prostaglandins, the requirements for COX-2 conversion are less rigid. An interesting example was recently reported in which it was shown that anandamide is a good substrate for COX-2 but not for COX-1 (Yu *et al.*, 1997; So *et al.*, 1998). The reaction was observed using either purified enzymes or intact cells. The major product produced by these models was the ethanolamide of PGE_2, a novel prostaglandin with no known biological function (Figure 14.3). The rate of COX-2 oxygenation of anandamide was slower in the presence of added arachidonic acid (So *et al.*, 1998).

PGE_2 ethanolamide did not compete with CP55,940 for binding to the CB_1 receptor at a concentration of $100 \mu M$ (Pinto *et al.*, 1994). The non-physiological ethanolamides of the PGA and PGB also showed no binding activity. It might be instructive to look at their binding activities with CB_2 or the appropriate PG

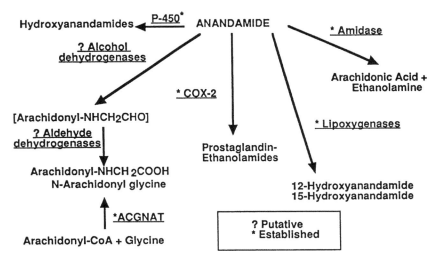

Figure 14.3 Metabolism of anandamide. The bioconversions shown are either taken from published reports (established), or, are proposed in this review (putative). ACGNAT; acyl-CoA:glycine N-acyltransferase (Burstein *et al.*, 2000).

receptors. An explanation for the lack of binding to CB_1 based on conformational analysis has been reported (Barnett-Norris *et al.*, 1998).

CANNABINOID ACIDS: INHIBITORS OF PROSTAGLANDIN SYNTHESIS

Metabolic origin

The *in vivo* metabolism of THC ultimately leads to a number of products containing a carboxyl group either at the 11 position (Figure 14.4) or at the end of the sidechain (Burstein, 1985; Harvey, 1987). The former is the predominant metabolite of THC and is notable for its relatively low affinity for the CB_1 receptor (Rhee *et al.*, 1997; Compton *et al.*, 1993; Yamamoto *et al.*, 1998) which is reflected in its lack of psychotropic activity in humans (Perez-Reyes, 1985). An interesting property of the acids is their inhibitory action on PG synthesis in cell culture models that suggests possible analgesic or anti-inflammatory activities similar to the NSAIDS (Burstein *et al.*, 1986). The structural basis for this profound change in pharmacological profile between THC and its hyroxylated metabolites on the one hand, and the CB acids on the other hand is not at all clear at this time. The identification of a high-affinity, saturable binding site for the acids would help to answer this question. The search for cannabinoid receptors other than CB_1 or CB_2 has thus far been unsuccessful, possibly due to the lack of an appropriate radio-labelled ligand.

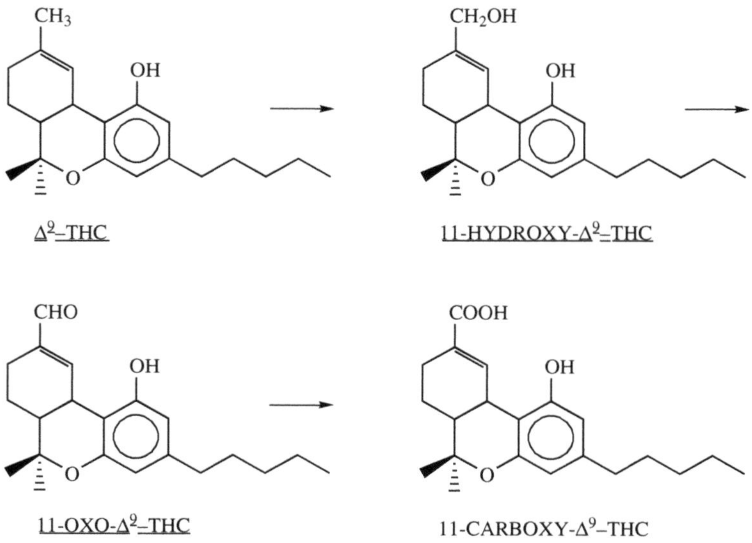

Figure 14.4 The principle route of metabolism for THC. The underlined structures show typical cannabimimetic properties.

Role of the acid metabolites in the actions of THC

Even though THC-11-oic acid binds weakly to the CB_1 receptor (Rhee *et al.*, 1997; Compton *et al.*, 1993; Yamamoto *et al.*, 1998), it effectively antagonizes the cataleptic effects of THC (Burstein *et al.*, 1987). In this respect it resembles the action of certain NSAIDs suggesting that it may be efficacious in NSAID models *in vivo*, and several reports have been published supporting this idea (Burstein *et al.*, 1988; Doyle *et al.*, 1990; Audette *et al.*, 1991; Audette and Burstein, 1990). A common property shared by the CB acids and the NSAIDs is their ability to inhibit prostaglandin synthesis providing a possible explanation for these observations on their actions.

There are interesting implications for the possible *in vivo* role of the acid as a regulator of some of the actions of THC. Since its levels in the body will increase with time following *Cannabis* exposure, it may limit or even reduce the psychotropic actions of the drug. On the other hand, the acid shares several of the pharmacological properties of THC. For example, in the mouse hot plate assay it shows activity comparable to THC (Burstein *et al.*, 1988) suggesting that some, if not most, of the anti-nociceptive action of THC might be due to its metabolite.

Ajulemic acid: a potent synthetic analog of THC-11-oic acid

The discovery of the properties of THC-11-oic acid raised the possibility that the long sought after goal of finding a non-psychoactive cannabinoid with useful therapeutic actions might be attained using it as a template molecule. A problem with THC-11-oic acid is its relatively low potency since in mice a dose of 20–40 mg/kg p.o. is generally required. The strategy employed in designing a synthetic analog (Burstein *et al.*, 1992) was taken from the well known principle that extending and branching of the sidechain will result in a significant increase in potency for most cannabinoids (Loev *et al.*, 1973). The molecular basis for this effect is not well understood, however, it may involve some form of hydrophobic interaction with the relevant binding site.

The application of this principle to the CB acids resulted in the molecule called ajulemic acid (Figure 14.1). As predicted, the analog has a pharmacological profile similar to the template molecule but at doses that are considerably lower (Burstein *et al.*, 1992; Dajani *et al.*, 1999; Zurier *et al.*, 1998). Interestingly, it shows an affinity for CB_1 comparable to that of THC (Rhee *et al.*, 1997) posing something of a dilemma since it lacks the psychotropic actions of THC. While ajulemic acid likewise inhibits COX-2 mediated PG synthesis, it is becoming apparent that its mechanism of action may be more complex. For example, like THC it stimulates the release of arachidonic acid in cell culture models (Burstein, unpublished data). What can be stated is that it is highly effective as both an analgesic and anti-inflammatory agent and shows no psychotropic action at therapeutic doses. Moreover, it has been subjected to a rigorous screening for toxic effects (S. Miller, personal communication) with negative results and was approved by the FDA for testing in human subjects. It has also been discovered that ajulemic acid has potent and selective anti proliferative effects on cancer cells (Recht *et al.*, personal communication). A preliminary *in vivo* study in a mouse model for brain cancer showed promise for its potential use in the clinic as a chemotherapeutic agent with generally low toxic side effects (Recht *et al.*, 2001).

N-arachidonylglycine (NAGly): a putative endogenous cannabinoid acid

The most studied pathway for the metabolism of anandamide involves the hydrolysis of the amide bond (Figure 14.3, Deutsch and Chin, 1993). This process has been postulated to be a possible mechanism for the physiological regulation of anandamide levels. Products are also produced through the actions of various lipoxygenases (Ueda *et al.*, 1995; Hampson *et al.*, 1995; Edgemond *et al.*, 1998) and, interestingly, anandamide is a good substrate for COX-2 giving rise to ethanolamide conjugates of PGE_2 (Yu *et al.*, 1997).

Preliminary data on the existence of an oxidative pathway involving the hydroxyl group of anandamide have been reported recently (Burstein *et al.*, 2000). The product of this metabolic route is the carboxyl derivative (Figure 14.3) that is, in essence, a conjugate of arachidonic acid and glycine (NAGly). Such a molecule could, of course, arise by a direct coupling of arachidonic acid and glycine and such a pathway is known to exist. NAGly appears to have a pharmacological profile different from anandamide and similar to the carboxyl metabolite of THC, namely, THC-11-oic acid. This raises the possibility that NAGly may be a novel endocannabinoid that acts through some mechanism other than activation of CB_1 or CB_2.

NON-CANNABINOID CONSTITUENTS OF *C. SATIVA*

In addition to its cannabinoid content, the marijuana plant contains a number of constituents with potential pharmacological activity (Fairbairn and Pickens, 1981). This may be the basis for claims by some medicinal users of the drug that greater benefit is derived from smoking marijuana as compared with the use of pure THC in the form of Marinol capsules. Several of these non-cannabinoid substances have been investigated and reported to have modulating effects on PG synthesis suggesting a biochemical basis for these claims.

Flavones

Two novel prenylated flavones were isolated from the ethanol extract of *Cannabis* and their structures elucidated (Barrett *et al.*, 1986). These were tested for cyclooxygenase inhibition in both seminal vesicle microsomes, a COX-1 preparation, and in rheumatoid synovial cells, a COX-2 preparation (Barrett *et al.*, 1985; Evans *et al.*, 1987). One of the compounds, cannflavin, was 30 times more potent than aspirin, however, it was less active than either indomethacin or dexamethasone. Thus, its presence should be considered in any study on the actions of the whole marijuana plant especially analgesic and anti-inflammatory effects.

Volatile oil constituents

Steam distillation of dried marijuana leaves gives a complex mixture of terpenes and other volatile oil components (Burstein *et al.*, 1975). These were separated by chromatographic methods and the fractions assayed for COX-1 inhibition. Several of the known components such as eugenol showed modest activity, and in addition, two new fractions were found to be active. In a subsequent study, one of these was identified as p-vinyl phenol (Burstein *et al.*, 1976). This molecule probably arises from the thermal decomposition of p-hydroxycinnamic acid, a known component of *Cannabis*.

Pyrolysis products

The phenolic cracking products of cannabidiol were isolated and tested for COX-1 inhibition (Spronck *et al.*, 1978). Several of the products showed activity, the most potent being 2-methylolivetol which was about 12 times more potent than cannabidiol but five times less active than indomethacin. Since cannabidiol is usually the most abundant cannabinoid in the plant, these pyrolysis products may add to the analgesic/anti-inflammatory effects of smoked *Cannabis*. It is probably safe to assume that THC also gives rise to bioactive products when subjected to pyrolysis conditions thus making the pharmacology of marijuana smoke a rather complex subject.

CONCLUSIONS

The information thus far available strongly suggests that eicosanoids have an important role in mediating many of the actions of the exocannabinoids. Much of the data were obtained before the discoveries of the cannabinoid receptors CB_1 and CB_2 with their different ligand specificities. Moreover, the importance of G-protein specificity in a particular mechanism was not fully appreciated at that time. The focus then was on explaining the psychotropic actions of cannabinoids, which are due mainly to CB_1 interaction. It may be, for example, that CB_2-prostaglandin interactions could explain other actions such as the effects of cannabinoids on the immune system. Perhaps a fresh look at possible mechanisms where eicosanoids have the dominant role would yield some useful hypotheses.

Another dimension to the subject of PG–CB interactions was added with the discovery that cannabinoid acids have important actions and are also inhibitors of eicosanoid synthesis. Since the acids have little or no psychotropic activity, they make attractive template molecules for the design of clinically useful drugs. This strategy has already showed promise with the discovery of ajulemic acid.

Finally, the intriguing similarities between the endocannabinoids and the eicosanoid family have raised some interesting questions. For example, is the regulation of their biosyntheses somehow interdependent? With regard to the whole subject of PG–CB interactions, it seems appropriate to quote the philosopher Goethe:

"It is easier to perceive error than to find truth,
for the former lies on the surface and is easily seen,
while the latter lies in the depth, where few are willing to search for it."
Johann Wolfgang Von Goethe

ACKNOWLEDGMENTS

This publication was made possible by grant numbers DA12178 and DA09017 from NIDA. Its contents are solely the responsibility of the author and do not necessarily represent the official views of the National Institute on Drug Abuse. The assistance of Pusha Karim in preparing this review is gratefully acknowledged.

REFERENCES

Audette, C. A. and Burstein, S. (1990) "Inhibition of leukocyte adhesion by the *in vivo* and *in vitro* administration of cannabinoids," *Life Sciences* **47**: 753–759.

Audette, C. A., Burstein, S. H., Doyle, S. A. and Hunter, S. A. (1991) "G-Protein mediation of cannabinoid-induced phospholipase activation," *Pharmacology, Biochemistry and Behavior* **40**: 559–563.

Ayalon, D., Nir, I., Cordova, T., Bauminger, S., Puder, M., Naor, Z., Kashi, R., Zor, U., Harell, A. and Lindner, H. R. (1977) "Acute effect of delta1-tetrahydrocannabinol on the hypothalamo-pituitary-ovarian axis in the rat," *Neuroendocrinology* **23**: 31–42.

Barnett-Norris, J., Guarnieri, F., Hurst, D. P. and Reggio, P. H. (1998) "Exploration of biologically relevant conformations of anandamide, 2-arachidonylglycerol, and their analogues using conformational memories," *Journal of Medicinal Chemistry* **41**: 4861–4872.

Barrett, M. L., Gordon, D. and Evans, F. J. (1985) "Isolation from Cannabis sativa L. of cannflavin a novel inhibitor of prostaglandin production," *Biochemical Pharmacology* **34**: 2019–2024.

Barrett, M. L., Scutt, A. M. and Evans, F. J. (1986) "Cannflavin A and B, prenylated flavones from *Cannabis sativa* L.," *Experientia* **42**: 452–453.

Bhattacharya, S. K. (1986) "Delta-9-tetrahydrocannabinol (THC) increases brain prostaglandins in the rat," *Psychopharmacology* **90**: 499–502.

Burstein, S. (1977) "Prostaglandins and *Cannabis* IV: a biochemical basis for therapeutic applications," in *The Therapeutic Potential of Marihuana*, edited by S. Cohen and R. C. Stillman, pp. 19–33. Plenum.

Burstein, S. (1985) "Biotransformations of the cannabinoids," in *Pharmacokinetic and Pharmacodynamics of Psychoactive Drugs*, edited by G. Barnett and N. Chiang, pp. 396–414. Biomed. Pubs.

Burstein, S. (1987) "Inhibitory and stimulatory effects of cannabinoids on eicosanoid synthesis," *NIDA Monograph* **79**: 158–172.

Burstein, S. (1991) "Cannabinoid-induced changes in eicosanoid synthesis by mouse peritoneal cells," in *Drugs of Abuse, Immunity and Immunodeficiency*, edited by H. Friedman, pp. 107–112. Plenum.

Burstein, S. (1992) "Eicosanoids as mediators of cannabinoid action," in *Marijuana/Cannabinoids*, edited by L. Murphy and A. Bartke. CRC Press.

Burstein, S. H. (1999) "The cannabinoid acids: non psychoactive derivatives with therapeutic potential," *Pharmacology and Therapeutics* **82**: 87–96.

Burstein, S. and Raz, A. (1972) "Inhibition of prostaglandin E2 biosynthesis by delta 1-tetrahydrocannabinol," *Prostaglandins* **2**: 369–374.

Burstein, S., Levin, E. and Varanelli, C. (1973) "Prostaglandins and *Cannabis* II. Inhibition of biosynthesis by the naturally occurring cannabinoids," *Biochemical Pharmacology* **22**: 2905–2910.

Burstein, S., Varanelli, C. and Slade, L. T. (1975) "Prostaglandins and *Cannabis* III. Inhibition of biosynthesis by essential oil components of marihuana," *Biochemical Pharmacology* **24**: 1053.

Burstein, S., Taylor, P., Turner, C. and El-Feraly, F. S. (1976) "Prostaglandins and *Cannabis* V. Identification of p-vinylphenol as a potent inhibitor of prostaglandin synthesis," *Biochemical Pharmacology* **25**: 2003.

Burstein, S. and Hunter, S. A. (1978) "Prostaglandins and *Cannabis* VI. Release of arachidonic acid from HeLa cells by delta 1-tetrahydrocannabinol and other cannabinoids," *Biochemical Pharmacology* **27**: 1275–1280.

Burstein, S. and Hunter, S. A. (1981a) "The biochemistry of the cannabinoids," *Pure Applied Pharmacological Sciences* **2**: 155–226.

Burstein, S. and Hunter, S. A. (1981b) "Prostaglandins and *Cannabis* VIII. Elevation of phospholipase A2 activity by cannabinoids in whole cells and subcellular preparations," *Journal of Clinical Pharmacology* **21**: 240S–248S.

Burstein, S., Hunter, S. A., Sedor, C. and Shulman, S. (1982a) "Prostaglandins and *Cannabis* IX. Stimulation of prostaglandin E2 synthesis in human lung fibroblasts by delta 1-tetrahydrocannabinol," *Biochemical Pharmacology* **31**: 2361–2365.

Burstein, S., Ozman, K., Burstein, E., Palermo, N. and Smith, E. (1982b) "Prostaglandins and *Cannabis* XI. Inhibition of delta 1-tetrahydrocannabinol-induced hypotension by aspirin," *Biochemical Pharmacology* **31**: 591–592.

Burstein, S., Hunter, S. A. and Ozman, K. (1983) "Prostaglandins and *Cannabis* XII. The effect of cannabinoid structure on the synthesis of prostaglandins by human lung fibroblasts," *Molecular Pharmacology* **23**: 121–126.

Burstein, S., Hunter, S. A., Ozman, K. and Renzulli, L. (1984) "Prostaglandins and *Cannabis* XIII. Cannabinoid-induced elevation of lipoxygenase products in mouse peritoneal macrophages," *Biochemical Pharmacology* **33**: 2653–2656.

Burstein, S., Hunter, S. A. and Renzulli, L. (1985) "Prostaglandins and *Cannabis* XIV. Tolerance to the stimulatory actions of cannabinoids on arachidonate metabolism," *The Journal of Pharmacology and Experimental Therapeutics* **235**: 87–91.

Burstein, S., Hunter, S. A., Latham, V. and Renzulli, L. (1986) "Prostaglandin and *Cannabis* XVI. Antagonism of 1-THC action by its metabolites," *Biochemical Pharmacology* **35**: 2553–2558.

Burstein, S., Hunter, S. A., Latham, V. and Renzulli, L. (1987) "A major metabolite of delta 1-tetrahydrocannabinol reduces its cataleptic effect in mice," *Experientia* **43**: 402–403.

Burstein, S. H., Hull, K., Hunter, S. A. and Latham, V. (1988) "Cannabinoids and pain responses: a possible role for prostaglandins," *The FASEB Journal* **2**: 3022–3026.

Burstein, S. H., Hull, K., Hunter, S. A. and Shilstone, J. (1989) "Immunization against prostaglandins reduces delta 1-tetrahydrocannabinol-induced catalepsy in mice," *Molecular Pharmacology* **35**: 6–9.

Burstein, S. H., Audette, C. A., Breuer, A., Devane, W. A., Colodner, S., Doyle, S. A. and Mechoulam, R. (1992) "Synthetic non-psychotropic cannabinoids with potent antiinflammatory, analgesic and leukocyte anti adhesion activities," *Journal of Medicinal Chemistry* **35**: 3135–3141.

Burstein, S., Budrow, J., Debatis, M., Hunter, S. A. and Subramanian, A. (1994) "Phospholipase participation in cannabinoid-induced release of free arachidonic acid," *Biochemical Pharmacology* **48**: 1253–1264.

Burstein, S., Young, J. K. and Wright, G. E. (1995a) "Relationships between eicosanoids and cannabinoids. Are eicosanoids cannabimimetic agents? COMMENTARY," *Biochemical Pharmacology* **50**: 1735–1742.

Burstein, S. H. and Hunter, S. A. (1995b) "Stimulation of anandamide biosynthesis in N-18TG2 neuroblastoma cells by THC," *Biochemical Pharmacology* **49**: 855–858.

Burstein, S. H., Rossetti, R. G., Yagen, B. and Zurier, R. B. (2000) "Oxidative metabolism of anandamide," *Prostaglandins & other Lipid Mediators* **61**: 29–41.

Chan, G. C., Hinds, T. R., Impey, S. and Storm, D. R. (1998) "Hippocampal neurotoxicity of Delta9-tetrahydrocannabinol," *The Journal of Neurosciences* **18**: 5322–5332.

Compton, D. R., Rice, K. C., De Costa, B. R., Razdan, R. K., Melvin, L. S., Johnson, M. R. and Martin, B. R. (1993) "Cannabinoid structure–activity relationships: correlation of receptor binding and *in vivo* activities," *The Journal of Pharmacology and Experimental Therapeutics* **265**: 218–226.

Coupar, I. M. and Taylor, D. A. (1982) "Alteration in the level of endogenous hypothalamic prostaglandins induced by delta 9-tetrahydrocannabinol in the rat," *British Journal of Pharmacology* **76**: 115–119.

Coupar, I. M. and Taylor, D. A. (1983) "Effect of delta 9-tetrahydrocannabinol on prostaglandin concentrations and fluid absorption rates in the rat small intestine," *The Journal of Pharmacy and Pharmacology* **35**: 392–394.

Dajani, E. Z., Larsen, K. R., Taylor, J., Dajani, N. E., Shahwan, T. G., Neeleman, S. D., Taylor, M. S., Dayton, M. T. and Mir, G. N. (1999) "1′,1′-Dimethylheptyl-delta-8-tetrahydrocannabinol-11-oic acid: a novel, orally effective cannabinoid with analgesic and anti-inflammatory properties," *The Journal of Pharmacology and Experimental Therapeutics* **31**: 291

Dalterio, S., Bartke, A., Roberson, C., Watson, D. and Burstein, S. (1978) "Direct and pituitary-mediated effects of delta9-THC and cannabinol on the testis," *Pharmacology of Biochemistry and Behavior* **8**: 673–678.

Dalterio, S., Bartke, A., Harper, M. J., Huffman, R. and Sweeney, C. (1981) "Effects of cannabinoids and female exposure on the pituitary-testicular axis in mice: possible involvement of prostaglandins," *Biology of Reproduction* **24**: 315–322.

Deutsch, D. G. and Chin, S. A. (1993) "Enzymatic synthesis and degradation of anandamide, a cannabinoid receptor agonist," *Biochemical Pharmacology* **46**(5):791–796.

Doyle, S. A., Burstein, S. H., Dewey, W. L. and Welch, S. P. (1990) "Further studies on the antinociceptive effects of delta 6-THC-7-oic acid," *Agents Actions* **31**: 157–163.

Edgemond, W. S., Hillard, C. J., Falck, J. R., Kearn, C. S. and Campbell, W. B. (1998) "Human platelets and polymorphonuclear leukocytes synthesize oxygenated derivatives of arachidonylethanolamide (anandamide): their affinities for cannabinoid receptors and pathways of inactivation," *Molecular Pharmacology* **54**: 180–188.

Ellis, E. F., Moore, S. F. and Willoughby, K. A. (1995) "Anandamide and delta 9-THC dilation of cerebral arterioles is blocked by indomethacin," *American Journal of Physiology* **269**: H1859–H1864.

Evans, A. T., Formukong, E. A. and Evans, F. J. (1987) "Actions of cannabis constituents on enzymes of arachidonate metabolism: anti-inflammatory potential," *Biochemical Pharmacology* **36**: 2035–2037.

Fairbairn, J. W. and Pickens, J. T. (1979) "The oral activity of delta′-tetrahydrocannabinol and its dependence on prostaglandin E2," *British Journal of Pharmacology* **67**: 379–385.

Fairbairn, J. W. and Pickens, J. T. (1980) "The effect of conditions influencing endogenous prostaglandins on the activity of delta′-tetrahydrocannabinol in mice," *British Journal of Pharmacology* **69**: 491–493.

Fairbairn ,J. W. and Pickens, J. T. (1981) "Activity of cannabis in relation to its delta′-trans-tetrahydro-cannabinol content," *British Journal of Pharmacology* **72**: 401–409.

Flower, R., Gryglewski, R., Herbaczynska-Cedro, K. and Vane, J. R. (1972) "Effects of anti-inflammatory drugs on prostaglandin synthesis," *Nature* **238**: 104–106.

Green, K. and Podos, S. M. (1974) "Antagonism of arachidonic acid-induced ocular effects by delta1-tetrahydrocannabinol," *Investigative Ophthalmology* **13**: 422–429.

Green, K. and Kim, K. (1976) "Interaction of adrenergic antagonists with prostaglandin E2 and tetrahydrocannabinol in the eye," *Investigative Ophthalmology* **15**: 102–111.

Green, K. and Kim, K. (1977) "Papaverine and verapamil interaction with prostaglandin E2 and delta9-Tetrahydrocannabinol in the eye," *Experimental Eye Research* **24**: 207–212.

Hampson, A. J., Hill, W. A., Zan-Phillips, M., Makriyannis, A., Leung, E., Eglen, R. M. and Bornheim, L. M. (1995) "Anandamide hydroxylation by brain lipoxygenase: metabolite structures and potencies at the cannabinoid receptor," *Biochimica et Biophysica Acta* **1259**: 173–179.

Harvey, D. J. (1987) "Mass spectrometry of the cannabinoids and their metabolites," *Biomed. Mass. Spectrom. Rev.* **6**: 135–229.

Hillard, C. J. and Bloom, A. S. (1983) "Possible role of prostaglandins in the effects of the cannabinoids on adenylate cyclase activity," *European Journal of Pharmacology* **91**: 21–27.

Howlett, A. C. (1984) "Inhibition of neuroblastoma adenylate cyclase by cannabinoid and nantradol compounds," *Life Sciences* **35**: 1803–1810.

Howlett, A. C., Fleming, R. M. (1984) "Cannabinoid inhibition of adenylate cyclase. Pharmacology of the response in neuroblastoma cell membranes," *Molecular Pharmacology* **26**: 532–538.

Howes, J. F. and Osgood, P. F. (1976) "Cannabinoids and the inhibition of prostaglandin synthesis," in *Marihuana: Chemistry, Biochemistry and Cellular Effects*, edited by G. Nahas, pp. 415–424. New York: Springer-Verlag.

Hunter, S. A., Burstein, S. and Sedor, C. (1984) "Stimulation of prostaglandin synthesis in WI-38 human lung fibroblasts following inhibition of phospholipid acylation by p-hydroxymercuribenzoate," *Biochimica et Biophysica Acta* **793**: 202–212.

Hunter, S. A., Burstein, S. and Renzulli, L. (1986) "Effects of cannabinoids on the activities of mouse brain lipases," *Neuro Chemical Research* **11**: 1273–1288.

Hunter, S. A., Audette, C. A. and Burstein, S. (1991) "Elevation of brain prostaglandin E2 levels in rodents by delta 1-tetrahydrocannabinol," *Prostaglandins Leukot Essent Fatty Acids*, **43**: 185–190.

Hunter, S. A. and Burstein, S. H. (1997) "Receptor mediation in cannabinoid stimulated arachidonic acid mobilization and anandamide synthesis," *Life Sciences* **60**: 1563–1573.

Jackson, D. M., Malor, R., Chesher, G. B., Starmer, G. A., Welburn, P. J. and Bailey, R. (1976) "The interaction between prostaglandin E1 and delta 9-tetrahydrocannabinol on intestinal motility and on the abdominal constriction response in the mouse," *Psychopharmacologia* **47**: 187–193.

Jordan, V. C. and Castracane, V. D. (1976) "The effect of reported prostaglandin synthetase inhibitors on estradiol-stimulated uterine prostaglandin biosynthesis *in vivo* in the ovariectomized rat," *Prostaglandins* **12**: 1073–1081.

Kaymacalan, S. and Turker, R. K. (1975) "The evidence of the release of prostaglandin-like material from rabbit kidney and guinea-pig lung by (–)-trans-delta9-tetrahydrocannabinol," *The Journal of Pharmacy and Pharmacology* **27**: 564–568.

Kelly, L. A. and Butcher, R. W. (1973) "The effects of delta 1-tetrahydrocannabinol on cyclic AMP levels in WI-38 fibroblasts," *Biochimica et Biophysica Acta* **320**: 540–544.

Kelly, L. A. and Butcher, R. W. (1979a) "Effects of delta 1-tetrahydrocannabinol (THC) on cyclic AMP metabolism in cultured human fibroblasts," *Progress in Clinical and Biological Research* **27**: 227–236.

Kelly, L. A. and Butcher, R. W. (1979b) "Effects of delta 1-tetrahydrocannabinol on cyclic AMP in cultured human diploid fibroblasts," *Journal of Cyclic Nucleotide Research* **5**: 303–313.

Laychock, S. G., Hoffman, J. M., Meisel, E. and Bilgin, S. (1986) "Pancreatic islet arachidonic acid turnover and metabolism and insulin release in response to delta-9-tetrahydrocannabinol," *Biochemical Pharmacology* **35**: 2003–2008.

Loev, B., Bender, P. E., Dowalo, F., Macko, E. and Fowler, P. J. (1973) "Cannabinoids: Structure–activity studies related to 1,2-dimethylheptyl derivatives," *Journal of Medicinal Chemistry* **16**: 1200–1206.

Martin, B. R. (1986) "Cellular effects of cannabinoids," *Pharmacology Reviews* **38**: 45–74.

Paton, W. D. M., Pertwee, R. G. and Temple, D. (1972) "The general pharmacology of cannabinoids," in *Cannabis and Its Derivatives*, edited by W. D. M. Paton and J. Crown, pp. 50–75. London: Oxford University Press.

Perez-Reyes M. (1985) "Pharmacodynamics of certain drugs of abuse," in *Pharmacokinetics and Pharmacodynamics of Psychoactive Drugs*, edited by G. Barnett, N. C. Chiang. Foster City: Biomedical Publishers.

Perez-Reyes, M., Burstein, S. H., White, W. R., McDonald, S. A. and Hicks, R. E. (1991) "Antagonism of marihuana effects by indomethacin in humans," *Life Sciences* **48**: 507–515.

Pestonjamasp, V. K. and Burstein, S. H. (1998) "Anandamide synthesis is induced by arachidonate mobilizing agonists in cells of the immune system," *Biochimica et Biophysica Acta* **1394**: 249–260.

Pinto, J. C., Potie, F., Rice, K. C., Boring, D., Johnson, M. R., Evans, D. M., Wilken, G. H., Cantrell, C. H. and Howlett, A. C. (1994) "Cannabinoid receptor binding and agonist activity of amides and esters of arachidonic acid," *Molecular Pharmacology* **46**: 516–178.

Pratt, P. F., Hillard, C. J., Edgemond, W. S. and Campbell, W. B. (1998) "N-arachidonylethanolamide relaxation of bovine coronary artery is not mediated by CB_1 cannabinoid receptor," *American Journal of Physiology* **274**: H375–H381.

Recht, L. D., Salmonsen, R., Rosetti, R., Jang, T., Pipia, G., Kubiatowski, T., Karim, P., Ross, A. H., Zurier, R., Litofsky, N. S. and Burstein, S. (2001) "Antitumor effects of ajulemic acid (CT3), a synthetic non-psychoactive cannabinoid," *Biochemical Pharmacology* **62**: 755–763.

Reichman, M., Nen, W. and Hokin, L. E. (1987) "Effects of delta 9-tetrahydrocannabinol on prostaglandin formation in brain," *Pharmacology* **32**: 686–690.

Rettori, V., Aguila, M. C., Gimeno, M. F., Franchi, A. M. and McCann, S. M. (1990) "*In vitro* effect of delta 9-tetrahydrocannabinol to stimulate somatostatin release and block that of luteinizing hormone-releasing hormone by suppression of the release of prostaglandin E2," *Proceedings of the National Academy of Sciences of the USA.* **87**: 10063–10066.

Rhee, M. H., Vogel, Z., Barg, J., Bayewitch, M., Levy, R., Hanus, L., Breuer, A. and Mechoulam, R. (1997) "Cannabinol derivatives: binding to cannabinoid receptors and inhibition of adenylylcyclase," *Journal of Medicinal Chemistry* **40**: 3228–3233.

Shivachar, A. C., Martin, B. R. and Ellis, E. F. (1996) "Anandamide- and delta9-tetra-hydrocannabinol-evoked arachidonic acid mobilization and blockade by SR141716A [N-(Piperidin-1-yl)-5-(4-chlorophenyl)-1-(2,4-dichlorophenyl)-4-methyl-1H-pyrazole-3-carboximidehyrochloride]," *Biochemical Pharmacology* **51**: 669–676.

So, O.-Y., Scarafia, L. E., Mak, A. Y., Callan, O. H. and Swinney, D. C. (1998) *The Journal of Biological Chemistry* **273**: 5801–5807.

Spronck, J. W., Lutein, M., Salemink, A. and Nugteren, H. (1978) "Inhibition of prostaglandin biosynthesis by derivatives of olivetol formed under pyrolysis of cannabidiol," *Biochemical Pharmacology* **27**: 607–608.

Ueda, N., Yamamoto, K., Yamamoto, S., Tokunaga, T., Shirakawa, E., Shinaki, H., Ogawa, M., Sato, T., Kudo, I., Inoue, K., Takizawa, H., Nagano, T., Hirobe, M. and Saito, H. (1995) "Lipoxygenase-catalyzed oxygenation of arachidonylethanolamide, a cannabinoid receptor agonist," *Biochimica et Biophysica Acta* **1254**: 127–134.

Wartmann, M., Campbell, D., Subramanian, A., Burstein, S. H. and Davis, R. J. (1995) "The MAP kinase signal transduction pathway is activated by the endogenous cannabinoid anandamide," *FEBS Letters* **359**: 133–136.

Yamaguchi, T., Shoyama, Y., Watanabe, S. and Yamamoto, T. (2001) "Behavioral suppression induced by cannabinoids is due to activation of the arachidonic acid cascade in rats," *Brain Research* **889**: 149–154.

Yamamoto, I., Kimura, T., Kamei, A., Yoshida, H., Watanabe, K., Ho, I. K. and Yoshimura, H. (1998) "Competitive inhibition of THC and its active metabolites for cannabinoid receptor binding," *Biological and Pharmaceutical Bulletin* **21**: 408–410.

Yu, M., Ives, D. and Ramesha, C. S. (1997) "Synthesis of PGE2 ethanolamide from anandamide by COX-2," *Journal of Biological Chemistry* **272**: 21181–21186.

Zurier, R. B., Rossetti, R. G., Lane, J. H., Goldberg, J. M., Hunter, S. A. and Burstein, S. H. (1998) "Dimethylheptyl-THC-11-oic acid: a non psychoactive antiinflammatory agent with a cannabinoid template structure," *Arthritis and Rheumatism* **41**: 163–170.

Chapter 15

Cannabinoid mediated signal transduction

Michelle Glass and Sean D. McAllister

ABSTRACT

Many scientists in the fields of pharmacology, physiology, and electrophysiology are focusing their efforts towards understanding the detailed mechanisms involved in cannabinoid receptor signal transduction. With cloned receptors, specific antagonists, and powerful approaches afforded to us through the use of molecular and cellular biology, an understanding of *in vivo* outcomes is being revealed by *in vitro* events. Like many other G-protein coupled receptors, cannabinoid receptors can activate a range of signal transduction pathways. This chapter reviews the studies to date that have investigated the type of G-proteins activated by cannabinoid receptors and the regions of these proteins involved in the interactions. We next consider the result of this activation on intracellular alterations in adenylate cyclase activity, modulation of ion channels, release of calcium from intracellular stores, and activation of transcription factors. This chapter also addresses the apparent constitutively active nature of cannabinoid receptors by discussing their susceptibility to inverse agonists. Receptor activation is a dynamic response which must be regulated by feedback mechanisms in order to achieve homeostasis within a physiological system. We consider these characteristics by relating *in vivo* tolerance to many recent *in vitro* studies that have investigated receptor desensitization, internalization, and up-regulation. Taken together the studies presented will demonstrate cannabinoid receptor mediated signal transduction to be a highly complex process which requires future investigation.

Key Words: cannabinoid, signal transduction, desensitization, internalization, G-protein

INTRODUCTION

Even prior to the discovery of the cannabinoid receptors, some indications of their signal transduction pathways had been determined. Early work demonstrated that Δ^9-THC could inhibit forskolin mediated cAMP accumulation in neuroblastoma cells and rat brain (Howlett and Fleming, 1984; Bidaut-Russell and Howlett, 1989) and that this effect was pertussis toxin sensitive, suggestive of a Gi/o protein mediated pathway. To date two subclasses of cannabinoid receptor have been isolated, CB_1, found primarily in the central nervous system (Matsuda *et al.*, 1990), and CB_2, located predominantly in the immune system (Munro *et al.*, 1993). The CB_1 and CB_2 receptors exhibit a low overall amino acid sequence identity (44%, with 68% in the transmembrane domains) but they share a common pharmacology and few of the available ligands distinguish between them. The sequences of the cannabinoid receptors were consistent with the seven transmembrane spanning domains characteristic of G-protein coupled receptors. The cloning of the cannabinoid receptors provided the tools for extensive study of the signal transduction of this family of receptors.

G-PROTEIN COUPLING

Cannabinoid receptors activate multiple intracellular signal transduction pathways. Figure 15.1 summarises the putative signal transduction pathways for the cannabinoid receptors that are discussed throughout this chapter. CB_1 and CB_2 receptor agonists inhibit forskolin-stimulated adenylate cyclase by activation of a pertussis toxin-sensitive G-protein (Felder *et al.*, 1995). However in transfected cells, CB_1 but not CB_2 receptors inhibit N- and P/Q-type calcium channels and activate inwardly rectifying potassium channels (Caulfield and Brown, 1992; Mackie and Hille, 1992; Felder *et al.*, 1995; Mackie *et al.*, 1995; Pan *et al.*, 1996). Inhibition of calcium channels and enhancement of inwardly rectifying potassium currents is pertussis toxin sensitive but independent of cAMP inhibition, suggesting that this inhibition is directly mediated by G-proteins (Mackie and Hille, 1992; Mackie *et al.*, 1995). An additional layer of complexity for the signaling of CB_1 receptors derives from their ability to stimulate cAMP formation under certain conditions, consistent with a possible G_s linkage of this receptor (Maneuf *et al.*, 1996; Glass and Felder, 1997; Bonhaus *et al.*, 1998).

When a G-protein coupled receptor is activated by an agonist, the rate of GDP/GTP exchange on the $G\alpha$ subunit increases favouring a GTP-bound G-protein. By using a radioactive non-hydrolysable analog of GTP ([γS^{35}]GTPγS), the formation of the activated G protein–GTPγS complex can be measured. Binding of GTPγS in native membrane fractions has been utilised to compare the ability of different cannabinoid agonists to activate G-proteins via the CB_1 receptor (Burkey *et al.*, 1997a; Breivogel *et al.*, 1998; Griffin *et al.*, 1998; Kearn *et al.*, 1999). These studies demonstrated that different cannabinoid agonists could produce different levels of GTPγS stimulation and led to the hypothesis that these chemically disparate agonists may direct receptor activation of selective G-proteins. Similar studies have been carried out in cells transfected with the CB_2 receptor (MacLennan *et al.*, 1998; Griffin *et al.*, 1999) however, this approach has not been utilized with CB_2 receptors in native membranes probably due to lower receptor levels. In cell lines and in native tissues the G-protein content is heterogeneous and undefined. In order to distinguish between different G-proteins, an *in situ* reconstitution approach was recently utilized (Glass and Northup, 1999). This involved stripping endogenous G-proteins from membranes expressing either the CB_1 or CB_2 receptors, and then adding back isolated G-proteins. As predicted by signal transduction studies CB_1 and CB_2 receptors did indeed show different abilities to activate the G-proteins. While both CB_1 and CB_2 receptors could activate G_i with similar apparent affinity, both receptors exhibited a lower apparent affinity for G_o with CB_2 being particularly inefficient in this interaction.

The confirmation that CB_1 and CB_2 receptors complex differentially to G-proteins may help to explain the differences in signal transduction pathways observed for these receptors. As inhibition of voltage gated Ca^{2+} channels is likely mediated via

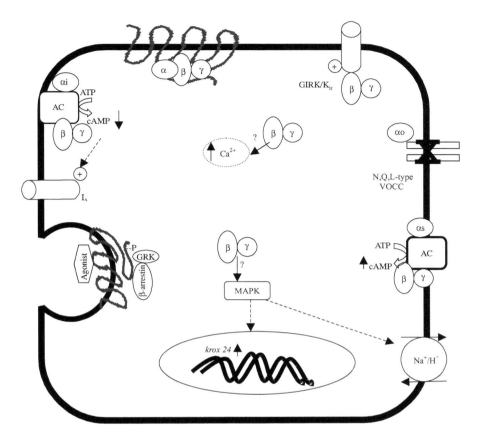

Figure 15.1 Schematic diagram of the proposed signal transduction pathways of the cannabinoid
receptors. Receptor activation leads to stimulation of multiple G-proteins, and intracellular pathways. Activation of Gi, leads to inhibition of adenylate cyclase, while a putative
activation of Gs may result in stimulation of adenylate cyclase. Inhibition of cAMP by CB_1
leads to stimulation of potassium A-current (I_A). Activation of Gαo probably directly
mediates cannabinoid CB_1 receptor inhibition of Ca^{2+} channels. G protein $\beta\gamma$ subunits
are likely to mediate multiple pathways including activation of inwardly rectifying potassium channels, activation of MAP kinase, and mobilization of intracellular calcium
(although this may not be a directly $\beta\gamma$ mediated effect). Depending on the adenylate
cyclase isoform present in the cell, $\beta\gamma$ subunits may also contribute to the inhibition or
stimulation of cAMP formation. Agonist binding to the CB_1 receptor leads to desensitization of the receptor via a GRK mediated phosphorylation pathway. Receptors may
also be internalised, via clatherin coated pits, by a pathway that involves the binding
of β-arrestin to the phosphorylated receptor. The CB_2 receptor has been demonstrated to produce only the inhibition of adenylate cyclase, activation of MAP-
kinase, and mobilization of intracellular calcium pathways.

Gα_o whereas activation of K^+ channels is probably via $\beta\gamma$ subunits derived from
either G_i or G_o (Gudermann *et al.*, 1997), the failure of CB_2 receptors to modulate
ion channels may be explained by its low affinity for G_o. It is also possible that $\beta\gamma$
subunits of differing composition have higher affinity for G_o vs. G_i (or CB_1 vs. CB_2),

and that these subunits differentiate the ability of cannabinoid receptors to activate K^+ channels. The *in situ* reconstitution approach demonstrated no coupling of either CB_1 or CB_2 receptors to G_t or G_q (Glass and Northup, 1999) and agrees with the majority of studies demonstrating cannabinoid agonists can not activate phospholipase C (Felder *et al.*, 1993, 1995). However, one recent study in which G-proteins were co-expressed with cannabinoid receptors suggests that CB_1 may be able to activate G_{11} and G_{16} (Ho *et al.*, 1999). The *in situ* findings are consistent with the regional distribution of the cannabinoid receptors and G-proteins. CB_1 receptors are localized to the brain and a few peripheral organs (Pertwee, 1997) while CB_2 receptors are localized primarily to immune cells. Both G_i and G_o are neuronally localized and have been co-localized with the CB_1 receptor in rat brain (Mukhopadhyay *et al.*, 2000), However, while G_i has been demonstrated to be present in immune cells, G_o is not abundant peripherally, suggesting that physiologically G_i is likely to be the G-protein encountered by CB_2.

Ligand-receptor-G-protein interactions

Studies on the CB_1 receptor suggest that different cannabinoid agonists can direct the interaction of CB_1 receptors with $G\alpha_i$ or $G\alpha_o$. Glass and Northup (1999) found that WIN55,212-2 and anandamide were full agonists in the activation of $G\alpha_i$ but only partial agonists in the activation of $G\alpha_o$. Consistent with the finding of partial agonism at both G_i and G_o proteins, Δ^9-THC has been demonstrated previously to produce either no activation or only partial activation of GTPγS binding in rat and mouse brain cerebellar membranes and slices (Sim *et al.*, 1995; Burkey *et al.*, 1997a,b; Griffin *et al.*, 1998). Anandamide also produced partial agonism at G_o in reconstitution assays (Glass and Northup, 1999) and has previously demonstrated sub-maximal activation of GTPγS in cerebellar and whole brain homogenates (Burkey *et al.*, 1997a; Griffin *et al.*, 1998; Kearn *et al.*, 1999). A recent study has demonstrated WIN55,212-2 to be more efficacious than other agonists tested in stimulating cAMP accumulation in CHO cells following pertussis toxin treatment (Bonhaus *et al.*, 1998). Previous studies have suggested that this pathway may be mediated by $G\alpha_s$ (Glass and Felder, 1997) thus agonist-receptor complexes may differ in their recognition of G_s in addition to G_i and G_o. The finding of agonist trafficking of the CB_1 receptor has broad therapeutic implications as it suggests that agonists could be designed which selectively target one intracellular pathway over another, thereby potentially separating therapeutic activities from other unwanted effects. However, it should be borne in mind that the physiological relevance of the difference in the ability of ligands to regulate G-protein signaling will depend on a combination of the number of receptors in the cell and the saturation properties of the effector molecules. Thus, if saturation of the second messenger response (e.g. inhibition of adenylate cyclase, enhancement of potassium channel conductance) requires full stimulation of G-protein, then partial efficacy would be visible if receptor number was limited. If however the maximal response can be generated by sub-maximal G-protein activation, then the difference between agonists may not be readily discernible. This provides a mechanism for explaining differences in responses observed in different brain regions, tissues and cell systems. For example, anandamide is a partial agonist in the activation of $G\alpha_o$, the G-protein thought to mediate calcium channel activation and was a partial agonist in the inhibition of calcium channels in N18 neuroblastoma cells (Mackie *et al.*, 1993). In contrast, anandamide produced full agonism of this effect in AtT20 cells which express higher receptor numbers (Mackie *et al.*, 1995). Thus, these findings are consistent with the observation that when receptor is limited the differences in response to particular agonists become detectable.

Receptor structure/activity relationships

The possibility that different agonists might induce different conformations of the CB_1 and CB_2 receptor is not entirely unpredicted given that the agonists have distinct chemical structures. Much work has been carried out to determine the critical sites of interaction of ligands and receptors, and receptors with G-proteins and information from these studies has provided target sites for cannabinoid receptor mutations (see Figure 15.2). Previous work with other G-protein coupled receptors has shown that amino acids in transmembrane region 3 (TM3) are critical for ligand binding (Baldwin, 1994). The first cannabinoid receptor mutation study therefore assessed the importance of a conserved lysine (K192) in TM3 of the CB_1 receptor (Song and Bonner, 1996; Chin *et al.*, 1998) (Figure 15.2). When the negative charge of K192 was removed by substitution with alanine (K192A) or glutamine (K192Q) or replaced with the positive charged glutamate (K192E) a variety of cannabinoid ligands including HU210, CP55,940, and anandamide were unable to recognise the mutant receptors. Interestingly, WIN55,212-2 bound to all the mutant receptors with equal affinity to the wild-type receptor, and this ligand was able to elicit full inhibition of adenylate cyclase. This suggested that while WIN55,212-2 competitively inhibited the binding of classical cannabinoids to the CB_1 receptor, its interaction with the receptor is not identical. The CB_2 receptor contains a lysine (K109) that is analogous to K192 in the CB_1 receptor. However, it was found that the mutant CB_2 receptor K109A expressed in HEK 293 cells exhibited binding and signal transduction profiles identical to the wild-type CB_2 receptor indicating a clear difference between CB_1 and CB_2 (Tao *et al.*, 1999). Another highly conserved sequence implicated in G-protein coupled receptor signally is the aspartate-asparagine-tyrosine (DRY) motif, at the junction of TM3 and intracellular loop 2 (Figure 15.2). Mutations of the CB_2 receptor in which the DRY sequence was consecutively changed to alanines (D130A, R131A, Y132A) suggested that D130 imparted a structure that was critical for ligand binding whereas Y132 appeared to play a role in CB_2 signal transduction (Rhee *et al.*, 2000).

A highly conserved amino acid among G-protein coupled receptors (GPCR) is an aspartic acid in TM2. Mutation of this amino acid in α_2 adrenergic receptors has resulted in loss of ligand recognition, cation selectivity and/or coupling to G-proteins (Ceresa and Limbird, 1994). As shown in Figure 15.2, in the human CB_1 receptor this TM2 aspartate corresponds to amino acid D163 and in the CB_2 this amino acid corresponds to the aspartate D80. In order to evaluate the role of this conserved aspartate residue in CB_1 and CB_2 receptors, Tao and Abood (1998) created a series of mutations. In both receptors the aspartate was replaced with an asparagine (D163N;D80N) or glutamate (D163E;D80E). Radioligand-binding

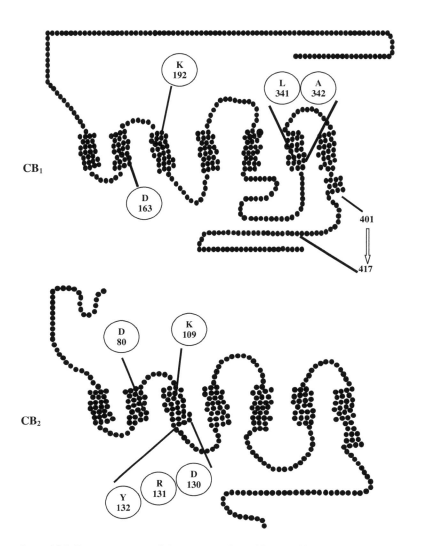

Figure 15.2 Representation of the regions of the CB_1 and CB_2 receptors deemed important for receptor signaling from mutational studies. Transmembrane domains (1–7) are represented from left to right. The amino acids involved in mutational analysis have been enlarged and the area corresponding to the juxtamembrane C-terminus fragment ($CB_1$401–417) is labeled.

analysis revealed normal ligand recognition for a variety of cannabinoid ligands with the exception of WIN55,212-2 for which both CB_1 receptor mutants displayed lowered affinity. However, unlike wildtype receptors, ligand binding to mutant receptors was not sensitive to nucleotide, and did not result in inhibition of cAMP accumulation. Further studies on the D163N mutant and its rat homolog (D164N), have confirmed this conclusion as both in AtT20 cells and oocytes the mutant receptor failed to couple efficiently to inwardly rectifying potassium (GIRK/K_{ir})

channels (McAllister *et al.*, 1999; Roche *et al.*, 1999). In addition to a highly conserved aspartate in TM2, GPCRs have a highly conserved asparagine in TM7 and studies with other GPCR have suggested that these two amino acids interact in the folded proteins (Zhou *et al.*, 1994), and that reciprocal mutations of these amino acids will return the receptor to normal functioning (Sealfon *et al.*, 1995). To test if this was the case in the CB_1 receptor Roche *et al.* (1999) produced a D164N/N394D mutant, however, coupling to GIRK/K_{ir} channels was not restored, suggesting that these amino acids do not interact in the CB_1 receptor. Overall, the study suggested that this aspartate residue is not generally important for ligand recognition in either cannabinoid receptor; however, it is required for communication with G proteins and appropriate signal transduction.

In contrast to the results of Tao and Abood (1998) the D163N mutant coupled to the cAMP cascade and Ca^{2+} channels in AtT20 cells (Roche *et al.*, 1999). Although, this contradiction appear surprising, it is not the first time the same mutation transfected into 2 different cell types produces different effects (Wang *et al.*, 1991; Surprenant *et al.*, 1992). These differences are thought to occur due to cell type variations in stoichiometry and class of G-protein families, second messenger systems, and other signaling moieties. That the same mutation produces apparently different effects in different cell lines emphasizes that one must be cautious when interpreting the outcomes of receptor mutation experiments. Moreover, when a single amino-acid mutation alters receptor signaling the tempting explanation would be that the residue served as part of the interaction point with the signaling moiety. However, modifications of a single amino acid may produce more global changes such as altering the orientation or folding of entire transmembrane regions. Some mutations may also alter the way the receptor is transported to or inserted into the plasma membrane making it less accessible to available signaling proteins (Keefer *et al.*, 1994; Saunders *et al.*, 1998). Indeed it has been demonstrated in many cases that single and double amino acid changes can profoundly alter normal receptor expression (Shire *et al.*, 1996; Tao *et al.*, 1999). Even with these caveats, the cannabinoid receptor mutational studies were largely consistent with previous work on other GPCRs which define amino acids in the second and third transmembrane domains as regulating receptor signaling and ligand selectivity.

In addition to mutagenesis, cannabinoid receptor interactions with G-proteins have been investigated using peptide-mimetics. Howlett *et al.* (1998) described the importance of the CB_1 receptor third intracellular loop and juxtamembrane carboxyl (C)-terminus in G-protein activation (Figure 15.2). Peptides corresponding to regions of the CB_1 receptor were tested for their ability to stimulate GTPγS binding or inhibit adenylate cyclase in rat brain and neuroblastoma cell membranes, respectively. They found that the juxtamembrane C-terminus fragment (CB_1 401-417) was a highly effective stimulator of GTPγS binding as well as an inhibitor of adenylate cyclase. Consistent with this finding, a CB_1 receptor mutant that was truncated past residue 418 still retained wild-type signal transduction (Jin *et al.*, 1999). Mukhopadhyay *et al.* (1999) went on to determine that Arg401 of CB_1 401-417 contributed primarily to the fragment's affinity for adenylate cyclase inhibition, however, further removal of charged residues up to $CB_1$408 did not negate the efficacy of the peptide. These results suggested that while essential for a high affinity interaction, the first seven amine (N)-terminal residues were not crucial for the efficacy of the fragment as $CB_1$408-417 could still form a structure that interacted with the adenylate cyclase cascade. Thus, it was not surprising that a CB_2 juxtamembrane C-terminal fragment, which only shared homology with CB_1 in the 401-406 region failed to inhibit adenylate cyclase. The most drastic reductions in efficacy for inhibition of adenylate cyclase activity were observed with mutations that neutralized the C- or N-terminal amino acids or constrained the conformation of the fragment by adding acetamido sulfhydryl blocking groups. A recent study by Mukhopadhyay et al. (2000), utilizing the technique of immunoprecipitation, demonstrated $CB_1$401-407 could directly interact with $G_{i/o}$ proteins agreeing with the previous findings that the effects of the fragment could be blocked by pertussis toxin.

It is beginning to appear likely that contact sites for different G-proteins on cytoplasmic receptor parts can be differentiated. Recently, Abadji et al. (1999) created a CB_1 mutant receptor in which leucine 341 and alanine 324 were replaced with the reciprocal amino acids (i.e. alanine 341 and leucine 342). In β_2-adrenergic and α_2-adrenergic receptors this type of amino acid substitution has resulted in a constitutively active receptor (Ren et al., 1993; Samama et al., 1993). These investigators found this CB_1 mutation had no effect on the $G_{i/o}$ mediated inhibition of adenylate cyclase but when they unmasked the putative G_s coupling with pertussis toxin treatment, they found a constitutive activation of the stimulatory pathway. Calandra et al. (1999) transfected CHO cells with CB_1 and a cAMP response element fused to the firefly luciferase coding region and found luciferase activity increased upon treatment with cannabimimetic compounds. As treatment with pertussis toxin actually enhanced CB_1 receptor stimulation of luciferase, it was suggested this effect was mediated through G_s. When a similar heterologous system was created with CB_2, luciferase activity could not be increased, consistent with the previous work suggesting that CB_2 does not couple to Gs (Glass and Felder, 1997). Since the CB_2 receptor failed to activate luciferase, CB_1/CB_2 chimeras were created in order to study the features involved in this CB_1 effect. Using chimeric constructs it was determined that the first and second TM regions of CB_1 were involved in transducing the signal to luciferase. Taken together the work of both these groups complements the ever growing literature suggesting distinct receptor regions can differentially target selective G-proteins.

INVERSE AGONISM AT CANNABINOID RECEPTORS

Several reports have suggested that the CB_1 receptor antagonists SR141716A may exhibit inverse agonist properties (Bouaboula et al., 1997; MacLennan et al., 1998; Pan et al., 1998). Inverse agonism differs from conventional antagonism, in that rather than possessing equivalent affinity for both the active and inactive receptor states, inverse agonists have a higher affinity for the inactive state, thereby inhibiting any spontaneous activity of the receptor. SR141716A has been reported to reduce basal GTPγS binding in membranes from cells transfected with CB_1 receptors (Landsman et al., 1997), however other studies in native membranes have failed to observe this (Breivogel et al., 1998; Kearn et al., 1999). In transfected cells, it has also been demonstrated that constitutively active CB_1 receptors regulating ion channels were sensitive to inverse agonists (Pan et al., 1998; McAllister et al., 1999).

These effects were not due to endogenous agonists, confirming that CB_1 receptors can be tonically active. In G-protein reconstitution studies CB_1 receptors exhibited spontaneous activation of both G_i and G_o which could be completely blocked by SR141716A, indicative of strong inverse agonism in *in vitro* systems (Glass and Northup, 1999). This demonstration of a strong pre-coupled receptor G protein state helps to explain recent observation that constitutively active CB_1 receptors could sequester $G_{i/o}$ proteins from other receptors (Vàsquez and Lewis, 1999).

Similarly inverse agonism has been reported for the CB_2 antagonist SR144528 (Bouaboula *et al.*, 1999b; Portier *et al.*, 1999). The physiological relevance of inverse agonism is not clear. It is difficult in whole animal systems to differentiate between the effects of blockade of constitutively active receptors (inverse agonism) and blockade of endogenous ligand (antagonism). Furthermore it is not known if the cannabinoid receptors are constitutively active under *in vivo* conditions. Theoretically at least, inverse agonists may provide some advantages over neutral antagonists therapeutically (Milligan *et al.*, 1995), such as in disorders resulting from constitutive activation of a receptor, or over-expression of a receptor. However, until a pure antagonist for the CB_1 receptor is developed, *in vivo* differences between these compounds remains purely speculative.

CANNABINOID RECEPTOR ACTIVATION OF ION CHANNELS

Activation of the CB_1 receptor can lead to alterations in the function of multiple ion channels (Figure 15.1). Most notable is the ability of cannabinoids to inhibit voltage dependent N- and Q-type calcium channel currents both in transfected cell lines and in cell lines that contain the native receptor (Caulfield and Brown, 1992; Mackie and Hille, 1992; Mackie *et al.*, 1995). More recently, it was demonstrated that cannabinoid CB_1 receptors can inhibit L-type Ca^{2+} current in cerebral arterial smooth muscle cells (Gebremedhin *et al.*, 1999). The effect of cannabinoids on calcium currents is pertussis toxin sensitive but not cAMP mediated, suggestive of direct G-protein activation, and indeed previous studies have suggested that inhibition of voltage gated Ca^{2+} channels is likely mediated via $G\alpha_o$ (see Gudermann *et al.*, 1997). Interestingly, one study has demonstrated that at agonist concentrations greater than 1 µM WIN55,212-2 can inhibit N and P/Q-type Ca^{2+} channels directly (Shen and Thayer, 1998). One possible consequence of a CB_1 mediated reduction in calcium influx is decreased activation of various calcium dependent intracellular processes. Consistent with this suggestion CB_1 receptor agonists have been demonstrated to inhibit depolarization induced synthesis of nitric oxide by neuronal nitric oxide synthase via inhibition of calcium influx through voltage-operated calcium channels in cerebellar granule cells (Hillard *et al.*, 1999).

Cannabinoid agonists can enhance the activity of G-protein coupled inwardly rectifying potassium channels (GIRK/K_{ir}) *in vitro* (Henry and Chavkin, 1995; Mackie *et al.*, 1995). This effect is pertussis toxin sensitive and likely to be mediated via $G_{i/o}$ $\beta\gamma$ subunits (see Gudermann *et al.*, 1997). Cannabinoid ligands have also been shown to enhance A-type potassium current in hippocampal cells, an effect which is indirectly mediated by inhibition of cAMP accumulation (Deadwyler *et al.*, 1995).

Cannabinoid receptors are located on the pre- and post-synaptic regions of multiple types of neurons (Herkenham *et al.*, 1991; Tsou *et al.*, 1998). If cannabinoids inhibit calcium channel currents pre-synaptically then transmitter release could be inhibited. Indeed it has recently been demonstrated that inhibition of N- and Q-type calcium channels by CB_1 receptor activation reduces the probability that neurotransmitter will be released in response to an action potential in hippocampal pyramidal neurons (Sullivan, 1999). Pre-synaptic inhibition of calcium channels provides a mechanism for studies that have shown cannabinoids to inhibit acetylcholine, noradrenaline, glutamate, and GABA release in hippocampal neurons (Gifford and Ashby, 1996; Shen *et al.*, 1996; Gessa *et al.*, 1997; Kathmann *et al.*, 1999; Katona *et al.*, 1999), noradrenaline release at sympathetic nerve terminals (Ishac *et al.*, 1996), and NMDA stimulated dopamine release in the striatum (Kathmann *et al.*, 1999). In the peripheral nervous system, pre-synaptic inhibition of transmitter release may also provide a mechanism for the ability of cannabinoids to inhibit electrically evoked contractions of guinea pig myenteric plexus-longitudinal muscle, mouse isolated vas deferens and urinary bladder (Pertwee and Griffin, 1995; Pertwee and Fernando, 1996; Coutts and Pertwee, 1997).

Activation of K_{ir} channel currents pre- or post-synaptically, by cannabinoid receptors, would result in hyperpolarization of the cell and this would make it more difficult for the cell to depolarize and produce action potentials (Mackie *et al.*, 1995). Finally, enhancement of A-type potassium conductance post-synaptically by cannabinoids would hyperpolarize the cell and result in suppression of repetitive cell firing (Deadwyler *et al.*, 1995). Taken together most evidence suggests cannabinoids inhibit cell function through their interactions with ion channels and this overall effect is consistent with observed physiological outcomes. However, recent studies have demonstrated that cannabinoids can decrease post-synaptic M-type potassium current in hippocampal CA1 neurons (Schweitzer, 2000) and enhance NMDA-elicited Ca^{2+} signals in cerebellar granule cells (Netzeband *et al.*, 1999), effects that would result in increased neuronal excitability.

CALCIUM MOBILIZATION AND ACTIVATION OF THE PHOSPHOLIPASE C/INOSITOL PHOSPHATE SYSTEM AND OTHER PATHWAYS

Many studies have addressed whether the cannabinoid ligands can mobilize intracellular calcium via activation of phospholipase C (PLC). Most studies have concluded that while the cannabinoid ligands can activate PLC, this effect is not receptor mediated. Early studies demonstrated that Δ^9-THC did not activate phosphoinositide-specific PLC in guinea-pig cerebral cortical slices (Reichman *et al.*, 1991). Studies in rat hippocampal cultured cells demonstrated that Δ^9-THC inhibited carbachol-induced formation of labelled inositol phosphates, however, this effect was pertussis toxin-insensitive, and cannabidiol, which is not a cannabinoid receptor agonist, produced the same effect with a slightly greater potency than Δ^9-THC. Further evidence that this effect was not cannabinoid receptor mediated came from Felder *et al.* (1992) who demonstrated that HU-210 had no detectable stimulatory effect on inositol phosphate production at concentrations up to 1000 times greater than its Ki, in either CHO or L cells expressing CB_1 receptors; similar results were subsequently obtained with anandamide (Felder *et al.*, 1993). Calcium imaging studies in CHO cells expressing CB_1 receptors demonstrated while many cannabimetic agents could result in increases in intracellular free calcium (Felder *et al.*, 1992, 1993), these effects lack stereo-selectivity, and occurred equivalently in CHO cells lacking the CB_1 receptor. Anandamide has been demonstrated to mobilize Ca^{2+} from caffeine-sensitive intracellular Ca^{2+} stores in human endothelial cells that functionally overlap in part with the internal stores mobilized by histamine, however, this effect was not pertussis toxin sensitive, or blocked by SR141716A at relevant doses (Mombouli *et al.*, 1999). This study also found that SR141716A at high doses could block the histamine mediated mobilization of intracellular stores suggesting that SR141716A may have non-specific inhibitory effects on this pathway (Mombouli *et al.*, 1999). As in earlier studies (Felder *et al.*, 1992; White and Hiley, 1998) anandamide did not induce capacitive Ca^{2+} entry in these cells, however, it may do so in DDT_1-MF-2 smooth muscle cells (Filipeanu *et al.*, 1997).

In contrast to these studies, extensive work has been carried out by Sugiura and colleagues examining CB_1 mediated transient increases in Ca^{2+} in NG108-15 cells which contain an endogenous CB_1 receptor. These studies have suggested that the increase in transient Ca^{2+} release seen is both CB_1 and G_i/G_o mediated (Sugiura *et al.*, 1996, 1997, 1999). Furthermore their studies have suggested differing efficacy profiles of cannabinoid ligands compared to other assays, with 2-arachidonyl glycerol (2AG) being the most efficacious and HU-210, CP55,940, Δ^9-THC, anandamide and WIN55,212-2 all being partial agonists (Sugiura *et al.*, 1999). A recent paper by Netzeband *et al.* (1999) is in agreement with a cannabinoid receptor mediated pathway that activates intracellular Ca^{2+} pools. This group reported that cannabinoids could enhance NMDA-elicited Ca^{2+} release in cerebellar granule cells. The enhancement was dependent on Ca^{2+} release from intracellular stores. However, in this case the effect could not be produced directly and was apparent only after the stimulation of the NMDA subtype of glutamate receptors. Cannabinoids are known for their modulatory role in the CNS and perhaps similar findings will be unmasked in other cells if this indirect approach is taken.

A possible pathway through which CB_1 receptors may activate PLC is through G_{11} (Ho *et al.*, 1999) a member of the G_q family. Alternatively, it has previously been demonstrated that $G_{i/o}$ $\beta\gamma$ subunits can stimulate PLCβ (Jin *et al.*, 1994) in NG108-15 cells, leading to an opioid receptor driven rapid increase in intracellular calcium concentrations, suggesting a possible mechanism through which CB_1 receptors might activate intracellular Ca^{2+} stores. The reasons for the discrepancies in this field are not clear, possibly, the non-responsive cells and tissues examined lack the relevant PLC isoform, or have different $\beta\gamma$ composition. However, this explanation would still not explain the differences in the relative agonist efficacy observed in these studies, as compared to other $\beta\gamma$ driven pathways such as inhibition of voltage-gated Ca^{2+} channels.

Similar discrepancies exist for CB_2 receptors. Results from CHO cells transfected with the CB_2 receptor indicated that these receptors do not activate phospholipases A_2, C or D, or mobilize intracellular calcium (Felder *et al.*, 1995; Slipetz *et al.*, 1995). However, a recent paper by Sugiura *et al.* (2000) demonstrated rapid transient increases in intracellular free Ca^{2+} concentrations in response to 2-AG in HL-60 cells that naturally express the CB_2 receptor. This effect was blocked by the CB_2 receptor antagonist SR144528, and was pertussis toxin sensitive. As for the CB_1 receptor, this group found that 2AG was the most potent agonist tested for this response, while anandamide was only a weak partial agonist. However in contrast to the CB_1 receptor, HU210 and CP55,940 were full agonists in this system, while Δ^9-THC was only a partial agonist.

Several studies have focused on the ability of the endogenous cannabinoid ligands to produce non-receptor mediated intracellular actions. DePetrocellis *et al.*, (1995) demonstrated that anandamide can modulate protein kinase C (PKC) activity *in vitro* by binding to the diacylglycerol regulatory site of this enzyme. Furthermore, there are numerous reports of cannabinoid-induced increases in free arachidonic acid accumulation, probably through activation of phospholipase A_2 (Reichman *et al.*, 1991; Felder *et al.*, 1992, 1993, 1995). These studies have demonstrated that the effects lack stereospecificity, and occur equipotently in cells not transfected with CB_1 and CB_2 receptors. Interestingly, astrocytes have been suggested to have a cannabinoid sensitive G-protein receptor distinct from the CB_1 receptor, through which cannabinoids can inhibit cAMP formation (Sagan *et al.*, 1999), as well as an anandamide mediated inhibition of gap junction conductance, that is not mimicked by other cannabinoid agonists, nor displays CB_1 antagonist sensitivity (Venance *et al.*, 1995). Recent studies have demonstrated CB_2 receptors to be present on C6 glioma cells (Galve-Roperh *et al.*, 2000) and microglia (Kearn and Hillard, 1997) suggesting that these receptors may account for some of the cannabinoid effects in the central nervous system.

MAP KINASE ACTIVATION AND REGULATION OF GENE EXPRESSION

In CHO cells, transfected with the CB_1 or CB_2 receptors it has been demonstrated that cannabinoids can increase the activity of mitogen-activated protein (MAP) kinase via a G-protein, but not cAMP dependent mechanism (Bouaboula *et al.*, 1995b, 1996). However, the pathways for activation of MAP-kinase appear to be different for each receptor as CB_2, but not CB_1 receptor stimulation of MAP-kinase can be attenuated by a PKC inhibitor, suggesting that it is PKC dependent (Bouaboula *et al.*, 1996). MAP-kinase activation by CB_1 has recently been demonstrated to lead to activation of the Na^+/H^+ exchanger NHE-1, a electroneutral transmembrane transporter involved in multiple cellular functions such as intracellular pH regulation and control of cell volume (Bouaboula *et al.*, 1999a). Furthermore, it is possible that MAP-kinase activation is an intermediate step in the cannabinoid receptor-mediated induction of multiple transcription factors. These include activation of *krox 24*, increased AP-1 DNA-binding activity and increased Fos-related-antigen (FRA) activity (Bouaboula *et al.*, 1995a; Glass and Dragunow, 1995; Porcella *et al.*, 1998). In addition to *krox 24*, *in vivo* activation of the transcription factors *c-fos*, and *c-jun* have been reported in rat cortical and striatal regions following intra-peritoneal administration of Δ^9-THC (Mailleux *et al.*, 1994). A recent study has demonstrated that activation of CB_2 receptors induces two chemokines involved in inflammatory disease, interleukin-8 and monocyte chemotactic protein-1 in the promyelocytic cell line HL60 which has an endogenous CB_2 receptor. The functional consequences of alterations in gene regulation via activation of cannabinoid receptors is still under investigation.

ACTIVITY DEPENDENT REGULATION OF CANNABINOID RECEPTORS

It is well established that tolerance develops to most pharmacological effects of cannabinoids *in vivo* after a period of chronic exposure (McMillan *et al.*, 1971; Pertwee, 1991; Pertwee *et al.*, 1993; Fan *et al.*, 1994), and are likely to be the response of activity dependent regulation of CB_1 receptors. Earlier studies looking at cannabinoid receptor density changes using ligand binding analysis gave conflicting results. Some investigators demonstrated decreases in the B_{max} of cannabinoid receptors after chronic treatment with Δ^9-THC (Oviedo *et al.*, 1993; Rodriguez de Fonseca *et al.*, 1994) whereas others reported no changes in receptor density (Abood *et al.*, 1993). Similar discrepancies have been noted with changes in mRNA levels (Fan *et al.*, 1996; Romero *et al.*, 1997). Differences in these studies may be related to the different tolerance paradigms utilized by the investigators. Furthermore, one study has suggested that CP55,940 can bind equivalently to internalized and membrane localized receptors (Rinaldi-Carmona *et al.*, 1998), suggesting ligand binding analysis may not be appropriate for the study of cannabinoid receptor regulation and may help to explain the previous discrepancies seen in tolerance studies.

The cellular mechanisms behind the production of tolerance of other G-protein coupled receptor systems have been studied extensively, and have been attributed to a combination of early and late events (Nestler, 1993). Early events include processes such as uncoupling of the receptor from second messenger responses through phosphorylation and receptor internalization followed by recycling of the receptor or degradation. Later events include alteration of the levels of protein involved in cell signaling through regulation of protein turnover, RNA translation, and gene transcription. With mechanisms such as these in mind, investigators began to focus on the molecular and cellular components of cannabinoid mediated tolerance. A majority of these studies have been carried out in transfected cell lines which have proven to be valuable systems with which to study the processes of cannabinoid receptor desensitization, internalization, and up-regulation.

Desensitization can be divided into two classes, homologous desensitization, caused by agonist activation of the receptor itself, and heterologous desensitisation in which the receptor is inactivated following the activation of other classes of receptors or intracellular pathways. In CB_1 transfected AtT20 cells and an oocyte expression system (Jin *et al.*, 1999), it was found that WIN55,212-2 produced marked desensitization in the ability of CB_1 agonists to activate inwardly rectifying potassium channels, which appeared to be mediated by G-protein coupled receptor kinase-3 (GRK3) and β-arrestin 2. A series of mutations indicated that S426 and S430, in the C-terminus region, were the targets for phosphorylation by GRK3 (Figure 15.3). In this study, it was also reported that a co-expressed opioid receptor responded normally after CB_1 desensitization, confirming the receptor specific nature of this effect.

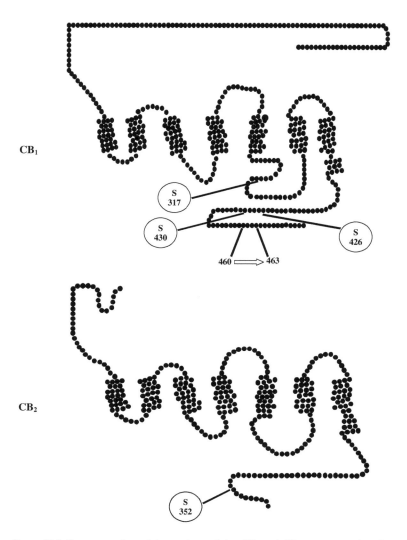

Figure 15.3 Representation of the regions of the CB₁ and CB₂ receptors thought to be involved in desensitization and internalization. Transmembrane domains (1–7) are represented from left to right. The amino acids involved in receptor desensitization have been enlarged and the residues involved in CB₁ receptor internalization (460–463) are labeled.

Garcia *et al.* (1998) investigated influences of cannabinoid receptor phosphorylation, in AtT20 cells transfected with CB₁, following PKC activation. As PKC is not thought to be activated by cannabinoid receptors in these cells this is a model of heterologous desensitization. When cells were pretreated with a PKC activator both agonist mediated effects on calcium channels and potassium channels were significantly decreased. Somatostatin receptors, which also couple the calcium channels and potassium channels in AtT20 cells, were only slightly effected by pretreatment with the PKC activator suggesting the observed effect was occurring at the level of the cannabinoid receptor, rather than a common downstream effector. Since, intracellular loop domains of GPCRs have been suggested to be phosphorylation sites for protein kinases these domains of CB_1 were evaluated for their role in the observed PKC effect (Garcia *et al.*, 1998). Mutational analysis suggested that phosphorylation of serine 317 was responsible for the attenuation of signaling produced by activation of PKC.

As with the CB_1 receptor, the CB_2 receptor has also been found to undergo the processes implicated in the production of desensitization. Bouaboula *et al.* (1999b) found that agonist treatment led to increased receptor phosphorylation at serine 352 whereas exposure to an inverse agonist decreased the amount of receptor in a phosphorylated state. Although this amino acid is located in the tail region of the CB_2 receptor, it shares no homology with residues that were targeted during CB_1 receptor desensitization. In the CB_2 transfected CHO cells, PKC and PKA inhibitors did not alter CB_2 receptor phosphorylation nor did pre-treatment with pertussis toxin. Bouaboula *et al.* (1999b) suggested that perhaps a GRK-type kinase, activated by non-pertussis toxin sensitive G-proteins could be responsible for the observed phosphorylation of CB_2.

Both CB_1 and CB_2 receptors are subject to the processes of internalization, sequestration, and up-regulation. Rapid internalization of CB_1 receptors has been observed after agonist exposure using immunofluorescence in CB_1 transfected CHO and AtT20 cells (Rinaldi-Carmona *et al.*, 1998; Hsieh *et al.*, 1999). Using cell flow cytometric analyses, Rinaldi-Carmona *et al.* (1998) were also able to show that an agonist caused receptor internalization whereas an inverse agonist caused receptor up-regulation and that these effects were related to changes in cannabinoid receptor function. The ability of agonists to induce internalization corresponded to agonist efficacy, with WIN55,212-2, CP55,940, and HU-210 causing rapid internalization, while agonists with lower intrinsic efficacy, such as methanadamide and Δ^9-THC, caused modest receptor internalization (Hsieh *et al.*, 1999). Hsieh *et al.* (1999) also demonstrated that CB_1 receptors are internalized via clathrin-coated pits and also provided evidence for receptor recycling. Recycling of CB_1 to the cell surface after short (20 min) but not long (90 min) agonist treatment was independent of new protein synthesis, but dependent on dephosphorylation.

In other GPCRs the C-terminus contains amino acid residues that play a role in the internalization of the receptor. Mutational analysis of the CB_1 receptor revealed residues 460–463 were critical for this process. As was previously noted, phosphorylation of S426 and S430 (tail region) or S317 (3rd IL) resulted in CB_1 receptor desensitization, however, these sites had no influence on receptor internalization (Garcia *et al.*, 1998; Jin *et al.*, 1999). Therefore, there is a distinct difference between the domains, in the C-terminal region of the CB_1 receptor, targeted during desensitization and internalization. Interestingly, internalization of the receptors was not affected when receptor G-protein signaling was disrupted using either pertussis toxin or cholera toxin, nor by a PKC activator. This suggested that G-protein activation was not necessary for cannabinoid receptor internalization. In contrast, Roche *et al.* (1999) showed that uncoupling of the G-protein/receptor interaction, with a mutation in the second transmembrane of the CB_1 receptor (D164N), blocked receptor down-regulation. Interestingly, the inverse agonist SR141716A has been demonstrated to cause a marked increase in receptor number on the cell surface (Rinaldi-Carmona *et al.*, 1998) suggesting that receptors may be recycled from an intracellular pool in response to prolonged inverse agonist exposure.

The CB_2 receptor can be seen to internalize or up-regulate in the presence of agonist or inverse agonists, respectively (Bouaboula *et al.*, 1999b). These findings could be correlated with cellular responses since pretreatment with the inverse agonist SR144528 (which would increase the amount of surface receptors available for stimulation) enhanced the response to agonist. Therefore both CB_2 and CB_1 share similarities in terms of these events. SR144528 also proved to be able to regenerate desensitized CB_2 receptors. As was observed with CB_1, this action on CB_2 was dependent on the dephosphorylation of internalized receptors and could be blocked with a phosphatase 2A inhibitor (Bouaboula *et al.*, 1999b; Hsieh *et al.*, 1999).

CONCLUSIONS

The last decade has providing exciting developments in the field of signal transduction of cannabinoid receptors, and cannabinoid science in general. The possibility that additional cannabinoid receptors exist has recently been strengthened with the creation of the CB_1 and the CB_1/CB_2 knockout mice (Ledent *et al.*, 1999; Zimmer *et al.*, 1999). In the $CB_1^{-/-}$ mice, even though most of the cannabinoid mediated behaviors were abolished there were still some effects present. Most notable was the presence of cannabinoid induced analgesia in the tail flick and tail immersion test. Knockout mice have also been used to implicate the existence of cannabinoid receptor subtypes that regulate vascular tone (Wagner *et al.*, 1998; Jàrai *et al.*, 1999; Ledent *et al.*, 1999). If subtypes exist, there is potential for explaining some of the discrepancies in cannabinoid mediated signal transduction and for the addition of new effector pathways. However, lack of specificity of the available ligands may also account for some of the observations in the knockout mice, with ligands potentially targeting non-cannabinoid receptor proteins (i.e. receptors or channels) such as vanniloid receptors (Zygmunt *et al.*, 1999; Smart *et al.*, 2000). Additional studies are needed to answer these important questions.

We see that both the CB_1 and the CB_2 receptors specifically target selective G-protein pools and activation of these results in modulation of multiple effector pathways. The tonically active nature of the receptor makes it susceptible to inverse agonism and perhaps provides us with the potential to produce physiological outcomes opposite of those seen with cannabinoid agonists. The receptors are influenced by a distinct feedback system which controls their activity and allows for the preservation of a homeostatic environment. This regulation may also contribute in part to cannabinoid tolerance that is observed *in vivo*.

We are continually gaining a greater insight to the intricacies of cannabinoid mediated signal transduction. This progress should open the door into an understanding of the physiological role of the cannabinoid system which in turn will enhance the therapeutic potential of this class of compounds.

ACKNOWLEDGMENTS

The authors wish to thank Dr's Mary Abood, Graeme Griffin and Christopher Kearn for critical reading of the manuscript.

REFERENCES

Abadji, V., Lucas-Lenard, J. M., Chin, C. N. and Kendall, D. A. (1999) "Involvement of the carboxyl terminus of the third intracellular loop of the cannabinoid CB_1 receptor in constitutive activation of G(s)," *Journal of Neurochemistry* **72**: 2032–2038.

Abood, M. E., Sauss, C., Fan, F., Tilton, C. L. and Martin, B. R. (1993) "Development of behavioral tolerance of Δ^9-THC without alteration of cannabinoid receptor binding or mRNA levels in the whole brain," *Pharmacology, Biochemistry and Behaviour* **46**: 575–579.

Baldwin, J. M. (1994) "Structure and function of receptors coupled to G-proteins," *Current Opinion in Cell Biology* **6**: 180–190.

Bidaut-Russell, M. and Howlett, A. C. (1989) "Opioid and cannabinoid analgesics both inhibit cyclic AMP production in the rat striatum," *Advances in Biosciences* **75**: 165–168.

Bonhaus, D. W., Chang, L. K., Kwan, J. and Martin, G. R. (1998) "Dual activation and inhibition of adenylyl cyclase by cannabinoid receptor agonists: Evidence for agonist specific trafficking of intracellular responses," *The Journal of Pharmacology and Experimental Therapeutics* **287**: 884–888.

Bouaboula, M., Bourrie, B., Rinaldi-Carmona, M., Shire, D., Le Fur, G. and Casellas, P. (1995a) "Stimulation of cannabinoid receptor CB_1 induces krox-24 expression in human astrocytoma cells," *Journal of Biological Chemistry* **270**: 13973–13980.

Bouaboula, M., Poinotchazel, C., Bourrie, B., Canat, X., Calandra, B., Rinaldi-Carmona, M., LeFur, G. and Casellas, P. (1995b) "Activation of mitogen-activated protein-kinase by stimulation of the central cannabinoid receptor CB_1," *Biochemical Journal* **312**: 637–641.

Bouaboula, M., Poinotchazel, C., Marchand, J., Canat, X., Bourrie, B., Rinaldi-Carmona, M., Calandra, B., LeFur, G. and Casellas, P. (1996) "Signaling pathway associated with stimulation of CB_2 peripheral cannabinoid receptor – Involvement of both mitogen-activated protein kinase and induction of Krox-24 expression," *European Journal of Biochemistry* **237**: 704–711.

Bouaboula, M., Perrachon, S., Milligan, L., Canat, X., Rinaldi-Carmona, M., Portier, M., Barth, F., Calandra, B., Pecceu, F., Lupker, J., Maffrand, J.-P. and Le Fur, G. (1997) "A selective inverse agonist for central cannabinoid receptor inhibits mitogen-activated protein kinase activation stimulated by insulin or insulin-like growth factor 1 – Evidence for a new model of receptor/ligand interactions," *Journal of Biological Chemistry* **272**: 22330–22339.

Bouaboula, M., Bianchini, L., McKenzie, F. R., Pouyssegur, J. and Casellas, P. (1999a) "Cannabinoid receptor CB_1 activates the Na^+/H^+ exchanger NHE-1 isoform via Gi-mediated mitogen activated protein kinase signaling transduction pathways," *FEBS Letters* **449**: 61–65.

Bouaboula, M., Dussossoy, D. and Casellas, P. (1999b) "Regulation of peripheral cannabinoid receptor CB_2 phosphorylation by the inverse agonist SR144528," *Journal of Biological Chemistry* **274**: 20397–20405.

Breivogel, C. S., Selley, D. E. and Childers, S. R. (1998) "Cannabinoid receptor agonist efficacy for stimulating [35S]GTPgS binding to rat cerebellar membranes correlates with agonist-induced decreases in GDP affinity," *Journal of Biological Chemistry* **273**: 16865–16873.

Burkey, T. H., Quock, R. M., Consroe, P., Ehlert, F. J., Hosohata, Y., Roeske, W. R. and Yamamura, H. I. (1997a) "Relative efficacies of cannabinoid CB_1 receptor agonists in the mouse brain," *European Journal of Pharmacology* **336**: 295–298.

Burkey, T. H., Quock, R. M., Consroe, P., Roeske, W. R. and Yamamura, H. I. (1997b) "Δ^9-THC is a partial agonist of cannabinoid receptors in the mouse brain," *European Journal of Pharmacology* **323**: R3–R4.

Calandra, B., Portier, M., Kernèis, A., Delpech, M., Carillon, C., Le Fur, G., Ferrara, P. and Shire, D. (1999) "Dual intracellular signalling pathways mediated by the human cannabinoid CB_1 receptor," *European Journal of Pharmacology* **374**: 445–455.

Caulfield, M. P. and Brown, D. A. (1992) "Cannabinoid receptor agonists inhibit Ca^{2+} current in NG108-15 neuroblastoma cells via a pertussis toxin sensitive mechanism," *British Journal of Pharmacology* **106**: 231–232.

Ceresa, B. P. and Limbird, L. E. (1994) "Mutation of an aspartate residue highly conserved among G-protein coupled receptors results in non-reciprocal disruption of alpha-2-adrenergic receptor-G-protein interactions," *Journal of Biological Chemistry* **269**: 29557–29564.

Chin, C. N., Lucas-Lenard, J., Abadji, V. and Kendall, D. A. (1998) "Ligand binding and modulation of cyclic AMP levels depend on the chemical nature of residue 192 of the human cannabinoid receptor 1," *Journal of Neurochemistry*, **70**: 366–373.

Coutts, A. A. and Pertwee, R. G. (1997) "Inhibition by cannabinoid receptor agonists of acetylcholine release from the guinea-pig mysenteric plexus," *British Journal of Pharmacology* **121**: 1557–1566.

Deadwyler, S. A., Hampson, R. E., Mu, J., Whyte, A. and Childers, S. (1995) "Cannabinoids modulate voltage sensitive potassium A-current in hippocampal neurons via a cAMP dependent process," *The Journal of Pharmacology and Experimental Therapeutics* **273**: 734–743.

DePetrocellis, L., Orlando, P. and DiMarzo, V. (1995) "Anandamide, an endogenous cannabinomimetic substance, modulates rat brain protein kinase C in vitro," *Biochemistry and Molecular Biology International* **36**: 1127–1133.

Fan, F., Compton, D. R., Ward, S., Melvin, L. and Martin, B.R. (1994) "Development of cross tolerance between Δ^9-THC, CP 55,940 and WIN 55,212," *The Journal of Pharmacology and Experimental Therapeutics* **271**: 1383–1390.

Fan, F., Tao, Q., Abood, M. and Martin, B. R. (1996) "Cannabinoid receptor down-regulation without alteration of the inhibitory effect of CP 55,940 on adenylyl cyclase in the cerebellum of CP 55,940-tolerant mice," *Brain Research* **706**: 13–20.

Felder, C. C., Veluz, J. S., Williams, H. L., Briley, E. M. and Matsuda, L. A. (1992) "Cannabinoid agonists stimulate both receptor and non-receptor mediated signal transduction pathways in cells transfected and expressing cannabinoid receptor clones," *Molecular Pharmacology* **42**: 838–845.

Felder, C. C., Briley, E. M., Axelrod, J., Simpson, J. T., Mackie, K. and Devane, W. A. (1993) "Anandamide, an endogenous cannabimetic eiconsanoid, binds to the cloned human cannabinoid receptor and stimulates receptor mediated signal transduction," *Proceedings of the National Academy of the Science, USA* **90**: 7656–7666.

Felder, C. C., Joyce, K. E., Briley, E. M., Mansouri, J., Mackie, K., Blond, O., Lia, Y., Ma, A. and Mitchell, R. L. (1995) "Comparison of the pharmacology and signal transduction of the human cannabinoid CB_1 and CB_2 receptors," *Molecular Pharmacology* **48**: 443–450.

Filipeanu, C. M., de Zeeuw, D. and Nelemans, S. A. (1997) "Delta-9-tetrahydrocannabinol activates $[Ca^{2+}]i$ increases partly sensitive to capacitative store refilling," *European Journal of Pharmacology* **336**: R1–R3.

Galve-Roperh, I., Sánchez, C., Cortés, M. L., Gómez del Pulgar, T., Izquierdo, M. and Guzmán, M. (2000) "Anti-tumoral action of cannabinoids: Involvement of sustained ceramide accumulation and extracellular signal-regulated kinase activation," *Nature Medicine* **6**: 313–319.

Garcia, D. E., Brown, S., Hille, B. and Mackie, K. (1998) "Protein kinase C disrupts cannabinoid actions by phosphorylation of the CB_1 cannabinoid receptor," *Journal of Neuroscience*, **18**: 2834–2841.

Gebremedhin, D., Lange, A. R., Campbell, W. B., Hillard, C. J. and Harder, D. R. (1999) "Cannabinoid CB_1 receptor of cat cerebral arterial muscle functions to inhibit L-type Ca^{2+} channel current," *American Journal of Physiology* **276**: h2085–h2093.

Gessa, G. L., Mascia, M. S., Casu, M. A. and Carta, G. (1997) "Inhibition of hippocampal acetylcholine release by cannabinoids reversal by SR141716A," *European Journal of Pharmacology* **327**: R1–R2.

Gifford, A. N. and Ashby, C. R. (1996) "Electrically-evoked acetylcholine-release from hippocampal slices is inhibited by the cannabinoid receptor agonist, WIN55, 212-2, and is potentiated by the cannabinoid antagonist, SR141716A," *The Journal of Pharmacology and Experimental Therapeutics* **277**: 1431–1436.

Glass, M. and Dragunow, M. (1995) "Selective induction of Krox 24 in striosomes by a cannabinoid agonist," *NeuroReport* **6**: 241–244.

Glass, M. and Felder, C. C. (1997) "Concurrent stimulation of cannabinoid CB_1 and dopamine D2 receptors augments cAMP accumulation in striatal neurons: Evidence for a Gs-linkage to the CB_1 receptor," *Journal of Neuroscience* **17**: 5327–5333.

Glass, M. and Northup, J. K. (1999) "Agonist selective regulation of G-proteins by cannabinoid CB_1 and CB_2 receptors," *Molecular Pharmacology* **56**: 1362–1369.

Griffin, G., Atkinson, P. J., Showalter, V. M., Martin, B. R. and Abood, M. E. (1998) "Evaluation of cannabinoid receptor agonists and antagonists using the guanosine-5'-O-2 [35guanosine-5'-O-2[35S]thio)-triphosphate binding assay in rat cerebellar membranes," *The Journal of Pharmacology and Experimental Therapeutics* **285**: 553–560.

Griffin, G., Wray, E. J., Tao, Q., McAllister, S. D., Rorrer, W. K., Aung, M., Martin, B. R. and Abood, M. E. (1999) "Evaluation of the cannabinoid CB_2 receptor-selective antagonist SR144528: further evidence for cannabinoid CB_2 receptor absence in the rat central nervous system," *European Journal of Pharmacology* **377**: 117–125.

Gudermann, T., Schöneberg, T. and Schultz, G. (1997) "Functional and structural complexity of signal transduction via G-protein coupled receptors," *Annual Reviews in Neuroscience* **20**: 399–427.

Henry, D. J. and Chavkin, C. (1995) "Activation of inwardly rectifying potassium channels (GIRK1) by co-expressed rat brain cannabinoid receptors in *Xenopus* oocytes," *Neuroscience Letters* **186**: 91–94.

Herkenham, M., Lynn, A. B., Johnson, M. R., Melvin, L. S., de Costa, B. R. and Rice, K. C. (1991) "Characterization and localization of cannabinoid receptors in the rat brain: a quantitative in vitro autoradiographic study," *Journal of Neuroscience* **11**: 563–583.

Hillard, C. J., Muthian, S. and Kearn, C. S. (1999) "Effects of CB_1 cannabinoid receptor activation on cerebellar granule cell nitric oxide synthase activity," *FEBS Letters* **459**: 277–281.

Ho, B. Y., Uezono, Y., Takada, S., Takase, I. and Izumi, F. (1999) "Coupling of the expressed cannabinoid CB_1 and CB_2 receptors to phospholipase C and G-protein coupled inwardly rectifying K+ channels," *Receptors and Channels* **6**: 363–374.

Howlett, A. C. and Fleming, R. M. (1984) "Cannabinoid inhibition of adenylate cyclase. Pharmacology of the response of neuroblastoma cell membranes," *Molecular Pharmacology* **26**: 532–538.

Howlett, A. C., Song, C., Berglund, B. A., Wilken, G. H. and Pigg, J. J. (1998) "Characterization of CB_1 cannabinoid receptors using receptor peptide fragments and site-directed antibodies," *Molecular Pharmacology* **53**: 504–510.

Hsieh, C., Brown, S., Derleth, C. and Mackie, K. (1999) "Internalization and recycling of the CB_1 cannabinoid receptor," *Journal of Neurochemistry* **73**: 493–501.

Ishac, E. J. N., Jiang, L., Lake, K. D., Varga, K., Abood, M. E. and Kunos, G. (1996) "Inhibition of exocytotic noradrenaline release by presynaptic cannabinoid CB_1 receptors on peripheral sympathetic nerves," *British Journal of Pharmacology* **118**: 2023–2028.

Jàrai, Z., Wagner, J. A., Varga, K., Lake, K. D., Compton, D. R., Martin, B. R., Zimmer, A. M., Bonner, T. I., Buckley, N. E., Mezey, E., Razdan, R. K., Zimmer, A. and Kunos, G. (1999) "Cannabinoid-induced mesenteric vasodilation through an endothelial site distinct from CB_1 or CB_2 receptors," *Proceedings of the National Academy of the Science, USA* **96**: 14136–14141.

Jin, W., Lee, N. M., Loh, H. H. and Thayer, S. A. (1994) "Opioids mobilize calcium from inositol 1,4,5-triphosphate-sensitive stores in NG108-15 cells," *Journal of Neuroscience* **14**: 1920–1929.

Jin, W., Brown, S., Roche, J. P., Hsieh, C., Celver, J. P., Kovoor, A., Chavkin, C. and Mackie, K. (1999) "Distinct domains of the CB_1 cannabinoid receptor mediate desensitization and internalization," *Journal of Neuroscience* **19**: 3773–3780.

Kathmann, M., Bauer, U., Schlicker, E. and Gothert, M. (1999) "Cannabinoid CB_1 receptor-mediated inhibition of NMDA and kainate-stimulated noradrenaline and dopamine release in the brain," *Naunyn-Schmiedeberg's Archives of Pharmacology* **359**: 466–470.

Katona, I., Sperlagh, B., Sik, A., Kafalvi, A., Vizi, E. S., Mackie, K. and Freund, T. F. (1999) "Presynaptically located CB_1 cannabinoid receptors regulate GABA release from axon terminals of specific hippocampal interneurons," *Journal of Neuroscience* **19**: 4544–4558.

Kearn, C. S. and Hillard, C. J. (1997) "Rat microglial cells express the peripheral-type cannabinoid receptor (CB_2) which is negatively coupled to adenylyl cyclase," Symposium on Cannabinoids, 1:61. Burlington Vt.:International Cannabinoid Research Society.

Kearn, C. S., Greenberg, M. J., DiCamelli, R., Kurzawak, K. and Hillard, C. J. (1999) "Relationship between ligand affinities for the cerebellar cannabinoid receptor CB_1 and the induction of GDP/GTP exchange," *Journal of Neurochemistry* **72**: 2379–2387.

Keefer, J. R., Kennedy, M. E. and Limbird, L. E. (1994) "Unique structural features important for stabilisation versus polarization of the alpha-2A-adrenergic receptor on the basolateral membrane of Madin-Darby canine kidney cells," *Journal of Biological Chemistry* **269**: 16425–16433.

Landsman, R. S., Burkey, T. H., Consroe, P., Roeske, W. R. and Yamamura, H. I. (1997) "SR141716A is an inverse agonist at the human cannabinoid CB_1 receptor," *European Journal of Pharmacology* **334**: r1–r2.

Ledent, C., Valverde, O., Cossu, C., Petitet, F., Aubert, L. F., Beslot, F., Bohme, G. A., Imperato, A., Pedrazzini, T., Roques, B. P., Vassart, G., Fratta, W. and Parmentier, M. (1999) "Unresponsiveness to cannabinoids and reduced addictive effects of opiates in CB_1 receptor knockout mice," *Science* **283**: 401–404.

Mackie, K. and Hille, B. (1992) "Cannabinoids inhibit N-type calcium channels in neuroblastoma-glioma cells," *Proceedings of the National Academy of the Science, USA* **89**: 3825–3829.

Mackie, K., Devane, W. A. and Hille, B. (1993) "Anandamide, an endogenous cannabinoid, inhibits calcium currents as a partial agonist in N18 neuroblastoma cells," *Molecular Pharmacology* **44**: 498–503.

Mackie, K., Lai, Y., Westenbroek, R. and Mitchell, R. (1995) "Cannabinoids activate an inwardly rectifying potassium conductance and inhibit Q-type calcium currents in AtT20 cells transfected with rat brain cannabinoid receptors," *Journal of Neuroscience* **15**: 6552–6561.

MacLennan, S. J., Reynen, P. H., Kwan, J. and Bonhaus, D. W. (1998) "Evidence for inverse agonism of SR141716A at human recombinant cannabinoid CB_1 and CB_2 receptors," *British Journal of Pharmacology* **124**: 619–622.

Mailleux, P., Verslype, M., Preud'homme, X. and Vanderhaeghen, J. J. (1994) "Activation of multiple transcription factor genes by tetrahydrocannabinol in rat forebrain," *NeuroReport* **5**: 1265–1268.

Maneuf, Y. P., Nash, J. E., Crossman, A. R. and Brotchie, J. M. (1996) "Activation of the cannabinoid receptor by Δ^9-THC reduces GABA uptake in the globus pallidus," *European Journal of Pharmacology* **308**: 161–164.

Matsuda, L. A., Lolait, S. J., Brownstein, M. J., Young, A. C. and Bonner, T. I. (1990) "Structure of cannabinoid receptor and functional expression of the cloned cDNA," *Nature* **346**: 561–564.

McAllister, S. D., Griffin, G., Satin, L. S. and Abood, M. E. (1999) "Cannabinoid receptors can activate and inhibit G-protein coupled inwardly rectifying potassium channels in a *Xenopus* oocyte expression system," *The Journal of Pharmacology and Experimental Therapeutics* **291**: 618–626.

McMillan, D. E., Dewey, W. L. and Harris, L. S. (1971) "Characteristics of tetrahydrocannabinol tolerance," *Annuals of the New York Academy of Science* **191**: 83–99.

Milligan, G., Bond, R. A. and Lee, M. (1995) "Inverse agonism: pharmacological curiosity or potential therapeutic strategy?" *Trends in Pharmacological Science* **16**: 10–13.

Mombouli, J. V., Schaeffer, G., Holzmann, S., Kostner, G. M. and Graier, W. F. (1999) "Anandamide induced mobilization of cytosolic Ca^{2+} in endothelial cells," *British Journal of Pharmacology* **126**: 1593–1600.

Mukhopadhyay, S., Cowsik, S. M., Lynn, A. M., Welsh, W. J. and Howlett, A. C. (1999) "Regulation of Gi by the CB_1 cannabinoid receptor c-terminal juxtamembrane region: structural requirements determined by peptide analysis," *Biochemistry* **38**: 3447–3455.

Mukhopadhyay, S., McIntosh, H. H., Houston, D. B. and Howlett, A. C. (2000) "The CB_1 cannabinoid receptor juxtamembrane C-terminal peptide confers activation to specific G-proteins in brain," *Molecular Pharmacology* **57**: 162–170.

Munro, S., Thomas, K. L. and Abu-Shaar, M. (1993) "Molecular characterization of a peripheral receptor for cannabinoids," *Nature* **365**: 61–65.

Nestler, E. J. (1993) "Cellular-responses to chronic treatment with drugs of abuse," *Critical Reviews in Neurobiology* **7**: 23–39.

Netzeband, J. G., Conroy, S. M., Parsons, K. L. and Gruol, D. L. (1999) "Cannabinoids enhance NMDA-elicited Ca^{2+} signals in cerebellar granule neurons in culture," *Journal of Neuroscience* **19**: 8765–8777.

Oviedo, A., Glowa, J. and Herkenham, M. (1993) "Chronic cannabinoid administration alters cannabinoid receptor binding in rat brain: a quantitative autoradiographic study," *Brain Research* **616**: 293–302.

Pan, X. H., Ikeda, S. R. and Lewis, D. L. (1996) "Rat brain cannabinoid receptor modulates N-type Ca^{2+} channels in a neuronal expression system," *Molecular Pharmacology* **49**: 707–714.

Pan, X., Ikeda, S. R. and Lewis, D. L. (1998) "SR 141716A Acts as an inverse agonist to increase neuronal voltage-dependent Ca^{2+} currents by reversal of tonic CB_1 cannabinoid receptor activity," *Molecular Pharmacology* **54**: 1064–1072.

Pertwee, R. G. (1991) "Tolerance to and dependence on psychotropic cannabinoids," in *The biological bases of drug tolerance and dependence*, edited by J. A. Pratt, pp. 231–263. New York: Academic Press.

Pertwee, R. G. (1997) "Pharmacology of cannabinoid CB_1 and CB_2 receptors," *Pharmacology and Therapeutics*, **74**: 129–180.

Pertwee, R. G. and Fernando, S. R. (1996) "Evidence for the presence of cannabinoid receptors in mouse urinary bladder," *British Journal of Pharmacology* **118**: 2056–2058.

Pertwee, R. G. and Griffin, G. (1995) "A preliminary investigation of the mechanisms underlying cannabinoid tolerance in the mouse vas deferens," *European Journal of Pharmacology* **272**: 67–72.

Pertwee, R., Stevenson, L. and Griffin, G. (1993) "Cross-tolerance between Δ^9-THC and the cannabimimetic agents, CP55,940, WIN 55,212 and anandamide," *British Journal of Pharmacology* **110**: 1483–1490.

Porcella, A., Gessa, G. L. and Pani, L. (1998) "Delta(9)-tetrahydrocannibinol increases sequence-specific AP-1 binding activity and Fos-related antigens in the rat brain," *European Journal of Neuroscience* **10**: 1743–1751.

Portier, M., Rinaldi-Carmona, M., Pecceu, F., Combes, T., Poinot-Chazel, C., Calandra, B., Barth, F., Le Fur, G. and Casellas, P. (1999) "SR 144528, an antagonist for the peripheral cannabinoid receptor that behaves as an inverse agonist," *The Journal of Pharmacology and Experimental Therapeutics* **288**: 582–589.

Reichman, M., Nen, W. and Hokin, L. E. (1991) "Δ^9-THC inhibits arachidonic acid acylation of phospholipids and triacylglycerols in guinea pig cerebral cortex slices," *Molecular Pharmacology* **40**: 547–555.

Ren, Q., Kurose, H., Lefkowitz, R. J. and Cotecchia, S. (1993) "Constitively active mutants of the alpha 2 adrenergic receptor," *Journal of Biological Chemistry* **268**: 16483–16487.

Rhee, M. H., Nevo, I., Levy, R. and Vogel, Z. (2000) "Role of the highly conserved Asp-Arg-Tyr motif in signal transduction of the CB_2 cannabinoid receptor," *FEBS Letters* **466**: 300–304.

Rinaldi-Carmona, M., Le Duigou, A., Oustric, D., Barth, F., Bouaboula, M., Carayon, P., Casellas, P. and Le Fur, G. (1998) "Modulation of CB_1 cannabinoid receptor functions after a long-term exposure to agonist or inverse agonist in the chinese hamster ovary cell expression system," *The Journal of Pharmacology and Experimental Therapeutics* **287**: 1038–1047.

Roche, J. P., Bounds, S., Brown, S. and Mackie, K. (1999) "A mutation in the second transmembrane region of the CB_1 receptor selectively disrupt G-protein signaling and prevents receptor internalization," *Molecular Pharmacology* **56**: 611–618.

Rodriguez de Fonseca, F., Gorriti, M. A., Fernandez-Ruiz, J. J., Palomo, T. and Ramos, J. A. (1994) "Down regulation of rat brain cannabinoid binding sites after chronic Δ^9-THC treatment," *Pharmacology, Biochemistry and Behaviour* **47**: 33–40.

Romero, J., Garcia-Palomero, J. G., Castro, L., Garcia-Gil, J. A., Ramos, J. J. and Fernandez-Ruiz, J. J. (1997) "Effects of chronic exposure to Δ^9-THC on cannabinoid receptor binding and mRNA levels in several rat brain regions," *Molecular Brain Research* **46**: 100–108.

Sagan, S., Venance, L., Torrens, Y., Cordier, J., Glowinski, J. and Giaume, C. (1999) "Anandamide and WIN 55,212-2 inhibit cyclic AMP formation through G-protein coupled receptors distinct from CB_1 cannabinoid receptors in cultured astrocytes," *European Journal of Neuroscience* **11**: 691–699.

Samama, P., Cotecchia, S., Costa, T. and Lefkowitz, R. J. (1993) "A mutation-induced activated state of the B2-adrenergic receptor," *Journal of Biological Chemistry* **268**: 4625–4636.

Saunders, C., Keefer, J. F., Bonner, C. A. and Limbird, L. E. (1998) "Targeting of G-protein coupled receptors to the basolateral surface of polarized renal epithelial cells involves multiple non-contiguous structural signals," *Journal of Biological Chemistry* **273**: 24196–24206.

Schweitzer, P. (2000) "Cannabinoids decrease the K^+ M-current in hippocampal CA_1 neurons," *Journal of Neuroscience* **20**: 51–58.

Sealfon, S. C., Chi, L., Ebersole, B. J., Rodic, V., Zhang, D., Ballesteros, J. A. and Weinstein, H. (1995) "Related contribution of specific helix 2 and 7 residues to conformational activation of the serotonin 5-HT2A receptor," *Journal of Biological Chemistry* **270**: 16683–16688.

Shen, M. X. and Thayer, S. A. (1998) "The cannabinoid agonist WIN 55,212-2 inhibits calcium channels by receptor-mediated and direct pathways in cultured rat hippocampal neurons," *Brain Research* **783**: 77–84.

Shen, M. X., Piser, T. M., Seybold, V. S. and Thayer, S. A. (1996) "Cannabinoid receptor agonists inhibit glutamatergic synaptic transmission in rat hippocampal cultures," *Journal of Neuroscience* **16**: 4322–4334.

Shire, D., Calandra, B., Delpech, M., Dumont, X., Kaghad, M., LeFur, G., Caput, D. and Ferrara, P. (1996) "Structural features of the central cannabinoid CB_1 receptor involved in the binding of the specific CB_1 antagonist SR141716A," *Journal of Biological Chemistry* **271**: 6941–6946.

Sim, L. J., Selley, D. E. and Childers, S. R. (1995) "In vitro autoradiography of receptor-activated G-proteins in rat brain by agonist-stimulated guanylyl 5'[gamma]35S-thio-triphosphate binding," *Proceedings of the National Academy of the Science, USA* **92**: 7242–7246.

Slipetz, D. M., O'Neill, G. P., Favreau, L., Dufresne, C., Gallant, M., Gareau, Y., Guay, D., Labelle, M. and Metters, K. M. (1995) "Activation of the human peripheral cannabinoid receptor results in inhibition of adenylyl cyclase," *Molecular Pharmacology* **48**: 352–361.

Smart, D., Gunthorpe, M. J., Jerman, J. C., Nasir, S., Gray, J., Muir, A. I., Chambers, J. K., Randall, A. D. and Davis, J. B. (2000) "The endogenous lipid anandamide is a full agonist at the human vanilloid receptor," *British Journal of Pharmacology* **129**: 227–230.

Song, Z. H. and Bonner, T. I. (1996) "A lysine residue of the cannabinoid receptor is critical for receptor recognition by several agonists but not WIN55212-2," *Molecular Pharmacology* **49**: 891–896.

Sugiura, T., Kodaka, T., Kondo, S., Tonegawa, T., Nakane, S., Kishimoto, S., Yamashita, A. and Waku, K. (1996) "2-Arachidonoylglycerol, a putative endogenous cannabinoid receptor ligand, induces rapid, transient elevation of intracellular free Ca2+ in neuroblastoma × glioma hybrid NG108-15 cells," *Biochemical and Biophysical Research Communication* **229**: 58–64.

Sugiura, T., Kodaka, T., Kondo, S., Nakane, S., Kondo, H., Waku, K., Ishima, Y., Watanabe, K. and Yamamoto, I. (1997) "Is the cannabinoid CB_1 receptor a 2-arachidonoylglycerol receptor? Structural requirements for triggering Ca2+ transient in NG108–15 cells," *Journal of Biochemistry* **122**: 890–895.

Sugiura, T., Kodaka, T., Nakane, S., Miyashita, T., Kondo, S., Suhara, Y., Takayama, H., Waku, K., Seki, C., Baba, N. and Ishima, Y. (1999) "Evidence that the cannabinoid CB_1 receptor is a 2-arachidonoylglycerol receptor," *Journal of Biological Chemistry* **274**: 2794–2801.

Sugiura, T., Kondo, S., Kishimoto, S., Miyashita, T., Nakane, S., Kodaka, T., Suhara, Y., Takayama, H. and Waku, K. (2000) "Evidence that 2-arachidonolyglycerol but not N-Palmitoylethanolamine or anandamide is the physiological ligand for the cannabinoid CB_2 receptor," *Journal of Biological Chemistry* **275**: 605–612.

Sullivan, J. M. (1999) "Mechanisms of cannabinoid receptor mediated inhibition of synaptic transmission in cultured hippocampal pyramidal neurons," *Journal of Neurophysiology* **82**: 1286–1294.

Surprenant, A., Horstman, D. A., Akbarali, H. and Limbird, L. E. (1992) "A point mutation of the a2-adrenoreceptor that blocks coupling to potassium but not calcium currents," *Science* **257**: 977–980.

Tao, Q. and Abood, M. E. (1998) "Mutation of a highly conserved aspartate residue in the second transmembrane domain of the cannabinoid receptors, CB_1 and CB_2, disrupts G-protein coupling," *The Journal of Pharmacology and Experimental Therapeutics* **285**: 651–658.

Tao, Q., McAllister, S. D., Andreassi, J., Nowell, K. W., Cabral, G. A., Hurst, D. P., Bachtel, K., Ekman, M. C., Reggio, P. H. and Abood, M. E. (1999) "Role of a conserved lysine residue in the peripheral cannabinoid receptor (CB_2): Evidence for subtype specificity," *Molecular Pharmacology* **55**: 605–613.

Tsou, K., Brown, S., Sanudo-Pena, M. C., Mackie, K. and Walker, J. M. (1998) "Immunohisto-chemical distribution of Cannabinoid CB_1 receptors in the rat central nervous system," *Neuroscience* **83**: 393–411.

Vàsquez, C. and Lewis, D. L. (1999) "The CB_1 cannabinoid receptor can sequester G-proteins, making them unavailable to couple to other receptors," *Journal of Neuroscience* **19**: 9271–9280.

Venance, L., Piomelli, D., Glowinski, J. and Giaume, C. (1995) "Inhibition by anandamide of gap junctions and intercellular calcium signalling in striatal astrocytes," *Nature* **376**: 590–594.

Wagner, J. A., Varga, K. and Kunos, G. (1998) "Cardiovascular actions of cannabinoids and their generation during shock," *Journal of Molecular Medicine* **76**: 824–836.

Wang, C. D., Buck, M. and Fraser, C. M. (1991) "Site-directed mutagenesis of alpha-2A-adrenergic receptors – identification of amino-acids involved in ligand-binding and receptor activation by agonists," *Molecular Pharmacology* **40**: 168–179.

White, R. and Hiley, C. R. (1998) "The actions of some cannabinoid receptor ligands in the rat isolated mesenteric artery," *British Journal of Pharmacology* **125**: 533–541.

Zhou, W., Flanagan, C., Ballesteros, J. A., Knovicka, K., Davidson, J. S., Weinstein, H., Millar, R. P. and Sealfon, S. C. (1994) "A reciprocal mutation supports helix 2 and helix 7 proximity in the gonadotrophin-releasing hormone receptor," *Molecular Pharmacology* **45**: 165–170.

Zimmer, A., Zimmer, A. M., Hohmann, A. G., Herkenham, M. and Bonner, T. I. (1999) "Increased mortality, hypoactivity, and hypoalgesia in cannabinoid CB_1 receptor knockout mice," *Proceedings of the National Academy of the Science, USA* **96**: 5780–5785.

Zygmunt, P. M., Petersson, J., Andersson, D. A., Chuang, H. H., Sorgard, M., Di Marzo, V., Julius, D. and Hogestatt, E. D. (1999) "Vanilloid receptors on sensory nerves mediate the vasodilator action of anandamide," *Nature* **400**: 452–457.

Chapter 16

Deregulation of membrane and receptor mediated signaling by THC – therapeutic implications

Gabriel G. Nahas, D. Harvey, K. M. Sutin,
H. Turndorf and R. Cancro

ABSTRACT

There has been much discussion over the last year (Select Committee on Science and Technology, 1998; Joy *et al.*, 1999; Nahas *et al.*, 1999) on the use of psychoactive cannabinoids of marijuana (THC) for medical purposes, particularly with respect to the treatment of neurological disorders (multiple sclerosis), glaucoma, and pain. However, little of this discussion appears to reflect the true nature of this drug that contains many active components, present in varying amounts. These compounds affect the body through a number of mechanisms that can, in some cases, produce opposing effects. Of particular importance, are the developing theories of signal transduction within the membrane and the emerging evidence that suggests that marijuana interferes with a basic regulation of cell function at the molecular level with unforeseen consequences. This paper reviews the molecular mechanisms by which THC produces its effects and presents a unified theory of membrane signaling transduction which could account for the therapeutic properties of the drug.

Key Words: cannabinoids, lipid ligand, membrane bilipid layer, G protein linked 7TM receptor, THC receptor, CB_1, CB_2, arachidonic acid, AEA, eicosanoid, (arichinodylethanolamine) neurotransmitters, allosteric molecular configuration, volume transmission, molecular signaling interactions, membrane receptors

CANNABINOIDS

Marijuana is a product of the plant Cannabis sativa L. that contains some sixty unique terpenoid-containing molecules known as cannabinoids. CBD, for example, inactivates certain isozymes of the cytochrome P-450 drug metabolizing enzyme system (Bornheim *et al.*, 1993) which, in turn, alters the relative amount of psychoactive metabolites produced from Δ^9-THC. It can be seen that marijuana from uncontrolled sources could produce inconsistent pharmacological effects. Furthermore, the high lipophilicity of THC (Garrett and Hunt, 1977; Leuschner *et al.*, 1986) produces a long half-life and results in substantial accumulation in body tissues after continuous use (Cridland *et al.*, 1983; Ellis *et al.*, 1985).

Cannabinoids and their metabolites (Harvey, 1991) are classified into two major categories, psychoactive and non-psychoactive, corresponding to particular chemical structures and pharmacological effects. The psychoactive cannabinoids, Δ^9-THC, Δ^8-THC (a synthetic analog) and their 11-hydroxy derivatives bind stereospecifically to unique receptors and have pharmacologic activity in nanomolar concentrations. The non-psychoactive cannabinoids, which consist of both natural compounds such as CBD and metabolites such as Δ^9-THC-11-oic acid, exert their actions at other sites. While not psychoactive, these latter compounds nevertheless possess other biological activities, and as first emphasized by Paton (Paton *et al.*, 1972), target the lipid bilayer. The membrane and its integral receptors closely interact; THC partitions into the membrane, alters membrane fluidity, (increases the molecular disorder of the bilipid layer) and affects membrane bound enzymes and receptors.

THE INTERACTION OF CANNABINOIDS WITH THE MEMBRANE LIPID BILAYER

Lawrence and Gill (Lawrence and Gill, 1975), in 1976, first demonstrated that THC at low concentrations (1 micromolar) increases the molecular disorder of the liposome. At higher concentration, the effects level off and do not increase as the molecular ratio of THC to lecithin is increased. The maximum degree of fluidization (disordering) produced by THC does not approach that required to produce anesthesia and THC has been designated a "partial anesthetic". The term incomplete could also be used. It was reported (Leuschner *et al.*, 1984) that the disordering effect of THC occurred *in vivo*, there was an equilibrium between the concentration of the drug in erythrocytes and plasma. A disordering effect in the membrane was also reported after chronic treatment of mice with ethanol (Wing *et al.*, 1982) which led the authors to suggest that tolerance might be associated with changes in the membrane lipid composition. By contrast, the non-psychoactive cannabinoids, CBN and CBD, produce an effect opposite to that of Δ^9-THC by decreasing the molecular disorder of the lipid bilayer (Lawrence and Gill, 1975) and one could expect that CBN and CBD would antagonize, in part, the effect of THC *in vivo*, a prediction that has been documented by experimental observations (Borgen and Davis, 1974). These results could contribute, in part, to the variability of observed effects in studies which use different cannabinoids.

EFFECTS OF CANNABINOIDS ON NEUROTRANSMITTER RECEPTORS

THC has been reported to affect the activity of several neurotransmitter receptor systems. THC, however, does not interact directly with the active site of the receptors, but rather causes an allosteric modification of the receptor, which in turn modifies its response to other agonists and antagonists. After THC administration, the effects of subsequent exposure to the receptor agonists acetylcholine (Domino, 1981; Gessa *et al.*, 1997) and opioids (Vaysse, 1987) and NMDA (Hampson *et al.*, 1998) are decreased, while the effects of catecholamines (Bloom *et al.*, 1978) and GABA are biphasic (Pryor *et al.*, 1977). Bloom and Hillard (1984) concludes that, since THC alters membrane fluidity, its effects on constituent receptors are due to changes in the membrane environment, which in turn, affect membrane receptors. "Although some differences exist in the interactions with specific proteins, the fact remains that psychoactive cannabinoids affect a broad range of membrane-bound receptors, which supports the concept of a direct influence on the membrane lipid bilayer, rather than an embedded receptor specific for Δ^9-THC on each of these very different receptor molecules to neurotransmitters."

INTERACTION OF THC WITH SPECIFIC RECEPTORS

As a result of these general interactions with membrane bound receptors, it was concluded that, while THC altered, in a specific fashion, the physicochemical organization of the membrane, this effect alone could not account for the stereospecific psychoactive properties of this cannabinoid. These are exerted at lower, nanomolar concentrations, and are more typical of drugs acting at specific receptors.

The first THC receptor (CB_1) was identified in rat brain (Devane *et al.*, 1988) and is a guanine nucleotide regulatory (G) protein linked (7 Trans membrane, 7TM) receptor. CB_1 Receptors in the brain are unevenly distributed, with highest concentrations in the globus pallidus, substantia nigra, cerebral cortex, striatum and the molecular layers of the cerebellum and hippocampus (Herkenham *et al.*, 1990). A second receptor (CB_2), has 44% sequence homology with the CB_1 receptor and is also a G protein-linked 7TM receptor (Munro *et al.*, 1993). Although the CB_2 receptors appear to be in the periphery, CB_1 receptors are found both centrally and peripherally (Pertwee, 1995).

I. Cannabinoids (tetrahydrocannabinoids)

Δ^9-THC

Δ^8-THC

Psychoactive Δ^9-THC, Δ^8-THC and hydroxy compounds and synthetic derivatives (nabilone, levonantradol)

CBD

CBN

Non psychoactive CBD CBN

II. Natural Ligands to the 7 transmembrane receptors (7TM) (G protein receptors) also called "cannabinoid" receptors (CB_1 and CB_2)

ARACHIDONYETHANOLAMIDE (AEA)

2-ARACHIDONOYL-GLYCEROL (2AC)

III. Synthetic ligands to the specific 7 transmembrane (7TM) AEA or 2AG receptor sites

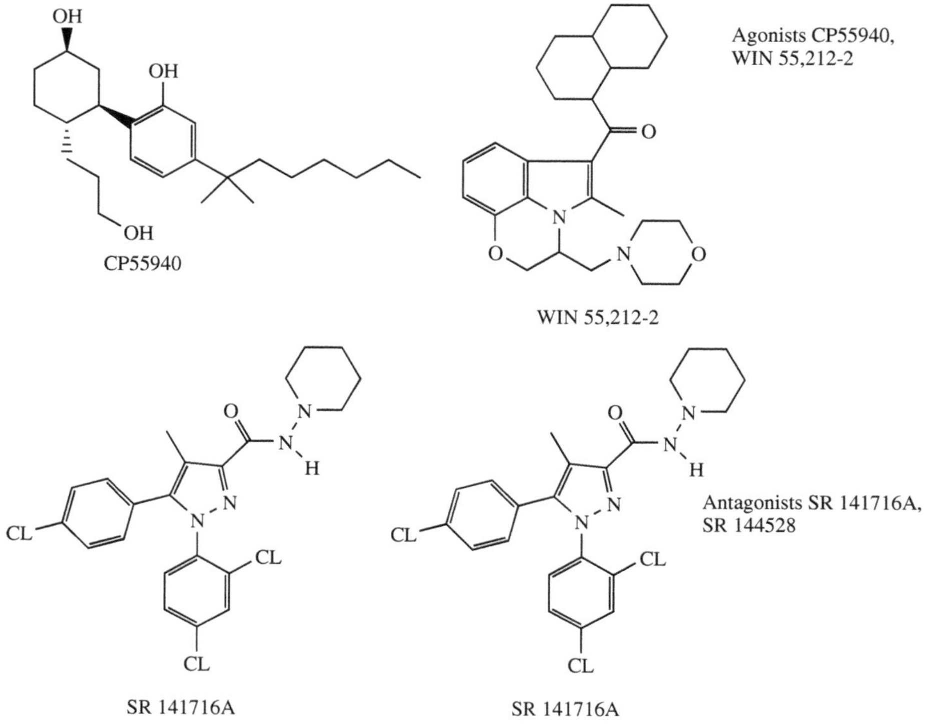

Figure 16.1 Structures of unrelated clinical compounds which bind to the cannabinoid receptor of the 7 transmembrane (7TM) G protein linked receptor.

FUNCTIONAL CORRELATES OF THC RECEPTOR BINDING AND SIGNALING ALTERATIONS

The binding of THC to its CB_1 or CB_2 G protein-7TM coupled receptors is associated with changes in functions of the brain, the immune system and the reproductive organs. In the brain, binding of THC to CB_1 receptors is associated with marked changes of sensory perception. These changes first described in 1845 by Moreau (Moreau, 1970), who emphasized alterations of visual, auditory, and "body image" perceptions. Subsequently, these self-reported alterations were correlated with functional markers used to estimate visual, auditory, and somatosensory perceptions: Cannabis intoxication induces an illusion of visual perception characterized by binocular depth inversion that is also observed in unmedicated persons with schizophrenia (Emrich *et al.*, 1997). It is also associated with a delay in the P300 response to auditory evoked potentials, a dysfunction that may persist for weeks after THC (Solowij *et al.*, 1995). Distortion of somatosensory perception reported by Moreau was also observed in a controlled clinical setting (Perez-Reyes *et al.*, 1991). Lowering of thermal pain perception has also been measured in heavy marijuana smokers (Clark *et al.*, 1981). Such perceptual alterations may be related to the persistent binding of THC to the CB_1 receptors in areas of the brain where sensory perceptions are transduced.

Binding of THC to peripheral CB_2 7TM receptors induces alterations of signaling in the immune system and the reproductive organs. The binding to CB_2 receptors on lymphocytes and macrophages impairs their signaling function (Cabral, 1999). Binding of THC to receptors on sperm cells and ovum interferes with the acrosomal reaction (Schuel *et al.*, 1994), fertilization and ovum implantation (Paria *et al.*, 1995). These signaling alterations could account for the clinical and experimental observations reporting a decreased spermatogenesis, decreased sperm motility and increase of abnormal forms of sperm in rodents and humans exposed to marijuana smoke or THC (Hembree *et al.*, 1999).

ENDOGENOUS MEMBRANE-DERIVED LIGANDS OF THE CANNABINOID RECEPTOR

The discovery of receptors with an affinity for THC suggested the occurrence of endogenous ligands for these receptors. The first such compound to be discovered was the lipid, *N*-arachidonylethanolamide (AEA) (Devane *et al.*, 1992), also named "anandamide" after the Sanskrit word "ananda" meaning bliss. Biosynthesis of AEA

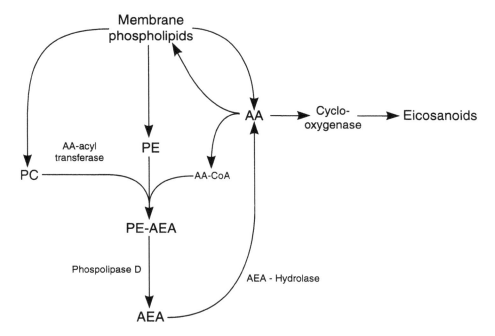

Figure 16.2 Biosynthesis and breakdown of AEA (AA = arachidonic acid, PE = phosphatidyleth-anolamine, PC = phosphatidylcholine, AEA = *N*-arachidonylethanolamide).

(Figure 16.2) is from membrane phosphatidylethanolamine which becomes esterified by free arachidonic acid or, by acyltransferase, from the 1-position of phosphatidylcholine (Di Marzo, 1998, 1999; Di Marzo et al., 1999). AEA is then released by the action of a phosphodiesterase (Sugiura et al., 1996a,b) identified in neuroblastoma cells as phospholipase D (Di Marzo et al., 1996). Degradation of AEA occurs within the membrane by the action of the amidase, anandamide amidohydrolase (Ueda et al., 1996; Bisogno et al., 1997a). AEA has a lower affinity for the 7TM receptor than THC, a much shorter duration of pharmacologic activity and it is rapidly eliminated like most signaling and second messenger molecules. It is not stored in cells, but accumulates in brain following hypoxia or cellular death (Schmid et al., 1995). Like THC, AEA has been found to fluidize the cell membrane (Bloom et al., 1997), and one might expect that AEA might play a role in the regulation of membrane fluidization and membrane function.

The full profile of AEA has activity yet to be elucidated and there are conflicting theories (Stanford and Mitchell, 1999). AEA, like THC may be a long-acting endothelium-derived vasorelaxant, known as endothelium-derived hyperpolarizing factor, or EDHF which is independent of NO. This mechanism has been discussed, although AEA has been shown to cause a dose dependent vasodilatation in the perfused rat mesenteric bed (Wagner et al., 1999), which appears to act through CB_1 or CB_2 receptors (Wagner et al., 1999; Niederhoffer and Szabo, 1999).

A second ligand of both CB_1 and CB_2 receptors has subsequently been identified as 2-arachidonyldiacylglycerol (2-AG). It was first isolated from canine gut (Mechoulam et al., 1995) and was later found in higher concentrations in brain cells (Sugiura et al., 1995; Stella et al., 1997). Its affinity for the receptors parallels that of AEA.

Two mechanisms for its biosynthesis have been proposed (Di Marzo, 1998; Di Marzo et al., 1999). The first involves cleavage of inositol phosphate from phosphatidylinositol by a phosphatidylinositol-specific phospholipase C followed by removal of the acyl-group from the 1 position. The second pathway involves the removal of the fatty acid from the 1-position of sn-arachidonic acid-containing diacylglycerides which, in turn, can originate from hydrolysis of triglycerides, phosphatidic acid, phosphatidylinositol or phosphatidylcholine.

EFFECT OF CANNABINOIDS ON PHOSPHOLIPID ENZYMES AND ARACHIDONIC ACID BIOSYNTHESIS

A phospholipid messenger system constitutes a major function of the lipid bilayer and involves signal-mediated hydrolysis of phospholipids within the membrane. The best characterized of these systems is that originating from arachidonic acid, which is released from phospholipids following activation by a phospholipase. Psychoactive cannabinoids and some of their non-psychoactive acid metabolites also release arachidonic acid from phosphatidylcholine in a dose-related manner (Hunter et al., 1985; Burstein et al., 1986). Arachidonic acid has been associated with biosynthesis of prostaglandins and other eicosanoids such as leukotrienes and platelet activation factors (PAF). Its metabolism has been shown to be inhibited by steroids (NSAIDs) (Burstein et al., 1986) and by the THC metabolite, THC-11-oic acid (Zurier et al., 1998). In direct contrast to the stimulation of prostaglandin production by THC induced arachidonic acid release, an inhibitory effect is produced by the nonpsychoactive THC metabolite THC-11-oic acid, that suppresses cyclooxygenase activity and inhibits prostaglandin production.

It is proposed that an alternative pathway of arachidonic acid metabolism, namely the formation of *N*-arachidonylethanolamine, is involved in a molecular signaling system of the cell (Zoli *et al.*, 1999). Protein receptor interaction with lipid ligands AEA and 2-AG would regulate signaling between boundary lipids and receptors or enzymes of the membrane. Boundary lipids surrounding the membrane proteins transduce signals from the AEA and 2-AG to the integral membrane receptors. The change of configuration of the 7TM receptor, caused by the lipid mediators, would produce allosteric changes of the protein, modulating or fine-tuning of membrane bound protein activity. According to this hypothesis, AEA and 2-AG, are indirect modulators of membrane and enzyme activity. The proposed mechanism of membrane signaling would fall within the category of volume transmission (VT) of intercellular communication.

MEMBRANE SIGNALING AND VOLUME TRANSMISSION

The terms wiring transmission (WT) and VT were introduced to provide a "systematic categorization of intracellular communication in the brain" (Zoli *et al.*, 1999). Wiring (voltage) transmission (WT) is a well defined quantifiable entity, while VT is not as clearly delineated. VT refers to hormonal messengers in brain extracellular space and cerebrospinal fluid. In the membrane bilayer signaling, VT might also occur and be more clearly defined. A signal transmission by a signaling lipid ligand occurring in the fluid milieu of the lipid bilayer will induce a volume change. This change in the lipid bilayer volume is imparted through VT of the lipid signaling molecule to the 7TM receptor protein, and in turn affects the volume of the receptor. A change in receptor volume will result in an *adiabatic* amplification of the original signal; the pressure volume relationships of the protein would be modulated by the recycling of the natural physiological lipid ligand AEA. Signaling through the lipid bilayer falls within the science of fluidics, which uses the technology of fluid pressure volume relationships of fluid dynamics to study signal amplification and transduction mechanisms.

THC deregulates this putative membrane signaling in three ways, (1) by altering the physicochemical organization of the boundary lipid bilayer which THC permeates, as a surrogate lipid signaling molecule; (2) by inducing or promoting AEA synthesis through release of arachidonic acid; and (3) by persistent binding to the CB receptors (Figure 16.3). The persistent deregulation by THC of this ubiquitous membrane signaling system is associated in a time and dose-related fashion with measurable function alterations of the brain and cerebellum, impairing visual, auditory, somatosensory perceptions, coordination, memory and consciousness (Perez-Reyes *et al.*, 1991; Clark *et al.*, 1981; Cabral, 1999; Schuel *et al.*, 1994). While binding to central and peripheral cannabinoid receptors and displacing their natural ligands, THC alters central and peripheral cardiovascular regulations inducing tachycardia (Trouve and Nahas, 1999) and vasodilatation (Stanford and Mitchell, 1999; Fulton and Quilley, 1998; Niederhoffer and Szabo, 1999).

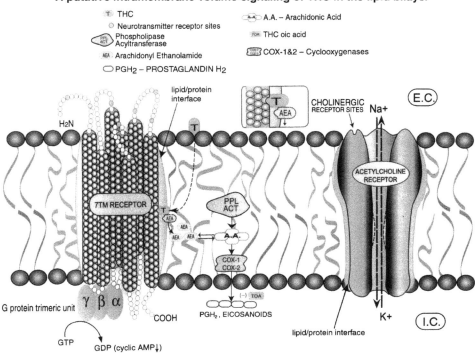

Figure 16.3 THC, a lipid molecule permeates the lipid bilayer of the membrane and disorders its fluidity, activating membrane bound enzymes phospholipids and acyl transferase. These enzymes control the biosynthesis of signaling lipid molecules, derived from membrane phospholipids, like arachidonic acid and its cascade of eicasanoids which include besides prostaglandins arachidonylethanolamide (AEA) or anandamide. AEA is the natural lipid ligand of the G protein linked 7 transmembrane domain receptor (7TM) a molecular switch which controls membrane signal transduction. THC is the only cannabinoid which also binds to same 7TM receptor and for which it has a greater affinity than AEA. It is proposed that THC would act as a "blocker" of AEA and impair its physiological function which is to regulate the 7TM receptor activity. THC alters in an alloteric fashion the responses of neurotransmitter and neuro modulator receptors (opiate, NMDA, 5HT, beta, muscarinic). These concurrent membrane molecular events have been correlated with multiple alterations of cellular function associated with THC administration and observed in brain synapses, gametes, gonads and blood vessels. Because of their common molecular mechanisms of action, the adverse effects of THC may not be dissociated from the therapeutic properties of the drug. It is proposed that the permeation of THC into the fluid milieu of the bilipid layer is associated with a volume change. This change in volume is imparted by "volume transmission" (VT) to the 7TM receptor protein which would in turn affect the volume of the receptor. This change in volume would result in an adiabatic amplification of the original signal transmitted by the lipid ligand to the receptor protein. (Copyright G. G. Nahas, 2000)

BIOSYNTHESIS OF MEMBRANE LIPID MEDIATORS

The fundamental nature of the AEA and 2-AG lipid mediators in controlling membrane signaling function is supported, in the case of AEA, by at least two biosynthetic mechanisms: direct enzyme-mediated esterification of phosphatidyl-ethanolamine by free arachidonic acid and a transesterification reaction from phosphatidylcholine (Figure 16.2). The first of these processes probably works to compensate, or fine-tune, the effects of the products of arachidonic acid oxidation (prostaglandins and the eicosanoid cascade), whereas the second, transferase-eicosanoid formation, could be a mechanism by which the cell produces AEA independently of eicosanoid production.

AEA and 2AG receptors have been discovered in primitive organisms such as the leech (Stefano *et al.*, 1997), *Mytillus* (Sepe *et al.*, 1998), *Hydra* (De Petrocellis *et al.*, 1999) and sea urchin (Bisogno *et al.*, 1997b) attesting to the fundamental and ubiquitous nature of this signaling mechanism. In organisms without, or with only a primitive nervous system, it appears to play a fundamental role in controlling behavior (De Petrocellis *et al.*, 1999) and may even predate the nervous system in unifying aggregations of single cells into a single functioning organism. Consequently, the AEA and 2-AG signaling pathways should be thought of as an important physiological control systems in contrast to the current emphasis on an "endogenous" cannabinoid system. We suggest that "cannabinoid" receptors should be renamed to reflect their *in vivo* function, rather than to their binding to exogenous cannabinoid ligands. Moreover, the CB$_1$ receptors have been shown to exhibit a significant degree of sequence homology with that of another phospholipid-derived bioactive mediator, lysophosphatidic acid (Piomelli *et al.*, 1998). This similarity suggests that all of these lipid-mediator receptors should be classified together rather than in separate groups, one of which, the "cannabinoid receptor" is classified as the result of interaction with an exogenous agonist which does not reflect its fundamental role.

PHYSIOLOGICAL IMPLICATIONS FOR MEDICINE AND THERAPEUTICS

Most therapeutic properties of cannabinoids have been attributed to natural psychoactive Δ^9-THC and to its 11-hydroxy metabolite which bind stereospecif-ically to the G-protein coupled receptors. A similar mode of action has been attributed to the synthetic analogs of Δ^9-THC, like nabilone and levonantradol and it is difficult or impossible to dissociate their therapeutic from their psychoactive effects because both these properties are mediated by the same molecular mechanism, an alteration of 7TM configuration. In addition, as discussed above, the highly lipo-philic THC, its non-psychoactive metabolites and its synthetic derivatives interact with the membrane lipid bilayer. The drugs also accumulate in the fatty tissues, maintaining blood levels during chronic administration, but making it difficult to titrate the precise therapeutic level in patients, especially in those with impaired drug metabolism. The persistence of this drug in lipid membranes induces a persistent state of neuronal membrane dysfunction, and continued use impairs the mental ability of the chronic user. The capability of THC to disrupt fundamental membrane signaling pathways must be considered when evaluating the full spectrum of marijuana's therapeutic potential.

In its smoked form, marijuana contains some 60 cannabinoids (Turner *et al.*, 1980) with variable agonist and antagonist properties. Smoke is toxic to the lung, impairs macrophage function and smoked marijuana thus cannabis is not a medically acceptable medium for THC delivery. Oral administration results in erratic uptake (Ohlsson *et al.*, 1980) and hepatic first-pass metabolism.

Analgesic effects

Over thousands of years of its consumption, pain relief and sedation were not the primary therapeutic properties attributed to marijuana. Cannabis preparations were recommended as general purpose medications to relieve a wide variety of ailments including asthma, urinary tract infection, inflammation, and headache. Opium has for several millennia been the medication of choice for the relief of pain and was prescribed preferentially over all other substances. This unique role was confirmed by Sydenham in the 17th century, when opium preparations (laudanum) became available in Europe. During the 19th century, after cannabis was introduced into the British and US medicinal formularies, opium remained the drug of choice for the relief of pain.

Since the discovery of Δ^9-THC, it has been claimed that this cannabinoid and its synthetic derivatives possess marked pain relieving properties, however, the role of marijuana and cannabinoids as analgesics has never been clearly defined in modern clinical medicine. Although THC and its synthetic derivatives have been shown to be effective in many antinociceptive tests in laboratory animals, results in man have been inconclusive or sometimes negative (Martin and Lichtman, 1998; Waser and Martin, 1999). A controlled clinical trial of smoked marijuana to evaluate the analgesic effect of the drug separately from its intoxicating effects concluded that marijuana enhanced pain perception (Clarke *et al.*, 1981). The authors point out that changes in reported pain probably reflect subjective sensation or reduced anxiety. In experimental studies, results are also inconsistent (Waser and Martin, 1999). Cannabinoids exert antalgic effects by interacting with the two major bio-chemical pathways that control pain, the opiate pathway of acute pain and the prostaglandin pathway of inflammatory pain. The opioid and prostaglandin pathways interact yet are separately targeted by the two major categories of pain mediation, opiates and NSAIDs which have well defined molecular mechanisms of action.

"Opiate pathway" (acute pain)

The effect of THC and of other psychoactive cannabinoids in relieving pain has been documented in some experimental studies on animals but not in controlled observations of marijuana smokers. The acute analgesic effect of THC and other psychoactive cannabinoids has been attributed to their interaction with the opioid-endorphin system by a mechanism yet to be determined. The euphoriant property of THC which cannot be measured with objective markers, may contribute to analgesia, and makes it difficult to assess the pain relieving properties of this drug. The combination of opiates or barbiturates and THC is also undesirable because their combination significantly increases the undesirable side-effects of both medications (Johnstone *et al.*, 1975). Marijuana smoking or THC administration are not suitable for the relief of acute pain; opiates remain the drugs of choice. Attempts to dissociate acute analgesic properties of THC from its unwanted psychoactive effects are problematic since these two properties appear to be related to the interaction of THC with the same 7TM receptor.

Inflammatory pain

Some non-psychoactive cannabinoids, especially metabolites of THC such as THC-11-oic acid and certain synthetic derivatives, inhibit cyclooxygenase-2 (COX-2) and, consequently, prostaglandin production. They are effective in relieving inflammatory pain in rodents (Zurier *et al.*, 1998) by the same mechanism as current anti-inflammatory agents (NSAIDS). This effect is independent of the CB receptors, and not associated with psychoactive effects. Cannabinoids increase arachidonic acid release, and exert a dual releasing and inhibiting effect on prostaglandin metabolism. A clinical comparison of non-psychoactive cannabinoids, such as THC-11-oic acid, with currently used COX-2 inhibitors in the treatment of inflammatory pain has not been performed. Cannabinoids might be considered weak analgesics (anodynes) that fail to directly control the major pathways of nociceptive pain.

Anesthetic action

THC possesses mild anesthetic properties as shown by its ability to increase barbiturate sleeping time in rodents and enhance sedation. THC also enhances the respiratory depression of opiates (Johnstone *et al.*, 1975) and alters cardiovascular function while increasing their side effects. Infiltration of tissues with local anesthetics and α_2 agonists offers another therapeutic intervention to relieve acute pain by a direct action on C fibers. THC does not possess a local anesthetic effect.

Antiemetic effects

The antiemetic effect of marijuana was first reported in patients who smoked marijuana and were receiving cancer chemotherapy. This effect was subsequently duplicated in patients who were prescribed THC or other psychoactive cannabinoids. The antiemetic property of marijuana was attributed to an effect of THC on its receptor since nonpsychoactive cannabinoids were not effective in relieving nausea. There are two main categories of antiemetic drugs with known mechanisms of action on the chemoreceptor trigger zone of the area postrema: the phenothiazines, (prochlorperazine) and substituted benzamides, (metoclopramide) which are D_2 (Dopamine) receptor antagonists and drugs such as ondansetron which are 5-HT$_3$ receptors antagonists (Brunton, 1996). THC affects indirectly and partially these specific 5-HT$_3$ and D_2 receptors and may be considered a limited antiemetic for mild to moderate chemotherapy; its administrations associated with doserelated and adverse psychoactive effects (Brunton, 1996).

Antiglaucoma effects

Psychoactive cannabinoids, either smoked or ingested, decrease intraocular pressure in patients with glaucoma. To be maintained, this effect requires repetitive daily exposure to marijuana smoking. THC may not be topically applied like other anti-emitic medications because of its ocular irritating property. The mechanism of action of cannabinoids on glaucoma appears to be mediated by their interaction with prostaglandins. Psychoactive and non-psychoactive cannabinoids stimulate arachidonic acid production and the eicosanoid cascade, including prostaglandin synthesis. At the same time, the non-psychoactive metabolite of THC, THC-11-oic acid, inhibits cyclooxygenase and biosynthesis of prostaglandins (Burstein *et al.*, 1986). THC may be considered an indirect antiglaucoma medication by the action of its metabolite on aqueous outflow pathways. The role of prostaglandins in glaucoma treatment has led to the development of the prostaglandin analog PGE_2-α-tromethamine that has proved to be highly effective in local application for the management of open angle glaucoma with minimal side effects (Camras *et al.*, 1996). Topical beta blockers (timolol) and carbonic anhydrase inhibitors like brinzolamide are also available for the effective management of glaucoma (Silver, 1998). The use of cannabinoids, with their dual stimulation and inhibition of prostaglandin production and their acute and chronic side effects, is not recommended in modern ophthalmology.

APPETITE STIMULATION

Open studies in cancer patients and healthy volunteers treated with marijuana smoking or with oral THC for appetite stimulation have reported conflicting results. Anecdotal information and case reports suggest that smoking marijuana is beneficial for treatment of AIDS patients with wasting syndrome. Studies comparing oral THC to megestrol acetate are inconclusive (Bennett and Bennett, 1999). In this case the most likely molecular mechanism of action of THC is related to the receptor mediated euphoriant effect of the drug.

Neurological disorders

These disorders involve many alterations of signal transmission in brain and spinal cord that are displayed by symptomatic disturbances of sensory and motor function. Present symptomatic medications attempt to target the major molecular causes of these ailments. Psychoactive cannabinoids are no substitute for these medications and their properties of destabilization of membrane and receptor regulating mechanisms may add to the poor prognosis of these ailments. For instance, THC and marijuana smoking have been used as potential therapy for multiple sclerosis patients with symptoms of spasticity, ataxia, tremor, pain and bladder dysfunction. Clinical studies are inconclusive and not supportive of benefits: ataxia and tremors appear to worsen and spasticity is not favorably affected (Francis, 1999). A similar lack of consistent therapeutic effects is also observed in studies of marijuana use for management of epilepsy and spasticity following spinal injury.

INTERACTIONS OF THC WITH OTHER DRUGS

The molecular mechanisms of THC on the lipid bilayer and its integral receptors are also targeted by other therapeutic and nontherapeutic drugs. The resulting interactions (Nahas, 1984; Sutin and Nahas, 1999) have been reported experimentally and clinically. THC increases the depressive effects of psychodepressants (alcohol, barbiturates, benzodiazepine anesthetics and opiates) and decreases the effects of stimulants such as amphetamines and cocaine. These clinical effects are not dose related and more often biphasic, first enhancing stimulation and subsequently depression or vice versa. The variability in individual responses of these interactions is also related to the genetic polymorphism of cytochrome P450 present in the brain, mainly the CYP2D6 family of alleles (Britto and Wedlund, 1992). The metabolism of THC (Watanabe *et al.*, 1988) by brain microsomal oxidation is a major source of pharmacokinetic variation and of variability of drug effect. The genetic polymorphism of P450 brain oxidative enzymes allows one to distinguish between extensive metabolizers (EMS) and poor metabolizers (PMS). This genetic polymorphism of P450 which determines THC metabolism and disposition, will affect the concurrent metabolism of other psychoactive drugs, the clinical response of different subjects and the occurrence of side effects. Because of such interactions THC should be only administered with caution in association with other therapeutic drugs (Ingelman-Sundberg *et al.*, 1999).

CONCLUSIONS

The therapeutic properties attributed to THC may be related to the simultaneous molecular interactions of this drug in the lipid bilayer and its integral 7TM receptors; similar mechanisms account for the drug's adverse effects which may not be dissociated from its therapeutic properties. Genetic polymorphism of brain P450 oxidative enzymes contributes to the variability of drug effects on different subjects.

Marijuana or THC do not qualify as modern effective medications comparable to those presently available or in the advanced process of development. Their reintroduction into modern Pharmacopoeia, from which cannabis preparations were eliminated in the first part of this century and replaced by active molecules targeted to specific receptors, appears problematic.

However, the experimental use of THC and its synthetic analogs in molecular physiology has provided invaluable information leading to a better understanding of membrane signaling. As a result, the relationship between allosteric receptor responsiveness, molecular configuration of proteins and regulations of cellular function has been reevaluated (Nahas *et al.*, 1999). Membrane enzymes and receptors are integrated in the lipid core of the membrane and have lateral mobility. Lipids may be considered the solvent of the protein (Makriyannis, 1995). The proposed hypothesis of volume transmission (Zoli *et al.*, 1999) of lipid ligands in the membrane deserves additional investigation. The major difficulty of such studies resides in the measurement of volume changes in the membrane and its receptors of the order of the Angstrom (Skou, 1992). The interaction of lipid ligands like THC or eicosanoids with protein molecules and their structural effects on protein configuration remains a most important area for future research (Janssen, 1999).

REFERENCES

Bennett, W. M. and Bennett, S. S. (1999) "Marihuana for AIDS Wasting," in *Marihuana and Medicine*, edited by G. Nahas, K. M. Sutin, D. J. Harvey and S. Agurell, pp. 717–721. Totowa: Humana Press.

Bisogno, T., Maurelli, S., Melck, D., De Petrocellis, L. and Di Marzo, V. (1997) "Biosynthesis, uptake, and degradation of anandamide and palmitoylethanolamide in leukocytes," *Journal of Biological Chemistry* **272**: 3315–3323.

Bisogno, T., Ventriglia, M., Milone, A., Mosca, M., Cimino, G. and Di Marzo, V. (1997) "Occurrence and metabolism of anandamide and related acyl-ethanolamides in ovaries of the sea urchin *Paracentrotus lividus*," *Biochimica et Biophysica Acta* **1345**: 338–348.

Bloom, A. S., Edgemond, W. S. and Moldvan, J. C. (1997) "Nonclassical and endogenous cannabinoids: effects on the ordering of brain membranes," *Neurochemical Research* **22**: 563–568.

Bloom, A. S. and Hillard, C. J. (1984) "Cannabinoids, neurotransmitter receptors and brain membranes," in *Marihuana '84*, edited by D. J. Harvey, pp. 217–231. Oxford: IRL Press.

Bloom, A. S., Johnson, K. M. and Dewey, W. L. (1978) "The effects of cannabinoids on body temperature and brain catecholamine synthesis," *Research Communications in Chemotheraphy, Pathology and Pharmacology* **20**: 51–57.

Borgen, L. A. and Davis, W. M. (1974) "Cannabidiol interaction with Δ^9-THC," *Research Communications in Chemotheraphy, Pathology and Pharmacology* **7**: 663–670.

Bornheim, L. M., Everhart, E. T., Li, J. and Correia, M. A. (1993) "Characterization of cannabidiol-mediated cytochrome P450 inactivation," *Biochemical Pharmacology* **45**: 1323–1331.

Britto, M. R. and Wedlund, P. J. (1992) "Cytochrome P-450 in the brain. Potential evolutionary and therapeutic relevance of localization of drug-metabolizing enzymes," *Drug Metabolism and Disposition* **20**: 446–450.

Brunton, L. L. (1996) "Agents affecting gastrointestinal water flux and motility: emesis and antiemetics; bile acids and pancreatic enzymes," in Goodman & Gilman' *The Pharmacological Basis of Therapeutics*, edited by J. Hardman and L. Limbird, pp. 917–936. New York: McGraw-Hill.

Burstein, S. H., Hunter, S. A., Latham, V. and Renzulli, L. (1986) "Prostaglandins and *Cannabis XVI*: Antagonism of Δ^1-THC action by its metabolites," *Biochemical Pharmacology* **35**: 2553–2558.

Cabral, G. A. (1999) "Marijuana and the Immune System," in *Marijuana and Medicine*, edited by G. G. Nahas, K. M. Sutin, D. Harvey and S. Agurell, pp. 317–325. Totowa: Humana Press.

Camras, C. B., Alm, A., Watson, P. and Stjernschantz, J. (1996) "Latanoprost, a prostaglandin analog; for glaucoma therapy. Efficacy and safety after 1 year of treatment in 198 patients. Latanoprost Study Groups," *Ophthalmology* **103**: 1916–1924.

Clark, W. C., Janal, M. N., Zeidenberg, P. and Nahas, G. G. (1981) "Effects of moderate and high doses of marihuana on thermal pain: a sensory decision theory analysis," *Journal of Clinical Pharmacology* **21**: 299S–310S.

Cridland, J. S., Rottanburg, D. and Robins, A. H. (1983) "Apparent half-life of excretion of cannabinoids in man," *Human Toxicology* **2**: 641–644.

De Petrocellis, L., Melck, D., Bisogno, T., Milone, A. and Di Marzo, V. (1999) "Finding of the endocannabinoid signalling system in hydra, a very primitive organism: Possible role in the feeding response," *Neuroscience* **92**: 377–387.

Devane, W. A., Dysarz, F. A., Johnson, M. R., Melvin, L. S. and Howlett, A. C. (1988) "Determination and characterization of a cannabinoid receptor in rat brain," *Molecular Pharmacology* **34**: 605–613.

Devane, W. A., Hanus, L., Breuer, A. *et al.* (1992) "Isolation and structure of a brain constituent that binds to the cannabinoid receptor," *Science* **258**: 1946–1949.

Di Marzo, V. (1998) "Biochemistry of the endogenous ligands of cannabinoid receptors," *Neurobiology of Disease* **5**: 384–404.

Di Marzo, V. (1999) "Biosynthesis and inactivation of endocannabinoids: relevance to their proposed role as neuromodulators," *Life Sciences* **65**: 645–655.

Di Marzo, V., De Petrocellis, L., Bisogno, T. and Meleck, D. (1999) "Metabolism of anadamide and 2-arachidonylglycerol: An historical overview and some recent developments," *Lipids* **34**: S319–S325.

Di Marzo, V., De Petrocellis, L., Sepe, N. and Buono, A. (1996) "Biosynthesis of anandamide and related acylethanolamides in mouse J774 macrophages and N18 neuroblastoma cells," *The Biochemical Journal* **316**: 977–984.

Domino, E. (1981) "Cannabinoids and the cholinergic system," *Journal of Clinical Pharmacology* **21** (suppl.): 249S–255S.

Ellis, G. M., Jr., Mann, M. A., Judson, B. A., Schramm, N. T. and Tashchian, A. (1985) "Excretion patterns of cannabinoid metabolites after last use in a group of chronic users," *Clinical Pharmacology and Therapeutics* **38**: 572–578.

Emrich, H. M., Leweke, F. M. and Schneider, U. (1997) "Towards a cannabinoid hypothesis of schizophrenia: cognitive impairments due to dysregulation of the endogenous cannabinoid system," *Pharmacology, Biochemistry and Behavior* **56**: 803–807.

Francis, G. S. (1999) "Marihuana use in Multiple Sclerosis," in *Marihuana and Medicine*, edited by G. G. Nahas, K. M. Sutin, D. J. Harvey and S. Agurell, pp. 631–637. Totowa: Humana Press .

Fulton, D. and Quilley, J. (1998) "Evidence against anadamide as the hyperpolarizing factor mediating the nitric oxide-independent coronary vasodilator effect of bradykinin in the rat," *The Journal of Pharmacology and Experimental Therapeutics* **286**: 1146–1151.

Garrett, E. R. and Hunt, C. A. (1977) "Pharmacokinetics of Δ^9-THC in dogs," *Journal of Pharmaceutical Sciences* **66**: 395–407.

Gessa, G. L., Mascia, M. S., Casu, M. A. and Carta, G. (1997) "Inhibition of hippocampal acetylcholine release by cannabinoids: reversal by SR 141716A," *European Journal of Pharmacology* **327**: R1–R2.

Hampson, A. J., Bornheim, L. M. and Scanziani, M. *et al.* (1998) "Dual effects of anandamide on NMDA receptor-mediated responses and neurotransmission," *Journal of Neurochemistry* **70**: 671–676.

Harvey, D. J. (1991) "Metabolism and pharmacokinetics of the cannabinoids," in *Biochemistry and Physiology of Substance Abuse*, edited by R. R. Watson, pp. 279-end. Boca Raton: CRC Press.

Hembree, W. C., Nahas, G. G., Zeidenberg, P. and Huang, H. F. S. (1999) "Changes in human spermatozoa associated with high dose marihuana smoking," in *Marihuana and Medicine*, G. G. Nahas, K. M. Sutin, D. J. Harvey and S. Agurell, pp. 367–378. Totowa: Humana Press.

Herkenham, M., Lynn, A. B., Little, M. D. *et al.* (1990) "Cannabinoid receptor localization in brain," *Proceedings of the National Academy of Sciences of the USA* **87**: 1932–1936.

Hunter, S. A., Burstein, S. and Renzulli, L. (1985) "Cannabinoid modulated phospholipase activities by mouse brain subcellular fractions," *Marihuana '84*, Oxford, England, IRL Press.

Ingelman-Sundberg, M., Oscarson, M. and Mclellan, R. A. (1999) "Polymorphic human cytochrome P450 enzymes: an opportunity for individualized drug treatment," *Trends in Pharmacological Sciences* **20**: 342–349.

Janssen, P. (1999) "How to Design the Ideal Drug for the Future," in *Marihuana and Medicine*, edited by G. Nahas, K. M. Sutin, D. Harvey, S. Agurell, pp. 327–331. New Jersey: Humana Press.

Johnstone, R. E., Lief, P. L., Kulp, R. A. and Smith, T. C. (1975) "Combination of Δ^9-THC with oxymorphone or pentobarbital: Effects on ventilatory control and cardiovascular dynamics," *Anesthesiology* **42**: 674–684.

Joy, J. E., Watson, S. J. and Benson, J. A. (1999) *Marijuana and Medicine: Assessing the Science Base*. Washington D. C.: Institute of Medicine, National Academy Press, 288pp.

Lawrence, D. K. and Gill, E. W. (1975) "The effects of D1-tetrahydrocannabinol and other cannabinoids on spin-labeled liposomes and their relationship to mechanisms of general anesthesia," *Molecular Pharmacology* **11**: 595–602.

Leuschner, J., Wing, D., Harvey, D. *et al.* (1984) "The partitioning of Δ^1-THC into erythrocyte membranes *in vivo* and its effect on membrane fluidity," *Experientia* **40**: 866–868.

Leuschner, J. T., Harvey, D. J., Bullingham, R. E. and Paton, W. D. (1986) "Pharmacokinetics of Δ^9-THC in rabbits following single or multiple intravenous doses," *Drug Metabolism and Disposition* **14**: 230–238.

Makriyannis, A. (1995) "The role of cell membranes in cannabinoid activity," in *Cannabinoid Receptors*, pp. 87–115. New York: Academic Press.

Martin, B. R. and Lichtman, A. H. (1998) "Cannabinoid transmission and pain perception," *Neurobiology of Disease* **5**: 447–461.

Mechoulam, R., Ben-Shabat, S., Hanus, L. *et al.* (1995) "Identification of an endogenous 2-monoglyceride, present in canine gut, that binds to cannabinoid receptors," *Biochemical Pharmacology* **50**: 83–90.

Moreau, J. J. (1970) *Hashish and Mental Health*. New York, Raven Press (reprinted).

Munro, S., Thomas, K. L. and Abu-Shaar, M. (1993) "Molecular characterization of a peripheral receptor for cannabinoids [see comments]," *Nature* **365**: 61–65.

Nahas, G., Sutin, K. M., Harvey, D. and Agurell, S. (1999) *Marihuana and Medicine*, Totowa, New Jersey: Humana Press, 826pp.

Nahas, G., Harvey, D., Sutin, K. M. and Agurell, S. (1999) "Receptor and nonreceptor membrane-mediated effects of THC and cannabinoids," in *Marihuana and Medicine*, edited by G. Nahas, K. M. Sutin, D. Harvey, S. Agurell, pp. 781–806. New Jersey: Humana Press.

Nahas, G. G. (1984) "Toxicology and pharmacology, excerpt," *Marihuana in Science and Medicine*, pp. 196–206, New York: Raven Press.

Niederhoffer, N. and Szabo, B. (1999) "Involvement of CB_1 cannabinoid receptors in the EDHF-dependent vasorelaxation in rabbits," *British Journal of Pharmacology* **126**: 1383–1386.

Ohlsson, A., Lindgren, J. E., Wahlen, A., Agurell, S., Hollister, L. E. and Gillespie, H. K. (1980) "Plasma Δ^9-THC concentration and clinical effects after oral and intravenous administration and smoking," *Clinical Pharmacology and Therapeutics* **28**: 409–416.

Paria, B. C., Das, S. K. and Dey, S. K. (1995) "The preimplantation mouse embryo is a target for cannabinoid ligand-receptor signaling," *Proceedings of the National Academy of Sciences of the USA* **92**: 9460–9464.

Paton, W. D. M., Pertwee, R. J. and Temple, D. M. (1972) "The general pharmacology of cannabinoids," in *Cannabis and its Derivatives: Pharmacology and Experimental Psychology*, edited by W. D. M. Paton and J. Crown, pp. 50–74. Oxford: Oxford University Press.

Perez-Reyes, M., Burstein, S. H., White, W. R., McDonald, S. A. and Hicks, R. E. (1991) "Antagonism of marihuana effects by indomethacin in humans," *Life Sciences* **48**: 507–515.

Pertwee, R. G. (1995) *Cannabinoid Receptors*. London: Academic Press.

Piomelli, D., Belttramo, M. and Guiffrida, A. (1998) "Endogenous cannabinoid signaling," *Neurobiology of disease* **5**: 462–473.

Pryor, G., Larsen, P., Carr, J., Carr, E. and Braude, M. (1977) "Interactions of Δ^9-THC with penobarbitol, ethanol and chlordiazepoxide," *Pharmacology Biochemistry and Behavior* **7**: 331–345.

Select Committee on Science and Technology. (1998) *Cannabis: The Scientific and Medical Evidence*, London, The Stationery Office: House of Lords, 54pp.

Sepe, N., De Petrocellis, L., Montanaro, F., Cimino, G. and Di Marzo, V. (1998) "Bioactive long chain N-acylethanolamines in five species of edible bivalve molluscs. Possible implications for mollusc physiology and sea food industry," *Biochimica et Biophysica Acta* **1389**: 101–111.

Schuel, H., Goldstein, E., Mechoulam, R., Zimmerman, A. and Zimmerman, S. (1994) "Anandamide (arachidonylethanolamide), a brain cannabinoid receptor agonist, reduces fertilizing capacity in sea urchins by inhibiting the acrosome reaction," *Proceedings the National Academy of Sciences of the USA* **91**: 7678–7682.

Schmid, P. C., Krebsbach, R. J., Perry, S. R., Dettmer, T. M., Maasson, J. L. and Schmid, H. H. O. (1995) "Occurence and postmortem generation of anandamide and other long-chain N-acylethanolamines in mammalian brain," *FEBS Letters* **375**: 117–120.

Silver, L. H. (1998) "Clinical efficacy and safety of brinzolamide (Azopt), a new topical carbonic anhydrase inhibitor for primary open-angle glaucoma and ocular hypertension Brinzolamide Primary Therapy Study Group," *American Journal of Ophthalmology* **126**: 400–408.

Skou, J. C. (1992) "The Na-K Pump," *News in Physiological Science* **7**: 95–100.

Solowij, N., Michie, P. T. and Fox, A. M. (1995) "Differential impairments of selective attention due to frequency and duration of cannabis use," *Biological Pshychiatry* **37**: 731–739.

Stanford, S. J. and Mitchell, J. A. (1999) "The prolonged phase of ATP-induced vasodilation in rat mesenteric vessels is mediated by a cannabinoid-like ligand," *British Journal of Pharmacology* **126**.

Stefano, G. B., Salzet, B. and Salzet, M. (1997) "Identification and characterization of the leech CNS cannabinoid receptor: coupling to nitric oxide release," *Brain Research* **753**: 219–224.

Stella, N., Schweitzer, P. and Piomelli, D. (1997) "A second endogenous cannabinoid that modulates long-term potentiation," *Nature* **388**: 773–778.

Sugiura, T., Kondo, S., Sukagawa, A. *et al.* (1996a) "Transacylase-mediated and phosphodiesterase-mediated synthesis of N-arachidonoylethanolamine, an endogenous cannabinoid-receptor ligand, in rat brain microsomes: Comparison with synthesis from free arachidonic acid and ethanolamine," *European Journal of Biochemistry* **240**: 53–62.

Sugiura, T., Kondo, S., Sukagawa, A. *et al.* (1996b) "Enzymatic synthesis of anandamide, an endogenous cannabinoid receptor ligand, through N-acylphosphatidylethanolamine pathway in testis: involvement of Ca(2+)-dependent transacylase and phosphodiesterase activities," *Biochemical and Biophysical Research Communications* **218**: 113–117.

Sugiura, T., Kondo, S., Sukagawa, A. *et al.* (1995) "2-Arachidonoylglycerol: a possible endogenous cannabinoid receptor ligand in brain," *Biochemical and Biophysical Research Communications* **215**: 89–97.

Sutin, K. M. and Nahas, G. G. (1999) "Physiological and Pharmacological Interactions of Marihuana (THC) with Drugs and Anesthetics," in *Marihuana and Medicine*, edited by G. G. Nahas, K. M. Sutin, D. J. Harvey and S. Agurell, pp. 253–271. Totowa: Humana Press.

Trouve, R. and Nahas, G. G. (1999) "Cardiovascular effects of marihuana and cannabinoids," in *Marihuana and Medicine*, edited by G. G. Nahas, K. M. Sutin, D. H. Harvey, S. Agurell, pp. 253–271. Totowa, N. J: Humana Press.

Turner, C. E., El Sohly, M. A. and Boeren, E. G. (1980) "Constituents of Cannabis sativa L. XVII, A review of the natural constituents," *Journal of Natural Products* **43**: 169–234.

Ueda, N., Kurahashi, Y., Yamamoto, K., Yamamoto, S. and Tokunaga, T. (1996) "Enzymes for anandamide biosynthesis and metabolism," *Journal of Lipid Mediators of Cell Signaling* **14**: 57–61.

Vaysse, P. J., Gardner, E. L. and Zukin, R. S. (1987) "Modulation of rat brain opioid receptors by cannabinoids," *The Journal of Pharmacology and Experimental Therapeutics* **241**: 534–539.

Wagner, J. A., Varga, K., Jarai, Z. and Kunos, G. (1999) "Mesenteric vasodilation mediated by endothelial anandamide receptors," *Hypertension* **33**: 429–434.

Waser, P. G. and Martin, A. (1999) "Analgesic and behavioral effects of THC derivatives in comparison with delta THC," in *Marihuana and Medicine*, edited by G. Nahas, M. K. Sutin, D. H. Harvey, S. Agurell, pp. 527–539. Totowa, N. J.: Humana press.

Watanabe, K., Tanaka, T., Yamamoto, I. and Yoshimura, H. (1988) "Brain microsomal oxidation of delta 8- and Δ^9-THC," *Biochemical and Biophysical Research Communications* **157**: 75–80.

Wing, D., Harvey, D., Hughes, J., Dunbar, P., McPherson, K. and Paton, W. (1982) "Effects of chronic ethanol administration on the composition of membrane lipids in the mouse," *Biochemical Pharmacology* **31**: 3431–3439.

Zoli, M., Jamsson, A., Sykova, E., Agnati, L. and Fuxe, K. (1999) "Volume transmission in CNS and its relevance for neuropsychopharmacology," *TIPS* **20**: 142–150.

Zurier, R. B., Rossetti, R. G., Lane, J. H., Goldberg, J. M., Hunter, S. A. and Burstein, S. H. (1998) "Dimethylheptyl-THC-11 oic acid: a nonpsychoactive antiinflammatory agent with a cannabinoid template structure," *Arthritis and Rheumatism* **41**: 163–170.

Chapter 17

Cannabinoid receptors: the relationship between structure and function

Patricia H. Reggio

ABSTRACT

This chapter reviews the current state of knowledge in the cannabinoid field concerning the relationship between cannabinoid (CB) receptor structure and function, with special emphasis upon computational models of the CB receptors. The cannabinoid CB_1 and CB_2 receptors are transmembrane proteins that belong to the G-protein coupled receptor (GPCR) family. The CB receptors bind four different structural classes of agonist ligands. Antagonists/inverse agonists of each receptor sub-type have also been identified. Recent biochemical studies have shown that the CB receptors can couple to more than one G-protein and signal to more than one effector system. These studies have led to the hypothesis that agonists induce different conformations of the CB receptors, which in turn distinguish between different G-proteins. Such agonist selective G-protein signaling is of potential therapeutic importance if ligands can be designed to regulate individual G-protein signaling pathways that are linked to specific pharmacological effects. Knowledge of CB receptor structure and the changes undergone by CB receptors upon ligand binding is, therefore, of fundamental importance.

The recent 2.8Å crystal structure of rhodopsin (Rho), a GPCR, in its inactive state has provided important information at an atomic level of resolution concerning the overall transmembrane helix bundle organization of GPCRs. While there is no experimental structure of the CB_1 or CB_2 receptors yet available, there is a growing body of literature from biophysical studies of other GPCRs that can be used in combination with structure–activity relationship (SAR) and mutation studies of the CB receptors to construct three dimensional computer models of the CB_1 and CB_2 receptors. These models can then be probed for the mechanism by which certain agonist structural classes may activate the CB receptors. The process of model construction and evaluation is illustrated here by models built in the Reggio lab and elsewhere.

Key Words: cannabinoid, CB_1, CB_2, signaling, molecular model, agonist selective coupling

THE CANNABINOID RECEPTORS AND THEIR LIGANDS

For centuries hashish and marijuana, both derived from the Indian hemp *Cannabis sativa* L., have been used for their medicinal, as well as their psychotomimetic effects. By definition, cannabinoids are the group of C_{21} compounds typical of and present in *Cannabis sativa* L., their carboxylic acids, analogs, and transformation products (Mechoulam and Gaoni, 1967). A surge of scientific interest in the cannabinoids followed Mechoulam's report that (−)-trans-Δ^9-tetrahydrocannabinol (Δ^9-THC, **1**) is the major psychopharmacologically active component of cannabis (Mechoulam and Gaoni, 1967; Gaoni and Mechoulam, 1964). The recognized CNS responses to preparations of Cannabis include alterations in cognition and memory, euphoria and sedation (Howlett, 1995). A multiple-evaluation paradigm of *in vivo* mouse assays is commonly employed to test for cannabimimetic effects. This paradigm includes assays for reduction in spontaneous activity, and the production of hypothermia, catalepsy, and antinociception (tail-flick assay) (Compton *et al.*, 1992). Like the opioids, cannabinoids inhibit electrically-evoked contractions of the mouse vas deferens (MVD) and the guinea pig ileum (GPI). Unlike the opioids, the MVD and GPI effects of the cannabinoids are not antagonized by naloxone (Pertwee, 1993; Pertwee *et al.*, 1992a,b).

Cannabinoid receptors

The demonstration that some of the effects of the cannabinoids are receptor medi-ated was first made by Allyn Howlett and co-workers, who developed membrane homogenate and tissue section binding assays for the characterization and localiza-tion of a cannabinoid receptor in brain using the potent radiolabeled ligand [^{3}H]CP55,940 (**2**) (Devane *et al.*, 1988). To date, two sub-types of the cannabinoid receptor, CB_1 and CB_2, have been identified. The cloning and expression of a complementary DNA from a rat cerebral cortex cDNA library that encoded the first cannabinoid receptor subtype (CB_1) was reported by L. A. Matsuda and co-workers (1990). The amino acid sequence of this receptor was found to be consistent with a tertiary structure typical of the G-protein coupled receptors (Matsuda *et al.*, 1990). Subsequently, the primary amino acid sequence of this same receptor (CB_1) in human brain was reported (Gerard *et al.*, 1991). The rat CB_1 receptor shares 97.3% sequence identity with the human CB_1 receptor with 100% identity within the transmembrane regions (Gerard *et al.*, 1991). An amino ter-minus variant CB_1 receptor has also been reported (Shire *et al.*, 1995; Rinaldi-Carmona *et al.*, 1996). In addition to being found in the central nervous system (CNS), mRNA for CB_1 has also been identified in testes (Gerard *et al.*, 1991).

The second cannabinoid receptor sub-type, CB_2, was derived from a human promyelocytic leukemia cell HL60 cDNA library (Munro *et al.*, 1993). The primary amino acid sequence of the CB_2 receptor is also consistent with a tertiary structure typical of G-protein coupled receptors (GPCRs). The human CB_2 receptor exhibits 68% identity to the human CB_1 receptor within the transmembrane regions, 44% identity throughout the whole protein (Munro *et al.*, 1993). The CB_2 receptor in both rat (Griffin *et al.*, 2000) and mouse (Shire *et al.*, 1996a) has been cloned as well. Sequence analysis of the coding region of the rat CB_2 genomic clone indicates 93% amino acid identity between rat and mouse and 81% amino acid identity between rat and human. Unlike the CB_1 receptor, which is highly conserved across human, rat and mouse, the CB_2 receptor is much more divergent (Figure 17.1). This has important implications for pharmacological studies, as CB_2 ligand affinities may be altered depending on the species employed in the study. Reports of such discrepancies have already been reported in the literature (Berglund *et al.*, 1998).

CNS responses to cannabinoid compounds are believed to be mediated largely by the CB_1 receptor. While Munro and colleagues found no CB_2 transcripts in

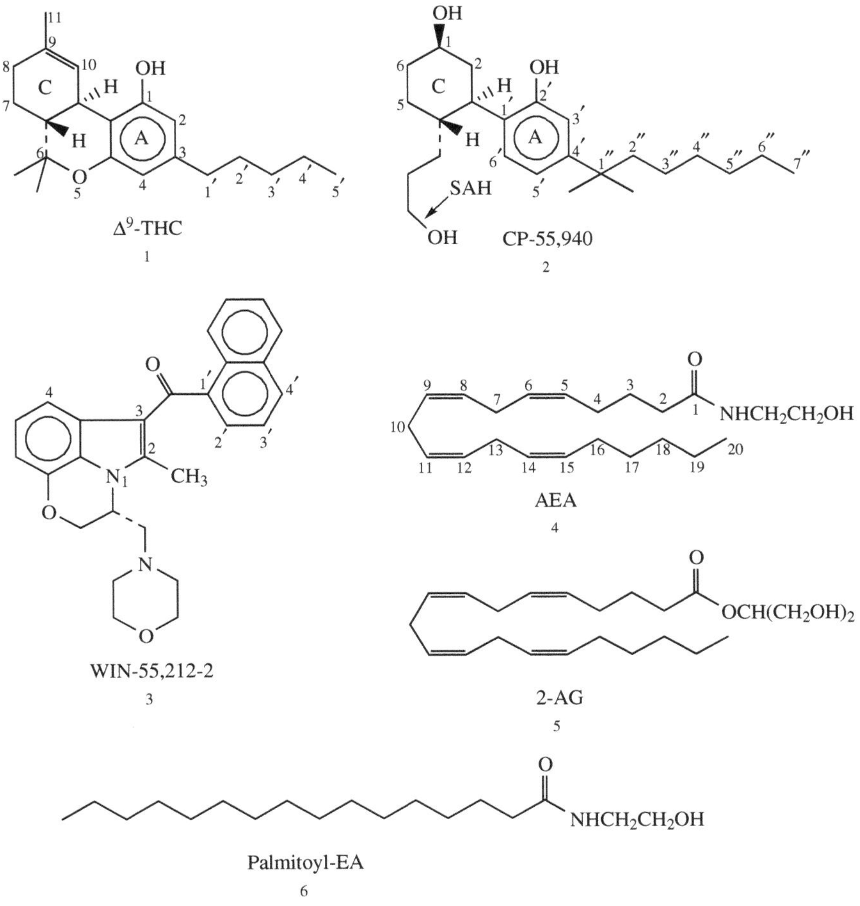

Figure 17.1 Cannabinoid receptor agonist structures and numbering systems.

brain tissue by either Northern analysis or *in situ* hybridization studies (Munro *et al.*, 1993), Skaper and colleagues have reported that cerebellar granule cells and cerebellum express genes encoding both the CB_1 and CB_2 receptors (Skaper *et al.*, 1996). CB_1 knockout mice have been produced. These mice show altered gene expression in striatal projection neurons. The mice display significantly elevated levels of Substance P, dynorphin, enkephalin and GAD 67 mRNAs in neurons of the two output pathways of the striatum that project to the substantia nigra and the globus pallidus (Steiner *et al.*, 1999). The mice display reduced locomotor activity, increased ring catalepsy and hypoalgesia in hot plate and formalin tests (Zimmer *et al.*, 1999).

The cannabinoid receptors activate multiple intracellular signal transduction pathways. CB_1 and CB_2 receptor agonists inhibit forskolin-stimulated adenylyl cyclase by activation of a pertussis toxin-sensitive G-protein (Felder *et al.*, 1995). CB_1 receptors can also stimulate cAMP formation under certain conditions, consistent with a possible G_s linkage in this receptor (Glass and Felder, 1997; Maneuf and Brotchie, 1997; Felder *et al.*, 1998). Rhee and co-workers have reported that activation of the CB_2 receptor can produce stimulation of cAMP formation (Rhee *et al.*, 1998). These investigators used co-transfection experiments between CB_1 and CB_2 and nine isoforms of adenylyl cyclase to show that both cannabinoid receptors inhibit the activity of adenylyl cyclase types I, V, VI, and VIII, whereas types II, IV, and VII were stimulated by cannabinoid receptor activation. The activity of adenylyl cyclase type IX was inhibited only marginally by cannabinoids and inhibition of adenylyl cyclase type III by cannabinoids was observed only when forskolin was used as a stimulant.

In heterologous cells, CB_1 but not CB_2 receptors inhibit N-, P-, and Q-type calcium channels and activate inwardly rectifying potassium channels (Caulfield and Brown, 1992; Mackie and Hille, 1992; Felder *et al.*, 1995; Mackie *et al.*, 1995; Pan *et al.*, 1996). Inhibition of calcium channels and enhancement of inwardly rectifying potassium currents is pertussis toxin-sensitive, but independent of cAMP inhibition, suggestive of a direct G-protein mechanism (Mackie and Hille, 1992; Mackie *et al.*, 1995). The CB_1 receptor also activates the MAP kinase cascade (Bouaboula *et al.*, 1995), and can activate phospholipase C via a G_α protein (Ho *et al.*, 1999).

Cannabinoid receptor agonists

The CB_1 receptor transduces signals in response to CNS-active constituents of Cannabis sativa, such as the classical cannabinoid $(-)-\Delta^9$-THC, (**1**) and to three other structural classes of ligands, the non-classical cannabinoids typified by CP55,940 (**2**) (Devane *et al.*, 1988; Melvin *et al.*, 1995), the aminoalkylindoles (AAIs) typified by WIN 55212-2 (**3**) (D'Ambra *et al.*, 1992; Ward *et al.*, 1991; Compton *et al.*, 1992) and the endogenous cannabinoids. The non-classical cannabinoids clearly share many structural features with the classical cannabinoids, e.g. a phenolic hydroxyl at C-1 (C2′), and alkyl side chain at C-3 (C-4′), as well as, the ability to adopt the same orientation of the carbocyclic ring as that in classical CBs (Reggio *et al.*, 1993). The AAIs, on the other hand, bear no obvious structural similarities with the classical/non-classical cannabinoids.

The first endogenous cannabinoid was isolated from porcine brain by Mechoulam and co-workers (Devane *et al.*, 1992). The endogenous cannabinoid ligands are unsaturated fatty-acid ethanolamides. The first identified ligand of this class was arachidonylethanolamide (AEA, also called anandamide), (**4**) (Devane *et al.*, 1992). Like other cannabinoid agonists, AEA produces a concentration dependent inhibition of the electrically-evoked twitch response of the mouse vas deferens (MVD) (Devane *et al.*, 1992), as well as, antinociception, hypothermia, hypomotility, and catalepsy in mice (Smith *et al.*, 1994). AEA inhibits forskolin-stimulated cAMP accumulation in CHO-HCR cells (Felder *et al.*, 1993) and exhibits higher affinity for the cannabinoid CB_1 receptor (K_i $CB_1 = 89 \pm 10$ nM) than for the CB_2 receptor (K_i $CB_2 = 371 \pm 102$ nM) (Showalter *et al.*, 1996).

sn-2-arachidonylglycerol (2-AG); (**5**) was isolated from intestinal tissue and shown to be a second endogenous CB ligand (CB$_1$ K$_i$ = 472±55 nM; CB$_2$ K$_i$ = 1400 ± 172 nM) (Mechoulam *et al.*, 1995). 2-AG has been found present in the brain at concentrations 170 times greater than anandamide (Stella *et al.*, 1997). Recently, Facci and colleagues reported that palmitoylethanolamide (**6**), a C16 saturated fatty acid ethanolamide behaves as an endogenous agonist of the CB$_2$ receptor on rat mast cells (RBL-2H3 cells) (Facci *et al.*, 1995). However, **6** was found to displace only 10% of [^{3}H] CP55,940 from human cloned CB$_2$ receptor at concentrations up to 10 µM (Showalter *et al.*, 1996). The origin of the divergence of affinities may be due to species differences between CB$_2$ in rat vs. human. At present, the identification of **6** as an endogenous cannabinoid remains controversial.

Cannabinoid receptor antagonists

The first CB$_1$ antagonist, SR 141716A (**7**) was developed by M. Rinaldi-Carmona and co-workers at Sanofi Recherche (1994). SR 141716A displays nanomolar CB$_1$ affinity (K$_i$ = 1.98±.13 nM), but very low affinity for CB$_2$. *In vitro*, SR 141716A antagonizes the inhibitory effects of cannabinoid agonists on both MVD contractions and adenylyl cyclase activity in rat brain membranes. SR 141716A also antagonizes the pharmacological and behavioral effects produced by CB$_1$ agonists after interperitoneal (IP) or oral administration (Rinaldi-Carmona *et al.*, 1994). Three other CB$_1$ antagonists have been reported, AM-630 (**8**) (Hosohata *et al.*, 1997a,b; Pertwee *et al.*, 1995), LY-320135 (**9**) (Felder *et al.*, 1998), and O-1184 (**10**) (Ross *et al.*, 1998).

The first CB$_2$ antagonist, SR 144528 (**11**), has been reported by M. Rinaldi-Carmona and co-workers at Sanofi Recherche (1998) (Figure 17.2). SR 144528 displays sub-nanomolar affinity for both the rat spleen and cloned human CB$_2$ receptors (K$_i$ = 0.60±0.13 nM). SR 144528 displays a 700-fold lower affinity for both the rat brain and cloned human CB$_1$ receptors.

Sub-type specific ligands

While there are several sub-type specific cannabinoid antagonists (see above), there are few sub-type specific cannabinoid agonists. The endogenous cannabinoids are usually CB$_1$ selective. Cannabinol (CBN) (**12**) was reported by Felder *et al.* (1995) to be 3.8-fold CB$_2$ selective. WIN 55,212-2 has been shown to be 19-fold selective for CB$_2$ (Felder *et al.*, 1995). Huffman *et al.* (1996) have shown that 1-deoxy-11-hydroxy-Δ^8-THC dimethylheptyl (**13**) has a 38-fold preference for CB$_2$. This analog also retains high CB$_1$ affinity (K$_i$ CB$_1$ = 1.2±0.1 nM; K$_i$ CB$_2$ = 0.032 ± 0.019 nM). 1-Deoxy-1′-1′-dimethylbutyl-Δ^8-THC was recently reported to have high CB$_2$ affinity (K$_i$ = 3.4±1.0 nM), 200-fold greater than for CB$_1$ (Huffman *et al.*, 1999). Huffman and co-workers have also shown that substitution of an n-propyl chain for the morpholino side chain of the AAI's to produce 1-propyl-2-methyl-3-(1-naphthoyl) indole (JWH-015) (**14**) yields a 28-fold CB$_2$ selective compound (Showalter *et al.*, 1996).

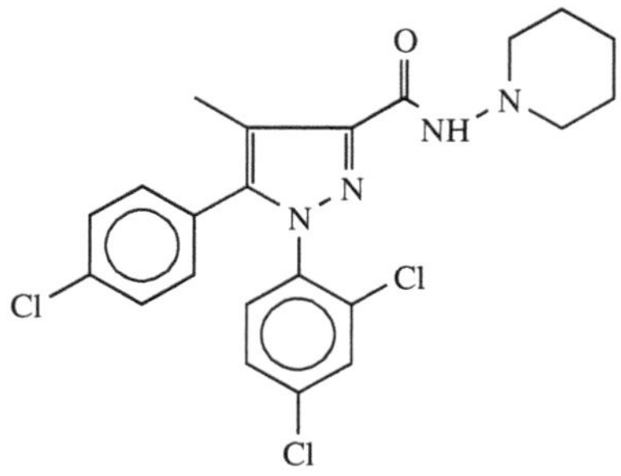

SR-141716A

7

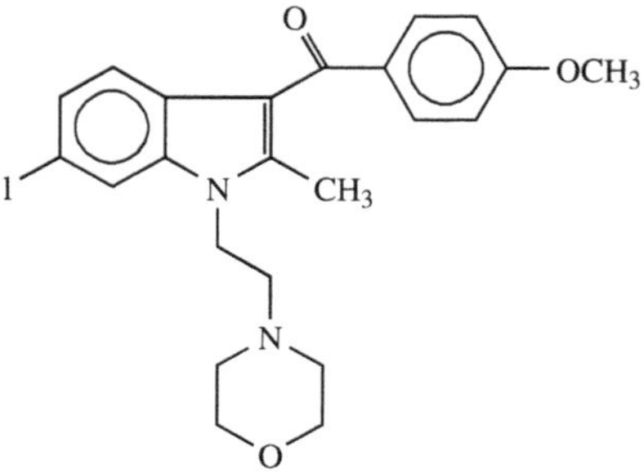

AM-630

8

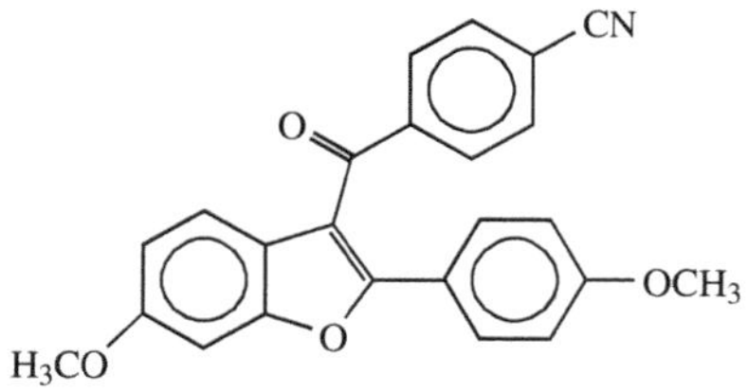

LY-320135

9

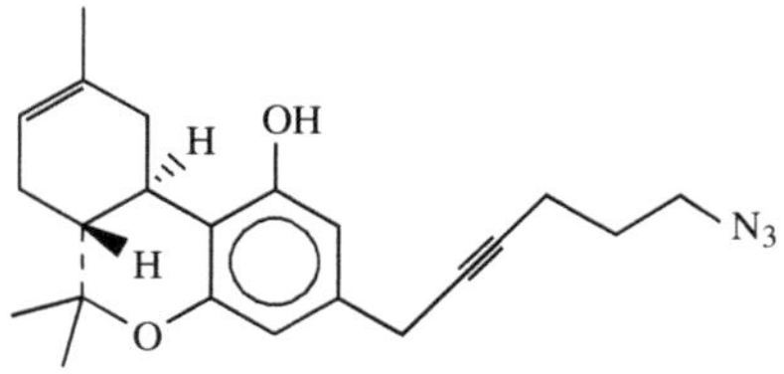

O-1184

10

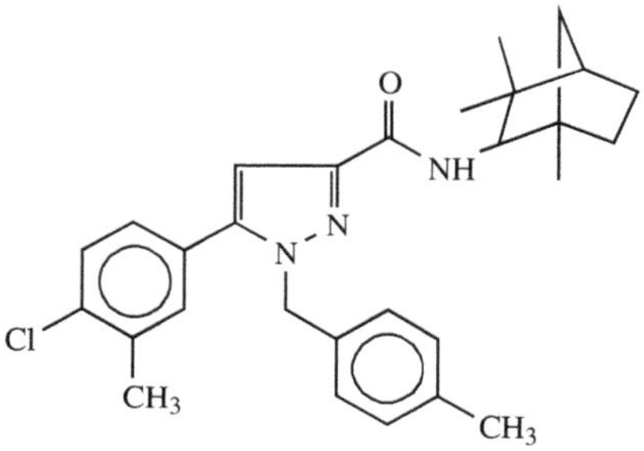

SR-144528

11

Figure 17.2 Cannabinoid receptor antagonist structures.

Endothelial receptor for anandamide

CB_1 knock-out mice along with CB_1/CB_2 knock-out mice were recently used by Jarai and co-workers (1999) to characterize an as-yet-unidentified endothelial receptor for anandamide, activation of which elicits NO-independent mesenteric vasodilation. Abnormal cannabidiol (Abn-CBD) (**15**) and cannabidiol (CBD) (**16**) have been reported to act as selective agonist and antagonist respectively of this receptor. The vasodilation produced by Abn-CBD can be inhibited by the CB_1 antagonist, SR 141716A (Jarai et al., 1999).

CANNABINOID SIGNALING

Models for GPCR interaction with G-protein

The traditional model for GPCR interaction with G-protein assumes that GPCRs exist in several dynamic states as reflected by changes in agonist affinity (Conklin and Bourne, 1993; Neubig and Sklar, 1993; Birnbauer et al., 1990; Samama et al., 1993). When an agonist binds, the receptor shifts to a state of higher affinity for agonists and induces a conformational change in the G-protein, triggering first the exchange of GDP for GTP bound to the G-protein α-subunit, and then triggering regulation by the GTP-bound α-subunit (α-GTP) and the free complex of $\beta\gamma$ subunits (Shiekh et al., 1996). Once the G-protein is released, the receptor reverts to a state of low affinity for agonists. Thus, not only does the receptor send information to the G-protein, but the G-protein feeds back information to the receptor inducing a conformational switch (Strader et al., 1994). This traditional two-state model for agonist action at GPCRs has been challenged by more complex models which accommodate, for example, agonists that can signal at one receptor through more than one G-protein, to two different effector systems (Leff et al., 1997). Studies in which levels of expression of certain GPCRs have been altered have resulted in a strong correlation of basal signal generation with GPCR levels (Samama et al., 1993). These results indicate strongly that total quiescence of a GPCR in the absence of ligand is not the norm. GPCR mutations which have produced constitutive activation (i.e. a state in which the receptor remains activated) have revealed that, even in the absence of ligand there must be a level of function of a GPCR (Lee et al., 1996). This has resulted in the development of ideas of conformational states of GPCRs in which equilibria must exist between the ground and the active states (R and R*). The binding of an agonist ligand shifts the equilibrium towards R*. The binding of an inverse agonist ligand shifts the equilibrium towards R. The binding of a (null) antagonist does not alter the equilibrium between R and R*, because it has equal affinity for both states. According to Milligan and Bond (1997), true antagonists may be very rare, but in practice it may be difficult to distinguish between an antagonist and a ligand with very weak positive or negative partial efficacy particularly at GPCRs in which the resting equilibrium is very heavily weighted towards R. The recent cannabinoid literature has begun to reveal the complexities discussed here.

Multiple affinity states of the CB receptors

Houston and Howlett (1998) detected multiple affinity states of the CB_1 receptor by agonist competition for [³H]SR 141716A binding to rat brain CB_1 receptors. Cannabinoid agonists, desacetylevonantradol (DALN) (17) and WIN 55212-2 both bound in two discrete affinity states (30% high affinity), but the ratios of the IC_{50}'s revealed distinct differences. Other affinity state differences included (1) differential Na^+ effects: Na^+ reduced the CB_1 affinity of DALN by 10-fold, but did not affect the affinity of WIN 55212-2; (2) GTP effects: a non-hydrolyzable GTP analogue decreased the fraction of high affinity WIN 55212-2 binding, but not that of DALN unless Na^+ was present. These investigators concluded that the differential modulation of CB_1-G-protein coupling by Na^+ and guanine nucleotides is dependent upon the agonist bound.

In their study of agonist induced $[^{35}S]GTP\gamma S$ binding in rat cerebellar membranes, Kearn et al. (1999) showed that ligand efficacy at the neuronal CB_1 receptor correlates with the ratio of ligand affinities for the active and inactive states of the receptor. These investigators found that in the absence of GDP, GTP or sodium, the majority of CB_1 receptors in cerebellar membranes were in the R* conformation. The addition of a non-hydrolyzable GTP analog shifted the equlibrium to favor R. These investigators concluded that in vivo their data suggest that the CB_1 receptor is capable of precoupling to G-proteins.

Under the incubation conditions used in their $[^{35}S]GTP\gamma S$ assay, Kearn and co-workers found 30% of CB_1 binding sites are in the R* state and 70% in R. The proportion of receptors in the high affinity state was 60–70% of the total for each agonist regardless of efficacy. The ligands of the CB_1 receptor exhibited different affinities for the active, G-protein coupled (R*) and inactive (R) states of the CB_1 receptor. Agonists such as WIN 55212-2 and CP55,940 had higher affinity for the precoupled (R*) state of the receptor, while SR 141716A had a slightly higher affinity for the uncoupled (R) state, in support of its designation as an inverse agonist (Bouaboula et al., 1997). CB_1 receptor agonists exhibited a continuum of efficacies as reflected by the E_{max} for $[^{35}S]GTP\gamma S$ binding. WIN 55212-2 and CP55,940 were found to be approximately equiefficacious. Δ^9-THC was found to have an efficacy 20–30% of WIN 55212-2 (Kearn et al., 1999). These results are consistent with results reported by Burkey and co-workers (1997a,b) and by Sim and co-workers (1996), although Petitet et al. (1997) reported that the E_{max} for Δ^9-THC was 88% of that for WIN 55212-2 and Griffin et al. (1998) found that Δ^9-THC increased $[^{35}S]GTP\gamma S$ binding to 57% of the E_{max} of WIN 55212-2. Kearn et al. (1999), Burkey et al. (1997b) and Selley et al. (1996) found the efficacy of AEA to be 50–65% of the E_{max} produced by WIN 55212-2. However, Griffin et al. (1998) reported AEA had no effect on $[^{35}S]GTP\gamma S$ binding.

Kearn and co-workers (1999) concluded that "since the proportion of receptors in each affinity state is a determinant of efficacy, the ability of CB_1 receptor agonists to initiate signaling could vary among brain regions and cell types depending upon the ratio of receptors in the R* and R states. In fact, regional differences in CB_1 receptor coupling and activation of GTP exchange have been demonstrated (Breivogel et al., 1997; Sim et al., 1995). A key determinant of CB_1-mediated signaling is, therefore, the G-protein distribution and the ratio of CB_1 receptors to G-proteins in a particular cell" (Kearn et al., 1999).

According to Glass and Northup (1999), the physiological relevance of the difference in the ability of ligands to regulate G-protein signaling will depend on a combination of the number of receptors in the cell, and the saturation properties of the effector molecules. Thus, if saturation of second messenger response (e.g. inhibition of adenylate cyclase, enhancement of K^+ conductance) requires full stimulation of G-protein, then partial efficacy would be visible if receptor number was limited. If, however, the maximal response can be generated by submaximal G-protein activation, then the difference between agonists may be readily discernable (Glass and Northup, 1999). This model, therefore, provides a mechanism for explaining the differences observed in potency of agonists in different cells and tissues. For example, Mackie and co-workers (1993) demonstrated that AEA was a partial agonist in the inhibition of calcium channels in N18 neuroblastoma cells, but they observed full agonism of AEA in AtT20 cells that express higher receptor number (Mackie *et al.*, 1995). Glass and Northrup (1999) demonstrated that AEA is a partial agonist in activation of $G_{\alpha o}$, the G-protein thought to mediate calcium channel activation. Their findings are consistent with the idea that when receptor number is limited, the differences in response to particular agonists become detectable.

Cannabinoid receptor selective G-protein coupling

Mukhopadhyay and co-workers (2000) have reported that the CB_1 receptor can exist as an SDS-resistant multimer. In 3-[(3-cholamidopropyl)dimethylammonio] propane-sulfonate (CHAPS) detergent, the CB_1 receptor exists in a complex with G-proteins of the $G_{i/o}$ family in the absence of exogenous agonists. A peptide derived from the CB_1 receptor juxtamembrane C-terminal domain, peptide CB_1 401–417 (see Activation of CB Receptors section below), autonomously activates $G_{i/o}$ proteins and competitively disrupted the CB_1 receptor association with $G_{\alpha o}$ and $G_{\alpha i3}$, but not with $G_{\alpha i1}$ or $G_{\alpha i2}$.

Many receptors signal through pertussis toxin-sensitive pathways. Physiological studies of cAMP and ion channel regulation have suggested that a single receptor type may couple to multiple distinct pertussis toxin-substrate-G-protein α subunits (Gudermann *et al.*, 1997; Prather *et al.*, 2000). Cannabinoid ligand binding studies (Houston and Howlett, 1998; Kearn *et al.*, 1999) and studies of the regulation of GTPγS binding (Burkey *et al.*, 1997a; Breivogel *et al.*, 1998; Griffin *et al.*, 1998; Kearn *et al.*, 1999) in membrane fractions have led to hypotheses of agonist–selective G-protein coupling (see above). However, diverse cell lines were used in these studies, with the likelihood of heterogeneity in G-protein available for coupling within each cell line. In order to circumvent this problem, Glass and Northup (1999) used recombinant expressed receptors *in situ* in membrane fractions from which extrinsic membrane proteins had been removed or inactivated by urea extraction. Although depleted of G-protein, the uncoupled receptors in this protocol remained fully functional for reconstitution with purified G-protein subunits. Glass and Northup (1999) examined the ability of receptors to catalyze the GDP-GTP exchange of G-protein with purified bovine brain G_i and G_o. Activation of CB_1 receptors produced high-affinity saturable interaction for both G_i and G_o. Agonist stimulation of CB_2 lead to high-affinity saturable interaction with G_i only. The ability to recognize G-proteins was selective because nonappropriate G-proteins such as G_q and G_t were not recognized by either CB_1 or CB_2.

Glass and Northup's finding that CB_2 couples only to G_i, while CB_1 couples to G_i or G_o is consistent with the regional distribution of the cannabinoid receptors and G-proteins (1999). CB_1 receptors are localized to the brain and a few peripheral organs (Pertwee, 1997). CB_2 receptors are localized primarily to immune cells. Although both G_i and G_o have been shown to be located in the CNS, only G_i has been demonstrated to be present in immune cells. So physiologically, G_i is likely to be the G-protein encountered by CB_2 and, therefore, the protein to which it will couple.

In addition to its cyclic AMP effects, the CB_1 receptor has also been shown to inhibit voltage gated Ca^{+2} channels and to activate inwardly rectifying K^+ channels (Caulfield and Brown, 1992; Mackie and Hille, 1992; Felder et al., 1995; Mackie et al., 1995; Pan et al., 1996). In contrast, the CB_2 receptor has not been found to modulate ion channels. The failure of CB_2 receptors to modulate ion channels may be explained by their low affinity for G_o (Glass and Northup, 1999). Inhibition of voltage gated Ca^{+2} channels has been proposed to be mediated via $G_{\alpha o}$, whereas activation of K^+ channels has been proposed to be via $\beta\gamma$ subunits derived from either G_i or G_o (Gudermann et al., 1997; Ho et al., 1999). It is also possible that $\beta\gamma$ subunits of differing composition have higher affinity for G_o vs. G_i (or CB_1 vs. CB_2), and that these subunits differentiate the ability of CB receptors to activate K^+ channels. This data would suggest that CB_1 receptor coupling to ion channels is G_o mediated and that the lower apparent affinity of CB_2 for G_o is sufficient to prevent regulation of ion channels (Glass and Northrup, 1999).

Cannabinoid agonist selective G-protein coupling

Prather and co-workers (2000) recently demonstrated that activation of cannabinoid receptors in rat brain by WIN 55212-2 produces a distinct pattern of G-protein activation that has similar percentage increases between several brain regions (hippocampus, striatum, amygdala and hypothalamus). The greatest amount of activation in these brain regions involved $G_{0\alpha1}$, followed by simular stimulation of $G_{0\alpha3}$ and $G_{0\alpha2}$. However, in cerebellum a greater overall percentage stimulation of G-proteins occurred than in the brain regions above and a slightly different pattern of activation was observed, with the most stimulaiton for $G_{0\alpha3}$ (212%) and similar levels of activation of $G_{0\alpha2}$ (161%) and $G_{0\alpha1}$ (132%).

Studies of WIN 55,212-2 activation in the cerebellum revealed that WIN 55212-2 activates G_α subunits with different potency (Prather et al., 2000). Although 0.25μM and 0.28 μM WIN 55212-2 was required to half-maximally activate $G_{0\alpha3}$ and $G_{0\alpha1}$, up to 11-fold greater concentrations were required to produce 50% activation of $G_{0\alpha2}$ (2.86 μM). WIN 55212-2 also activated $G_{i\alpha1}$ and $G_{i\alpha2}$, requiring 0.1 μM or 0.62 μM to produce half-maximal effects. The authors conclude that "... data from this study suggest that cannabinoid receptors may produce distinct intracellular signals by activation of a specific pattern of G-proteins responsible for regulation of a unique blend of intracellular effectors in a concentration-dependent manner".

Glass and Northup (1999) reported that for the interaction of CB_1 with G_i, HU-210 (**18**), WIN 55212-2 and AEA all elicited maximal activation, whereas Δ^9-THC (56±6%) caused only partial activation. For the interaction of CB_1 with G_o, only HU-210 affected maximal stimulation with AEA, WIN 55212-2 and

Δ^9-THC stimulating between 60 and 75% compared with HU-210. For interaction of CB_2 receptors with G_i, Glass and Northup (1999) reported that HU-210 was the only compound that demonstrated maximal activation. WIN 55212-2 (64%), AEA (42%) and Δ^9-THC (44%) all initiated sub-maximal levels of G-protein activation. In order to account for these differences, Glass and Northrup (1999) proposed that HU-210 stabilizes a conformation of the CB_1 receptor that can fully activate G_i or G_o, whereas WIN 55212-2 must induce a different conformation, one that can fully activate G_i, but only partially activate G_o. These results demonstrate that a particular ligand can induce a receptor conformation that is maximally active in stimulating one G-protein. Glass and Northup (1999) propose that this finding clearly suggests that ligands may be designed that are fully selective for one G-protein pathway over another.

A recent study has demonstrated specificity among CB_1 receptor agonists in their relative abilities to activate G_s vs. G_i coupled transduction pathways (Bonhaus *et al.*, 1998). In CHO cells expressing human CB_1 receptors, Bonhaus and co-workers found that CB receptor agonists inhibited forskolin-stimulated cAMP accumulation in a rank order identical to their CB_1 affinities (HU-210 > CP 55,940 > Δ^9-THC > WIN 55212-2 > anandamide). Δ^9-THC was a partial agonist whereas, CP 55,940 and WIN 55212-2 were full agonists. In cells pretreated with pertussis toxin, in the presence of forskolin, these same CB receptor agonists, in the same rank order of potency as above, stimulated cAMP accumulation, albeit with potencies 5- to 10-fold less than they inhibited its production above. WIN 55212-2, however, was the only full agonist in the G_s-linked assay. Thus, agonist–receptor complexes may differ in their recognition of G_s in addition to G_i and G_o.

Mutations affecting cannabinoid signaling

Among the highly conserved residues in GPCRs is an aspartic acid residue in TMH2 at position 2.50 (D163 in human CB_1; D164 in rat CB_1). Mutation of this residue in various GPCRs has had a variety of effects, including decreasing agonist affinity (e.g. see Chakrabarti *et al.*, 1997), blocking coupling to KIR current (e.g. see Surprenant *et al.*, 1992), decreasing phosphoinositide hydrolysis (e.g. see Sealfon *et al.*, 1995), blocking inhibition of cAMP production (Chakraborti *et al.*, 1997) and eliminating allosteric receptor modulation by sodium ions (Parent *et al.*, 1996). This conserved aspartate is thought to interact with an asparagine (N7.49) in TMH 7. Many of the actions resulting from mutation of the aspartate are thought to be due to the disruption of this interaction (Zhou *et al.*, 1994). Studies of the GnRH and 5HT-2A receptors have shown that exchanging these two residues results in a receptor that functions like WT (Zhou *et al.*, 1994; Sealfon *et al.*, 1995).

The importance of D2.50 in the cannabinoid CB_1 receptor has been the subject of two separate studies which led to different results. Tao and Abood (1998) reported that in HEK293 cells, the D2.50(163)N/E in human CB_1 and corresponding D2.50(80)N/E mutations in (human) CB_2 led to an unaltered binding profile for CP55,940, AEA and Δ^9-THC. Binding of SR 141716A in CB_1 was unaltered as well.

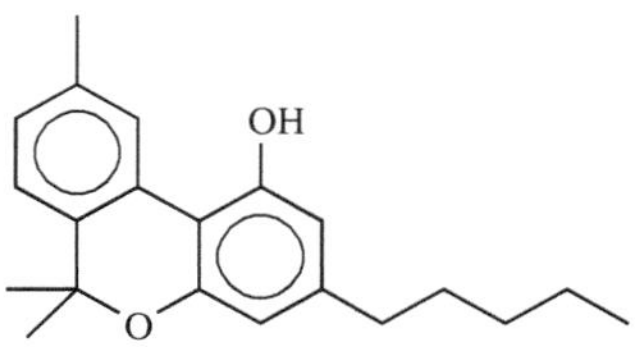

CBN
12

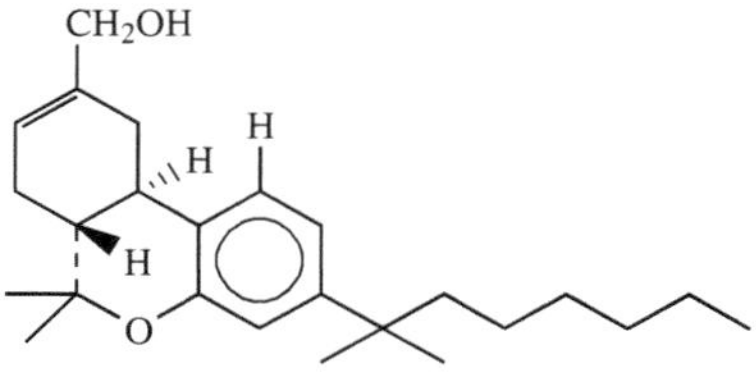

1-deoxy-11-OH-Δ^8-THC DMH
13

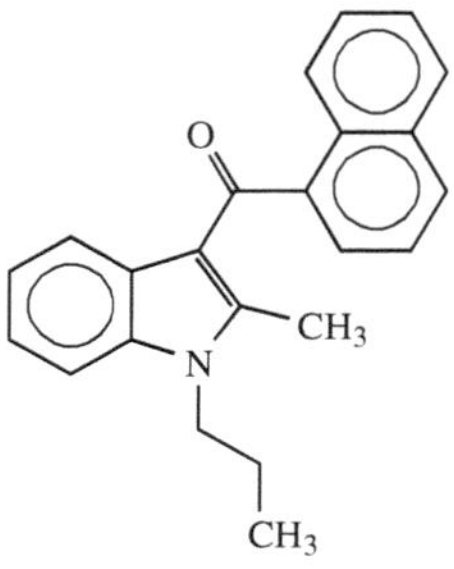

JWH-015
14

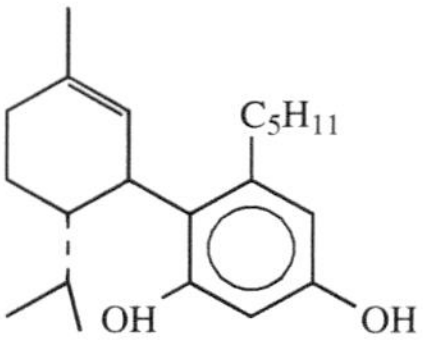

Abn-CBD
15

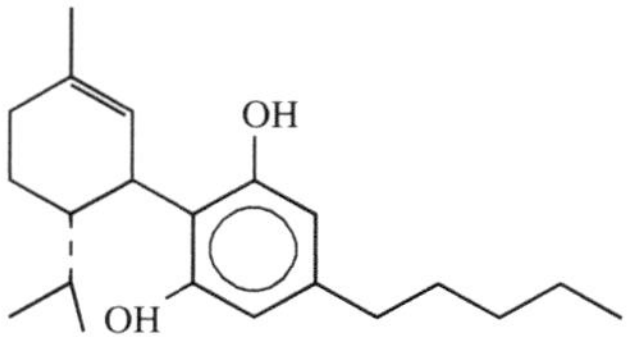

CBD
16

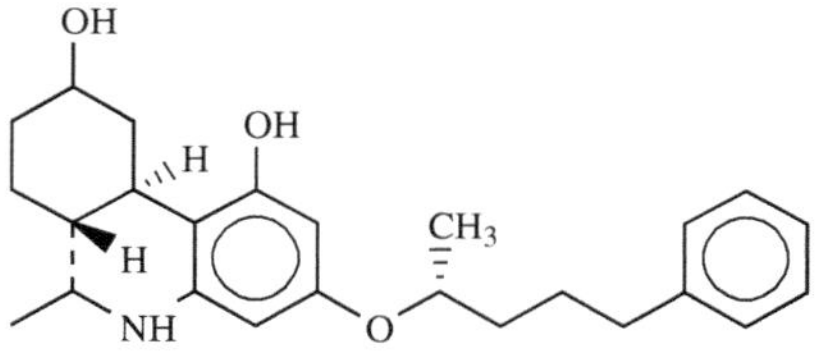

DALN
17

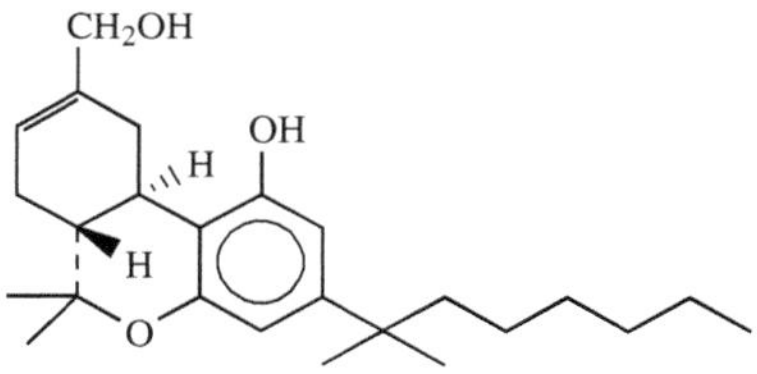

HU-210
18

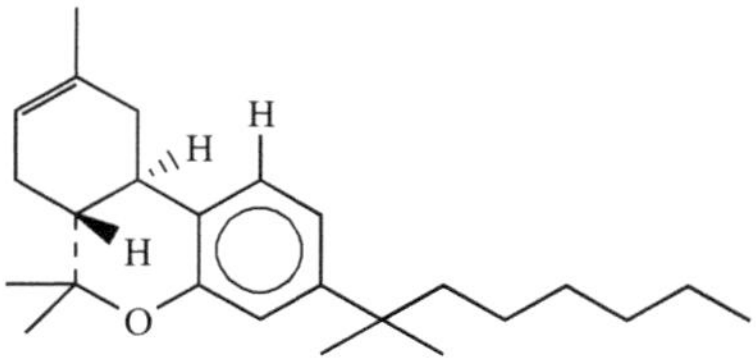

deoxy-Δ^8-THC DMH
19

Figure 17.3 Other cannabinoid structures referenced in the text.

However, the affinity of WIN 55212-2 was attenuated significantly in CB_1, but not in CB_2. Studies examining inhibition of cAMP accumulation showed reduced effects of cannabinoid agonists in the mutated receptors. These investigators concluded that D2.50 was not generally important for ligand recognition; however, it was required for communication with G-proteins and signal transduction.

In a separate study in AtT20 cells, Roche and co-workers (1999) found that a CB_1 D2.50(164)N mutant bound WIN 55212-2 with an affinity matching WT CB_1. The D2.50(164)N mutant inhibited cAMP and Ca^{2+} currents with a potency and efficacy equivalent to WT as well. However, this mutant did not couple to the potentiation of inwardly rectifying potassium channel (KIR) currents and prevented internalization of the receptor after exposure to agonist. Despite the fact that the D2.50(164)N mutant did not internalize, it was still found capable of activating p42/44 MAP kinase. In addition, the double revertant mutation D2.50(164)N/N7.49(394)D did not produce a receptor that internalized.

Roche and co-workers (1999) have suggested that these disparate results for mutation of amino acid 2.50 have precedent in the α_2-adrenergic receptor literature. Here the D2.50(79)N mutation in AtT20 cells had no effect on coupling to inhibition of adenylyl cyclase (Surprenant *et al.*, 1992), but in chinese hamster ovary (CHO) cells, the D2.50(79)N mutation disrupted coupling of the receptor to inhibition of adenylyl cyclase (Wang *et al.*, 1991). Roche *et al.* (1999) point out that the disparities in their results and those of Tao and Abood are likely due to the cell types used for expression of the receptor. The cell types used may provide different ratios of G-protein families that couple to the CB_1 receptor, provide different forms of adenylyl cyclase, or localize the receptors in different manners.

Calandra and co-workers (1999) provided evidence for dual coupling of the CB_1 receptor to the classical pertussis toxin-sensitive Gi/Go inhibitory pathway and to a pertussis toxin-insensitive adenylyl cyclase stimulatory pathway initiated with low quantities of agonist in the absence of any co-stimulant. These investigators used CB_1/CB_2 chimeric constructs that maintained high affinity for CP55940 to investigate the functionality of the chimeras. The cAMP stimulatory property of CB_1 was found to be conserved in all chimeras that contained the first two intracellular loops (I-1 and I-2) of this receptor. However, these investigators were not able to delineate the role of each more precisely since the cannabinoid CB_1/CB_2 receptor chimeras with connections in the second and third TM regions were not translocated into the plasma membrane (Shire *et al.*, 1996b).

Inverse agonism produced by cannabinoid antagonists

There is growing evidence in the cannabinoid literature that SR 141716A (**7**) actually functions as a CB_1 inverse agonist. Bouaboula *et al.* (1997) reported that CHO cells transfected with human CB_1 receptor exhibit high constitutive activity at both levels of mitogen-activated protein kinase (MAPK) and adenylyl cyclase. Guanine nucleotides enhanced the binding of SR 141716A, a property of inverse agonists. These authors propose the existence of an R^- state in which the inverse agonist promotes or stabilizes an inactive receptor/G-protein complex. Lewis and co-workers (Pan *et al.*, 1998) demonstrated constitutive activity of CB_1 receptors in inhibiting Ca^{2+} currents that was not due to endogenous agonist, confirming that CB_1 receptors can be tonically active. These investigators reported that SR 141716A acts as an inverse agonist to increase neuronal voltage – dependent Ca^{+2} currents by reversal of tonic CB_1 receptor activity. In addition, these investigators reported that SR 141716A behaves as a null antagonist in a CB_1 K3.28(192)A mutant receptor. AM630 (**8**) has also recently been reported to function as an inverse agonist at the human CB_1 receptor (Landsman *et al.*, 1998). SR 141716A also has been reported to reduce basal GTPγS binding in membrane from cells with CB_1 receptors (Landsman *et al.*, 1997). However, other studies have failed to observe this effect or have observed this effect only at high drug concentrations (Breivogel *et al.*, 1998; Kearn *et al.*, 1999; Sim-Selley *et al.*, 2001). The high drug concentrations needed to see this effect led both Sim-Selley and co-workers (2001) and Breivogel and co-workers (1998) to conclude that these concentrations preclude the determination that the effect is mediated exclusively by the CB_1 receptor. Glass and Northrup (1999) reported spontaneous activation of both G_i and G_o by CB_1 receptors that could be enhanced by additional magnesium ion. This spontaneous activity was completely blocked by SR 141716A, indicating strong inverse agonism. AM630 has also been reported to function as an inverse agonist (Landsman *et al.*, 1998). Portier and co-workers (1999) recently demonstrated that the CB_2 receptor in CHO cells is constitutively active. This activity could be reduced by SR 144528, thus providing evidence that SR 144528 also functions as an inverse agonist.

MODELING THE CANNABINOID RECEPTORS

It is clear from the previous section that recent biochemical studies of the CB receptors suggest that these receptors may exist in several dynamical states in order to couple to different G-proteins. These states may depend on the conformation stabilized by a particular agonist. It is also clear that the label of agonist or partial agonist cannot be applied to any specific ligand for its interaction with all G-proteins, but may be a function of the types and proportions of G-proteins available within a particular cell line or preparation. An understanding of the origins of these complexities at a molecular level requires not only a knowledge of the structures of the CB_1 and CB_2 receptors in their inactive (R) and active (R*) states, but also a knowledge of possible gradations in the R* state recognized by various G-proteins.

The recent 2.8 Å crystal structure of the GPCR, rhodopsin in its inactive state has provided important information at an atomic level of resolution concerning the overall helix bundle organization in GPCRs (Palczewski *et al.*, 2000). While there is no experimental structure of the CB_1 or CB_2 receptors yet available, there is a growing body of literature from biophysical studies of other GPCRs that can be used in combination with structure-activity relationship (SAR) and mutation studies of the CB receptors to construct three dimensional computer models of the CB_1 and CB_2 receptors in both their inactive (R) and activated (R*) states. Computer models of these receptors and their states can prove to be a very important tool for hypothesis generation and testing. In this section, we review what is known about the structures and functions of G-protein coupled receptors (GPCRs) in general, how computer models of the CB receptors have been built and what is known about cannabinoid receptor structure based on mutation and computer modeling experiments.

Both the CB_1 and CB_2 receptors belong to the super family of G-protein coupled receptors (GPCRs). These receptors are membrane proteins that serve as a very important link through which cellular signal transduction mechanisms are activated. It has been reported that approximately 80% of known hormones and neurotransmitters work through coupling with GPCRs (Kobilka, 1992). Much of what was originally thought about the structure of GPCRs was based on the solved structure of bacteriorhodopsin (BR), an integral membrane protein from *Halobacterium halobium* (Henderson *et al.*, 1990). The analogy between the known structure of BR and the proposed structure of the GPCRs was based upon the functional and structural relation of the G-protein coupled visual pigment, rhodopsin, to BR and the homology in the amino acid sequences of rhodopsin and other GPCRs (Findlay and Eliopoulus, 1990).

By analogy with BR, GPCRs were thought to possess an extracellular N terminus, seven α-helical transmembrane (TM) regions, with intervening loops extending intra- and extracellularly, and an intracellular C terminus. The alpha helices were thought to be oriented approximately perpendicular to the membrane and arranged to form a closed bundle. Largely hydrophilic amino acids that are part of a TM helix (TMH) were thought to face the protein interior, while largely hydrophobic amino acids were thought to face the exterior of the protein away from the center of the bundle (Donnelly and Cogdell, 1993). For many GPCRs, such as the cationic neurotransmitters, the ligand binding pocket has been proposed to be within the transmembrane bundle (Findlay and Eliopoulus, 1990). For other GPCRs, such as peptide receptors which have larger ligands, the ligand binding sites are thought to be extracellular loops (Xie *et al.*, 1990) or to be a combination of TM residues and extracellular loop residues (Greenwood *et al.*, 1997).

Some molecular modeling studies of GPCRs (see Mahmoudian, 1997; Hibert *et al.*, 1991; Teeter *et al.*, 1994; Hutchins, 1994; Westkaemper and Glennon, 1993), but not all (Zhang and Weinstein, 1993; Findlay and Donnelly, 1995; Ballesteros and Weinstein, 1995; Paterlini *et al.*, 1997; Sealfon *et al.*, 1997; Pogozheua *et al.*, 1997; Strahs and Weinstein, 1997; Laakkonen *et al.*, 1996), have been based on the structure of BR. Some of these studies have used BR as a direct template, while others have used the BR structure as a departure point. The use of BR as a direct template for GPCR modeling is difficult to justify because BR is not coupled to G-proteins and its sequence shows none of the distinctive patterns of the GPCR family (Zhang and Weinstein, 1994). In fact, its sequence has very low homology with GPCR sequences (Pardo *et al.*, 1992). Rhodopsin, on the other hand, is a GPCR that displays the characteristic sequence patterns of this family of receptors.

The recent 2.8 Å resolution crystal structure of bovine rhodopsin (Palczewski *et al.* 2000) along with the earlier 5 Å electron cryomicroscopy results for bovine rhodopsin (Krebs *et al.*, 1998) and 7.5 Å resolution results for frog rhodopsin confirm the validity of the seven transmembrane alpha-helix model for GPCRs (Unger *et al.*, 1997; Baldwin *et al.*, 1997). However, it is clear from the early projection maps of rhodopsin, and the 2.8 Å crystal structure that the arrangement of the helix bundle in rhodopsin differs from that of BR both in helix tilt and in helix orientation (Unger *et al.*, 1997; Baldwin *et al.*, 1997; Krebs *et al.*, 1998; Palczewski *et al.*, 2000).

The crystal and cryoelectron microscopy structures of rhodopsin have taken many years (Palczewski *et al.*, 2000; Krebs *et al.*, 1998; Unger *et al.*, 1997; Schertler *et al.*, 1993, 1995; Unger *et al.*, 1995). It is not likely that a structure with well defined degrees of certainty will be available for other GPCRs, including the cannabinoid receptors, in the forseeable future. Because of this, molecular models based on experimental data and theoretical considerations represent one way to begin an exploration of structure–function relations for GPCRs such as the cannabinoid receptors. Such models have been developed in the literature for many GPCRs, including the cannabinoid receptors. Clearly, computational models of these receptors cannot be considered on par with structures obtained from direct experimental methods such as X-ray crystallography, NMR, or, for the case of membrane proteins, cryoelectron microscopy. However, such computational models when linked in an iterative fashion with SAR and molecular biological experiments can serve as hypothesis generators and can contribute to an understanding of the molecular pharmacology of GPCRs such as the cannabinoid receptors at an atomic level of detail (Ballesteros and Weinstein, 1995).

In the following section, the development of CB_1 and CB_2 receptor models in the Reggio lab is described. This section is followed by a review of the cannabinoid receptor mutation literature and its use in the validation of 3D cannabinoid receptor models. In the discussion of receptor residues which follows, the amino acid numbering scheme proposed by Ballesteros and Weinstein (1995) is used. In this numbering system, the most highly conserved residue in each transmembrane helix (TMH) is assigned a locant of .50. This number is preceded by the TMH number and may be followed in parentheses by the sequence number. All other residues in a TMH are numbered relative to this residue. In this numbering system, the most highly conserved residue in TMH 2 of the human CB_1 receptor is D2.50(163). The residue that immediately precedes it is A2.49(162).

Development of preliminary cannabinoid receptor models

Results from Biophysical studies of GPCRs have provided information and techniques that can be used to construct a preliminary GPCR model. These techniques have been used in the Reggio lab to construct models of the CB_1 and CB_2 receptors (Bramblett *et al.*, 1995; Huffman *et al.*, 1996; Song *et al.*, 1999; Tao *et al.*, 1999). Helix nets of the CB_1 and CB_2 receptors are included in Figures 17.4 and 17.5.

The TMH ends of the CB_1 receptor were initially identified using a Fourier Transform analysis of periodicity (in hydrophobicity and variability) in the primary amino acid sequence of the CB_1 receptor and a sub-set of other GPCRs with high homology to CB_1 (Bramblett *et al.*, 1995). Variability moment vectors for each seven amino acid window in the sequence were used to delineate the arc of each TMH that faced lipid. The seven transmembrane (TM) alpha helices were arranged into a bundle consistent with that predicted for rhodopsin (Rho) (Baldwin, 1993), as well as consistent with their predicted lipid exposure. This analysis also revealed the TMH ends and orientations for the other receptors in the sequence alignment including the CB_2 receptor.

Initial models of the transmembrane domains of the CB_1 and CB_2 receptors were refined using a convergence of methods which relied on sequence similarities

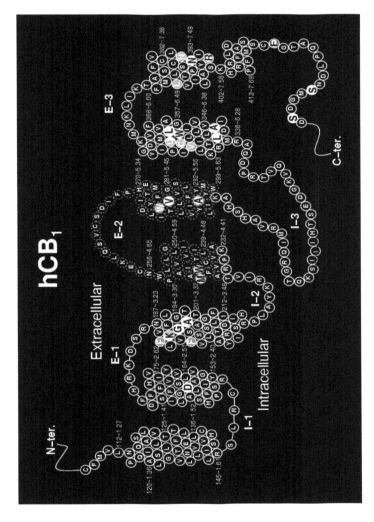

Figure 17.4 Helix net representation of the human CB₁ receptor sequence. Yellow filled circles denote sites that have been mutated (see Song and Bonner, 1996; Chin *et al.*, 1998, 1999; Tao and Abood, 1998, 1999; Song *et al.*, 1999; Roche *et al.*, 1999; Abadji *et al.*, 1999; Jin *et al.*, 1999; Mukhopadhyay *et al.*, 1999). Blue filled circles denote residues identified by Mahmoudian (1997) to be involved in Δ⁹-THC binding. Green filled circles denote residues identified in Tao *et al.* (1999) to be involved directly in CP55,940 binding. Magenta filled circles denote residues identified in Song *et al.* (1999) to be involved directly in WIN 55212-2 binding. Purple dashed circles denote residues involved in chimera studies reported by Shire *et al.* (1996b,1999). White circles denote the segments of the I-3 loop and C-terminus that Howlett *et al.* (1998b) have reported to be involved in G-protein coupling. The residues with white outlines and black letters within this span have been suggested by Mukhopadhyay *et al.* (1999) to be of importance in G-protein docking. Circles filled with two colors serve the dual roles connoted by those two colors described above. (*See Color plate 3*)

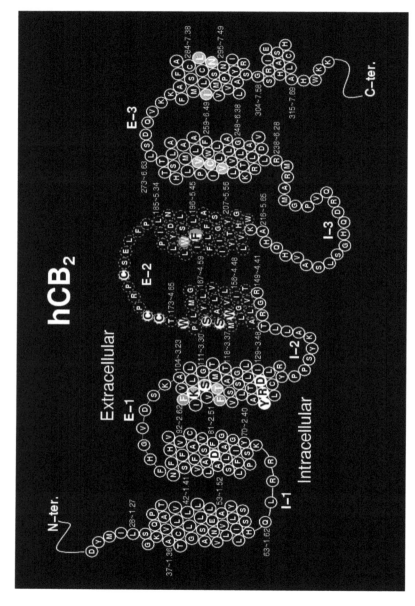

Figure 17.5 Helix net representation of the human CB$_2$ receptor sequence. Yellow filled circles denote sites that have been mutated (see Tao and Abood, 1998; Song *et al.*, 1999; Tao *et al.*, 1999; Rhee *et al.*, 2000a, 2000b; Gouldson *et al.*, 2000). Green filled circles denote residues reported by Tao *et al.* (1999) to be involved directly in CP55,940 binding. Magenta filled circles denote residues reported in Song *et al.* (1999) to be involved directly in WIN 55212-2 binding. Purple dashed circles denote residues involved in chimera studies reported by Shire *et al.* (1999). Blue filled circles denote residues reported by Gouldson *et al.* (2000) to be involved in SR 144528 binding. Circles filled with two colors serve the dual roles connoted by those two colors described above. (*See Color plate 4*)

between GPCRs (Ballesteros and Weinstein, 1995), on photo-affinity labeling (Nakanishi *et al.*, 1995), spin-labeling (Altenbach *et al.*, 1996), NMR (Han and Smith, 1995), mutation (Zhou *et al.*, 1994; Cohen *et al.*, 1994), scanning cysteine accessibility (Fu *et al.*, 1996; Javitch *et al.*, 1995a, 1995b, 1998, 1999, 2000), metal ion cross linking (Sheikh *et al.*, 1996), electron cryomicroscopy (Unger *et al.*, 1997) and other studies of GPCRs. Each TMH bundle was energy minimized using Amber with its all atom force field parameters (Pearlman *et al.*, 1995).

The models were revised upon the publication of the 2.8 Å crystal structure of bovine rhodopsin (Rho), in its inactive (R) state (Palczewski *et al., 2000*). Both the CB receptors and Rho belong to the same sub-family of GPCRs and can therefore be expected to bear many structural similarities, particularly on their intracellular sides. Because the original CB_1 and CB_2 models were built using a wide variety of experimental evidence as discussed above, the models compared very well with the Rho structure in helix orientation, helix tilt, helix packing and relative helix heights. Key interactions present in the original CB models were found also to be present in the Rho structure.

We were able, however, to capitalize on new information provided by the Rho structure in order to update our CB_1 and CB_2 models. The Rho structure was used to adjust helix ends, particularly in TMH 1 (see Figure 17.4, CB_1 Helix Net which illustrates the new TMH 1 ends) and to modify the conformation of our TMH 7 intracellular to the proline at position 7.50. **It is important to note, however, that the revised CB models are not identical to the Rho structure and should not be expected to be so. Clearly, at the very minimum, the CB models should reflect the consequences of sequence variations between Rho and CB_1/CB_2. This will be most evident in TMHs in which the sequences diverge with respect to the location of helix bending residues, such as proline, and the location of helix breaking/bending motifs such as the GG motif.** The CB receptors lack a proline in TMH 1 (Rho P1.48(53), CB_1/CB_2 L1.48), lack one of the two prolines in TMH 4 (Rho P4.59(170), CB_1/CB_2 L4.59), and lack the proline in TMH 5 (Rho P5.50(215), CB_1/CB_2 L5.50). The CB receptors also differ from Rho in the occurrence of GG motifs that can create kinks in TMHs when the motif faces lipid. The CB receptors lack the GG motif in TMH 2 (Rho G2.56(89), G2.57(90), CB_1 I2.56(169), F2.57(170), CB_2 V2.56(86), F2.57(87)). However, there is a GG motif in TMH 3 of CB_1 that faces lipid, that may have an effect on the conformation of TMH 3 (G3.31(194), G3.32(195)). A final important difference between Rho and the CB receptors is a key disulfide bridge. In Rho, a TMH 3 residue Cys 3.25(110) is engaged in a disulfide bridge with a Cys residue in the E2 loop. The 2.8 Å structure of Rho shows that this causes the E2 loop to dip down in the binding site crevice above (extracellular to) 11-*cis*-retinal. There is no corresponding Cys residue in TMH 3 of the CB receptors. However, there is a Cys residue at the extracellular end of TMH 4, and a Cys near the middle of the E2 loop in the CB receptors. Recent mutation results (Gouldson *et al.*, 2000) for CB_2 suggest that a bridge between these two Cys residues (C4.66(174), C179 in E2 loop) may exist, but further work is needed to prove the existence of this bridge. As the result of this important difference between Rho and the CB receptors, the binding site crevice around TMHs 3-4-5 is likely to be different.

The preliminary CB_1 and CB_2 TMH bundles prepared here were used as the starting point for the construction of models for the inactive (R) and active (R*) forms of each CB receptor sub-type as described in the following sections.

Creation of cannabinoid R and R* models

It is clear from the biochemical results on CB_1 and CB_2 coupling to G-proteins discussed above that a single model of each receptor sub-type will not be sufficient to probe the molecular pharmacology of the CB receptors. At the very minimum, both R and R* models are necessary. To this end, the biophysical literature can be used to delineate the basic features necessary for such models. There is a growing body of evidence in the literature that activation of GPCRs is accompanied by rigid domain motions of certain helices in the transmembrane bundle. Site-directed spin labeling (SDSL) studies of rhodopsin (Rho) conducted by Farrens and co-workers (1996) revealed that the cytoplasmic ends of Helices C and F (Helices 3 and 6) of Rho are in proximity and that changes in their interaction upon Rho light activation suggest a rigid body movement of these helices relative to one another. From an intracellular view, these authors suggest that TMH F (TMH 6) undergoes a clockwise rotation upon activation. In their fluorescence spectroscopy study of the β_2 adrenergic receptor using the cysteine reactive fluorophore IANBD, Gether and co-workers (1997) obtained similar results. Their results suggested that the intracellular end of TMH 6 may kink towards TMH 3 in the inactive/ground state (R) of the β_2 adrenergic receptor and that this end of TMH 6 may move away from the intracellular portion of TMH 3 upon activation (R*). Using fluoresence studies of labeled cysteines at the cytoplasmic end of TMH 6 in the β_2 adrenergic receptor, Jensen and co-workers (2001) concluded that activation is accompanied by a movement of the cytoplasmic part of TMH 6 away from the receptor core and upwards towards the lipid bilayer. Most recently, Ballesteros and co-workers (2001) have provided evidence for the existence of an interaction which constrains the relative mobility of the cytoplasmic end of TMH 3 and TMH 6 in the inactive state of the β_2 adrenergic receptor. These investigators proposed that the highly conserved R3.50(131) at the cytoplasmic end of TMH 3 interacts with E6.30(268) at the cytoplasmic end of TMH 6 and that this interaction may constitute a common switch governing activation of many rhodopsin-like GPCRs.

Most GPCRs contain the CWXP sequence motif in TMH 6. Recent evidence in the literature points to the importance of C6.47 and W6.48, both part of the CWXP motif, to the conformational changes that TMH 6 may undergo. In their substituted cysteine accessibility method (SCAM) study of a constitutively activated β_2 adrenergic receptor mutant, Javitch and co-workers (1997) found that upon activation of the β_2 adrenergic receptor, Cys 6.47(285) became accessible within the binding site crevice. These authors suggested that a rotation and/or tilting of TMH 6 was associated with activation of the β_2 receptor.

Lin and Sakmar (1996) reported that perturbations in the environment of W6.48 (265) of Rho (along with W3.41(126)) occur during the conformational change concomitant with receptor activation. Spectroscopic experiments of Rho suggest that the g^+ conformation of W6.48 is prevalent when Rho is in its inactive/ground state (R), while the trans conformation of W6.48 is prevalent in the activated state (R*).

The inactive/ground (R) form of CB_1 and CB_2

Clearly, the conformation of TMH 6 is key to the creation of R and R* models. The Conformational Memories (CM) method (Guarnieri and Weinstein, 1996) was recently used to explore the conformation of TMH 6 in the CB receptors in unbound and ligand complexed states (Ballesteros et al., 2000a; Reggio et al., 2000). The conserved Pro6.50 of the CWXP motif in TMH 6 of CB_1 was found to generate a kink in the α-helical structure that behaved as a flexible hinge. Two major families of TMH 6 conformers were found. In the context of a 3D model of the CB_1 receptor, the major cluster contained more kinked conformations that would bring the intracellular portion of TMH 6 in proximity to TMH 3, and thus were used to select an inactive state conformation for TMH 6 in analogy with experimental data on rhodopsin (Farrens et al., 1996) and the β_2 adrenergic receptor (Gether et al., 1997; Jensen et al., 2001; Ballesteros et al., 2001) . Less kinked conformations, which formed another cluster, were used to choose an active state conformation of TMH 6 (see below). CM calculations for TMH 6 in CB_2 revealed that while the major cluster for CB_2 TMH 6 is kinked, it does not exhibit the range of motion seen for the CB_1 TMH 6. The larger range of motion of the TMH 6 kink in CB_1 vs. CB_2 may be related to the ability of CB_1 to couple to a wider range of G-proteins than does CB_2 (Reggio et al., 2000). The TMH 6 from the preliminary bundle was replaced in both the CB_1 and CB_2 bundles with appropriate new TMH 6s. Each TMH 6 was oriented so that C6.47 was near the TMH 6–7 interface, but not accessible from within the binding site crevice to be consistent with Javitch et al. (1997) SCAM studies.

The literature cited above concerning TMHs 3 and 6 clearly indicates that the intracellular ends of these two helices are in close proximity in the inactive state (Farrens et al., 1996; Javitch et al., 1997). This would suggest a TMH 3/TMH 6 direct interaction. Ballesteros et al. (1998) have proposed that in the inactive state, the highly conserved residue at the intracellular end of TMH 3, R3.50 forms an ionic interaction with D3.49. Ballesteros and co-workers (2001) also recently described an interaction between R3.50 and E6.30 in the β_2 adrenergic receptor that forms an "ionic lock" to stabilize the receptor in its inactive conformation. Kosugi et al. (1998) have reported that a negatively charged amino acid at position 6.30(564) is important for maintaining the inactive conformation of the lutropin/choriogonadotropin receptor. Amongst the rhodopsin/β adrenergic sub-family of GPCRs, position 6.30 is D or E. In the CB receptors, residue 6.30 is an aspartic acid (D). Interaction between D6.30 and R3.50 in the cannabinoid receptors may be the interaction that brings TMHs 3 and 6 in proximity at their intracellular sides. We, therefore, have hypothesized by analogy with the β_2 adrenergic receptor that a R3.50/D6.30 interaction may be the "ionic lock" that stabilizes the CB receptors in their inactive states.

Constitutive activation is produced by destabilization of the inactive/ground (R) state of a receptor. Abadjii et al. (1999) reported that a L6.33 (341)A/A6.34 (342)L double mutation of CB_1 results in constitutive activation for coupling with G_s. In the Reggio model, residue 6.34 points towards R3.50. The enlargement of residue 6.34 from alanine to leucine creates steric interference for the R3.50/D6.30

interaction and may therefore interfere with the "ionic lock" that stabilizes the inactive conformation of CB_1. This result supports the choice of the R3.50/D6.30 interaction as a stabilizer of the inactive/ground state of the CB receptors.

The (R*) form of CB_1 and CB_2

The first major change to the preliminary CB_1 and CB_2 models which was made to create active (R*) bundles of CB_1 and CB_2 was the placement in the bundle of a representative structure from the major cluster of TMH 6 conformers calculated using Conformational Memories for CB_1 and CB_2 (Ballesteros *et al.*, 2000a; Reggio *et al.*, 2000). Each conformer is less kinked and therefore is not bent toward TMH 3 at the intracellular end as was the TMH 6 conformer chosen for the inactive bundle. The second major change made to create an R* model was to orient TMH 6 so that C6.47 was rotated counter-clockwise from an extracellular view into the binding cavity to be consistent with the literature (Javitch *et al.*, 1997).

Finally, it has been reported that when Rho is activated (i.e. passes from the dark state to the photoactivated Meta II state), two protons are taken up. Proton uptake is blocked in a E3.49Q mutation of Rho, suggesting that protonation of E3.49 (analogous to D3.49 in CB_1/CB_2) may account for one of the adsorbed protons (Arnis *et al.*, 1994). Rhee and co-workers (2000a) reported that D3.49(130) was essential for the capacity of the CB_2 receptor to bind CB agonists. Mutation of the corresponding residue in the β_2 adrenergic receptor to Asn, (i.e. D3.49(130)N) to mimic the protonated state and to Ala (i.e. D3.49(130)A) to fully remove the functionality of the side chain produced constitutive receptor activation (Rasmussen *et al.*, 1999). Furthermore, in the D3.49(130)N mutant, C6.47 (a residue that is not accessible from within the binding site crevice in WT β_2), became accessible to methanethiosulfonate ethylammonium (MTSEA), a charged, sulfhydryl-reactive reagent. This is consistent with a counterclockwise rotation of TMH 6 (from extracellular view) upon activation such that C6.47 is now exposed in the binding site crevice. The constitutive activity of the D3.49(130)N mutant also provides structural evidence linking charge-neutralizing mutations of D3.49 of the DRY motif to the overall conformation of the receptor. In the CB_1 and CB_2 R* (active) bundles, the charge on D3.49 was neutralized to be consistent with results reported above. The neutralization of charge on D3.49 and the counter-clockwise (extracellular view) movement of TMH 6 breaks the interaction of the D3.49/R3.50/D6.30 patch.

CB RECEPTOR MUTATION/CHIMERA STUDIES: IMPLICATIONS FOR CB RECEPTOR MODELS

One means of validation of a receptor model is consistency with mutation/chimera data. However, results of mutation/chimera studies need to be interpreted carefully for a mutation may indirectly alter a distant binding site by causing a conformational change in the receptor as opposed to altering a site of interaction directly (Perlman *et al.*, 1994). As discussed by Ward, Timms, and Fersht (1990), the retention of binding equal to wild type (WT) of at least some ligands of a receptor is one way of assuring that a gross structural change of the receptor has not occurred upon mutation. These authors also suggest that a mutation that affects ligand binding, but not signal transduction is an alternate way to test if a gross structural change has occurred. Figures 17.4 and 17.5 identify the amino acids in the CB_1 and CB_2 receptors (respectively) that have been mutated to date, as well as regions shown to be important via chimera experiments.

Mutations suggesting pharmacophore separation

Some experimental evidence points to a separation of pharmacophores for the AAIs from that of the other three classes of cannabinoid agonist ligands. Song and Bonner (1996) have reported that a K3.28(192)A mutation of CB_1 results in no loss of affinity or efficacy for WIN 55212-2 (3), but greater than 1,000-fold loss in affinity and efficacy for HU-210(18), CP55,940(2) and anandamide (4). Song and Bonner's results have been confirmed and have been extended by Chin and co-workers who reported that only the conservative mutation K3.28(192)R allowed retention of CP55,940 binding, while WIN 55212-2 retained affinity for K3.28(192)R, K3.28(192)Q and K3.28(192)E mutants (Chin et al., 1998). These results prompted Reggio and co-workers (1998) to pursue the hypothesis of a separate binding site for AAIs at CB_1 and prompted Tong et al.'s (1998) development of an independent COMFA model for AAIs binding at CB_1. These same results combined with modeling results (see above) prompted Dutta et al. (1997) to suggest that an independent AAI pharmacophore might be necessary. Although anandamide (AEA) is largely structurally dissimilar from the classical and non-classical cannabinoids, the K3.28(192)A mutation results suggest that AEA may share at least one interaction site with the classical and non-classical cannabinoids (i.e. K3.28(192)). This result supports the possibility of a common pharmacophore (and common binding site) for the classical/non-classical/endogenous cannabinoids. Such a pharmacophore has been developed by B. F. Thomas et al. (1996) and also by Tong et al. (1998) (see above).

The second extracellular loops (i.e. E-2 loops) of CB_1 and CB_2 share a sequence motif CSXXFP (see Figures 17.4 and 17.5). These loops differ by two amino acids in length. In addition, the CB_2 loop is quite rich in proline residues (four prolines in CB_2 vs. one proline in CB_1), suggesting that these loops may differ substantially in their preferred conformations. Shire et al. (1996b) have reported that exchange of the E-2 loops between the CB_1 (residues 256–273) and CB_2 (residues 173–188) receptors eliminated CP55,940 binding in both mutants. However, binding of SR 141716A in the CB_1/CB_2 (E-2) chimera was normal. One interpretation of this result is that key residues for CP55,940 are located on the E-2 loops in each receptor sub-type. Another interpretation is that exchange of these two loops produced a change in helix orientations within the transmembrane helix bundle that perturbed a CP55,940 binding site within each receptor sub-type. This later interpretation would seem to be more plausible considering the relative richness in proline residues in the E-2 loop of CB_2, which would presumably promote a very different loop conformation than seen in the E-2 CB_1 loop.

Shire and co-workers also have used chimeric CB_1 receptor/CB_2 receptor constructs to identify the region encompassed by the fourth and fifth transmembrane helices (from residues 4.42 to 5.63) as being critical for binding of both SR 141716A and SR 144528 (Shire *et al.*, 1999).

Mutations revealing sub-type selectivity

In addition to Song and Bonner's CB_1 K3.28(192)A mutation (1996), other mutation studies of the CB receptors have revealed important differences between the binding pocket of AAIs and other cannabinoid agonist classes. WIN 55212-2 has been reported to possess a 19-fold higher selectivity for CB_2 over CB_1 (Felder *et al.*, 1995). Using the CB_1 and CB_2 models developed in the Reggio lab, Song *et al.* have recently demonstrated in CB_1 V5.46(282)F and CB_2 F5.46(197)V mutation studies that the documented CB_2 selectivity of **3** may be due to the presence of an additional aromatic residue F5.46(197) in the CB_2 receptor (Song *et al.*, 1999). These mutations had no effect upon the binding of HU-210 (**18**), CP55,940 (**2**), or anandamide (**4**) and normal signal transduction was maintained in the mutants.

Receptor chimera studies of the CB_1 and CB_2 receptors conducted by D. Shire *et al.* (1996) have demonstrated that the region delimited by the fourth and fifth transmembrane domains of the CB_1 receptor (residues 233 to 299) were crucial for the binding of SR 141716A, but not CP55,940 and, that this same region in the CB_2 receptor (residues 150 to 214) was crucial for the binding of SR 144528 and WIN 55212-2 (Shire *et al.*, 1999). Figures 17.4 and 17.5 illustrate the range of amino acids in the CB_1 and CB_2 receptors, respectively, that were considered in this chimera study.

Unlike K3.28(192) in CB_1, the analogous CB_2 residue K3.28(109) does not appear to be critical for ligand recognition. Tao *et al.* (1999) recently reported that mutation of K3.28(109) in the CB_2 receptor to alanine had no effect on ligand binding and signal transduction. These authors suggest that CP55,940 may bind in a different orientation in WT CB_1 and CB_2 (see Proposed Binding Sites for Classical/Non-Classical Cannabinoids below).

Gouldson and co-workers (2000) recently reported a CB_2 receptor mutational analysis of TMH 4 and the E-2 loop. S4.53(161) and S4.57(165) were each mutated to their corresponding residue in CB_1 (i.e. A4.53 and A4.57 in CB_1). Each mutation showed little effect on the affinity and efficacy of agonists, CP 55940 and WIN 55212-2, but produced a dramatic loss in SR 144528 affinity (>1000-fold) and a concomitant loss in the ability of SR 144528 to antagonize the effect of CP55940 in a corticotrophin releasing factor luminescence assay. These authors propose that S4.53 and S4.57 have a direct hydrogen bonding interaction with SR 144528 in WT CB_2. However, because serine residues have been shown to bend alpha helices (Ballesteros *et al.*, 2000b), it is possible that the S4.53(161)A and S4.57(165)A mutations altered the conformation of TMH 4 in CB_2, causing steric interference with the SR 144528 binding site, but not with the binding sites of the agonists tested. This latter explanation is being explored in the Reggio lab at present.

Gouldson and co-workers (2000) also probed the importance of cysteine residues in the E-2 loop of CB_2 through mutation. C174S and C179S mutations resulted in loss of CP55940, WIN 55212-2 and SR 144528 binding, while a C175S mutation resulted in WT binding for CP55940, an eight-fold reduction in WIN

55212-2 binding, and a loss of SR 144528 binding. A reason for the loss of binding in the C174S and C179S mutants remains unclear, however, Gouldson and co-workers did document that the receptor had been translocated to the cell surface. The authors suggest that the effect of the C175S mutation on SR 144528 binding may be the result of alternative di-sulfide bridge formation or that there is a sulfur-π interaction with SR 144528 which is lost in the C175S mutation.

Proposed binding sites for classical/non-classical cannabinoids

Mahmoudian (1997) used the AUTODOCK program to identify a binding site for Δ^9-THC within the pore formed by the transmembrane helix bundle of his model (based upon BR) using the orientation of retinal in the BR bundle as a starting point. Figure 17.4 includes the residues identified by Mahmoudian to be in the Δ^9-THC binding site. The ligand was found to bind to a hydrophobic site which consisted of M4.49(240), W4.50(241), A5.57(293), W6.48(356), L6.51(359) and L6.52(360). The phenolic hydroxyl group formed a hydrogen bond with the carboxy group of A3.34(198). The docked position of Δ^9-THC was consistent with an area of steric repulsion behind the C-ring of Δ^9-THC and the result that steric bulk of the C-3 side chain contributes to increased binding affinity.

Reggio and co-workers (Tao *et al.*, 1999) have used their CB_1/CB_2 receptor models, as well as models of CB_1 K3.28A and CB_2 K3.28A mutants to probe possible differences in the binding site of CP55,940 in CB_1 vs. CB_2. In both wild type (WT) CB_1 and CB_2, the alkyl side chain of CP55,940 resided in a hydrophobic binding pocket formed by residues on Helices (TMHs) 6 and 7 (V6.43(351), C6.47(355), L6.51(359), L7.44(388) in CB_1 and (V6.43(253), C6.47(257), V6.51(261), L7.41(287) and I7.44(290)) in CB_2. In WT CB_1, ring A of CP55,940 (**2**) was near TMH 3, while the cyclohexyl ring (Ring C) pointed toward TMH 5. The C-2' phenolic hydroxyl formed a hydrogen bond with K3.28(192), the C-1 hydroxyl formed a hydrogen bond with W5.43(279) and the SAH group formed a hydrogen bond with N7.45(389). V3.32(196) created a region of steric interference behind the C-1 hydroxyl. In WT CB_2, the phenyl ring of CP55,940 was near TMH 7, while the cyclohexyl ring was between TMHs 3 and 7 pointing towards TMH 2. Hydrogen bonding interactions involved the SAH group with K3.28(109), the C-2' phenolic hydroxyl with N7.45(291) and the C-1 hydroxyl group with both S3.31(112) and T3.35(116) (Tao *et al.*, 1999).

In the CB_1 K3.28(192)A mutant, the loss of the lysine interaction with CP55,940 resulted in the cyclohexyl ring (Ring C) angling deeper into the pocket. The C-1 hydroxyl/W5.43 (279) hydrogen bond was lost as a result. Only one amino acid, N7.45(389) interacted with CP55,940 producing a hydrogen bond with the phenolic hydroxyl and with the SAH group. In contrast, in the CB_2 K3.28(109)A mutant, three nearly linear hydrogen bonds with S3.31(112), T3.35(116) and N7.45(291) were retained (Tao *et al.*, 1999). To test this proposed binding site in the CB_2 K3.28(109)A mutant, a double CB_2 mutant, K3.28(109)A/S3.31(112)G was prepared. This mutant retained binding only for WIN 55212-2 and a related structure, JWH-015. This result was found to be consistent with the proposed molecular model (Tao *et al.*, 1999).

These results were found to be consistent with the CB_1 K3.28(192)A mutation results reported by Song and Bonner (1996) and Chin *et al.* (1998) which showed a dramatic loss of affinity and efficacy for CP55,940 in the CB_1 K3.28(192)A mutant. These results also are consistent with CB_2 K3.28(109)A mutation results reported by Tao *et al.* (1999) which showed no loss of affinity or efficacy for CP55,940 in the CB_2 K3.28(109)A mutant .

Deoxy-Δ^8-THC-DMH analogs

Huffman et al. (1996) reported that removal of the phenolic hydroxyl of HU-210 resulted in a CB_2 selective compound. Docking studies of deoxy-Δ^8-THC analogs (**13** and **19**) in the Reggio model of the CB_1 receptor revealed that in the absence of a C-1 phenolic hydroxyl, K3.28(192) can hydrogen bond with the 11-hydroxyl group of deoxy-HU-210 (i.e. 1-deoxy-11-OH-Δ^8-THC-DMH), (**13**) and with the pyran oxygen (O-5) in deoxy-Δ^8-THC DMH (**19**), while the C-3 side chain maintains a hydrophobic interaction with hydrophobic residues in the TMH 6/TMH 7 region. However, modeling also revealed that the ligand repositioning necessary to allow K3.28(192) to interact with the pyran oxygen of deoxy-Δ^8-THC DMH (**19**) causes a decrease in the number of alkyl side chain atoms that can maintain an interaction with the hydrophobic binding pocket. This repositioning would cause a deoxy-Δ^8-THC analog which possesses only a pentyl side chain to be unable to extend to reach a TMH 6/7 hydrophobic binding pocket. This may be the reason why the K_i for deoxy-Δ^8-THC at CB_1 is greater than 10,000 nM (Huffman *et al.*, 1999).

Proposed aminoalkylindole binding sites

The Reggio lab CB_1 and CB_2 receptor transmembrane helix bundles have been used to identify amino acids responsible for the CB_2 selectivity of WIN 55212-2 (Song *et al.*, 1999). One distinguishing feature of these CB_1 and CB_2 models is the hypothesis that aminoalkylindole agonists interact primarily with Helices 3, 4 and 5, while traditional, non-traditional and endogenous cannabinoids are hypothesized to interact with helices 3, 5, 6, and 7 (see above). The ability of each structural class to displace another class (as demonstrated in radioligand binding assays) is hypothesized to be due to spatial overlap between binding sites, but not necessarily the sharing of the same primary interaction sites.

Huffman *et al.* (1994) have shown that the morpholino group of the AAIs can be replaced by an alkyl chain without loss of CB_1 affinity or efficacy. Kumar *et al.* (1995) have shown that rigid AAI analogs that lack a carbonyl oxygen still exhibit high CB_1 affinity and efficacy. These results suggest that hydrogen bonding may not be a key interaction of AAIs at CB_1. Instead, because the AAIs are aromatic ligands, Reggio *et al.* (1998) hypothesized that aromatic stacking interactions may be important for AAI binding at the CB_1 receptor. A receptor region (TMH 3-4-5) rich in aromatic residues was identified as a likely AAI interaction region. This binding region is consistent with the report by Shire *et al.* (1999) that the TMH 4-E2 loop-TMH 5 region of the CB_1 receptor contains amino acids important for WIN 55212-2 binding. Docking studies in a CB_1 transmembrane helix bundle revealed that WIN 55212-2 (in its S-trans conformation) can participate in aromatic stacking interactions with F3.25(189), W5.43(279) and F3.36(200) and that WIN 55212-3, the inactive stereoisomer of WIN 55212-2, will not fit in this binding site due to a steric clash with TMH 5. In CB_2, WIN 55212-2 (in its S-trans conformation), can participate in aromatic stacking interactions with F3.36(117), W5.43(194), F3.25(106) and F5.46(197). On the whole, **3** participates in greater aromatic interactions in CB_2. This is largely due to an additional aromatic residue in the binding pocket F5.46(197), a residue unique to CB_2. The importance of F5.46 (197) to the CB_2 selectivity of WIN 55212-2 has been confirmed by mutation studies (Song *et al.*, 1999). Proposed binding sites are included in Figures 17.4 and 17.5.

In their study of residues on TMH 3 that may be involved in the CB_2 selectivity of WIN 55212-2, Chin and co-workers (1999) mutated CB_1 residues, G3.31(195) and A3.34(198) to their corresponding residues in CB_2, i.e., G3.31(195)S and A3.34(195)M. The A3.34(195)M mutation in CB_1 could not be distinguished from WT. The G3.31(195)S mutation in CB_1, on the other hand, resulted in a four-fold increase in WIN 55212-2 affinity. The authors hypothesize that the serine at 3.31 in CB_2 can hydrogen bond with WIN 55212-2 and thereby enhance its affinity. This interpretation of the results is in opposition to the hypothesized AAI binding site proposed by Song and co-workers (1999) discussed above, because residue 3.31 does not face TMHs 4-5, but faces TMH 2 instead. It is possible that the four-fold gain in affinity documented by Chin *et al.* (1999) for the CB_1 G3.31(195)S mutant results from conformational differences between TMH 3 in CB_1 vs. CB_2. Because serine residues have been shown to bend alpha helices (Ballesteros *et al.* 2000b), it is possible that the G3.31(195)S mutation altered the conformation of TMH 3 in CB_1, causing a slight change in the TMH 3-4-5 AAI binding pocket. The small enhancement in affinity (4 fold) reported by Chin *et al.* (1999) is more consistent with a subtle difference in TMH 3 conformation which improves an existing interaction, than with the gain of a hydrogen bonding interaction between two uncharged partners. Fersht (1988) estimates that such a hydrogen bond is worth about 1.8 kcal/mol, which would translate into a larger (~20 fold) gain in affinity.

Mutations of functional importance

Jin and co-workers (1999) reported that truncation of the CB_1 receptor at residue 7.73(418) almost completely abolished desensitization, but did not affect agonist activation of GRK-3. In contrast, truncation at residues 439 and 460 did not significantly affect GRK-3- and β-arrestin 2-dependent desensitization. A deletion mutant (Δ418-439) did not desensitize, indicating residues within this region are important for GRK-3- and β-arr 2-mediated desensitization. Phosphorylation in this region was likely involved in desensitization, because mutation of either of two putative phosphorylation sites (S7.81(426)A or S7.85(430)A) significantly attenuated desensitization (Jin *et al.*, 1999).

As discussed previously, working with stably transfected human CB_1 and CB_2 receptors in HEK293 cells, Tao and Abood (1998) reported that a highly conserved aspartic acid residue D2.50(163) in CB_1 and D2.50(80) in CB_2 is required for communication with G-proteins and signal transduction. Mutation of this residue in each receptor to glutamic acid (E) or asparagine (N), did not disrupt high affinity agonist binding with the exception of WIN 55212-2 in the CB_1 D2.50(163)E and D2.50(163)N mutants.

In contrast to these results, Roche and co-workers (1999) found that in AtT20, a rat CB_1 D2.50(164)N mutant bound WIN 55212-2 with an affinity matching WT CB_1. This mutant inhibited cAMP and Ca^{+2} currents with a potency and efficacy equal to WT. Because in other receptors a non-functional D2.50N mutant can be rescued by a concomitant N7.49D mutation, Roche and co-workers also performed a N7.49(394)D mutation. However, activation of the D2.50(164)N/N7.49(394)D mutant did not potentiate KIR (inwardly rectifying potassium channel) current, nor did this double mutant internalize. Roche and co-workers (1999) showed that D2.50(164) (which corresponds to D2.50(163) in the human CB_1 sequence) is necessary for potentiation of KIR current and internalization of the receptor, but not necessary for agonist binding, inhibition of cAMP production, inhibition of calcium ion currents, or activation of p42/44 MAP kinase. Furthermore, these investigators showed that CB_1 receptor internalization is not necessary for MAP kinase activation.

One of the most conserved sequence motifs across GPCRs (including the CB receptors) is the E/DRY motif at the intracellular end of TMH 3. This motif has been shown to be of functional importance in many GPCRs. Rhee and co-workers (2000a) used transiently transfected COS cells to study the DRY motif in CB_2. These investigators found that D3.49(130) was essential for the capacity of the receptor to bind cannabinoid agonists and that Y3.51(132) has a role in receptor downstream signaling. Surprisingly, mutation of R3.50(133) to A only partially reduced cannabinoid induced inhibition of adenylate cyclase.

Rhee and co-workers (2000b) recently explored the importance of W4.50(158) and W4.64(172) in the CB_2 receptor. These investigators found that only the conservative mutation W4.50(158)F retained WT binding and signaling activities. On the otherhand, mutation to tyrosine or alanine resulted in loss of ligand binding capacity. Mutation of W4.64(172) to other aromatic amino acids (i.e. Y or F), retained cannabinoid binding and signaling (inhibition of adenylate cyclase), whereas removal of the aromatic side chain by mutation to alanine or leucine, eliminated agonist binding. Results for mutation at position 4.64(172) are very consistent with the role of residue W4.64 in Rho (Palczewski et al., 2000). Here an aromatic cluster is formed by W175 at the extracellular end of TMH 4 and F203 and Y207 near the extracellular end of TMH 5 in Rho. This aromatic cluster in Rho may serve to stabilize the relative positions of TMHs 4 and 5 on their extracellular sides. It is very likely that in CB_2, the W4.64(172) mutations to non-aromatic residues reported by Rhee and co-workers (2000b) resulted in a significant change in the binding pocket due to the re-positioning of TMHs 4 and 5 as the result of the mutation. This may explain why mutation to non-aromatic amino acids at position W4.64(172) resulted in loss of ligand binding.

Activating mutations or peptides of the CB receptors

It has been suggested that activating mutations release the native receptor from an inactive conformation (i.e. its R form) (Kjelsburg *et al.*, 1992). Several groups have suggested that specific interactions among residues in the helical bundle of GPCRs are involved in constraining native receptors in inactive conformations (Robinson *et al.*, 1992; Scheer *et al.*, 1996; Groblewski *et al.*, 1997; Lin *et al.*, 1997; Ballesteros *et al.* 2001). One study of W6.48 in the Thyrotropin-Releasing Hormone receptor showed that an aromatic cluster on Helices 5 and 6 constrains the receptor in an inactive conformation (Colson *et al.*, 1998). In fact, many studies have documented that mutation of key residues in TMH 6 of the GPCRs results in the constitutive activation of these receptors [see Kosugi *et al.*, 1998 and references therein]. Abadjii and co-workers (1999) mutated two residues near the intracellular loop 3 (I3)-TMH 6 border of CB_1 L6.33(341)A and A6.34(342)L. This resulted in agonist independent enhancement of cAMP levels that could be inhibited by treatment with the CB_1 specific inverse agonist, SR 141716A. Studies of the effect of cholera toxin (CTX) vs. pertussis toxin (PTX) on adenylyl cyclase accumulation produced upon the binding of CP 55,940 to WT and the L6.33(341)A/A6.34(342)L mutant suggested that the mutant couples to G_s more than WT. Further experiments revealed that the mutant displays partial constitutive activity with G_s, not G_i, its primary coupling agent.

The movement of transmembrane helices (particularly TMHs 3 and 6) that accompanies the R to R* transition (see above) may produce movement of the intracellular loops. In fact, lines of evidence point to the importance of certain intracellular loops in activating G-proteins (i.e. the second and third intracellular loops, as well as the carboxyl terminus, see Strader *et al.*, 1994). A site-directed spin-labeling study of rhodopsin, for example, has indicated that photo-activation results in patterns of structural changes that can be interpreted in terms of movements of helices that extend into the IL-3 (TMH 5–6) loop (Altenbach *et al.*, 1996). Initial studies with the β-adrenergic receptor demonstrated that removal of residues from either end of the third intracellular domain (i.e. loop between TMH 5 and TMH 6) uncoupled the receptor from G_s (Dixon *et al.*, 1987). Thus, the picture which is emerging is that the R → R * transition is accompanied by movements of transmembrane helices. These movements, in turn, affect the positions/conformations of intracellular loops. Resultant changes in the conformation of key loop and C-terminal regions then permit receptor interaction with G-protein.

Recent work reported by Howlett *et al.* (1998a) has shown that the juxta-membrane portion of the carboxyl terminus of the CB_1 receptor (residues 401–417 in rat CB_1, which corresponds to 400–416 in human CB_1) can directly activate $G_{i/o}$ proteins in a pertussis toxin sensitive manner and that a peptide from the amino-side IL3 (residues 301–317 in rat CB_1, which corresponds to 300–316 in human CB_1) also may interact with G_i proteins leading to inhibition of adenylate cyclase. Figure 17.4 includes both of these regions of the CB_1 sequence. Data from the rat CB_1 401–417 peptide suggest that this peptide can associate with the G-protein at the same site as does the receptor. Fourier Transform analysis of the periodicity in the human CB_1 sequence reported by Bramblett *et al.* (1995) identified the

403–413 segment in the carboxyl terminus to be an intracellular helical segment (See Figure 17.4). Howlett and Mukhopadhyay have reported that in CD studies under solvent conditions that would mimic an aqueous or a hydrophobic environment, the rat CB_1 401–417 peptide maintained a random conformation. The anionic detergent sodium dodecyl sulfate promoted a more helical conformation, which may mimic an interaction of the peptide with phospholipid head groups or with a negatively charged patch on an associated protein. These authors concluded that conformational changes in the juxtamembrane C-terminal region of CB_1 may modulate G-protein interactions (Howlett *et al.*, 1998b; Mukhopadhyay *et al.*, 1999).

Mukhopadhyay and co-workers (1999) studied the sequence and structural requirements of this C-terminal juxtamembrane region 7.56(401) to 7.72(417) of the rat CB_1 receptor (which corresponds to 7.56(400) to 7.72(416) in human CB_1) that previously had been shown to be capable of coupling to G_i in the absence of the CB_1 receptor (Howlett *et al.*, 1998a). Through truncation and mutation of Arg401 to norleucine, these investigators found that R7.56(401) (rat CB_1 sequence number) was of primary importance in determining apparent affinity for G_i protein. These investigators suggested that R7.56(401) may be required for docking with the G-protein. Mukhopadhyay and co-workers (1999) also showed that the single charged residue on the C-terminal side E7.72(417) (rat sequence number) can be truncated or neutralized by Leu substitution with some loss of activity. These investigators concluded that E7.72 may participate in docking with G-protein, as well.

CONCLUSIONS

New experimental evidence reviewed here shows that cannabinoid agonists can signal at a particular cannabinoid receptor through more than one G-protein to more than one effector system. While computer models of the CB_1 and CB_2 receptors have proven their utility in identifying amino acids integral to the binding of certain ligand structural classes, there is much that has yet to be addressed in these models. The proposal made by Glass and Northup (1999) that individual CB agonists may stabilize the CB_1 receptor in different active state conformations and that each resultant receptor conformation may then be selectively recognized by specific G-proteins is an intriguing hypothesis. The development of both R and R* models of each CB receptor sub-type described here is the first step towards testing Glass and Northup's hypothesis. The next step will be to determine through experiment and modeling which amino acids are key to the production of activation by each CB agonist ligand structural class. The separation of cannabinoid agonist pharmacophores shown by mutation studies reviewed here clearly suggests that the mechanism by which the aminoalkylindoles activate the CB receptors may be quite different from that of the other three agonist ligand structural classes. Such differences may ultimately be found to explain the agonist selective G-protein coupling documented by Prather and co-workers (2000), Glass and Northup (1999), Bonhaus and co-workers (1998) and others.

ACKNOWLEDGEMENTS

The author wishes to acknowledge the technical assistance of Beverly Brookshire and Dow Hurst in preparing this manuscript. Financial support from the National Institute on Drug Abuse (Grant DA-03934) is also gratefully acknowledged.

REFERENCES

Abadjii, V., Lucas-Lenard, J. M., Chin, C. and Kendall, D. A. (1999) "Involvement of the carboxyl terminus of the third intracellular loop of the cannabinoid CB_1 receptor in constitutive activation of G_S," *Journal of Neurochemistry* **72**: 2032–2038.

Altenbach, C., Yang, K., Farrens, D. L., Farahbakhsh, Z. T., Khorana, H. G. and Hubbell, W. L. (1996) "Structural features and light-dependent changes in the cytoplasmic interhelical E-F loop region of rhodopsin: A site directed spin-labeling study," *Biochemistry* **35**: 12470–12478.

Arnis, S., Fahmy, K., Hofman, K. P. and Sakmar, T. P. (1994) "A conserved carboxylic acid group mediates light-dependent proton uptake and signaling by rhodopsin," *The Journal of Biological Chemistry* **269**: 23879–23881.

Baldwin, J. (1993) "The probable arrangement of the helices in G-protein coupled receptors," *The EMBO Journal* **12**: 1693–1703.

Baldwin, J. M., Schertler, G. F. X. and Unger, V. M. (1997) "An alpha-carbon template for the transmembrane helices in the rhodopsin family of G-protein coupled receptors," *Journal of Molecular Biology* **272**: 144–164.

Ballesteros, J. A. and Weinstein, H. (1995) "Integrated methods for the construction of three dimensional models and computational probing of structure function relations in G-protein coupled receptors," in *Methods in Neuroscience*, edited by P. M. Conn and S. C. Sealfon, pp. 366–428.

Ballesteros, J., Kitanovic, S., Guarnieri, F., Davies, P., Fromme, B. J., Konvicka, K., Chi, L., Millar, R. P., Davidson, J. S., Weinstein, H. and Sealfon, S. C. (1998) "Functional microdomains in G-protein coupled receptors," *The Journal of Biological Chemistry* **273**: 10445–10453.

Ballesteros, J. A., Norris, J. B., Guarnieri, F., Hurst, D. P. and Reggio, P. H. (2000a) "The importance of the helix 6 $\beta XX\beta$ motif to ligand binding and activation of the cannabinoid receptors," *Biophysical Journal* **78**: 40A.

Ballesteros, J. A., Deupi, X., Olivella, M., Haaksma, E. E. J. and Pardo, L. (2000b) "Serine and threonine residues bend α-helices in the $\chi_1 = g^-$ conformation," *Biophysical Journal* **79**: 2754–2760.

Ballesteros, J. A., Jensen, A. D., Liapakis, G., Rasmussen, S. G. F., Shi, L., Gether, U. and Javitch, J. A. (2001) "Activation of the β_2 adrenergic receptor involves disruption of an ionic lock between the cytoplasmic ends of transmembrane segments 3 and 6," *The Journal of Biological Chemistry* **276**: 29171–29177.

Berglund, B. A., Boring, D. L., Wilken, G. H., Makriyannis, A. and Howlett, A. C. (1998) "Structural requirements for arachidonylethanolamide interaction with CB_1 and CB_2 cannabinoid receptors: Pharmacology of the carbonyl and ethanolamide groups," *Prostaglandins Leukotrienes and Essential Fatty Acids* **59**: 111–118.

Birnbaumer, L., Abramowitz, J. and Brown, A. M. (1990) "Receptor effector coupling by G-proteins," *Biochimica et Biophysica Acta* **1031**: 163–224.

Bonhaus, D. W., Chang, L. K., Kwan, J. and Martin, G. R. (1998) "Dual activation and inhibition of adenylyl cyclase by cannabinoid receptor agonists: Evidence for agonist-specific trafficking of intracellular responses," *The Journal of Pharmacology and Experimental Therapeutics* **287**: 884–888.

Bouaboula, M., Poinot-Chazel, C., Bourrié, B., Canat, X., Calandra, B., Rinaldi-Carmona, M., LeFur, G. and Casellas, P. (1995) "Activation of mitogen activated protein kinases by stimulation of the central cannabinoid receptor CB_1," *The Biochemical Journal* **312**: 637–641.

Bouaboula, M., Perrachon, S., Milligan, L., Carrat, X., Rinaldi-Carmona, M., Portier, M., Barth, F., Calandra, B., Pecceu, F., Lupker, J., Maffrand, J.P., Lefur, G. and Casellas, P. (1997) "A selective inverse agonist for central cannabinoid receptor inhibits mitogen-activated protein kinase activation stimulated by insulin or insulin-like growth factor 1. Evidence for a new model of receptor/ligand interactions," *The Journal of Biological Chemistry* **272**: 22330–22339.

Bramblett, R. D., Panu, A. M., Ballesteros, J. A. and Reggio, P.H. (1995) "Construction of a 3D model of the cannabinoid CB_1 receptor: Determination of helix ends and helix orientation," *Life Science* **56**: 1971–1982.

Breivogel, C., Sim, L. and Childers, S. (1997) "Regional differences in cannabinoid receptor/G-protein coupling in rat brain," *The Journal of Pharmacology and Experimental Therapeutics* **282**: 1632–1642.

Breivogel, C. S., Selley, D. E. and Childers, S. R. (1998) "Cannabinoid receptor agonist efficacy for stimulating [^{35}S]GTPγS binding to rat cerebellar membranes correlates with agonist-induced decreases in GDP affinity," *The Journal of Biological Chemistry* **273**: 16865–16873.

Burkey, T. H., Quock, R. M., Consroe, P., Ehlert, F. J., Hosohata, Y., Roeske, W. R. and Yamamura, H. I. (1997a) "Relative efficacies of cannabinoid CB_1 receptor agonists in the mouse brain," *European Journal of Pharmacology* **336**: 295–298.

Burkey, T., Quock, R., Consroe, P., Roeske, W. and Yamamura, H. (1997b) "Delta-9-tetrahydrocannabinol is a partial agonist of cannabinoid receptors in mouse brain," *European Journal of Pharmacology* **323**: R3–R4.

Calandra, B., Portier, M., Kerneis, A., Delpech, M., Carillon, C., Le Fur, G., Ferrara, P. and Shire, D. (1999) "Dual intracellular signaling pathways mediated by the human cannabinoid CB_1 receptor," *European Journal of Pharmacology* **374**: 445–455.

Caulfield, M. P. and Brown, D. A. (1992) "Cannabinoid receptor agonists inhibit Ca^{2+} current in NG108-15 neuroblastoma cells via a pertussis toxin sensitive mechanism," *British Journal of Pharmacology* **106**: 231–232.

Chakrabarti, S., Yang, W., Law, P. Y. and Loh, H. H. (1997) "The mu-opioid receptor down regulates differently from the delta-opioid receptor: Requirements of a high affinity receptor/G-protein complex formation," *Molecular Pharmacology* **52**: 105–113.

Chin, C., Lucas-Lenard, J., Abadjii, V. and Kendall, D. A. (1998) "Ligand binding and modulation of cyclic AMP levels depend on the chemical nature of residue 192 of the human cannabinoid receptor 1," *Journal of Neurochemistry* **70**: 366–373.

Chin, C.-N., Murphy, J. W., Huffman, J. W. and Kendall, D. A. (1999) "The third trans-membrane helix of the cannabinoid receptor plays a role in the selectivity of aminoalkylindoles for CB_2, peripheral cannabinoid receptor," *The Journal of Pharmacology and Experimental Therapeutics* **291**: 837–844.

Cohen, G. B., Oprian, D. D. and Rao, V. R. (1994) "Rhodopsin mutation G90D and a molecular mechanism for congenital night blindness," *Nature* **367**: 639–641.

Colson, A. O., Perlman, J. H., Jinsi-Parimoo, A., Nussenzveig, D. R., Osman, R. and Gershengorn, M. C. (1998) "A hydrophobic cluster between transmembrane helices 5 and 6 constrains the thyrotropin-releasing hormone receptor in an inactive conformation," *Molecular Pharmacology* **54**: 968–978.

Compton, D. R., Gold, L. H., Ward, S. J., Balster, R. L. and Martin, B. R. (1992) "Aminoalkylindole analogs: Cannabimimetic activity of a class of compounds structurally distinct from

Δ^9-tetrahydrocannabinol," *The Journal of Pharmacology and Experimental Therapeutics* **263**: 1118–1126.

Conklin, B. R. and Bourne, H. R. (1993) "Structural elements of G_α subunits that interact with $G_{\beta\gamma}$, receptors, and effectors," *Cell* **73**: 631–641.

D'Ambra, T. E., Estep, K. G., Bell, M. R., Eissenstat, M. A., Josef, K. A., Ward, S. J., Haycock, D. A., Baizman, E. R., Casiano, F. M., Beglin, N. C., Chippari, S. M., Grego, J. D., Kullnig, R. K. and Daley, G. T. (1992) "Conformationally restrained analogues of pravadoline: Nanomolar potent, enantioselective, (aminoalkyl) indole agonists of the cannabinoid receptor," *Journal of Medicinal Chemistry* **35**: 124–135.

Devane, W. A., Hanus, L., Breuer, A., Pertwee, R. G., Stevenson, L. A., Griffin, G., Gibson, D., Mandelbaum, A., Etinger, A. and Mechoulam, R. (1992) "Isolation and structure of a brain constituent that binds to the cannabinoid receptor," *Science* **258**: 1946–1949.

Devane, W. A., Dysarz III, F. A., Johnson, M. R., Melvin, L. S. and Howlett, A. C. (1988) "Determination and characterization of a cannabinoid receptor in rat brain," *Molecular Pharmacology* **34**: 605–613.

Dixon, R. A. F., Sigal, I. S., Rands, E., Register, R. B., Candelore, M. R., Blake, A. D. and Strader, C. D. (1987) "Ligand binding to the β-adrenergic receptor involves its rhodopsin-like core," *Nature* **326**: 73–77.

Donnelly, D. and Cogdell, R. J. (1993) "Predicting the point at which transmembrane helices protrude from the bilayer: A model of the antenna complexes from photosynthetic bacteria," *Protein Engineering* **6**: 629–635.

Dutta, A. K., Ryan, W., Thomas, B. F., Singer, M., Compton, D. R., Martin, B. R. and Razdan, R. K. (1997) "Synthesis, pharmacology and molecular modeling of novel 4-alkyloxy indole derivatives related to cannabimimetic aminoalkyl indoles (AAIs)," *Bioorganic and Medicinal Chemistry* **5**: 1591–1600.

Facci. L., Toso, R. D., Romanello, S., Buriani, A., Skaper, S. D. and Leon, A. (1995) "Mast cells express a peripheral cannabinoid receptor with differential sensitivity to anandamide and palmitoylethanolamide," *Proceedings of the National Academy of Sciences of the USA* **92**: 3376–3380.

Farrens, D. L., Altenbach, C., Yang, K., Hubbell, W. L. and Khorana, H. G. (1996) "Requirement of rigid-body motion of transmembrane helices for light activation of rhodopsin," *Science* **274**: 768–770.

Felder, C. C., Briley, E. M., Axelrod, J., Simpson, J. T., Mackie, K. and Devane, W. A. (1993) "Anandamide, an endogenous cannabimimetic eicosanoid, binds to the cloned human cannabinoid receptor and stimulates receptor-mediated signal transduction," *Proceedings of the National Academy of Sciences of the USA* **90**: 7656–7660.

Felder, C. C., Joyce K. E., Briley, E. M., Mansouri, J., Mackie, K., Blond, O., Lai, Y., Ma, A. L. and Mitchell, R. L. (1995) "Comparison of the pharmacology and signal transduction of the human CB_1 and CB_2 receptors," *Molecular Pharmacology* **48**: 443–450.

Felder, C. C., Joyce, K. E., Briley, E. M., Glass, M., Mackie, K. P., Fahey, K. J., Culinan, G. J., Hunden, D. C., Johnson, D. W., Chaney, M. O., Koppel, G. A. and Brownstein, M. (1998) "LY320135, a novel cannabinoid CB_1 receptor antagonist, unmasks coupling of the CB_1 receptor to stimulation of cAMP accumulation," *The Journal of Pharmacology and Experimental Therapeutics* **284**: 291–297.

Fersht, A. R. (1988) "Relationships between apparent binding energies measured in site-directed mutagenesis experiments and energetics of binding and catalysis," *Biochemistry* **27**: 1577–1580.

Findlay, J. and Eliopoulus, E. (1990) "Three-dimensional modelling of G-protein-linked receptors," *Trends in Pharmacological Science* **11**: 492–499.

Findlay, J. B. C. and Donnelly, D. (1995) "GT-Pases in biology," edited by B. Dicky, and L. Birnbaumer. Heidelberg: Springer-Verlag.

Fu, D., Ballesteros, J. A., Weinstein, H., Chen, J. and Javitch, J. A. (1996) "Residues in the seventh membrane-spanning segment of the dopamine D2 receptor accessible in the binding-site crevice," *Biochemistry* **35**: 11278–11285.

Gaoni, Y. and Mechoulam, R. (1964) "Hashish III: The isolation, structure, and partial synthesis of an active constituent of hashish," *Journal of the American Chemical Society* **86**: 1646–1647.

Gerard, C. M., Mollereau, C., Vassart, G. and Parmentier, M. (1991) "Molecular cloning of a human brain cannabinoid receptor which is also expressed in testis," *The Biochemical Journal* **279**: 129–134.

Gether, U., Lin, S., Ghanouni, P., Ballesteros, J. A., Weinstein, H. and Kobilka, B. K. (1997) "Agonists induce conformational changes in transmembane domains III and VI of the β_2 adrenoceptor," *The EMBO Journal* **16**: 6737–6747.

Glass, M. and Felder, C. C. (1997) "Concurrent stimulation of cannabinoid CB_1 and dopamine D2 receptors augments cAMP accumulation in striatal neurons: Evidence for a G_s-linkage to the CB_1 receptor," *The Journal of Neuroscience* **17**: 5327–5333.

Glass, M. and Northup, J. K. (1999) "Agonist selective regulation of G-proteins by cannabinoid CB_1 and CB_2 receptors," *Molecular Pharmacology* **56**: 1362–1369.

Gouldson, P., Calandra, B., Legoux, P., Kerneis, A., Rinaldi-Carmona, M., Barth, F., LeFur, G., Ferrara, P. and Shire, D. (2000) "Mutational analysis and molecular modeling of the antagonist SR 144528 binding site on the human cannabinoid CB_2 receptor," *European Journal of Pharmacology* **401**: 17–25.

Greenwood, M. T., Hukovic, N., Kuman, U., Panetta, R., Hjorth, S. A., Srikant, C. B. and Patel, Y. C. (1997) "Ligand binding pocket of the human somatostatin receptor 5: Mutational analysis of the extracellular domains," *Molecular Pharmacology* **52**: 807–814.

Griffin, G., Atkinson, P. J., Showalter, V. M., Martin, B. R. and Abood, M. E. (1998) "Evaluation of cannabinoid receptor agonists and antagonists using the guanosine-5′-O- (3-[^{35}S] thio]-triphosphate binding assay in rat cerebellar membranes," *The Journal of Pharmacology and Experimental Therapeutics* **285**: 553–560.

Griffin, G., Tao, Q. and Abood, M. E. (2000) "Cloning and pharmacological characterization of the rat CB_2 cannabinoid receptor," *The Journal of Pharmacology and Experimental Therapeutics* **292**: 886–894.

Groblewski, T., Maigret, B., Larguier, R., Lombard, C., Bonnafous, J. C. and Marie, J. (1997) "Mutation of Asn^{111} in the third transmembrane domain of the AT_{1A} angiotensin II receptor induces its constitutive activation," *The Journal of Biological Chemistry* **272**: 1822–1826.

Guarnieri, F. and Weinstein, H. (1996) "Conformational memories and the exploration of biologically relevant peptide conformations: An illustration for gonadotropin-releasing hormone," *Journal of the American Chemical Society* **118**: 5580–5589.

Gudermann, T., Schöneberg, T. and Schultz, G. (1997) "Functional and structural complexity of signal transduction via G-protein coupled receptors," *Annual Review of Neuroscience* **20**: 399–427.

Han, M. and Smith, O. (1995) "Structural model of rhodopsin and bathorhodopsin based on NMR constraints," *Biophysical Journal* **68**: A21.

Henderson, R., Baldwin, J. M., Ceska, T. A., Zemlin, F., Beckmann, E. and Downing, K. H. (1990) "Model for the structure of bacteriorhodopsin based on high-resolution electron cryo-microscopy," *Journal of Molecular Biology* **213**: 899–929.

Hibert, M. F., Trumpp-Kallmeyer, S., Brunvels, A. and Hoflack, J. (1991) "Three-dimensional models of neurotransmitter G-binding protein coupled receptors," *Molecular Pharmacology* **40**: 8–15.

Ho, B. Y., Uezono, Y., Takada, S., Takase, I. and Izumi, F. (1999) "Coupling of the expressed cannabinoid CB_1 and CB_2 receptors to phospholipase C and G-protein coupled inwardly rectifying K^+ channels," *Receptors and Channels* **6**: 363–374.

Hosohata, K., Quock, R. M., Hosohata, Y., Burkey, T. H., Makriyannis, A., Consroe, P., Roeske, W. R. and Yamamura, H. I. (1997a) "AM630 is a competitive cannabinoid receptor antagonist in the guinea pig brain," *Life Science* **61**: PL115–PL118.

Hosohata, Y., Quock, R. M., Hosohata, K., Makriyannis, A., Consroe, P., Roeske, W. R. and Yamamura, H. I. (1997b) "AM630 antagonism of cannabinoid-stimulated [^{35}S] GTP-γS binding in mouse brain," *European Journal of Pharmacology* **321**: R1–R2.

Houston, D. B. and Howlett, A. C. (1998) "Differential receptor-G-protein coupling evoked by dissimilar cannabinoid receptor agonists," *Cell Signal* **10**: 667–674.

Howlett, A. (1995) "Pharmacology of cannabinoid receptors," *Annual Review of Pharmacology and Toxicology* **35**: 607–634.

Howlett, A. C., Song, C., Berglund, B. A., Wilken, G. H. and Pigg, J. J. (1998a) "Characterization of CB$_1$ cannabinoid receptors using receptor peptide fragments and site-directed antibodies," *Molecular Pharmacology* **53**: 504–510.

Howlett, A. C., Mukhopadhyay, S., Cossik, S. M. and Welsh, W. J. (1998b) "Interaction of the CB$_1$ cannabinoid receptor with G-proteins: Importance of the C-terminal region," *Archives of Pharmacology* **358**: 2121.

Huffman, J. W., Dai, D., Martin, B. R. and Compton, D. R. (1994) "Design, synthesis and pharmacology of cannabimimetic indoles," *Biorganic and Medicinal Chemistry Letters* **4**: 563–566.

Huffman, J. W., Yu, S., Showalter, V., Abood, M. E., Wiley, J. L., Compton, D. R., Martin, B. R., Bramblett, R. D. and Reggio, P. H. (1996) "Synthesis and pharmacology of a very potent cannabinoid lacking a phenolic hydroxyl with high affinity for the CB$_2$ receptor," *Journal of Medicinal Chemistry* **39**: 3875–3877.

Huffman, J. W., Yu, S., Liddle, J., Wiley, J. L., Abood, M., Martin, B. R. and Aung, M. M. (1999) "3-(1′-1′-Dimethylbutyl)-1-deoxy-Δ^8-THC and related compounds: Synthesis of selective ligands for the CB$_2$ receptor," *Biorganic and Medicinal Chemistry* **7**: 2905–2914.

Hutchins, C. (1994) "Three-dimensional models of the D1 and D2 dopamine receptors," *Endocrine Journal* **2**: 7–23.

Jarai, Z., Wagner, J. A., Varga, K., Lake, K. D., Compton, D. R., Martin, B. R., Zimmer, A. M., Bonner, T. I., Buckley, N. E., Mezey, E., Razdan, R. K., Zimmer, A. and Kunos, G. (1999) "Cannabinoid-induced mesenteric vasodilation through an endothelial site distinct from CB$_1$ or CB$_2$ receptors," *Proceedings of the National Academy of Sciences of the USA* **96**: 14136–14141.

Javitch, J. A., Fu, D., Chen, J. and Karlin, A. (1995a) "Mapping the binding-site crevice of the dopamine D2 receptor by the substituted-cysteine accessibility method," *Neuron* **14**: 825–831.

Javitch, J. A., Fu, D. and Chen, J. (1995b) "Residues in the fifth membrane-spanning segment of the dopamine D2 receptor exposed in the binding-site crevice," *Biochemistry* **34**: 16433–16439.

Javitch, J. A., Fu, D., Liapakis, G. and Chen, J. (1997) "Constitutive activation of the β_2 adrenergic receptor alters the orientation of its sixth membrane-spanning segment," *The Journal of Biological Chemistry* **272**: 18546–18549.

Javitch, J. A., Ballesteros, J. A., Weinstein, H. and Chen, J. (1998) "A cluster of aromatic residues in the sixth membrane-spanning segment of the dopamine D2 receptor is accessible in the binding-site crevice," *Biochemistry* **37**: 998–1006.

Javitch, J. A., Ballesteros, J. A., Chen, J., Chiappa, V. and Simpson, M. M. (1999) "Electrostatic and aromatic microdomains within the binding site crevice of the D2 receptor: Contributions of the second membrane-spanning segment," *Biochemistry* **38**: 7961–7968.

Javitch, J. A., Shi, L., Simpson, M. M., Chen, J., Chiappa, V., Visiers, I., Weinstein, H. and Ballesteros, J. A. (2000) "The fourth transmembrane segment of the dopamine D2 receptor:

accessibility in the binding site crevice and position in the transmembrane bundle," *Biochemistry* **39**: 12190–12199.

Jensen, A. D., Guarnieri, F., Rasmussen, S. G. F., Asmar, F., Ballesteros, J. A., and Gether, G. (2001) "Agonist-induced conformational changes at the cytoplasmic side of TM 6 in the β_2 adrenergic receptor mapped by site-selective fluorescent labeling," *The Journal of Biological Chemistry* **276**: 9279–9290.

Jin, W., Brown, S., Roche, J. P., Hsieh, C., Celver, J. P., Kovoor, A., Chavkin, C. and Mackie, K. (1999) "Distinct domains of the CB_1 cannabinoid receptor mediate desensitization and internalization," *Journal of Neuroscience* **19**: 3773–3780.

Kearn, C. S., Greenberg, M. J., DiCamelli, R., Kurzawak, K. and Hillard, C. J. (1999) "Relationship between ligand binding affinities for the cerebellar cannabinoid receptor CB_1 and the induction of GDP/GTP exchange," *Journal of Neurochemistry* **72**: 2379–2387.

Kjelsberg, M. A., Cotecchia, S., Ostrowski, J., Caron, M. G. and Lefkowitz, R. J. (1992) "Constitutive activation of the $\alpha_{1\beta}$ adrenergic receptor by all amino acids substitutions at a single site: Evidence for a region which constrains receptor activation," *The Journal of Biological Chemistry* **267**: 1430–1433.

Kobilka, B. (1992) "Adrenergic receptors as models for G-protein coupled receptors," *Annual Review of Neuroscience* **51**: 87–114.

Kosugi, S., Mori, T. and Shenker, A. (1998) "An anionic residue at position 564 is important for maintaining the inactive conformation of human lutropin/choriogonadotropin receptor," *Molecular Pharmacology* **53**: 894–901.

Krebs, A., Villa, C., Edwards, P. C. and Schertler, G. F. X. (1998) "Characterization of an improved two-dimensional p 22_12_1 crystal from bovine rhodopsin," *Journal of Molecular Biology* **282**: 991–1003.

Kumar, V., Alexander, M. D., Bell, M. R., Eissenstat, M. A., Casiano, F. M., Chippari, S. M., Haycock, D. A., Lutinger, D. A., Kuster, J. E., Miller, M. S., Stevenson, J. I. and Ward, S. J. (1995) "Morpholinoalkylindenes as antinociceptive agents: Novel cannabinoid receptor agonists," *Biorganic and Medicinal Chemistry Letters* **5**: 381–386.

Laakkonen, L., Li, W., Perlman, J. H., Guarnieri, F., Osman, R., Moeller, K. D. and Gershengorn, M. C. (1996) "Restricted analogues provide evidence of a biologically active conformation of thyrotropin-releasing hormone," *Molecular Pharmacology* **49**: 1092–1096.

Landsman, R. S., Burkey, T. H., Consroe, P., Roeske, W. R. and Yamamura, H. I. (1997) "SR 141716A is an inverse agonist at the human cannabinoid CB_1 receptor," *European Journal of Pharmacology* **334**: R1–R2.

Landsman, R. S., Makriyannis, A., Deng, H., Consroe, P., Roeske, W. R. and Yamamura, H. I. (1998) "AM630 is an inverse agonist at the human cannabinoid CB_1 receptor," *Life Sciences* **62**: PL109–PL113.

Lee, T. W., Wise, A., Cotecchia, S. and Milligan, C. (1996) "A constitutively active mutant of the α-1B-adrenergic receptor can cause greater agonist-dependent down regulation of the G-protein G_9 alpha and G_{11} alpha than the wild-type receptor," *The Biochemical Journal* **320**: 78–86.

Leff, P., Scaramillini, C., Law, C. and McKechnie, K. (1997) "A three state model of agonist action," *Trends in Pharmacological Sciences* **18**: 355–362.

Lin, S. W. and Sakmar, T. P. (1996) "Specific tryptophan UV-absorbance changes are probes of the transition of rhodopsin to its active state," *Biochemistry* **35**: 11149–11159.

Lin, Z. L., Shenker, A. and Pearlstein, R. (1997) "A model of the lutropin/choriogonadotropin receptor: Insights into the structural and functional effects of constitutively activating mutations," *Protein Engineering* **10**: 501–510.

Mackie, K. and Hille, B. (1992) "Cannabinoids inhibit N-type calcium channels in neuroblastoma-glioma cells," *Proceedings of the National Academy of Sciences of the USA* **89**: 3825–3829.

Mackie, K., Devane, W. A. and Hille, B. (1993) "Anandamide, an endogenous cannabinoid, inhibits calcium currents as a partial agonist in N18 neuroblastoma cells," *Molecular Pharmacology* **44**: 498–503.

Mackie, K., Lai, Y., Westenbroek, R. and Mitchell, R. (1995) "Cannabinoids activate an inwardly rectifying potassium conductance and inhibit Q-type calcium currents in AtT20 cells transfected with rat brain cannabinoid receptors," *The Journal of Neuroscience* **15**: 6552–6561.

Mahmoudian, M. (1997) "The cannabinoid receptor: Computer-aided molecular modeling and docking of ligand," *Journal of Molecular Graphics and Modelling* **15**: 149–153.

Maneuf, Y. P. and Brotchie, J. M. (1997) "Paradoxical action of cannabinoid WIN 55,212-2 in stimulated and basal cyclic AMP accumulation in rat globus pallidus slices," *British Journal of Pharmacology* **120**: 1397–1398.

Matsuda, L. A., Lolait, S. J., Brownstein, M. J., Young, A. C. and Bonner, T. I. (1990) "Structure of a cannabinoid receptor and functional expression of the cloned cDNA," *Nature* **346**: 561–564.

Mechoulam, R. and Gaoni, Y. (1967) "Recent advances in the chemistry of hashish," *Fortschr. Chem. Org. Naturst.* **25**: 175–213.

Mechoulam, R., Ben-Shabat, S., Hanus, L., Ligumsky, M., Kaminski, N. E., Schatz, A. R., Gopher, A., Almoy, S., Martin, B. R., Compton, D. R., Pertwee, R. G., Griffin, G., Bayewitch, M., Barg, J. and Vogel, Z. (1995) "Identification of an endogenous 2-monoglyceride, present in canine gut, that binds to cannabinoid receptors," *Biochemical Pharmacology* **50**: 83–90.

Melvin, L. S., Milne, G. M., Johnson, M. R., Wilken, G. H. and Howlett, A. C. (1995) "Structure-activity relationships defining the ACD-tricyclic cannabinoids: Cannabinoid receptor binding and analgesic activity," *Drug Design and Discovery* **13**: 155–166.

Milligan, G. and Bond, R. A. (1997) "Inverse agonism and the regulation of receptor number," *Trends in Pharmacological Sciences* **18**: 468–474.

Mukhopadhyay, S., Cowsik, S.M., Lynn, A. M., Welsh, W. J. and Howlett, A. C. (1999) "Regulation of G_i by the CB_1 cannabinoid receptor C-terminal juxtamembrane region: Structural requirements determined by peptide analysis," *Biochemistry* **38**: 3447–3455.

Munro, S., Thomas, K. L. and Abu-Sharr, M. (1993) "Molecular characterization of a peripheral receptor for cannabinoids," *Nature* **365**: 61–65.

Nakanishi, K., Zhang, H., Lerro, K. A., Takekuma, S., Yamamoto, T., Lien, T. H., Sastry, L., Baek, D.-J., Moquin-Pattey, C., Boehrn, M. F., Derguini, F. and Gawinowicz, M. A. (1995) "Photoaffinity labeling of rhodopsin and bacteriorhodopsin," *Biophysical Chemistry* **56**: 13–22.

Neubig, R. R. and Sklar, L. A. (1993) "Subsecond modulation of formyl peptide-linked guanine nucleotide-binding proteins by guanosine 5′-O-(3-thio)triphosphate in permeabilized neutrophils," *Molecular Pharmacology* **43**: 734–740.

Palczewski, K., Kumasaka, T., Hori, T., Behnke, C. A., Motoshima, H., Fox, B. A., LeTrong, I., Teller, D. C., Okada, T., Stenkamp, R. E., Yamamoto, M. and Miyano, M. (2000) "Crystal structure of rhodopsin: A G-protein coupled receptor," *Science* **289**: 739–745.

Pan, X. H., Ikeda, S. R. and Lewis, D. L. (1996) "Rat brain cannabinoid receptor modulates N-type Ca^{2+} channels in a neuronal expression system," *Molecular Pharmacology* **49**: 707–714.

Pan, X., Ikeda, S. R. and Lewis, D. L. (1998) "SR 141716A acts as an inverse agonist to increase neuronal voltage-dependent Ca^{2+} currents by reversal of tonic CB_1 receptor activity," *Molecular Pharmacology* **54**: 1064–1072.

Pardo, L., Ballesteros, J. A., Osman, R. and Weinstein, H. (1992) "On the use of transmembrane domains of bacteriorhodopsin as a template for modeling the three-dimensional structure of guanine nucleotide-binding regulatory protein coupled receptors," *Proceedings of the National Academy of Sciences of the USA* **89**: 4009–4012.

Parent, J. L., LeGouill, C., Rola-Pleszczynski, M. and Stankova, J. (1996) "Mutation of an aspartate at position 63 in the human platelet-activating factor receptor augments binding affinity but abolishes G-protein-coupling and inositol phosphate production," *Biochemical and Biophysical Research Communications* **219**: 968–975.

Paterlini, G., Portoghese, P. S. and Ferguson, D. M. (1997) "Molecular simulation of dynorphin A-(1–10) binding to extracellular loop 2 of the κ-opioid receptor. A model for receptor activation," *Journal of Medicinal Chemistry* **40**: 3254–3262.

Pearlman, D. A., Case, D. A., Caldwell, J., Siebel, G. L., Singh, C., Weiner, P. and Kollman, P. A. (1995) "Amber 4.0," *Department of Pharmaceutical Chemistry*, University of California, San Francisco.

Perlman, J. H., Thaw, C. N., Laakkonen, L., Bowers, C. Y., Osman, R. and Gershengorn, M. C. (1994) "Hydrogen bonding interaction of thyrotropin-releasing hormone (TRH) with transmembrane tyrosine 106 of the TRH receptor," *The Journal of Biological Chemistry* **269**: 1610–1613.

Pertwee, R. G., Stevenson, L. A. and Griffin, G. (1992a) "Cross-tolerance between delta-9-tetrahydrocannabinol and the cannabimimetic agents CP55,940, WIN 55212-2 and anandamide," *British Journal of Pharmacology* **105**: 980–984.

Pertwee, R. G., Stevenson, L. A., Elrick, D. B., Mechoulam, R. and Corbett, A. D. (1992b) "Inhibitory effects of certain enantiomeric cannabinoids in the mouse vas deferens and the myenteric plexces preparation of guinea pig small intestine," *British Journal of Pharmacology* **105**: 980–984.

Pertwee, R. (1993) "The evidence for the existence of cannabinoid receptors," *General Pharmacology* **24**: 811–824.

Pertwee, R., Griffin, G., Fernando, S., Li, X., Hill, A. and Makriyannis, A. (1995) "AM630, a competitive cannabinoid antagonist," *Life Science* **56**: 1949–1955.

Pertwee, R. G. (1997) "Pharmacology of cannabinoid CB_1 and CB_2 receptors," *Pharmacology and Therapeutics* **74**: 129–180.

Pogozheua, I. D., Lomizi, A. L. and Mosberg, H. I. (1997) "The transmembrane 7-α-bundle of rhodopsin: Distance geometry calculations with hydrogen bonding constraints," *Biophysical Journal* **70**: 1963–1985.

Portier, M., Rinaldi-Carmona, M., Pecceu, F., Combes, T., Poinot-Chazel, C., Calandra, B., Barth, F., LeFur, G. and Casellas, P. (1999) "SR 144528, an antagonist for the peripheral cannabinoid receptor that behaves as an inverse agonist," *The Journal of Pharmacology and Experimental Therapeutics* **288**: 582–589.

Prather, P. L., Martin, N. A., Breivogel, C. S. and Childers, S. R. (2000) "Activation of cannabinoid receptors in rat brain by WIN 55212-2 produces coupling to multiple G-protein α-subunits with different potencies," *Molecular Pharmacology* **57**: 1000–1010.

Rasmussen, S. G. F., Jensen, A. D., Liapakis, G., Ghanouni, P., Javitch, J. A. and Gether, U. (1999) "Mutation of a highly conserved aspartic acid in the β_2 adrenergic receptor: Constitutive activation, structural instability, and conformational rearrangement of transmembrane segment 6," *Molecular Pharmacology* **56**: 175–184.

Reggio, P. H., Panu, A. M. and Miles, S. (1993) "Characterization of a region of steric interference at the cannabinoid receptor using the active analog approach," *Journal of Medicinal Chemistry* **36**: 1761–1771.

Reggio, P. H., Basu-Dutt, S., Barnett-Norris, J., Castro, M. T., Hurst, D. P., Seltzman, H. H., Roche, M. J., Gilliam, A. F., Thomas, B. F., Stevenson, L. A., Pertwee, R. G. and Abood, M. E. (1998) "The bioactive conformation of aminoalkylindoles at the cannabinoid CB_1 and CB_2 receptors: Insights gained from E and Z naphthylidene indenes," *Journal of Medicinal Chemistry* **41**: 5177–5187.

Reggio, P., Norris, J., Ballesteros, J., Guarnieri, F. and Hurst, D.P. (2000) "The importance of the helix 6 βXXβ motif to ligand binding and activation of the cannabinoid receptors,"

2000 Symposium on the Cannabinoids, International Cannabinoid Research Society, Burlington, VT, p. 4.

Rhee, M.-H., Bayewitch, M., Avidor-Reiss, T., Levy, R. and Vogel, Z. (1998) "Cannabinoid receptor activation differentially regulates the various adenylyl cyclase isozymes," *Journal of Neurochemistry* **71**: 1525–1534.

Rhee, M.-H., Nevo, I., Levy, R. and Vogel, Z. (2000a) "Role of the highly conserved Asp-Arg-Tyr motif in signal transduction of the CB_2 cannabinoid receptor," *FEBS Letters* **466**: 300–304.

Rhee, M.-H., Nevo, I., Bayewitch, M. L., Zagoory, O., and Vogel, Z. (2000b) "Functional role of tryptophan residues in the forurth transmembrane domain of the CB_2 cannabinoid receptor," *Journal of Neurochemistry* **75**: 2485–2491.

Rinaldi-Carmona, M., Barth, F., Héaulme, M., Shire, D., Calandra, B., Congy, C., Martinez, S., Maruani, J., Néliat, G., Caput, D., Ferrara, P., Soubrié, P., Brelière, J.-C. and LeFur, G. (1994) "SR 141716A, a potent and selective antagonist of the brain cannabinoid receptor," *FEBS Letters* **350**: 240–244.

Rinaldi-Carmona, M., Calandra, B., Shire, D., Bouaboula, M., Oustric, D., Barth, F., Casellas, P., Ferrara, P. and LeFur, G. (1996) "Characterization of two cloned human CB_1 cannabinoid receptor isoforms," *The Journal of Pharmacology and Experimental Therapeutics* **278**: 871–878.

Rinaldi-Carmona, M., Barth, M., Millan, J., Derocq, J.M., Casellas, P., Congy, C., Oustric, D., Sarran, M. Bouboula, M., Calandra, B., Poutier, M., Shire, D., Breliére, J.C. and LeFur, G. (1998) "SR 144528, the first potent and selective antagonist of the CB_2 cannabinoid receptor," *The Journal of Pharmacology and Experimental Therapeutics* **284**: 644–650.

Robinson, P. R., Cohen, G. B., Zhukovsky, E. A. and Oprian D. D. (1992) "Constitutively active mutants of rhodopsin," *Neuron* **9**: 719 -725.

Roche, J. P., Bounds, S., Brown, S. and Mackie, K. (1999) "A mutation in the second transmembrane region of the CB_1 receptor selectively disrupts G-protein signaling and prevents receptor internalization," *Molecular Pharmacology* **56**: 611–618.

Ross, R. A., Brockie, H. C., Fernando, S. R., Saha, B., Razdan, R. K. and Pertwee, R. G. (1998) "Comparison of cannabinoid binding sites in guinea-pig forebrain and small intestine," *British Journal of Pharmacology* **125**: 1345–1351.

Samama, P., Cotechia, S., Costa, T. and Lefkowitz, R. J. (1993) "A mutation-induced activated state of the β_2 adrenergic receptor extending the ternary complex model," *The Journal of Biological Chemistry* **268**: 4625–4636.

Scheer, A., Fanelli, F., Costa, T., DeBenedetti, P. G. and Cotecchia, S. (1996) "Constitutively active mutants of the α_{1B} – adrenergic receptor: Role of highly conserved polar amino acids in receptor activation," *The EMBO Journal* **15**: 3566–3578.

Schertler, G. F. X., Villa, C. and Henderson, R. (1993) "Projection structure of rhodopsin," *Nature* **362**: 770–772.

Schertler, G. F. X., Unger, V. M. and Hargrave, P. A. (1995) "The structure of rhodopsin obtained by cryo-electron microscopy to 8Å resolution," *Biophysical Journal* **68**: A330.

Sealfon, S. C., Chi, L., Ebersole, B. J., Rodic, V., Zhang, D., Ballesteros, J. A. and Weinstein, H. (1995) "Related contribution of specific helix 2 and 7 residues to conformational activation of the serotonin 5-HT2A receptor," *The Journal of Biological Chemistry* **270**: 16683–16688.

Sealfon, S. C., Weinstein, H. and Millar, R. P. (1997) "Molecular mechanisms of ligand receptor interaction with the gonadotropin-releasing hormone receptor," *Endocrine Reviews* **18**: 180–205.

Selley, D., Stark, S., Sim, L. and Childers, S. (1996) "Cannabinoid receptor stimulation of guanosine-5′-O-(3-[35S]thio) triphosphate binding in rat brain membranes," *Life Sciences* **59**: 659–668.

Sheikh, S. P., Zuyaga, T. A., Lichtarge, O., Sakmar, T. P. and Bourne, H. R. (1996) "Rhodopsin activation blocked by metal ion-binding sites linking transmembrane helices C and F," *Nature* **383**: 347–350.

Shire, D., Carillon, C., Kaghad, M., Calandra, B., Rinaldi-Carmona, M., LeFur, G., Caput, D. and Ferrara, P. (1995) "An amino-terminal variant of the central cannabinoid receptor resulting from alternative splicing," *The Journal of Biological Chemistry* **270**: 3726–3731; *Erratum* (1996) **271**: 33706b.

Shire, D., Calandra, B., Rinaldi-Carmona, M., Oustric, D., Pessèque, B., Bonnin-Cabanne, O., LeFur, G., Caput, D. and Ferrara, P. (1996a) "Molecular cloning, expression and function of the murine CB_2 peripheral cannabinoid receptor," *Biochimica et Biophysica Acta* **1307**: 132–136.

Shire, D., Calandra, B., Delpech, M., Dumont, X., Kaghad, M., LeFur, G., Caput, D. and Ferrar, P. (1996b) "Structural features of the central cannabinoid CB_1 receptor involved in the binding of the specific CB_1 antagonist SR 141716A," *The Journal of Biological Chemistry* **271**: 6941–6946.

Shire, D., Calandra, B., Bouaboula, M., Barth, F., Rinaldi-Carmona, M., Casellas, P. and Ferrara, P. (1999) "Cannabinoid receptor interactions with the antagonists SR 141716A and SR 144528," *Life Sciences* **65**: 627–635.

Showalter, V. M., Compton, D. R., Martin, B. R. and Abood, M. E. (1996) "Evaluation of binding in a transfected cell line expressing a peripheral cannabinoid receptor (CB_2): Identification of cannabinoid receptor subtype selective ligands," *The Journal of Pharmacology and Experimental Therapeutics* **278**: 989–999.

Sim, L., Selley, D. and Childers, S. (1995) "*In vitro* autoradiography of receptor-activated G-proteins in rat brain by agonist-stimulated guanylyl 5'-[gamma-[35S]thio]-triphosphate binding," *Proceedings of the National Academy of Sciences of the USA* **92**: 7242–7246.

Sim-Selley, L. J., Brunk, L. K. and Selley, D. E. (2001) "Inhibitory effects of SR 141716 on G-protein activation in rat brain," *European Journal of Pharmacology* **414**: 135–143.

Sim, L. J., Hampson, R. E., Deadwyler, S. A. and Childers, S. R. (1996) "Effects of chronic treatment with delta-9-tetrahydrocannabinol on cannabinoid-stimulated [35S] GTPγS autoradiography in rat brain," *The Journal of Neuroscience* **16**: 8057–8066.

Skaper, S. D., Buriani, A., DalToso, R., Petrelli, L., Romanello, S., Facci, L. and Leon, A. (1996) "The ALIAmide palmitoylethanolamide and cannabinoids, but not anandamide, are protective in a delayed postglutamate paradigm of excitotoxic death in cerebellar granule neurons," *Proceedings of the National Academy of Sciences of the USA* **93**: 3984–3989.

Smith, P. B., Compton, D. R., Welch, S. P., Razdan, R. K., Mechoulam, R. and Martin, B. R. (1994) "The pharmacological activity of anandamide, a putative endogenous cannabinoid, in mice," *The Journal of Pharmacology and Experimental Therapeutics* **270**: 219–227.

Song, Z. H. and Bonner, T. I. (1996) "A lysine residue of the cannabinoid receptor is critical for receptor recognition by several agonists, but not WIN 55212," *Molecular Pharmacology* **49**: 891–896.

Song, Z. H., Slowey, C.-A., Hurst, D. P. and Reggio, P. H. (1999) "The difference between the CB_1 and CB_2 cannabinoid receptors at position 5.46 is crucial for the selectivity of WIN 55,212-2 for CB_2," *Molecular Pharmacology* **56**: 834–840.

Stella, N., Schweitzer, P. and Piomelli, D. (1997) "A second endogenous cannabinoid that modulates long-term potentiation," *Nature* **388**: 773–778.

Strader, C. D., Fong, T. M., Tota, M. R. and Dixon, R. A. F. (1994) "Structure and function of G-protein coupled receptors," *Annual Review of Biochemistry* **63**: 101–132.

Strahs, D. and Weinstein, H. (1997) "Comparative modeling and molecular dynamics studies of the δ, κ, and μ opioid receptors," *Protein Engineering* **10**: 1019–1038.

Steiner, H., Bonner, T. I., Zimmer, A. M., Kitai, S. T. and Zimmer, A. (1999) "Altered gene expression in striatal projection neurons in CB_1 cannabinoid receptor knockout mice," *Proceedings of the National Academy of Sciences of the USA* **96**: 5786–5790.

Surprenant, A., Horstman, D. A., Akbarali, H. and Limbird, L. E. (1992) "A point mutation on the α_2-adrenoreceptor that blocks coupling to potassium but not calcium channels," *Science* **257**: 977–980.

Tao, Q. and Abood, M. E. (1998) "Mutation of a highly conserved aspartate residue in the second transmembrane domain of the cannabinoid receptors, CB_1 and CB_2, disrupts G-protein coupling," *The Journal of Pharmacology and Experimental Therapeutics* **285**: 651–658.

Tao, Q., McAllister, S. D., Andreassi, J., Nowell, K. W., Cabral, G. A., Hurst, D. P., Reggio, P. H. and Abood, M. E. (1999) "Role of a conserved lysine residue in the CB_2 cannabinoid receptor: Evidence for subtype specificity," *Molecular Pharmacology* **55**: 605–613.

Teeter, M. M., Froimowitz, M., Stec, B. and Durand, C. J. (1994) "Homology modeling of the dopamine D_2 receptor and its testing by docking agonists and tricyclic antagonists," *Journal of Medicinal Chemistry* **37**: 2874–2888.

Thomas, B. F., Adams, I. B., Mascarella, S. W., Martin, B. R. and Razdan, R. K. (1996) "Structure–activity analysis of anandamide analogs: Relationship to a cannabinoid pharmacophore," *Journal of Medicinal Chemistry* **39**: 471–479.

Tong, W., Collantes, E. R. and Welsh, W. J. (1998) "Derivation of a pharmacophore model for anandamide using constrained conformational searching and comparative molecular field analysis," *Journal of Medicinal Chemistry* **41**: 4207–4215.

Unger, V. M., Hargrave, P. A. and Schertler, G. F. X. (1995) "Localization of the transmembrane helices in the three-dimensional structure of frog rhodopsin," *Biophysical Journal* **68**: A21.

Unger, V. M., Hargrave, P. A., Baldwin, J. M. and Schertler, G. F. X. (1997) "Arrangement of rhodopsin transmembrane α-helicies," *Nature* **389**: 203–206.

Wang, C. D., Buck, M. A. and Fraser, C. M. (1991) "Site-directed mutagenesis of alpha 2A-adrenergic receptors: Identification of amino acids involved in ligand binding and receptor activation by agonists," *Molecular Pharmacology* **40**: 168–179.

Ward, S. J., Baizman, E., Bell, M., Childers, S., D'Ambra, T., Eissenstat, M., Estep, K., Haycock, D., Howlett, A., Luttinger, D., Miller, M. and Pacheco, M. (1991) "Amino-alkylindoles (AAIs): A new route to the cannabinoid receptor?" *Proceedings of the 1990 Committee on Problems of Drug Dependence, NIDA Monograph* **105**: 425–426.

Ward, W. H. J., Timms, D. and Fersht, A. R. (1990) "Protein engineering and the study of structure–function relationships in receptors," *Trends in Pharmacological Sciences* **17**: 280–284.

Westkaemper, R. B. and Glennon, R. A. (1993) "Molecular graphics models of members of the 5-HT2 Subfamily: 5-HT2A, 5-HT2B, and 5-HT2C receptors," *Medicinal Chemistry Research* **3**: 317–334.

Xie, Y. B., Wang, H. and Segaloff, D. L. (1990) "Extracellular domain of lutropin/choriogonadotropin receptor expressed in transfected cells binds choriogonadotropin with high affinity," *The Journal of Biological Chemistry* **265**: 21411–21414.

Zhang, D. and Weinstein, H. (1993) "Signal transduction by a 5-HT_2 receptor: A mechanistic hypothesis from molecular dynamics simulations of the three-dimensional model of the receptor complexed to ligands," *Journal of Medicinal Chemistry* **36**: 934–938.

Zhang, D. and Weinstein, H. (1994) "Polarity conserved positions in transmembrane domains of G-protein coupled receptors and bacteriorhodopsin," *FEBS Letters* **337**: 207–212.

Zhou, W., Flanagan, C., Ballesteros, J. A., Konvicka, K., Davidson, J. S., Weinstein, H., Millar, R. P. and Sealfon, S. C. (1994) "A reciprocal mutation supports helix 2 and helix 7 proximity in the gonadotropin-releasing hormone receptor," *Molecular Pharmacology* **45**: 165–170.

Zimmer, A., Zimmer, A. M., Hohmann, A. G., Herkenham, M. and Bonner, T. I. (1999) "Increased mortality, hypoactivity and hypoalgesia in cannabinoid CB_1 receptor knockout mice," *Proceedings of the National Academy of Sciences of the USA* **96**: 5780–5785.

Chapter 18

Endocannabinoid proteins and ligands

Sonya L. Palmer, Atmaram D. Khanolkar and Alexandros Makriyannis

ABSTRACT

During the last decade, the field of cannabinoid biology has observed important advances that have propelled it to the forefront of biomedical research. These new developments have also provided an opportunity to examine the physiological and biochemical events related to the use and abuse of cannabis as well as elucidating the biological role of the endogenous cannabinoid ligands (endocannabinoids). The biological targets for endocannabinoids include the two known cannabinoid receptors (CB_1 and CB_2), the enzyme anandamide amidase (ANAase), and the carrier protein referred to as the anandamide transporter (ANT).

The identification of arachidonylethanolamide (anandamide, AEA) and more recently, 2-arachidonylglycerol (2-AG) as endogenous cannabinoids has been an important development in cannabinoid research which has led to the identification of two proteins associated with cannabinoid physiology in addition to the CB_1 and CB_2 receptors. These proteins are anandamide amidase (ANAase), an enzyme responsible for the hydrolytic breakdown of anandamide and 2-arachidonylglycerol and the anandamide transporter (ANT), a carrier protein involved in the transport of anandamide across the cell membrane. Evidence obtained so far suggests that these two proteins, in combination, are responsible for the termination of the biological actions of anandamide. Also, the discovery of anandamide has led to the development of a novel class of more selective agents possessing somewhat different pharmacological properties than the cannabinoids. A number of such analogs have now been reported many of which possess markedly improved cannabinoid receptor affinities and metabolic stabilities when compared to those of the parent ligand. Generally, anandamide and all known analogs exhibit significant selectivities with high affinities for the CB_1 receptor and modest to very low affinities for CB_2. Within the last two or three years, pharmacological and biochemical studies have confirmed initial speculations that anandamide is either a neuromodulator or neurotransmitter and have significantly advanced our understanding of cannabinoid biochemistry. This summary seeks to define the pharmacology of endocannabinoids and to focus on the structure–activity relationships (SAR) of anandamide for the CB_1 cannabinoid receptor.

Key Words: anandamide, endocannabinoids, cannabinoid receptors, anandamide amidase, anandamide transporter

INTRODUCTION

Cannabis (marijuana), the mixture of natural cannabinoids found in *Cannabis sativa*, is one of the most frequently used drugs among recreational substance abusers.

The medicinal uses of marijuana have been recognized for many centuries, but the isolation and structure elucidation of the active ingredient in cannabis (Gaoni and Mechoulam, 1964), (−)-Δ^9-tetrahydrocannabinol **1**, (-)-Δ^9-THC, Figure 18.1), afforded the initial opportunity to investigate the pharmacological properties of cannabinoids. These terpenoid compounds have been demonstrated to exert analgesic, antiemetic and anticonvulsive effects, hyperactivity, hypothermia, lowering of intraocular pressure and immunosupression (Abood and Martin, 1992; Dewey, 1986; Hollister, 1986). Some of these effects are due to the direct interaction with cannabinoid receptors (Pertwee, 1993). Currently, two cannabinoid receptors have been identified. The CB_1 receptor was first discovered in mammalian brain (Devane et al., 1988) and the CB_2 receptor was found in the periphery and in immune cells such as B and T lymphocytes (Munro et al., 1993; Brower, 2000). Evidence for a third cannabinoid receptor subtype has started to emerge (Jarai et al., 1999). Direct evidence for the existence of the CB_1 receptor was demonstrated using membrane homogenates and tissue section binding assays for the characterization and localization of a cannabinoid receptor in the brain using the potent radiolabeled ligand [^{3}H] CP-55,940 (Devane et al., 1988). The [^{3}H] CP-55,940 binding site was found to be saturatable, and to have high affinity and specificity for agonist ligands (Devane et al., 1988). Herkenham and his coworkers first used autoradiographic techniques to reveal a heterogeneous distribution of the CB_1 receptor throughout the brain (Herkenham et al., 1990). The binding pattern was conserved across several mammalian species, including humans, with the greatest abundance of sites in the basal ganglia, hippocampus, and cerebellum (Bidaut-Russell et al., 1990; Herkenham et al., 1990).

ENDOCANNABINOID LIGANDS

In 1992, an arachidonic acid derivative, anandamide (**2**, AEA, arachidonylethanolamide, Figure 18.1), was first identified as an endogenous ligand for CB_1 (Devane et al., 1992). AEA is lipophilic and highly unsaturated with four non-conjugated cis double bonds, sensitive to both oxidation and hydrolysis, and difficult to isolate (Di Marzo and Fontana, 1995). Anandamide, originally isolated from porcine (Devane et al., 1992) and bovine (Johnson et al., 1993) brains, has been demonstrated to bind to the central cannabinoid receptor (CB_1) with a rather moderate affinity (K_i 61 nM) and has a low affinity for the CB_2 receptor (K_i 1930 nM) (Lin et al., 1998). AEA has also been found to inhibit both forskolin-stimulated adenylyl cyclase activity (Childers et al., 1994; Vogel et al., 1993) and voltage-dependent N-type calcium channels (Mackie et al., 1993). In addition, this endogenous cannabinoid produces the characteristic in vivo effects of cannabinoids such as antinociception, hypothermia, analgesia, and catalepsy in mice (Crawley et al., 1993; Fride and Mechoulam, 1993; Smith et al., 1994) and rats (Adams et al., 1995a,b). These findings strongly suggest a role for anandamide as an endogenous ligand involved in the modulation of behavior, memory, cognition, and pain perception.

Since the discovery of AEA, a number of reports have been published concerning the biochemical, pharmacological and behavioral properties of anandamide showing that this novel neuromodulator is a cannabinoid agonist (Fride and Mechoulam,

1993). Thus, like other cannabimimetic agents, anandamide was shown to modulate cAMP levels (Vogel *et al.*, 1993), inhibit N-type calcium currents through a pertussis toxin-sensitive pathway (Mackie *et al.*, 1993) and also to inhibit the electrically evoked twitch response of the mouse vas deferens (Pertwee *et al.*, 1992, 1995).

Shortly after the discovery of anandamide, two other endogenous unsaturated fatty acid ethanolamides were also isolated and shown to be cannabinoid receptor agonists (Hanus *et al.*, 1993). These are docosatetraenylethanolamide (**3**) and homo-γ-linolenylethanolamide (**4**). Their chemical structures along with that of anandamide are shown in Figure 18.1. Like anandamide, these two agents have specific binding affinities for the CB_1 receptor, and inhibited both forskolin-stimulated adenylyl cyclase and the mouse vas deferens twitch response, although the K_is and IC_{50}s were higher than anandamide (Barg *et al.*, 1995; Pertwee *et al.*, 1994). These findings suggest that there is a family of anandamides with cannabimimetic and neuromodulatory properties in the central nervous system (CNS). However, the polyunsaturated fatty acyl moiety of these amides appears to be essential for their binding to the CB_1 receptor (Felder *et al.*, 1993; Vogel *et al.*, 1993).

More recently, 2-arachidonylglycerol (**5**, 2-AG, Figure 18.1), isolated from intestinal tissue, was shown to be another endogenous cannabinoid (Mechoulam *et al.*, 1995; Stella, Schweitzer and Piomelli, 1997) present in brain in concentration approximately 170 times higher than that of anandamide. 2-arachidonylglycerol is a monoglyceride found to bind to both the CB_1 and CB_2 receptors (Di Marzo *et al.*, 1998; Mechoulam *et al.*, 1996; Stella *et al.*, 1997). It produced the typical effects of Δ^9-THC, including antinociception, immobility, immunomodulation, and inhibition of electrically evoked contractions of the mouse vas deferens (Lee *et al.*, 1995; Mechoulam *et al.*, 1995; Sugiura *et al.*, 1996a,b,c; Sugiura *et al.*, 1995). Another endogenous agonist for both CB_1 and CB_2 receptors is mead ethanolamide (**6**, eicosatrienoic acid, Figure 18.1). It inhibited forskolin-stimulated adenylyl cyclase activity with similar potency to that of anandamide, and inhibited N-type Ca^{2+} currents with a lower efficiency than anandamide (Priller *et al.*, 1995). Recent studies indicate that 2-AG induces a rapid increase in intracellular free calcium ions in HL-60 cells that express the cannabinoid receptor CB_2 (Sugiura *et al.*, 2000). In a recent publication it was shown that the effect of 2-AG was blocked by pretreatment of the cells with SR144528, a CB_2 receptor-specific antagonist, but not with SR141716A, a CB_1 receptor-specific antagonist, indicating that only the CB_2 receptor is involved in this cellular response. Work by Sugiura *et al.* suggests that the CB_2 receptor is originally a 2-arachidonoylglycerol receptor (Sugiura *et al.*, 2000).

Other endogenous ligands for CB_2 receptors were also discovered. It has been reported that palmitylethanolamide (**7**, PEA, Figure 18.1), an anti-inflammatory compound, interacts with the peripheral CB_2 cannabinoid receptor on mast cells, producing the same important non-psychotropic effects of cannabinoids such as anti-inflammatory and immunosuppressant activities (Facci *et al.*, 1995). However, more recently, evidence suggested that PEA may not interact primarily with CB_2 receptors, bringing into debate the cannabimimetic role of PEA (Lambert and Di Marzo, 1999; Lambert *et al.*, 1999). At present, the role of PEA is not yet fully elucidated. In general, these families of lipid modulators of the cannabinoid system are structurally dissimilar from the plant-derived cannabinoids. For this purpose they have been designated, by some authors, as endocannabinoids.

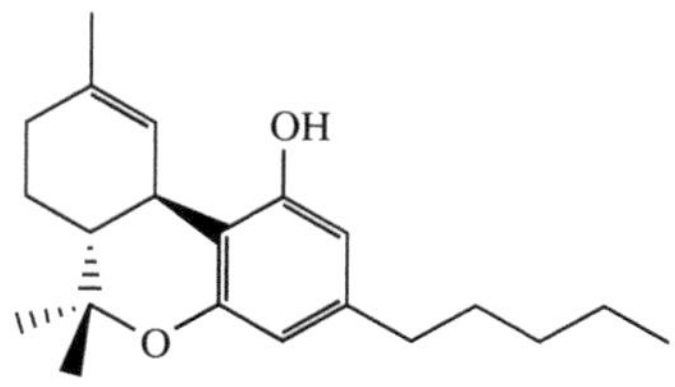

(1) (-)-Δ^9-Tetrahydrocannabinol (Δ^9-THC)

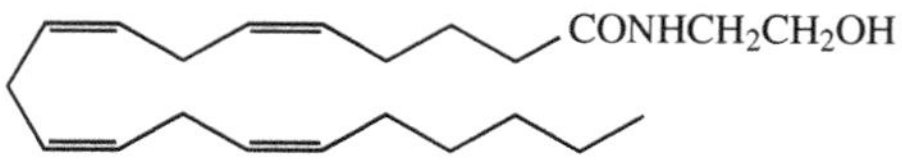

(2) Anandamide (AEA)

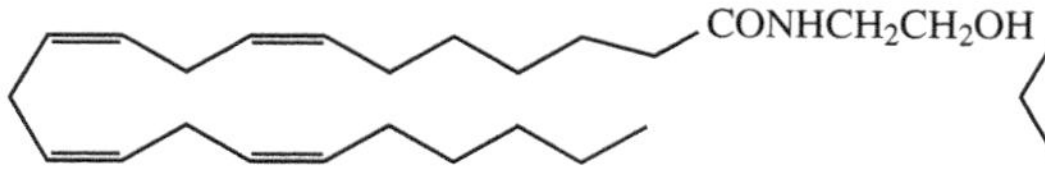

(3) Docosatetraenylethanolamide

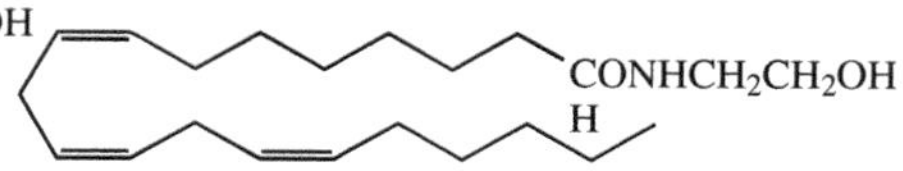

(4) Homo-γ-linolenylethanolamide

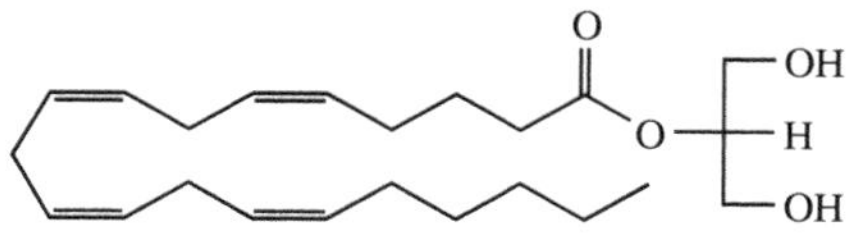

(5) 2-Arachidonylglycerol (2-AG)

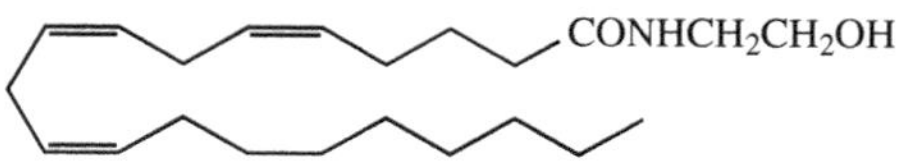

(6) Mead Ethanolamide

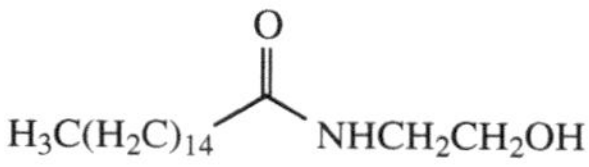

(7) Palmitylethanolamide

Figure 18.1 Natural and endogenous cannabinoid ligands.

PHARMACOLOGICAL EFFECTS OF ANANDAMIDE

The pharmacological effects of anandamide and its analogs *in vivo* and at the cellular level, as summarized in Table 18.1, have been assessed and compared to the actions of the naturally occurring THC (Berdyshev *et al.*, 1996; Hillard and Campbell, 1997; Jarbe *et al.*, 1998b; Mechoulam *et al.*, 1986; Schmid *et al.*, 1997; Sulcova *et al.*, 1998; Welch, 1997) in order to elucidate whether anandamide is a true cannabinoid receptor agonist. Hypothermia, antinociception, and hypomotility effects of anandamide were tested on mice and rats in immobility and open field studies (Fride and Mechoulam, 1993; Jarbe *et al.*, 1998a,b; Stein *et al.*, 1996). These effects are the common psychotropic effects caused by cannabinoids at low doses. It was thus shown that the pharmacological effects caused by anandamide require higher doses, have more rapid onset, and shorter durations of action when compared to those induced by THC (Smith *et al.*, 1994). In addition, anandamide also produces an endothelium-derived vasorelaxation effects which has

Table 18.1 Summary of endocannabinoid functional properties

Pharmacological actions	References
Hypothermia	(Crawley *et al.*, 1993; Fride and Mechoulam, 1993)
Antinociceptive	(Fride and Mechoulam, 1993; Smith *et al.*, 1994; Adams *et al.*, 1995a,b; Stein *et al.*, 1996)
Vasorelaxant	(Pate *et al.*, 1995; Randall *et al.*, 1996)
Hypotensive	(Varga *et al.*, 1995)
Cataleptic	(Fride and Mechoulam, 1993; Smith *et al.*, 1994)
Immunomodulatory effects	(Schwarz *et al.*, 1994; Lee *et al.*, 1995)
Inhibition of locomotor and rearing activity	(Fride and Mechoulam, 1993; Smith *et al.*, 1994; Romero *et al.*, 1995; Stein *et al.*, 1996)
Enhancement of muscimol-induced catalepsy	(Wickens and Pertwee, 1993)
Effects of ACTH and corticosterone secretion	(Weidenfeld *et al.*, 1994; Wenger *et al.*, 1995)
Inhibition of sperm acrosome reaction	(Schuel *et al.*, 1994)

Actions at the cellular level	References
Displacement of specific THC agonists	(Felder *et al.*, 1993; Vogel *et al.*, 1993; Sugiura *et al.*, 1996a,b,c)
Inhibition of adenylyl cyclase	(Van der Kloot, 1994; Barg *et al.*, 1995; Mechoulam *et al.*, 1996)
Inhibition of N-type calcium channels	(Felder *et al.*, 1993; Mackie *et al.*, 1993)
Activation of phospholipase A2	(Felder *et al.*, 1993)
Release of intracellular calcium channels	(Felder *et al.*, 1993; Venance *et al.*, 1995)
Reversal inhibition of L-type calcium channels	(Johnson *et al.*, 1993)
Inhibition of voltage-gated potassium channels	(Mackie *et al.*, 1995; Poling *et al.*, 1996)
Regulation of focal adhesion kinase	(Derkinderen *et al.*, 1996)
Modulation of protein kinase C	(De Petrocellis *et al.*, 1995)
Stimulation of MAP kinase	(Wartmann *et al.*, 1995)
Inhibition of twitch response	(Pertwee *et al.*, 1993; Pertwee *et al.*, 1994)
Inhibition of Q-type calcium channels	(Mackie *et al.*, 1995)

been reported by Randall *et al.* These effects were found to be inhibited by the highly selective CB_1 antagonist, SR141716A, indicating these effects are CB_1 receptor mediated (Randall *et al.*, 1996). Receptor binding studies on anandamide were performed in NG18 neuroblastoma cells and the chinese hamster ovary (CHO) cells transfected with the CB_1 receptor (Felder *et al.*, 1993; Vogel *et al.*, 1993). In these cells, anandamide displaced specific cannabinoid agonists from their binding sites, inhibited the formation of cyclic AMP and reduced N-type calcium current with K_is and IC_{50}s in the medium to high nanomolar range. These effects were not found in non-transfected cells. The inhibition of both adenylyl cyclase activity and N-type calcium current, which are the two typical cannabinoid receptor signaling functions by anandamide, were prevented by pertussis toxin pretreatment (Felder *et al.*, 1993; Vogel *et al.*, 1993). Furthermore, anandamide has been reported to play a role in the regulation of brain protein kinase C.

Anandamide was also found to demonstrate some cannabinoid actions which may not be receptor mediated, such as the stimulation of phospholipase A2 and the release of intracellular calcium (Felder *et al.*, 1993). These effects are observed only in the mid to high micromolar range (Mackie *et al.*, 1993). Additional studies suggest that anandamide inhibits shaker-related, voltage-gated, potassium channels (Poling *et al.*, 1996). Again, this action of anandamide was pertussis toxin insensitive, not coupled to a cannabinoid receptor, and could possibly be related to phospholipase D modulation (Poling *et al.*, 1996).

FUNCTIONS OF ENDOCANNABINOIDS AT THE CELLULAR LEVEL

Elucidating the cellular action of endocannabinoids can assist in the understanding and discovery of the functional significance of cannabinoid receptors in the neural systems as well as uncovering the therapeutic usefulness of cannabinoids. The known cellular actions of cannabinoids include; effects on ion channels, inhibition of adenylyl cyclase, stimulation of mitogen-activated protein kinase, and the inhibition of phospholipase C which leads to arachidonic acid (AA) accumulation. Studies have show that both CB_1 and CB_2 receptors are coupled to the inhibition of adenylyl cyclase and the stimulation of mitogen-activated protein kinase through pertussis toxin-sensitive G proteins (Caulfied *et al.*, 1993; Felder *et al.*, 1992; Mackie and Hille, 1992; Schatz *et al.*, 1992). However, the CB_1 receptor mediates several ion channels while the CB_2 receptor has not yet been shown to modulate the activity of ion channels. The observed cellular endocannabinoid effects are mediated, at least in part, by cannabinoid receptors (Di Marzo and Fontana, 1995; Felder *et al.*, 1992; Makriyannis, 1995; Makriyannis and Rapaka, 1990; White and Tansik, 1980) and are summarized in Table 18.1.

CANNABINOID EFFECTS ON ORGAN SYSTEMS

Understanding the effects of cannabinoids on the organ systems can help us to explore the pharmacological and potential therapeutic value of these compounds. Cannabinoids affect the functions of a wide variety of organ systems. Of these, the central nervous, immune, and cardiovascular systems have received particular attention.

Central nervous system (CNS)

There have been several studies aimed at elucidating the effects of natural cannabinoids on CNS. The cerebral blood flow and cerebral metabolic rate are two parameters of brain activity. Cannabinoid-induced changes in these parameters represent a change in brain function (Mathew and Wilson, 1993). Initial exposure to Δ^9-THC reduces global cerebral blood flow, while prolonged exposure produces increased global cerebral blood flow mainly in the frontal and left temporal regions (Mathew and Wilson, 1992, 1993; Mathew *et al.*, 1992). It has also been shown that the cerebral metabolic rate is increased by exposure to Δ^9-THC (Margulies and Hammer, 1991; Volkow and Fowler, 1993).

The observed effects with natural cannabinoids have sparked interest in delineating the role of endogenous cannabinoids in the CNS. Recent studies designed to evaluate the role of AEA in the CNS revealed that AEA affects body temperature and locomotor activity in rats, although with a shorter duration of action relative to Δ^9-THC (McGregor *et al.*, 1998). AEA has also been reported to be released extracellularly in the brain of rats via activation of membrane receptors such as the dopamine D2 receptors (Giuffrida *et al.*, 1999). Giuffrida *et al.* measured the release of AEA in the dorsal striatum of mobile rats by microdialysis and gas chromatography/mass spectrometry. Neural activity stimulated the release of AEA, but not that of 2-AG (Giuffrida *et al.*, 1999). Additionally, AEA release was increased eight fold when the rats were treated with a D2-like (D2, D3, D4) dopamine receptor agonist. Subsequently, the increase in AEA release in the dorsal striatum was suppressed using a D2-like receptor antagonist while administration of a D1-like (D1, D5) receptor agonist ultimately had no such effect on the levels of AEA released in the brain. These results suggest that functional interactions between endocannabinoid and dopaminergic systems may contribute to striatal signaling (Giuffrida *et al.*, 1999). The fact that AEA behaves as a neuromodulator by interacting with dopamine receptors may provide further insight into diseases such as Parkinson's disease and Tourette's Syndrome which occur due to abnormalities in striatal neuromodulation and provide new approach for therapeutic intervention.

Immune system

Δ^9-THC decreases the weight of lymphoid organs and high doses of cannabinoids affected the function of the stem cells and decreased the size of the spleen in rodents (Munson and Fehr, 1983). Cannabinoids can also affect the morphology of macrophages (Cabral *et al.*, 1991), phagocytic and spreading ability (Lopez-Cepero *et al.*, 1986; Spector and Lancz, 1991), superoxide production (Sherman *et al.*, 1991) tumor necrosis factor (Fisher-Stenger *et al.*, 1993; Zheng *et al.*, 1992), and interleukin release (Klein and Friedman, 1990; Shivers *et al.*, 1994). The effects of cannabinoids on B lymphocyte production and T lymphocyte production were investigated. Cannabinoids modulate the production of lymphocytes, however the clinical relevance of these effects is unclear (Dax *et al.*, 1989; Nahas and Ossweman, 1991; Wallace *et al.*, 1988). It has been suggested that AEA plays a role in the immune response to cannabinoids and bacterial endotoxins (Pestonjamasp and Burstein, 1998).

Cardiovascular system

Cannabinoids, as well as the endogenous ligand AEA, can produce tachycardia and orthostatic hypotension through the activation of the CB_1 cannabinoid receptors and lead to reduced platelet aggregation (Merritt *et al.*, 1980; Schaefer *et al.*, 1979). Administration of high doses of cannabinoids may result in pink eye due to dilation of blood vessels and increased heart rate with a concomitant peripheral vasodilation (Dewey, 1986).

AEA has been reported to produce a dose-dependent decrease in systemic blood pressure of anesthesized guinea pigs.(Calignano *et al.*, 1997) These effects can be suppressed by treatment with the CB_1 cannabinoid receptor antagonist SR 141716A

[N-(piperidin-1-yl)-5-(4-chlorophenyl)-1-(2,4-dichlorophenyl)-4-methyl-1H-pyra-zole-3-carboxamide x HCl]. Subsequently, the observed vasodepression induced by anandamide was significantly potentiated and prolonged by the anandamide transport inhibitor, N-(4-hydroxyphenyl) arachidonylethanolamide (AM404) (Calignano et al., 1997). These results suggest that anandamide transport participates in terminating the vascular actions of AEA.

Recently Jarai et al. have examined the potential of 2-AG to elicit hypotension. 2-AG was metabolically unstable but an analog of 2-AG caused hypotension in mice. The cardiovascular effects of 2-AG may be produced by a metabolite through a noncannabinoid mechanism, but the CB_1 receptor-mediated cardiovascular effects of a stable 2-AG ligand leaves open the possibility that endogenous 2-AG may elicit cardiovascular effects through CB_1 receptors (Jarai et al., 2000).

Several endocannabinoids, including AEA, have been found to be vasodilators but the mechanism by which they exert their actions is not fully elucidated. It was reported by Zygmunt et al. that the cardiovascular effects of anandamide may be mediated through the activation of vanilloid receptors (VR) on perivascular sensory nerves and not through cannabinoid receptors (Zygmunt et al., 1999). Other evidence also suggests that anandamide is an antagonist of vanilloid receptors (Smart et al., 2000). However, these postulates regarding the actions of anandamide through the vanilloid receptors remain to be validated.

Reproductive system

Natural cannabinoids such as Δ^9-THC, have been known to exert inhibitory effects on the reproductive system. The discovery of endocannabinoids brings into question their possible role in neuroendocrine regulation of reproduction. Wenger and colleagues studied the effects of AEA on reproduction in rats (Wenger et al., 1997) and showed that anandamide decreases serum luteinizing hormone (LH) and prolactin (PRL) levels in both male and female rats. Recently, the presence of cannabinoid receptors in human sperm was reported (Schuel et al., 1999) and AM356, a metabolically stable analog of anandamide, was shown to inhibit the acrosome reaction in sub-nanomolar doses (Schuel et al., 2000).

BIOSYNTHESIS AND METABOLISM OF ANANDAMIDE

Anandamide amidase (ANAse)

Although the data obtained to date indicate that anandamide is a neuromodulator, its exact physiological role has not yet been fully elucidated. A distinguishing feature of anandamide is that it undergoes facile enzyme catalyzed hydrolysis as demonstrated in rat brain homogenates and intact neurons (Desarnaud et al., 1995; Deutsch and Chin, 1993; Di Marzo et al., 1994). This membrane-bound enzyme, known as anandamide amidase (AEAase), anandamide amidohydrolase (ANAse), or fatty acid amidohydrolase (FAAH), has been characterized by several groups and partially purified from mammalian brains (Desarnaud et al., 1995; Di Marzo et al., 1994; Qin et al., 1997; Ueda et al., 1995a,b) and cultured neuroblastoma cells

(Maurelli *et al.*, 1995a,b). Anandamide amidase catalyzes the hydrolysis of ananda-mide to arachidonic acid and ethanolamine (Desarnaud *et al.*, 1995; Di Marzo *et al.*, 1994; Hillard *et al.*, 1995; Katayama *et al.*, 1997; Lang *et al.*, 1996, 1999; Omeir *et al.*, 1995; Ueda *et al.*, 1995a,b) and has a high affinity for arachidonic acid (Desarnaud *et al.*, 1995; Deutsch and Chin, 1993; Hillard *et al.*, 1995; Lang *et al.*, 1996, 1999; Omeir *et al.*, 1995; Paria *et al.*, 1996). Its presence in the brain was shown to correlate well with the distribution of the CB_1 receptors. Highest activity was found in the hippocampus, cerebellum, and cerebral cortex, while the lowest activity occurred in the striatum, brain stem, and white matter (Fisher-Stenger *et al.*, 1993; Liu *et al.*, 2000; Zheng *et al.*, 1992). There is evidence suggesting that ANAse, which exhibits substrate selectivity for anandamide when compared to other unsaturated fatty acid ethanolamides, is an intracellular enzyme with a special role in terminating the biological activity of anandamide (Desarnaud *et al.*, 1995; Di Marzo *et al.*, 1994; Ueda *et al.*, 1995a). The intracellular localization of ANAse is supported by studies with subcellular membrane fractions (Hillard *et al.*, 1995), by its deduced amino acid sequence (Cravatt *et al.*, 1996), and by the fact that ananda-mide hydrolysis takes place after this protein has been accumulated in cells (Di Marzo *et al.*, 1994). Ethanolamine, the other hydrolysis product, was found to exist on both sides of the cell membrane. In the extracellular domain it exists in the free form while intracellularly it is either in the free or in the esterified form as phosphatidylethanolamine (PE). On the other hand, arachidonic acid is immediately re-acylated and is converted into membrane phospholipids such as phosphati-dylserine, phosphatidylinositol, phosphatidylethanolamine, and phosphatidylcholine (PS, PI, PE, PC, Figure 18.3). The affinity of anandamide for ANAse was studied using partially purified membrane preparations (Desarnaud *et al.*, 1995; Hillard *et al.*, 1995; Ueda *et al.*, 1995a). However, definitive evidence characterizing the enzyme's selectivity was reported by Lang *et al.* (1999). These studies evaluated the ANAse catalyzed hydrolysis of a series of anandamide analogs in order to determine the structural requirements for substrate specificity and enzyme selectiv-ity. It was found that arachidonamide is a highly selective substrate with K_m values in the μM range. The enzyme was observed not only to be substrate selective but to be stereoselective as well. Recently, ANAse reported as fatty acid amide hydrolase, was isolated, molecularly cloned (Cravatt *et al.*, 1996; Giang and Cravatt, 1997) and expressed from rat liver plasma membranes (Cravatt *et al.*, 1996). The molecular characterization of human and mouse fatty acid hydrolase was also reported (Giang and Cravatt, 1997; Thomas *et al.*, 1997).

The anandamide transporter (ANT)

Recently, evidence was reported for the high-affinity carrier mediated transport of anandamide in neurons and astrocytes (Beltramo *et al.*, 1997; Di Marzo *et al.*, 1994). In 1994, the presence of a rapid saturatable process of anandamide accumulation in neuronal cells was demonstrated and attributed to a transmembrane carrier (Di Marzo *et al.*, 1994). This transmembrane anandamide transporter (ANT) pro-tein appears to be involved in terminating the biological actions of anandamide by reuptake of extracellular anandamide into the cells followed by AAH-mediated intracellular hydrolysis.

In 1997, further evidence to support the presence of the anandamide transporter came from the laboratories of Makriyannis and Piomelli in which drug inhibitors of AEA transport were developed and their pharmacological properties investigated in rat neurons and astrocytes (Beltramo et al., 1997). It was shown that radioactively labeled anandamide is avidly taken up in rat astrocytes and neurons and this uptake is temperature dependent, substrate selective, high affinity, and saturable (Beltramo et al., 1997; Di Marzo et al., 1994; Hillard et al, 1997). These observations regarding anandamide uptake are consistent with carrier-mediated transport. The criteria for the presence of such transport include time and temperature-dependence as well as high substrate affinity and selectivity (Beltramo et al., 1997; Hillard et al., 1997; Piomelli et al., 1999). Strong support for the existence of an anandamide transporter has also come from the development of a compound, N-(4-hydroxyphenyl)-arachidonylethanolamide (AM404), which selectively inhibits anandamide transport competitively (Beltramo et al., 1997; Piomelli et al., 1999). Using this inhibitor, it was demonstrated that carrier mediated transport is involved in the inactivation of anandamide (Beltramo et al., 1997; Calignano et al., 1997; Piomelli et al., 1999). The discovery of AM404 will help in understanding the physiological functions of anandamide and may serve as a molecular prototype for development of future generation ANT ligands.

Anandamide biosynthesis

Anandamide is released from neurons upon depolarization by a mechanism that is not fully elucidated (Devane et al., 1992) and is rapidly inactivated. AEA biosynthesis is thought to take place via a Ca^{2+}-activated mechanism requiring phospholipase D-mediated hydrolysis of the phospholipid precursor, N-arachidonyl phosphatidylethanolamine (NAPE, Figure 18.2). This normal constituent of neuronal membranes (Cadas et al., 1997; Cadas et al., 1996a,b; Di Marzo et al., 1994; Sugiura et al., 1996b) is obtained by N-acylation of membrane phosphatidylethanolamine. Microsomal phosphodiesterase (PLD) is stimulated or inhibited, respectively, by low or high concentrations of calcium ions (Natarajan et al., 1984). After release, anandamide binds with high affinity to cannabinoid receptors, mimicking many cannabinoid actions in vitro and in vivo (Devane et al., 1992; Fride and Mechoulam, 1993; Howlett, 1995).

Anandamide metabolism

The observation that anandamide can be biosynthesized and released by central neurons also provides evidence that it may act as a neurotransmitter or neuromodulator (Self, 1999). In order to terminate its signal rapidly, such a mediator must be inactivated rapidly. There are two possible routes for anandamide metabolism once it is inside the cell. One is anandamide amidase mediated hydrolysis of anandamide. The second pathway is the oxidation of anandamide by enzymes of the arachidonic acid cascade.

Anandamide inactivation can occur through uptake and degradation mechanisms (Di Marzo and Fontana, 1995; Di Marzo et al., 1994). The degradation of anandamide, involves the selective transmembrane transport of extracellular AEA by the

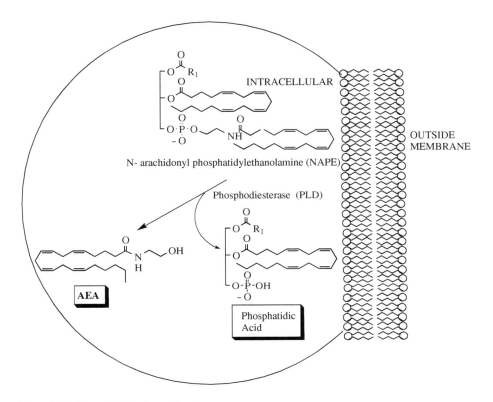

Figure 18.2 Biosynthesis of anandamide.

ANT transporter into the cell and the intracellular degradation of anandamide catalyzed by AAH (Figure 18.3) (Di Marzo *et al.*, 1994). After the release of newly formed anandamide (Figure 18.2) into the extracellular space, where it may activate CB_1 receptors, anandamide is thought to be removed from its sites of action by the ANT and intracellularly hydrolyzed by membrane-bound ANAse (Figure 18.3). Arachidonic acid produced during the ANAse reaction may be rapidly reincorporated into phospholipids and is unlikely to undergo further metabolism.

Oxidative metabolism of anandamide

It was reported that many of the enzymes which catalyze the oxidation of arachidonic acid (AA) to eicosanoids seem to have a broad substrate specificity (Takahashi *et al.*, 1993) and may also recognize anandamide. This possibility leads to an alternative route for anandamide metabolism (Figure 18.3), which may be used by cells not only to inactivate the mediator, but also to generate metabolites possibly active either at the cannabinoid receptor or at other extra- and intra-cellular targets. Therefore this process may potentially produce a novel class of lipids, the eicosanoid ethanolamides, prostaglandins (PG), Leukotrienes (LT), or hydroxyeicosatetraenoic

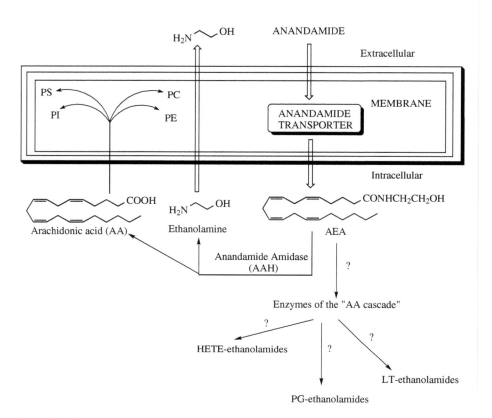

Figure 18.3 Metabolism of anandamide.

(HETE) ethanolamides, which may act as both primary and secondary messengers (Di Marzo and Fontana, 1995).

Several experiments have been conducted to support this metabolic pathway (Bornheim *et al.*, 1993, 1995; Hampson *et al.*, 1995; Ueda, 1995; Ueda *et al.*, 1995a,b; Wise *et al.*, 1994) and have shown that anandamide can be hydroxylated by the 11-, 12-, or 15-lipoxygenases to produce 11-, 12-, or 15-hydroxyananda-mide (Hampson *et al.*, 1995; Ueda, 1995; Ueda *et al.*, 1995a,b). Pharmacological studies on the ability of anandamide and its metabolites to inhibit the mouse vas deferens twitch response have shown that all three metabolites are less active than anandamide. A P450-dependent hydroxylase found in mouse hepatic microsomal cytochrome, catalyzed the oxidation of anandamide to at least 20 metabolites whose chemical structures and biological functions are not yet fully understood (Bornheim *et al.*, 1993, 1995). In addition, a novel enzyme, polyenoic fatty acid isomerase, was found to mediate the conversion of both AA and anandamide into their conjugate (5Z, 7Z, 9E)-triene derivatives. Recently, cyclooxygenase-2 was found to be responsible for the formation of prostaglandin E_2 ethanolamide from anandamide (Yu *et al.*, 1997).

PHARMACOLOGICAL AND THERAPEUTIC EFFECTS OF ENDOCANNABINOIDS

Cannabis has been used in the treatment of many disease conditions for several centuries. However, it was not until the 20th century that the cannabinoids were extensively investigated to assess their therapeutic potential. The major therapeutic uses of cannabinoids are discussed below.

Analgesia (antinociceptive)

Cannabinoids and opioids have historically been used in combination as a pain therapy. Cannabinoids have significant analgesic properties (Segal, 1986). The finding that several new cannabinoids with high analgesic activity also have high affinity for the CB_1 cannabinoid receptor (Johnson et al., 1988) suggests that the analgesic effects of cannabinoids are coupled with this cannabinoid receptor (Abood and Martin, 1992). In addition, the antinociceptive effects of cannabinoids are detectable in the same concentration range that produces marked changes in motor activity and hypothermia (Compton et al., 1992). The similarity in dose-dependence clearly indicates that these different functions are mediated by a common cannabinoid receptor.

Recently, Calignano et al. reported that anandamide and palmitylethanolamide may contribute to the control of pain transmission within the central nervous system (CNS) and peripherally (Calignano et al., 1998). They showed that anandamide attenuates the pain behavior produced by chemical damage to cutaneous tissue by interacting with CB_1 receptors. PEA exerts a similar effect by interacting with peripheral CB_2 receptors. When administered together, AEA and PEA act synergistically, to reduce pain responses in mice 100 times more potently than does each compound alone. These results indicate that CB_1 and CB_2 receptors may participate in the intrinsic control of pain initiation and that anandamide and PEA may mediate this effect.

Walker et al. carried out a series of in vivo studies in rats which also supports the role of endogenous AEA in pain modulation. In these studies the periaqueductal gray (PAG) region of the rat brain was electrically stimulated to produce a CB_1 mediated analgesia. When the rat brain is stimulated, there is an observed increase in the levels of anandamide in the PAG as evaluated by LCMS. The rats were also injected subcutaneously with the irritant formalin in the hindpaws inducing peripheral pain which also resulted in the increase of AEA in the PAG. These studies suggest that AEA plays a role in cannabinergic pain suppression (Walker et al., 1999).

Additionally, cannabinoids have been studied for their potential in the treatment of severe acute and chronic pain (Brower, 2000; Walker et al., 1999). Recent animal research indicates that cannabinoids control pain by direct interaction with the receptors on the rostral ventromedial medulla (RVM) of the brain, which modulates pain perception. Cannabinoids act as analgesics by preventing pain perception via the activation of RVM cells which function to suppress pain signaling (Klein et al., 1998). The analgesic activity of cannabinoids has prompted much research designed to explore their therapeutic potential.

Anticonvulsant effects

The anticonvulsant uses of cannabis were reported several decades ago (Loewe and Goodman, 1947). Subsequent investigations in various animal species have confirmed and increased our understanding of this action. Turkanis and Karler, 1985 reported that cannabinoids induced a reduction of cortical-evoked response and spinal monosynaptic reflexes which corresponded to a decrease in neurotransmission. They postulated several possible biochemical mechanisms for the effect, including altered neurotransmitter release, altered transmitter equilibrium potential, and drug receptor interactions. So far, the mechanism of this action remains unclear.

Antiglaucoma

Glaucoma is a disease of the eye which is characterized by an increase in intraocular pressure (Lemberger, 1980). Many reports have shown that Δ^9-THC and other synthetic cannabinoids can lower intraocular pressure (Lemberger, 1980). Intraocular pressure is a reflection of the balance between the formation and the outflow of the aqueous humor. Δ^9-THC can decrease fluid production and increase the total outflow facility of the eye (Adler and Geller, 1986). This may be one possible mechanism for the action. Because cannabinoids lower intraocular pressure, they have a therapeutic potential in the treatment of glaucoma. Unfortunately, simultaneously, they show CNS and cardiovascular effects as well. Recently, Buchwald and collegue (Klein *et al.*, 1998) performed a SAR study designed to generate cannabinoid like anti-glaucoma agent which have local effects, but no systemic effect. The preliminary pharmacological study indicate that there is a good possibility for the development of atopical anti-glaucoma agent. Thus cannabinoids are still being explored as therapeutic agents for glaucoma.

Antiemetic effect in cancer chemotherapy

Cannabinoids can prevent the nausea and vomiting induced by cancer chemotherapy (Dewey, 1986). Both clinical and animal studies indicate that certain cannabinoids have therapeutic potential as antiemetic agents. Vomiting is the expulsion of contents of the gut, largely by forces generated by the respiratory muscles (Levitt, 1986). Cannabinoids can affect cerebral function, above the level of the vomiting reflex (Steele, 1980). Therefore, cannabinoids may suppress vomiting through descending inhibitory connections to the lower brain stem centers (Levitt, 1986). However, other possible mechanisms have been investigated (Levitt, 1986) and no agreement about the mode of action has been reached.

Antineoplastic

It has been reported that cannabinoids have some antitumor effect (Hollister, 1986). This may be due to the immunosuppressant effect of cannabinoids and their ability to prevent the synthesis of nucleic acids.

Galve-Roperh *et al.* has reported evidence to support the use of cannabinoid agonists in the treatment of malignant gliomas (Galve-Roperh *et al.*, 2000) which are a rare, usually fatal, form of brain cancer (Piomelli, 2000). *In vivo* studies in wistar rats inoculated with glioma cells indicated that Δ^9-THC and the synthetic agonist WIN-55,212-2 slowed the progress of the glioma and in some cases completely eradicated the cancer as evaluated by gadolinium enhanced magnetic resonance (Galve-Roperh *et al.*, 2000). Additional studies in glioma cell culture indicated that Δ^9-THC signals apoptosis by a pathway that involves cannabinoid receptors (Galve-Roperh *et al.*, 2000). Other synthetic agonists such as WIN-55,212-2, CP-55,940, and HU-210 also induced glioma cell death at even lower concentrations than Δ^9-THC (Galve-Roperh *et al.*, 2000). These findings open new avenues for the therapeutic use of cannabinoids as antineoplastic agents.

SAR OF ENDOCANNABINOIDS FOR CANNABINOID RECEPTORS, ANANDAMIDE AMIDASE AND THE ANANDAMIDE TRANSPORTER

Cannabinoids produce a wide range of pharmacological effects, which include anticonvulsant, antiglaucoma, antiemetic, antianxiety, analgesic, and antiasthmatic activity. The structure–activity relationships of cannabinoids and anandamide analogs indicate that their pharmacological activities are significantly influenced by their chemical structures and stereochemical features.

The chemical structure of anandamide can be divided into two major molecular fragments: (a) a polar ethanolamido head group; and (b) a hydrophobic arachidonyl chain (Figure 18.4). The polar head group comprises a secondary amide functionality with an *N*-hydroxyalkyl substituent while the hydrophobic fragment is a non-conjugated *cis* tetraolefinic chain and an n-pentyl tail reminiscent of the lipophilic side chain found in the classical cannabinoids. A number of anandamide analogs have been synthesized and tested for their biological activities. These efforts have resulted in the development of several potent metabolically stable analogs some of which are important pharmacological tools useful in elucidating the physiological role of anandamide. The emerging SAR picture is that all of the anandamide analogs developed to date have considerable selectivity for the CB_1 receptor. Indeed, all known arachidonylethanolamides are primarily CB_1-selective ligands and bind poorly to the peripheral CB2 receptor (Munro *et al.*, 1993). Therefore, the following discussion summarizing the SAR of anandamides will focus on the CB_1 receptor.

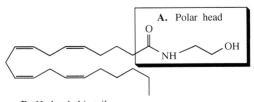

Figure 18.4 Structural features of anandamide.

The polar ethanolamido head group

Structural modifications of the N-hydroxyethyl group of anandamide

Structural modifications leading to the one carbon homolog of the N-hydroxypropyl analog increases CB_1 receptor affinity. However, further extension, with or without branching produce a decrease in binding affinity (Pinto *et al.*, 1994; Sheskin *et al.*, 1997). Thus, it appears that a three carbon chain separating the amido NH group from the terminal OH is an optimal requirement for a favorable ligand–receptor interaction.

The hydroxyl group is not a necessary requirement for receptor affinity and potency. N-alkyl analogs such as N-ethyl, N-propyl, and N-butyl, all show good receptor affinity. Indeed, N-(n-propyl)arachidonamide has a threefold higher CB_1 affinity than anandamide while the n-butyl homolog has about equal affinity (Pinto *et al.*, 1994). N-allyl and N-propargyl substitutions lead to high CB_1 affinity (Lin *et al.*, 1998) as does substitution of the hydroxyl group with a halogen such as F and Cl (Adams *et al.*, 1995a,b; Lin *et al.*, 1998). It, thus, appears that the presence of a H-bonding group in the headgroup of arachidonamides is not essential for a favorable ligand–receptor interaction.

One of the shortcomings of anandamide as a ligand is its facile *in vivo* and *in vitro* enzymatic degradation. It was, thus, important to develop analogs that are resistant to the hydrolytic actions of ANAse. Abadji *et al.* (1999), have reported four chiral anandamide analogs substituted with a methyl group at the C-1' or the C-2' positions. The rationale behind the design was to slow down the enzymatic hydrolysis by increasing steric hindrance around the amido group. Of these, the 1'-R-methyl isomer (AM356; R-methanandamide) showed 4 times higher CB_1 affinity than anandamide while exhibiting excellent metabolic stability (Figure 18.5). This analog is now being used as an important pharmacological tool in cannabinoid research. Interestingly, 1'-*R*-methanandamide was a poorer substrate for ANAse than its less potent *S*-enantiomer. It appears that the active site for the CB_1 receptor and ANAse have opposite absolute stereochemical requirements for ligand recog-

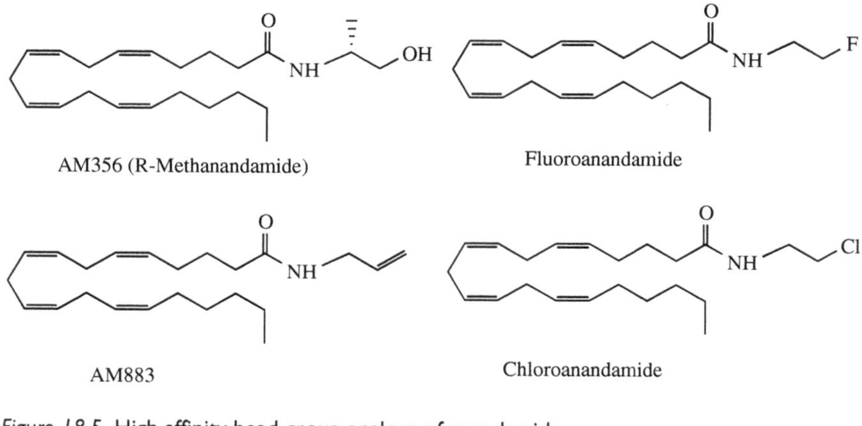

AM356 (R-Methanandamide)

Fluoroanandamide

AM883

Chloroanandamide

Figure 18.5 High affinity head group analogs of anandamide.

nition. Introduction of larger alkyl groups e.g. ethyl, isopropyl has a detrimental effect on the CB_1 affinity (Adams *et al.*, 1995a).

Substitution of the 2-hydroxyethyl group of AEA with a phenolic group results in decreased affinity for CB_1 (Khanolkar *et al.*, 1996). However, *N*-(*o*-hydroxy) phenylarachidonamide (AM403) was found to be an excellent substrate for the amidohydrolase (Lang *et al.*, 1999) while a second phenolic analog, *N*-(*p*-hydroxy) phenylarachidonamide (AM404), was found to be an inhibitor for the recently discovered (Beltramo *et al.*, 1997) anandamide transporter (ANT).

Arachidonamide and arachidonic acid esters (methyl, ethyl, propyl) do not exhibit significant affinity for CB_1 (Pinto *et al.*, 1994). Cyclization of the head group of AEA into an oxazoline ring also diminishes receptor affinity (Lin *et al.*, 1998).

Modifications of the amido group

Modifications of the polar head group of anandamide through the replacement of the keto O by S results in reduced affinity for CB_1. Thus, both thioanandamide and thio-R-methanandamide bind weakly to CB_1 and have no significant biological activity (Lin *et al.*, 1998).

The SAR indicates that the amide group in AEA must be secondary. Primary amides e.g. arachidonamide as well as tertiary amides e.g. *N*-methylanandamide do not effectively bind the CB_1 receptor (Lin *et al.*, 1998; Pinto *et al.*, 1994; Sheskin *et al.*, 1997).

Reversing the position of the carbonyl and the NH groups increases receptor affinity. These anandamides, designated as retro-anandamides (Figure 18.6) which were developed in our laboratory have high affinity for CB_1 and exhibit exceptional stability with regard to hydrolysis by ANAse (Lin *et al.*, 1998).

The hydrophobic arachidonyl chain

Importance of cis-olefinic bonds for cannabimimetic activity

Drastic structural modifications of the arachidonyl component such as complete saturation or replacement of the double bonds with triple bonds result in complete loss of CB_1 affinity (Sheskin *et al.*, 1997). Furthermore, ethanolamides of partially unsaturated fatty acids such as linoleic (two double bonds) and oleic (one double bond) acids exhibit considerably diminished affinity for CB_1 and low cannabimimetic activity (Lin *et al.*, 1998; Sheskin *et al.*, 1997). From these results it can be argued that the presence of the four cis olefinic bonds is optimal for activity. Prostaglandins

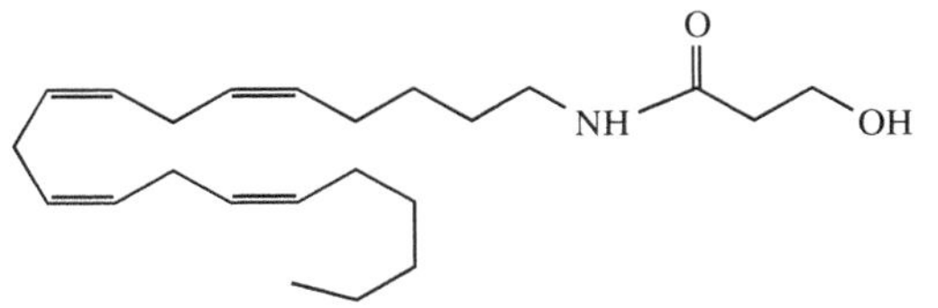

Figure 18.6 Retro-anandamide, a metabolically stable anandamide analog.

and related analogs which can be considered as conformationally rigid arachidonic acid analogs, do not bind to CB_1 (Pinto *et al.*, 1994). Their inability to interact with the receptor may be due to the conformational restriction imposed by the five-member carbocyclic ring which leads to preferred conformations that are incongruent with those of arachidonylethanolamide and its analogs. The lack of affinity of prostaglandin analogs for CB_1 could also be due to the presence of hydroxyl and/or keto substituents whose regio- and stereochemical features may destabilize their interactions with the receptor.

Introduction of a methyl group or *gem*-dimethyl group at the C-2 position results in metabolically stable analogs with concomitant increase in CB_1 affinity as in the case of C-1' methylation (Makriyannis and Goutopoulos; unpublished results).

Tail n-pentyl group modifications

Anandamide behaves as a partial agonist in the biochemical and pharmacological tests used to characterize cannabimimetics. Although there is no apparent structural similarity between the classical cannabinoids and anandamide, there is considerable evidence suggesting that these two classes of cannabimimetic agents bind similarly to the CB_1 active site. Chemical and computational data indicate that arachidonic acid, the parent fatty acid of anandamide, favors a bent or looped conformation in which the carbonyl group is proximal to the C14-C15 olefinic bond. This is suggested by the highly regiospecific intramolecular epoxidation of arachidonyl peracid (Corey *et al.*, 1984) and the facile macrolactonization of the C20 hydroxyl methyl arachidonate (Corey *et al.*, 1983). The above experimental results are corroborated by molecular dynamics calculations (Rich, 1993) which indicate that indeed a bent conformation is thermodynamically favorable. In the case of arachidonylethanolamides, molecular modeling studies (Barnett-Norris *et al.*, 1998; Piomelli *et al.*, 1999) have shown that anandamide and other fatty acid ethanolamides and esters also prefer a hairpin conformation. Additional data (Thomas *et al.*, 1996; Tong *et al.*, 1998) indicate that such a bent conformation is capable of mimicking the three-dimensional structure of tetrahydro-and hexahydrocannabinols. However, it is unclear whether the hairpin conformation is also the conformation at the CB_1 receptor active site. Recent biophysical work provides evidence for a more extended alignment for the C20 chain of anandamide in the membrane bilayer (Makriyannis and Tian; unpublished results) and suggest alternative CB_1 pharmacophoric conformations.

According to the SAR data for classical and non-classical cannabinoids, a 1,1-dimethylheptyl (DMH) side chain is optimal for activity. It was thus predicted

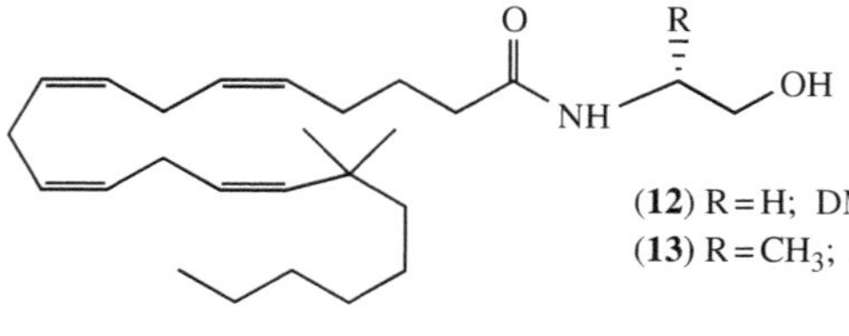

(**12**) R = H; DMH-anandamide
(**13**) R = CH₃; DMH-R-methanandamide

Figure 18.7 Dimethylheptyl anandamides.

that a similar substitution in anandamide should lead to analogs with increased CB_1 receptor affinity and potency. This postulate was validated by work showing that dimethylheptyl anandamide analogs exhibited a marked increase in receptor affinity and *in vivo* potency when compared to their *n*-pentyl counterparts (Makriyannis and Kawakami; unpublished results) (Seltzman *et al.*, 1997).

ANANDAMIDE AMIDASE (ANAse) SUBSTRATES AND INHIBITORS

Structural requirements of a substrate for ANAse

The hydrolysis of several long chain fatty acid amides by rat brain microsomal ANAse was studied in some detail (Desarnaud *et al.*, 1995; Lang *et al.*, 1996, 1999; Qin *et al.*, 1998). These studies have served to identify the structural features required within ANAse ligands for active site recognition and subsequent catalytic hydrolysis as summarized below:

1. Arachidonamide and *N*-(*O*-hydroxyphenyl) arachidonamide are found to be the best substrates for AAH with relative rates of hydrolysis approximately twice that of anandamide. Enzymatic hydrolysis of the *N*-(*O*-hydroxylphenyl) arachidonamide analog presumably is facilitated by anchimeric assistance due to the phenolic hydroxyl group (Lang *et al.*, 1999).
2. Introduction of a methyl group in the C2, C1', or C2' positions of anandamide leads to metabolically stable analogs (Khanolkar *et al.*, 1996; Lang *et al.*, 1999). The resistance to hydrolysis among these analogs can be attributed to the increased steric hindrance around the carbonyl group. A noteworthy example of this class of substrates is R-1'-methylanandamide (AM356); a metabolically stable analog exhibiting four-fold higher affinity for CB_1 than anandamide.
3. ANAse exhibits considerable stereoselectivity. In the case of methanandamides, R-1'-methanandamide and S-2'-methanandamide are approximately 10- and 19-times more stable than their corresponding enantiomers (Abadji *et al.*, 1994; Lang *et al.*, 1999).
4. The enzyme appears to favor anandamide substrates in which the nitrogen of the ethanolamido head group is either primary or secondary. Analogs containing a tertiary amide nitrogen are poor enzyme substrates (Lang *et al.*, 1999).
5. Anandamide analogs with H-bonding and/or an electronegative head group substituents, such as hydroxyl and fluoro, are better substrates for the amidase than their unsubstituted counterparts (Lang *et al.*, 1999).
6. Other fatty acid ethanolamides containing fewer than four cis non-conjugated double bonds are not as good substrates as anandamide (Desarnaud *et al.*, 1995; Lang *et al.*, 1999). This may be attributed to a requirement for a hairpin conformation, discussed earlier, for substrate recognition at the ANAse active site. Arguably this bent conformation is best accommodated by the presence of four cis non-conjugated double bonds as in anandamide. Such a conformation is thermodynamically less favored in the cis-diene linolenyl and cis-ene oleyl analogs and much less so in the fully saturated palmityl

analog thus decreasing the ability of these substrates to recognize the ANAse catalytic site.

7   The second endogenous cannabinoid ligand, 2-arachidonylglycerol (2-AG), a meso ester and its racemic isomer, α-arachidonylglycerol are excellent substrates for ANAse (Goparaju *et al.*, 1998; Lang *et al.*, 1999) and are hydrolyzed about 4 times faster than anandamide. It is thus possible that ANAse may also be involved in the biological inactivation of 2-AG.

ANAse inhibitors

Shortly after the existence of ANAse was demonstrated, a number of inhibitors of anandamide hydrolysis such as fatty acid trifluoromethylketones, α-keto esters and α-keto amides were reported (Koutek *et al.*, 1994). These first generation inhibitors are the transition-state analogs and form a stable tetrahedral intermediate with a putative serine residue at the enzyme active site. Of these, arachidonyl trifluoromethyl ketone (ATFMK) was found to be the most potent. With regard to irreversible inhibition, PMSF was the first compound shown to inhibit the hydrolysis of ANAse through a covalent mechanism. However, because of its low activity and limited selectivity, this compound is not an optimal ANAse inhibitor. We have, thus, developed a series of saturated fatty acid (C12 to C20) sulfonyl fluorides as specific and highly potent irreversible inhibitors of ANAse (Deutsch *et al.*, 1997a; Deutsch and Makriyannis, 1997). Moreover, these sulfonyl fluorides were shown to react irreversibly with the CB_1 receptor. Thus, AM374 (palmitylsulfonyl fluoride, Figure 18.8) was approximately 20 times more potent as an enzyme inhibitor than PMSF and 50 times more potent than arachidonyl TFMK in preventing the hydrolysis of anandamide in brain homogenates. In cultured cells, AM374 was over 1000 times more effective than PMSF in inhibiting AAH activity. These sulfonyl fluoride analogs generally showed decreasing affinities for the CB_1 receptor as the chain length increased; thus, C12 sulfonyl fluoride had an IC_{50} of 18 nM while C20 sulfonyl fluoride showed an IC_{50} of 78 μM. AM374 is the most successful inhibitor as it demonstrates the highest selectivity for ANAse versus the CB_1 receptor. Recently methyl arachidonyl fluorophosphonate was reported as a potent irreversible inhibitor of ANAse (Deutsch *et al.*, 1997a,b). However, this analog has lower affinity and selectivity for ANAse than the corresponding sulfonyl fluorides.

Hydrolysis of anandamide by ANAse has also been studied in the presence of arachidonic acid congeners and other cannabimimetics (Lang *et al.*, 1999). Arachidonyl alcohol and anandamine did not show significant ANAse inhibition while arachidonyl aldehyde exhibited significant inhibition. This inhibitory activity may be attributed to a covalent attachment of the aldehyde group at the enzyme active site through the formation of a tetrahedral intermediate. Arachidonic acid also has an inhibitory effect on anandamide hydrolysis by ANAse.

Boger *et al.* have recently reported a series of heterocyclic AAH inhibitors of which several analogs possess more potent activity than their corresponding trifluoromethyl ketones (Lynn and Herkenham, 1994). Of these heterocyclic fatty acid ketone analogs **8–10** (Figure 18.8) were shown to be powerful ANAse inhibitors with K_i values of 17 nM, 20 nM and 1 nM, respectively. These Ki values reach activities similar to the sulfonyl fluorides.

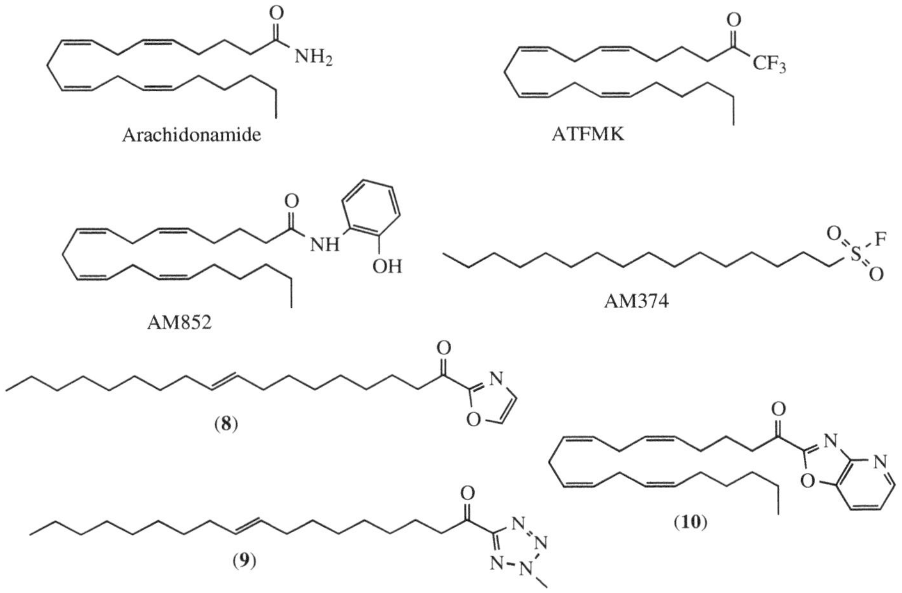

Figure 18.8 Substrates and inhibitors of anandamide amidohydrolase (AAH).

Our studies have also shown that certain cannabimimetic ligands although structurally unrelated to anandamide, nevertheless, are able to recognize and competitively inhibit ANAse. For example, classical cannabinoids such as (–)-Δ^8-THC, cannabidiol and cannabinol were shown to inhibit anandamide hydrolysis (Lang *et al.*, 1999; Watanabe *et al.*, 1996). This inhibitory ability is reduced with cannabinoids carrying additional OH groups and longer side chains such as the classical cannabinoids (–)-11-OH-DMH-Δ^8-THC and the non-classical cannabinoids CP-55,940 and CP-55,244. No enzyme inhibition was observed with a cannabinoid ligand in which the phenolic hydrogen was substituted with a methyl group as in OMe-Δ^8-THC. Of the remaining cannabimimetic structural prototypes, SR141716A, a cannabinoid antagonist, exhibited AAH inhibitory activity while the aminoalkylindole agonist WIN-55,212-2 did not show significant inhibition.

ANANDAMIDE TRANSPORTER (ANT) SUBSTRATES AND INHIBITORS

N-(4-hydroxyphenyl)arachidonamide (AM404), was shown to inhibit anandamide reuptake in neurons and astrocytes and was used as a pharmacological tool for characterization of the anandamide transporter (ANT) (Beltramo *et al.*, 1997).

The SAR of ANT ligands has been described in a recent publication (Piomelli *et al.*, 1999) in which the structural requirements for recognition and translocation are highlighted.

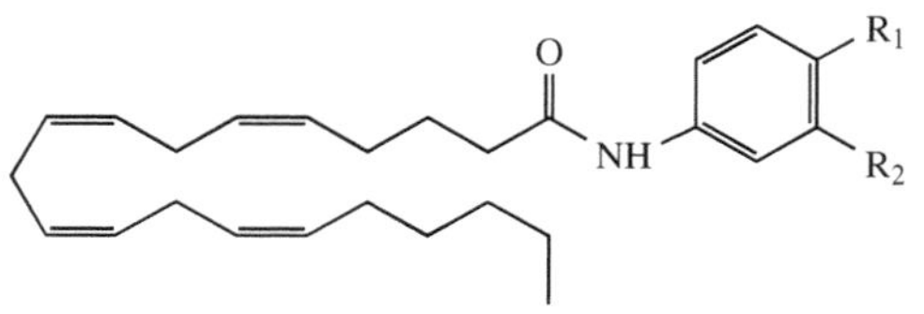

R$_1$ = OH; R$_2$ = H (AM404)
R$_1$ = H; R$_2$ = OH (AM403)

Figure 18.9 Anandamide transport inhibitors.

Unlike the CB$_1$ receptor and ANAse, the presence of a secondary arachidonyl amido group is not a requirement for transport. For example, arachidonyl esters such as 2-arachidonylglycerol (2-AG), a second endogenous cannabinoid, also exhibit high affinity for ANT and are translocated across the cell membrane. These results suggest that ANT may be involved in terminating the biological activity of both groups of endocannabinoids.

The ANT shows considerable stereoselectivity. Thus, S-methanandamide is 4 times more effective than R-methanandamide in inhibiting anandamide transport. Interestingly, the absolute stereochemical preference for the 1'-methylananda- mides of ANT is similar to that of ANAse but opposite to that of CB$_1$.

The studies find that a polar head group is necessary for favorable ligand- transporter interaction. Although there is no requirement for the polar group to be a H-bonding donor, analogs with OH-containing headgroups are somewhat more potent than the corresponding hydrogen bond accepting alkyl ethers. Comparative SAR among the anandamide analogs shows that the presence of an OH in the headgroup leads to better ANAse substrate but is not a requirement for interaction with CB$_1$.

Ligand recognition by the transporter requires at least one cis double bond located at the midpoint of the carbon chain. Fully saturated hydrophobic chains and trans double bond containing fatty acid amides are not recognized. Con- versely, to be translocated across the membrane via the transporter, ligands must have all four cis double bonds. Fatty acid ethanolamides containing fewer than four double bonds are recognized but not transported or transported very slowly.

CONCLUSIONS

The discovery of anandamide and 2-arachidonylglycerol as two families representing the endocannabinoids has accelerated the pace of research in this field and brought a better understanding of the molecular basis of cannabinoid activity. Cannabinoid research has taken major strides towards the goal of understanding the molecular mechanism of cannabinoid action. A characteristic common feature of the above two principal endocannabinoid ligands is the arachidonyl side chain. We now know that structurally related arachidonic acid derivatives are capable of interacting selectively with all of the known cannabinoid proteins, namely, the CB$_1$ receptor, the metabolizing enzyme ANAse and the ANT. Conformational studies suggest that when in solution, the arachidonic acid moiety assumes a hairpin conformation in which the two ends are brought into close proximity by folding at the chain mid-point. However, the studies also show that the rotational barrier around the mid-chain single bonds are relatively small and allow the arachidonic acid moiety to assume other low energy conformations. An intriguing question that remains to be answered is whether all three cannabimimetic sites have similar pharmacophoric requirements for anandamide and 2-AG or whether each of these proteins has its own requirement with the arachidonic acid chain assuming distinctive pharmacophoric conformations for each of the active sites.

Currently, all of the known cannabimimetic sites (CB_1 and CB_2, ANAse, ANT) can be considered as potential therapeutic targets for developing useful medications in the treatment of a variety of ailments including drug addiction, pain and neurodegenerative disorders. Also available are a number of ligands (receptor-selective agonists/antagonists, enzyme inhibitors, transport inhibitors) to serve as research tools for exploring the cannabinoid system and its role in the modulation of behavior, memory, cognition and pain perception. This is remarkable progress in light of the fact that only about a decade ago the sites of action of cannabinoids had not been identified and their molecular mechanism of action was unknown. Arguably, the future of endocannabinoid research is very promising with the full therapeutic potential of cannabimimetic agents being realized in the not too distant future.

ACKNOWLEDGMENTS

This work was supported by Grants DA-3801, DA-152, DA-7215, DA-9158 (A. M.) and DA-00355 (A. D. K.) from the National Institute on Drug Abuse.

REFERENCES

Abadji, V., Lin, S., Taha, G., Griffin, G., Stevenson, L. A., Pertwee, R. G. and Makriyannis, A. (1994) "*R*-Methanandamide: a chiral novel anandamide possessing higher potency and metabolic stability," *Journal of Medicinal Chemistry* **37**: 1889–1893.

Abood, M. E. and Martin, B. R. (1992) "Neurobiology of marijuana abuse," *Trends in Pharmacological Sciences* **13**: 201–206.

Adams, I. B., Ryan, W., Singer, M., Razdan, R. K., Compton, D. R. and Martin, B. R. (1995a) "Pharmacological and behavioral evaluation of alkylated anandamide analogs," *Life Sciences* **56**: 2041–2048.

Adams, I. B., Ryan, W., Singer, M., Thomas, B. F., Compton, D. R. Razdan, R. K. and Martin, B. R. (1995b) "Evaluation of cannabinoid receptor binding and *in vivo* activities for anandamide analogs," *Journal of Pharmacology and Experimental Therapeutics* **273**: 1172–1181.

Adler, M. W. and Geller, E. B. (1986) In *Cannabinoids as Therapeutic Agents*, edited by R. Mechoulam, pp. 52–67. Boca Raton: CRC Press.

Barg, J., Fride, E., Hanus, L., Levy, R., Matus Leibovitch, N., Heldman, E., Bayewitch, M., Mechoulam, R. and Vogel, Z. (1995) "Cannabinomimetic behavioral effects of and adenylate cyclase inhibition by two new endogenous anandamides," *European Journal of Pharmacology* **287**: 145–152.

Barnett-Norris, J., Guarnieri, F., Hurst, D. P. and Reggio, P. H. (1998) "Exploration of biologically relevant conformations of anandamide, 2-arachidonylglycerol and their analogues using conformational memories," *Journal of Medicinal Chemistry* **41**: 4861–4872.

Beltramo, M., Stella, N., Calignano, A., Lin, S. Y., Makriyannis A. and Piomelli, D. (1997) "Functional role of high-affinity anandamide transport, as revealed by selective inhibition," *Science* **277**: 1094–1097.

Berdyshev, E. V., Boichot, E. and Lagente, V. (1996) "Anandamide – a new look on fatty acid ethanolamides," *Journal of Lipid Mediators and Cell Signalling* **15**: 49–67.

Bidaut-Russell, M., Devane, W. A. and Howlett, A. C. (1990) "Cannabinoid receptors and modulation of cyclic AMP accumulation in the rat brain," *Journal of Neurochemistry* **55**: 21.

Bornheim, L. M., Kim, K. Y., Chen, B. and Correia, M. A. (1993) "The effect of cannabidiol on mouse hepatic microsomal cytochrome P450-dependent anandamide metabolism," *Biochemical and Biophysical Research Communications* **197**: 740–746.

Bornheim, L. M., Kim, K. Y., Chen, B. and Correia, M. A. (1995) "Microsomal cytochrome P450-mediated liver and brain anandamide metabolism," *Biochemical Pharmacology* **50**: 677–686.

Brower, V. (2000) "New paths to pain relief: A better understanding of the mechanisms by which pain signals are relayed in the nervous system is paving the way for novel treatments," *Nature Biotechnology* **18**: 387–391.

Cabral, G. A., Stinnet, A. L., Bailey, J., Ali, S. F., Paule, M. G., Scallet, A. C. and Slikker, W. (1991) "Chronic marijuana smoke alters alveolar macrophage morphology and protein expression," *Journal of Pharmacology, Biochemistry and Behavior* **40**: 643–649.

Cadas, H., di Tomaso, E. and Piomelli, D. (1997) "Occurrence and biosynthesis of endogenous cannabinoid precursor, N-arachidonoyl phosphatidylethanolamine, in rat brain," *Journal of Neuroscience* **17**: 1226–1242.

Cadas, H., Gaillet, S., Beltramo, M., Venance, L. and Piomelli, D. (1996a) "Biosynthesis of an endogenous cannabinoid precursor in neurons and its control by calcium and cAMP," *Journal of Neuroscience* **16**: 3934–4392.

Cadas, H., Schinelli, S. and Piomelli, D. (1996b) "Membrane localization of N-acylphosphatidylethanolamine in central neurons: studies with exogenous phospholipases," *Journal of Lipid Mediators and Cell Signalling* **14**: 63–70.

Calignano, A., La Rana, G., Beltramo, M., Makriyannis, A. and Piomelli, D. (1997) "Potentiation of anandamide hypotension by the transport inhibitor, AM404," *European Journal of Pharmacology* **337**: R1–R2.

Calignano, A., La Rana, G., Giuffrida, A. and Piomelli, D. (1998) "Control of pain initiation by endogenous cannabinoids," *Nature* **394**: 277–281.

Caulfied, M. P., Rinaldi, M. and Carayon, P. (1993) *European Journal of Biochemistry* **214**: 173.

Childers, S. R., Sexton, T. and Roy, M. B. (1994) "Effects of anandamide on cannabinoid receptors in rat brain microsomes," *Biochemical Pharmacology* **47**: 711–715.

Compton, D. R., Gold, L. H., Ward, S. J., Balster, R. L. and Martin, B. R. (1992) *Journal of Pharmacology Experimental Therapeutics* **263**: 1118.

Corey, E. J., Cashman, J., Kantner, S. S. and Wright, S. W. (1984) "Rationally designed, potent competitive inhibitors of leukotriene biosynthesis," *Journal of American Chemical Society* **106**: 1503–1504.

Corey, E. J., Iguchi, S., Albright, J. O. and De, B. (1983) "Studies on the conformational mobilities of arachidonic acid facile macrolactonization of 20-hydroxyarachidonic acid," *Tetrahedron Letters* **24**: 37–40.

Cravatt, B. F., Giang, D. K., Mayfield, S. P., Boger, D. L., Lerner, R. A. and Gilula, N. B. (1996) "Molecular characterization of an enzyme that degrades neuromodulatory fatty-acid amides," *Nature* **384**: 83–87.

Crawley, J. N., Corwin, R. L., Robinson, J. K., Felder, C. C., Devane, W. A. and Axelrod, J. (1993). "Anandamide, an endogenous ligand of the cannabinoid receptor, induces hypomotility and hypothermia *in vivo* in rodents," *Pharmacology, Biochemistry and Behavior* **46**: 967–972.

Dax, E. M., Pilotte, N. S., Alder, W. H., Nagel, J. E. and Lange, W. R. (1989) "The effects of 9-ene-tetrahydrocannabinol on hormone release and immune function," *Journal of Steroid Biochemistry* **34**: 263–270.

Derkinderen, P., Toutant, M., Burgaya, F., Le Bert, M., Siciliano, J. C., de Franciscis, V., Gelman, M. and Girault, J. A. (1996) "Regulation of neuronal form of focal adhesion kinase by anandamide," *Science* **273**: 1719–1722.

DePetrocellis, L., Orlando, P. and Di Marzo, V. (1995) "Anandamide, and endogenous cannabinomimetic substance, modulates rat brain protein kinase C *in vitro*," *Biochemistry Molecular Biology International* **36**: 1127–1133.

Desarnaud, F., Cadas, H. and Piomelli, D. (1995) "Anandamide amidohydrolase activity in rat brain microsomes. Identification and partial characterization," *Journal of Biological Chemistry* **270**: 6030–6035.

Deutsch, D. G. and Chin, S. A. (1993) "Enzymatic synthesis and degradation of anandamide, a cannabinoid receptor agonist," *Biochemical Pharmacology* **46**: 791–796.

Deutsch, D. G., Lin, S., Hill, W. A. G., Morse, K. L., Salehani, D., Arreaza, G., Omeir, R. L. and Makriyannis, A. (1997a) *Biochemistry and Biophysics Research Communications* **231**: 217–221.

Deutsch, D. G. and Makriyannis, A. (1997) "Inhibitors of anandamide breakdown," *NIDA Research Monograph* **173**: 65–84.

Deutsch, D. G., Omeir, R., Arreaza, G., Salehani, D., Prestwich, G. D., Huang, Z. and Howlett, A. (1997b) "Methyl arachidonyl fluorophosphonate: a potent irreversible inhibitor of anandamide amidase," *Biochemical Pharmacology* **53**: 255–260.

Devane, W. A., Dysarz, F. A. I., Johnson, R. M., Melvin, L. S. and Howlett, A. C. (1988) "Determination and characterization of a cannabinoid receptor in rat brain," *Molecular Pharmacology* **34**: 605–613.

Devane, W. A., Hanus, L., Breuer, A., Pertwee, R. G., Stevenson, L. A., Griffin, G., Gibson, D., Mandelbaum, A., Etinger, A. and Mechoulam, R. (1992) "Isolation and structure of a brain constituent that binds to the cannabinoid receptor [see comments]," *Science* **258**: 1946–1949.

Dewey, W. (1986) "Cannabinoid pharmacology," *Pharmacological Reviews* **38**: 151–178.

Di Marzo, V., Bisogno, T., Sugiura, T., Melck, D. and De Petrocellis, L. (1998) "The novel endogenous cannabinoid 2-arachidonoylglycerol is inactivated by neuronal- and basophil-like cells: connections with anandamide," *Biochemical Journal* **331**: 15–19.

Di Marzo, V. and Fontana, A. (1995) "Anandamide, an endogenous cannabinomimetic eicosanoid: Killing two birds with one stone," *Prostoglandins Leukotriens and Essential Fatty Acids* **53**: 1–11.

Di Marzo, V., Fontana, A., Cadas, H., Schinelli, S., Cimino, G., Schwartz, J. C. and Piomelli, D. (1994) "Formation and inactivation of endogenous cannabinoid anandamide in central neurons," *Nature* **372**: 686–691.

Facci, L., Dal Toso, R., Romanello, S., Buriani, A., Skaper, S. D. and Leon, A. (1995) "Mast cells express a peripheral cannabinoid receptor with differential sensitivity to anandamide and palmitoylethanolamide," *Proceedings of the National Academy of Sciences of the USA* **92**: 3376–3380.

Felder, C. C., Briley, E. M., Axelrod, J., Simpsom, J. T., Mackie, K. and Devane, W. A. (1993) "Anandamide, an endogenous cannabimimetic eicosanoid, binds to the cloned human cannabinoid receptor and stimulates receptor-mediated signal transduction," *Proceedings of the National Academy of Sciences of the USA* **90**: 7656–7660.

Felder, C. C., Veluz, J. S., Williams, H. L., Briley, E. M. and Matsuda, L. A. (1992) "Cannabinoid agonists stimulate both receptor- and non-receptor-mediated signal transduction pathways in cells transfected with and expressing cannabinoid receptor clones," *Molecular Pharmacology* **42**: 838–845.

Fisher-Stenger, K., Pettit, D. A. D. and Cabral, G. A. (1993) "Delta-9-tetrahydrocannabinol inhibition tumor necrosis factor-alpha: suppression of post-translational events," *Journal of Pharmacology and Experimental Therapeutics* **267**: 1558.

Fride, E. and Mechoulam, R. (1993) "Pharmacological activity of the cannabinoid receptor agonist, Anandamide, a brain constituent," *European Journal of Pharmacology* **231**: 313–314.

Galve-Roperh, I., Sachez, C., Cortes, M. L., Gomez del Pulger, T., Izquierdo, M. and Guzman, M. (2000) "Anti-tumoral action of cannabinoids: Involvement of sustained ceramide accumulation and extracellular signal-regulated kinase activation," *Nature Medicine* **6**: 313–319.

Gaoni, Y. and Mechoulam, R. (1964) "Isolation, structure and partial synthesis of an active constituent of hashish," *Journal of the American Chemical Society* **86**: 1646–1647.

Giang, D. K. and Cravatt, B. F. (1997) "Molecular characterization of human and mouse fatty acid amide hydrolases," *Proceedings of the National Academy of Sciences of the USA* **94**: 2238–2242.

Giuffrida, A., Parsons, L. H., Kerr, T. M., Rodriguez de Fonseca, F., Navarro, M. and Piomelli, D. (1999) "Dopamine activation of endogenous cannabinoid signaling in dorsal striatum," *Nature Neuroscience* **2**: 358–363.

Goparaju, S. K., Ueda, N., Yamamguchi, H. and Yamamoto, S. (1998) "Anandamide amidohydrolase reacting with 2-arachidonylglycerol, another cannabinoid receptor ligand," *FEBS Letters* **422**: 69–73.

Hampson, A. J., Hill, W. A., Zan Phillips, M., Makriyannis, A., Leung, E., Eglen, R. M. and Bornheim, L. M. (1995) "Anandamide hydroxylation by brain lipoxygenase: metabolite structures and potencies at the cannabinoid receptor," *Biochimica et Biophysica Acta* **1259**: 173–179.

Hanus, L., Gopher, A., Almog, S. and Mechoulam, R. (1993) "Two new unsaturated fatty acid ethanolamides in brain that bind to the cannabinoid receptor," *Journal of Medicinal Chemistry* **36**: 3032–3034.

Herkenham, M., Lynn, A. B., Little, M. D., Johnson, M. R., Melvun, L. S., Decosta, B. R. and Rice, K. C. (1990) "Cannabinoid receptor localization in brain," *Proceedings of the National Academy of Sciences of the USA* **87**: 1932.

Hillard, C. J. and Campbell, W. B. (1997) "Biochemistry and pharmacology of arachidonylethanolamide, a putative endogenous cannabinoid," *Journal of Lipid Research* **38**: 2383–2398.

Hillard, C. J., Edgemond, W. S., Jarrahian, A. and Campbell, W. B. (1997) "Accumulation of N-arachidonoylethanolamine (anandamide) into cebeller granule cells occurs via facilitated transport," *Journal of Neurochemistry* **69**: 631–638.

Hillard, C. J., Wilkison, D. M., Edgemond, W. S. and Campbell, W. B. (1995) "Characterization of the kinetics and distribution of N-arachidonylethanolamine (anandamide) hydrolysis by rat brain," *Biochimica et Biophysica Acta* **1257**: 249–256.

Hollister, L. E. (1986) "Health aspects of cannabis," *Pharmacological Reviews* **38**: 1–20.

Howlett, A. C. (1995) "Pharmacology of Cannabinoid Receptors," *Annual Reviews in Pharmacology Toxicology* **35**: 607–634.

Jarai, Z., Wagner, J. A., Goparaju, S. K., Wang, L., Razdan, R. K., Sugiura, T., Zimmer, A. M., Bonner, T. I., Zimmer, A. and Kunos, G. (2000) "Cardiovascular effects of 2-arachidonoyl glycerol in anesthetized mice," *Hypertension* **35**: 679–684.

Jarai, Z., Wagner, J. A., Varga, K., Lake, K. D., Compton, D. R., Martin, B. R., Zimmer, A. M., Bonner, T. I., Buckley, N. E., Mezey, E., Razdan, R. K., Zimmer, A. and Kunos, G. (1999) "Cannabinoid-induced mesenteric vasodilation through an endothelial site distinct from CB$_1$ or CB$_2$ receptors," *Proceedings of the National Academy of Sciences of the USA* **96**: 14136–14141.

Jarbe, T. U., Lamb, R. J., Makriyannis, A., Lin, S. and Goutopoulos, A. (1998a) "Delta-9-THC training dose as a determinant for *R*-methanandamide generalization in rats," *Psychopharmacology* **140**: 519–522.

Jarbe, T. U., Sheppard, R., Lamb, R. J., Makriyannis, A., Lin, S. and Goutopoulos, A. (1998b) "Effects of delta-9-tetrahydrocannabinol and (R)-methanandamide on open-field behavior in rats," *Behavioural Pharmacology* **9**: 169–174.

Johnson, D. E., Heald, S. L., Dally, R. D. and Janis, R. A. (1993) "Isolation, identification, and synthesis of an endogenous arachidonic amide that inhibits calcium channel antagonist 1,4-dihydropyridine binding," *Prostograndins, Leukotrienes, Essential Fatty Acids* **48**: 429–437.

Johnson, M., Devane, W., Howlett, A., Melvin, L. and Milne, G. (1988) "Structural studies leading to the cannabinoid binding site," *NIDA Research Monograph Series* **90**: 129–135.

Katayama, K., Ueda, N., Kurahashi, Y., Suzuki, H., Yamamoto, S. and Kato, I. (1997) "Distribution of anandamide amidohydrolase in rat tissues with special reference to small intestine," *Biochimica et Biophysica Acta* **1347**: 212–218.

Khanolkar, A. D., Abadji, V., Lin, S., Hill, W. A. G., Taha, G., Abouzid, K., Meng, Z., Fan, P. and Makriyannis, A. (1996) "Head group analogs of arachidonylethanolamide, the endogenous cannabinoid," *Journal of Medicinal Chemistry* **39**: 4515–4519.

Klein, T. W. and Friedman, H. (1990). In *Drugs of Abuse and Immune Function*, edited by R. R. Watson, p. 87. Boca Raton: CRC Press.

Klein, T. W., Newton, C. and Friedman, H. (1998) "Cannabinoid receptors and immunity," *Immunology Today* **19**: 373–381.

Koutek, B., Prestwich, G. D., Howlett, A. C., Chin, S. A., Salehani, D., Akhavan, N. and Deutsch, D. G. (1994) "Inhibitors of arachidonoyl ethanolamide hydrolysis," *Journal of Biological Chemistry* **269**: 22937–22940.

Lambert, D. M. and Di Marzo, V. (1999) "The palmitoylethanolamide and oleamide enigmas: are these two fatty acid amides cannabimimetic?" *Current Medicinal Chemistry* **6**: 757–773.

Lambert, D. M., DiPaolo, F. G., Sonveaux, P., Kanyonyo, M., Govaerts, S. J., Hermans, E., Bueb, J., Delzenne, N. M. and Tschirhart, E. J. (1999) "Analogues and homologues of N-palmitoylethanolamide, a putative endogenous CB(2) cannabinoid, as potential ligands for the cannabinoid receptors," *Biochimica et Biophysica Acta* **1440**: 266–274.

Lang, W., Qin, C., Hill, W. A. G., Lin, S., Khanolkar, A. D. and Makriyannis, A. (1996) "High-performance liquid chromatographic determination of anandamide amidase activity in rat brain microsomes," *Analytical Biochemistry* **238**: 40–45.

Lang, W., Qin, C., Lin, S., Khanolkar, A. D., Goutopoulos, A., Fan, P., Abouzid, K., Meng, Z., Biegel, D. and Makriyannis, A. (1999) "Substrate specificity and stereoselectivity of rat brain microsomal anandamide amidohydrolase," *Journal of Medicinal Chemistry* **42**: 896–902.

Lee, M., Yang, K. H. and Kaminski, N. E. (1995) "Effects of putative cannabinoid receptor ligands, anandamide and 2-arachidonyl-glycerol, on immune function in B6C3F1 mouse splenocytes," *Journal of Pharmacology and Experimental Therapeutics* **275**: 529–536.

Lemberger, L. (1980) "Potential therapeutic usefulness of marijuana," *Annual Reviews in Pharmacology and Toxicology* **20**: 151–172.

Levitt, M. (1986) In *Cannabinoids as Therapeutic Agents*, edited by R. Mechoulam, pp. 73–79. Boca Raton: CRC Press.

Lin, S., Khanolkar, A. D., Fan, P., Goutopoulos, A., Qin, C., Papahadjis, D., and Makriyannis, A. (1998) "Novel analogues of arachidonylethanolamide (anandamide): affinities for the CB1 and CB2 cannabinoid receptors and metabolic stability," *Journal of Medicinal Chemistry* **41**: 5353–5361.

Liu, J., Gao, B., Mirshahi, F., Sanyal, A. J., Khanolkar, A., Makriyannis, A. and Kunos, G. (2000) "Functional CB1 cannabinoid receptors in human vascular endothelial cells," *Biochemical Journal* **346**: 835–840.

Loewe, S. and Goodman, L. S. (1947) *Federal Proceedings* **6**: 352.

Lopez-Cepero, M., Friedman, M., Klein, T. and Friedman, H. J. (1986) *Journal of Leukocyte Biology* **39**: 679.

Lynn, A. B. and Herkenham, M. (1994) "Localization of cannabinoid receptors and nonsaturable high-density cannabinoid binding sites in peripheral tissue of the rat: implications for receptor-mediated immune modulation by cannabinoids," *Journal of Pharmacology and Experimental Therapeutics* **268**: 1612–1623.

Mackie, K., Devane, W. A. and Hille, B. (1993) "Anandamide, an endogenous cannabinoid, inhibits calcium currents as partial agonist in N18 neuroblastoma cells," *Molecular Pharmacology* **44**: 498–503.

Mackie, K. and Hille, B. (1992) *Proceedings of the National Academy of Sciences of the USA* **106**: 231.

Mackie, K., Lai, Y., Westenbroek, R. and Mitchell, R. (1995) "Cannabinoids activate an inwardly rectifying potassium conductance and inhibit Q-type calcium currents in AtT cells transfected with rat brain cannabinoid receptor," *Journal of Neuroscience* **15**: 6552–6561.

Makriyannis, A. (1995) "The role of cell membranes in cannabinoid activity," in *The Cannabinoid Receptors*, edited by R. G. Pertwee, pp. 87–116. New York: Academic Press.

Makriyannis, A., and Rapaka, R. S. (1990) "The molecular basis of cannabinoid activity," *Life Sciences* **47**: 2173–2184.

Margulies, J. E. and Hammer, R. P. (1991) "Delta-9-tetrahydrocannabinol alters cerebral metabolism in a biphasic, dose-dependent manner in rat brain," *European Journal of Pharmacology* **202**: 373–378.

Mathew, R. J. and Wilson, W. H. (1992) In *Marijuana/Cannabinoids: Neurobiology and Neurophysiology*, edited by L. Murphy and A. Bartke, p. 337. Boca Raton: CRC Press.

Mathew, R. J. and Wilson, W. H. (1993) "Acute changes in cerebral blood flow after smoking marijuana," *Life Sciences* **52**: 757–767.

Mathew, R. J., Wilson, W. H., Humphreys, D. F., Lowe, J. V. and Wiethe, K. E. (1992) *Journal of Cerebral Blood Flow Metabolism* **12**: 750.

Maurelli, S., Bisogno, T., Petrocellis, L. D., Lucci, A. D., Marino, G. and Marzo, V. D. (1995) "Two novel classes of neuroactive fatty acid amides are substrates for mouse neuroblastoma anandamide amidohydrolase," *FEBS Letters* **377**: 82–86.

McGregor, I. S., Arnold, J. C., Weber, M. F., Topple, A. N. and Hunt, G. E. (1998) "A comparison of delta 9-THC and anandamide induced c-fos expression in the rat forebrain," *Brain Research* **802**: 19–26.

Mechoulam, R., Ben Shabat, S., Hanus, L., Fride, E., Vogel, Z., Bayewitch, M. and Sulcova, A.E. (1996) "Endogenous cannabinoid ligands-chemical and biological studies," *Journal of Lipid Mediators and Cell Signalling* **14**: 45–49.

Mechoulam, R., Ben-Shabat, S., Hanus, L., Ligumsky, M., Kaminski, N. E., Schatz, A. R., Gopher, A., Almog, S., Martain, B. R. and Comton, D. R. (1995) "Identification of an endogenous 2-monoglyceride, present in canine gut, that binds to cannabinoid receptors," *Biochemical Pharmacology* **50**: 83–90.

Mechoulam, R., Hanus, L. and Martin, B. R. (1986) *Cannabinoibs as Therapeutic Agents*, pp. 21–159. Boca Raton: CRC Press.

Merritt, J., Crawford, W., Alexander, P., Anduze, A. and Gelbart, S. (1980) "Effect of marihuana on intraocular and blood pressure in glaucoma," *Ophthalmology* **87**: 222.

Munro, S., Thomas, K. L. and Abu Shaar, M. (1993) "Molecular characterization of a peripheral receptor for cannabinoids," *Nature* **365**: 61–65.

Munson, A. E. and Fehr, K. O. (1983) In *Cannabis and Health Hazard*, edited by K. O. Fehr and H. Kalant, p. 257. Toronto.

Nahas, G. G. and Ossweman, E. F. (1991) In *Drugs of Abuse, Immunity, and Immunodeficiency*, edited by S. Spector and T. W. Klein, p. 25. New York: Plenum Press.

Natarajan, V., Schmid, P. C. and Schmid, H. H. O. (1984) *Biochimica et Biophysica Acta* **878**: 32.

Omeir, R. L., Chin, S., Hong, Y., Ahern, D. G. and Deutsch, D. G. (1995) "Arachidonoyl ethanolamide-[1,2-14C] as a substrate for anandamide amidase," *Life Sciences* **56**: 1999–2005.

Paria, B. C., Deutsch, D. D. and Dey, S. K. (1996) "The uterus is a potential site for anandamide synthesis and hydrolysis: differential profiles of anandamide synthase and hydrolase activities in the mouse uterus during the periimplantation period," *Molecular Reproduction and Development* **45**: 183–192.

Pate, D. W., Jarvinen, K., Urtti, A., Jarho, P. and Jarvinen, T. (1995) "Ophthalmic arachido-nylethanolamide decreases intraocular pressure in normotensive Rabbits," *Current Eye Research* **14**: 791–797.

Pertwee, R. G., Stevenson, L. A. and Griffin, G. (1993) "Cross-tolerance between delta 9-tetrahydrocannabinol and the cannabimimetic agents, CP 55,940, WIN 55,212-2 and anandamide," *British Journal of Pharmacology* **110**: 1483–1490.

Pertwee, R. (1993) "The evidence for the existence of cannabinoid receptors," *General Pharmacology* **24**: 811–824.

Pertwee, R., Griffin, G., Hanus, L. and Mechoulam, R. (1994) "Effects of two endogenous fatty acid ethanolamides on mouse vasa deferentia," *European Journal of Pharmacology* **259**: 115–120.

Pertwee, R. G., Fernando, S. R., Griffin, G., Abadji, V. and Makriyannis, A. (1995) "Effect of phenylmethylsulfonylfluoride on the potency of anandamide as an inhibitor of electrically evoked contractions in two isolated tissue preparations," *European Journal of Pharmacology* **272**: 73–78.

Pertwee, R. G., Stevenson, L. A., Elrick, D. B., Mechoulam, R. and Corbett, A. D. (1992) "Inhibitory effects of certain enantiomeric cannabinoids in the mouse vas deferens and myenteric plexus preparation of guinea-pig small intestine," *British Journal of Pharmacology* **105**: 980–984.

Pestonjamasp, V. K. and Burstein, S. H. (1998) "Anandamide synthesis is induced by arachi-donate mobilizing agonists in cells of the immune system," *Biochimica et Biophysica Acta* **1394**: 249–260.

Pinto, J. C., Poti, F., Rice, K. C., Boring, D., Johnson, M. R., Evans, D. M., Wilken, G. H., Cantrell, C. H. and Howlett, A. C. (1994) "Cannabinoid receptor binding and agonist activity of amides and esters of arachidonic acid," *Molecular Pharmacology* **46**: 516–522.

Piomelli, D. (2000) "Pot of Gold for Glioma Therapy," *Nature Medicine* **6**: 255–256.

Piomelli, D., Beltramo, M., Glasnapp, S., Lin, S. Y., Goutopoulos, A., Xie, X. Q. and Makriy-annis, A. (1999) "Structural determinants for recognition and translocation by the anandamide transporter," *Proceedings of the National Academy of Sciences of the USA* **96**(10): 5802–5807.

Poling, J. S., Rogawski, M. A., Salem, N., Jr. and Vicini, S. (1996) "Anandamide, an endo-genous cannabinoid, inhibits Shaker-related voltage-gated K+ channels," *Neuropharmacology* **35**: 983–991.

Priller, J., Briley, E. M., Mansouri, J., Devane, W. A., Mackie, K. and Felder, C. C. (1995) "Mead ethanolamide, a novel eicosanoid, is an agonist for the central (CB1) and peripheral (CB2) cannabinoid receptors," *Molecular Pharmacology* **48**: 288–292.

Qin, C., Hill, W. A. G., Goutopoulos, A., Lin, S. and Makriyannis, A. (1997) "Purification of anandamide amidase by affinity chromatography," Paper presented at the *Proceedings of the International Cannabinoid Research Society*, Program and abstract, 17pp, Stone Mountain, GA.

Qin, C., Lin, S., Land, W., Goutopoulos, A., Pavlopoulos, S., Mauri, F. and Makriyannis, A. (1998) "Determination of anandamide amidase activity using ultraviolet-active amine derivatives and reverse-phase high-performance liquid chromatography," *Analytical Biochemistry* **261**: 8–15.

Randall, M. D., Alexander, S. P., Bennett, T., Boyd, E. A., Fry, J. R., Gardiner, S. M., Kemp, P. A., McCulloch, A. I. and Kendall, D. A. (1996) "An endogenous cannabinoid as an endothelium-derived vasorelaxant," *Biochemical and Biophysical Research Communications* **229**: 114–120.

Rich, M. R. (1993) "Conformational analysis of arachidonic and related fatty acids using molecular dynamics simulations," *Biochimica et Biophysica Acta* **1178**: 87–96.

Romero, J., Garcia, L., Cebeira, M., Zadrony, D., Fernandez-Ruiz, J. J. and Ramos, J. A. (1995) "The endogenous cannabinoid receptor ligand, inhibits the motor behavior: role of nigrostriatal dopaminergic neurons," *Life Science* **56**: 2033–2040.

Schaefer, C., Brackett, D., Gunn, C. and Dubowski, K. J. (1979) *Oklahoma State Medical Association* **72**: 435.

Schatz, A. R., Kessler, F. K. and Kaminski, N. E. (1992) "Inhibition of adenylate cyclase by delta-9-tetrahydrocannabinol in mouse spleen cells: apotentil mechanism for cannabinoid-mediated immunosuppression," *Life Sciences* **51**: 25–30.

Schmid, P. C., Kuwae, T., Krebsbach, R. J. and Schmid, H. H. (1997) "Anandamide and other N-acylethanolamines in mouse peritoneal macrophages," *Chemistry and Physics of Lipids* **87**: 103–110.

Schuel, H., Goldstein, E., Mechoulam, R., Zimmerman, A. M. and Zimmerman, S. (1994) "Anandamide (arachidonylethanolamide), a brain cannabinoid receptor agonist, reduces sperm fertilizing capacity in sea urchins by inhibiting the acrosome reaction," *Proceedings of the National Academy of Science* **91**: 7678–7682.

Schuel, H., Chang, M. C., Burkman, L. J., Picone, R. P., Makriyannis, A., Zimmerman, A. M. and Zimmerman, S. (1999) "Cannabinoid receptors in soerm," in *Marihuana and Medicine*, edited by G. G., Nahas, K. M. Sutin and S. Agurell, pp. 335–345, Humana Press.

Schuel, H., Burkman, L. J., Lippes, J., Picone, R. P., Makriyannis, A., Mahony, M., Giuffrida, A. and Piomelli, D. (2000) "An endocannabinoid-signal system regulates sperm," *International cannabinoid research society symposium*, Hunt Valley, Maryland.

Schwarz, H., Blanco, F. J. and Lotz, M. (1994) "Anandamide an endogenous cannabinoid receptor agonist inhibits lymphocyte proliferation and induces apoptosis," *Journal of Neuroimmunology* **55**: 107–115.

Segal, M. (1986) In *Cannabinoids as Therapeutic Agents*, edited by R. Mechoulam, p. 115. Boca Raton: CRC Press.

Self, D. W. (1999) "Anandamide: a candidate neurotransmitter heads for the big leagues," *Nature Neuroscience* **2**: 303–304.

Seltzman, H. H., Fleming, D. N., Thomas, B. F., Gilliam, A. F., McCallion, D. S., Pertwee, R. G., Compton, D. R. and Martin, B. R. (1997) "Synthesis and pharmacological comparison of dimethylheptyl and pentyl analogs of anandamide," *Journal of Medicinal Chemistry* **40**: 3626–3634.

Sherman, M. P., Roth, M. D., Gong, H. and Tashkin, D. P. (1991) "Marijuana smoking, pulmonary function and lung macrophage oxidant release," *Pharmacology, Biochemistry and Behaviour* **40**: 663–669.

Sheskin, T., Hanus, L., Slager, J., Vogel, Z. and Mechoulam, R. (1997) "Structural requirements for binding of anandamide-type compounds to the brain cannabinoid receptor," *Journal of Medicinal Chemistry* **40**: 659–667.

Shivers, S. C., Newton, C., Friedman, H. and Klein, T. W. (1994) *Life Sciences* **54**: 1281.

Smart, D., Gunthorpe, M. J., Jerman, J. C., Nasir, S., Gray, J., Muir, A. I., Chambers, J. K., Randall, A. D. and Davis, J. B. (2000) "The endogenous lipid anandamide is a full agonist at the human vanilloid receptor," *British Journal of Pharmacology* **129**: 227–230.

Smith, P. B., Compton, D. R., Welch, S. P., Razdan, R. K., Mechoulam, R. and Martin, B. R. (1994). "The pharmacological activity of anandamide, a putative endogenous cannabinoid, in mice," *Journal of Pharmacology and Experimental Therapeutics* **270**: 219–227.

Spector, S. and Lancz, G. (1991). "Suppression of human macrophage function in vitro by delta-9-tetrahydrocannabinol," *Journal Leukocyte Biology* **50**: 423–426.

Steele, N. (1980). *Cancer Treatment Report* **64**: 219.

Stein, E. A., Fuller, S. A., Edgemond, W. S. and Campbell, W. B. (1996) "Physiological and behavioural effects of the endogenous cannabinoid, arachidonylethanolamide (anandamide), in the rat," *British Journal of Pharmacology* **119**: 107–114.

Stella, N., Schweitzer, P. and Piomelli, D. (1997) "A second endogenous cannabinoid that modulates long-term potentiation," *Nature* **388**: 773–778.

Sugiura, T., Kodaka, T., Kondo, S., Tonegawa, T., Nakane, S., Kishimoto, S., Yamashita, A. and Waku, K. (1996a) "2-arachidonoylglycerol, a putative endogenous cannabinoid receptor ligand, induces rapid, transient elevation of intracellular free Ca^{2+} in neuroblastoma × glioma hybrid NG108-15 cells," *Biochemical and Biophysical Research Communications* **229**: 58–64.

Sugiura, T., Kondo, S., Kishimoto, S., Miyashita, T., Nakane, S., Kodaka, T., Suhara, Y., Takayama, H. and Waku, K. (2000) "Evidence that 2-arachidonoylglycerol but not N-palmitoylethanolamine or anandamide is the physiological ligand for the cannabinoid CB2 receptor. Comparison of the agonistic activities of various cannabinoid receptor ligands in HL-60 cells," *Journal of Biological Chemistry* **275**: 605–612.

Sugiura, T., Kondo, S., Sukagawa, A., Nakane, S., Shinoda, A., Itoh, K., Yamashita, A. and Waku, K. (1995) "2-arachidonoylglycerol: a possible endogenous cannabinoid receptor ligand in brain," *Biochemical and Biophysical Research Communications* **215**: 89–97.

Sugiura, T., Kondo, S., Sukagawa, A., Tonegawa, T., Nakane, S., Yamashita, A. and Waku, K. (1996b) "Enzymatic synthesis of anandamide, an endogenous cannabinoid receptor ligand, through N-acylphosphatidylethanolamine pathway in testis: involvement of Ca(2+)-dependent transacylase and phosphodiesterase activities," *Biochemical and Biophysical Research Communications* **218**: 113–117.

Sugiura, T., Kondo, S., Sukagawa, A., Tonegawa, T., Nakane, S., Yamashita, A. and Waku, K. (1996c) "N-arachidonoylethanolamine (anandamide), an endogenous cannabinoid receptor ligand, and related lipid molecules in the nervous tissues," *Journal of Lipid Mediators and Cell Signalling* **14**: 51–56.

Sulcova, E., Mechoulam, R. and Fride, E. (1998) "Biphasic effects of anandamide," *Pharmacology, Biochemistry and Behavior* **59**: 347–352.

Takahashi, Y., Glasgow, W. C., Suzuki, H., Taketani, Y., Yamamoto, S., Anton, M., Kuhn, H. and Brash, A. R. (1993) "Investigation of the oxygenation of phospholipids by the porcine leukocyte and human platelet arachidonate 12-lipoxygenase," *European Journal of Biochemistry* **218**: 165–171.

Thomas, B. F., Adam, I. B., Mascarella, S. W., Martin, B. R. and Razdan, R. K. (1996) "Structure–activity of anandamide analogues: relationship to a cannabinoid pharmacophore," *Journal of Medicinal Chemistry*, **39**: 471–479.

Thomas, E. A., Cravatt, B. F., Danielson, P. E., Gilula, N. B. and Sutcliffe, J. G. (1997) "Fatty acid amide hydrolase, the degradative enzyme for anandamide and oleamide, has selective distribution in neurons within the rat central nervous system," *Journal of Neuroscience Research* **50**: 1047–1052.

Tong, W., Collantes, E. R., Welsh, W. J., Berglund, B. A. and Howlett, A. C. (1998). "Derivation of a pharmacophore model of anandamide using constrained conformational searching and comparative molecular field analysis," *Journal of Medicinal Chemistry* **41**: 4207–4215.

Turkanis, S. A. and Karler, R. (1985) In *Marijuana '84'*, edited by D. J. Harvey. Oxford: IRL Press.

Ueda, N. (1995) "Oxygenation of arachidonylethanolamide (anandamide) by lipoxygenases," *Advances in Prostaglandin, Thromboxane and Leukotriene Research* **23**: 163–165.

Ueda, N., Kurahashi, Y., Yamamoto, S. and Tokunaga, T. (1995a) "Partial purification and characterization of the porcine brain enzyme hydrolyzing and synthesizing anandamide," *Journal of Biological Chemistry* **270**: 23823–23827.

Ueda, N., Yamamoto, K., Yamamoto, S., Tokunaga, T., Shirakawa, E., Shinkai, H., Ogawa, M., Sato, T., Kudo, I. and Inoue, K. (1995b) "Lipoxygenase-catalyzed oxygenation of arachidonylethanolamide, a cannabinoid receptor agonist," *Biochimica et Biophysica Acta* **1254**: 127–134.

Van der Kloot, W. (1994) "Anandamide a naturally-occurring agonist of the cannabinoid receptor, blocks adenylate cyclase at the frog neuromuscular junction," *Brain Research* **649**: 181–184.

Varga, K., Lake, K., Martin, B. R. and Kunos, G. (1995) "Novel antagonist implicates the CB_1 cannabinoid receptor in the hypotensive action of anandamide," *European Journal of Pharmacology* **278**: 279–283.

Venance, L., Piomelli, D., Glowinski, J. and Giaume, C. (1995) "Inhibition by ananda-mide of gap junctions and intercellular calcium signaling in striatal astrocytes," *Nature* **376**: 590–594.

Vogel, Z., Barg, J., Levy, R., Saya, D., Heldman, E. and Mechoulam, R. (1993). "Anand-amide, a brain endogenous compound, interacts specifically with cannabinoid receptors and inhibits adenylate cyclase," *Journal of Neurochemistry* **61**: 352–355.

Volkow, N. D. and Fowler, J. S. (1993) In *Cannabis: Physiopathology, Epidemiology, Detection*, edited by C. G. Nahas and C. Latour, p. 21. Boca Taton: CRC Press.

Walker, J. M., Huang, S. M., Strangman, N. M., Tsou, K. and Sanudo-Pena, M. C. (1999) "Pain modulation by release of the endogenous cannabinoid anandamide," *Proceedings of the National Academy of Sciences of the USA* **96**: 12198–12203.

Wallace, J. M., Tashkin, D. P., Oishi, J. S. and Barbers, R. G. (1988) "Peripheral blood lymphocyte subpopulations and mitogen responsiveness in tobacco and marijuana smokers," *Journal of Psychoactive Drugs* **20**: 9–14.

Wartmann, M., Campbell, D., Subramanian, A., Burstein, S. H. and Davis, R. J. (1995) "The MAP kinase signal transduction pathway is activated by the endogenous cannabinoid anandamide," *FEBBS LETTER* **359**: 133–136.

Watanabe, K., Kayano, Y., Matsunaga, T., Yamamoto, I. and Yoshimura, H. (1996) "Inhib-ition of anandamide amidase activity in mouse brain microsomes by cannabinoids," *Biological and Pharmaceutical Bulletin* **19**: 1109–1111.

Weidenfeld, J., Feldman, S. and Mechoulam, R. (1994) "Effect of the brain constituent anandamide, a cannabinoid receptor agonist, on the hypothalamopituitary adrenal axis in the rat," *Neuroendocrinology* **59**: 110–112.

Welch, S. P. (1997) "Characterization of anandamide-induced tolerance: comparison to delta 9-THC-induced interactions with dynorphinergic systems," *Drug and Alcohol Dependence* **45**: 39–45.

Wenger, T., Fragkakis, G., Giannikou, P., Probonas, K. and Yiannikakis, N. (1997) "Effects of anandamide on gestation in pregnant rats," *Life Sciences* **60**: 2361–2371.

Wenger, T., Toth, B. E. and Martin, B. R. (1995) "Effects of anandamide (endogen cannab-inoid) on anterior pituitary hormone secretion in adult ovariectomized rats," *Life Science* **56**: 2057–2063.

White, H. L. and Tansik, R. L. (1980) "Effects of delta-9-tetrahydrocannabinol and canna-bidiol on phospholipase and other enzymes regulating arachidonate metabolism," *Prostag-landins Med.* **4**: 409–417.

Wickens, A. P. and Pertwee, R. J. (1993) "Delta-9-tetrahydrocannabinoil and anandamide enhance the ability of muscimol to induce catalepsy in the globus pallidus of rats," *European Journal of Pharmacology* **250**: 205–208.

Wise, M. L., Hamberg, M. and Gerwick, W. H. (1994) "Biosynthesis of conjugated triene-containing fatty acids by a novel isomerase from the red marine alga *Ptilota filicina*," *Biochemistry* **33**: 15223–15232.

Yu, M., Ives, D. and Ramesha, C. S. (1997) "Synthesis of prostaglandin E2 ethanolamide from anandamide by cyclooxygenase-2," *Journal of Biological Chemistry* **272**: 21181–21186.

Zheng, Z. M., Specter, S. and Friedman, H. (1992) "Inhibition by delta-9-tetrahydrocanna-binol of tumor necrosis alfa production by mouse and human macrophages," *International Journal of Immunopharmacology* **14**: 1445–1452.

Zygmunt, P. M., Petersson, J., Andersson, D. A., Chuang, H., Sorgard, M., Di Marzo, V., Julius, D. and Hogestatt, E. D. (1999) "Vanilloid receptors on sensory nerves mediate the vasodilator action of anandamide," *Nature* **400**: 452–457.

Chapter 19

Electrophysiological actions of marijuana

Paul Schweitzer

ABSTRACT

Marijuana alters brain physiology by activating specific cannabinoid receptors (CB_1) to modulate neuronal activity. Electrophysiological data obtained with extracellular and intracellular recording techniques indicate that activation of CB_1 influences several neuronal properties. However, full (synthetic) and partial (natural) CB_1 ligands appear to differentially affect some aspects of neurotransmission, a feature to consider when interpreting cannabinoid actions.

The full CB_1 ligand WIN 55,212-2 markedly decreases excitatory and inhibitory synaptic transmission, whereas the natural cannabinoids 2-arachidonylglycerol and anandamide have partial or no effects on these parameters, and Δ^9-tetrahydrocannabinol may increase inhibitory transmission. All cannabinoids, however, consistently decrease long-term potentiation of synaptic transmission, a form of synaptic plasticity associated with learning and memory processes. Endogenously formed cannabinoids appear to have a minimal tonic influence on basal neurotransmission, but play an important role in restricting the potentiation process and may therefore influence synaptic plasticity. Cannabinoids also act on several ion channels at the postsynaptic level to modulate intrinsic neuronal properties. Cannabinoids inhibit Ca^{2+} channels and affect sustained and transient K^+ conductances, with concomitant excitatory and inhibitory consequences. These postsynaptic effects involve transduction mechanisms such as G-proteins and protein kinase A, but eicosanoids are unlikely candidates.

Thus, cannabinoids acting at CB_1 have dual effects on both synaptic transmission and intrinsic neuronal properties, and elicit excitatory and inhibitory effects at the pre- and postsynaptic levels. It now appears that marijuana interferes with an endogenous system of neurotransmitters and has a complex mode of action on brain neurons.

Key Words: cannabinoid, electrophysiology, neuron, synaptic, ion channel

INTRODUCTION

The major active constituent of marijuana, Δ^9-tetrahydrocannabinol (THC), is a cannabinoid compound (Mechoulam *et al.*, 1970). Cannabinoids have powerful psychoactive properties and alter many central physiological processes, such as cognition, locomotion, and nociception (Ameri, 1999). It has now become clear that cannabinoid effects are mediated through specific receptors (CB_1) in mammalian brain (Pertwee, 1997). The CB_1 receptor is one of the most abundantly expressed of the neuronal receptors, with the highest distribution observed in hippocampus, basal ganglia, and cerebellum (Herkenham *et al.*, 1991), areas likely to be involved in marijuana's known effects on cognition and motor control.

The discovery of specific cannabinoid receptors led to the isolation of endogenously formed ligands, the arachidonate derivatives arachidonylethanolamide (anandamide, AEA) and 2-arachidonylglycerol (2-AG), both found in brain (Di Marzo *et al.*, 1994; Stella *et al.*, 1997). Exogenously applied endocannabinoids mimic the effects obtained with synthetic cannabinoids, but with lower efficacy. Indeed, AEA and 2-AG have a relatively low affinity for CB_1 receptors and are partial agonists (Felder and Glass, 1998). Similarly, THC is a partial agonist at CB_1. The development of specific cannabinoid receptor agonists and antagonists, such as the full CB_1 agonist WIN 55,212-2 (WIN-2) and the selective CB_1 antagonist SR 141716 (SR1), has provided powerful pharmacological tools and boosted cannabinoid research.

The existence of specific cannabinoid receptors and of endogenous ligands for these receptors suggests that marijuana affects brain function by interacting with a naturally occuring physiological system of endogenous transmitters. However, the cellular mechanisms of action contributing to the variety of central actions of cannabinoids remain poorly understood, and a primary challenge is to identify the signaling system underlying these effects. This chapter will present electrophysiological data obtained with extracellular and intracellular recordings using the hippocampal slice preparation, and the ongoing research on the electrophysiological actions of marijuana and cannabinoids will be discussed.

METHODS

The hippocampal slice

The high CB_1 density observed in hippocampus combined with its important role in cognitive processes designate this structure as a prime target to investigate. Numerous studies aimed at understanding learning and memory have been conducted in hippocampus (Shen *et al.*, 1994). Hippocampal neurophysiology has been extensively studied and a wealth of data is available on its network organization, neuronal population, and neuronal properties. Thus, cannabinoid studies can be largely focused on a pharmacological approach. *In vitro* slice preparations such as the hippocampal slice (Figure 19.1) offer technical advantages and control over experimental variables (visualization of recording site, drug concentration applied, preservation of native neuronal network), whereas remote effects (e.g. cardiovascular) are eliminated. In general, the experimental approach for electrophysiological studies in slices relies on the analysis of changes in electrical signals during responses to drug superfusion onto slices placed in a recording chamber. Extracellular and intracellular recording techniques are used to investigate the alteration of synaptic activity and plasticity, and postsynaptic membrane properties.

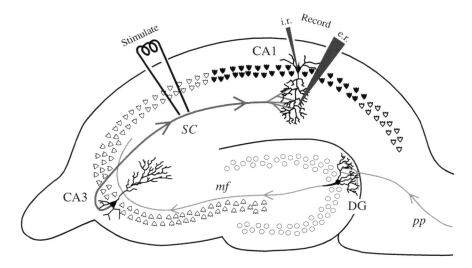

Figure 19.1 The hippocampal slice. Diagram of a transverse hippocampal slice. The stimulating electrode was placed at the Schaffer collaterals (*SC*), a fiber tract that projects from CA_3 to CA_1, to evoke synaptic responses in CA_1. Extracellular recordings (e.r.) were performed in CA_1 stratum radiatum to monitor dendritic fields of excitatory postsynaptic potentials. Intracellular recordings (i.r.) were obtained from the soma of CA_1 pyramidal neurons. DG: dentate gyrus; *mf*: mossy fibers; *pp*: perforant path.

Recording techniques

Slice preparation

Standard recording techniques in adult rat hippocampal slices (350 μm thick) were used as described previously (Stella *et al.*, 1997; Schweitzer, 2000). The slices were submerged and superfused at a constant rate in gassed artificial cerebrospinal fluid (ACSF) of the following composition in mM: NaCl, 130; KCl, 3.5; NaH_2PO_4, 1.25; $MgSO_4$, 1.5; $CaCl_2$, 2.0; $NaHCO_3$, 24; glucose, 10. Drugs were dissolved in 0.1% DMSO.

Extracellular recordings

Extracellular fields of excitatory postsynaptic potentials (fEPSP) were recorded with a glass micropipette (3M NaCl) placed in CA_1 stratum radiatum (Figure 19.1). To evoke synaptic activity, the Schaffer collaterals were stimulated with a bipolar tungsten electrode delivering constant current pulses that elicited a 40% maximal response. Long-term potentiation was elicited by applying 2 trains (20 sec apart) of high frequency stimulations (100 Hz) at basal intensity (40%). Moderate stimulations paradigms were also tested with the use of short trains (0.1, 0.2, or 0.5 sec, i.e. 10, 20, or 50 stimulations). Voltage records were acquired and analyzed with software.

Intracellular recordings

Single-electrode voltage-clamp studies were performed using sharp micropipettes (3M KCl) to penetrate CA_1 pyramidal neurons. Tetrodotoxin (1 µM) was added to the ACSF to block action potentials and synaptic transmission. Current and voltage records were acquired by D/A sampling and analyzed using software. To observe I_M, neurons were depolarized to potentials around −45 mV and hyperpolarizing commands were applied. The various problems (for example, space-clamp) associated with voltage-clamping of neurons with extended processes are discussed elsewhere (Halliwell and Adams, 1982).

CANNABINOIDS MODULATE SYNAPTIC ACTIVITY

Excitatory synaptic transmission

Fast excitatory synaptic transmission at neuronal synapses is mediated by the release of glutamate. Glutamate elicits synaptic responses by activating α-amino-3-hydroxy-5-methyl-4-isoxazolepropionate (AMPA) receptors, the principal determinant of the depolarizing response, and N-methyl-D-aspartate (NMDA) receptors, which allow calcium entry into the cell.

Extracellular fields of excitatory postsynaptic potentials (fEPSPs) were recorded in CA_1 stratum radiatum upon single stimulation of the Schaffer collaterals, a fiber tract that projects from CA_3 to CA_1. Cannabinoid effects were investigated by superfusing the endocannabinoid 2-AG and the aminoalkyndole WIN-2 onto the slices. Addition of 30 µM 2-AG in the superfusate had a small effect to decrease fEPSPs by 6 ± 5% (Figure 19.2). Superfusion of the full CB_1 agonist WIN-2 (2 µM) elicited a much stronger effect and decreased fEPSP by 60 ± 2% (Figure 19.2). The cannabinoid effects were completely prevented by the CB_1 antagonist SR_1 (1 µM), supporting a receptor-mediated action. Note that the time-to-peak from onset for maximum effect was rather long (20 min) and could reflect the difficulty for the drug to penetrate into the slice, despite the use of DMSO as a vehicle. Another possibility is the occurrence of slow transduction mechanisms.

These results indicate that cannabinoids decrease synaptically evoked responses via CB_1. The natural CB_1 ligand 2-AG, however, was ineffective when compared to WIN-2. Although 2-AG is reportedly a partial agonist at CB_1, endogenous inactivation mechanisms described for AEA (transport and degradation) may also play a role to regulate 2-AG activity. Yet, the large difference of effect observed between 2-AG and WIN-2 despite the use of saturating concentrations suggests that receptor activation or binding properties may differ.

Experiments performed in hippocampus and cerebellum have shown that the full CB_1 agonist WIN-2 markedly diminishes glutamatergic synaptic transmission at AMPAR (Shen et al., 1996; Lévénès et al., 1998; Sullivan, 1999). This effect involves the inhibition of Ca^{2+} channels at a presynaptic site, with a consequent decrease of glutamate release. On the other hand, 2-AG had little effect on synaptic transmission in our hands. The other endocannabinoid, AEA, acts at CB_1 to partially decrease synaptic responses, but has also been shown to increases NMDAR-mediated transmission via a non-CB_1 route in hippocampal slices (Ameri

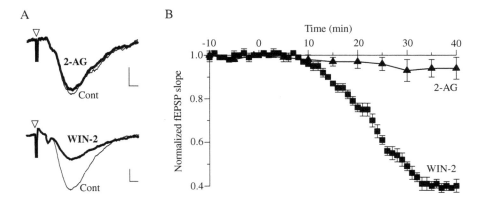

Figure 19.2 Cannabinoids decrease basal synaptic responses. (A) Example of extracellular voltage recordings showing fEPSPs (downward deflection; superimposed for each condition) obtained in CA$_1$ stratum radiatum upon stimulation of the Schaffer Collaterals (triangle indicates stimulation artifact). fEPSPs were recorded in absence of drug (control; thin line) and after a 40 min exposure to a CB$_1$ agonist (thick line). The amplitude of the synaptic response was unaffected by 30 μM 2-AG (top), but decreased by 2 μM WIN-2 (bottom). Calibration: 0.2 mV, 2 ms. (B) Graph average showing the cannabinoids effects on fEPSP overtime (normalized to pre-drug level; drugs were applied at t = 0). 2-AG had a small effect, but WIN-2 greatly decreased fEPSP.

et al., 1999; Hampson *et al.*, 1998). The active substance of marijuana, THC, does not affect synaptic responses in hippocampal slices, although it acts as a partial CB$_1$ agonist to decrease AMPAR-mediated currents between cultured neurons (Hampson *et al.*, 1998; Shen and Thayer, 1999). Overall, results obtained on neurons in slices and cultures indicate that full CB$_1$ agonists strongly decrease excitatory synaptic transmission by about 50%, whereas AEA, 2-AG and THC have a partial or no effect.

Inhibitory synaptic transmission

Inhibitory synaptic transmission is principally mediated by the release of γ-aminobutyric acid (GABA) which activates GABA$_A$ and GABA$_B$ receptors to transiently hyperpolarize neurons. Cannabinoids have been shown to affect inhibitory GABA transmission, but the effects of THC appear to differ from those of WIN-2, a feature reminiscent of the alteration of excitatory glutamatergic transmission by partial and full CB$_1$ agonists.

Thus, THC increases the GABA response in CA$_1$ hippocampus and GABA levels in globus pallidus, an effect believed to occur via inhibition of GABA uptake by THC (Coull *et al.*, 1997; Maneuf *et al.*, 1996). On the other hand, WIN-2 decreases GABA release in hippocampal slices and GABA$_A$ (but not GABA$_B$) receptor-mediated synaptic currents in CA$_1$ pyramidal neurons (Katona *et al.*, 1999; Hoffman and Lupica, 2000). The lack of WIN-2 effect on GABA$_B$ responses has been attributed to the occurrence of distinct inhibitory terminals innervating GABA$_A$ and GABA$_B$

receptors. WIN-2 also decreases GABA$_A$ synaptic currents in striatal neurons, substantia nigra pars reticulata, and rostral medulla (Szabo *et al.*, 1998; Chan *et al.*, 1998; Vaughan *et al.*, 1999). The mechanism implicated in the decrease of inhibitory transmission by WIN-2 appears to involve the inhibition of presynaptic Ca^{++} channels to diminish GABA release, in a manner similar to the cannabinoid-elicited decrease of glutamate release. So far, available data tend to indicate that full CB$_1$ agonists may decrease, whereas THC may increase, inhibitory GABA responses in brain neurons.

Long-term potentiation

Long-term potentiation of synaptic transmission, a long-lasting increase of synaptic strength, is the leading experimental model for the synaptic changes that may underlie learning and memory (Malenka and Nicoll, 1999). Long-term potentiation is observed as the increase of an evoked synaptic response that can last hours, and is induced by rapidly delivering a high frequency stimulation (tetanus) to the synaptic afferences of the recorded region. The stimulation paradigm usually comprises 2 trains delivered 20 seconds apart, each train consisting of 100 shocks delivered in 1 second. Because cannabinoids acting at CB$_1$ impair memory, experiments were conducted to investigate cannabinoid effects on the induction of long-term potentiation elicited in CA$_1$ by stimulating the Schaffer Collaterals.

In control (untreated) slices, delivery of a tetanus strongly increased the fEPSP that remained potentiated to 156 ± 6% (expressed as % of pre-tetanus values) 60 min post-delivery (Figure 19.3A,B). In slices pretreated with the cannabinoid 2-AG, long-term potentiation was completely prevented: the fEPSP transiently increased,

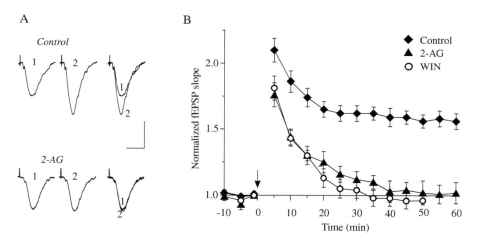

Figure 19.3 Cannabinoids prevent long-term potentiation. (A) fEPSP recordings prior to delivery of a tetanus (1) and 60 min post-delivery (2) (superimposed on the right) in control or in the presence of 30 μM 2-AG. 2-AG completely prevented long-term potentiation. Calibration: 0.4 mV, 10 ms. (B) Graph average of fEPSP slopes over time. Tetanus delivery (arrow; t = 0) elicited long-term potentiation in control slices, but pretreatment with 30 μM 2-AG or 2 μM WIN-2 completely prevented the potentiation process. Drugs were applied 20 to 40 min before tetanus and throughout the experiments.

then decayed back to baseline values within 40 min ($102 \pm 8\%$ 60 min post-delivery; Figure 19.3). This effect was CB_1-mediated, because addition of SR_1 prior to 2-AG in the superfusate permitted the establishment of a potentiation comparable to control ($152 \pm 9\%$). The effects of WIN-2 were then assessed. Similarly to 2-AG, potentiation could not be established in slices pre-treated with WIN-2: the fEPSP was $96 \pm 6\%$ 60 min after delivery of the tetanus (Figure 19.3B), an effect prevented by SR_1. Thus, long-term potentiation was consistently prevented by CB_1 agonists.

Several studies conducted on hippocampal slices have shown that cannabinoids, including THC, prevent the induction of long-term potentiation by activating CB_1 (Nowicky *et al.*, 1987; Terranova *et al.*, 1995; Stella *et al.*, 1997). Because WIN-2 decreases AMPAR-mediated responses, further experiments showed that the WIN-2-induced decrease of glutamate release diminishes postsynaptic depolarization, thus failing to relieve the Mg^{++} block of postsynaptic NMDA receptors and preventing the induction of the potentiation process (Misner and Sullivan, 1999). Such mechanism, however, might be insufficient to explain the effect of 2-AG or THC on long-term potentiation, because these two cannabinoids had little effect on basal synaptic transmission (Stella *et al.*, 1997; Hampson *et al.*, 1998).

Summary

Synthetic and natural cannabinoids differentially affect synaptic responses. Experiments performed in slices or cultures indicate that WIN-2 markedly decreases excitatory and inhibitory synaptic transmission. However, 2-AG and AEA have a partial or no effect, and THC may increase inhibitory transmission in brain. On the other hand, the results obtained on long-term potentiation are consistent among synthetic and natural ligands. Full and partial CB_1 agonists prevent the potentiation process, suggesting that degradation mechanisms or drug penetration are insufficient to interpret the large difference of effect observed on synaptic transmission (Figure 19.4).

THC, 2-AG and AEA are partial agonists and have a rather low potency at CB_1. WIN-2, on the other hand, is a full agonist and believed to be 50 to 100 times more potent than THC, and 200 to 1000 times more potent than 2-AG or AEA. Receptor trafficking and internalization mechanisms are likely to play a major role in cannabinoid pharmacology: for example, CB_1 receptors are rapidly internalized by WIN-2, whereas THC is not effective at all (Hsieh *et al.*, 1999). Thus, different mechanisms regulating neuronal activity could be affected according to the degree of receptor/ligand interaction at CB_1.

ENDOGENOUSLY FORMED CANNABINOIDS AND SYNAPTIC ACTIVITY

Formation of endocannabinoids upon neural activity

Because exogenously applied cannabinoids decreased synaptic activity, the next question was to reveal whether endogenously formed cannabinoids affect synaptic

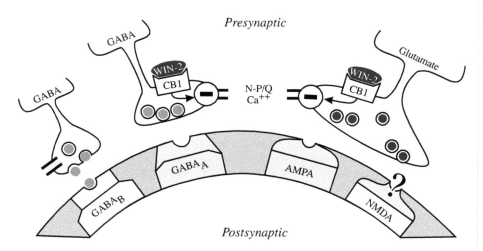

Figure 19.4 Cannabinoid modulation of synaptic transmission. Schematic summarizing cannabinoid actions on synaptic transmission. The full CB_1 agonist WIN-2 acts on presynaptic terminals to decrease the release of glutamate onto AMPA receptors and the release of GABA on $GABA_A$ (but not $GABA_B$) receptors. These actions are believed to occur via inhibition of N-P/Q types Ca^{2+} channels. THC, however, reportedly increases GABAergic transmission and has a partial or no effect on glutamatergic transmission. Although WIN-2 is likely to also decrease glutamate release onto NMDA receptors, AEA has been shown to augment NMDA transmission in a CB_1-independent manner.

transmission and potentiation. In the first place, it was necessary to establish the presence of endogenous CB_1 ligands in hippocampal slices, and if neural activity could influence their formation. AEA and 2-AG were isolated in peripheral tissue after the discovery of CB_1. AEA had been detected in cultured brain neurons (Di Marzo *et al.*, 1994), but the existence of 2-AG in brain still had to be demonstrated.

In collaboration with Drs. Stella and Piomelli, we investigated the occurrence of 2-AG in brain and hippocampus (Stella *et al.*, 1997). Analyses of lipid extracts from rat whole brains by gas chromatography/mass spectrometry showed that 2-AG was present in brain and about 170 times more abundant than AEA. The hippocampal slice preparation was then used to determine if neural activity increased the formation of endocannabinoids. Stimulation of the Schaffer collaterals resulted in a 4-fold increase of the basal level of 2-AG (Figure 19.5). The levels of AEA and 1-palmitylglycerol remained unchanged, indicating a selective effect on 2-AG. The stimulation-induced augmentation of 2-AG was prevented by addition of tetrodotoxin in the superfusate to block action potentials and neural activity, or removal of Ca^{++} to prevent neurotransmitter release.

These results indicated that 2-AG is found in brain and hippocampus and can act as an endogenous cannabinoid ligand. Fiber tract stimulation selectively increased its formation in a depolarization- and calcium-dependent manner, consistent with a neurotransmitter role. Other experiments conducted in striatum have revealed that neural activity in this brain region selectively enhances the

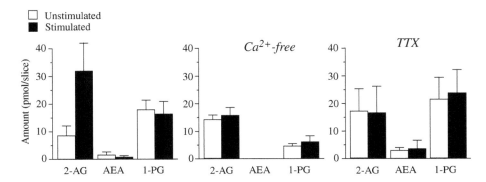

Figure 19.5 Neural activity augments 2-AG formation in hippocampal slices. GC/MS analyses of lipid extracts from hippocampal slices that received an electrical stimulation of the Schaffer Collaterals (stimulated) or no stimulation (unstimulated). In a standard bathing solution (left), delivery of a tetanus selectively increased the formation of 2-AG, without affecting the levels of AEA and 1-palmitylglycerol (1-PG). The increased formation of 2-AG was prevented in a nominally Ca^{2+}-free bathing solution to prevent transmitter release (middle), or when tetrodotoxin was added to the superfusate to block neuronal activity (right).

release of AEA instead of 2-AG (Giuffrida *et al.*, 1999). The selective formation of these two transmitters indicates that they may be produced under different physiological conditions.

Basal transmission

A possible role of endogenously formed cannabinoids to modulate neuronal activity was then investigated. To address this issue, various stimulations paradigms were used to elicit synaptic responses in untreated slices (control) or slices pre-treated with the CB_1 antagonist SR_1 (i.e. does a CB_1 antagonist, by preventing endocannabinoid activation of CB_1, facilitate synaptic activity).

The addition of SR_1 in the superfusate slightly augmented fEPSP by $6 \pm 2\%$, suggesting that endocannabinoids may tonically activate CB_1 to decrease excitatory neurotransmission. Although SR_1 reportedly possesses inverse agonist properties, the detection of substantial amounts of 2-AG and AEA in hippocampal slices in basal conditions is consistent with these results. Furthermore, the lack of effect of SR_1 alone on synaptic transmission in other preparations (Lévénès *et al.*, 1998; Szabo *et al.*, 1998) supports the idea that SR_1 did not elicit an inverse agonist effect, but rather prevented surrounding endocannabinoids to activate CB_1.

Long-term potentiation

A typical paradigm used to elicit long-term potentiation is the delivery of a tetanus consisting of 2 trains of 100 stimulations (see Figure 19.3). This paradigm elicited a long-term potentiation that reached $156 \pm 6\%$ 60 min post-delivery. Because exogenous cannabinoids prevent induction of long-term potentiation by activating

CB_1, one would expect the potentiation process to be greater in the presence of a CB_1 antagonist. When the tetanus was administered in the presence of SR_1, the synaptic response was potentiated to 152 ± 9%, a value similar to that observed in absence of SR_1 (Figure 19.6). Thus, pretreatment of the slices with SR_1 did not increase long-term potentiation elicited with 2 trains of 100 stimulations. We hypothesized that the tetanus paradigm could be too intense and prevent the observation of modulatory effects elicited by endocannabinoids. For example, such long trains (100 shocks, twice) may release vast amounts of glutamate that would elicit the maximum level of potentiation, preventing any further augmentation or facilitation to occur (a "ceiling" effect). Another possibility is that long trains could "overpower" the network and obliterate subtle cellular mechanisms that may otherwise occur.

To test this possibility, experiments were conducted using moderate stimulation paradigms: instead of 100 shocks delivered twice, shorter trains of 10, 20, or 50 shocks (10S, 20S, 50S) were delivered once. Delivery of 10S transiently increased fEPSPs, which rapidly returned to basal levels, an effect unaffected by the presence of SR_1 in the superfusate (Figure 19.6). Delivery of 20S in control conditions also induced a transient increase of fEPSPs, which again rapidly returned to basal levels. However, delivery of 20S in the presence of SR_1 elicited a sustained increase of synaptic responses that remained potentiated (113±3% of basal) 20 min post-delivery. A subsequent train consisting of 50S only induced a limited potentiation in control condition (108±4% 60 min post-delivery), but application of 50S in the presence of SR_1 elicited a sustained increase of fEPSPs that remained largely potentiated to 132±7% 60 min post-delivery (Figure 19.6).

Thus, the use of brief trains indeed revealed the influence of endocannabinoids on synaptic potentiation. Delivery of 10S in the presence of SR_1 was not sufficient to induce potentiation, incidentally ruling out a non-specific effect of SR_1. On the other hand, twice 100 stimulations appeared too strong to permit the regulation of

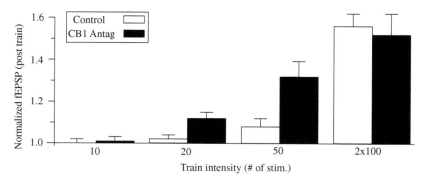

Figure 19.6 Blockade of CB_1 facilitates potentiation. Hippocampal slices were untreated (control, open bars), or pretreated with the CB_1 antagonist SR1 (filled bars) to block CB_1 and prevent its activation by endogenously formed cannabinoids. The graph shows the average level of potentiation reached at each condition after delivering trains of stimulations of various intensities: 10, 20, 50, or twice 100 shocks. Potentiation was facilitated with 20 and 50 shocks, but unchanged with 2 trains of 100 shocks.

the potentiation process by endocannabinoids, whereas 20S or 50S greatly facilitated the potentiation process in slices pretreated with SR_1. The CB_1 antagonist therefore permitted a robust synaptic potentiation by preventing activation of CB_1 by endogenously formed cannabinoids. These results indicate that endocannabinoids may restrict the enhancement of synaptic activity in hippocampus, and may therefore modulate synaptic plasticity in normal brain functioning.

Summary

Endogenously formed cannabinoids appear to have a minimal influence on basal neuronal activity. However, synaptic potentiation elicited with moderate trains of stimulations in the presence of a CB_1 antagonist was greatly facilitated, suggesting that endocannabinoids restrict the potentiation process and may serve as a "brake" to regulate synaptic plasticity. Because GABA-mediated mechanisms modulate potentiation elicited with moderate but not intense tetani (Wigström and Gustafsson, 1983; Chapman *et al.*, 1998), the results suggest a possible action of endocannabinoids on the GABA system to restrict synaptic potentiation. The endocannabinoid system may serve as an inhibitory feedback mechanism in hippocampus, as postulated in striatum (Giuffrida *et al.*, 1999).

CANNABINOIDS AFFECT K$^+$ AND CA^{++} CONDUCTANCES

In central neurons, few studies have investigated cannabinoid actions at the postsynaptic level and available data have been obtained in hippocampus, where cannabinoids affect sustained and transient K^+ conductances, as well as Ca^{++} conductances.

K$^+$ Conductances

Potassium conductances control the transmembrane flow of K^+ to set the membrane potential and are therefore important determinants of neuronal activity and excitability. Sustained (non-inactivating) K^+ conductances, such as the M-current (I_M), have a predominant influence on the intrinsic excitability of neurons because they remain activated overtime and tonically modulate neuronal excitability. Transient (inactivating) K^+ conductances, such as the A- and D-currents (I_A and I_D), are activated during fluctuations of the membrane potential and readily inactivate with respect to their inactivation kinetics.

Cannabinoid effects on sustained conductances were investigated by performing intracellular voltage-clamp recordings of CA_1 pyramidal neurons in the hippocampal slice preparation. Generation of current–voltage relationships indicated that superfusion of WIN-2 elicited an inward steady-state current in the depolarized range, an effect prevented by pretreatment of the slices with SR_1 (Figure 19.7A). The properties and current-voltage profile of the cannabinoid effect were consistent with a selective decrease of I_M. The I_M amplitude was directly assessed by using a specific voltage protocol (Figure 19.7B). These experiments indicated that WIN-2 decreased I_M in a concentration-dependent manner

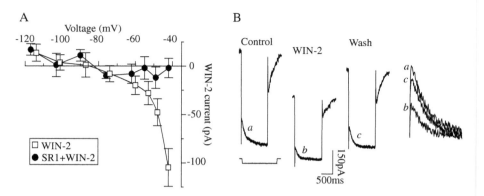

Figure 19.7 Cannabinoids selectively decrease I_M. (A) Net currents elicited by WIN-2 in absence or presence of SR_1, obtained by performing current–voltage relationships on voltage-clamped CA_1 pyramidal neurons. WIN-2 elicited a voltage-dependent inward current that reversed at -85 mV (open squares), an effect prevented by $SR1$ (filled circles). (B) Current recordings of I_M relaxations obtained from a voltage-clamped neuron held at -44 mV and subjected to a 15 mV hyperpolarizing step. Addition of WIN-2 in the superfusate elicited an I_M decrease, with recovery on washout of WIN-2. I_M relaxations magnified and superimposed on the right for comparison.

(EC50 of $0.6 \mu M$), with a maximum inhibition of $45 \pm 3\%$. The cannabinoid-induced I_M decrease was prevented by SR_1 but unaffected by the muscarinic receptor antagonist atropine. Conversely, the cholinergic agonist carbamylcholine decreased I_M in the presence of SR_1, indicating that cannabinoid and muscarinic receptor activation independently diminish I_M. These results show that cannabinoids specifically decrease I_M via CB_1, without affecting other sustained K^+ conductances.

The I_M is the only K^+ current that both activates below the action potential threshold and does not inactivate, readily controlling neuronal activity (Marrion, 1997). Because I_M opposes membrane depolarization, decreasing I_M is an excitatory mechanism that favors increased firing of action potentials and prolonged bursting activity. Consistent with this effect, cannabinoids reinforce bursting activity in CA_1 and increase neuronal firing rate and bursting activity in the ventral tegmentum and substantia nigra *in vivo* (Xue *et al.*, 1993; French *et al.*, 1997).

Cannabinoids also affect transient K^+ currents activated by depolarization in cultured hippocampal neurons, where WIN-2 concurrently increases I_A and decreases I_D (Mu *et al.*, 1999). The activation of these conductances generates a delay in the discharge of action potentials by slowing membrane repolarization for tens of milliseconds (I_A) to several seconds (I_D). The alteration of I_A and I_D by cannabinoids may also alter synaptic input and therefore modulate synaptic integration. Whereas the augmentation of I_A is inhibitory, the diminution of I_M and I_D has excitatory consequences, pointing to dual effects of cannabinoids on intrinsic neuronal excitability.

Ca²⁺ conductances

Calcium conductances mediate calcium influx in response to membrane depolarization and can regulate numerous intracellular processes such as secretion, neurotransmission, and gene expression. Voltage-gated Ca^{2+} channels of the N- and P/Q-types play an important role because they initiate neurotransmission at most fast synapses, and inhibition of presynaptic N- and P/Q-type Ca^{2+} channels reduces the probability of neurotransmitter release.

Cannabinoids have been shown to reduce barium currents passing through N- and P/Q type Ca^{2+} channels via activation of a Gi/o-protein in cultured hippocampal neurons (Twitchell *et al.*, 1997). Although those experiments had to be conducted on cultures grown for a short period, other studies indirectly point at the inhibition of such channels by cannabinoids to decrease synaptic transmission. Thus, the diminution of N- and P/Q Ca^{2+} conductances by cannabinoids is believed to be the mechanism by which cannabinoids presynaptically decrease glutamate and GABA release on hippocampal neurons (Sullivan, 1999; Hoffman and Lupica, 2000). However, other presynaptic mechanisms may occur. The cannabinoid-induced decrease of neurotransmitter release appeared to be Ca^{2+}-independent in periaqueductal gray neurons (Vaughan *et al.*, 2000), and cannabinoids did not affect voltage-dependent Ca^{2+} currents recorded in striatal neurons (Szabo *et al.*, 1998).

Other studies using coexpression or transfection of CB_1 in non-neuronal systems also showed that cannabinoids may activate an inwardly rectifying K^+ conductance and decrease sodium currents, but such effects have not been reported in central neurons.

Transduction mechanisms

Multiple second messenger pathways have been associated with CB_1 (Pertwee, 1997). The inhibition of cyclic AMP production by CB_1 activation has been widely reported and implicates the coupling of CB_1 with an inhibitory G-protein ($G_{i/o}$) to inhibit adenylate cyclase. The modulation of I_A and I_D by cannabinoids occurs via a $G_{i/o}$-protein and inhibition of adenylate cyclase, with subsequent modulation of cyclic AMP and protein kinase (Deadwyler *et al.*, 1995; Mu *et al.*, 1999).

Cannabinoids also alter intracellular Ca^{2+} levels in various ways. NMDA-elicited Ca^{2+} signals may be decreased in hippocampal slices (Hampson *et al.*, 1998), or increased in cerebellar neurons where a mechanism involving phospholipase C and Ca^{2+} release from inositol triphosphate-sensitive Ca^{2+} stores has been postulated (Netzeband *et al.*, 1999). Although the mechanisms of I_M modulation remain elusive, intracellular Ca^{2+} levels appear to play a key role in the decrease of I_M by various transmitters (Marrion, 1997). Interestingly, bradykinin inhibits I_M in ganglion neurons via phospholipase C and Ca^{2+} release from inositol triphosphate-sensitive Ca^{2+} stores (Cruzblanca *et al.*, 1998), and the phospholipase C/inositol triphosphate system has been involved in the muscarinic-induced decrease of I_M in CA_1 pyramidal neurons (Dutar and Nicoll, 1988). It is thus possible that cannabinoids decrease I_M via activation of phospholipase C and enhancement of Ca^{2+} release from inositol triphosphate-sensitive Ca^{2+} stores.

Cannabinoids reportedly release arachidonic acid (Hunter and Burstein, 1997), a molecule produced upon endocannabinoid degradation. Arachidonic acid and its metabolites, the eicosanoids, are potent signaling molecules implicated in several forms of neuromodulation. However, eicosanoids augment I_M in CA_1 pyramidal neurons (Schweitzer *et al.*, 1990), an effect opposite to those of cannabinoids. Furthermore, arachidonic acid also decreases I_A and increases long-term potentiation in hippocampus (Keros and McBain, 1997; Williams *et al.*, 1989), whereas cannabinoids increase I_A and decrease long-term potentiation. Thus, cannabinoids and eicosanoids act on similar targets in hippocampus, but in an opposite direction.

Summary

Cannabinoids act on several channels to modulate neuronal activity at the postsynaptic level. Cannabinoids inhibit N- and P/Q-type Ca^{2+} channels, which consequently diminishes Ca^{2+} entry into synaptic terminals and decreases neurotransmitter release. The diminution of K^+ conductances indicates that cannabinoids also directly increase intrinsic neuronal activity. In contrast with the decrease of glutamatergic transmission, the diminution of I_M and I_D by cannabinoids has excitatory consequences and may amplify synaptic input, perhaps a compensatory mechanism. The postsynaptic effects on ion conductances appear to involve second messengers such as the protein kinase A system. The reported augmentation of intracellular calcium levels by cannabinoids, which contrasts with their inhibition of Ca^{2+} channels, could also play a role to mediate cannabinoid actions, whereas eicosanoids are unlikely candidates. The modulation of various postsynaptic neuronal properties in both excitatory and inhibitory directions point at complex mechanisms of cannabinoid action (Figure 19.8).

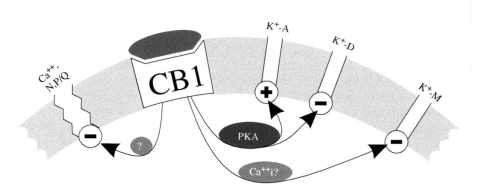

Figure 19.8 Cannabinoid modulation of intrinsic membrane properties. Schematic summarizing cannabinoid actions on K^+ and Ca^{2+} conductances. CB_1 agonists decrease the K^+ M- and D-currents, consequently increasing neuronal activity and excitability. But cannabinoids also increase the K^+ A-current, a transient inhibitory action, and inhibit Ca^{2+} channels, consequently decreasing the release of neurotransmitters. The cannabinoid modulation of the A- and D-currents has been shown to be mediated by cyclic AMP and protein kinase A mechanisms, and the M-current decrease could possibly occur via alteration of intracellular Ca^{2+} levels.

FUNCTIONAL IMPLICATIONS

Impairment of memory

One of the highest CB_1 densities is found in the hippocampus, a brain structure associated with learning and memory processes (Zola-Morgan and Squire, 1993). Acute consumption of marijuana impairs cognitive and performance tasks, including memory, learning, and attention, and such effects are believed to occur via activation of CB_1 (Ameri, 1999). Long-term potentiation of synaptic transmission, a form of synaptic plasticity, is an electrophysiological model believed to underlie the cellular mechanisms by which memories are formed and stored. Although the relevance of long-term potentiation with memory is still controversial, the prevention of this phenomenon by cannabinoids is consistent with the alteration of cognitive processes by THC. Whether the modulation of synaptic potentiation by natural cannabinoids implicates glutamatergic or GABAergic transmission remains to be clarified.

Epileptiform activity

Epilepsy is a brain disorder characterized by seizures, i.e. excessive and synchronous firing of CNS neurons that disrupts normal neurotransmission. Cannabinoids are believed to affect temporal lobe epilepsy, one of the most common forms of this disease that involves the hippocampus and other limbic structures (Ameri, 1999). Research performed before the identification of specific receptors showed that THC has both anticonvulsant and convulsant actions (Martin, 1986). AEA and WIN-2 have been shown to attenuate both stimulus-triggered and spontaneously occurring epileptiform activity elicited by omission of Mg^{2+} in hippocampal slices (Ameri *et al.*, 1999). Although the exact cellular mechanisms implicated in these effects are unknown, the anticonvulsant effect was tentatively attributed to the cannabinoid inhibition of glutamate release. On the other hand, cannabinoids reduce I_M and consequently favor increased firing of action potentials and prolonged bursting activity. Thus, consistent with the alteration of M-channel expression in some form of epilepsy (Biervert *et al.*, 1998), the cannabinoid inhibition of I_M could play a role in the reported convulsant action of THC.

Interaction with the opiate system

There is increasing evidence for physiological interactions between the cannabinoid and opiate systems. The most important cannabinoid–opioid interactions described so far involve antinociception and, to a lesser extent, drug reinforcement (Manzanares *et al.*, 1999). Previous work has shown that the endogenous κ opioid receptor agonist dynorphin decreases synaptic transmission and prevents long-term potentiation (Wagner *et al.*, 1993), and we recently showed that I_M is increased by dynorphin acting at κ opioid receptors in CA_1 hippocampus (Madamba *et al.*, 1999). Interestingly, the interaction of opioid and cannabinoid receptor agonists on antinociceptive mechanisms implicates κ receptors in other brain structures (Manzanares *et al.*, 1999). Similar interactions may take place in hippocampus, where both cannabinoids and κ opioids modulate synaptic transmission, long-term potentiation, and I_M.

CONCLUSION

Neurophysiology research is rapidly progressing to further uncover the mode of action of marijuana and cannabinoids. At the cellular level, electrophysiological data indicate that THC and cannabinoids have complex effects and affect a wide range of neuronal properties at the pre- and postsynaptic levels. Thus, cannabinoids depress neuronal activity via inhibition of Ca^{2+} channels to decrease excitatory neurotransmission at glutamatergic synapses, increase the K^+ I_A, and may increase inhibitory neurotransmission at GABAergic synapses. But cannabinoids also augment neuronal activity by decreasing the K^+ I_M and I_D, and elicit disinhibitory effects associated with decreased GABAergic transmission.

Few studies, however, have assessed the electrophysiological effects of the natural CB_1 ligands THC, 2-AG and AEA. The interpretation of the results is further complicated by the partial agonist properties of these ligands at CB_1. Available data suggest that synthetic and natural CB_1 agonists may differ in their modulation of neuronal activity. It should be considered that the effects of full CB_1 agonists on excitatory and inhibitory synaptic transmission might not readily compare to those of natural CB_1 ligands such as THC. Thus, the physiological consequences of CB_1 activation by these two classes of ligands may diverge.

A variable degree of CB_1 activation elicited by different concentrations of ligands, as well as the cellular interaction between THC and endogenous cannabinoids, could favor particular mechanisms to alter normal brain physiology or pathophysiological conditions. The opposite excitatory and inhibitory actions elicited by cannabinoids may provide a delicate balance to fine tune neuronal activity and influence central physiological processes.

ACKNOWLEDGEMENTS

This work was supported by the National Institute on Drug Abuse.

REFERENCES

Ameri, A. (1999) "The effects of cannabinoids on the brain," *Progress in Neurobiology* **58**: 315–348.

Ameri, A., Wilhelm, A. and Simmet, T. (1999) "Effects of the endogeneous cannabinoid, anandamide, on neuronal activity in rat hippocampal slices," *British Journal of Pharmacology* **126**: 1831–1839.

Biervert, C., Schroeder, B. C., Kubisch, C., Berkovic, S. F., Propping, P., Jentsch, T. J. *et al.* (1998) "A potassium channel mutation in neonatal human epilepsy," *Science* **279**: 403–406.

Chan, P. K., Chan, S. C. and Yung, W. H. (1998) "Presynaptic inhibition of GABAergic inputs to rat substantia nigra pars reticulata neurones by a cannabinoid agonist," *NeuroReport* **9**: 671–675.

Chapman, C. A., Perez, Y. and Lacaille, J. C. (1998) "Effects of GABA$_A$ inhibition on the expression of long-term potentiation in CA1 pyramidal cells are dependent on tetanization parameters," *Hippocampus* **8**: 289–298.

Coull, M. A., Johnston, A. T., Pertwee, R. G. and Davies, S. N. (1997) "Action of Δ-9-tetrahydrocannabinol on GABA-A receptor-mediated responses in a grease-gap recording preparation of the rat hippocampal slice," *Neuropharmacology* **36**: 1387–1392.

Cruzblanca, H., Koh, D. S. and Hille, B. (1998) "Bradykinin inhibits M current via phospholipase C and Ca^{2+} release from IP$_3$-sensitive Ca^{2+} stores in rat sympathetic neurons," *Proceedings of the National Academy of Sciences of the USA* **95**: 7151–7156.

Deadwyler, S. A., Hampson, R. E., Mu, J., Whyte, A. and Childers, S. R. (1995) "Cannabinoids modulate voltage sensitive potassium A-current in hippocampal neurons' *via* a cAMP-dependent process," *Journal of Pharmacology and Experimental Therapeutics* **273**: 734–743.

Di Marzo, V., Fontana, A., Cadas, H., Schinelli, S., Cimino, G., Schwartz, J.-C. *et al.* (1994) "Formation and inactivation of endogenous cannabinoid anandamide in central neurons," *Nature* **372**: 686–691.

Dutar, P. and Nicoll, R. A. (1988) "Classification of muscarinic responses in hippocampus in terms of receptor subtypes and second-messenger systems: electrophysiological studies in vitro," *Journal of Neuroscience* **8**(11): 4214–4224.

Felder, C. C. and Glass, M. (1998) "Cannabinoid receptors and their endogenous agonists," *Annual Review of Pharmacology and Toxicology* **38**: 179–200.

French, E. D., Dillon, K. and Wu, X. (1997) "Cannabinoids excite dopamine neurons in the ventral tegmentum and substantia nigra," *NeuroReport* **8**: 649–652.

Giuffrida, A., Parsons, L. H., Kerr, T. M., De Fonseca, F. R., Navarro, M. and Piomelli, D. (1999) "Dopamine activation of endogenous cannabinoid signaling in dorsal striatum," *Nature Neuroscience* **2**: 358–363.

Halliwell, J. V. and Adams, P. R. (1982) "Voltage-clamp analysis of muscarinic excitation in hippocampal neurons," *Brain Research* **250**: 71–92.

Hampson, A. J., Bornheim, L. M., Scanziani, M., Yost, C. S., Gray, A. T., Hansen, B. M. *et al.* (1998) "Dual effects of anandamide on NMDA receptor-mediated responses and neurotransmission," *Journal of Neurochemistry* **70**: 671–676.

Herkenham, M., Lynn, A. B., Johnson, M. R., Melvin, L. S., De Costa, B. R. and Rice, K. C. (1991) "Characterization and localization of cannabinoid receptors in rat brain: a quantitative *in vitro* autoradiographic study," *Journal of Neuroscience* **11**: 563–583.

Hoffman, A. F. and Lupica, C. R. (2000) "Mechanisms of cannabinoid inhibition of GABA$_A$ synaptic transmission in the hippocampus," *Journal of Neuroscience* **20**: 2470–2479.

Hsieh, C., Brown, S., Derleth, C. and Mackie, K. (1999) "Internalization and recycling of the CB$_1$ cannabinoid receptor," *Journal of Neurochemistry* **73**: 493–501.

Hunter, S. A. and Burstein, S. H. (1997) "Receptor mediation in cannabinoid stimulated arachidonic acid mobilization and anandamide synthesis," *Life Sciences* **60**: 1563–1573.

Katona, I., Sperlágh, B., Sík, A., Käfalvi, A., Vizi, E. S., Mackie, K. *et al.* (1999) "Presynaptically located CB$_1$ cannabinoid receptors regulate GABA release from axon terminals of specific hippocampal interneurons," *Journal of Neuroscience* **19**: 4544–4558.

Keros, S. and McBain, C. J. (1997) "Arachidonic acid inhibits transient potassium currents and broadens action potentials during electrographic seizures in hippocampal pyramidal and inhibitory interneurons," *Journal of Neuroscience* **17**: 3476–3487.

Lévénès, C., Daniel, H., Soubrié, P. and Crépel, F. (1998) "Cannabinoids decrease excitatory synaptic transmission and impair long-term depression in rat cerebellar Purkinje cells," *Journal of Physiology* **510**: 867–879.

Madamba, S. G., Schweitzer, P. and Siggins, G. R. (1999) "Dynorphin selectively augments the M-current in hippocampal CA1 neurons by an opiate receptor mechanism," *Journal of Neurophysiology* **82**: 1768–1775.

Malenka, R. C. and Nicoll, R. A. (1999) "Long-term potentiation – A decade of progress?" *Science* **285**: 1870–1874.

Maneuf, Y. P., Nash, J. E., Crossman, A. R. and Brotchie, J. M. (1996) "Activation of the cannabinoid receptor by delta-9-tetrahydrocannabinol reduces gamma-aminobutyric acid uptake in the globus pallidus," *European Journal of Pharmacology* **308**: 161–164.

Manzanares, J., Corchero, J., Romero, J., Fernández-Ruiz, J. J., Ramos, J. A. and Fuentes, J. A. (1999) "Pharmacological and biochemical interactions between opioids and cannabinoids," *Trends in Pharmacological Sciences* **20**: 287–294.

Marrion, N. V. (1997) "Control of M-current," *Annual Review of Physiology* **59**: 483–504.

Martin, B. R. (1986) "Cellular effects of cannabinoids," *Pharmacological Reviews* **38**: 45–74.

Mechoulam, R., Shani, A., Edery, H. and Grunfeld, Y. (1970) "Chemical basis of hashish activity," *Science* **169**: 611–612.

Misner, D. L. and Sullivan, J. M. (1999) "Mechanism of cannabinoid effects on long-term potentiation and depression in hippocampal CA1 neurons," *Journal of Neuroscience* **19**: 6795–6805.

Mu, J., Zhuang, S. Y., Kirby, M. T., Hampson, R. E. and Deadwyler, S. A. (1999) "Cannabinoid receptors differentially modulate potassium A and D currents in hippocampal neurons in culture," *Journal of Pharmacology and Experimental Therapeutics* **291**: 893–902.

Netzeband, J. G., Conroy, S. M., Parsons, K. L. and Gruol, D. L. (1999) "Cannabinoids enhance NMDA-elicited Ca^{2+} signals in cerebellar granule neurons in culture," *Journal of Neuroscience* **19**: 8765–8777.

Nowicky, A. V., Teyler, T. J. and Vardaris, R. M. (1987) "The modulation of long-term potentiation by delta-9-tetrahydrocannabinol in the rat hippocampus, *in vitro*," *Brain Research Bulletin* **19**: 663–672.

Pertwee, R. G. (1997) "Pharmacology of cannabinoid CB_1 and CB_2 receptors," *Pharmacological Therapeutics* **74**: 129–180.

Schweitzer, P. (2000) "Cannabinoids decrease the K^+ M-current in hippocampal CA1 neurons," *Journal of Neuroscience* **20**: 51–58.

Schweitzer, P., Madamba, S. G. and Siggins, G. R. (1990) "Arachidonic acid metabolites as mediators of somatostatin-induced increase of neuronal M-current," *Nature* **346**: 464–467.

Shen, M. X., Piser, T. M., Seybold, V. S. and Thayer, S. A. (1996) "Cannabinoid receptor agonists inhibit glutamatergic synaptic transmission in rat hippocampal cultures," *Journal of Neuroscience* **16**: 4322–4334.

Shen, M. X. and Thayer, S. A. (1999) "Delta-9-tetrahydrocannabinol acts as a partial agonist to modulate glutamatergic synaptic transmission between rat hippocampal neurons in culture," *Molecular Pharmacology* **55**: 8–13.

Shen, Y., Specht, S. M., De Saint Ghislain, I. and Li, R. (1994) "The hippocampus: a biological model for studying learning and memory," *Progress in Neurobiology* **44**: 485–496.

Stella, N., Schweitzer, P. and Piomelli, D. (1997) "A second endogenous cannabinoid that modulates long-term potentiation," *Nature* **388**: 773–778.

Sullivan, J. M. (1999) "Mechanisms of cannabinoid-receptor-mediated inhibition of synaptic transmission in cultured hippocampal pyramidal neurons," *Journal of Neurophysiology* **82**: 1286–1294.

Szabo, B., Dörner, L., Pfreundtner, C., Nörenberg, W. and Starke, K. (1998) "Inhibition of gabaergic inhibitory postsynaptic currents by cannabinoids in rat corpus striatum," *Neuroscience* **85**: 395–403.

Terranova, J. P., Michaud, J. C., Le Fur, G. and Soubrié, P. (1995) "Inhibition of long-term potentiation in rat hippocampal slices by anandamide and WIN 55212-2: Reversal by SR141716 A, a selective antagonist of CB_1 cannabinoid receptors," *Naunyn-Schmiedebergs Archives of Pharmacology* **352**: 576–579.

Twitchell, W., Brown, S. and Mackie, K. (1997) "Cannabinoids inhibit N- and P/Q-type calcium channels in cultured rat hippocampal neurons," *Journal of Neurophysiology* **78**: 43–50.

Vaughan, C. W., Connor, M., Bagley, E. E. and Christie, M. J. (2000) "Actions of cannabinoids on membrane properties and synaptic transmission in rat periaqueductal gray neurons in vitro," *Molecular Pharmacology* **57**: 288–295.

Vaughan, C. W., McGregor, I. S. and Christie, M. J. (1999) "Cannabinoid receptor activation inhibits GABAergic neurotransmission in rostral ventromedial medulla neurons *in vitro*," *British Journal of Pharmacology* **127**: 935–940.

Wagner, J. J., Terman, G. W. and Chavkin, C. (1993) "Endogenous dynorphins inhibit excitatory neurotransmission and block LTP induction in the hippocampus," *Nature* **363**: 451–454.

Wigström, H. and Gustafsson, B. (1983) "Facilitated induction of hippocampal long-lasting potentiation during blockade of inhibition," *Nature* **301**: 603–604.

Williams, J. H., Errington, M. L., Lynch, M. A. and Bliss, T. V. P. (1989) "Arachidonic acid induces a long-term activity-dependent enhancement of synaptic transmission in the hippocampus," *Nature* **341**: 739–742.

Xue, B. G., Belluzzi, J. D. and Stein, L. (1993) "*In vitro* reinforcement of hippocampal bursting by the cannabinoid receptor agonist (-)-CP-55,940," *Brain Research* **626**: 272–277.

Zola-Morgan, S. and Squire, L. R. (1993) "Neuroanatomy of memory," *Annual Review of Neuroscience* **16**: 547–563.

Chapter 20

The vascular pharmacology of endocannabinoids

Michael D. Randall, David Harris and David A. Kendall

ABSTRACT

Endogenous cannabinoids (endocannabinoids), which were first identified in the central nervous system, exert cardiovascular actions. The prototypic endocannabinoid, anandamide, which is derived from arachidonic acid, is a vasodilator in the resistance vasculature. However, the mechanisms of vasorelaxation to endocannabinoids are currently unclear but may involve both endothelium-dependent and independent pathways. To date, the mechanisms proposed for the vasorelaxant actions of anandamide have included the release of endothelial autacoids, activation of myoendothelial gap junctions, activation of the sodium pump, activation of potassium channels, inhibition of calcium channels, and activation of vanilloid receptors leading to the release of sensory neurotransmitters. The vasodilator actions of endocannabinoids have been implicated in the hypotension associated with both septic and hemorrhagic shock, but their physiological significance remains to be determined.

Key Words: endocannabinoids, anandamide, vasorelaxation, endothelium, gap junctions, hyperpolarization

ENDOCANNABINOIDS

In 1992 the first endocannabinoid, anandamide (*N*-arachidonoylethanolamine), which is the ethanolamide of arachidonic acid, was isolated from the porcine brain (Devane *et al.*, 1992). It was shown both to occupy cannabinoid receptors and to mimic the functional effects of Δ^9-tetrahydrocannabinol. Anandamide is the prototype of a family of *N*-acylethanolamines and other polyunsaturated *N*-acylethanolamines (Hanus *et al.*, 1993), which have similar effects via activation of G protein-linked cannabinoid (CB_1 or CB_2) receptors. To date, CB_1 receptors have been found predominantly in the brain and peripheral nervous system, and the CB_2 receptors appear to be exclusive to immune tissues.

CARDIOVASCULAR EFFECTS OF ENDOCANNABINOIDS

Actions *in vivo*

One of the therapeutic indications for cannabis has been as an antihypertensive, although developments in this direction have been stifled by the stigma associated with cannabis. The relatively limited literature concerning the cardiovascular effects of exogenous cannabinoids contains variable observations with both vasodilator and vasoconstrictor actions being reported (Stark and Dews, 1980). In one of the most recent reports, Niederhoffer and Szabo (1999) demonstrated that CB_1-receptor agonists (CP 55940 and WIN 55212-2) cause presynaptic inhibition of sympathetic activity leading to hypotension. By contrast Vidrio et al. (1996) reported that the cannabinoid agonist HU210, on administration to both conscious and anesthetized rats, caused prolonged bradycardia and hypotension, which was not mediated through a sympatholytic action. Similarly, in isolated atrial and rat mesenteric vessel preparations, Lay et al. (2000) have demonstrated that neither synthetic cannabinoid ligands nor anandamide influence sympathetic neurotransmission.

Given the recent interest in endocannabinoids, attention has now turned towards their actions on the cardiovascular system. In this respect, exogenous anandamide, causes bradycardia (with secondary hypotension) and a transient pressor effect which is followed by a longer lasting depressor effect in urethane-anesthetized rats (Varga et al., 1995; Lake et al., 1997). This depressor effect was believed to be mediated by CB_1 receptor-dependent inhibition of sympathetic tone via a presynaptic mechanism, as the effect was independently attenuated by cervical spinal transection, α-adrenoceptor blockade and cannabinoid receptor blockade. This sympatholytic action is greater in spontaneously hypertensive rats compared to normotensive controls, perhaps reflecting the higher level of sympathetic tone in the former (Lake et al., 1997). The pressor component of the response to anandamide was not sensitive to cannabinoid receptor blockade, perhaps reflecting a non CB_1 receptor-mediated response (Lake et al., 1997). By contrast, Stein et al. (1996) reported that, although anandamide caused bradycardia in conscious rats, it caused a transient depressor response, followed by a longer pressor phase, and only at high doses was there delayed hypotension. In mice, both anandamide and synthetic cannabinoid receptor agonists cause biphasic hypotension, which is thought to be solely CB_1-receptor mediated as the responses are absent in CB_1-receptor knockout mice (Ledent et al., 1999).

The cardiovascular actions of endocannabinoids are complicated by their rapid metabolism. This is especially apparent for 2-arachidonoyl glycerol (2-AG) which is particularly unstable and is believed to be the natural ligand for CB_1-receptors. When 2-AG was applied exogenously to mice it caused CB_1-receptor-independent effects via arachidonic acid metabolites (Jarai et al., 2000). However, when a stable analogue of 2-AG was used, the hypotensive effects were mediated via CB_1-receptors. The rapid metabolism of 2-AG might therefore point to the actions of endocannabinoids, in particular 2-AG being localized to their site of production, with arachidonic acid metabolites contributing at more distant sites.

Consistent with efficient local metabolism of endocannabinoids, Calignano et al. (1997) found that the hypotension caused by anandamide in guinea-pigs (which occurs independently of the autonomic nervous system but is mediated via CB_1 receptors) was potentiated by inhibition of anandamide reuptake. Furthermore, in isolated mesenteric vessels inhibition of the cannabinoid transporter with AM404 or bromocresol green also enhanced vasorelaxation to anandamide (Harris et al., 1998; Figure 20.1). These observations provided functional evidence that this system terminates the cardiovascular actions of cannabinoids. The actions of the

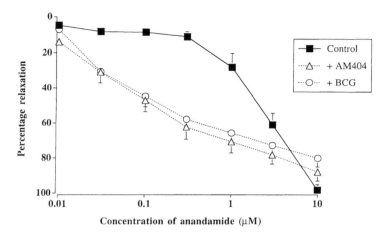

Figure 20.1 Vasorelaxation to anandamide in the rat isolated perfused mesenteric arterial bed. The graph shows that in the presence of either AM404 (3 µM) or bromocresol green (30 µM), both of which block the reuptake of anandamide, the potency of anandamide is increased. Data shown as mean ± S.E.M.

transporter might not be confined to terminating the actions of endocannabinoids as it has been found in rabbit mesenteric vessels that inhibition of the transporter actually attenuates the actions of anandamide (Chaytor *et al.*, 1999). This led to the suggestion that the cannabinoid transporter might, in some circumstances, function to allow anandamide access to intracellular sites, where it exerts vascular effects.

Vascular actions of endocannabinoids

Anandamide was first shown to be a vasodilator in the rat cerebral vasculature, but these effects were sensitive to indomethacin, suggesting that cannabinoids may cause relaxation through the stimulation of the metabolism of arachidonic acid (Ellis *et al.*, 1995). Dependence on prostanoids was also found for the vasorelaxant effects of Δ^9-tetrahydrocannabinol. In the rat isolated mesenteric and coronary vasculatures, anandamide is a vasodilator (Randall *et al.*, 1996; Randall and Kendall, 1997; Figure 20.2). In mesenteric arterial vessels (Randall *et al.*, 1996; Randall *et al.*, 1997; Plane *et al.*, 1997; White and Hiley, 1997), and the coronary vasculature (Randall and Kendall, 1997), anandamide induces relaxation in the presence of blockers of both nitric oxide synthase and cyclooxygenase, and also in the absence of the endothelium (Randall *et al.*, 1996; White and Hiley, 1997; Figure 20.2). Accordingly in these vessels it acts independently of endothelial autacoids. Early studies reported that responses to anandamide were abolished by high extracellular potassium, and proposed that endocannabinoids might act via hyperpolarization (Randall *et al.*, 1996).

The relaxant effects of anandamide show tissue selectivity, as it does not relax conduit vessels such as rat carotid arteries (Holland *et al.*, 1999) or the rat aorta

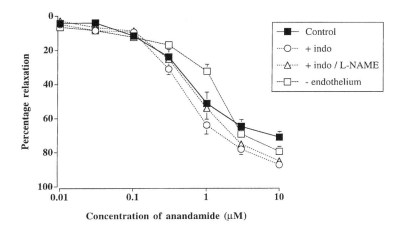

Figure 20.2 Shows that in the rat isolated perfused mesenteric arterial bed that vasorelaxation to anandamide is unaffected by inhibition of nitric oxide synthesis (with 300 μM L-NAME), inhibition of cyclooxygenase (with 10 μM indomethacin) or removal of the endothelium. Data shown as mean ± S.E.M.

(Darker *et al.*, 1998). Indeed, it may be that the actions of endocannabinoids are localized to the resistance vasculature.

Endocannabinoids and EDHF

In 1996, it was proposed that an endocannabinoid might represent an endothelium-derived hyperpolarising factor (EDHF) (Randall *et al.*, 1996). This was based on the observations that anandamide was an endothelium-independent vasorelaxant, which appeared to act via a hyperpolarizing mechanism, and that EDHF responses were sensitive to the CB_1-receptor antagonist, SR 141716A. This proposal, although not generally accepted, generated substantial interest in the vascular actions of endocannabinoids. The finding that SR 141716A opposed EDHF-mediated responses was confirmed by some (White and Hiley, 1997; Hewitt *et al.*, 1997; Rowe *et al.*, 1998; Sedhev *et al.*, 1998) but not all groups (Plane *et al.*, 1997; Zygmunt *et al.*, 1997; Chataigneau *et al.*, 1998; Pratt *et al.*, 1998). In particular Chataigneau *et al.* (1998) reported that SR 141716A opposed EDHF-mediated hyperpolarization in some preparations, but ascribed this effect to the antagonist interacting with potassium channels rather than CB receptors. Pratt *et al.* (1998) provided evidence that SR 141716A interferes with arachidonic acid metabolism, and this could, potentially, explain the effects against EDHF, which they proposed to be a non-cannabimimetic metabolite of arachidonic acid. Further insight into the actions of SR 141716A came from the work of Chaytor *et al.* (1999). In their study they reported that high concentrations of SR 141716A did oppose EDHF responses, but that this was due to the antagonist acting via inhibition of myoendothelial gap junctions. Interestingly, this finding added further weight to the contention that EDHF-type relaxations could be explained by myoendothelial gap junctions mediating heterocellular communication between the endothelium and vascular smooth muscle, which may or may not also involve a humoral factor.

Where are endocannabinoids produced in the vasculature?

If endocannabinoids are to be significant regulators of vascular function then they must be produced in areas associated with the cardiovascular system. The proposal that endocannabinoids might be endothelium-derived autacoids was supported by the finding that cultured rat renal endothelial cells contain anandamide, together with synthase and amidase activities (Deutsch *et al.*, 1997). In addition, cultured human umbilical vein endothelial cells release the endocannabinoid, 2-arachidonylglycerol, on stimulation with a calcium ionophore (Sugiura *et al.*, 1998). However, bovine coronary endothelial cells do not produce endocannabinoids, and furthermore, these cells metabolize exogenous anandamide, possibility via a cytochrome P450 monooxygenase to vasoactive metabolites (Pratt *et al.*, 1998).

Sensory nerves have been proposed as another potential site of production and release. In this respect Ishioka and Bukoski (1999) demonstrated that nerve-dependent calcium-induced relaxation of rat mesenteric vessels was blocked by SR141716A, with the suggestion that an endocannabinoid was released from the nerves, and mediated the vasorelaxation.

In the context of pathophysiology, Wagner *et al.* (1997) demonstrated in a rat model of hemorrhagic shock that activated macrophages released anandamide. Similarly in endotoxic shock the synthesis of 2-AG in platelets is increased and anandamide is only detectable in macrophages after exposure to lipopolysaccharide (Varga *et al.*, 1998). *In vitro,* mouse J774 macrophages also release both 2-AG and anandamide, and participate in their degradation (Di Marzo *et al.*, 1999). These findings certainly point to the genesis of endocannabinoids in blood cells, which is enhanced in shock, and contributes towards the cardiovascular sequelae.

Mechanisms of vasorelaxation for endocannabinoids

The original findings suggested that anandamide might act via a hyperpolarizing mechanism (Randall *et al.*, 1996). This proposal was confirmed electrophysiologically by the demonstration that anandamide causes hyperpolarization or repolarization of vascular smooth muscle, but in both cases this effect was independent of cannabinoid CB_1 receptors (Plane *et al.*, 1997, Chaitaigneau *et al.*, 1998). The hyperpolarization was also found to be endothelium-dependent (Chataigneau *et al.*, 1998; Zygmunt *et al.*, 1997), with the implication that anandamide acted via the release of EDHF (Figure 20.3). The latter study also provided evidence that anandamide acted via inhibition of calcium mobilization in vascular smooth muscle cells, without direct effects on potassium conductance.

In rat mesenteric vessels, anandamide-induced relaxation is sensitive to non-specific potassium channel blockers, including cytochrome P450 inhibitors (Randall *et al.*, 1997). In isolated mesenteric arterial segments the relaxation to anandamide was blocked by selective inhibitors of large conductance calcium-activated K^+-channels (charybdotoxin and iberiotoxin; Plane *et al.*, 1997). Furthermore, in similar mesenteric vessels, the anandamide-induced relaxation was insensitive to

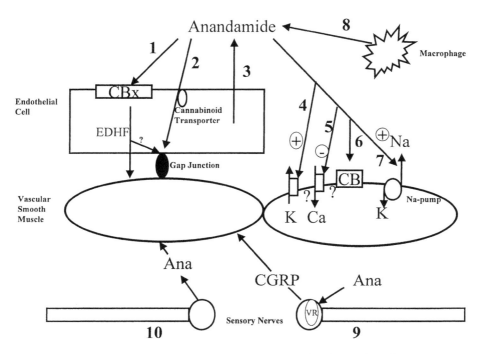

Figure 20.3 Summary diagram of the vascular pharmacology of endocannabinoids, incorporating the various putative sites of synthesis, and proposed mechanisms and sites of action. (1) The proposal that anandamide (ana) acts on novel endothelial cannabinoid receptors (CBx) to elicit the release of the endothelium-derived hyperpolarizing factor (EDHF) (Wagner *et al.*, 1999; Jarai *et al.*, 1999). (2) The suggestion that anandamide gains access to the endothelium via the transporter and acts via gap junctions (Chaytor *et al.*, 1999). (3) The proposal that the endothelium releases endocannabinoids (Randall *et al.*, 1996; Deutsch *et al.*, 1997; Sugiura *et al.*, 1998). The actions of anandamide via: (4) activation of the potassium channels (Randall *et al.*, 1996; Plane *et al.*, 1997; Randall *et al.*, 1997; Randall and Kendall, 1998; Chataigneau *et al.*, 1998; Ishioka and Bukoski, 1999); (5) inhibition of voltage-operated calcium channels (Gebremedhin *et al.*, 1999) (both 4 and 5 via coupling (6) to CB-receptors); (7) activation of the sodium pump (Figure 20.4). (8) Illustrates the release of anandamide from macrophages in shock (Wagner *et al.*, 1997; Varga *et al.*, 1998). (9) The proposal that anandamide acts via vanilloid receptors (VR) to release CGRP from sensory nerves (Zygmunt *et al.*, 1999). (10) The proposal that sensory nerves release anandamide (Ishioka and Bukoski, 1999).

the combination of charybdotoxin and apamin (White and Hiley, 1997). By contrast in the perfused mesenteric arterial bed the combination of charybdotoxin and apamin abolished relaxation to anandamide, although neither agent alone affected the responses (Randall and Kendall, 1998). In the guinea-pig carotid artery, the anandamide-induced hyperpolarization, which was insensitive to charybdotoxin plus apamin, was blocked by the ATP-sensitive potassium channel inhibitor, glibenclamide (Chataigneau *et al.*, 1998). By contrast, glibenclamide does not affect anandamide-induced relaxation in the rat mesentery (Randall *et al.*, 1997; White and Hiley, 1997). In conclusion some, but not all, studies point to the involvement of K^+-channels, at some stage, in the vasorelaxant actions of anandamide (Figure 20.3).

The endothelium-dependent hyperpolarization and sensitivity, in some cases, to K^+-channel inhibitors, raised the possibility that anandamide might act in an endothelium-dependent manner via the release of EDHF. However, in this respect initial studies indicated that vasorelaxation was preserved following removal of the endothelium (Randall et al., 1996; White and Hiley, 1997; Figure 20.2). By contrast this was not found to be the case in all blood vessels. In bovine coronary vessels anandamide certainly causes endothelium-dependent relaxation by metabolism to cytochrome P450 metabolites of arachidonic acid (Pratt et al., 1998). In rabbit mesenteric vessels anandamide acts partly in an endothelium-dependent manner following uptake into the endothelial cells, where it appears to promote gap junctional opening, leading to smooth muscle relaxation (Chaytor et al., 1999). Endothelial uptake may also be important in allowing anandamide to have intracellular effects, such as raising cytosolic calcium (Mombouli et al., 1999). This action itself might trigger endothelial-mediated responses, such as the release of EDHF.

In rat mesenteric vessels, Wagner et al. (1999) identified a small endothelial component of relaxation to anandamide which was SR 141716A-sensitive but not mediated by CB_1-receptors. This led to them to propose that there is a novel endothelial cannabinoid receptor (Figure 20.3). One alternative explanation for this could be that SR141716A was acting to inhibit responses to anandamide via inhibition of gap junctions (Chaytor et al., 1999). Further work in this area has indicated that a neurobehaviorally inactive cannabinoid, abnormal cannabidiol, causes SR141716A-senstive mesenteric vasodilatation which is also blocked by cannabidiol (Jarai et al., 1999). From these findings it was proposed that cannabidiol was an antagonist of this novel endothelial cannabinoid receptor, which is coupled to EDHF release (Figure 20.3).

Under some circumstances anandamide has been shown to act via release of endothelium-derived nitric oxide (Deutsch et al., 1997), although in many instances (see Randall et al., 1996; White and Hiley, 1997; Jarai et al., 1999) the responses to anandamide are insensitive to inhibition of nitric oxide synthase.

The possibility that vasorelaxation to anandamide might involve EDHF release led us to investigate further the pharmacology of vasorelaxation to anandamide. To this end the effects of gap junction inhibitors and ouabain, the sodium pump inhibitor which also inhibits gap junctional communication, were investigated against responses to anandamide (Figure 20.4). In this respect we were unable to demonstrate any endothelial-dependence of vasorelaxant responses to anandamide. However, the vasorelaxation was sensitive to gap junction inhibitors (18α-glycyrrhetinic acid and ouabain) which also block the sodium pump but was unaffected by agents which are selective for gap junctions (carbenoxolone and palmitoleic acid). This has raised the possibility that the sodium pump may at some stage be involved in vasorelaxation to anandamide independently of any contribution of EDHF.

Vascular smooth muscle calcium channels have also been proposed to be the target for endocannabinoids (Gebremedhin et al., 1999; Figure 20.3). In feline cerebral vessels it was shown that both endocannabinoids and synthetic

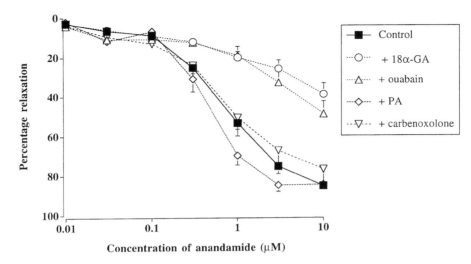

Figure 20.4 Vasorelaxation to anandamide in the rat isolated perfused mesenteric arterial bed in the presence of the combined gap junction and sodium pump inhibitors (1mM ouabain and 100 μM 18α-glycyrrhetinic acid (18α-GA)) and pure gap junction inhibitors (50 μM palmitoleic acid (PA) and 100 μM carbenoxolone). These data support the proposal that anandamide may act via activation of the sodium pump but not via gap junction activation. Data shown as mean ± S.E.M.

cannabinoids act via G-protein coupled CB_1-receptors to cause inhibition of voltage-sensitive calcium channels, leading to vasodilatation.

One of the most recent proposals for the vascular actions of anandamide has been that anandamide is a vanilloid agonist. In support of this, Zygmunt *et al.* (1999) reported that vasorelaxant responses to anandamide (but not 2-AG, palmitylethanolamide or synthetic cannabinoid receptor agonists) were essentially abolished by pre-treatment with capsaicin to deplete the sensory nerves, especially of calcitonin gene-related peptide (CGRP) (Figure 20.3). In addition, the responses to anandamide were sensitive to the vanilloid antagonist capsazepine and, also, CGRP receptor antagonism. The conclusion from this study was that anandamide evoked the release of transmitters from sensory nerves leading to vasorelaxation. In support of this, similar observations have been made with the analog of anandamide, methanandamide (Ralevic *et al.*, 2000). By contrast, Harris *et al.* (2000) have reported that the responses to anandamide were only partly sensitive to capsaicin pre-treatment in the presence of a functional NO system. In the presence of NO synthase blockade vasorelaxation due to anandamide was insensitive to capsaicin pretreatment and thus does not occur via sensory nerves. Accordingly the activation of sensory nerves by anandamide may only explain part of the actions anandamide, and only under some circumstances. Indeed, the fact that the hypotensive action of anandamide is absent in mice lacking CB_1 receptors (Ledent *et al.*, 1999) certainly suggests that any action via vanilloid receptors is only of minor importance.

Vascular cannabinoid receptors

The vascular actions of endocannabinoids suggests that there may be vascular cannabinoid receptors, which may either fall into the classical CB_1/CB_2 classification or represent a new subtype. To date the sensitivity of vasorelaxant responses to CB-receptor antagonists has been controversial with some studies indicating that the responses are opposed by SR 141716A (Randall et al., 1996; White and Hiley, 1997) and others demonstrating that they are insensitive to this antagonist (Plane et al., 1997). Indeed, as described above, sensitivity to SR 141716A might reflect non-cannabinoid receptor actions (Pratt et al., 1998; Chaytor et al., 1999), although Jarai et al. (1999) have suggested the presence of a novel vascular CB-receptor.

Using reverse transcriptase–polymerase chain reaction we have identified a gene product in mesenteric resistance arterioles and cerebral micro-vessels which may suggest that a CB_1-like transcript is present in rat and human resistance vessels suggesting local CB_1 receptor expression (Darker et al., 1998; Randall et al., 1999). Using immunohistochemistry, antibody staining for CB_1-like immunoreactivity was associated with both the endothelium and smooth muscle in mesenteric vessels but not in the thoracic aorta (which does not relax to anandamide) (Randall et al., 1999).

Cannabinoid CB_1 receptors have also been localized to cat cerebral arterial smooth muscle (Gebremedhin et al., 1999). In their study it was demonstrated that feline vascular smooth muscle contained CB_1-receptors together with cDNA showing very close homology to that associated with neuronal CB_1-receptors.

Endocannabinoids and pathophysiology

A role for endocannabinoids in hypotension associated with hemorrhagic shock has been advanced (Wagner et al., 1997; Figure 20.3). In this respect, in a rat model of hemorrhagic shock, the accompanying hypotension was reversed by the cannabinoid receptor antagonist, SR 141716A, whilst activated macrophages were found to release anandamide. Similarly in endotoxic shock the syntheses of 2-AG in platelets and anandamide in macrophages are increased (Varga et al., 1998). It is conceivable that the activated blood cells could also stimulate the release of endocannabinoids from the endothelium or other vascular sites, contributing further towards the hypotension. The release of anandamide by central neurones under hypoxic conditions, leading to improved blood flow and protection against ischaemia has also been advanced as a pathophysiological role for anandamide (Gebremedhin et al., 1999).

CONCLUDING REMARKS

Endocannabinoids exert powerful vasorelaxant effects. However, the precise mechanism of action of vasorelaxation is uncertain. Whether these responses have an endothelium-dependent component varies between species and between vascular sites. The responses may involve gap junctional communication, actions on the sodium pump, release of endothelial autacoids, inhibition of voltage-operated calcium channels or non-cannabinoid actions on vanilloid receptors leading to release of neurotransmitters from sensory nerves (Figure 20.3). Indeed there may be several targets for the endocannabinoids and experimentally (or in pathologically) removing one target may be compensated by action at another site.

The fact that endocannabinoids have vascular effects raises the question of their relevance. As discussed above there is some emerging evidence pointing to their participation in shock but what roles, if any, they play in normal physiology remains to be established.

ACKNOWLEDGMENTS

We thank the British Heart Foundation for financial support. DH holds an MRC Studentship.

REFERENCES

Calignano, A., La Rana, G., Beltramo, M., Makriyannis, A. and Piomelli, D. (1997) "Potentiation of anandamide hypotension by transport inhibitor, AM404," *European Journal of Pharmacology* **337**: R1–R2.

Chataigneau, T., Feletou, M., Thollon, C., Villeneuve, N., Vilaine, J. P., Duhault, J. *et al.* (1998) "Cannabinoid CB1 receptor and endothelium-dependent hyperpolarization in guinea-pig carotid, rat mesenteric and porcine coronary arteries," *British Journal of Pharmacology* **123**: 968–974.

Chaytor, A. T., Martin, P. E. M., Evans, W. H., Randall, M. D. and Griffth, T. M. (1999) "The endothelial component of cannabinoid-induced relaxation in rabbit mesenteric artery depends on gap junctional communication," *The Journal of Physiology* **520**: 539–550.

Darker, I. T., Millns, P. J., Selbie, L., Randall, M. D., S-Baxter, G. and Kendall, D. A. (1998) "Cannabinoid (CB$_1$) receptor expression is associated with mesenteric resistance vessels but not thoracic aorta in the rat," *British Journal of Pharmacology* **125**: 95P.

Deutsch D. G., Goligorsky M. S., Schmid P. G., Krebsbach R. J., Schmid H. H. O., Das S. K. *et al.* (1997) "Production and physiological actions of anandamide in the vasculature of the rat kidney," *The Journal of Clinical Investigation* **100**: 1538–1546.

Devane W. A., Hanus, L., Breuer, A., Pertwee, R. G., Stevenson, L. A., Griffin, G. *et al.* (1992) "Isolation and structure of a brain constituent that binds to the cannabinoid receptor," *Science* **258**: 1946–1949.

Di Marzo, V., Bisogno, T., DePetrocellis, L., Melck, D., Orlando, P., Wagner, J. A. *et al.* (1999) "Biosynthesis and inactivation of the endocannabinoid 2-arachidonoylglycerol in circulating and tumoral macrophages," *European Journal of Biochemistry* **264**: 258–267.

Ellis, E. F., Moore, S. F. and Willoughby, K. A. (1995) "Anandamide and Δ^9-THC dilation of cerebral arterioles is blocked by indomethacin," *American Journal of Physiology* **269**: H1859–H1864.

Gebremedhin, D., Lange, A. R., Campbell, W. B., Hillard, C. J. and Harder, D. R. (1999) "Cannabinoid CB$_1$ receptor of cat cerebral arterial muscle functions to inhibit L-type Ca^{2+} channel current," *American Journal of Physiology* **45**: H2085–H2093.

Hanus L., Gopher, A., Almog, S. and Mechoulam, R. (1993) "Two new unsaturated fatty acid ethanolamides in brain that bind to the cannabinoid receptor," *Journal of Medicinal Chemistry* **36**: 3032–3034.

Harris, D., Kendall, D. A. and Randall, M. D. (1998) "Effects of AM404, a cannabinoid reuptake inhibitor, on EDHF-mediated relaxations in the rat isolated mesentery," *British Journal of Pharmacology* **125**: 15P.

Harris, D., Ralevic, V., Kendall, D. A. and Randall, M. D. (2000) "Capsaicin and ruthenium red antagonism of anandamide-induced relaxation is blocked by L-NAME in the rat mesenteric arterial bed," *British Journal of Pharmacology* (in press).

Hewitt, N., Plane, F. and Garland, C. J. (1997) "Bioassay of EDHF in the rabbit isolated femoral artery," *British Journal of Pharmacology* **122**: 122P.

Holland, M., Challis, R. A., Standen, N. B. and Boyle, J. P. (1999) "Cannabinoid CB_1 receptors fail to cause relaxation, but couple via Gi/Go to inhibition of adenylyl cyclase in carotid artery smooth muscle," *British Journal of Pharmacology* **128**: 597–604.

Ishioka, N. and Bukoski, R. D. (1999) "A role for N-arachidonylethanolamine (anandamide) as the mediator of sensory nerve-dependent Ca^{2+}-induced relaxation," *The Journal of Pharmacology and Experimental Therapeutics* **289**: 245–250.

Jarai, Z., Wagner, J. A., Goparaju, S. K., Wang, L., Razadan, R. K., Sugiura, T. *et al.* (2000) "Cardiovascular effects of 2-arachidonoylglycerol in anaesthetized mice," *Hypertension* **35**: 679–684.

Jarai, Z., Wagner, J. A., Varga, K., Lake, K., Compton, D. R., Martin, B. R. *et al.* (1999) "Cannabinoid-induced mesenteric vasodilation through an endothelial site distinct from CB_1 or CB_2 receptors," *Proceedings of the National Academy Sciences of the USA* **96**: 14136–14141.

Lake, K., Martin, B. R., Kunos, G. and Varga, K. (1997) "Cardiovascular effects of anandamide in anaesthetized and conscious normotensive rats," *Hypertension* **29**: 1204–1210.

Lay, L., Angus, J. A. and Wright, C. E. (2000) "Pharmacological characterisation of cannabinoid CB_1 receptors in the rat and mouse," *European Journal of Pharmacology* **391**: 151–161.

Ledent, C., Valverde, O., Cossu, C. Petitet, F., Aubert, L. F., Beslot, F. *et al.* (1999) "Unresponsiveness to cannabinoids and reduced additive effects of opiates in CB_1 receptor knockout mice," *Science* **283**: 401–404.

Mombouli, J. V., Schaeffer, G., Holzmann, S., Kostner, G. M. and Graier, W. F. (1999) "Anandamide-induced mobilization of cytosolic Ca^{2+} in endothelial cells," *British Journal of Pharmacology* **126**: 1593–1600.

Niederhoffer, N. and Szabo, B. (1999). "Effect of the cannabinoid receptor agonist WIN 55212-2 on sympathetic cardiovascular regulation," *British Journal of Pharmacology* **126**: 457–466.

Plane, F., Holland, M., Waldron, G. J., Garland, C. J. and Boyle, J. P. (1997) "Evidence that anandamide and EDHF act via different mechanisms in rat isolated mesenteric arteries," *British Journal of Pharmacology* **121**: 1509–1512.

Pratt, P. F., Hillard, C. J., Edgemond, W. S. and Campbell, W. B. (1998) "N-arachidonylethanolamide relaxation of bovine coronary artery is not mediated by CB_1 cannabinoid receptor," *American Journal of Physiology* **274**: H375–H381.

Ralevic, V., Kendall, D. A., Randall, M. D., Zygmunt, P. M., Movahed, P. and Högestatt, E. D. (2000) "Vanilloid receptors on capsaicin-sensitive nerves mediate relaxation to methanandamide in the rat isolated mesenteric bed," *British Journal of Pharmacology* (in press).

Randall M. D., Alexander, S. P. H., Bennett, T., Boyd, E. A., Fry, J. R., Gardiner, S. M. *et al.* (1996) "An endogenous cannabinoid as an endothelium-derived vasorelaxant," *Biochemical and Biophysical Research Communications* **229**: 114–120.

Randall, M. D., Harris, D., Darker, I. T., Millns, P. J. and Kendall, D. A (1999). "Endocannabinoids: endothelium-derived vasodilators," in *Endothelium-dependent hyperpolarizations*, edited by P. M. Vanhoutte, pp. 149–155. Amsterdam: Harcourt.

Randall, M. D. and Kendall, D. A. (1997) "The involvement of an endogenous cannabinoid in EDHF-mediated vasorelaxation in the rat coronary vasculature," *European Journal of Pharmacology* **335**: 205–209.

Randall, M. D. and Kendall, D. A. (1998) "Anandamide and endothelium-derived hyperpolarizing factor act via a common vasorelaxant mechanism in rat mesentery," *European Journal of Pharmacology* **346**: 51–53.

Randall, M. D., McCulloch, A. I. and Kendall, D. A. (1997). "Comparative pharmacology of endothelium-derived hyperpolarizing factor and anandamide in rat isolated mesentery," *European Journal of Pharmacology* **333**: 191–197.

Rowe, D. T. D., Garland, C. J. and Plane, F. (1998) "Multiple pathways underlie NO-independent relaxation to the calcium ionophore A23187 in the rabbit isolated femoral artery," *British Journal of Pharmacology* **123**: 1P.

Sedhev, J., Garland, C. J. and Plane, F. (1998) "Further characterization of the mediator of NO-independent dilatation to acetylcholine in the rat isolated perfused mesenteric bed," *British Journal of Pharmacology* **123**: 2P.

Stark, P. and Dews, P. B. (1980) "Cannabinoids. II. Cardiovascular effects," *The Journal of Pharmacology and Experimental Therapeutics* **214**: 131–138.

Stein, E. A., Fuller, S. A., Edgemond, W. S. and Campbell, W. B. (1996) "Physiological and behavioural effects of the endogenous cannabinoid, arachidonylethanolamine (anandamide), in the rat," *British Journal of Pharmacology* **119**: 107–114.

Sugiura, T., Kodaka, T., Nakane, S., Kishimoto, S., Kondo, S. and Waku, K. (1998) "Detection of an endogenous cannabimimetic molecule, 2-arachidonoylglycerol, and cannabinoid CB_1 receptor mRNA in human vascular cells: is 2-arachidonoylglycerol a possible vasomodulator," *Biochemical and Biophysical Research Communications* **243**: 838–843.

Varga, K., Lake, K., Martin, B. R. and Kunos, G. (1995) "Novel antagonist implicates CB_1 cannabinoid receptor in the hypotensive action of anandamide," *European Journal of Pharmacology* **278**: 279–283.

Varga, K., Wagner, J. A., Bridgen, D. T. and Kunos, G. (1998) "Platelet- and macrophage-derived endogenous cannabinoids are involved in endotoxin-induced hypotension," *FASEB Journal* **12**: 1035–1044.

Vidrio, H., Sanchez-Salvatori, M. A. and Medina, M. (1996). "Cardiovascular effects of $(-)$-11-OH-Δ^8-tetrahydrocannabinol-dimethylheptyl in rats," *Journal of Cardiovascular Pharmacology* **28**: 332–336.

Wagner, J. A., Varga, K., Ellis, E. F., Rzigalinski, B. A., Martin, B. R. and Kunos, G. (1997) "Activation of peripheral CB1 cannabinoid receptors in haemorrhagic shock," *Nature* **390**: 518–521.

Wagner, J. A., Varga, K., Jarai, Z. and Kunos, G. (1999) "Mesenteric vasodilation mediated by endothelial anandamide receptors," *Hypertension* **33**: 429–434.

White, R. and Hiley, C. R. (1997) "A comparison EDHF-mediated and anandamide-induced relaxations in the rat isolated mesenteric artery," *British Journal of Pharmacology* **122**: 1573–1584.

Zygmunt, P. M., Högestatt, E. D., Waldeck, K., Edwards, G., Kirkup, A. J. and Weston, A. H. (1997) "Studies on the effects of anandamide in the rat hepatic artery," *British Journal of Pharmacology* **122**: 1679–1686.

Zygmunt, P. M., Petersson, J., Anderssson, D. A., Chuang, H-h., Sorgard, M., Di Marzo, V. *et al.* (1999) "Vanilloid receptors on sensory nerves mediate the vasodilator action of anandamide," *Nature* **400**: 452–457.

Chapter 21

The cannabinoid receptors and their interactions with synthetic cannabinoid agonists and antagonists

David Shire, Paul Gouldson, Bernard Calandra, Marielle Portier, Monsif Bouaboula, Francis Barth, Murielle Rinaldi-Carmona, Pierre Casellas, Gérard Le Fur and Pascual Ferrara

ABSTRACT

Here we describe the characteristics of the cannabinoid receptor subtypes, CB_1 and CB_2, and what we know of their interactions with certain ligands. CB_1 and CB_2 are members of the superfamily of 7 transmembrane domain (7TM) receptors that transduce intracellular signals via heterotrimeric GTP-binding proteins. Like many other 7TM receptors the cannabinoid receptors have been the subject of intense research with a view to designing highly specific and potent ligands that may turn out to be effective medicaments for alleviating human ailments. A detailed knowledge of the structure of both ligand and receptor and how they interact may be useful in this respect. Although the two human receptors share only 44% overall identity, ranging from 35% to 82% in the TM regions, it is only recently that subtype specific agonists have been found, those hitherto available showing no or little subtype selectivity. In contrast, two highly specific antagonists, SR 141716A for the CB_1 receptor and SR 144528 for the CB_2 receptor, besides being promising drug candidates, are also proving to be excellent tools for investigating the structural and functional characteristics of the cannabinoid receptors. We describe experiments with mutated receptors undertaken to try to discover which amino acid residues contact ligands. We also consider the constitutive activity of CB_1 and CB_2 when they are overexpressed in heterologous systems and the classical inverse agonist properties exhibited by SR 141716A and SR 144528 with their respective target receptors.

Key Words: structure, mutagenesis, ligand–receptor interactions, inverse agonism

THE CANNABINOID RECEPTORS

The first cannabinoid receptor to be identified, CB_1, was isolated from a rat brain cDNA library a decade ago (Matsuda *et al.*, 1990) and the human equivalent was cloned within a year (Gérard *et al.*, 1991). CB_1 is found principally in the central nervous system, but also occurs in many peripheral regions. Three years after the cloning of CB_1 a second subtype, CB_2, was discovered fortuitously in differentiated human myeloid cells by a PCR-based strategy (Munro *et al.*, 1993). CB_2 is found mainly in the spleen and in cells of the immune system. The human CB_1 receptor gene, designated *Cnr1*, has been localized on chromosome 6 at 6q14-q15 (Hoehe *et al.*,

1991) and the mouse CB_1 receptor gene on proximal chromosome 4 (Stubbs *et al.*, 1996). The CB_2 receptor gene, designated *Cnr2* is located on the distal end of mouse chromosome 4 and on human chromosome 1p36 (Valk *et al.*, 1997). The coding region of both CB_1 and CB_2 occurs in a single exon, which also contains a short 5' leader sequence and an extensive 3'-untranslated sequence. A truncated human CB_1 receptor having a 61 amino acid deletion within the amino terminus resulting from an alternative splicing event, was described at the DNA level (Shire *et al.*, 1995) but the receptor has not been detected *in vivo* and, without this confirmation, its existence remains questionable. Current knowledge of cannabinoid gene structure and tissue distribution was excellently reviewed in 1997 by Matsuda (Matsuda, 1997). Much remains to be done at the molecular level concerning cannabinoid promoter structure and regulation of mRNA expression and role played by the long 3' untranslated regions, which contain several polyadenylation signals.

Mammalian CB_1 receptor coding sequences are presently available for human, hCB_1, (Gérard *et al.*, 1991), rat, rCB_1, (Matsuda *et al.*, 1990), mouse, mCB_1, (Chakrabarti *et al.*, 1995) and cat, cCB_1, receptors (see SwissProt: CB1R_FELCA), together with a partial bovine (Pfister-Genskow *et al.*, 1997) sequence and also two related sequences from the puffer fish (Yamaguchi *et al.*, 1996). The puffer fish sequences show only 73.9% and 60.1% identity with the hCB_1 receptor sequence and the encoded proteins have not been characterized for cannabinoid binding and activity. Recently a newt receptor, nCB_1, has been sequenced (see Genbank: AF 181,894) and partial leech sequences have been published (Stefano *et al.*, 1997). An alignment of all the complete receptor amino acid sequences with a consensus sequence at fully conserved positions reveals a remarkable preservation of the CB_1 receptor primary structure, most of the few differences occurring in the extremities (Figure 21.1). The rCB_1 and mCB_1 receptors differ in length from the cCB_1 and hCB_1 receptors by an amino acid insertion in the amino terminus, but otherwise show only 12 and 13 amino acid differences, respectively, with the hCB_1 receptor, the differences being primarily in the amino and carboxyl ends of the receptors. The cCB_1 receptor sequence has 18 residues that differ from those of the hCB_1 receptor, the differences also being mainly in the extremities. The cross-species preservation of the CB_1 receptor, almost perfect in the helical bundles, is remarkable and is particularly evident if one considers the DNA coding sequences in the various species (not shown). For example, compared to the over 1400 nucleotides encoding the human sequence, the rat and mouse sequences have 134 and 133 differences respectively, less than 10% of which result in amino acid changes. This remarkably conserved receptor primary sequence across species, together with the conserved tissue localization recorded by Herkenham (Herkenham, 1995), would indicate an important and universal role for the CB_1 receptor. An intransigent structural requirement could well explain the difficulties we have encountered with the expression of some mutated CB_1 receptors (Shire *et al.*, 1996a).

The conservatism of the CB_1 receptor sequence contrasts with the variability seen with the CB_2 receptor. Full sequence information is available for the human, hCB_2, (Munro *et al.*, 1993) mouse, mCB_2, (Shire *et al.*, 1996b) and rat, rCB_2, (Griffin *et al.*, 2000) receptors. The hCB_2 and mCB_2 receptors share only 82.2% amino acid identity; the mouse sequence is shorter by 13 residues at the carboxyl terminus and has 57 other differences distributed throughout the receptor (Figure 21.2).

```
hCB₁  MKSILDGLADTTFRTITTDLLYVGSNDIQYEDIKGDMASKLGYFPQKFPLTSFRGSPFQEKMTAGDNPQL.VPA.DQVNITEFYNKSLSSFKENEENIQCGE  100
cCB₁  MKSILDGLADTTFRTITTDLLYVGSNDIQYEDIKGDMASKLGYFPQKFPLTSFRGSPFQEKMTAGDNSQL.VPA.DQVNITEFYNKSLSSYKENEENIQCGE  100
mCB₁  MKSILDGLADTTFRTITTDLLYVGSNDIQYEDIKGDMASKLGYFPQKFPLTSFRGSPFQEKMTAGDNSPL.VPAGDTTNITEFYNKSLSSFKENEDNIQCGE  101
rCB₁  MKSILDGLADTTFRTITTDLLYVGSNDIQYEDIKGDMASKLGYFPQKFPLTSFRGSPFQEKMTAGDNSPL.VPAGDTTNITEFYNKSLSSFKENEENIQCGE  101
nCB₁  MKSILDGLADTTFRTITTDLLYMGSNDVQYEDTKGEMASKLGYFPQKLPLSSFRRDHSPDKMTIGDDNLLSFYPLDQFNVTEFFNRSVSTFKENDDNLKCGE  102
Cons  MKSILDGLADTTFRTITTDLLY*GSND*QYED-KG-MASKLGYFPQK-PL*SFR-----*KMT-GD---L-----D--N*TEF*N*S*S**KEN**N*-CGE

                               I                                          II
hCB₁  NFMDIECFMVLNPSQQLAIAVLSLTGTFTVLENLLVLCVILHSRSLRCRPSYHFIGSLAVADLLGSVIFVYSFIDFHVFHRKDSRNVFLFKLGGVTASF  200
cCB₁  NFMDMECFMILNPSQQLAIAVLSLTGTFTVLENLLVLCVILHSRSLRCRPSYHFIGSLAVADLLGSVIFVYSFVDFHVFHRKDSPNVFLFKLGGVTASF  200
mCB₁  NFMDMECFMILNPSQQLAIAVLSLTGTFTVLENLLVLCVILHSRSLRCRPSYHFIGSLAVADLLGSVIFVYSFVDFHVFHRKDSPNVFLFKLGGVTASF  201
rCB₁  NFMDMECFMILNPSQQLAIAVLSLTGTFTVLENLLVLCVILHSRSLRCRPSYHFIGSLAVADLLGSVIFVYSFVDFHVFHRKDSPNVFLFKLGGVTASF  201
nCB₁  NFMDMECFMILTASQQLIIAVLSLTGTFTVLENFLVLCVILQSRTLRCRPSYHFIGSLAVADLLGSVIFVYSFLDFHVFHRKDSSNVFLFKLGGVTASF  202
Cons  NFMD*ECFM-L-SQQL*IIAVLSLTGTFTVLEN-LVLCVIL-SR*LRCRPSYHFIGSLAVADLLGSVIFVYSF*DFHVFHRKDS-NVFLFKLGGVTASF

       III                       IV                                              V
hCB₁  TASVGSLFLTAIDRYISIHRPLAYKRIVTRPKAVVAFCLMWTIAIVIAVLPLLGWNCEKLQSVCSDIFPHIDETYLMFWIGVTSVLLLFIVYAYMYILWK  300
cCB₁  TASVGSLFLTAIDRYISIHRPLAYKRIVTRPKAVVAFCLMWTIAIVIAVLPLLGWNCKKLQSVCSDIFPLIDETYLMFWIGVTSVLLLFIVYAYMYILWK  300
mCB₁  TASVGSLFLTAIDRYISIHRPLAYKRIVTRPKAVVAFCLMWTIAIVIAVLPLLGWNCKKLQSVCSDIFPLIDETYLMFWIGVTSVLLLFIVYAYMYILWK  301
rCB₁  TASVGSLFLTAIDRYISIHRPLAYKRIVTRPKAVVAFCLMWTIAIVIAVLPLLGWNCKKLQSVCSDIFPLIDETYLMFWIGVTSVLLLFIVYAYMYILWK  301
nCB₁  TASVGSLFLTAIDRYISIHRPLAYKRIVTRTKAVIAFCVMWTIAIIIAVLPLLGWNCKKLSVCSDIFPLIDENYLMFWIGVTSILLLFIVYAYMYILWK  302
Cons  TASVGSLFLTAIDRYISIHRPLAYK*IVTR-KAV*AFC*MWTIAI*IAVLPLLGWNC-KL-SVCSDIFP-IDE-YLMFWIGVTS*LLLFIVYAY*YILWK

                                                       VI                                     VII
hCB₁  AHSHAVRMIQRGTQKSIIIHTSEDGKVQVTRPDQARMDIRLAKTLVLILVVLIICWGPLLAIMVYDVFGKMNKLIKTVFAFCSMLCLLNSTVNPIIYALR  400
cCB₁  AHIHAVRMIQRGTQKSIIIHTSEDGKVQVTRPDQARMDIRLAKTLVLILVVLIICWGPLLAIMVYDVFGKMNKLIKTVFAFCSMLCLLNSTVNPIIYALR  400
mCB₁  AHSHAVRMIQRGTQKSIIIHTSEDGKVQVTRPDQARMDIRLAKTLVLILVVLIICWGPLLAIMVYDVFGKMNKLIKTVFAFCSMLCLLNSTVNPIIYALR  401
rCB₁  AHSHAVRMIQRGTQKSIIIHTSEDGKVQVTRPDQARMDIRLAKTLVLILVVLIICWGPLLAIMVYDVFGKMNKLIKTVFAFCSMLCLLNSTVNPIIYALR  401
nCB₁  AHSHAVRMLQRGTQKSIIIHTSEDGKVQITRPEQTRMDIRLAKTLVLILVVLIICWGPLLAIMVYDVFGKMNNPIKTVFAFCSMLCLMDSTVNPIIYALR  402
Cons  AH-HAVRM*QRGTQKSIIIHTSEDGKVQ*TRP*Q-RMDIRLAKTLVLILVVLIICWGPLLAIMVYDVFGKMN--IKTVFAFCSMLCL*-STVNPIIYALR

hCB₁  SKDLRHAFRSMFPSCEGTAQPLDNSMGDSDCLHKHANNAASVHRAAESCIKSTVKIAKVTMSVSTDTSAEAL  472
cCB₁  SKDLRHAFRSMFPSCEGTAQPLDNSMGDSDCLHKHANNTANVHRAAENCIKNTVQIAKVTISVSTNTSAKAL  472
mCB₁  SKDLRHAFRSMFPSCEGTAQPLDNSMGDSDCLHKHANNTASMHRAAESCIKSTVKIAKVTMSVSTDTSAEAL  473
rCB₁  SKDLRHAFRSMFPSCEGTAQPLDNSMGDSDCLHKHANNTASMHRAAESCIKSTVKIAKVTMSVSTDTSAEAL  473
nCB₁  SQDLRHAFLEQCPPCEGTSQPLDNSM.ESDCQHRHGNNAGNVHRAAENCIKSTVKIAKVTMSVSTETSGEAV  473
Cons  S-DLRHAF----P-CEGT-QPLDNSM-*SDC-H*H-NN---*HRAAE-CIK-TV-IAKVT*SVST-TS--A*
```

Figure 21.1 Alignment of all known full-length CB₁ sequences. Human, cat, mouse, rat and newt sequences are shown by the prefixes h, c, m, r, and n. Gaps are represented by a dot. In the consensus (Cons) sequence a dash indicates residues of different classes, an asterisk residues of the same class. The canonic glycosylation sites are shown in bold characters. The lines labeled I–VII indicate the putative transmembrane domains.

```
                                            I                                      II
hCB₂  MEECWVTEIANGSKDGLDSNPMKDYMILSGPQKTAVAVLCTLLGLLSALENVAVLYLILSSHQLRRKPSYLFIGSLAGADFLASVVFACSFVNFHVFHGV  100
mCB₂  MEGCRETEVTNGSNGGLEFNPMKEYMILSSGQQIAVAVLCTLMGLLSALENMAVLYIILSSRRLRRKPSYLFISSLAGADFLASVIFACNFVIFHVFHGV  100
rCB₂  MEGCRELELTNGSNGGLEFNPMKEYMILSDAQQIAVAVLCTLMGLLSALENVAVLYLILSSQRLRRKPSYLFIGSLAGADFLASVIFACNFVIFHVFHGV  100
Cons  ME-C---E*-NGS--GL*-NPMK*YMILS--Q---AVAVLCTL-GLLSALEN-AVLY*ILSS--LRRKPSYLFI-SLAGADFLASV*FAC-FV-FHVFHGV

          III                       IV
hCB₂  DSKAVFLLKIGSVTMTFTASVGSLLLTAIDRYLCLRYPPSYKALLTRGRALVTLGIMWVLSALVSYLPLMGWTCCPRPCSELFPLIPNDYLLSWLLFIAF  200
mCB₂  DSNAIFLLKIGSVTMTFTASVGSLLLTAVDRYLCLCYPPTYKALVTRGRALVALCVMWVLSALISYLPLMGWTCCPSPCSELFPLIPNDYLLGWLLFIAI  200
rCB₂  DSRNIFLLKIGSVTMTFTASVGSLLLTAVDRYLCLCYPPTYKALVTRGRALVALGVMWVLSALISYLPLMGWTCCPSPCSELFPLIPNDYLLGWLLFIAI  200
Cons  DS--*FLLKIGSVTMTFTASVGSLLLTA-DRYLCL-YPP*YKAL*TRGRALV-L--MWVLSAL*SYLPLMGWTCCP-PCSELFPLIPNDYLL-WLLFIA-

          V                                      VI                            VII
hCB₂  LFSGIIYTYGHVLWKAHQHVASLSGHQDRQVPGMARMRLDVRLAKTLGLVLAVLLICWFPVLALMAHSLATTLSDQVKKAFAFCSMLCLINSMVNPVIYA  300
mCB₂  LFSGIIYTYGYVLWKAHRHVATLAEHQDRQVPGIARMRLDVRLAKTLGLVLAVLICWFPALALMGHSLVTTLSDQVKEAFAFCSMLCLVNSMVNPIIYA  300
rCB₂  LFSGIIYTYGYVLWKAHVASLTEHLDRQVPGIARMRLDVRLAKTLGLVMAVLLICWFPALALMGHSLVTTLSDKVKEAFAFCSMLCLVNSMVNPIIYA  300
Cons  LFSGIIYTYG*VLWKAH-HVA*L--H-DRQVPG-ARMRLDVRLAKTLGL-AVLLICWFP*LALM-HSL-TTLSD-VK-AFAFCSMLCL*NSMVNP*IYA

hCB₂  LRSGEIRSSAHHCLAHWKKCVRGLGSEAKEEAPRSSVTETEADGKITPWPDSRDLDLSDC  360
mCB₂  LRSGEIRSAAQHCLIGWKKYLQGLGPEGKEEGPRSSVTETEADVKTT               347
rCB₂  LRSGEIRSAAQHCLTGWKKYLQGLGSEGKEEAPKSSVTETEAEVKTTTGPGSRTPGCSNC  360
Cons  LRSGEIRS-A-HCL--WKK-*-GLG-E-KEE-P*SSVTETEA*-K-T-------------
```

Figure 21.2 Alignment of all known full-length CB₂ sequences. Human, mouse and rat sequences are shown by the prefixes h, m, and r. In the consensus (Cons) sequence a dash indicates residues of different classes, an asterisk residues of the same class. The canonic glycosylation site is shown in bold characters. The lines labeled I–VII indicate the putative transmembrane domains.

The mCB$_2$ and rCB$_2$ sequences are 91.2% identical. The hCB$_2$ and rCB$_2$ are 81.2% identical and are of the same length. The hCB$_1$ and hCB$_2$ receptors share only 44% overall identity, rising to 68% if the highly disparate extremities are disregarded and varying between 35% and 82% in the transmembrane regions. A schematic diagram of a typical 7TM receptor, representative of what we believe to be the general structure of CB$_1$ and CB$_2$ is shown in Figure 21.3a. The unfolded forms of hCB$_1$ and hCB$_2$ is depicted in Figure 21.3b, the white circles representing the subtype-specific residues. It can be clearly seen that, apart from the EL and IL regions and the extremities of the receptors, most of the differences between the two receptors are found in TMs 4 and 5. As described below, this region has been revealed to be of importance for the subtype specificity of certain ligands.

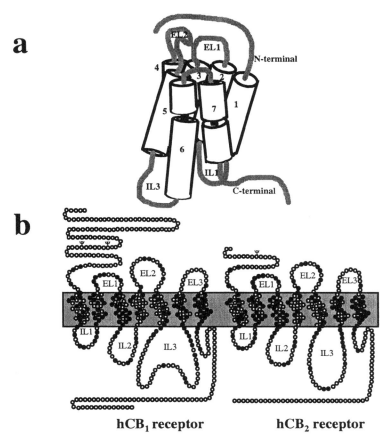

Figure 21.3 The human cannabinoid receptors. (a) A general representation of a 7TM G protein-coupled receptor, with the transmembrane regions shown as cylinders, linked by three extracellular loops (EL) and three intracellular loops (IL). (b) An unfolded representation of the two human receptors, with conserved residues shown as black circles, subtype-specific residues shown as white circles. The position of the glycosylation sites is shown as Ψ. The gray rectangle represents the membrane lipid bilayer.

CANNABINOID LIGANDS

Here we present a very small selection of the hundreds of cannabinoid ligands, representing different chemical series, that have been synthesized by medicinal chemists over the years. The selection includes the natural cannabinoids and the ligands that have most contributed to our current knowledge of the cannabinoid receptors.

Agonists

Classical cannabinoids

Although marijuana (*Cannabis sativa* L.) has been used since time immemorial for recreational and therapeutic purposes, it was only in 1964 that its main psychotropic constituent was identified by Mechoulam and Gaoni (1965) as Δ^9-tetrahydrocannabinol (Δ^9-THC) (Figure 21.4). Many other classical cannabinoids based on the Δ^9-THC dibenzopyran structure have subsequently been developed with the aim of improving the non-specificity and only moderate affinities ($K_i \sim 30$–$60\,nM$) and efficacy of Δ^9-THC. Among the best of these are derivatives of Δ^8-THC in which the n-pentyl C3 side-chain has been replaced by a dimethylheptyl and a hydroxyl group at position 11, as in HU-210 (Figure 21.4) and other derivatives lacking the phenolic OH group, such as L759 633, L759 656, JWH-133 and JWH-139 (Figure 21.4). HU-210 is highly potent but not specific ($K_i = 0.73\,nM$ for CB_1 and $0.22\,nM$ for CB_2), whereas the L and JWH compounds show specificity for CB_2 (affinity ratios CB_2/CB_1 160–800) with K_is for CB_2 in the nanomolar range (Pertwee, 1999).

Endogenous cannabinoids

In 1992 the first endogenous compound that bound to cannabinoid receptors was isolated from porcine brain (Devane et al., 1992). This was found to be an arachidonic acid derivative, arachidonoylethanolamide, named as anandamide (Figure 21.4). Anandamide is present in significant levels in the brain and spleen. In binding experiments with cloned receptors, it showed higher affinity for CB_1 ($K_i = 89\,nM$) than for CB_2 ($K_i = 371\,nM$) (Showalter et al., 1996). Anandamide is readily degraded by fatty acid amide hydrolase and has to be studied in the presence of an enzyme inhibitor such as phenylmethylsulfonyl fluoride. A second endogenous cannabinoid agonist, *sn*-2 arachidonylglycerol (Figure 21.4) was isolated from canine gut (Mechoulam et al., 1995) and has also been detected in rat brain (Stella et al., 1997). Two other ethanolamide derivatives have been identified as endogenous cannabinoids and these discoveries have led to the synthesis of many anandamide analogs designed to improve the affinity, selectivity and metabolic stability of anandamide. Among these, two show good selectivity for CB_1, methanandamide (Figure 21.4) (CB_1, $K_i = 20\,nM$, CB_2, $K_i = 815\,nM$) (Khanolkar et al., 1996) and fluoroanandamide (in which the OH of anandamide is replaced by F, CB_1, $K_i = 9.6\,nM$, CB_2, $K_i = 324\,nM$) (Showalter et al., 1996). A satu-

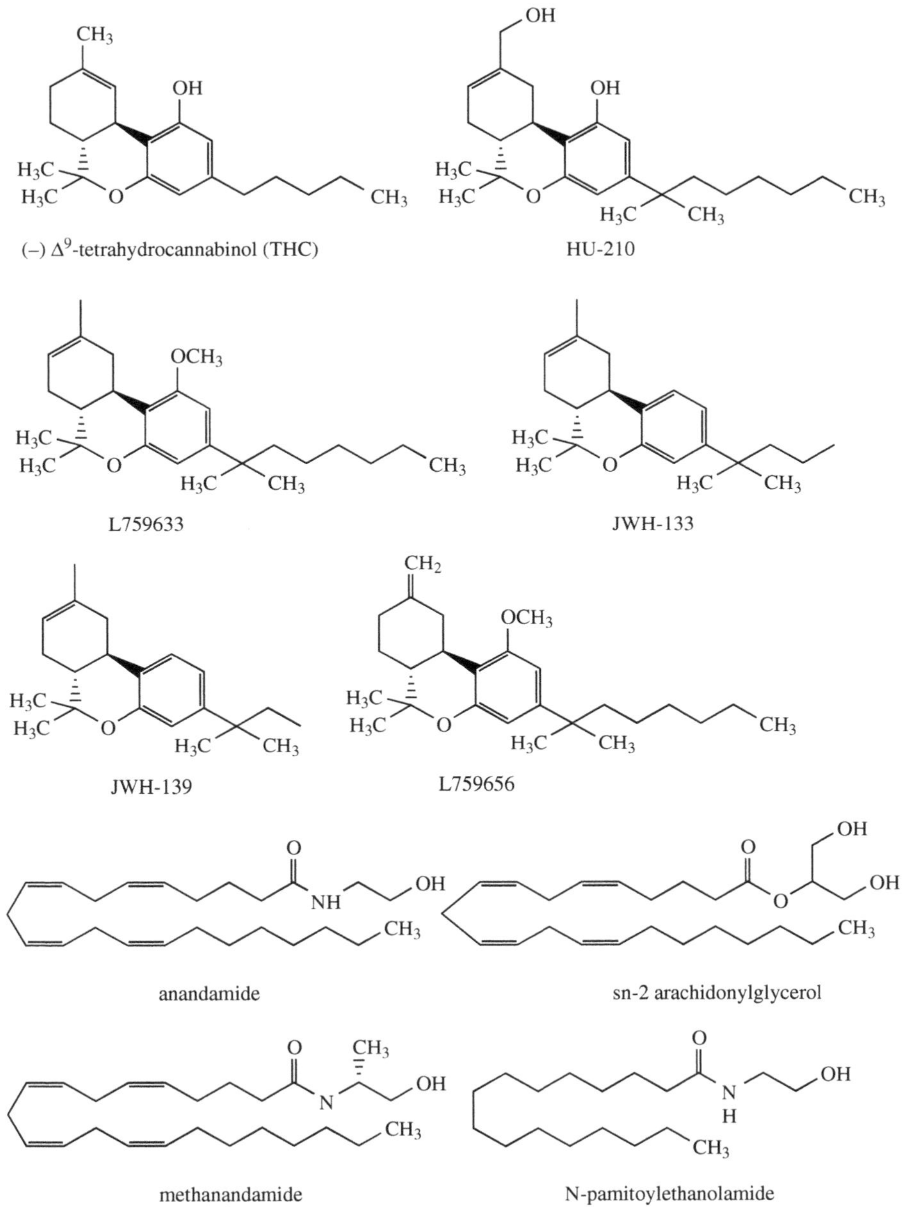

Figure 21.4 Structures of classical cannabinoids, anandamide and analogs.

rated arachidonic acid derivative, N-palmitoylethanolamide (Figure 21.4) was claimed to bind with high affinity ($IC_{50} = 1.0$ nM) to CB_2 receptors in mast cells (Facci *et al.*, 1995), but this has not been confirmed and indeed it proved to have poor affinity for cloned CB_2 (Showalter *et al.*, 1996; Griffin *et al.*, 2000).

Non-classical cannabinoids and aminoalkylindoles

Bicyclic or tricyclic THC analogs lacking the pyran ring are often referred to as non-classical cannabinoids, the most prominent of which is CP 55,940 (Figure 21.5) (Johnson and Melvin, 1986). CP 55,940 and the radiolabeled version [^{3}H]CP 55,940 are still the most widely used cannabinoid ligands, the latter being instrumental in the discovery of CB_1 in the brain (Devane *et al.*, 1988). CP 55,940 is nonspecific, has a high affinity ($K_i = 0.6$ nM) (Showalter *et al.*, 1996) and is highly potent.

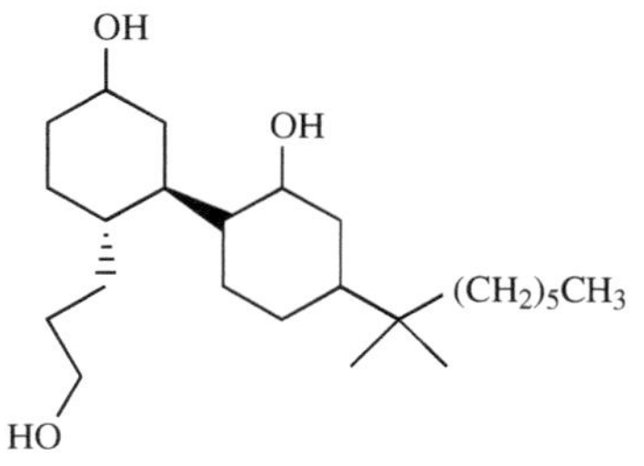

CP 55,940

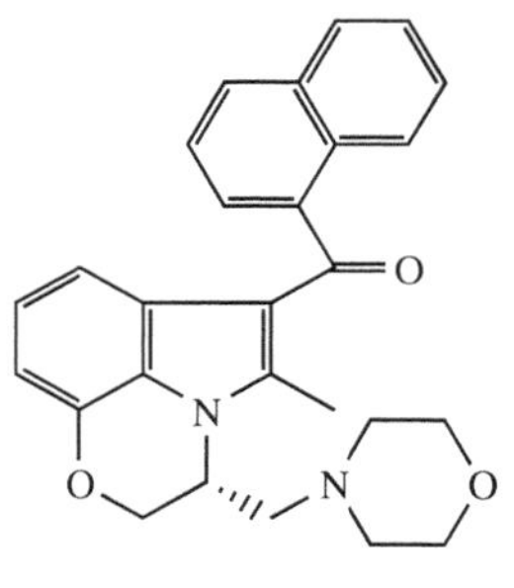

WIN 55,212-2

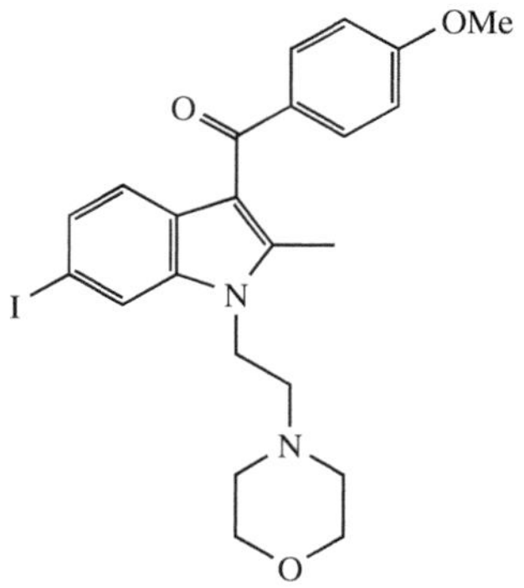

AM-630

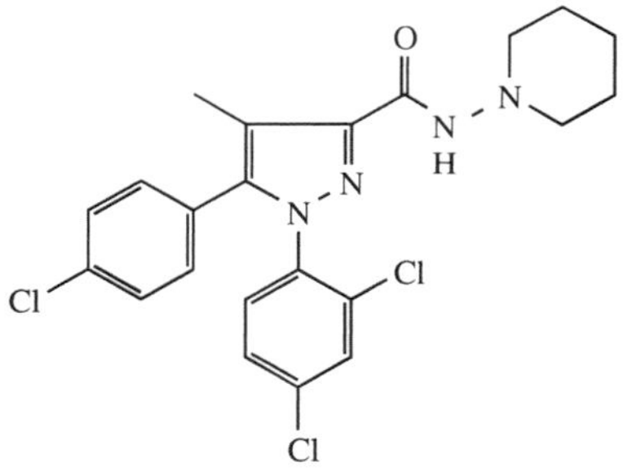

SR 141716A

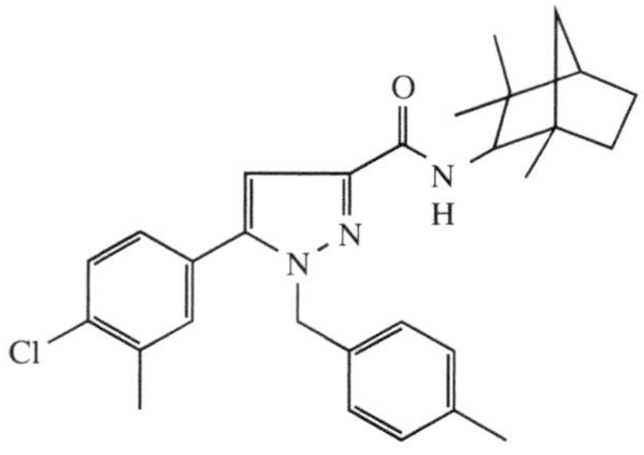

SR 144528

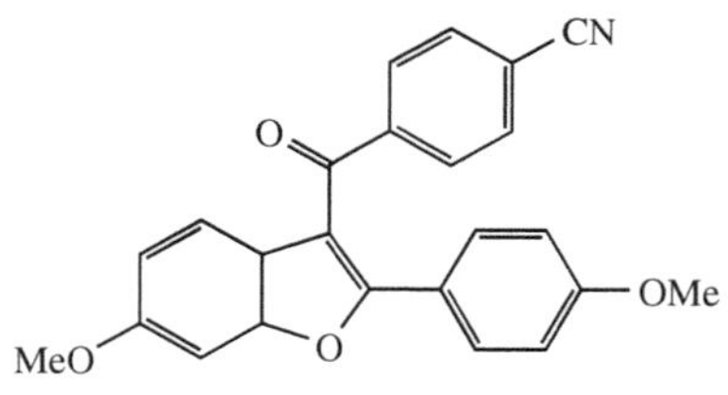

LY 320135

Figure 21.5 Structures of some prominent agonists and antagonists.

The aminoalkylindoles represent another important class of agonists; the most widely used is WIN 55212-2 (Figure 21.5) (Pacheco *et al.*, 1991), since it has high affinity for both receptor subtypes, with moderate selectivity for human CB_2 ($K_i = 0.3$ nM) over human CB_1 ($K_i = 1.9$ nM) expressed in CHO cell lines (Showalter *et al.*, 1996). Even more CB_2-selective aminoalkylindoles have been described in the literature and in patent applications recently reviewed by Barth (1998).

Antagonists

Cannabinoid antagonists have potentially interesting therapeutic applications as appetite suppressants and in the treatment of psychotic dysfunction and memory disorders. As yet, no effective antagonist has been developed from classical or non-classical cannabinoid structures, among the aminoalkylindole class of molecules or the fatty acid anandamidelike series. In a new structural series of 1,5-diphenylpyrazole derivatives described in a Sanofi patent in 1994, one molecule out of the 194 presented was chosen for development based on its potency and selectivity for CB_1 and its oral activity. SR 141716A (Figure 21.5) is a selective CB_1 antagonist (CB_1, $K_i = 5.6$ nM; CB_2, $K_i => 1000$ nM) (Rinaldi-Carmona *et al.*, 1994) that prevents the characteristic effects produced by cannabinoids both *in vitro* and *in vivo* and has intrinsic effects attributable to disruption of endogenous cannabinoid tone or inverse agonist properties (see below and a review by Pertwee) (Pertwee, 1999). Since its discovery SR 141716A has proven to be an invaluable tool for investigating cannabinoid activities.

A second pyrazole derivative, SR 144528 (Figure 21.5), was also selected from numerous similar compounds for its oral activity and especially for its specificity as an antagonist for the CB_2 receptor (CB_1, $K_i = 437$ nM; CB_2, $K_i = 0.6$ nM) (Rinaldi-Carmona *et al.*, 1998). Given the preferential distribution of CB_2 in cells of the immune system, SR 144528 has potential applications as an immunomodulator. It is also proving an useful tool for investigating the properties of CB_2, for example in demonstrating the involvement of CB_2 during B-cell differentiation (Carayon *et al.*, 1998).

Representative of another chemical series of analogs, LY320135 (Figure 21.5) is a selective CB_1 antagonist, having greater than 70-fold higher affinity for CB_1 than for CB_2. The K_i values for LY320135 at CB_1 and CB_2, stably expressed in cell lines, were 224 nM and >10 μM, respectively (Felder *et al.*, 1995). AM-630 (Figure 21.5) has a complex pharmacological profile in that it is a selective antagonist for hCB_2 expressed in CHO cells and a weak partial agonist for hCB_1, ($K_i = 31.2$ nM for hCB_2, $K_i = 5152$ nM for hCB_1), although depending on the CB_1 preparation or assay used, AM-630 can behave as an agonist, an antagonist or an inverse agonist (Pertwee, 1999).

CANNABINOID LIGAND–CANNABINOID RECEPTOR INTERACTIONS

Structural features of the cannabinoid receptors

Many studies have been undertaken to determine the nature of ligand–cannabinoid receptor interactions at the molecular level. The clear objective is to arrive at more potent and more subtype specific ligands, associating therapeutic effectiveness with a minimum of undesirable psychoactive and other side-effects. Most of the past efforts have concentrated on classical structure–activity relationships to determine the cannabinoid pharmacophore using different series of classical and non-classical cannabinoids. This has resulted in a three-point contact model for agonist–receptor interactions and a general picture of the architecture of the binding site (see Reggio (1999) for a recent review).

Since the cloning of the two receptor subtypes, efforts have also been directed towards pinpointing the actual amino acid residues on the receptors implicated in ligand binding, an approach that has produced satisfactory results with many other 7TM receptors. In the absence of 3D crystal structures the determination of binding sites relies on the use of a number of techniques, including cross-linking/degradation studies, computer model building and various mutational analysis procedures. Apart from the 3D crystal structure of bacteriorhodopsin, which is not coupled to G-proteins, such structures are not yet available for 7TMs that are G-protein coupled. Fortunately, the 2D electron cryomicroscopic structure of the G-protein coupled bovine rhodopsin receptor has been obtained at 5 Å resolution (Krebs et al., 1998) and this has allowed more realistic models to be built for many other class A (rhodopsin-like) 7TM receptors. Most physical and functional analyses of 7TMs are in accord with the rhodopsin crystallographic structure, but various other structures have come to light, particularly multimeric receptor forms, truncated receptors and splice variants, some of which are perfectly functional. The cannabinoid receptors are members of the class A rhodopsin-like 7TM receptors, but form a distinct subclass of lipid-binding receptors. As yet, no cross-linking studies have been made with the cannabinoid receptors to analyze binding sites, although irreversible probes have been developed by the Makryannis laboratory (Guo et al., 1994). Therefore, attempts to resolve the ligand binding sites on the CB_1 and CB_2 receptors depend heavily on computer models (Bramblett et al., 1995; Huffman et al., 1996; Mahmoudian, 1997). Such models have to be refined gradually by an iterative procedure based on data from binding and functional assays with mutated receptors.

All known cannabinoids, including the endogenous ones, are small nonpeptide molecules, but the existence of others cannot be excluded. In all other class A 7TMs studied to date, small nonpeptide molecules bind within the helix bundles, with relatively few interactions in the extracellular loop (EL) and amino terminal regions (Strader et al., 1994). An outstanding feature of the CB_1 receptor is its exceptionally long amino terminal region (Figure 21.1), remarkable for a class A 7TM. Why the receptor has such a long extracellular tail is a mystery. Nearly the whole of it, 89 residues, incidentally containing two glycosylation sites, can be removed without affecting receptor expression, binding of known ligands or biological activity (Rinaldi-Carmona et al., 1996). Despite their low identity in the TMs, particularly in TM1, TM4 and TM5 (Figures 21.3 and 21.6), the hCB_1 and hCB_2 receptors bind many potent agonists, such as CP 55,940, with similar affinities. Subtype-specific agonists have been reported (see Barth (1998) for a recent review) and may provide interesting tools in the future for investigating binding sites. In the past, as a result of their absence of selectivity, agonists have not proved useful for ascertaining binding site contacts in the cannabinoid receptors, but there are exceptions as described below. Clues based on results from mutational analysis of other 7TMs must be treated with caution, because the cannabinoid receptors are members of a subfamily characterized by structural features not found in the majority of 7TMs. In particular, they lack the EL1/TM3 cysteine found in most 7TMs, but contain two conserved EL2 cysteines. An otherwise highly conserved TM5 proline residue is absent in the subfamily, which may have an important effect on the structure of TM5. Other 7TMs that share the characteristic cannabinoid receptor features are the EDG series of receptors (Lynch and Im, 1999), the natural ligands for which, sphingosine 1-phosphate and sphingosylphosphorylcholine, are lipids like the endogenous cannabinoid receptor ligands.

Mutated receptors and binding site models

Chimeric CB_1/CB_2 receptors

The most prominent ligands used for investigating fine details of interactions with the cannabinoid receptors are CP 55,940, WIN 55212-2 and SR 141716A (Figure 21.5), because of their high affinity and especially their availability as tritiated molecules. Despite the structural difference between CB_1 and CB_2, [^{3}H]CP 55,940 binds equally well to the two receptors and activates them with similar efficacy and potency. It was striking to find that [^{3}H]CP 55,940 also failed to discriminate any of a series of chimeric receptors we constructed in which cognate CB_1 and CB_2 receptor regions were interchanged (Shire et al., 1996a). Several chimeras, although expressed in the cell, failed to reach the plasma membrane (Shire et al., 1996a), but those that did bound to [^{3}H]CP 55,940 equally well regardless of the proportion of each component. This appeared to indicate that the CB_1 and CB_2 receptors contained amino acid residues that were probably common to both receptors, arranged in such a way as to form the high affinity [^{3}H]CP 55,940 binding site, but mutation

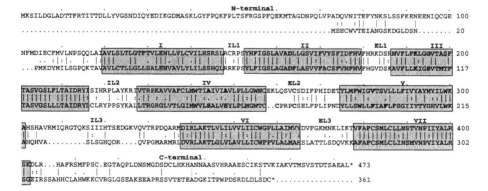

Figure 21.6   A comparison of hCB$_1$ and hCB$_2$. Vertical lines join identical residues, two similar residues, one dot more distant residues. Gaps introduced to produce the best alignment are shown as dots. The gray rectangles represent the putative transmembrane regions. The mutated positions described in the text are shown in bold characters.

studies (see below) contradict this supposition. Furthermore, the chimeric receptors were functional, since they have been shown to transduce a CP 55,940-mediated signal to G proteins (unpublished data).

Although the chimeric constructs provided no clues as to the CP 55,940 binding site, they permitted us to use [^{3}H]CP 55,940 as a universal ligand for competition binding assays. The wild-type hCB$_1$ and hCB$_2$ receptors have different binding affinities for WIN 55212-2 (IC$_{50}$ 70 nM and 3 nM, respectively) and for SR 141716A (IC$_{50}$ 6 nM and > 1000 nM, respectively) (Shire *et al.*, 1996a) and SR 144528 (IC$_{50}$ > 1000 nM and 2.5 nM, respectively). These differences may reflect recognition of distinct subtype specific amino acid residues present in the two receptors. As CB$_2$ receptor regions replaced those of the CB$_1$ receptor in the chimeras, so the competition binding affinities of these three ligands with respect to [^{3}H]CP 55,940 were modified (Shire *et al.*, 1996a an unpublished data). The results led us to the conclusion that the TM4-EL2-TM5 region of both receptors contained residues critical for the binding of WIN 55212-2 and the SR compounds. Using a chimeric receptor in which TM3 of CB$_1$ was replaced by that of CB$_2$ it was recently shown that TM3 contained residues important for WIN 55212-2 binding and activity (Chin *et al.*, 1999). The subsequent point mutation of Gly^{195} to the serine found in CB$_2$ (Figure 21.6) enhanced WIN 55212-2 binding and activity (Chin *et al.*, 1999). The reciprocal mutation in CB$_2$, serine to glycine, has not yet been done to confirm the result.

A model for the interaction between SR 144528 and CB$_2$

Single mutations of the conserved cysteine residues in EL2 of CB$_2$, Cys^{174} and Cys^{179} to serine (Figure 21.6) resulted in correctly translocated receptors as evidenced by immunofluorescence measurements, however the receptors failed to bind any of the ligands tested (Shire *et al.*, 1996a an unpublished data). Rather than concluding that the conserved EL2 cysteines are directly involved in ligand interactions, it is possible that they play some important structural role in both the CB$_1$ and CB$_2$ receptors. CB$_2$ has a third cysteine in this region, Cys^{175} (Figure 21.6), which we also mutated to serine, with interesting results. The wild-type binding affinity of CP 55,940 was unaffected (C175S, IC$_{50}$ = 0.2 nM; wild-type, IC$_{50}$ = 1.1 nM). However, the IC$_{50}$ for WIN 55212-2 in competition with [^{3}H]CP 55,940 fell from 2.8 nM for wild-type to 23.4 nM for the mutated receptor. In a functional assay we have developed based on a reporter gene directing firefly luciferase expression (Calandra *et al.*, 1999), we found that CP 55,940 inhibition of luciferase induction in the mutant receptor remained at wild-type levels, whereas the inhibition elicited by WIN 55212-2 fell in accord with its loss in binding affinity. At the same time, SR 144528 binding and activity was completely lost (submitted for publication). Two CB$_2$ specific serine residues in TM4 (Figure 21.6) were also found to be crucial contact points for SR 144528. Their separate mutation to the CB$_1$ specific residue, alanine, had no effect on CP 55,940 or WIN 55212-2 binding and activity, which SR 144528 at 10^{-6} M failed to antagonize. Based on these results we propose a docking model for SR 144528 on CB$_2$ (Figure 21.7). According to this model, the antagonist interacts with CB$_2$ through hydrogen bonds with the two serines in TM4 and a threonine in

TM3, together with a number of hydrophobic and aromatic interactions with residues in TMs 3, 4 and 5.

A model for the interaction between WIN 55212-2 and CB_2

The SR 144528 binding model has several points in common with that for WIN 55212-2 on CB_2 recently proposed by Song *et al.* (1999). WIN 55212-2 has selectivity for CB_2 over CB_1 with a K_i ratio of CB_1/CB_2 of 19 (Felder *et al.*, 1995). Song *et al.* (1999) showed that the selectivity could be attributed to a single aromatic residue in TM5 of CB_2, Phe[197] (Figure 21.6). Its replacement by the CB_1 residue, valine, reduced binding 14-fold to the wild-type CB_1 level. The reciprocal mutation in CB_2, valine to phenylalanine, had the inverse effect. Based on this result and taking previous work into consideration, the authors proposed a model for the WIN 55212-2–CB_2 interaction in which the ligand interacted with three shared CB_1 and CB_2 aromatic residues, together with the CB_2-specific Phe[197] (Song *et al.*, 1999).

The WIN 55212-2 binding site is evidently different from those of other agonists, as underscored by its properties with the receptors mutated in other domains. For example, in a study of CHO cell lines expressing CB_1 receptor D163N and D163E TM2 mutants it was found that binding of CP 55,940, SR

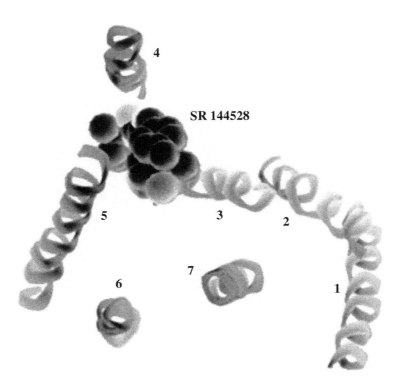

Figure 21.7   SR 144528 docked into CB_1. The receptor is viewed from the extracellular side, with the transmembrane regions organized in an anti-clockwise orientation.

141716A, anandamide, Δ^9-THC and (–)-11-OH-Δ^9-THC was largely unaffected compared to the wild-type CB_1 receptor, but that the WIN 55212-2 binding affinity was reduced about 100-fold (Tao and Abood, 1998). The cognate mutations in the CB_2 receptor had little effect on agonist binding. In both mutated CB_1 and CB_2 receptors the inhibition of forskolin-induced cAMP accumulation mediated by the agonists was reduced.

Other mutational studies

The role of a conserved CB_1/CB_2 lysine in TM3 is particularly interesting. Lys^{192} of the CB_1 receptor (Figure 21.6) has been studied by two groups (Song and Bonner 1996; Chin *et al.*, 1998). In HEK293 cells stably expressing the K192A mutated CB_1 receptor [^{3}H]WIN 55212-2 binding was reduced by about 50%, while other agonists, anandamide, CP 55,940 and HU-210 failed to compete with [^{3}H]WIN 55212-2. Only WIN 55212-2 could induce activity with the mutated receptor (Song and Bonner, 1996). Similar results with CP 55,940 and WIN 55212-2 were obtained with K192L, K192Q and K192E CB_1 receptor mutants stably expressed in CHO cells, whereas K192R resembled the wild-type receptor, leading the authors to conclude that a basic residue is required in this position (Chin *et al.*, 1998). The precipitous drop in binding and activity of all the agonists apart from WIN 55212-2 when Lys^{192} was replaced by a neutral or negative amino acid residue is generally accepted to be a result of the suppression of direct ligand contacts with Lys^{192}. Remarkably, in view of almost identical binding properties of CP 55,940 to CB_1 and CB_2, mutation of the cognate lysine in the CB_2 receptor, Lys^{109}, gave a receptor having essentially wild-type CB_2 receptor binding properties (McAllister *et al.*, 1997; Tao *et al.*, 1999). The conclusion is that CP 55,940 binds to structurally different binding sites with fortuitously similar affinity and/or the lysine in the two subtypes is involved in different structural arrangements of the receptors. Modeling studies by Tao *et al.* (1999) revealed a difference in orientation of CP 55,940 in CB_1 versus CB_2. In these studies, Lys^{109} in CB_2 was found not to be important for CP 55,940 binding, but that a subtype specific hydrogen bonding cluster in TM3 provided by Ser^{112} and Thr^{116} was important. Loss of binding not only of CP 55,940, but also of anandamide and Δ^9-THC, in the double mutant K109AS112G (Figure 21.6) provided evidence for the model. WIN 55212-2 was again the exception, binding well to the doubly mutated receptor, but its ability to antagnoize forskolin stimulated cAMP accumulation was severely impaired and immunofluorescence studies revealed a problem with receptor translocation and/or compartmentation. The single S122G mutant was not described. The mutagenesis results fully support recent quantitative structure–activity relationship models using CP 55,940 and a series of aminoalkylindoles (Shim *et al.*, 1998), from which it was concluded that the different classes of cannabinoids exhibited unique interactions within the CB_1 receptor, but that the binding sites may partly overlap. This conclusion can obviously be extended to the CB_2 receptor.

To summarize, good experimental evidence for the interaction of the agonists WIN 55212-2 and CP 55,940 and the antagonist SR 144528 with CB_2 has led to docking models for these two ligands. In addition, a docking model for CP 55,940 with CB_1 has also been proposed. In the future we can expect to see propositions for the interactions of other important cannabinoids with the two receptors, which, together with the knowledge gleaned from structure–activity relations, should lead to a clear understanding of these interactions at a molecular level.

CONSTITUTIVELY ACTIVE CANNABINOID RECEPTORS AND INVERSE AGONISM OF SR 141716A AND SR 144528

Several years before the cannabinoid receptors were cloned, Howlett showed that cannabimimetic drugs inhibited adenylyl cyclase activity through the mobilization of the pertussis toxin-sensitive G_i protein (Howlett et al., 1986). Both CB_1 and CB_2 transduce agonist stimulation through the mediation of the α-subunit of the heterotrimeric G_i protein, but CB_1 can also couple to $G_s\alpha$ to stimulate adenylyl cyclase activity, at least in heterologous expression systems (Glass and Felder, 1997; Calandra et al., 1999). Indeed, recent data from in situ reconstitution experiments showed that different agonists induce different conformations of CB_1, which in turn can distinguish between different G-proteins (Glass and Northup, 1999). To further complicate the picture, a further signaling pathway for both CB_1 and CB_2 leads to an enhancement of mitogen-activated protein kinase (MAPK) (Bouaboula et al., 1995a; Bouaboula et al., 1996) and the immediate-early gene Krox24 (Glass and Dragunow 1995; Bouaboula et al., 1995b).

CB$_1$ stably expressed in Chinese hamster ovary (CHO) cells exhibits high constitutive activity towards basal levels of MAPK and repression of adenylyl cyclase activity; SR 141716A has been shown to act as an inverse agonist by inhibiting these responses in a dose-response manner (Bouaboula et al., 1997). Similarly, in CHO cells doubly transfected with CB_2 and a reporter gene, inhibition of adenylyl cyclase activity and induction of Krox24 was observed (Portier et al., 1999). SR 144528 was shown to inhibit both activities in a dose–response manner, thereby exhibiting inverse agonism. A further example of the inverse agonism of SR 144528 was shown by its inhibition of the phosphorylation and internalization of autoactivated CB_2, leading to an upregulation of the receptor at the cell surface (Bouaboula et al., 1999). Additionally, in the presence of guanine nucleotides, binding of the SR compounds is enhanced, also a characteristic property of inverse agonists, contrasting with the reduction in binding seen with agonists (Bouaboula et al., 1997; Portier et al., 1999).

MAPK phosphorylation and activation in CHO cells also follows treatment with insulin or with insulin-like growth factor (IGF1). Most surprisingly, when CHO cells stably expressing the CB_1 receptor were pretreated with SR 141716A before stimulation with insulin or IGF1, MAPK activation was completely inhibited (Bouaboula et al., 1997). The effect was not seen in control CHO cells, showing that the effect was CB_1-mediated. In contrast, MAPK activity stimulated by basic fibroblast growth factor (FGF-b) was unaffected. Insulin and IGF1, but not FGF-b, are known to stimulate MAPK through the pertussis toxin-sensitive G_i protein/ phosphoinositide-3' kinase pathway. G_i is therefore a common intermediary to both the cannabinoid- and growth factor-induced MAPK activation. Pertussis toxin treatment completely abrogated both pathways and also the direct stimulation of G_i by a small synthetic peptide, the mastoparan analog Mas-7. Pretreatment of the CB_1 receptor-expressing CHO cell line with SR 141716A completely blocked the direct stimulation of G_i and hence MAPK by Mas-7. This implied that the interaction of SR 141716A with the CB_1 receptor resulted in sequestration of the G_i proteins in an inactive GDP-bound form, demonstrating an unexpected role for SR 141716A. In another series of experiments it was shown that CB_1 expressed in superior cervical ganglion neurons by microinjection of hCB_1 cDNA abolished the Ca^{2+} current inhibition induced by norepinephrine and somatostatin, a pathway mediated by $G_{i/o}$; SR 141617A treatment enhanced the Ca^{2+} current through inverse agonist activity (Vasquez and Lewis, 1999). It was also reported that both the active and inactive SR 141716A-treated states of CB_1 can sequester $G_{i/o}$ proteins from a common pool. Cannabinoid receptors thus have the potential to prevent other $G_{i/o}$-coupled receptors from transducing their biological signals. If such effects occur *in vivo* they may be of considerable importance.

CONCLUDING REMARKS

We have tried to show that although our knowledge of the cannabinoid receptors and their interactions with their respective ligands has considerably progressed in recent years, it is evident that much remains to be learned. With regard to the topics covered in this chapter, we know nothing, for example, about the promoters and how the expression of the receptor genes is regulated. We have constructed computer models of the receptors, but lack direct evidence to prove their validity. Mutational studies have led to docking models for several synthetic ligands, but no models have yet been proposed for the classical cannabinoids and the endogenous ligands, although extensive structure-activity relationship studies have provided a robust pharmacophore. Finally, we have touched on the extremely complex subject of signal transduction and inverse agonism, studied in *in vitro* systems, but it is clear that *in vivo* the situation will prove to be even more complicated.

REFERENCES

Barth, F. (1998) "Cannabinoid receptor agonists and antagonists," *Expert Opinion of Therapeutic Patents* **8**: 301–313.

Bouaboula, M., Bourrié, B., Rinaldi-Carmona, M., Shire, D., Le Fur, G. and Casellas, P. (1995b) "Stimulation of cannabinoid receptor CB1 induces *krox-24* expression in human astrocytoma cells," *Journal of Biological Chemistry* **270**: 13973–13980.

Bouaboula, M., Dussossoy, D. and Casellas, P. (1999) "Regulation of peripheral cannabinoid receptor CB_2 phosphorylation by the inverse agonist SR 144528 – Implications for receptor biological responses," *Journal of Biological Chemistry* **274**: 20397–20405.

Bouaboula, M., Perrachon, S., Milligan, L., Canat, X., Rinaldi-Carmona, M., Portier, M., Barth, F., Calandra, B., Pecceu, F., Lupker, J., Maffrand, J. P., Le Fur, G. and Casellas, P. (1997) "A selective inverse agonist for central cannabinoid receptor inhibits mitogen-activated protein kinase activation stimulated by insulin or insulin-like growth factor 1 – Evidence for a new model of receptor/ligand interactions," *Journal of Biological Chemistry* **272**: 22330–22339.

Bouaboula, M., Poinot-Chazel, C., Bourrié, B., Canat, X., Calandra, B., Rinaldi-Carmona, M., Le Fur, G. and Casellas, P. (1995a) "Activation of mitogen-activated protein kinases by stimulation of the central cannabinoid receptor CB1," *Biochemical Journal* **312**: 637–641.

Bouaboula, M., Poinot-Chazel, C., Marchand, J., Canat, X., Bourrié, B., Rinaldi-Carmona, M., Calandra, B., Le Fur, G. and Casellas, P. (1996) "Signaling pathway associated with stimulation of CB$_2$ peripheral cannabinoid receptor – Involvement of both mitogen-activated protein kinase and induction of Krox-24 expression," *European Journal of Biochemistry* **237**: 704–711.

Bramblett, R. D., Panu, A. M., Ballesteros, J. A. and Reggio, P. H. (1995) "Construction of a 3D model of the cannabinoid CB1 receptor: Determination of helix ends and helix orientation," *Life Sciences* **56**: 1971–1982.

Calandra, B., Portier, M., Kerneis, A., Delpech, M., Carillon, C., Le Fur, G., Ferrara, P. and Shire, D. (1999) "Dual intracellular signaling pathways mediated by the human cannabinoid CB1 receptor," *European Journal of Pharmacology* **374**: 445–455.

Carayon, P., Marchand, J., Dussossoy, D., Derocq, J. M., Jbilo, O., Bord, A., Bouaboula, M., Galiegue, S., Mondiere, P., Penarier, G., Fur, G. L., Defrance, T. and Casellas, P. (1998) "Modulation and functional involvement of CB2 peripheral cannabinoid receptors during B-cell differentiation," *Blood* **92**: 3605–3615.

Chakrabarti, A., Onaivi, E. S. and Chaudhuri, G. (1995) "Cloning and sequencing of a cDNA encoding the mouse brain-type cannabinoid receptor protein," *DNA Sequence* **5**: 385–388.

Chin, C. N., Lucas-Lenard, J., Abadji, V. and Kendall, D. A. (1998) "Ligand binding and modulation of cyclic AMP levels depend on the chemical nature of residue 192 of the human cannabinoid receptor 1," *Journal of Neurochemistry* **70**: 366–373.

Chin, C. N., Murphy, J. W., Huffman, J. W. and Kendall, D. A. (1999) "The third transmembrane helix of the cannabinoid receptor plays a role in the selectivity of aminoalkylindoles for CB2, peripheral cannabinoid receptor," *Journal of Pharmacology and Experimental Therapeutics* **291**: 837–844.

Devane, W. A., Dysarz, F. A., III, Johnson, M. R., Melvin, L. S. and Howlett, A. C. (1988) "Determination and characterization of a cannabinoid receptor in rat brain," *Molecular Pharmacology* **34**: 605–613.

Devane, W. A., Hanus, L., Breuer, A., Pertwee, R. G., Stevenson, L. A., Griffin, G., Gibson, D., Mandelbaum, A., Etinger, A. and Mechoulam, R. (1992) "Isolation and structure of a brain constituent that binds to the cannabinoid receptor," *Science* **258**: 1946–1949.

Facci, L., Dal Toso, R., Romanello, S., Buriani, A., Skaper, S. D. and Leon, A. (1995) "Mast cells express a peripheral cannabinoid receptor with differential sensitivity to anandamide and palmitoylethanolamide," *Proceedings of the National Academy of Sciences of the USA* **92**: 3376–3380.

Felder, C. C., Joyce, K. E., Briley, E. M., Mansouri, J., Mackie, K., Blond, O., Lai, Y., Ma, A. L. and Mitchell, R. L. (1995) "Comparison of the pharmacology and signal transduction of the human cannabinoid CB$_1$ and CB$_2$ receptors," *Molecular Pharmacology* **48**: 443–450.

Gérard, C. M., Mollereau, C., Vassart, G. and Parmentier, M. (1991) "Molecular cloning of a human cannabinoid receptor which is also expressed in testis," *Biochemical Journal* **279**: 129–134.

Glass, M. and Dragunow, M. (1995) "Induction of the Krox 24 transcription factor in striosomes by a cannabinoid agonist," *Neuroreport* **6**: 241–244.

Glass, M. and Felder, C. C. (1997) "Concurrent stimulation of cannabinoid CB1 and dopamine D2 receptors augments cAMP accumulation in striatal neurons – evidence for a G$_s$ linkage to the CB1 receptor," *Journal of Neuroscience* **17**: 5327–5333.

Glass, M. and Northup, J. K. (1999) "Agonist selective regulation of G-proteins by cannabinoid CB$_1$ and CB$_2$ receptors," *Molecular Pharmacology* **56**: 1362–1369.

Griffin, G., Tao, Q. and Abood, M. E. (2000) "Cloning and pharmacological characterization of the rat CB$_2$ cannabinoid receptor," *Journal of Pharmacology and Experimental Therapeutics* **292**: 886–894.

Guo, Y., Abadji, V., Morse, K. L., Fournier, D. J., Li, X. and Makriyannis, A. (1994) "(−)-11-hydroxy-7′-isothiocyanato-1′,1′-dimethylheptyl-Delta^8-THC: A novel, high-affinity irreversible probe for the cannabinoid receptor in the brain," *Journal of Medicinal Chemistry* **37**: 3867–3870.

Herkenham, M. (1995) in *Cannabinoid receptors*, edited by R. G. Pertwee, pp. 145–146. London: Academic Press.

Hoehe, M. R., Caenazzo, L., Martinez, M. M., Hsieh, W. T., Modi, W. S., Gershon, E. S. and Bonner, T. I. (1991) "Genetic and physical mapping of the human cannabinoid receptor gene to chromosome 6q14-q15," *New Biologist*, **3**: 880–885.

Howlett, A. C., Qualy, J. M. and Khachatrian, L. L. (1986) "Involvement of G$_i$ in the inhibition of adenylate cyclase by cannabimimetic drugs," *Molecular Pharmacology* **29**: 307–313.

Huffman, J. W., Yu, S., Showalter, V., Abood, M. E., Wiley, J. L., Compton, D. R., Martin, B. R., Bramblett, R. D. and Reggio, P. H. (1996) "Synthesis and pharmacology of a very potent cannabinoid lacking a phenolic hydroxyl with high affinity for the CB2 receptor," *Journal of Medicinal Chemistry* **39**: 3875–3877.

Johnson, M. R. and Melvin, L. S. (1986) "The discovery of non-classical cannabinoid analgesics," in *Cannabinoids as Therapeutic Agents*, edited by R. Mechoulam, pp. 121–145. Boca Raton: CRC Press.

Khanolkar, A. D., Abadji, V., Lin, S. Y., Hill, W. A. G., Taha, G., Abouzid, K., Meng, Z. X., Fau, P. S. and Makriyannis, A. (1996) "Head group analogs of arachidonylethanolamide, the endogenous cannabinoid ligand," *Journal of Medicinal Chemistry* **39**: 4515–4519.

Krebs, A., Villa, C., Edwards, P. C. and Schertler, G. F. X. (1998) "Characterization of an improved two-dimensional p22121 crystal from bovine rhodopsin," *Journal of Molecular Biology* **282**: 991–1003.

Lynch, K. R. and Im, D. S. (1999) "Life on the edge," *Trends in Pharmacological Sciences* **20**: 473–475.

Mahmoudian, M. (1997) "The cannabinoid receptor: computor-aided molecular modeling and docking of ligand," *Journal of Molecular Graphics and Modelling* **15**: 149–153.

Matsuda, L. A. (1997) "Molecular aspects of cannabinoid receptors," *Critical Reviews in Neurobiology* **11**: 143–166.

Matsuda, L. A., Lolait, S. J., Brownstein, M. J., Young, A. C. and Bonner, T. I. (1990) "Structure of a cannabinoid receptor and functional expression of the cloned cDNA," *Nature* **346**: 561–564.

McAllister, S. D., Tao, Q., Reggio, P. H. and Abood, M. E. (1997) "A conserved lysine residue in the third transmembrane domain of the CB1 and CB2 receptor mediates different functional roles," *1997 Symposium on the Cannabinoids, Stone Mountain, Georgia, International Cannabinoid Research Society*, 49pp.

Mechoulam, R., Ben-Shabat, S., Hanus, L., Ligumsky, M., Kaminski, N. E., Schatz, A. R., Gopher, A., Almog, S., Martin, B. R., Compton, D. R., Pertwee, R. G., Griffin, G., Beyewitch, M., Barg, J. and Vogel, Z. (1995) "Identification of an endogenous 2-monoglyceride, present in canine gut, that binds to cannabinoid receptors," *Biochemical Pharmacology* **50**: 83–90.

Mechoulam, R. and Gaoni, Y. (1965) "Hashish IV. The isolation and structure of cannabinolic cannabidiolic and cannabigerolic acids," *Tetrahedron* **21**: 1223–1229.

Munro, S., Thomas, K. L. and Abu-Shaar, M. (1993) "Molecular characterization of a peripheral receptor for cannabinoids," *Nature* **365**: 61–65.

Pacheco, M., Childers, S. R., Arnold, R., Casiano, F. and Ward, S. J. (1991) "Aminoalkylindoles: actions on specific G-protein-linked receptors," *Journal of Pharmacology and Experimental Therapeutics* **257**: 170–183.

Pertwee, R. G. (1999) "Pharmacology of cannabinoid receptor ligands," *Current Medicinal Chemistry* **6**: 635–664.

Pfister-Genskow, M., Weesner, G. D., Hayes, H., Eggen, A. and Bishop, M. D. (1997) "Physical and genetic localization of the bovine cannabinoid receptor (Cnr1) Gene to bovine chromosome 9," *Mammalian Genome* **8**: 301–302.

Portier, M., Rinaldi-Carmona, M., Pecceu, F., Combes, T., Poinot-Chazel, C., Calandra, B., Barth, F., Le Fur, G. and Casellas, P. (1999) "SR 144528, an antagonist for the peripheral cannabinoid receptor that behaves as an inverse agonist," *Journal of Pharmacology and Experimental Therapeutics* **288**: 582–589.

Reggio, P. H. (1999) "Ligand–ligand and ligand-receptor approaches to modeling the cannabinoid CB1 and CB2 receptors: Achievements and challenges," *Current Medicinal Chemistry* **6**: 665–683.

Rinaldi-Carmona, M., Barth, F., Héaulme, M., Shire, D., Calandra, B., Congy, C., Martinez, S., Maruani, J., Néliat, G., Caput, D., Ferrara, P., Soubrié, P., Brelière, J.-C. and Le Fur, G. (1994) "SR141716A, a potent and selective antagonist of the brain cannabinoid receptor," *FEBS Letters* **350**: 240–244.

Rinaldi-Carmona, M., Barth, F., Millan, J., Derocq, J. M., Casellas, P., Congy, C., Oustric, D., Sarran, M., Bouaboula, M., Calandra, B., Portier, M., Shire, D., Breliére, J. C. and Le Fur, G. (1998) "SR 144528, the first potent and selective antagonist of the CB2 cannabinoid receptor," *Journal of Pharmacology and Experimental Therapeutics* **284**: 644–650.

Rinaldi-Carmona, M., Calandra, B., Shire, D., Bouaboula, M., Oustric, D., Barth, F., Casellas, P., Ferrara, P. and Le Fur, G. (1996) "Characterization of two cloned human CB1 cannabinoid receptor isoforms," *Journal of Pharmacology and Experimental Therapeutics* **278**: 871–878.

Shim, J. Y., Collantes, E. R., Welsh, W. J., Subramaniam, B., Howlett, A. C., Eissenstat, M. A. and Ward, S. J. (1998) "Three-dimensional quantitative structure-activity relationship study of the cannabimimetic (aminoalkyl)indoles using comparative molecular field analysis," *Journal of Medicinal Chemistry* **41**: 4521–4532.

Shire, D., Calandra, B., Delpech, M., Dumont, X., Kaghad, M., Le Fur, G., Caput, D. and Ferrara, P. (1996a) "Structural features of the central cannabinoid CB1 receptor involved in the binding of the specific CB1 antagonist SR 141716A," *Journal of Biological Chemistry* **271**: 6941–6946.

Shire, D., Calandra, B., Rinaldi-Carmona, M., Oustric, D., Pessegue, B., Bonnin-Cabanne, O., Le Fur, G., Caput, D. and Ferrara, P. (1996b) "Molecular cloning, expression and function of the murine CB2 peripheral cannabinoid receptor," *Biochimica et Biophysica Acta: Gene Structure and Expression* **1307**: 132–136.

Shire, D., Carillon, C., Kaghad, M., Calandra, B., Rinaldi-Carmona, M., Le Fur, G., Caput, D. and Ferrara, P. (1995) "An amino terminal variant of the central cannabinoid receptor resulting from alternative splicing," *Journal of Biological Chemistry* **270**: 3726–3731.

Showalter, V. M., Compton, D. R., Martin, B. R. and Abood, M. E. (1996) "Evaluation of binding in a transfected cell line expressing a peripheral cannabinoid receptor (CB2): Identification of cannabinoid receptor subtype selective ligands," *Journal of Pharmacology and Experimental Therapeutics* **278**: 989–999.

Song, Z. H. and Bonner, T. I. (1996) "A lysine residue of the cannabinoid receptor is critical for receptor recognition by several agonists but not WIN55212-2," *Molecular Pharmacology* **49**: 891–896.

Song, Z. H., Slowey, C. A., Hurst, D. P. and Reggio, P. H. (1999) "The difference between the CB_1 and CB_2 cannabinoid receptors at position 5.46 is crucial for the selectivity of WIN55212-2 for CB_2," *Molecular Pharmacology* **56**: 834–840.

Stefano, G. B., Salzet, B. and Salzet, M. (1997) "Identification and characterization of the leech CNS cannabinoid receptor: Coupling to nitric oxide release," *Brain Research* **753**: 219–224.

Stella, N., Schweitzer, P. and Piomelli, D. (1997) "A second endogenous cannabinoid that modulates long-term potentiation," *Nature* **388**: 773–778.

Strader, C. D., Fong, T. M., Tota, M. R., Underwood, D. and Dixon, R. A. F. (1994) "Structure and function of G protein-coupled receptors," *Annual Reviews of Biochemistry* **63**: 101–132.

Stubbs, L., Chittenden, L., Chakrabarti, A. and Onaivi, E. (1996) "The gene encoding the central cannabinoid receptor is located in proximal mouse chromosome 4," *Mammalian Genome* **7**: 165–166.

Tao, Q. and Abood, M. E. (1998) "Mutation of a highly conserved aspartate residue in the second transmembrane domain of the cannabinoid receptors, cb1 and cb2, disrupts g-protein coupling," *Journal of Pharmacology and Experimental Therapeutics* **285**: 651–658.

Tao, Q., McAllister, S. D., Andreassi, J., Nowell, K. W., Cabral, G. A., Hurst, D. P., Bachtel, K., Ekman, M. C., Reggio, P. H. and Abood, M. E. (1999) "Role of a conserved lysine residue in the peripheral cannabinoid receptor (CB$_2$): Evidence for subtype specificity," *Molecular Pharmacology* **55**: 605–613.

Valk, P. J. M., Hol, S., Vankan, Y., Ihle, J. N., Askew, D., Jenkins, N. A., Gilbert, D. J., Copeland, N. G., De Both, N. J., Löwenberg, B. and Delwel, R. (1997) "The genes encoding the peripheral cannabinoid receptor and α-L-fucosidase are located near a newly identified common virus integration site," *Journal of Virology* **71**: 6796–6804.

Vasquez, C. and Lewis, D. L. (1999) "The CB$_1$ cannabinoid receptor can sequester G-proteins, making them unavailable to couple to other receptors," *Journal of Neuroscience* **19**: 9271–9280.

Yamaguchi, F., Macrae, A. D. and Brenner, S. (1996) "Molecular cloning of two cannabinoid type 1-like receptor genes from the puffer fish, *Fugu rubripes*," *Genomics* **35**: 603–605.

Chapter 22

Cannabinoids as analgesics

*J. Michael Walker, Nicole M. Strangman
and Susan M. Huang*

ABSTRACT

Although cannabis-based preparations have been used for centuries to treat pain, the biological basis for such pain-ameliorative effects of cannabinoids was unknown until recently. In animal studies, cannabinoids suppress pain behavior, noxious stimulus-evoked immediate early gene *c-fos* expression in the spinal cord, and noxious stimulus-evoked neuronal response. The effects on tactile sensitivity are selective for pain since cannabinoids were without effect on non-nociceptive neurons in the spinal cord and thalamus. The suppression of noxious stimulus-evoked responses are mediated by cannabinoid receptors through multiple sites of action in the brain, the spinal cord, and the periphery. Cannabinoids appear to be effective in both physiological (or acute) pain and clinical (or chronic) pain. Furthermore, studies in endocannabinoids have revealed that they serve a role in pain modulation. Advances such as these offer hope for new pharmacotherapies for pain, particularly in conditions that remain unresolved through current treatments.

Key Words: cannabinoids, analgesia, pain, endocannabinoid

HISTORY

Historical accounts suggest that the medicinal use of cannabis was once widespread both culturally and geographically. Among other things, cannabis was appreciated for its analgesic and anesthetic actions. The Chinese pharmacopeia of 2800 B.C. describes the use of cannabis as an analgesic (Iversen, 2000). Cannabis preparations were reportedly used in 200 A.D. by the Chinese physician Hua T'o to anesthetize patients before abdominal surgery (Li, 1974). Documentations regarding the medicinal use of cannabis through various times since then exist in civilizations such as ancient Greece and Rome, India, Argentina, and in Europe.

Nearly all evidence for the ancient medicinal use of cannabis is literary. However, there is at least one possible piece of physical evidence of cannabis use in the ancient Middle East. Scientists unearthed a tomb near Jerusalem dating to the 4th century AD which contained the skeletons of a teenage girl and a 40-week fetus (Zias *et al.*, 1993). GC MS and NMR spectroscopy of carbonized matter found in the skeleton's abdominal region revealed the presence of Δ^6-THC, a by-product of two chemicals that are formed when marijuana is burned. Although this finding is difficult to interpret with any great certainty, the team of scientists hypothesized that marijuana smoke was administered to the girl in order to ease what the skeleton's small cervix suggested would be very difficult child labor.

Due to the rampant medical and recreational consumption of cannabis by the 19th century, the Indian government solicited an investigation of the use of the hemp plant. After hearing testimony from numerous witnesses, including medical officers and private native medical practitioners, the resulting report of the Indian Hemp Drugs Commission of 1893–1894 (Commission, 1969) concluded: "*Cannabis indica* must be looked upon as one of the most important drugs of Indian Materia Medica". Although the drug was prescribed for a large number of medical afflictions, the report stated that "...one of the commonest uses is for the relief of pain..." including toothache, labor pain, dysmenorrhoea, neuralgia, stomach pain, headache, cramps, and neuralgia.

By contrast, the therapeutic potential of cannabis was little appreciated by Western physicians until the mid 19th century. The first record of cannabis' medicinal properties in America appeared in a homeopathic journal, the *American Provers' Union*, in 1839. However, recognition of its therapeutic potential was largely due to the efforts of British physician W. B. O'Shaughnessy, who systematically characterized the effects of cannabis extracts in animals and humans and documented their safe use for treatment of tetanus, cholera, epilepsy and rheumatism (Mikuriya, 1969; O'Shaughnessy, 1838). O'Shaughnessy's studies stimulated Western physicians to undertake their own investigations of cannabis' medicinal properties.

The results of such investigations were often positive, and approximately between 1840 and 1900, cannabis became quite popular with Western physicians. Around 1854, hemp products began to appear in drug catalogs in the United States. In 1860, a report by the Committee on *Cannabis indica* of the Ohio State Medical Society validated cannabis' clinical efficacy with respect to conditions including stomach pain, rheumatism, labor pain and neuralgia (Mikuriya, 1969; Grinspoon, 1977). The 1890 Dispensatory of the United States of America lists *Cannabis indica* for its pain-relieving properties and notes its failure to diminish appetite or cause constipation. Further evidence of the increasing Western medical faith in cannabis are the numerous clinical reports of cannabis' analgesic efficacy in the late 19th century. An article by Reynolds (1890), physician to the Queen of England (who studied cannabis' medicinal properties for many years) underscores the effectiveness of cannabis in treating migraine, neuralgia, and dysmenorrhoea.

In many cases, cannabis preparations were favorably compared to opiates. Some case reports even suggest that cannabis is a more effective analgesic than morphine (Mattison, 1891). Potency aside, numerous reports emphasize the low toxicity of cannabis preparations. Others note their failure to produce opiate side-effects such as nausea (Hare, 1887). Reynolds (1890) even commented that cannabis eases the nausea, vomiting and constipation that accompany certain medical afflictions.

Nevertheless, many of these reports note the variable potency of cannabis preparations, their instability, and differences in patient sensitivity to their effects (Anonymous, 1882; O'Shaughnessy, 1838; Fox, 1897; Hare, 1887). These factors complicated titration. Furthermore, the compounds' poor gastrointestinal absorption slowed their action, and their hydrophobicity limited their potential routes of administration – a factor that became particularly detrimental with the advent of the syringe (Snyder, 1971). These problems, and the development of analgesics such as aspirin and barbiturates, led to a decline in the medicinal use of cannabis in the mid 20th century (Mikuriya, 1969). The decisive event in the drug's clinical demise was the Marijuana Tax Act in 1937, which banned growing, distributing or possessing cannabis, effectively arresting marijuana research and medicinal use in the United States. In 1941, cannabis was removed from the U.S. Pharmacopoeia and National Formulary (Grinspoon, 1977). Nevertheless, as reviewed recently by Russo (1998), many authorities continued to herald cannabis' efficacy in treating migraine well into the 20th century.

Admittedly, by today's standards the historical literature is decidedly unscientific, lacking the necessary controls and mired in cultural issues that bring to question their objectivity. Several major developments in the latter part of the 20th century greatly facilitated a more scientific approach to the evaluation of cannabis' medical potential. These included isolation of the primary psychoactive constituents of marijuana: Δ^8-THC and Δ^9-THC (Gaoni and Mechoulam, 1964), the development of even more potent and more readily dissolved synthetic ligands, and, ultimately, discovery of endogenous cannabinoid ligands and cannabinoid receptors (Devane *et al.*, 1988, 1992).

BRIEF PHYSIOLOGY OF PAIN

In nociception (the neural processes that lead to the sensation of pain), primary sensory neurons and dorsal horn nociceptive neurons play a fundamental role. Physiological nociception is characterized by its acute nature and serves a protective function, while clinical nociception is distinguished by plastic changes in these neural circuits that lead to a prolonged enhancement of their excitability.

In the absence of injury, intense noxious stimuli trigger the activation of small diameter primary sensory neurons (Aδ and C fibers) that have their receptive endings in the skin and the peripheral organs. This event subsequently activates two categories of spinal nociceptive neuron: spinal wide dynamic range (which encode stimuli ranging from non-noxious to noxious) and nociceptive-specific neurons (which encode only noxious stimuli). The firing of these spinal nociceptive neurons leads to the activation of nociceptive circuits in the brain and ultimately, pain. In contrast, innocuous stimuli activate only large diameter primary sensory neurons (Aβ fibers), which, via a pathway involving both non-nociceptive neurons and wide dynamic range neurons in the spinal cord and central somatosensory circuits, signal only an innocuous sensation.

CANNABINOIDS AND PAIN

Cannabinoids suppress the pain behavior evoked by noxious stimuli of various types including thermal, mechanical and chemical (tail flick, hot plate: Buxbaum, 1972; Bloom *et al.*, 1977; Jacob *et al.*, 1981; Lichtman and Martin, 1991a,b; Martin *et al.*, 1996, Yaksh and Rudy, 1976; mechanical pinch tests: Sofia *et al.*, 1973; chemical: Sofia *et al.*, 1973; Formukong *et al.*, 1988). Typically, cannabinoids were comparable to opiates both in their potency and their efficacy (Bloom *et al.*,

1977; Jacob *et al.*, 1981). These findings are interesting and important, but more work was needed to make a convincing case for cannabinoid analgesia, because at appropriate doses cannabinoids suppress the motor system, a potential confound in most tests of experimental pain (Martin *et al.*, 1996). This, together with the emerging possibility that endogenous cannabinoids may mediate endogenous pain suppression, led to studies that examined the neural basis of cannabinoid analgesia.

Suppression of noxious stimulus-evoked expression of *c-fos* in spinal cord by cannabinoids

Hunt *et al.* (1987) demonstrated that noxious stimuli induce the expression of the immediate early gene *c-fos* in the spinal dorsal horn, and this response is diminished by analgesics such as morphine (Presley *et al.*, 1990). This laboratory first studied the effects of cannabinoids on the neural processing of pain by examining their effects on spinal *c-fos* expression induced by injections of dilute formalin in the hindpaw (Tsou *et al.*, 1995). The cannabinoid agonist WIN 55,212-2 markedly suppressed this effect. The mediation of the actions of WIN 55,212-2 by cannabinoid receptors was indicated by the dose-dependency of the effect, its reduced magnitude in cannabinoid-tolerant animals, and the lack of efficacy of the cannabinoid receptor-inactive enantiomer WIN 55,212-3.

Cannabinoid suppression of responses to noxious stimuli in spinal wide dynamic range and nociceptive specific neurons

The suppression of pain-induced expression of an immediate early gene by a cannabinoid agonist suggested that cannabinoids suppress the spinal processing of nociceptive messages, but many questions remained that were better addressed using neurophysiological methods. One of these was, whether cannabinoids suppress noxious stimulus-evoked firing of wide dynamic range and nociceptive specific neurons. These studies were carried out in urethane-anesthetized rats using the cannabinoid agonists WIN 55,212-2 and CP 55,940. Three types of noxious stimuli were used: noxious pressure applied to regions of the ipsilateral hind paw corresponding to the receptive field of the recorded neuron (Hohmann *et al.*, 1995), noxious thermal stimuli applied by a Peltier thermode (Hohmann *et al.*, 1998; Hohmann and Walker, 1999), or painful C-fiber strength electrical stimulation (Strangman and Walker, 1999). Low doses (62.5 µg/kg to 500 µg/kg, i.v.) of the cannabinoid agonists produced a profound inhibition of noxious stimulus-evoked firing of wide dynamic range (Figure 22.1) and nociceptive specific spinal dorsal horn neurons (Hohmann *et al.*, 1995; Hohmann and Walker, 1999; Strangman and Walker, 1999). These effects of cannabinoids are mediated by cannabinoid receptors (Figure 22.2) since pretreatment competitive cannabinoid CB_1 receptor antagonist SR141716A blocked the effect of the agonist and the inactive enantiomer (WIN 55,212-3) failed to alter the nociceptive responses.

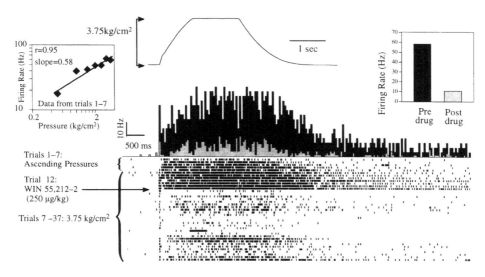

Figure 22.1 Example of inhibition of evoked activity in a WDR neuron by the cannabinoid WIN 55,212-2. The responses of the neuron to mechanical pressure were examined during 37 trials corresponding to each row of dots in the raster plot (top row = trial 1); each dot represents the time of occurrence of a single action potential relative to the stimulus onset. Trials 1 to 7 consisted of applications of increasingly strong mechanical stimulation ranging from non-noxious to noxious levels (0.5, 1, 1.5, 2, 2.5, 3, 3.75 kg/cm²). The concomitant increases in density of dots under the stimulus in the first 7 rows are indicative of the increasingly strong response of the neuron. LEFT (INSET): The mean firing rates of the neuron during a graded series of stimulations are plotted (log-log coordinates) against the applied pressure. The neuron's systematic change in responsiveness was the basis for classifying this cell as a WDR neuron. CENTER: The noxious stimulus illustrated by the pressure waveform (top center) was administered every 2 min for trials 7 to 37. Trials 8–12 constituted baseline trials; after trial 12 (arrow), WIN 55,212-2 (250 mg/kg, i.v.) was administered. A marked decrease in the responsiveness of the neuron is indicated by the sharply decreased density of dots in subsequent rows of the raster plot. RIGHT (INSET): Comparison of the mean firing rate during the stimulus for the 5 baseline trials to the firing rate during the stimulus for the first 10 post-injection trials illustrating, approximately, an 82% decrease in responsiveness. The black peristimulus time histogram between the raster plot and the pressure waveform represents the baseline firing rate prior to injection, whereas the gray peristimulus time histogram represents the firing rate for the first 10 postinjection trials.

Cannabinoid analgesia or anesthesia?

In order to distinguish between analgesia and anesthesia, the question of whether cannabinoid agonists alter the responses of non-nociceptive neurons to mild stimuli was examined (Hohmann and Walker, 1999; Martin *et al.*, 1996). In non-nociceptive neurons, maximum firing rates are achieved at non-noxious levels of stimulation. In contrast to the effects on nociceptive neurons, cannabinoid agonists did not alter the evoked activity of non-nociceptive neurons in the spinal cord or the ventroposterolateral nucleus of the thalamus (Figure 22.3) These findings

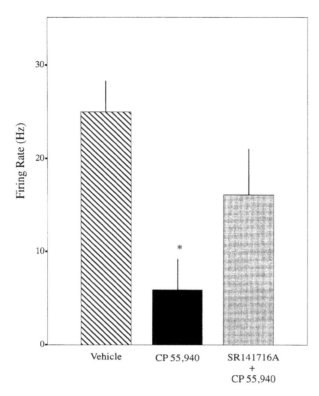

Figure 22.2 Pretreatment with SR141716A, the competitive antagonist for the central cannabinoid receptor (CB_1), blocks the suppression of noxious heat-evoked activity by the classical cannabinoid CP 55,940 (125 mg/kg i.v.). *Significant difference from antagonist pretreatment group, p < 0.05.

suggest that cannabinoids suppress the reactions of nociceptive neurons selectively, hence producing analgesia, not anesthesia.

Cannabinoid suppression of nociceptive responses in the ventroposterolateral thalamus

The classical ascending pain transmission pathway is the lateral spinothalamic tract, which terminates in the ventroposterolateral nucleus of the thalamus, an area that contains many nociceptive neurons (Kenshalo *et al.*, 1980; Guilbaud *et al.*, 1977). Therefore, the responses of neurons in this nucleus provide another measure of the efficacy of nociceptive neurotransmission. The effects of cannabinoids in this area of the brain are very similar to those observed in the spinal cord (Martin *et al.*, 1996). As shown in Figure 22.4, cannabinoids and morphine produced similar effects shifting the stimulus–response function downward. At higher doses, the cannabinoid flattened the stimulus–response function such that the neurons entirely lost their former ability to encode the strength of noxious stimuli in their firing rates. Hence

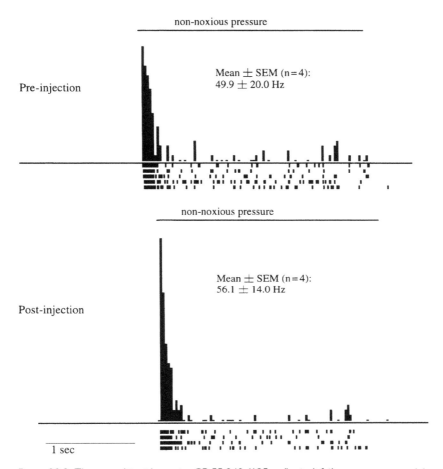

Figure 22.3 The cannabinoid agonist CP 55,940 (125 μg/kg i.v.) fails to suppress activity evoked by non-noxious pressure in a non-nociceptive mechanosensitive cell recorded in the lumbar dorsal horn. A mild pressure stimulus was applied for 3 sec as indicated. TOP: Pre-injection levels of evoked activity. BOTTOM: Evoked activity following administration of CP 55,940. The raster plot shows the time of occurrence of each action potential relative to the stimulus onset. Successive stimulation trials are represented by the horizontal rows of dots in the raster plot. The peristimulus time histogram, shown in black, summarizes the response to the stimulus pre- and post-drug. A similar experiment in the VPL yielded very similar results.

cannabinoids dose-dependently suppress neuronal responses to noxious stimuli, and diminish the neurons' ability to encode pain.

Relationship between effects on nociceptive neurons and behavior

If the neurophysiological changes produced by cannabinoids account for their suppression of pain behavior, one would expect to find the same potency and time course

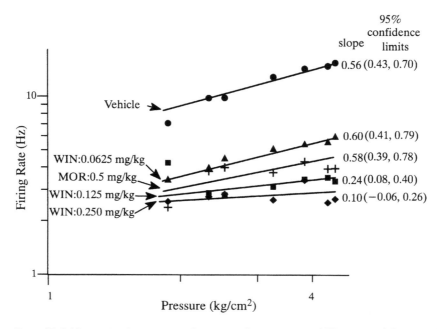

Figure 22.4 Mean stimulus-response functions of a nociceptive VPL neuron following administration of (●) vehicle, WIN 55,212-2 [(▲) 0.0625, (■) 0.125, or (♦) 0.25 mg/kg] or
(+) morphine. The lowest dose of the drug (0.0625 mg/kg) reduced the overall firing
but did not alter the slope of the stimulus–response function. Morphine (0.5 mg/kg)
showed a similar effect. Significant decreases in slope occurred at higher doses of
WIN 55,212-2 (0.125 and 0.25 mg/kg). Post-injection slope values and confidence
limits for estimation of b are shown to the right of each regression line. Note that
the confidence intervals of the slope for the dose of 0.250 mg/kg includes 1 indicating
that the firing rate of the neuron no longer encoded the level of applied pressure.

for the changes in both physiology and behavior. The time course of the suppression
of the thalamic neurophysiological responses was virtually identical to the time course
of suppression of the behavioral reaction (Figure 22.5, Martin *et al.*, 1996). Likewise,
the potency of the cannabinoid agonist in suppressing thalamic nociceptive neuronal
responses was virtually identical to its potency in suppressing thermal nocifensor
(tail flick) reflexes. These findings, along with the established role in pain transmission
of this neuronal population, provide strong support for the conclusion that cannabinoids effectively and selectively suppress neuronal responses to pain.

SITES OF ACTION FOR CANNABINOID ANALGESIA

Role of descending modulation

To address the question of the sites of action of cannabinoids, one approach was to
administer the drug by the intracerebroventricular (i.c.v.) route. Microinjections

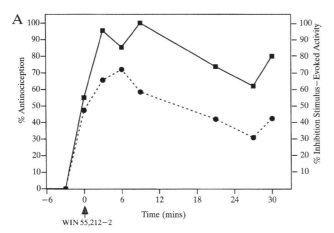

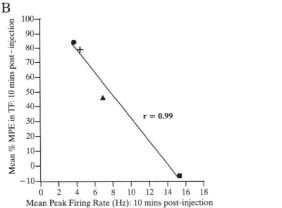

Figure 22.5 (A) Time-course of the antinociceptive and electrophysiological effects of WIN 55, 212-2 (0.25 mg/kg, i.v.). Antinociception was assessed as the pressure at which rats (lightly anesthetized) exhibited a withdrawal response to a mechanical stimulus that increased in intensity over time as described in the text. Data are presented as percent antinociception. For electrophysiology experiments (in separate animals under surgical anesthesia), the same mechanical stimulus was used to apply pressure to the contralateral hind paw while stimulus-evoked activity was recorded from individual neurons in the VPL. The effects of WIN 55,212-2 are presented as % inhibition of stimulus-evoked activity relative to pre-injection values. Note that the (●) inhibition of paw withdrawal (as % antinociception) parallels the (■) inhibition of stimulus-evoked activity in VPL neurons. (B) The relationship between inhibition of noxious stimulus evoked activity and inhibition of tail flick reflex were determined using regression analysis. For each procedure, (■) vehicle or WIN 55,212-2 [(▲) 0.0625, (+) 0.125, or (●) 0.25 mg/kg] was administered to separate groups of animals. Tail flick data (awake animals) are presented as mean % MPE during the ten min following injection. Electrophysiology data are presented as mean peak firing rate during the ten minutes following injection. The high correlation (r = 0.99) is indicative of a relationship between the behavioral and the electrophysiological responses.

of 20 µg WIN 55,212-2 suppressed the noxious stimulus-evoked responses of WDR neurons in the spinal cord (Hohmann *et al.*, 1995). This finding indicated that cannabinoids suppress spinal nociceptive processing in part by supraspinal descending influences. This conclusion is consistent with the behavioral suppression of tail flick reflex via the same route. Moreover, Lichtman and colleagues (1991b) observed that spinal administration of the α_2 receptor antagonist yohimbine markedly reduced the analgesia induced by systemically administered cannabinoids. In light of the crucial role of bulbospinal noradrenergic pathways in the modulation of spinal pain responses (Proudfit, 1988), the actions of yohimbine are also suggestive of a descending modulatory action of cannabinoids. Microinjection of cannabinoids in numerous brain sites that are involved in pain processing (Martin *et al.*, 1993, 1995, 1998) revealed that cannabinoids are active in at least seven brain areas. Among these, it is notable that the cannabinoids produce analgesia following microinjections in the dorsal and lateral periaqueductal gray, the amygdala, and the rostral ventral medulla. These interconnected structures are part of a well-established circuit that mediates, in part, the descending pain-suppressive influences of opiates (Liebeskind *et al.*, 1976). A study published subsequently also provided neurophysiological evidence that indeed this pathway is involved in cannabinoid analgesia (Meng *et al.*, 1998).

Spinal cannabinoid action

Besides their actions in the brain, cannabinoids act directly in the spinal cord to produce analgesia. This was first observed by Yaksh and Rudy (1976) and studied by Welch's group (Welch and Stevens, 1992) and others (Lichtman *et al.*, 1991a). Work from this laboratory demonstrated that topical spinal administration of cannabinoids suppresses the stimulus-evoked activity of nociceptive neurons (Hohmann *et al.*, 1998). The presence of CB_1 receptor mRNA in the dorsal root ganglion suggests that some primary afferent neurons have presynaptic CB_1 receptors (Hohmann and Herkenham, 1998) which could mediate the spinal effects of cannabinoids. Cannabinoid CB_1 receptor mRNA is abundant in the spinal dorsal horn suggesting an alternative site of action on spinal interneurons or projection neurons. Transection of the spinal cord above the level of recording virtually abolished the suppression of nociceptive neurons normally produced by intravenously administered cannabinoids (Hohmann and Walker, 1999). This indicates that either the descending modulatory influences are the dominant mechanism or that the main effect of spinally administered cannabinoids is a presynaptic action on descending tracts inactivated by spinal transection.

Peripheral cannabinoid action

Hargreaves' group (Richardson *et al.*, 1998c) demonstrated that cannabinoids also produce analgesia by peripheral actions. This finding is in line with the presence of cannabinoid receptor mRNA in the dorsal root ganglion, since the receptors formed from the mRNA would be expected to travel to the distal as well as the proximal terminals of these pseudo-unipolar neurons. Furthermore, very low doses of anandamide injected intradermally inhibit the release of a pro-inflammatory factor (calcitonin gene related peptide) from distal nerve endings, and the hyper-algesia consequent to intradermal injections of carrageenan. These effects were absent when the same dose of anandamide was applied to the contralateral paw to control for any systemic effects of the drug. This peripherally-mediated analgesia results from activation of the CB_1 receptor subtype (as expected from the neuronal localization of the receptor), since the effect could be blocked with the selective CB_1 receptor antagonist SR141716. Some recent evidence also suggests a role of CB_2 receptors in peripheral cannabinoid analgesia (Calignano *et al.*, 1998).

CANNABINOIDS AND CHRONIC PAIN/CENTRAL SENSITIZATION

Early work by Kosersky *et al.* (1973) was perhaps the first indication of an effect of cannabinoids in a chronic pain model. Later work from this laboratory using the formalin test (Tsou *et al.*, 1995), and from Bennet's group who examined neuro-pathic pain (Herzberg *et al.*, 1997) have provided confirmation and extensions of the earlier investigations. The study on neuropathic pain revealed that canna-binoids show greater potency in this model than in tests of physiological pain block-ing allodynia and hyperalgesia at doses that do not produce any side effects in rodents.

   Little is known about the effects of cannabinoids on processes that lead to hyperexcitability of nociceptive neurons following prolonged noxious stimulation. Windup, the increasingly strong response of spinal nociceptive neurons over a series of rapidly repeated noxious electrical stimuli, produces central sensitization (Mendell, 1966; Woolf *et al.*, 1988, 1991, 1996; Li *et al.*, 1999). Intravenous injection of WIN 55,212-2, but not the inactive enantiomer, WIN 55,212-3, significantly decreased the windup of spinal wide dynamic range and nociceptive-specific neurons as compared to vehicle (Strangman and Walker, 1999). At low doses (0.25 mg/kg) the effect is highly selective in reducing windup with no effect on the initial C-fiber response of cells (Figure 22.6). These findings indicate that cannabinoids are capable of blocking central sensitization, a result that suggests promise for cannabinoids as therapeutic agents in clinical pain syndromes which typically involve this type of pathophysiological process.

ROLE OF ENDOGENOUS CANNABINOIDS IN PAIN MODULATION

Studies that employ the selective cannabinoid agonists, such as those discussed above, established the consequence of cannabinoid receptor activation for nocicep-tive processing and pain. Work done with selective antagonists not only helped to confirm the actions of the agonists, but to shed light on the endogenous canna-binoid system when administered alone in appropriate physiological systems. Work

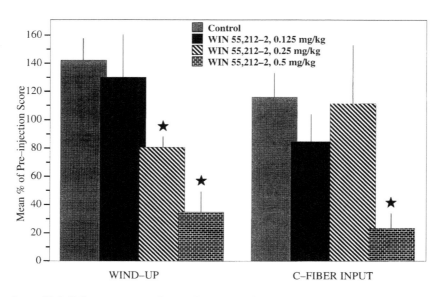

Figure 22.6 Differences in wind-up and the acute-C-fiber response between WIN 55,212-2 and control groups during the first 5 post-injection trials (*p < 0.05). There were no significant differences between the groups' wind-up or acute C-fiber responses during the pre-injection period. However, i.v. WIN 55,212-2, at the dose of 0.25 mg/kg (n = 7) and 0.5 mg/kg (n = 4) but not 0.125 mg/kg (n = 7), significantly decreased the mean wind-up score as compared to vehicle. Only the highest dose of WIN 55,212-2, 0.5 mg/kg, significantly reduced the acute response to C-fiber activation as compared to controls.

from this laboratory showed that blocking the cannabinoid CB_1 receptor with the antagonist SR141716A produces hyperalgesia in the formalin test (Strangman *et al.*, 1998), suggesting an endogenous analgesic tone maintained by endocannabinoids. The finding confirmed those of Richardson *et al.* (1997, 1998a,b) who found that this cannabinoid antagonist, injected intrathecally, produced hyperalgesia, and that the same effect occurs with spinal CB_1 receptor knockdown. Furthermore, the CB_1 antagonist blocks the analgesia produced by electrical stimulation of the dorsal periaqueductal gray (PAG; Walker *et al.*, 1999). These pro-nociceptive actions of the antagonist are reasonable evidence for an antinociceptive action of one or more endocannabinoids.

Even though agonists and antagonists are extremely useful pharmacological agents and provide information on the likely functional roles of the system, they do not by themselves settle the question on the roles of endogenous cannabinoids for pain. In both cases, there is no information on the identity of the one or more possible endocannabinoids. Moreover, in the case of the cannabinoid antagonists that have been commonly used, SR 141716A and SR 144528, there have been reports on their acting as inverse agonists (e.g. Bouaboula *et al.*, 1997; Landsman *et al.*, 1997; Rinaldi-Carmona *et al.*, 1998). Though cases of inverse agonism were typically identified in *in vitro* systems and maybe unlikely to occur to a large extent *in vivo*, more reassurance would be achieved if methods that are independent of these compounds were used to verify the existence of endogenous cannabinoids.

In order to directly address the questions regarding the role of endocannabinoids that were made inferentially from the actions of an antagonist, this laboratory took the further step of measuring endocannabinoids in the PAG using microdialysis (Walker *et al.*, 1999). This method permits collection of neurotransmitters/modulators from the extracellular space, and is therefore an indicator of the release of these modulators. Using liquid chromatography/mass spectrometry, it was established that the analgesia-producing electrical stimulation or injections of the chemical irritant formalin into the hindpaws of anesthetized rats, induced the release of anandamide in the PAG (Figure 22.7). Thus it appears that either pain itself, or electrical stimulation leads to the release of anandamide, which acts on cannabinoid receptors in the PAG to inhibit nociception.

As well as actions in the central nervous system, recent work also indicates an action of endocannabinoids in the periphery. Consistent with the presence of cannabinoid receptors in the peripheral nerve and their transport to the distal endings (Richardson *et al.*, 1998c), work by Calignano *et al.* (1998) demonstrated that endocannabinoids acting in the periphery may modulate pain responses.

SUMMARY AND CLINICAL IMPLICATIONS

As of 1990, virtually nothing was known about the neural mechanisms of cannabinoid analgesia. The discovery of cannabinoid receptors and the putative endogenous ligands suggested that endogenous cannabinoids play an important role in the nervous system, with one potential function being the modulation of pain sensitivity. This led to renewed interest in cannabinoid analgesia and the mechanisms by which it occurs. While there is much to be learned, it is now clear that (1) cannabinoids selectively suppress nociceptive neurotransmission; (2) synthetic cannabinoids are equal to morphine in potency and efficacy; (3) the effects are mediated by descending modulatory tracts that overlap with those affected by morphine, by direct actions in spinal cord, and by actions in the periphery; (4) cannabinoids suppress central sensitization, a mechanism that may explain the high efficacy of these compounds in models of chronic pain; and (5) endogenous cannabinoids may have influence on pain sensitivity and nociceptive responses.

More than 52,000,000 Americans (20% of the population) live in areas that have passed ballot initiatives legalizing the medical use of marijuana, despite the federal ban on the drug. The conflict between local and federal laws led to funding of a study on the clinical potential of cannabinoids, which was conducted by the National Academy of Science's Institute of Medicine (IOM). This and a similar study by NIH held pain pharmacotherapy among the prime candidates for medical marijuana. The IOM team found that second to its use as an antiemetic agent, the most frequent use reported by medical marijuana users is for analgesia. Safer alternatives are needed, due to the health risks associated with smoked marijuana. A better understanding of cannabinoid analgesia may reveal appropriate indications for cannabinoid use and lead to cannabinoid-based pharmacotherapies for pain that avoid the risks associated with inhaling smoke.

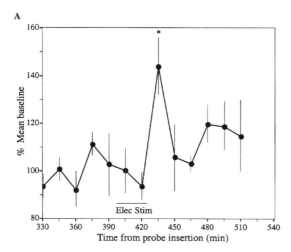

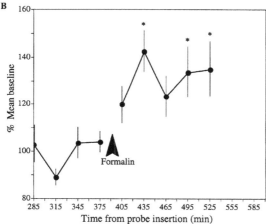

Figure 22.7 Stimulation of the release of anandamide in the PAG of the rat using electrical depolarization or pain. (A) Increased extracellular levels of anandamide following electrical stimulation of the periaqueductal gray matter (PAG) in urethane-anesthetized rats. Following the establishment of stable baseline values, electrical stimulation (monopolar 0.1 msec/1 mA, 60 Hz, 5 sec trains, 5 sec rest) was delivered for 30 min. Microdialysis samples were collected in 15 min intervals and analyzed using HPLC with detection by mass spectrometry, via selected ion monitoring mode at molecular weight 348.3 ($N = 5$, $p < 0.05$, repeated measures analysis of variance). * Significantly different from baseline average via post-hoc test ($p < 0.05$). The delay in the measurement presumably reflects the time needed to produce sufficient overflow of anandamide in the extracellular space to achieve recovery by nucrodialysis. (B) Increased extracellular levels of anandamide in the PAG following induction of prolonged pain in urethane-anesthetized rats. Following establishment of a stable baseline, formalin solution was injected subcutaneously in both hind paws (4%, 150 μl). Samples shown span 30 min intervals ($N = 6$; $p < 0.001$, repeated measures analysis of variance).

Although the literature on the antinociceptive effects of cannabinoids in man is replete with contradictions (e.g. Noyes *et al.*, 1975a,b; Raft *et al.*, 1977), several elements of cannabinoid pharmacology suggest significant clinical potential: (1) Cannabinoids have shown efficacy in the treatment of chronic pain in humans (Noyes *et al.*, 1975a,b); (2) Progress has been made in the synthesis of better, less lipophilic, synthetic cannabinoids such as WIN 55,212-2 and CP 55,940 (D'Ambra *et al.*, 1992; Compton *et al.*, 1992a,b). (3) Endogenous cannabinoids are a logical target for novel pain pharmacotherapies (e.g. reuptake blockers, breakdown inhibitors, or biosynthetic precursors). (4) Cannabinoids possess very low toxicity. There have been no deaths unequivocally attributable to overdose with cannabinoids (reviewed by Harris *et al.*, 1977), an unsurprising finding in light of the low levels of cannabinoid receptors in brainstem areas that control the respiratory and cardiovascular systems (Herkenham *et al.*, 1991). Further work on the neurobiology of cannabinoids offers hope for new pharmacotherapies for pain, especially in instances where opiates lack efficacy or produce intolerable side effects.

REFERENCES

Anonymous (1882) "*Cannabis indica* – Value in facial neuralgia, etc," *Proceedings of the American Pharmaceutical Association* **30**: 500–501.

Bloom, A. S., Dewey, W. L., Harris, L. S. and Brosius, K. K. (1977) "9-nor-9β-hydroxyhexahydrocannabinol a cannabinoid with potent antinociceptive activity: Comparisons with morphine," *The Journal of Pharmacology and Experimental Therapeutics* **200**: 263–270.

Bouaboula, M., Perrachon, S., Milligan, L., Canat, X., Rinaldi-Carmona, M., Portier, M., Barth, F., Calandra, B., Pecceu, F., Lupker, J., Maffrand, J.-P., Le-Fur, G. and Casellas, P. (1997) "A selective inverse agonist for central cannabinoid receptor inhibits mitogen-activated protein kinase activation stimulated by insulin or insulin-like growth factor 1," *Journal of Biological Chemistry* **272**: 22330–22339.

Buxbaum, D. M. (1972) "Analgesic activity of Δ^9-tetrahydrocannabinol in the rat and mouse," *Psychopharmacology* **25**: 275–280.

Calignano, A., La Rana, G., Giuffrida, A. and Piomelli, D. (1998) "Control of pain initiation by endogenous cannabinoids," *Nature* **394**: 277–281.

Commission, I. H. D. (1969) *Marijuana: Report of the India Hemp Drugs Commission 1893–1894*, Silver Spring, Thos. Jefferson Publishing Co.

Compton, D. R., Johnson, M. R., Melvin, L. S. and Martin, B. R. (1992a) "Pharmacological profile of a series of bicyclic cannabinoid analogs: classification as cannabimimetic agents," *The Journal of Pharmacology and Experimental Therapeutics* **260**: 201–209.

Compton, D. R., Gold, L. H., Ward, S. J., Balster, R. L. and Martin, B. R. (1992b) "Aminoalkylindole analogs: cannabimimetic activity of a class of compounds structurally distinct from delta 9-tetrahydrocannabinol," *The Journal of Pharmacology and Experimental Therapeutics* **263**: 1118–1126.

D'Ambra, T. E., Estep, K. G., Bell, M. R., Eissenstat, M. A., Josef, K. A., Ward, S. J., Haycock, D. A., Baizman, E. R., Casiano, F. M., Beglin, N. C., Chippari, S. M., Grego, J. D., Kullnig, R. K. and Daley, G. T. (1992) "Conformationally restrained analogues of pravadoline: Nanomolar potent enantioselective (aminoalkyl) indole agonists of the cannabinoid receptor," *Journal of Medicinal Chemistry* **35**: 124–135.

Devane, W. A., Dysartz III, F. A., Johnson, M. R., Melvin., L. S. and Howlett, A. C. (1988) "Determination and characterization of a cannabinoid receptor in rat brain," *Molecular Pharmacology* **34**: 605–613.

Devane, W. A., Hanus, L., Breuer, A., Pertwee, R. G., Stevenson, L. A., Griffin, G., Gibosn, D., Mandelbaum A., Etinger, A. and Mechoulam, R. (1992) "Isolation and structure of a brain constituent that binds to the cannabinoid receptor," *Science* **258**: 1946–1949.

Formukong, E. A., Evans, A. T. and Evans, F. J. (1988) "Analgesic and antiinflammatory activity of constituents of *Cannabis sativa* L.," *Inflammation* **12**: 361–371.

Fox, R. H. (1897) "Headaches: A study of some common forms with especial reference to arterial tension and to treatment," *The Lancet* **II**: 307–309.

Gaoni, Y. and Mechoulam, R. (1964) "Isolation, structure and partial synthesis of an active constituent of hashish," *Journal of the American Chemical Society* **86**: 1646–1647.

Grinspoon, L. (1977) *Marihuana Reconsidered*. Cambridge and London: Harvard University Press.

Guilbaud, G., Caille, D., Besson, J. M. and Benelli, G. (1977) "Single units activities in ventral posterior and posterior group thalamic nuclei during nociceptive and non nociceptive stimulations in the cat," *Archieves Italiennes de Biologie* **115**: 38–56.

Hare, H. A. (1887) "Clinical and physiological notes on the action of *Cannabis indica*," *Therapeutic Gazette* **11**: 225–228.

Harris, L. S., Dwewy, W. L. and Razdan, R. K. (1977) "*Cannabis*: Its chemistry pharmacology and toxicology," in *Drug Addiction II*, edited by W. R. Martin, pp. 372–429. New York: Springer-Verlag.

Herkenham, M., Lynn, A. B., Johnson, M. R., Melvin, L. S., De Costa, B. R. and Rice, K. C. (1991) "Characterization and localization of cannabinoid receptors in rat brain: A quantitative in vitro autoradiographic study," *The Journal of Neuroscience* **11**: 563–583.

Herzberg, U., Eliav, E., Bennett, G. J. and Kopin, I. J. (1997) "The analgesic effects of R(+)-WIN 55,212-2 mesylate a high affinity cannabinoid agonist, in a rat model of neuropathic pain," *Neuroscience Letters* **221**: 157–160.

Hohmann, A. G. and Herkenham, M. (1998) "Regulation of cannabinoid and mu opioid receptors in rat lumbar spinal cord following neonatal capsaicin treatment," *Neuroscience Letters* **252**: 13–16.

Hohmann, A. G., Martin, W. J., Tsou, K. and Walker, J. M. (1995) "Inhibition of noxious stimulus-evoked activity of spinal cord dorsal horn neurons by the cannabinoid WIN 55, 212-2," *Life Sciences* **56**: 2111–2119.

Hohmann, A. G., Tsou, K. and Walker, J. M. (1998) "Cannabinoid modulation of wide dynamic range neurons in the lumbar dorsal horn of the rat by spinally administered WIN55212-2," *Neuroscience Letters* **257**: 1–4.

Hohmann, A. G. and Walker, J. M. (1999) "Cannabinoid suppression of noxious heat-evoked activity in wide dyanamic range neuron in the lumbar dorsal horn of the rat," *Journal of Neurophysiology* **81**: 575–583.

Hunt, S. P., Pini, A. and Evan, G. (1987) "Induction of c-fos-like protein in spinal cord neurons following sensory stimulation," *Nature* **328**: 632–634.

Iversen, L. L. (2000) *The science of marijuana*. New York, Oxford University Press.

Jacob, J., Ramabadran, K. and Campos-Medeiros, M.A. (1981) "Pharmacological analysis of levonantradol antinociception in mice," *Journal of Clinical Pharmacology* **21**: 327S–333S.

Kenshalo, D. R. Jr., Giesler, G. J. Jr., Leonard, R. B. and Willis, W. D. (1980) "Responses of neurons in primate ventral posterior lateral nucleus to noxious stimuli," *Journal of Neurophysiology* **43**: 1594–1614.

Kosersky, D. S., Dewey, W. L. and Harris, L. S. (1973) "Antipyretic analgesic and anti-inflammatory effects of delta 9-tetrahydrocannabinol in the rat," *European Journal of Pharmacology* **24**: 1–7.

Landsman, R. S., Burkey, T. H., Consroe, P., Roeske, W. R. and Yamamura, H. I. (1997) "SR141716A is an inverse agonist at the human cannabinoid CB1 receptor," *European Journal of Pharmacology* **334**: R1–R2.

Li, H. (1974) "An archaeological and historical account of *Cannabis* in China," *Economic Botany* **28**: 437–448.

Li, J., Simone, D. A. and Larson, A. A. (1999) "Windup leads to characteristics of central sensitization," *Pain* **79**: 75–82.

Lichtman, A. H. and Martin, B. R. (1991a) "Spinal and supraspinal components of cannabinoid-induced antinociception," *The Journal of Pharmacology and Experimental Therapeutics* **258**: 517–523.

Lichtman, A. H. and Martin, B. R. (1991b) "Cannabinoid-induced antinociception is mediated by a spinal α_2-noradrenergic mechanism," *Brain Research* **559**: 309–314.

Liebeskind, J. C., Geisler, G. J. and Urca, G. (1976) "Evidence pertaining to an endogenous mechanism of pain inhibition in the central nervous system," in *Sensory Functions of the Skin in Primates*, edited by I. Zotterman, pp. 561–573. Pargamon Press.

Martin, W. J., Hohman, A. G. and Walker, J. M. (1996) "Inhibition of noxious stimulus-evoked activity in the ventral posterolateral nucleus of the thalamus by the cannabinoid WIN 55,212-2 correlation of electrophysiological effects with antinociceptive actions," *The Journal of Neuroscience* **16**: 6601–6611.

Martin, W. J., Lai, N. K., Patrick, S. L., Tsou, K. and Walker, J. M. (1993) "Antinociceptive actions of WIN 55,212-2 following intraventricular administration in rats," *Brain Research* **629**: 300–304.

Martin, W. J., Patrick, S. L., Coffin, P. O., Tsou, K. I. and Walker, J. M. (1995) "An examination of the central sites of action of cannabinoid-induced antinociception in the rat," *Life Sciences* **56**: 2103–2110.

Martin, W. J., Tsou, K. and Walker, J. M. (1998) "Cannabinoid receptor-mediated inhibition of the rat tail-flick reflex after microinjection into the rostral ventromedial medulla," *Neuroscience Letters* **242**: 33–36.

Mattison, J. B. (1891) "*Cannabis indica* as an anodyne and hypnotic," *The St. Louis Medical and Surgical Journal* **61**: 265–271.

Mendell, L. M. (1966) "Physiological properties of unmyelinated fiber projections to the spinal cord," *Experimental Neurology* **16**: 316–332.

Meng, I. D., Manning, B. H., Martin, W. J. and Fields, H. L. (1998) "An analgesia circuit activated by cannabinoids," *Nature* **395**: 381–383.

Mikuriya, T. H. (1969) "Historical aspects of *Cannabis sativa* in western medicine," *Comm. Prob. Drug. Depend.* **3**: 6121–6133.

Noyes, R. Jr., Brunk, S. F., Baram, D. A. and Canter, A. (1975a) "Analgesic effect of delta-9-tetrahydrocannabinol," *Journal of Clinical Pharmacology* **1**: 139–143.

Noyes, R. Jr., Brunk, S., Avery, D. H. and Canter, A. (1975b) "The analgesic properties of delta-9-tetrahydrocannabinol and codeine," *Clinical Pharmacology and Therapeutics* **18**: 84–89.

O'Shaughnessy, R. (1838) "On the preparations of the Indian hemp, or gunjah," *Trans. med. phys. Soc. Bengal* 71–102.

Presley, R. W., Menetrey, D., Levine, J. D. and Basbaum, A. I. (1990) "Systemic morphine suppresses noxious-evoked Fos protein-like immunoreactivity in the rat spinal cord," *The Journal of Neuroscience* **10**: 323–335.

Proudfit, H. K. (1988) "Pharmacologic evidence for the modulation of nociception by noradrenergic neurons," in *Progress in Brain Research*, edited by H. L. Fields and J. M. Besson, **77**: 357–370. Elsevier Science Publishers B V.

Raft, D., Gregg, J., Ghia, J. and Harris, L (1977) "Effects of intravenous tetrahydrocannabinol on experimental and surgical pain. Psychological correlates of the analgesic response," *Clinical Pharmacology and Therapeutics* **21**: 26–33.

Reynolds, J. R. (1890) "Therapeutical uses and toxic effects of *Cannabis indica*," *Lancet* **1**: 637–638.

Richardson, J. D., Aanonsen, L. and Hargreaves, K. M. (1997) "SR141716A a cannabinoid receptor antagonist, produces hyperalgesia in untreated mice," *European Journal of Pharmacology* **319**: R3–R4.

Richardson, J. D., Aanonsen, L. and Hargreaves, K. M. (1998a) "Antihyperalgesic effects of spinal cannabinoids," *European Journal of Pharmacology* **345**: 145–153.

Richardson, J. D., Aanonsen, L. and Hargreaves, K. M. (1998b) "Hypoactivity of the spinal cannabinoid system results in NMDA-dependent hyperalgesia," *The Journal of Neuroscience* **18**: 451–457.

Richardson, J. D., Kilo, S. and Hargreaves, K. M. (1998c) "Cannabinoids reduce hyperalgesia and inflammation via interaction with peripheral CB1 receptors," *Pain* **75**: 111–119.

Rinaldi-Carmona, M., Barth, F., Millan, J., Derocq, J. M., Casellas, P., Congy, C., Oustric, D., Sarran, M., Bouaboula, M., Calandra, B., Portier, M., Shire, D., Breliere, J. C. and Le Fur, G. L. (1998) "SR 144528, the first potent and selective antagonist of the CB2 cannabinoid receptor," *The Journal of Pharmacology and Experimental Therapeutics* **284**: 644–650.

Russo, E. (1998) "*Cannabis* for migraine treatment: the once and future prescription? An historical and scientific review," *Pain* **76**: 3–8.

Snyder, S. H. (1971) *Uses of Marijuana*. New York: Oxford University Press.

Sofia, R. D., Nalepa, S. D., Harakal, J. J. and Vassar, H. B. (1973) "Anti-edema and analgesic properties of Δ^9-tetrahydrocannabinol (THC)," *The Journal of Pharmacology and Experimental Therapeutics* **186**: 646–655.

Strangman, N. M., Patrick, S. L., Hohmann, A. G., Tsou, K. and Walker, J. M. (1998) "Evidence for a role of endogenous cannabinoids in the modulation of acute and tonic pain sensitivity," *Brain Research* **813**: 323–328.

Strangman, N. M. and Walker, J. M. (1999) "The Cannabinoid WIN 55, 212-2 inhibits the activity-dependent facilitation of spinal nociceptive responses," *Journal of Neurophysiology* **82**: 472–477.

Tsou, K., Lowitz, K. A., Hohman, A. G., Martin, W. J., Hathaway, C. B., Bereiter, D. A. and Walker, J. M. (1995) "Suppression of noxious stimulus-evoked expression of fos-like immunoreactivity in rat spinal cord by a selective cannabinoid agonist," *Neuroscience* **70**: 791–798.

Walker, J. M., Huang, S. M., Strangman, N. M., Tsou, K. and Sañudo-Peña, M. C. (1999) "Pain modulation by release of the endogenous cannabinoid anandamide," *Proceedings of the National Academy of Sciences (USA)* **96**: 12198–12203.

Welch, S. P. and Stevens, D. L. (1992) "Antinociceptive activity of intrathecally administered cannabinoids alone and in combination with morphine in mice," *The Journal of Pharmacology and Experimental Therapeutics* **262**: (1992) 10–18.

Woolf, C. J., Thompson, S. W. and King, A. E. (1988) "Prolonged primary afferent induced alterations in dorsal horn neurones, an intracellular analysis in vivo and in vitro," *The Journal of Physiology (Paris)* **83**: 255–266.

Woolf, C. J. and Thompson, S. W. (1991) "The induction and maintenance of central sensitization is dependent on N-methyl-D-aspartic aid receptor activation implications for the treatment of post-injury hypersensitivity states," *Pain* **44**: 293–299.

Woolf, C. J. (1996) "Windup and central sensitization are not equivalent," *Pain* **66**: 105–108.

Yaksh, T. L. and Rudy, T. A. (1976) "Chronic cathertization of the subarachnoid space," *Physiology and Behaviour* **17**: 1031–1036.

Zias, J., Stark, H., Sellgman, J., Levy, R., Werker, E., Breuer, A. and Mechoulam, R. (1993) "Early medical use of *Cannabis*," *Nature* **363**: 215.

Chapter 23

Effects of acute and chronic cannabinoids on memory: from behavior to genes

*Robert E. Hampson, Elena Grigorenko
and Sam A. Deadwyler*

ABSTRACT

Δ^9-tetrahydrocannabinol (Δ^9-THC) and associated ligands for the cannabinoid (CB_1) receptor produce a profound effect on memory processes both in humans and animals. These effects are largely due to influences on the hippocampus, and indeed, effects of cannabinoids on hippocampal neurons have been extensively studied *in vitro*. In rats performing a spatial delayed-nonmatch-to-Sample (DNMS) task, Δ^9-THC and the potent CB_1 receptor agonist WIN 55,212–2 (WIN-2) produce deficits in performance over long trial delays. These deficits can be accounted for by the hippocampal neurons' failure to strongly encode information critical to performance of the task. These effects are strikingly similar to complete removal of the hippocampus, thereby eliminating its role in short-term memory. Chronic exposure to cannabinoids causes development of tolerance to the behavioral and electrophysiological deficits, and is accompanied by biochemical changes as well as functional genetic changes in hippocampal neurons. Wide spread changes in functional gene expression appear across 21 days of chronic cannabinoid administration. Many of these genes have known roles with respect to their relation to cannabinoid receptor initiated processes. Given that the cannabinoid receptor system is one of the most ubiquitous G-protein coupled neural systems in hippocampus, not to mention the mammalian brain, it is likely that these changes reflect essential alterations in the ability of these neurons to process critical information underlying short-term memory.

Key Words: short-term memory, hippocampus, delayed-nonmatch-to-sample, neuron, electrophysiology, functional gene expression, cDNA microarray

HISTORY OF CANNABINOID EFFECTS ON MEMORY AND BEHAVIOR

Cannabinoids and hippocampus

The first indication that there was a receptor in the brain for Δ^9-tetrahydrocannabinol [Δ^9-THC], the psychoactive ingredient in marijuana or *Cannabis*, came from structure activity (SAR) studies of cannabimimetic compounds that were developed in the late 1980s by Pfizer Inc (Devane *et al.*, 1988). These compounds (of which CP 55940 is the most studied) were extremely potent and produced effects similar to Δ^9-THC on assays of adenylyl cyclase inhibition and cAMP accumulation (Howlett *et al.*, 1986).

Another compound WIN 55,212-2, an aminoalkylindole with equal or higher potency to CP 55940, was developed by Susan Ward and colleagues at Sterling Winthrop (Ward *et al.*, 1991). These reports were followed closely by the serendipitous cloning of the cannabinoid receptor from a large set of G-protein related receptor genes (Matsuda *et al.*, 1990). It was determined that the cannabinoid receptor had a high degree of homology to other G-protein coupled receptors found in brain. Shortly afterward, an endogenous cannabinoid receptor ligand "anandamide" was isolated and purified by Devane *et al.* (1992) which mimicked, to a lesser degree, the actions of the potent synthetic cannabinoids. With the discovery of the endogenous ligand and receptor, it was not long before an antagonist to the receptor was developed by Sanofi Reserch Inc, SR 141716A, which proved highly effective for the CB_1 receptor subtype (Rinaldi-Carmona *et al.*, 1994). A second receptor subtype, CB_2, found primarily in the periphery (thymus and tonsils) and not in brain, has since been cloned in both rats and humans (Munro *et al.*, 1993). Sanofi has also developed a specific antagonist, SR 144528, for the CB_2 subtype (Rinaldi-Carmona *et al.*, 1998). It has also become apparent that possibly more than one "endocannabinoid" (endogenous substance) exists as indicated by studies on 2-arachidonylethenolglyseride (Calignano *et al.*, 1998). Brain areas dense in both receptor and message for the cannabinoid receptor include substantia nigra, globus pallidus, cerebellum and hippocampus (Herkenham *et al.*, 1990, 1991). It should be noted that the CB_1 receptor is quite ubiquitous and is as dense as glutamate or $GABA_A$ receptor subtypes in regions such as neocortex and thalamus and exceeds the concentration of other known G-protein coupled receptors in brain by much as ten fold (Childers and Deadwyler, 1996; Childers and Breivogel, 1998).

Physiological effects of cannabinoids on hippocampal neurons

For such a newly discovered receptor system, the effects of cannabinoids on hippocampal neurons have been surprisingly well studied *in vitro*. The first demonstration of the effects of cannabinoids on NG108 cells in culture was a blockade of N-type calcium channels (Caulfield and Brown, 1992; Mackie and Hille, 1992). Shortly thereafter cannabinoid receptor mediated effects on potassium A-current in cultured hippocampal neurons were demonstrated and shown to be cAMP dependent (Deadwyler *et al.*, 1993, 1995a). Subsequently, a reduction in inward rectifier current was also demonstrated in cultured hippocampal cells (Twitchell *et al.*, 1997) which mimicked the effect displayed in oocytes following injections of CB_1 and GIRK1 mRNA (Henry and Chavkin, 1995; Priller *et al.*, 1995). Cannabinoids suppress glutamatergic transmission in cultured hippocampal neurons and calcium mediated postsynaptic spontaneous discharges in a receptor specific manner (Shen *et al.*, 1996; Shen and Thayer, 1998). Recently, cannabinoids have been tested on adult hippocampal neurons. In hippocampal slices cannabinoids and endocannabinoids have been shown to depress synaptic transmission and selective interfere with long-term potentiation (Stella *et al.*, 1997; Terranova *et al.*, 1995). Although there is anatomic evidence for CB_1 receptor localization on projection neurons to the dentate gyrus as evidenced by high density of receptors in the molecular layer

(Herkenham, 1991), involvement of selective populations of GABAergic interneurons is the more likely mechanism of control of hippocampal cell firing by cannabinoids (Tsou et al., 1998; Katona et al., 1999; Hoffman and Lupica, 2000), suggesting that cannabinoid receptor mediated physiological actions in hippocampus are likely to involve local circuit mechanisms as one means of exerting influence over relatively large numbers of pyramidal cells.

Effects of cannabinoids on short-term memory

Miller and Branconnier (1983) after reviewing the literature on effects of acute marijuana exposure in humans, concluded that the most prominent cognitive impairment was a disruption in short-term memory. Subsequent studies have confirmed this finding in both acute as well as long-term marijuana users (Chait and Pierri, 1992; Hall et al., 1994; Hall and Solowij, 1998). Recent imaging studies suggest that such cognitive effects on memory are associated with altered blood flow and metabolic changes in cannabinoid receptor dense brain regions of subjects exposed to cannabinoids (Fowler and Volkow, 1998). The high density of cannabinoid receptors in hippocampus implicates this system in short-term memory effects and has provided the rationale for investigations in animal models of short-term memory.

Acute exposure to cannabinoids shows marked dose-dependent impairment in rodent studies of short-term memory, whether administered systemically (Heyser et al., 1993) or directly into the hippocampus (Lichtman et al., 1995; Lichtman and Martin, 1996). The nature of this deficit with respect to delayed-match-(DMS) and nonmatch-to-sample (DNMS) performance has been analyzed extensively in comparison to animals with selective removal of the hippocampus using ibotenic acid lesions (Heyser et al., 1993; Hampson et al., 1999). A remarkable similarity exists between behavioral deficits in DNMS in hippocampectomized animals and animals given acute doses of Δ^9-THC (Hampson and Deadwyler, 1998a). As shown in Figure 23.1a this deficit in performance is delay-dependent with performance at short delays relatively unaffected. The deficit is completely reversed within 24 hours and performance is normal (solid circles in Figure 23.1a are the same animals tested 24 hours later). Like hippocampectomized rats, cannabinoid treated rats show a selective deficiency in processing of *Sample* phase information (Hampson et al., 1996; Hampson and Deadwyler, 1999a). The deficit is selectively blocked by SR 141716A, the CB_1 receptor antagonist.

Initial investigations in this laboratory of single neuron activity in hippocampus showed a selective reduction in discharge in the Sample but not the Match phase of the trial in cannabinoid treated rats (Heyser et al., 1993). Hippocampal neurons appear to lose the ability to effectively "encode" Sample information such that on long delay trials animals are *at risk* for increased errors due to temporal decay of the information or proactive interference from prior trials (Hampson and Deadwyler, 1996a). Thus cannabinoids induced a decrease in encoding "strength" in ensembles of hippocampal neurons. Firing in the Match phase however, is not affected, indicating that the deficit in encoding does not result from pharmacologically induced suppression in neuron firing (see Heyser et al., 1993), or generalized behavioral effects of the drug on arousal or motivational factors (see Campbell et al., 1986).

SIMILARITIES BETWEEN CANNABINOID EFFECTS AND HIPPOCAMPAL LESIONS

Hippocampal removal and "residual" memory

One of the longest standing controversies with respect to the role of the hippocampus in behavior is whether hippocampal activity is involved in the processing of spatial or nonspatial information (Cohen and Eichenbaum, 1993; Deacon and Rawlins, 1995; O'Keefe and Nadel, 1978; Nadel, 1991; Muller, 1996; Murray and Mishkin, 1998). In past years, this distinction has existed primarily in the rodent literature, but evidence is rapidly accumulating to extend this controversy to nonhuman primate and even human studies (Tulving and Markowitsch, 1997; Vargha-Khadem *et al.*, 1997; Squire and Zola, 1997; Wagner *et al.*, 1998; Brewer *et al.*, 1998; Stern *et al.*, 1996). We recently examined this issue by assessing the effects on short-term memory in two operant versions of delay tasks, matching (DMS) and nonmatching (DNMS) to sample in rats with selective ibotenate lesions of the hippocampus (HIPP) and in rats with lesions that encroached on retrohippocampal (HCX) structures (Hampson *et al.*, 1999). In HIPP lesioned animals DMS/DNMS performance was not affected on trials with very short (< 6.0 sec) delays, but as the delay was increased, the effects of the lesion became more debilitating. Performance in HIPP lesioned animals was characterized by: (1) severely limited retention of sample information (6.0 sec); and (2) marked proactive influences from preceding trials. HIPP lesioned animals had truncated capacity to utilize within-trial sample information, HCX animals had the same deficits as HIPP lesioned animals but also exhibited significant impairment on trials with <10 sec delays due to a strong *recency effect* in which performance was disrupted as a function of the temporal proximity to the previous trial (Hampson *et al.*, 1999). The findings suggest that an intact retrohippocampal region is important for protecting information from interference during the early portion (0–10 sec) of the delay interval.

Effects of cannabinoids on DNMS performance

As reported previously, delay-dependent performance in the delayed-*match*-to-sample task was impaired following exposure to Δ^9-THC (Heyser *et al.*, 1993). The same is true for animals performing the delayed-*nonmatch*-to-sample task (Figure 23.1a). Administration of Δ^9-THC (1.5 mg/kg, i.p.) produced a highly significant decrease in performance ($F_{(5,257)} = 18.11$, $p < 0.001$) at all delays >5 sec. This delay-dependent deficit was blocked when the animals were injected with the competitive cannabinoid receptor antagonist SR 141716A (1.5 mg/kg, i.p.). Animals received the antagonist 10 minutes before receiving an injection of Δ^9-THC (1.5 mg/kg), and the behavioral session was started 10 minutes afterwards. Injection of only the antagonist alone produced no changes from vehicle-only (pluronic) sessions. Δ^9-THC effects on DNMS were receptor-specific, and contributed primarily to an increase in delay-dependent errors (LDEs).

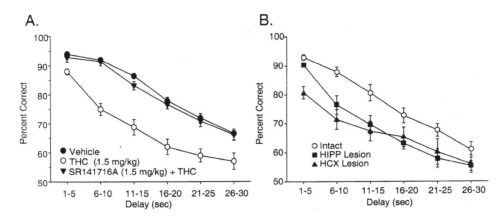

Figure 23.1 Cannabinoid effects on delayed-nonmatch-to-sample (DNMS) performance. A. DNMS performance curves following exposure to Δ^9-THC (1.5 mg/kg). DNMS trials (n = 13 animals, > 200 trials each) were sorted by length of delay, in increments of 5 sec, and are plotted as mean±S.E.M. percent correct DNMS trials. Control (vehicle) session performance is depicted by the filled circles, and performance during Δ^9-THC sessions by the open circles. The effects of the CB_1 receptor antagonist SR 141716A (1.5 mg/kg) to block the 1.5 mg/kg dose of Δ^9-THC are shown by the filled triangles. B. DNMS performance following ibotenate lesions of hippocampus (HIPP, n = 6, filled squares) is shown compared to behavior of nonlesioned animals (Intact, n = 12, open circles). Note similarity in effects of THC to hippocampal lesion. Additional short-delay deficit produced by lesions that encroached on extrahippocampal regions (HCX, n = 6) is shown by filled triangles.

Effects of ibotenate lesions of the hippocampus on DNMS behavior

Given that the DNMS task involved short-term, trial-specific memory, and that this memory was impaired by cannabinoids, the necessity for an intact hippocampus in performance of the DNMS task was investigated in rats which received ibotenate lesions of the hippocampus (Hampson *et al.*, 1999). Performance was assessed following the lesion for at least 15 days, after which retraining and further testing was attempted. Animals with total loss of hippocampal tissue with no encroachment on surrounding limbic structures showed delay dependent deficits in the DNMS task (Figure 23.1b). Mean performance on the DNMS task following the lesion was significantly below prelesion levels ($F_{2,327} = 18.90$, $p < 0.001$) and remained there for the duration of the initial 15 day test period. However, performance on a 0 sec delay was not affected by the lesion. Performance by lesioned animals was not uniform for both levers, suggesting a strategy that resulted in improved performance, but did not involve actual encoding of Sample information. Thus the animals did not require an intact hippocampus to perform the task as long as there was no delay between the Sample and Nonmatch responses. A subsequent, longer exposure to the 1–30 sec delay version of both tasks revealed significant improvement relative to the initial test, but in no case was complete recovery of prelesion performance observed. In animals that received lesions that were not strictly confined to the hippocampus (i.e. including subiculum and entorhinal cortex – HCX group in Figure 23.1b), performance was decreased at short delay intervals as well (Hampson et al., 1999).

Both hippocampal lesions and exposure to cannabinoids alter sequential dependency in the DNMS task

The similarity in behavioral effects of both cannabinoids (Figure 23.1a) and hippocampal lesions (Figure 23.1b) suggested that cannabinoid receptors in the hippocampus may "shut down" information processing in the hippocampus in much the same manner as a "reversible" hippocampal lesion (Hampson and Deadwyler, 1998b). The proactive interference or sequential dependency shown in Figure 23.2a appeared to involve a hippocampal "protection" from proactive interference in normal animals, these same measures of proactive interference were examined both for lesioned animals and animals exposed to Δ^9-THC (Figure 23.2b). Following ibotenate lesions of the hippocampal cell fields, the carryover of Nonmatch information occurred with equal likelihood following all trial delays (Figure 23.2c and 23.2d). This reflected an increase in the number of trials with the "delay-dependent" type of error (even at short delays, see Figure 23.1b) produced by the fact that Nonmatch information was carried over to the next trial. Administration of Δ^9-THC produced the same effect on proactive interference as hippocampal lesions (Figure 23.2b) (Hampson and Deadwyler, 1998b). Since the delay-dependent errors were previously demonstrated to correspond to trials in which Sample phase information was weakly or poorly encoded by the hippocampus, and both lesions and cannabinoids impair that encoding, the increase in these errors is attributed to the lack of the hippocampal encoding of Sample information. The increase in the frequency of carryover from Nonmatch to Sample may thus indicate a facilitation of proactive influences from the prior trial under such circumstances.

HIPPOCAMPAL ENSEMBLE ENCODING OF SAMPLE PHASE INFORMATION IS DISRUPTED BY CANNABINOIDS

Role of hippocampus in short term memory – DNMS

Recent reports from this laboratory described the results of population statistical analyses applied to ensembles of simultaneously-recorded hippocampal CA_1 and CA_3 pyramidal cells (Deadwyler et al., 1996; Hampson and Deadwyler, 1996a, 1998a, 1999b; Deadwyler and Hampson, 1997). Ensemble neural activity was analyzed by canonical discriminant analysis (CDA) which indicated that the major sources of variance in ensemble activity corresponded to encoding of task phase, lever position, and trial performance. Significant variance sources represented by 5 discriminant functions, were extracted, contributing 80% of total variance. The largest proportion of variance (41%, discriminant function 1, DF1) discriminated Sample from

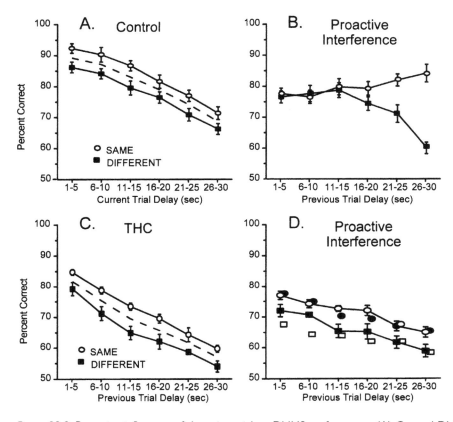

Figure 23.2 Proactive influences of the prior trial on DNMS performance. (A) Control DNMS trials were sorted according to current trial delay as in Figure 1, and as well as the SAME (unfilled circles) vs. DIFFERENT (filled squares) Sample lever position on the previous trial. Mean ± S.E.M. percent correct is shown for each combination of delay and trial type. Dashed line: overall mean performance irrespective of prior trial. (B) DNMS trials sorted by *previous trial delay* and same vs. different trial type. Note that under Control conditions, proactive interference occurred only when previous trial delay was >15 sec. (C) Proactive interference following exposure to Δ^9-THC. DNMS trials were sorted as in A during Δ^9-THC sessions. Overall performance was reduced (dashed line = overall mean), but there was no change in proactive influence on current trial. (D) DNMS trials sorted as in B above, showed altered influence of *previous trial delay* by Δ^9-THC. Opposite shaded symbols (filled circles and unfilled squares) depict the same data following HIPP lesions as in Figure 23.1.

Nonmatch response. Two additional functions, DF2 and DF3, together accounted for 24% of variance, and discriminated the type of behavior (i.e. leverpress vs. nose-poke, inter- vs. intratrial). A critically important source of variance was represented by DF4, which accounted for 11% of variance and encoded *position* of the Sample or Nonmatch responses. Finally, DF5 accounted for 8% of variance, and discriminated trial type as a function of only the Sample response position. Since these DF1, DF4 and DF5 *discriminated* between two possibilities of phase, position or trial type, the concomitant *encoding* of this information by the ensemble was represented by the discriminant scores for those functions. Success or failure to respond with the appropriate response on a given trial was shown to be directly related to the "strength" of the DF scores and consequently the pattern of firing of neurons within the ensemble.

Canonical discriminant analyses of ensemble activity following exposure to Δ^9-THC revealed the same 5 sources of variance as before, with the exception that the contribution of lever position (DF4) was reduced from 11% to 5%, and the mean discriminant scores for that root were reduced as well (Hampson and Deadwyler, 2000). In fact, DF4 scores in the Sample phase were reduced from a mean of 0.76 ± 0.12 on Vehicle sessions to a mean of 0.37 ± 0.22 on Δ^9-THC sessions ($p < 0.001$), while scores in the Nonmatch phase were not significantly altered (VEH: 1.27 ± 0.14, Δ^9-THC: 1.12 ± 0.18). The decrease in strength of encoding was accompanied by a proportional 60% increase in the number of total trials with reduced DF scores, and a significant increase (from 21% to 34% of total trials) in long delay error trials. Thus, as with behavioral measures of DNMS performance, it appears that information encoded during the Sample phase is most dependent on an intact hippocampus, and it is Sample information encoding that is influenced by Δ^9-THC.

Cannabinoid effects on DNMS information processing

Since Δ^9-THC was previously shown to reduce firing peaks in the Sample phase of the task, with only minimal reduction in Nonmatch phase firing (Heyser *et al.*, 1993), the effects of Δ^9-THC on ensemble firing during the DNMS task were examined to determine if encoding of Sample phase information was compromised by the reduction in amplitude of Sample firing. As shown in Figure 23.3, firing during the Nonmatch (recognition) phase of the DNMS task firing was only slightly attenuated, whereas firing during the Sample (encoding) phase was reduced by over 50% (arrows, Figure 23.3). Thus the principal effect of cannabinoids on hippocampal neurons in DMS tasks was to decrease Sample encoding as was shown previously (Heyser *et al.*, 1993). Figure 23.4 shows the DNMS encoding trial sequence from a recently completed study (Hampson and Deadwyler, 2000). The sequence following exposure to cannabinoids (bold dashed lines) differs from the sequence under Control conditions. The bold arrows show a lack of strong encoding, and hence an increase of 10% in LDE's and miscodes. This effect is consistent with the overall 65% decrease in Sample encoding as shown above (Figure 23.3). An analysis of the information content of Sample ensembles (Hampson and Deadwyler, 1996b) showed nearly 60% reduction in information encoded following exposure to cannabinoids. The results confirm the relationship between strength of ensemble code and performance (Figure 23.4) since the reduction in Sample encoding strength produced a proportional increase in DNMS errors.

The distribution of cell types within hippocampus was studied in 4 animals (39 neurons) under control conditions, and following administration of WIN 55, 212-2 (0.35 mg/kg). Figure 23.5 shows examples of single trial firing patterns of

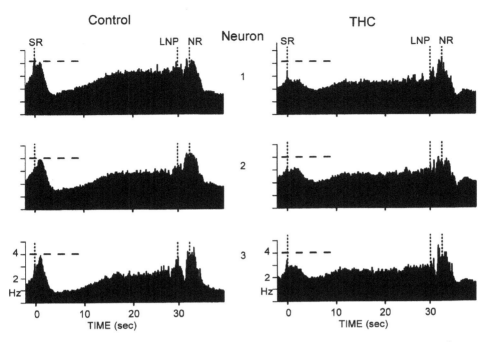

Figure 23.3 Three simultaneously recorded hippocampal neurons Control (vehicle) and Δ^9-THC sessions. Neurons 1–2 were recorded from CA_3, and neuron 3 from CA_1. The trial-based histograms (TBHs) depict summed firing rates of each neuron on 30 sec delay trials (n > 100 trials). Each Control TBH (left) shows characteristic firing rate increases at Sample response (SR), and Nonmatch response (NR) as well as prior to the last nosepoke (LNP) during the delay interval. TBHs at right were recorded following exposure to Δ^9-THC (1.5 mg/kg). Dashed horizontal line depicts SR peak firing rate during Control session, note suppression of SR and delay firing. Firing rate scale (Hz) is shown on the TBHs at bottom.

the three principal cell types during the Sample and Nonmatch phases. The *Sample-only cell* (Figure 23.5, top) which normally exhibited peak firing on Left and Right Sample and not during the Nonmatch phase, did not increase firing at all after injection of WIN-2 (Figure 23.5, bottom). In contrast, firing of the *Nonmatch-only cell* (Figure 23.5, second from top) was unaffected by WIN-2. Finally, the *Trial-type cell* (Figure 23.5, center) which fired during both Sample and Nonmatch phases for a given trial type (i.e. left or right), showed a selective loss of Sample, but not Nonmatch, firing following exposure to WIN-2. All cells resumed control firing correlates when recorded the next day following vehicle only injections. Results for tested functional cell types showed that of 37 neurons that were active during the Sample phase under vehicle conditions, only 9 were active in WIN-2 sessions (Table 23.1). In comparison, of 43 neurons that fired during the Nonmatch phase, 29 remained active in cannabinoid recording sessions (Table 23.1) (Hampson and Deadwyler, 2000).

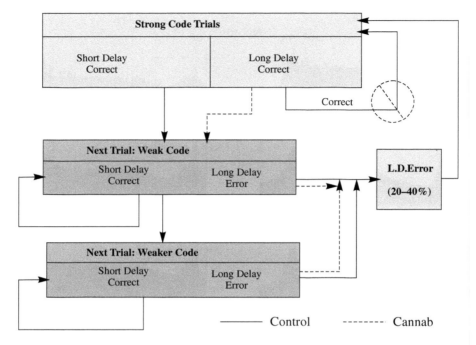

Figure 23.4 DNMS behavioral cascade dependent on strength of sample encoding. Behavioral on Control trials (solid lines) starts with strong encoding of SR. Result will be a correct trial irrespective of delay, however if delay is short, following trial receives a weaker encoding of SR. As code weakens through successive trials, long delays produce errors (L.D. Errors, 20% of total trials), and subsequently reset the cascade to strong encoding. Correct trials at short delays successively weakens the SR code, eventually leading to an error at short delays. Cannabinoid exposure disrupts the cascade (dashed "No" symbol) by producing weakened SR codes irrespective of prior trial outcome (dashed lines). This increases the number of delay errors (to 40%), and accelerates the weakening of the SR code.

Cannabinoids affect encoding strength but not encoding functions

The loss of Sample firing correlates described above (Figure 23.5) confirmed an earlier report that exposure to Δ^9-THC reduced firing of hippocampal cells during the Sample phase of DMS, but not the Nonmatch phase (Heyser *et al.*, 1993). In terms of the behavioral cascade, this means that more trials show weak encoding (red arrow, Figure 23.4b), and consequently, LDEs and miscodes are increased. The frequency distributions of ensemble Sample phase firing rates (Hampson and Deadwyler, 2000) reveals that cannabinoids shift the distribution toward weaker codes (lower rates). Comparison with the vehicle (control) distribution shows that a larger number of trials were "at risk" and the mean encoding strength was

Table 23.1 Cannabinoid effects on hippocampal functional cell types

Functional Cell Type	Number of Cells		% of Cells
	Vehicle	Cannabinoid	Suppressed
Sample Only	11	3	73
Nonmatch Only	11	9	18
Left Only	6	1	83[†]
Right Only	7	2	71[†]
Left-Sample	7	2	71
Right-Sample	6	1	83
Left-Nonmatch	8	12*	13[‡]
Right-Nonmatch	11	15**	0

* 5 cells shifted from Left-Only to Left-Nonmatch.
** 5 cells shifted from Right-Only to Right-Nonmatch.
[†] 5 cells each ceased firing as Left-Only or Right-Only cells.
[‡] 1 Left-Nonmatch cell was suppressed.

reduced 25% following cannabinoid exposure. The cannabinoid plot terminates at a lower overall percentage (60%) of correct trials than control due to the increased number of trials with lower encoding strengths. This suggests that the ensemble is unable to generate the appropriate Nonmatch pattern (code) when Sample information is reduced or absent.

CHRONIC EXPOSURE TO CANNABINOIDS ATTENUATES EFFECTS ON SHORT-TERM MEMORY

A striking paradox uncovered in our initial investigation of chronic cannabinoid treatment on short-term memory was the marked attenuation in the acute disruptive effects that occurred over a 20–30 day period of daily exposure to high doses (10mg/kg) of Δ^9-THC (Deadwyler et al., 1995b). It is well known that the half-life of Δ^9-THC in rats is quite long (10–100 days), and that repeated daily dosing leads to accumulation of the drug in fatty tissue and re-release of active metabolites as well as primary compound (Dewey, 1986). Remarkably, animals initially unable to even locomote due to the severe catalepsy induced by the large (10 mg/kg) dose of Δ^9-THC (Abood et al., 1993; Abood and Martin, 1992), showed systematic improvement in DMS performance and finally complete "tolerance" to the same large dose after 35 days of exposure. The curves in Figure 23.6 show a predominantly delay-dependent effect of Δ^9-THC on days 5 and 16, and fewer correct trials even at the shortest delays. By day 35, DMS/DNMS performance 1 hour after Δ^9-THC (10 mg/kg) administration was not significantly different from control.

This led to a series of investigations into the basic mechanisms of "tolerance" of the cannabinoid system in collaboration with Drs. Laura Sim, Chris Breivogel and Stephen Childers, using in vitro autoradiographic method for visualizing CB_1 receptor stimulated binding of GTPγS. This method demonstrated that the coupling of the receptor to the G-protein pool was reduced in hippocampus following chronic drug exposure (Sim et al., 1996). Animals subjected to the chronic dose regimen

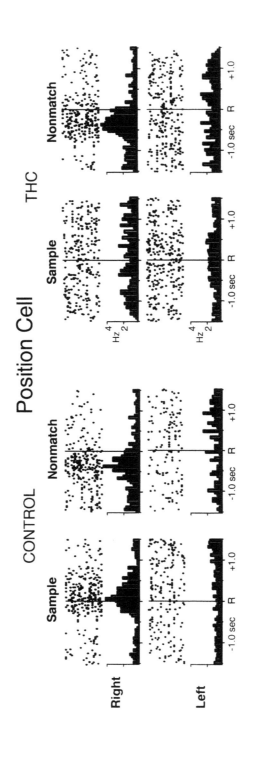

Trial-Type Cell

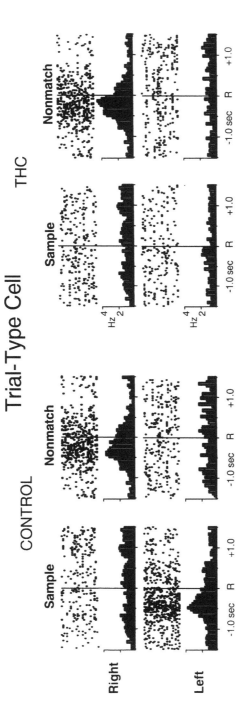

Figure 23.5 Rastergrams and perievent histograms (PEHs) depicting change in hippocampal functional cell types (FCTs) following exposure to cannabinoid. Two FCTs that fire during both the Sample and Nonmatch phases are shown. Rastergrams illustrate firing on single trials with a tick mark each time the cell fired an action potential. Successive lines depict 20 successive trials. PEHs below illustrate average firing across 100 trials for the same ± 1.5 sec interval around the sample or nonmatch response as the rastergrams. (A) Under Control (vehicle) conditions, right *Position-only cell* fires during both Sample and Nonmatch phases, but only at the right, and not the left response position. Right panel (Cannab.) shows loss of Sample firing but not nonmatch firing following exposure to cannabinoid (WIN-2, 0.35 mg/kg). (B) Left *Trial-type cell* fires only to left sample and right nonmatch – the two response that comprise a left trial. The cell never fired on right sample or left nonmatch (the opposite trial type). Left sample firing was suppressed following cannabinoid exposure (right panel), but right nonmatch firing was unaffected.

showed down regulation of receptor number as well as receptor desensitization in the form of reduced GTPγS binding in hippocampus within 7–14 days (Sim *et al.*, 1996; Breivogel *et al.*, 1999). We have pursued these studies by characterizing molecular changes in the cannabinoid system such as CB_1 receptor message (Zhuang *et al.*, 1998), as well as other genes in hippocampus (Kittler *et al.*, 2000) using the same chronic exposure paradigm. It is clear that there is a high degree of correspondence between behavioral tolerance and the ability for cannabinoid receptors to stimulate G-protein binding.

THE MOLECULAR BASIS OF TOLERANCE TO DRUGS OF ABUSE

Tolerance is defined as reduced drug efficacy following repeated exposure and has been described for many different abused substances including opiates and marijuana. Tolerance to opiates may also contribute to their addictive nature, and has been intensely investigated for many years (Hoffman and Tabakoff, 1989; Compton *et al.*, 1990; Adams and Martin, 1996; Self, 1998). We are only beginning to understand alterations in gene expression that accompanies the development of tolerance to abused substances.

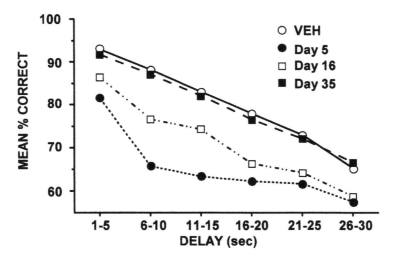

Figure 23.6 Development of behavioral tolerance to chronic cannabinoid exposure. Delay curves (as in Figure 23.1) reflect DNMS performance at successive daily intervals over the treatment period (daily 10 mg/kg Δ^9-THC). Control performance (VEH, unfilled circles) reflects mean ($\pm$ S.E.M.) DNMS performance (n = 6 animals) during a session with only vehicle injection, immediate preceding onset of daily Δ^9-THC injections. Daily curves (Days 5, 16, and 35) reflect mean performance for a single session on that day of chronic exposure. Performance on Day 35 reflects complete behavioral tolerance to the daily exposure to Δ^9-THC.

Both opioid and cannabinoid receptors are coupled to G-proteins, thus genetic changes that occur with tolerance may reveal biochemical mechanisms that become altered. One model of chronic heroin abuse suggests that opioid receptors are reduced in numbers or internalized rapidly (Pak *et al.*, 1996; Sternini *et al.*, 1996). Internalization of opioid receptors could occur via receptor phosphorylation of G-protein receptor kinases promoting receptor downregulation (Freedman and Lefkowitz, 1996). A similar demonstration of CB_1 receptor internalization and recovery during short (20 min) vs. longer (90 min) exposure times to CB_1 receptor agonists, has been shown by Hsieh *et al.* (1999). Another potential level at which tolerance can occur involves the efficacy of coupling between receptors and their G-proteins. Changes in coupling could be mediated either by receptor phosphorylation or by regulation of G-protein subunit expression which changes the primary coupling between G-proteins (Terwilliger *et al.*, 1991; Van Vilet *et al.*, 1993; Sim *et al.*, 1996). Sim *et al.* (1996) utilized [^{35}S]GTPgS *in vitro* autoradiography for analysis of coupling of CB_1 receptor with G-proteins during chronic exposure to Δ^9-THC and showed a reduced ability of CB_1 receptor to bind [^{35}S]GTPgS. Such uncoupling after chronic Δ^9-THC treatment may affect the function of ion channels (Mackie and Hille, 1992; Shen *et al.*, 1996; Deadwyler *et al.*, 1993) and other cellular processes normally linked to cannabinoid receptors (Martin *et al.*, 1993; Glass and Felder, 1997; Howlett, 1998; Lichtman and Martin, 1996; Calignano *et al.*, 1998).

Functional gene changes in drug abuse research

Evaluating the effects of drugs of abuse on gene expression profiles in the brain represents a novel strategy for investigating the underlying causes of drug addiction. This approach is supported in part by the observations that drugs such as cocaine, amphetamine and other psychostimulants induce expression of immediate early genes in specific brain regions (Hope *et al.*, 1994; Chen *et al.*, 1997). The ability of drugs to produce long-term changes in gene transcription represents a mechanism whereby cellular functions can remain substantially altered when drug taking ceases. Molecular changes within individual neurons following chronic drug administration or withdrawal have been studied with both differential display and subtractive hybridization approaches (Douglass *et al.*, 1995; Walker and Sevarino, 1995; Couceyro *et al.*, 1997; Ennulat and Cohen, 1997; Wang *et al.*, 1997). However, only cDNA technology allows monitoring of the expression of literally thousands of genes simultaneously and provides a format for identifying genes that might be *at risk* (see Nature Genetics, 1999). This approach has been referred to as functional genomics, and uses gene sequences taken from a cDNA library which are PCR amplified and arrayed by robotic assistance on glass slides or nylon filters (Schena *et al.*, 1995, 1996; Trower, 1997; Adryan *et al.*, 1999). The microarrays serve as gene targets for hybridization to cDNA probes prepared from RNA samples of brain tissue of treated and untreated subjects. This technique is now widely applied for identification of unique gene expression patterns associated with different types of pathology (Heller *et al.*, 1997; de Waard *et al.*, 1999; Debouck and Goodfellow, 1999), as well as drug abuse studies (Nestler, 1997).

Chronic exposure to cannabinoids results in marked changes in CB_1 mRNA expression profiles in rat brain

Several lines of evidence indicate that adaptive changes occur in the brain in response to repeated cannabinoid administration (Abood *et al.*, 1993; Heyser *et al.*, 1993; Romero *et al.*, 1995, 1997; Fan *et al.*, 1996; Aceto *et al.*, 1996; Adams and Martin, 1996; Solowij, 1998). Significant behavioral tolerance to memory disruptive effects of Δ^9-THC was demonstrated after 21–35 days (Deadwyler *et al.*, 1995b). It was later determined that "coupling" of the CB_1 receptor to $G_{i/o}$-proteins, as measured by receptor mediated GTPgS binding was decreased after such chronic exposure (Sim *et al.*, 1996). More recent findings showed that this change occurred within 3–7 days (Brievogel *et al.*, 1999). Using semi-quantitative RT-PCR analyses it was also demonstrated that pronounced changes occurred in the expression of CB_1 receptor mRNA over 21 days of Δ^9-THC treatment with different time courses in hippocampus, cerebellum and striatum (Zhuang *et al.*, 1998). Large changes in CB_1 receptor message occurred primarily after marked receptor desensitization and down regulation (Brievogel *et al.*, 1999). In hippocampus and cerebellum CB_1 receptor desensitization appeared to precede the alterations in CB_1 receptor mRNA while in striatum changes took longer. However, despite the heterogeneity of CB_1 mRNA changes across the three brain structures, CB_1 message levels returned to control (vehicle) values within 21 days in all regions even though animals were still exposed to relatively high doses (10 mg/kg) of Δ^9-THC. Data from primary large scale cDNA microarray screens indicated marked changes in the expression of other genes also occurred during the development of Δ^9-THC tolerance in subsets of these same animals.

DNA array designs and technical implications

The most attractive feature of the rapidly developed cDNA microarray technology is that it is possible to obtain expression profiles of thousands genes in a single experiment (Schena *et al.*, 1995, 1996; DeRisi *et al.*, 1996; Duggan *et al.*, 1999). This approach uses a potent feature of DNA-the sequence complementarity of two strands. The technology is based on hybridization between nucleic acids, one of which is immobilized on a glass or nylon membrane (Southern *et al.*, 1999). Individually arrayed, thousands of gene-specific fragments are simultaneously probed with labeled cDNA derived from mRNA pools of "control" or "treated" tissue. Direct comparisons of radioactive or fluorescent signal determines the *differential level* of transcript expression between animals for all genes that have been selected on the array. Such profiles can indicate expression "motives" as well as statistical differences (Chen *et al.*, 1998; Khan, 1998; Yuh *et al.*, 1998). Commonly used arrays employ isotope-based detection with cDNA fragments spotted onto positively charged nylon membranes at high density via a robotic arraying system. The membranes are then hybridized with radioactive mRNA probes from brain tissue. Fabrication of arrays and evaluation of their reliability with respect to cDNA spotting is the most critical step in microarray procedures. Several commercially available arrays are currently available which can offer up to 50% of human or mouse genes arrayed for large-scale profiling (Genomic Systems, St. Louis, MO) or arrayed genes arranged into specific functional classes (Clontech, Palo Alto, CA). These arrays include internal controls that allow determination of sensitivity and normalization across several arrays. A variety of approaches have been proposed to improve the uniformity of the hybridization signal on the array. They include changes in hybridization mixtures to increase the contact of labeled molecules with immobilized probes (Duggan *et al.*, 1999), post-hybridizational amplification using either detectable molecules (Chen *et al.*, 1998) or using small hybridization volumes (Bertucci *et al.*, 1999). Thus, cDNA microarray technology is now an affordable, reliable "tool" providing new perspectives for analyses of molecular mechanisms of tolerance to cannabinoids.

Application of cDNA microarrays to analyze effects of chronic cannabinoids

Arrays employed in collaboration with Glaxo Wellcome utilized Poly(A)$^+$RNA from male rat brains as a template for cDNA library construction with oligo-dT primers. Two 10 cm by 12 cm cDNA arrays each containing 12,228 clones in duplicate were constructed. The Glaxo Wellcome proprietary differential gene expression analysis program (DGENT), determined minimal variation (<5%) between filters in the amount of template DNA spotted at a particular coordinate (Figure 23.7). Arrays probed with the M13 internal oligonucleotide for normalization purposes were compared to the hybridization profiles of labeled probes mRNA derived from hippocampus and cerebellum of both Δ^9-THC or saline treated animals (n = 2 animals per exposure time point). Results from two independent large scale arrays from hippocampal mRNA of vehicle or Δ^9-THC treated tissues were averaged. Results are shown in the Table 23.2.

The selection criteria for differentially expressed clones or "hits" were determined as follows: (1) a minimum 1.5-fold (i.e. 50%) difference in intensity for a particular cDNA locus between vehicle and Δ^9-THC treated animals and (2) consistency of intensity between duplicate cDNA clones in the same array (Patten *et al.*, 1998). Only clones which met these two criteria were subjected to further sequencing and verification tests using RNA dot blot analyses. The sequences were compared to EMBL and Genbank databases using the BLAST program. Probes from tissues of vehicle or Δ^9-THC treated animals hybridized to an average of only 11,225 or roughly 40% of the clones on the array (Figure 23.7). This indicated that not all the clones derived from the whole rat brain cDNA library were represented and expressed in mRNA isolated from hippocampus. This was true for Δ^9-THC-treated as well as vehicle animals. In hippocampus, a total of 29 genes were altered (up or downregulated) by chronic Δ^9-THC treatment. Analysis of the hybridization pattern of the complex probes derived from hippocampal and cerebellar mRNA from Δ^9-THC treated animals revealed a total of 104 upregulated genes and 56 downregulated genes in hippocampus and 91 upregulated genes and 49 downregulated genes in cerebellum derived across all animals and time points of the chronic treatment regimen. Taking into account redundant hits of the same clone, a total of 36 in hippocampus and 35 in cerebellum were identified as altered by either acute or chronic Δ^9-THC treatment (Table 23.2).

Of the total 36 and 35 respective gene changes differentially expressed in hippocampus and cerebellum, 17 were commonly expressed in both tissues and

Table 23.2 List of genes differentially expressed in rat hippocampus and cerebellum following 24 hours, 7 days and 21 days of Δ^9-THC exposure

Clone/Cellular function	Accession number	Hippocampus			Cerebellum		
		24H	7D	21D	24H	7D	21D
Metabolism							
Fructose-biphosphate aldolase	M12919	1.6 ↓*	1.5 ↓	1.7 ↓			
Glyceraldehyde-3-phosphate dehydrogenase	M17701			1.8 ↓			
Cytochrome oxidase	S79304	2.0 ↑ (n=15)	1.7 ↑ (n=7)	2.8 ↑ (n=43)	3.0 ↑ (n=11)	2.1 ↑ (n=9)	2.0 ↑ (n=40)
Starch synthase	U48227	2.6 ↑			2.1 ↑		
α-Enolase	X02610		2.3 ↑				
Malate dehydrogenase	M29462			2.0 ↓			
Phospholipid glutathione peroxidase	X82679			1.8 ↓			
Cell adhesion/Growth factors							
Laminin	X58531	2.3 ↑			2.1 ↑		
NCAM	X06564	3.2 ↑					
Growth-associated protein ST2	Y07519	1.6 ↑					
Cystatin-related protein	S57980			1.8 ↑			2.0 ↑
Telencephalin	U89893				5.1 ↑		
Structural and Cytoskeletal							
SC1 protein	U27562	1.7 ↑	1.5 ↑	1.9 ↑	3.1 ↑	2.0 ↑	2.5 ↑
Myelin proteolipid protein	M11185	1.9 ↓	2.2 ↑ (n=2)			1.8 ↑	2.5 ↑
Myelin basic protein	K00512		1.8 ↓ (n=2)	2.0 ↓ (n=6)		2.5 ↓ (n=2)	
Tau microtibule associated protein	X79321						2.0 ↓
β-tubilin	U08342		1.6 ↑				
Myelin basic protein	M25889					2.0 ↓	2.1 ↓
Mitochondrial genome Genes coding for 16S rRNA; tRNAs specific for leucine and phenylalanine	V00681		1.8 ↓ (n=11)	3.5 ↓ (n=7)		2.5 ↓ (n=18)	2.1 ↓ (n=15)
Receptors/Transporters							
Angiotensin AT1 receptor	S66402		1.6 ↓	2.2 ↑			2.0 ↑
Sodium channel 1	U57352				7.5 ↑		
Peptide-histidine transporter	AB000280						6.7 ↓
Brain lipid binding protein	X82679					2.5 ↓	2.3 ↓
Signal transduction/Receptors							
Vibrator critical region	U96726		2.5 ↓				
Receptor for hyaloronidan-mediated motility	U87983		2.0 ↓				
Proteosomal ATPase	D83521			8.6 ↓			
Calmodulin	M17069	1.6 ↓	2.6 ↑ (n=3)	1.8↓	1.8↓		
Prostaglandin D synthase	J04488	1.8 ↑ (n=11)			2.4 ↑ (n=7)		

14-3-3 protein, regulator of PKC	D17447	1.7 ↑			2.3 ↑		
PKU beta subunit	AB004885			2.5 ↑			
Transferrin	D38380	3.0 ↓			1.7↓		
Transcription factors							
Elongation factor	X61043				1.9 ↑		
Histon H1 gene	J03482		3.9↓				
Elongation factor-1	L10339		2.1↓	2.2 ↓			
Elongation factor	X63561	2.5 ↑	3.1↑	1.7 ↓	1.6 ↑	2.1 ↑	
Elongation factor	Z11531				1.9 ↑		
Transcription/Translation Machinery							
ATP-dependent RNA helicase	AA995214	2.5↓	1.8 ↓				
Ubiquitin conjugating enzyme	P51966				2.4 ↑		
SnRNP-associated polypeptide	J05497					1.9 ↑	
Guanine-nucleotide releasing protein	L10336					2.7 ↓	
Protein folding/Degradation							
HSP 70				1.8↓ (n = 2)		1.7↓	2.3↓
Chaperonin containing TCP-1	Z31553				2.7 ↑		
Ubiquitin-containing enzyme	P51966				2.4 ↑		
Polyubiquitin	D17296	2.3 ↓ (n = 4)	1.7↑				
Others							
Brain-specific mRNA	M19861		1.8↓	1.6↓			
Acidic 82 kD protein	U15552		2.0 ↑				
Brain-specific small nuclear RNA	M29340		1.5 ↓		2.1↑		
KIAA0275 gene	D87465				1.8 ↑		
Glu-Pro dipeptide repeat protein mRNA	U40628						2.5↑
Notch 4	U43691		1.6↓				3.0↓
Tetratricopeptide repeat gene	D84296						1.9 ↓
ID repeat gene	U25468				2.1 ↓		
T8G15 genomic clone	B12172			2.1 ↑			1.6 ↑

*Expression level was measured as ratio of Δ^9-THC treated/control samples for upregulated (↓) genes and control/Δ^9-THC treated samples for downregulated (↑) genes. Genes commonly expressed in both tissues are in italics.

showed similar changes at the same time points during the course of Δ^9-THC treatment. The response to the initial high acute dose (10 mg/kg) of Δ^9-THC (24 h time point) produced changes in the expression of 15 and 19 genes in hippocampus and cerebellum, respectively, eight of which were expressed in both tissues. The increased expression of certain cell adhesion and growth factors in hippocampus (NCAM) and cerebellum (telencephalin), by the acute dose of Δ^9-THC, did not persist through the first 7 days of Δ^9-THC treatment (Table 23.2). These changes could be mediated by either non-specific stress reactions to the high dose of Δ^9-THC triggering the expression of immediate early genes such as *c-fos* and *knox* (McGregor *et al.*, 1998) and/or cannabinoid receptor-mediated decreased levels of cAMP (Howlett *et al.*, 1991; Felder *et al.*, 1992; Childers and Deadwyler, 1996)

Figure 23.7 Rat hippocampal gene expression monitored on a rat brain cDNA microarray. Each array contained 24,576 spots, representing 12,288 cDNAs in duplicate. 33^P-labeled first strand cDNA probes prepared from control and 21 days Δ^9-THC treated rat hippocampus were hybridized onto individual arrays. Differentially regulated genes are viewed as a pair at a time (cDNA spots are circled) in close-up images of the differentially expressed cDNA clones. Note left and right images are identical clones. Encircled: 1: transthyretin; 2: prostaglandin D synthase; 4: myelin basic protein.

shown to affect mRNA transcription mechanisms (Herring *et al.*, 1998). The changes could also have resulted from severe catalepsy produced by this high dose of Δ^9-THC (Bloom *et al.*, 1978; Gough and Olley, 1978; Abood and Martin, 1992; Deadwyler *et al.*, 1995a). Interestingly, other genes maintained altered expression over the entire time of Δ^9-THC exposure, which implies their possible involvement in cannabinoid receptor mediated signaling processes. For instance SC1 (hevin), a matricellular protein, cell surface receptors and other molecules such as cytokines and proteases (Soderling *et al.*, 1997), were elevated (in both hippocampus and cerebellum) by a single injection of Δ^9-THC and maintained over the 21 day time-course of Δ^9-THC treatment. This is the first assessment of differential gene expression corresponding to acute and chronic Δ^9-THC treatment reported in the literature (Kittler *et al.*, 2000).

Rapid and significant increases in the level of arachidonic acid have been shown to result from acute exposure to Δ^9-THC (Shivachar *et al.*, 1996). In agreement with this, we found an increased level of prostaglandin D synthase (PDS) expression in both hippocampus and cerebellum after the acute Δ^9-THC injection (Figure 23.8). Prostaglandin D2, a compound recently shown to induce sedation and sleep in rats, is the end product of the reaction of PDS with its substrate arachidonic acid (Hayaishi, 1997; Urade *et al.*, 1996). Thus, accumulation of prostaglandin D2 in brain has been shown to be sensitive to high levels of cannabinoids.

By day 7, animals no longer exhibited cataleptic reactions to the large dose (10 mg/kg) of Δ^9-THC and exhibited normal locomotor behavior after the injection, however major deficits in cognitive function produced by daily injections of this dose are still present (Heyser *et al.*, 1993; Deadwyler *et al.*, 1995). Changes in gene expression detected at this time point were therefore less likely to be related to activation of immediate early genes. This was supported by the fact that the pattern of gene expression after 7 days was different from that following an acute injection. Expression levels of the majority of affected genes in hippocampus were downregulated (12 out of 21) at this time point, while changes in cerebellum showed a more symmetric trend (5 upregulated and 6 downregulated). Expression levels of five of these genes (cytochrome oxidase, SC1, myelin basic protein, myelin proteolipid protein, elongation factor and HSP70) were altered in the same way in both cerebellum and hippocampus. The expression of myelin proteolipid protein was significantly elevated in hippocampus and cerebellum at day 7 of Δ^9-THC treatment, while the expression of myelin basic protein was downregulated at days 7 and 21 (Table 23.2) in both brain regions. The level of HSP70 expression was also decreased in cerebellum at days 7 and 21 of Δ^9-THC treatment and in hippocampus at day 21. Recent studies have shown that HSP70 participates in the folding of myelin basic protein (Aquino *et al.*, 1996; 1998) and it has been reported to have a high affinity binding site for Δ^9-THC (Nye *et al.*, 1988). It is therefore possible that these changes reflect alterations in myelination processes over the timecourse of Δ^9-THC treatment.

Increased expression of calmodulin mRNA in hippocampus at day 7 of Δ^9-THC treatment (Table 23.2) coincided with the time course of changes reported for CB_1 receptor mRNA expression (Zhuang *et al.*, 1998) and cannabinoid induced changes in activation of Ca^{2+}/calmodulin-stimulated adenylyl cyclase (Hutchenson *et al.*, 1998). The similarity in time course of changes of these proteins suggests activation of a common cellular signaling mechanism: perhaps in order to compensate for receptor

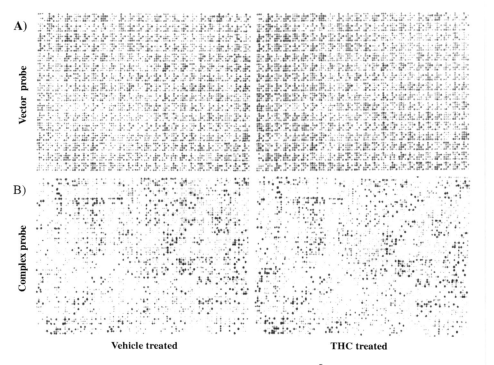

Figure 23.8 Chronic cannabinoid effects on gene expression. 33^P- labeled first strand cDNA probes prepared from Control (vehicle) and 21 days Δ^9-THC (THC) treated rat hippocampus were hybridized onto individual arrays. Differentially regulated genes are viewed as a pair at a time (cDNA spots are circled) in close-up images of the differentially expressed cDNA clones. Note left and right images are identical clones. Encircled: 1: calmodulin; 2: prostaglandin D synthase; 3: myelin basic protein.

desensitization in hippocampus (Sim *et al.*, 1996). In contrast, there were no changes in the expression of calmodulin in cerebellum after 7 days of Δ^9-THC treatment.

Behavioral tolerance to 21 day exposure to Δ^9-THC is accompanied by continued CB_1 receptor desensitization similar to 7 days of Δ^9-THC treatment (Zhuang *et al.*, 1998). As might be expected, the largest number of differentially expressed genes were revealed at this time point, 18 in hippocampus and 15 in cerebellum, and 9 genes were commonly expressed in both tissues (Table 23.2). The expression level of genes encoding brain specific lipid binding protein, myelin basic protein and mitochondrial 16S RNA, were downregulated in hippocampus and cerebellum (same as 7 days). Changes in expression of calmodulin and elongation factor were reversed (downregulated) compared to 7 days (Table 23.2; Figure 23.8).

Verification of large scale cDNA microarray screen by RNA dot blot analyses

RNA dot blots of hippocampal mRNA isolated from a different set of animals (Δ^9-THC treated and control, n = 3 per group) were obtained to confirm those genes identified in the large scale analysis. Blots were hybridized with 15 identi-fied candidate clones from the primary screen. Even if a change in gene expres-sion was detected at only one of three time points by the primary cDNA array screen, we examined (by dot blot) it's expression level at all three time points (24 hr, 7 and 21 days) of Δ^9-THC treatment in order to further verify the temporal profile extracted from the original screen. The results of RNA dot-blot analysis showed exceptions to only 4 of the 15 results of the large scale screen (angiotensin receptor: 7 days of Δ^9-THC treatment, sodium channel, glutathione peroxidase and SC1: 24 hour of Δ^9-THC treatment). The direction and magnitude of expression of the other 11 clones were consistant with the results of primary cDNA screen.

SUMMARY

The effect of cannabinoids on short-term memory in rats performing the DNMS task is quite similar to the effects of complete hippocampal removal: (1) behavioral deficits are produced at long delays; (2) proactive influences between trials are no longer "protected" by hippocampus; and (3) hippocampal neurons fail to strongly encode Sample phase information. The result of these deficits is a behavioral cascade in which the hippocampus is rarely "engaged" in strongly encoding information critical to the completion of the task. Canna-binoids thus mimic a "temporary" hippocampal lesion, that is fully reversed in 24–48 hours.

A critical exception to the similarity between cannabinoid effects and hippocam-pal lesion is the development of tolerance to chronic cannabinoid exposure. Both behavioral and electrophysiological measures show a return to Control levels within 21 to 35 days of chronic exposure. Biochemical and molecular biological measures have shown a similar timecourse for decoupling the receptor from its associated G-protein (Sim *et al.*, 1998; Brievogel *et al.*, 1999), and for alterations in mRNA expression for the CB$_1$ receptor (Zhuang *et al.*, 1998). Finally, chronic cannabinoid exposure produces functional gene changes of interest to understand the mechanisms of the above effects on memory.

Many of the above genes identified as differentially expressed in the large scale primary and secondary screen following acute and/or chronic Δ^9-THC treatment, have known roles with respect to their relation to cannabinoid receptor initiated processes. However, it should be pointed out that to date that the cannabinoid receptor system is one of the most ubiquitous G-protein cou-pled neural systems in mammalian brain (Oviedo *et al.*, 1993; Brievogel and Childers, 1998). There are many actions and processes that have not been fully investigated both at the cellular and systems level with respect to cannabinoid receptor involvement (Howlett, 1998). The above widespread changes in gene expression provide important information about the complexity of cellular alterations involved in the effects of acute and chronic cannabinoid exposure and the development of tolerance to the short-term memory deficits produced by cannabinoid drugs.

REFERENCES

Abood, M. E. and Martin, B. R. (1992) "Neurobiology of marijuana abuse," *Trends in Pharmacological Sciences* **13**: 201–206.

Abood, M. E., Sauss, C., Fan. F., Tilton, C. L. and Martin, B. R. (1993) "Development of behavioral tolerance to Δ^9-THC without alteration of cannabinoid receptor binding or mRNA levels in whole brain," *Pharmacology, Biochemistry and Behavior* **46**: 575–579.

Aceto, M. D., Scates, S. M., Lowe, J. A. and Martin, B. R. (1996) "Dependence on Δ^9-THC studies on precipitated and abrupt withdrawal," *The Journal of Pharmacology and Experimental Therapeutics* **27**: 1290–1295.

Adams, I. B. and Martin, B. R. (1996) "*Cannabis*: pharmacology and toxicology in animals and humans," *Addiction* **91**: 1585–1614.

Adryan, B., Carlguth, V. and Decker, H. J. (1999) "Digital image processing for rapid analysis of differentially expressed transcripts on high-density cDNA arrays," *Biotechniques* **26**: 1174–1179.

Aquino, D. A., Lopez, C. and Farooq, M. (1996) "Antisense oligonucleotide to the 70-kDa heat shock cognate protein inhibits synthesis of myelin basic protein," *Neurochemical Research* **21**: 417–422.

Aquino, D. A., Peng, D., Lopez, C. and Farooq, M. (1998) "The constitutive heat shock protein 70 is required for optimal expression of myelin basic protein during differentiation of oligodendrocytes," *Neurochemical Research* **23**: 413–420.

Bertucci, F., Bernard, K., Loriod, B. and Chang, Y. C. (1999) "Sensitivity issues in DNA array-based measurements and performance of nylon microarrays for small samples," *Human Molecular Genetics* **8**: 1715–1722.

Bloom, A. S., Johnson, K. M. and Dewey, W. L. (1978) "The effects of cannabinoids on body temperature and brain catecholamine synthesis," *Research Communications in Chemotherapy, Pathology and Pharmacology* **20**: 51–57.

Breivogel, C. S. and Childers, S. R. (1998) "The functional neuroanatomy of brain cannabinoid receptors," *Neurobiology of Disease* **5**: 417–431.

Breivogel, C. S., Childers, S. R., Deadwyler, S. A., Hampson, R. E., Vogt, L. J. and Sim-Selley, L. J. (1999) "Chronic Δ^9-THC treatment produces a time-dependent loss of cannabinoid receptors and cannabinoid receptor-activated G-proteins in rat brain," *Journal of Neurochemistry* **73**: 2447–2459.

Brewer, J. B., Zhao, Z., Desmond, J. E., Glover, G. H. and Gabrieli, J. D. (1998) "Making memories, brain activity that predicts how well visual experience will be remembered [see comments]," *Science* **281**: 1185–1187.

Calignano, A., La Rana, G., Giuffrida, A. and Piomelli, D. (1998) "Control of pain initiation by endogenous cannabinoids," *Nature* **394**: 277–281.

Campbell, K. A., Foster, T. C., Hampson, R. E. and Deadwyler, S. A. (1986) "Δ^9-THC differentially affects sensory-evoked potentials in the rat dentate gyrus," *The Journal of Pharmacology and Experimental Therapeutics* **239**: 936–940.

Caulfield, M. P. and Brown, D. A. (1992) "Cannabinoid receptor agonists inhibit Ca current in NG108–15 neuroblastoma cells via a pertussis toxin-sensitive mechanism," *British Journal of Pharmacology* **106**: 231–232.

Chait, L. D. and Pierri, J. (1992) "Effects of smoked marijuana on human performance: a critical review," in *Marijuana/Cannabinoids, Neurobiology and Neurophysiology*, edited by L. Murphy, A. Bartke, pp. 387–423. Boca Raton, FL: CRC Press.

Chen, J., Kelz, M. B., Hope, B. T., Nakabeppu, Y. and Nestler, E. J. (1997) "Chronic Fos related antigens, stable variants of FosB induced in brain by chronic treatments," *The Journal of Neuroscience* **17**: 4933–4941.

Chen, J. J., Wu, R., Yang, P. C., Hyang, J. P., Sher, Y. P., Han, M. H., Kao, W. C., Lee, P. J., Chiu, T. F., Chang, F., Chu, Y. W., Wu, C. W. and Peck, K. (1998) "Profiling expression patterns and isolating differentially expressed genes by cDNA microarray system with colorimetry detection," *Genomics* **51**: 313–324.

Chen Y., Dougherty, E. R. and Bittner, M. L (1997) "Ratio-based decisions and the quantitative analysis of cDNA microarray images," *Journal of Biomedical Optics* **2**: 364–374.

Childers, S. R. and Breivogel, C. S. (1998) "Cannabis and endogenous cannabinoid systems," *Drug and Alcohol Dependence* **51**: 173–187.

Childers, S. R. and Deadwyler, S. A. (1996) "Role of cyclic AMP in the actions of cannabinoid receptors," *Biochemical Pharmacology* **52**: 819–827.

Cohen, N. J. and Eichenbaum, H. (1993) "Memory, amnesia, and the hippocampal system," Cambridge, M. A., M. I. T. Press.

Compton, D. R., Dewey, W. L. and Martin, B. R. (1990) "Cannabis dependence and tolerance production," *Adv. Alcohol. Subst. Abuse* **9**: 129–147.

Couceyro, P., Mohammed, S., McCoy, M., Goldberg, S. R. and Kuhar, M. J. (1997) "Cocaine self-administration alters brain NADH dehydrogenese mRNA levels," *Neuroreport* **8**: 2437–2441.

Deacon, R. M. and Rawlins, J. N. (1995) "Serial position effects and duration of memory for nonspatial stimuli in rats," *Journal of Experimental Psychology and Animal Behavior Processes* **21**: 285–292.

Deadwyler, S. A., Bunn, T. and Hampson, R. E. (1996) "Hippocampal ensemble activity during spatial delayed-nonmatch-to-sample performance in rats," *The Journal of Neuroscience* **16**: 354–372.

Deadwyler, S. A. and Hampson, R. E. (1997) "The significance of neural ensemble codes during behavior and cognition," in *Annual Review of Neuroscience*, edited by W. M. Cowan, E. M. Shooter, C. F. Stevens, R. F. Thompson, pp. 217–244. Palo Alto, C. A. Annual Reviews, Inc.

Deadwyler, S. A., Hampson, R. E., Bennett, B. A., Edwards, T. A., Mu, J., Pacheco, M. A., Ward, S. J. and Childers, S. R. (1993) "Cannabinoids modulate potassium current in cultured hippocampal neurons," *Receptors and Channels* **1**: 121–134.

Deadwyler, S. A., Hampson, R. E., Mu, J., Whyte, A. and Childers, S. R. (1995a) "Cannabinoids modulate voltage-sensitive potassium A-current in hippocampal neurons via a cAMP-dependent process," *The Journal of Pharmacology and Experimental Therapeutics* **273**: 734–743.

Deadwyler, S. A., Heyser, C. J. and Hampson, R. E. (1995b) "Complete adaptation to the memory disruptive effects of delta-9-THC following 35 days of exposure," *Neur. Res. Comm.* **17**: 9–18.

Debouck, C. and Goodfellow, P. N. (1999) "DNA microarrays in drug discovery and development," *Nature Genetics* **21**: 48–50.

DeRisi, J., Penland, L., Brown, P. O., Bittner, M. L., Meltzer, P. L., Ray, M., Chen, Y., Su, Y. A. and Trent, J. M. (1996) "Use of a cDNA microarray to analyze gene expression patterns in human cancer," *Nature Genetics* **14**: 457–460.

Devane, W. A., Dysarz, F. A., Johnson, M. R., Melvin, L. S. and Howlett, A. C. (1988) "Determination and characterization of a cannabinoid receptor in rat brain," *Molecular Pharmacology* **34**: 605–613.

Devane, W. A., Hanus, L., Breuer, A., Pertwee, R. G., Stevenson, L. A., Griffin, G., Gibson, D., Mandelbaum, A., Etinger, A. and Mechoulam, R. (1992) "Isolation and structure of a brain constituent that binds to the cannabinoid receptor," *Science* **258**: 1946–1949.

de Waard, V., van der Berg, B. M., Veken, J. and Schultz-Heinkenbrok, R. (1999) "Serial analysis of gene expression to assess the endothelial cell response to an atherogenic stimulus," *Gene* **226**: 1–8.

Dewey, W. L. (1986) "Cannabinoid pharmacology," *Pharmacological Reviews* **38**: 151–178.

Douglass, J., McKinzie, A. M. and Couceyro, P. (1995) "PCR differential display identifies a rat brain mRNA that is transcriptionally regulated by cocaine and amphetamine," *The Journal of Neuroscience* **75**: 2471–2481.

Duggan, D. J., Bittner, M., Chen, Y., Meltzer, P. and Trent, J. M. (1999) "Expression profiling using cDNA microarrays," *Nature Genetics* **21**: 10–14.

Ennulat, D. J. and Cohen, B. M. (1997) "Multiplex differential display identifies a novel zinc-finger protein repressed during withdrawal form cocaine," *Molecular Brain research* **49**: 299–302.

Fan, F., Tao, Q., Abood, M. and Martin, B. R. (1996) "Cannabinoid receptor down regulation without alteration of the inhibitory effect of CP 55,940 on adenylyl cyclase in the cerebellum of CP 55,940-tolerant mice," *Brain Research* **706**: 13–20.

Felder, C. C., Veluz, J. S., Williams, H. L., Briley, E. M. and Matsuda, L. A. (1992) "Cannabinoid agonists stimulate both receptor- and non-receptor-mediated signal transduction pathways in cells transfected with and expressing cannabinoid receptor clones," *Molecular Pharmacology* **42**: 838–845.

Freedman, N. J. and Lefkowitz, R. J. (1996) "Desensitization of G protein-coupled receptors," *Recent Progress in Hormone Research* **51**: 319–351.

Fowler, J. S. and Volkow, N. D. (1998) "PET imaging studies in drug abuse. [Review] [77 refs]," *Journal of Toxicology – Clinical Toxicology* **36**: 163–174.

Glass, M. and Felder, C. C. (1997) "Concurrent stimulation of cannabinoid CB_1 and dopamine D2 receptors augments cAMP accumulation in striatal neurons, evidence for a Gs linkage to the CB_1 receptor," *The Journal of Neuroscience* **17**: 5327–5333.

Gough, A. L. and Olley, J. E. (1978) "Catalepsy induced by intrastriatal injection of Δ^9-THC and 11-OH-Δ^9-THC in the rat," *Neuropharmacology* **17**: 137–144.

Hall, W. and Solowij, N. (1998) "Adverse effects of *Cannabis*," *Lancet* **352**: 1611–1616.

Hall, W., Solowij, N. and Lemon, J. (1994) "The Health and Psychological Consequences of Cannabis Use," *Canberra, Australian Government Public Service*.

Hampson, R. E., Byrd, D. R., Konstantopoulos, J. K., Bunn, T. and Deadwyler S. A. (1996) "Delta-9-tetrahydrocannabinol influences sequential memory in rats performing a delayed-nonmatch-to-sample task," *Society for Neuroscience and Proceedings of National Academy of Sciences* **22**: 1131.

Hampson, R. E. and Deadwyler, S. A. (2000) "Cannabinoids reveal the necessity of hippocampal neural encoding for short-term memory in rats," *The Journal of Neuroscience*.

Hampson, R. E. and Deadwyler, S. A. (1996a) "Ensemble codes involving hippocampal neurons are at risk during delayed performance tests," *Proceedings of the National Academy of Sciences of the USA* **93**: 13487–13493.

Hampson, R. E. and Deadwyler, S. A. (1996b) "LTP and LTD and the encoding of memory in small ensembles of hippocampal neurons," in *Long-term Potentiation*, Vol. 3, edited by M. Baudry, J. Davis, pp. 199–214. Cambridge, MA: MIT Press.

Hampson, R. E. and Deadwyler, S. A. (1998a) "Methods, results and issues related to recording neural ensembles," in *Neuronal Ensembles, Strategies for Recording and Decoding*, edited by H. Eichenbaum, J. Davis, pp. 207–234. New York, Wiley.

Hampson, R. E. and Deadwyler, S. A. (1998b) "Role of cannabinoid receptors in memory storage," *Neurobiology of Disease, Experimental Neurology, Part B* **5**: 474–482.

Hampson, R. E. and Deadwyler, S. A. (1999a) "Cannabinoids, hippocampal function and memory," *Life Sciences* **65**: 715–723.

Hampson, R. E. and Deadwyler, S. A. (1999b) "Pitfalls and problems in the analysis of neuronal ensemble recordings during performance of a behavioral task," in *Methods for Simultaneous Neuronal Ensemble Recordings*, edited by M. Nicolelis, pp. 229–248. New York: Academic Press.

Hampson, R. E., Jarrard, L. E. and Deadwyler, S. A. (1999) "Effects of ibotenate hippocampal and extrahippocampal destruction on delayed-match and -nonmatch-to-sample behavior in rats," *The Journal of Neuroscience* **19**: 1492–1507.

Hayaishi, O. (1997) "Prostaglandin D synthase, beta-trace and sleep," *Advances in Experimental Medicine and Biology* **433**: 347–350.

Heller, R., Schena, M., Chai, A., Shalon, D., Bedilion, T., Gilmore, J., Wooley, D. E. and Davis, R. W. (1997) "Discovery and analysis of inflammatory disease-related genes using cDNA microarrays," *PNAS* **94**: 2150–2155.

Henry, D. J. and Chavkin, C. (1995) "Activation of inwardly rectifying potassium channels (GIRK1) by co-expressed rat brain cannabinoid receptors in *Xenopus* oocytes," *Neuroscience Letters* **186**: 91–94.

Herkenham, M. (1991) "Characterization and localization of cannabinoid receptors in brain, an in vitro technique using slide-mounted tissue sections," *NIDA Research Monograph* **112**: 129–145.

Herkenham, M., Lynn, A. B., Johnson, M. R., Melvin, L. S., de Costa, B. R. and Rice, K. C. (1991) "Characterization and localization of cannabinoid receptors in rat brain, A quantitative in vitro autoradiographic study," *The Journal of Neuroscience* **11**: 563–583.

Herkenham, M., Lynn, A. B., Little, M. D., Johnson, M. R., Melvin, L. S., de Costa, B. R. and Rice, K. C. (1990) "Cannabinoid receptor localization in brain," *Proceedings of the National Academy of Sciences of the USA* **87**: 1932–1936.

Herring, A. C., Koh, W. S. and Kaminski, N. E. (1998) "Inhibition of the cyclic AMP signaling cascade and nuclear factor binding to CRE and kappaB elements by cannabinol, a minimally CNS-active cannabinoid," *Biochemical Pharmacology* **55**: 1013–1023.

Heyser, C. J., Hampson, R. E. and Deadwyler, S. A. (1993) "The effects of delta-9-THC on delayed match to sample performance in rats, alterations in short-term memory produced by changes in task specific firing of hippocampal neurons," *The Journal of Pharmacology and Experimental Therapeutics* **264**: 294–307.

Hoffman, A. F. and Lupica, C. R. (2000) "Mechanisms of cannabinoid inhibition of GABA-A synaptic transmission in the hippocampus," *The Journal of Neuroscience* **20**: 2470–2479.

Hoffman, P. L. and Tabakoff, B. (1989) "Mechanisms of alcohol tolerance," *Alcohol* **24**: 251–254.

Hope, B. T., Nye, H. K., Kelz, M. B., Self, D. W., Ladarola, M. K., Nakabeppu, Y., Duman, R. S. and Nestler, E. J. (1994) "Induction of long-lasting AP-1 complex composed of altered fos-like proteins in brain by chronic cocaine and other chronic treatments," *Neuron* **13**: 1235–1244

Howlett, A. C. (1998) "The CB$_1$ cannabinoid receptor in the brain. [Review] [60 refs]," *Neurobiology of Disease* **5**: 405–416.

Howlett, A. C., Champion, T. M., Wilken, G. H. and Mechoulam, R. (1991) "Stereochemical effects of 11-OH-delta-8-tetrahydrocannabinol-dimethylheptyl to inhibit adenylate cyclase and bind to the cannabinoid receptor," *Neuropharmacology* **29**: 161–165.

Howlett, A. C., Qualy, J. M. and Khachatrian, L. L. (1986) "Involvement of Gi in the inhibition of adenylate cyclase by cannabimimetic drugs," *Molecular Pharmacology* **29**: 307–313.

Hsieh, C., Brown, S., Derleth, C. and Mackie, K. (1999) "Internalization and recycling of the CB$_1$ cannabinoid receptor," *Journal of Neurochemistry* **73**: 493–501.

Hutchenson, D. M., Tzavara, E. T., Smadja, C., Valjent, E., Roqies, B. P., Hanoune, J. and Maldonaldo, R. (1998) "Behavioral and biochemical evidence for signs of abstinence in mice chronically treated with delta-9-cannabinol," *British Journal of Pharmacology* **125**: 1567–1577.

Katona, I., Sperlagh, B., Sik, A., Kafalvi, A., Vizi, E. S., Mackie, K. and Freund, T. F. (1999) "Presynaptically located CB$_1$ cannabinoid receptors regulate GABA release from axon terminals of specific hippocampal interneurons," *Journal of Neuroscience* **19**: 4544–4558.

Khan, J. (1998) "Gene expression profiling of alveolar rhadomyosarcoma with cDNA microarrays," *Cancer Research* **53**: 5009–5013.

Kittler, J. T., Grigorenko, E. V., Clayton, C., Zhuang, S. Y., Bundey, S. C., Trower, M. M., Wallace, D., Hampson, R. and Deadwyler, S. (2000) "Large-scale analysis of gene expression changes during acute and chronic exposure to delta 9-THC in rats," *Physiological Genomics* **3**: 175–185.

Lichtman, A. H., Dimen, K. R. and Martin, B. R. (1995) "Systemic or intrahippocampal cannabinoid administration impairs spatial memory in rats," *Psychopharmacology (Berliner)* **119**: 282–290.

Lichtman, A. H. and Martin, B. R. (1996) "Delta 9-tetrahydrocannabinol impairs spatial memory through a cannabinoid receptor mechanism," *Psychopharmacology (Berliner)* **126**: 125–131.

Mackie, K. and Hille, B. (1992) "Cannabinoids inhibit N-type calcium channels in neuroblastoma-glioma cells," *Proceedings of the National Academy of Sciences of the USA* **89**: 3825–3829.

Martin, W. J., Lai, N. K., Patrick, S. L., Tsou, K. and Walker, J. M. (1993) "Antinociceptive actions of cannabinoids following intraventricular administration in rats," *Brain Research* **629**: 300–304.

Matsuda, L. A., Lolait, J., Brownstein, M. J., Young, A. C. and Bonner, T. I. (1990) "Structure of a cannabinoid receptor and functional expression of the cloned cDNA," *Nature* **346**: 561–564.

McGregor, I. S., Arnold, J. C., Weber, M. F., Topple, A. N. and Hunt, G. E. (1998) "A comparison of delta 9-THC and anadamide induced c-fos expression in the rat forebrain," *Brain Research* **802**: 19–26.

Miller, L. L. and Branconnier, R. J. (1983) "*Cannabis*, effects on memory and the cholinergic limbic system," *Psychological Bulletin* **93**: 441–456.

Muller, R. (1996) "A quarter of a century of place cells. [Review] [50 refs]," *Neuron* **17**: 813–822.

Munro, S., Thomas, K. L. and Abu-Shaar, M. (1993) "Molecular characterization of a peripheral receptor for cannabinoids [see comments]," *Nature* **365**: 61–65.

Murray, E. A. and Mishkin, M. (1998) "Object recognition and location memory in monkeys with excitotoxic lesions of the amygdala and hippocampus," *The Journal of Neuroscience* **18**: 6568–6582.

Nadel, L. (1991) "The hippocampus and space revisited," *Hippocampus* **1**: 221–229.

Nestler, E. J. (1997) "Molecular mechanisms of opiate and cocaine addiction," *Current Opinion in Neurobiology* **7**: 713–719.

Nye, J. S., Voglmaier, S., Martenson, R. E. and Snyder, S. H. (1988) "Myelin basic protein is an endogenous inhibitor of the high-affinity cannabinoid binding site in brain," *Journal of Neurochemistry* **50**: 1170–1180.

O'Keefe, J. A. and Nadel, L. (1978) *The Hippocampus as a Cognitive Map*. Oxford: Clarendon Press.

Oviedo, A., Glowa, J. and Herkenham, M. (1993) "Chronic cannabinoid administration alters cannabinoid receptor binding in brain, a quantitative autoradiographic study," *Brain Research* **616**: 293–302.

Pak, Y. S., Kouvelas, A., Schneider M. A., Rasmussen, J., O'Dowd, B. F. and George, S. R. (1996) "Agonist-induced functional desensitization of the mu opioid receptor is mediated by loss of membrane receptors rather than uncoupling from G-protein," *Molecular Pharmacology* **50**: 1214–1222.

Patten, C., Clayton, C. L., Blakemore, S. J., Trower, M. K., Wallace, D. M. and Hagan, R. M. (1998) "Identification of two novel diurnal genes by screening of a rat brain cDNA library," *Neuroreport*, pp. 3935–3941.

Priller, J., Briley, E. M., Mansouri, J., Devane, W. A., Mackie, K. and Felder, C. C. (1995) "Mead ethanolamide, a novel eicosanoid, is an agonist for the central (CB_1) and peripheral (CB_2) cannabinoid receptors," *Molecular Pharmacology* **48**: 288–292.

Rinaldi-Carmona, M., Barth, F., Heaulme, M., Shire, D., Calandra, B., Congy, C., Martinez, S., Marauni, J., Neliat, G., Caput, D., Ferrara, P., Soubrie, P., Breliere, J. C. and LeFur, G. (1994) "SR 141716A, a potent and selective antagonist of the brain cannabinoid receptor," *FEBS Letters* **350**: 240–244.

Rinaldi-Carmona, M., Barth, F., Millan, J., Derocq, J. M., Casellas, P., Congy, C., Oustric, D., Sarran, M., Bouaboula, M., Calandra, B., Pertier, M., Shires, D., Breliere, J. C. and LeFur, G. L. (1998) "SR 144528, the first potent and selective antagonist of the CB$_2$ cannabinoid receptor," *The Journal of Pharmacology and Experimental Therapeutics* **284**: 644–650.

Romero, J., Garcia, L., Fernendez-Ruiz, J. J., Cebeira, M. and Ramos, J. A. (1995) "Changes in rat cannabinoid binding sites after acute or chronic exposure to their endogenous agonist, anandamide or to Δ^9-tetrahydrocannabinol," *Pharmacology, Biochemistry and Behavior* **51**: 731–737.

Romero, J., Garcia-Palomero, E., Castro, J. G., Garcia-Gil, L., Ramos, J. A. and Fernendez-Ruiz, J. J. (1997) "Effects of chronic exposure to delta-9-tetracannbinol on cannabinoid receptor binding and mRNA levels in several rat brain regions," *Molecular Brain Research* **46**: 100–108.

Schena, M., Shalon, D., Davis, R. W. and Brown, P. O. (1995) "Quantitative monitoring of gene expression patterns with a complementary DNA microarray," *Science* **270**: 467–470.

Schena, M., Shalon, D., Heller, R., Chai, A., Brown, P. O. and Davis, R. W. (1996) "Parallel human genome analysis, Microarray-based expression monitoring of 1000 genes," *PNAS* **93**: 10614–10619.

Self, D. W. (1998) "Neural substrates of drug craving and relapse in drug addition," *Annals of Medicine* **30**: 379–389.

Shen, M., Piser, T. M., Seybold, V. S. and Thayer, S. A. (1996) "Cannabinoid receptor agonists inhibit glutamatergic synaptic transmission in rat hippocampal cultures," *The Journal of Neuroscience* **16**: 4322–4334.

Shen, M. and Thayer, S. A. (1998) "Cannabinoid receptor agonists protect cultured rat hippocampal neurons from excitotoxicity," *Molecular Pharmacology* **54**: 459–462.

Shivachar, A. C., Martin, B. R. and Ellis, E. F. (1996) "Anandamide- and delta-9-tetrahydro-cannabinol-evoked arachidonic acid mobilization and blockade by SR141716A [N-(piperi-din-1-yl)-5-(4-chlorophenyl)-1-(2,4-dichlorophenyl)-4-methyl-1H-pyrazol-3-carboximide hydrochloride]," *Biochemical Pharmacology* **51**: 669–676.

Sim, L. J., Hampson, R. E., Deadwyler, S. A. and Childers, S. R. (1996) "Effects of chronic treatment with delta-9-tetrahydrocannabinol on cannabinoid stimulated [35S]GTP-gamma-S autoradiography in rat brain," *The Journal of Neuroscience* **16**: 8057–8066.

Soderling, J. A., Reed, M. J., Corsa, A. and Sage, E. H. (1997) "Cloning and expression of murine SC1, a gene product homologous to SPARC," *Journal of Neurochemistry and Cytochemistry* **45**: 823–835.

Solowij, N. (1998) "*Cannabis* and Cognitive functioning," Cambridge University Press, UK.

Southern, E., Mir, K. and Schchepinov, M. (1999) "Molecular interactions on microarrays," *Nature Genetics* (suppl) **21**: 5–10.

Squire, L. R. and Zola, S. M. (1997) "Amnesia, memory and brain systems. Philosophical Transactions of the Royal Society of London – Series B," *Biological Sciences* **352**: 1663–1673.

Stella, N., Schweitzer, P. and Piomelli, D. (1997) "A second endogenous cannabinoid that modulates long-term potentiation," *Nature* **388**: 773–778.

Stern, C. E., Corkin, S., Gonzalez, R. G., Guimaraes, A. R., Baker, J. R., Jennings, P. J., Carr, C. A., Sugiura, R. M., Vedantham, V. and Rosen, B. R. (1996) "The hippocampal formation participates in novel picture encoding, evidence from functional magnetic resonance imaging," *Proceedings of the National Academy of Sciences of the USA* **93**: 8660–8665.

Sternini, C., Spann, M., Anton, B., Keith, D. E., Bunnett, N. W., von Zastrow, M., Evans, C. and Brecha, N. C. (1996) "Agonist-selective endocytosis of mu opioid receptor by neurons in vivo," *PNAS* **93**: 9241–9246.

Terranova, J. P., Michaud, J. C., Le Fur, G. and Soubrie, P. (1995) "Inhibition of long-term potentiation in rat hippocampal slices by anandamide and WIN55212-2, reversal by SR

141716 A, a selective antagonist of CB$_1$ cannabinoid receptors," *Naunyn–Schmiedeberg's Archives of Pharmacology* **352**: 576–579.

Terwilliger, R. Z., Better-Johnson, D., Sevarino, K. A., Crain, S. M. and Nestler, E. J. (1991) "A general role for adaptations on G-proteins and the cyclic AMP system in mediating the chronic actions of morphine and cocaine on neuronal function," *Brain Research* **548**: 100–110.

Trower, M. K. (1997) "Gene expression profiling using high-density cDNA arrays," *Microbial and Comparative Genomics* **2**: 236–240.

Tsou, K., Brown, S., Sanudo-Pena, M. C., Mackie, K. and Walker, J. M. (1998) "Immunohistochemical distribution of cannabinoid CB$_1$ receptors in the rat central nervous system," *Neuroscience* **83**: 393–411.

Tulving, E. and Markowitsch, H. J. (1997) "Memory beyond the hippocampus," *Current Opinion in Neurobiology* **7**: 209–216.

Twitchell, W., Brown, S. and Mackie, K. (1997) "Cannabinoids inhibit N- and P/Q-type calcium channels in cultured rat hippocampal neurons," *Journal of Neurophysiology* **78**: 43–50.

Urade, Y., Hayashi, O., Marsumura, H. and Watanabe, K. (1996) "Molecular mechanism of sleep regulation by prostaglandin D2. J Lipid Mediat," *Cell Signal* **14**: 71–82.

Van Vilet, B. J., Van Riijswiik, A. L., Wardeh, G., Mulder, A. H. and Schoffermeer, A. N. (1993) "Adaptive changes in the number of Gs and Gi proteins underlie adenylyl cyclase sensitization in morphine-treated rat striatal neurons," *European Journal of Pharmacology* **245**: 23–29.

Vargha-Khadem, F., Gadian, D. G., Watkins, K. E., Connelly, A., Van Paesschen, W. and Mishkin, M. (1997) "Differential effects of early hippocampal pathology on episodic and semantic memory," *Science* **277**: 376–380.

Wagner, A. D., Schacter, D. L., Rotte, M., Koutstaal, W., Maril, A., Dale, A. M., Rosen, B. R. and Buckner, R. L. (1998) "Building memories, remembering and forgetting of verbal experiences as predicted by brain activity," *Science* **281**: 1188–1191.

Walker, J. R. and Sevarino, K. A. (1995) "Regulation of cytochrome c oxidase subunit mRNA and enzyme activity in rat brain reward regions during withdrawal from chronic cocaine," *Journal of Neurochemistry* **64**: 497–502.

Wang, X. B., Funada, M., Imai, Y., Revay, R. S., Ujike, H., Vandenbergh, D. J. and Uhl, G. R. (1997) "rGbeta 1, a psychostimulant-regulated gene essential for establishing cocaine sensitization," *The Journal of Neuroscience* **15**: 5993–6000.

Ward, S. J., Baizman, E., Bell, M., Childers, S., D'Ambra, T., Eissenstat, M., Estep, K., Haycock, D., Howlett, A., Luttinger, D. *et al.* (1991) "Aminoalkylindoles (AAIs), a new route to the cannabinoid receptor?" *NIDA Research Monograph* **105**: 425–426.

Yuh, C. H., Bolouri, H. and Davidson, E. H. (1998) "Genomic cis-regulatory logic, experimental and computational analysis of sea urchin gene," *Science* **279**: 1896–1902.

Zhuang, S.-Y., Kittler, J., Grigorenko, E. V., Kirby, M. T., Sim, L. J., Hampson, R. E., Childers, S. R. and Deadwyler, S. A. (1998) "Effects of long-term exposure to delta-9-THC on the expression of cannabinoid (CB$_1$) receptor mRNA in different rat brain regions," *Molecular Brain Research* **62**: 141–149.

Chapter 24

Adverse effects of marijuana

John R. Hubbard

ABSTRACT

Marijuana is a complex substance with many potential physical and neuropsychiatric adverse effects. Acute physical effects include tachycardia, decreased task performance, and reduced cerebral blood flow. These effects may be particularly dangerous if the user is driving a car, operating machinery, or has certain pre-existing medical problems (such as cardiovascular disease). Acute neuropsychiatric effects may include paranoia, changes in libido, altered time and sensory perceptions, and others. Chronic marijuana use may lead to adverse effects on the respiratory system (due to tar, carbon monoxide, carcinogens and other chemicals), reproduction system, motivation, memory and other systems. Chemical dependence to cannabinoids may insidiously develop and marijuana use can be a "gateway" to use of other substances of potential abuse. Although human studies on marijuana have many limitations, adverse effects of marijuana is of considerable clinical and social importance.

Key Words: marijuana, cannabinoids, chemical dependence, amotivation syndrome, task performance, neuropsychiatric effects, cardiovascular effects, reproduction system, respiratory effects

INTRODUCTION

Marijuana is typically used for recreational purposes to achieve a dream-like state or "high". Peer pressure may motivate use not only in adolescents and teenagers, but in sub-populations of adults as well. In some cases, people use marijuana in attempts at self-medication for anxiety and other emotional problems, nausea (such as from medications to fight cancer), chronic pain, and for other purposes (Hubbard *et al.*, 1999). The use of marijuana peaked in the 1960s, but is still very high (Hubbard *et al.*, 1999), because marijuana is the most commonly used illicit drug in the United States and is used in some states for medical purposes, understanding its potential adverse effects is of significant clinical and social interest.

Adverse effects of marijuana may occur with or without the knowledge of the user. For example, long-term use of marijuana may cause changes too slowly to be noticed or to be certain of the cause. Even with acute use, an adverse effect may not be obvious to the user because they are "high" or because the effect occurs significantly after the intoxicating effect is over. Also an adverse effect may not be observed because it is internal (such as brain blood flow or hormonal changes) or because a measurement instrument is needed to observe the change. In one survey 10–15% of chronic cannabis users noticed adverse effects to marijuana (Halibo *et al.*, 1971). Others have reported that 40–60% of marijuana users have undesired side effects (Smart and Adlaf, 1982). In this chapter we will review scientific and clinical information on the adverse effects of marijuana and related cannabinoid containing substances.

LIMITATIONS OF HUMAN STUDIES ON MARIJUANA

Marijuana has been an area of significant social and scientific interest for many years, yet many questions remain unanswered or not fully answered. In part, this is due to the inherent difficulties of human research on any illegal substance and in part it is due to special characteristics of marijuana. For example, marijuana has a very long half-life making the cause and effect less obvious, and marijuana is not a specific chemical, but rather a substance derived from the cannabis sativa plant that contains numerous active and inert chemicals. In general, studies on acute exposure to marijuana are better controlled than studies on long-term use. However most acute exposure studies using human subjects were done decades ago. Many of these older studies may underestimate adverse effects of current marijuana preparations since marijuana is now generally much more potent than that used in the past. Investigations on the adverse effects of chronic marijuana are very important, but most of these studies are limited by necessity to naturalistic and retrospective designs.

Results and interpretation of investigations of marijuana therefore depend on many factors that vary between studies. Some of these differences include:

(a) the content of the active chemicals in the marijuana preparation used [particularly the major psychoactive chemical Δ^9-tetrahydrocannabinol (THC)]
(b) acute vs. chronic exposure
(c) frequency of use
(d) concomitant use of other alcohol or drugs of abuse
(e) psychological and biological differences between subjects
(f) controlled vs. uncontrolled investigations
(g) prospective vs. retrospective investigation
(h) time(s) of adverse effect measurement (i.e. too late, too early, or optimal time to observe the change)

ACUTE ADVERSE EFFECTS OF MARIJUANA

Many acute adverse effects of marijuana have been reported after recent use as shown in Table 24.1. Some of these effects include headache, dry mouth, decreased coordination, tachycardia, changes in pulmonary functioning, altered body tem-

Table 24.1 Acute adverse physical effects of marijuana

Dry mouth	Tachycardia
Decreased coordination/task performance	Changes in pulmonary functioning
Altered body temperature	Reduced muscle strength
Appetite	Decreased cerebral blood flow

References 1, 4–9, 13–20

perature, reduced muscle strength, increased appetite, decreased cerebral blood flow, task performance and others (Hubbard *et al.*, 1999).

Acute neuropsychiatric adverse effects to marijuana have been reported. Some of these side effects include anxiety, paranoia, hallucinations, time perception distortion, sensory (color/sound), altered libido, derealization, poor memory, decreased motivation and others (Table 24.2).

The potency of the preparation, the rate and duration of exposure, use of other drugs or alcohol, the setting of use, the psychological state of the individual and many other factors all influence the occurrence and severity of the acute adverse effect(s).

Task performance

Marijuana exposure can reduce physical motor performance. This has been demonstrated using many different measurements such as hand-eye coordination, tracking ability, body sway, reaction time, muscle strength tests and others (Ashton, 1999). The effect of marijuana on coordination lasts considerably longer than on the feeling of intoxication (Hubbard *et al.*, 1999; Losken *et al.*, 1996). The prolonged disruptive effect of marijuana on motor skills may be particularly important with regard to automobile driving, or in operating other vehicles or machinery (Klonoff, 1974; Smiley, 1986; Peck *et al.*, 1986; Gold, 1994). For example, driving on an obstacle course was shown to be more difficult during cannabis intoxication (Peck *et al.*, 1986). After an initial increase in motor activity, subjects often have ataxia, poor coordination, and psychomotor retardation (Ashton, 1999).

Performance of simulated airplane operating skills has also been shown to be reduced in pilots both during, and many hours after, marijuana intoxication (Gold, 1994; Janowsky *et al.*, 1976; Hollister, 1998). For example, in a study by Leirer and Yesavag, (1991) marijuana exposure worsened simulated landings one hour after smoking a 19 mg THC marijuana cigarette. In addition, some deficiencies

Table 24.2 Acute psychiatric adverse effects of marijuana

Paranoia	Depersonalization
Anxiety	Altered time perception
Dysphoria	Worsened memory
Hallucinations	Altered motivation
Changes in libido	Possible increased suicidal ideation
Derealization	Sensory perception

References 1, 3, 4, 9, 13, 15, 16, 19, 21–25, 27, 28

were noted 24 h later even though the pilots were unaware of their decreased performance (Leirer and Yesavag, 1991).

Task performance may be diminished with marijuana use not only due to the physical effects, but by the additional effect of marijuana on sensory perception (Ashton, 1999). For example, time, space, color and sound experiences may be altered (Ashton, 1999). Even after low dose exposure to marijuana subjects often overestimate the amount of time that has gone by.

Human position emission tomography (PET) studies show that cannabinoids change frontal lobe, parietal lobe, and cerebellum lobe metabolism of glucose for several hours (Nahas and Latour, 1992). These changes in metabolism may be related to the effects of marijuana on task performance (Nahas and Latour, 1992).

Cardiovascular

Marijuana has many affects on the cardiovascular system, such as increasing heart rate and cardiac work load (Hubbard et al., 1999). Tachycardia appears to be primarily due to decreased vagel tone (Clark et al., 1974). The cardiac output has been reported to be increased up to 30% (Ashton, 1999). Peripheral vasodilatation and postural hypotension have also been reported (Ashton, 1999).

The acute effect of marijuana on stimulating heart rate and cardiac output may be dangerous to some users due to increased myocardial demand, especially those with preexisting cardiovascular disease (Hubbard et al., 1999; Ashton, 1999; Schuckit, 1989; Hollister, 1988; Lu et al., 1993; Benowitz and Jones, 1975; Gottschalk et al., 1977). Marijuana has been shown to decrease cardiac oxygen and yet to increase myocardial demand in patients with angina (Gottschalk et al., 1977; Aronow and Cassidy, 1974). A 18 mg THC cigarette reduced exercise time to develop angina by 48% in cardiac patients (Gottschalk et al., 1977; Aronow and Cassidy, 1974). Heart attacks and transient ischemic attacks have been reported even in young healthy marijuana users (Ashton, 1999).

Neuropsychiatric

The human brain contains numerous cannabinoid receptors which may in part account for the many neuropsychiatric effects of marijuana (Hubbard et al., 1999). Some common neuropsychiatric side effects of marijuana include paranoia, anxiety, dysphoria, aggressiveness, hallucinations, changes in libido, derealization, depersonalization, altered time perception, worsened short-term memory, altered motivation, possible increased suicidal ideation (Hubbard et al., 1999; Smart and Adlaf, 1982; Ashton, 1999; Nahas and Latour, 1992; Schuckit, 1989; Hollister, 1988; Hubbard et al., 1993; Gottschalk et al., 1977; Nahas, 1977; Weil, 1970; Tunving, 1985). Most of the acute adverse neuropsychiatric effects appear to be anxiety reactions (Tunving, 1985). Sedation often occurs after the initial feeling of intoxication (Ashton, 1999).

The detrimental effect of marijuana on short-term memory is well known (Nahas and Latour, 1992). Interestingly, however, it can persist long after cessation of use. For example, in 1989 Schwartz et al. (Schwartz et al., 1989) reported short-term memory deficits in middle-class youths (median age 16 and matched for age and intelligence) who were dependent on cannabis both initially and after six weeks of controlled abstinence.

Paranoia, panic reactions and anxiety are unpleasant effects that can occur with marijuana exposure especially in those with a history of psychiatric disturbances (Weil, 1970; Hubbard *et al.*, 1993). Acute psychosis, melancholia and manic episodes have been described for many years (Tunving, 1985). Rapid thoughts, often considered by the user to be "profound" may occur, and confusion can develop at high doses (Ashton, 1999).

Patients with Schizophrenia are at increased risk of marijuana-induced psychosis (Gold, 1994; Nahas and Latour, 1992). In a study by Hubbard *et al.* (1993), patients with Schizophrenia and Bipolar disorders reported being particularly prone to paranoia with marijuana use.

ADVERSE EFFECTS OF CHRONIC MARIJUANA ABUSE

Evidence for adverse effects from chronic marijuana is difficult to establish in both research studies and anecdotally because of the numerous uncontrolled variables over a long period of time. Overall, however, available data suggests that long-term marijuana abuse may adversely effect behavior, mental functioning, the cardiovascular system, immune system, respiratory system, reproductive system and others as discussed below (Tables 24.3 and 24.4). In addition, chronic abuse of marijuana can lead to a clinical state of cannabis dependence.

Neuropsychiatric

Although there is no clear demonstration of structural brain damage caused by marijuana, evidence suggests that cannabis has neurotoxic effects on the animal and human brains. The hippocampus appears to be particularly susceptible to marijuana exposure, which may help explain apparent deficits in memory and learning skills (Ashton, 1999; Gold, 1994; Janowsky *et al.*, 1976; Nahas and Latour, 1992; Tunving, 1985). In some cases, changes resemble that of accelerated brain aging (Hollister, 1998). The changes tend to be slow and subtle, which may not be readily noticed by the users without repeated measurements (Gold, 1994; Hollister, 1998). Memory and learning deficits were not initially found in a cohort of heavy marijuana users in Costa Rica, yet abnormalities in short-term memory were observed 10 years later (Nahas and Latour, 1992; Page *et al.*, 1988). Other studies have shown effects on short-term memory as well (Nahas and Latour, 1992).

Table 24.3 Neuropsychiatric effects of chronic marijuana use

Memory deficits	Cognitive deficits/Learning deficits
Worsening of Psychosis	Judgement problem
Low motivation	Lower math/verbal testing
Chemical dependence	

References 1, 4, 9, 10, 13, 25, 30–32, 46

Table 24.4 Physical effects of chronic marijuana use

Alteration of sex hormones	Risk to patients with coronary artery disease
Fetal decrease in weight and length	Worsening emphysema/bronchitis
Duration of labor	Squamous cell metaplasia
Immune system deficits	Weight gain

References 4, 9–11, 13, 25, 32, 35–37, 41–44

In a study of 23 chronic cannabis users compared to non-using controls, differences in judgement, communication, verbalization and compromised were noted using the Wechsler Adult Intelligence Scale (Gold, 1994; Corrion, 1990). In a separate investigation of 144, 12th grade chronic marijuana users and 72 non-users (matched for IQ in the 4th grade), marijuana users scored poorer on verbal, math and memory testing (Ashton, 1999; Black and Ghanesia, 1993). Other studies did not notice differences (Gold, 1994). Studies on prenatal exposure to marijuana are preliminary but suggest possible development of subtle cognitive deficits in children (Walker *et al.*, 1999).

In addition to causing acute decompensation of patients with schizophrenia, high associations between marijuana use and schizophrenia have been noted (Hollister, 1998; Gersten, 1980). For example, a sixfold increase in the diagnosis of schizophrenia was reported in marijuana users in a study of about 55,000 Swedish military recruits (Andreasean *et al.*, 1987). Flashbacks of sensations similar to the original drug exposure have been reported, but are rare (Halibo *et al.*, 1971; Ashton, 1999). They most often occur weeks to months after heavy use (Ashton, 1999).

Concerns about marijuana abuse leading to an "amotivational syndrome" have been an area of considerable clinical concern and scientific debate. Low motivation and drive is often observed in marijuana abusers, however, it is uncertain if use of the drug leads to this condition, if people with lower drive tend to use marijuana, or both. Apparent low motivation may also be due to frequent states of intoxication (Ashton, 1999).

Reproduction system, hormone system and the fetus

Many investigations suggest that marijuana has adverse effects on the reproduction system and may be harmful to the fetus (Nahas and Latour, 1992; Witorsch *et al.*, 1995). By necessity, most studies on the reproductive system have been done in animals, however, human data has been collected as well. For example, marijuana appears to alter the menstrual cycle and decrease ovulation (Ashton, 1999; Nahas and Latour, 1992; Witorsch *et al.*, 1995). This may be due to effects of cannabinoids on sex hormones (Witorsch *et al.*, 1995). While acute marijuana exposure tends to decrease prolactin levels, chronic use may increase prolactin and lead to gynecomastia in males and galactorrhea in females (Ashton, 1999). Duration of labor may also be affected by marijuana use (Gold, 1994; Martin and Hubbard, 2000).

Epidemiological investigations suggest that marijuana may affect fetus weight gain, growth in length, and possibly behavioral characteristics of the child (Ashton,

1999; Gold, 1994; Martin and Hubbard, 2000). In a large study of 1,200 mothers, the children of the 10% that had a positive urinalysis for marijuana had reduced birth weight by 79 g, averaged shorter length by 0.5 cm, and had smaller head size (Gersten, 1980; Zuckerman *et al.*, 1989). Children of mothers who used marijuana during pregnancy were reported to more frequently have deficits in language skills, visual perception tasks, memory and attention at four years of age (Ashton, 1999; Janowsky *et al.*, 1976). High-speed computerized voice analysis of newborns showed voice abnormalities in babies of Jamaican women who smoked marijuana compared to controls (Lester and Dreher, 1989). Also, cannabinoids have been reported to cause hypotonicity, lethargy, tremor, and increased startle in newborns (Walker *et al.*, 1999).

In males, cannabinoids are anti-androgenic and appears to decrease sperm mobility, sperm count and may alter sperm shape (Ashton, 1999). Decreased testosterone has also been noted in marijuana users (Witorsch *et al.*, 1995). Effects on fertility are uncertain (Ashton, 1999).

Cancer

Tissue culture and Ames tests showed that marijuana smoke may be mutagenic (Nahas and Latour, 1992). However, teratological effects have not been clearly demonstrated. In a 204 pair case-controlled study Robinson *et al.* (1989) reported a 10 times increase in the risk of non-lymphoblastic leukemia in children of mothers who used marijuana just before or during pregnancy. Leukemia was not increased by other drugs such as alcohol or tobacco (Nahas and Latour, 1992; Robinson *et al.*, 1989). Children of marijuana smoking mothers were also reported to have a threefold greater risk of rhabdomyosarcoma (Janowsky *et al.*, 1976). With regard to the respiratory system, squamous cell hyperplasia has been shown even in young adults (20–26 years old) who are heavy hashish users (Tennani and Guerny, 1980).

Immune system

Numerous changes in the immune system to marijuana and THC have been reported in *in vitro* and *in vivo* systems (Ashton, 1999; Hollister, 1998). However, the clinical medical significance of these alterations is not certain (Ashton, 1999; Hollister, 1998). Marijuana use did not appear to alter progression of AIDS in a study of nearly 5,000 HIV-positive homosexual men (Ashton, 1999; Hollister, 1998; Kaslow *et al.*, 1989).

Cardiovascular system (CVS)

The effect of chronic marijuana use on the CVS has not been well demonstrated (Hollister, 1998). However, some clinicians report concern over the potential dangers of the large quantities of carbon monoxide in cannabis smoke (Ashton, 1999). Possible dangers of marijuana use on the CVS may be more important in the elderly, and those with hypertension, coronary artery disease, and/or other cardiovascular diseases (Janowsky *et al.*, 1976).

Respiratory system

Smoking marijuana is believed to have several dangerous effects on the respiratory system. Like tobacco smoke, cannabis smoke contains both tar carbon monoxide and carcinogens (Gold, 1994; Nahas and Latour, 1992). Unlike cigarettes, marijuana joints do not have filters, and smokers often inhale deeply and keep the smoke in their lungs for extended time to get the full euphoric effect. Marijuana smoke has about three times more tar and five to six times more carbon monoxide than cigarette smoke (Ashton, 1999; Janowsky et al., 1976; Walker et al., 1999; Wu et al., 1988). Chronic use of marijuana has been associated with worsening emphysema, bronchitis, airway obstruction and squamous cell metaplasia (Ashton, 1999; Nahas and Latour, 1992). Case reports of chronic marijuana users developing large lung bullae are also of potential concern (Johnson et al., 2000).

Weight gain

Users of marijuana generally agree that marijuana increases appetite (Hollister, 1970). However, many regular marijuana smokers may not think about the effect of marijuana on their weight. In a study of 10 controls (non-users), 12 "casual" users, and 15 "heavy" marijuana users, significant weight gain was found in both casual and heavy marijuana users after 3 weeks of use (Greenberg et al., 1976). Casual users (use of about 12 times per month) gained 2.8 lbs on average, heavy users (use of about 42 times per month) 3.7 lbs, and non-users gained only 0.2 lbs in the 21-day interval. Weight gain did not appear to be water retention. Weight gain over a longer period of time may be of significant importance to some patients, especially those with diabetes or other health related problems.

Chemical dependence

Dependence to cannabinoids appears to occur in humans and develop slowly with increased risk at higher doses and frequency of use (Gold, 1994; Martin and Hubbard, 2000). Compulsive use has been reported antidotally and in survey studies (Ashton, 1999). Like other drugs of abuse, cannabinoids has been shown to stimulate release of dopamine in neuroanatomical reward centers of the brain (Gardner and Lowinson, 1991). Cannabis dependence is estimated to be about 4% of the population (Gold, 1994; Martin and Hubbard, 2000). Tolerance to marijuana often develops and craving and compulsive urges to use are often reported. In clinical practice, patients often report being surprised to discover how difficult it can be to stop marijuana use. Some people resume marijuana use after initial attempts to quit despite significant personal consequences at home, school or work and yet many deny that they have a chemical dependence problem.

Although the Diagnostic and Statistical Manual of Mental Disorders, 4th ed. (DSM-IV) does not include cannabinoid withdrawal as a diagnostic category, withdrawal symptoms from cannabis have been reported in human and animal studies (Ashton, 1999; Tunving, 1985). For example, withdrawal symptoms have been observed in a laboratory setting when cannabis was used every day for 10 days or more and then abruptly stopped (Tunving, 1985). Anxiety, tremor, irritability, perspiration, nausea, muscle cramps, and insomnia are some of the reported withdrawal symptoms (Hubbard *et al.*, 1999; Ashton, 1999; Tunving, 1985). More often withdrawal symptoms do not occur or are mild due to the long half-life of THC (Hubbard *et al.*, 1999). In chronic users withdrawal symptoms are reported to occur in about 16–29% of subjects (Ashton, 1999).

It has also been a great concern that marijuana is a "gateway" drug to other illicit substances of abuse for some people (Hubbard *et al.*, 1999; Gold, 1994). For example, it was reported in 1985 that one of the best predictors of cocaine use in adolescents is marijuana use (Gold, 1994; Kandel *et al.*, 1985). The common gateway pattern of drug abuse generally begins with legal substances such as alcohol and cigarettes prior to marijuana use.

CONCLUSIONS

Overall, it appears that marijuana has potential for significant acute and long-term impact on the mental and physical health of users. Some adverse effects of marijuana are disturbing to the user (such as anxiety and paranoid states), while others (such as effects on memory and motivation) may not be noticed because they develop slowly and the association with marijuana use is not recognized. Marijuana does not appear to be directly life threatening (unlike alcohol, opiates and many other drugs of abuse) but may increase risk of harm during intoxication (such as during driving or doing other potentially dangerous activities).

The potential short-term and long-term risks of marijuana should be considered prior to medical use or for recreational purposes. Clearly this is an area where more research on the potential benefits and adverse effects is of significant social, scientific, and medical interest.

ACKNOWLEDGMENT

I would like to thank Michelle Gooch for her help in the preparation of this manuscript.

REFERENCES

Andreasean, S., Allebeck, P., Engstrom, A. and Rydbery, V. (1987) "*Cannabis* and Schizophrenia: a longitudinal study of Swedish concepts," *Lancet* **2**: 1483–1485.

Aronow, W. S. and Cassidy, J. (1974) "Effect of marijuana and placebo-marijuana smoking on Angina Pectoris," *The New England Journal of Medicine* **291**: 65–67.

Ashton, C. H. (1999) "Adverse effects of cannabis and cannabinoids," *British Journal of Anesthesia* **83**: 637–649.

Benowitz, N. L. and Jones, R. T. (1975) "Cardiovascular effects of prolonged delta-9-tetrahydrocannabinol ingestion," *Clinical Pharmacology Therapeutics* **18**: 287–297.

Black, R. I. and Ghanesia, M. M. (1993) "Effect of chronic marijuana use on human cognitive," *Psychopharmacology* **110**: 219–228.

Clark, C. S., Greene, C., Karry, G. W., Macconnell, K. L. and Milsteam, S. L. (1974) "Cardio-vascular effects of marijuana in man," *Can. J. Physical.* **52**: 706–719.

Corrion, J. L. (1990) "Mental performance in long-term heavy cannabis use; a preliminary report," *Psychological Reports* **67**: 947–952.

Gardner, E. L. and Lowinson, J. H. (1991) "Marijuana's interaction with brain reward systems: update 1991," *Pharmacology, Biochemistry and Behavior* **40**: 571–580.

Gersten, S. P. (1980) "Long-term adverse effects of brief marijuana usage," *The Journal of Clinical Psychiatry* **41**: 60–61.

Gold, M. S. (1994) "Marijuana," in: *Principles of Addiction*, American Society of Addiction Medicine, Inc., Maryland, Section II, Chap 8, pp. 1–5.

Gottschalk, L. A., Aronow, W. S. and Prakash, R. (1977) "Effect of marijuana and placebo-marijuana smoking on physiological state and on psychophysiological cardiovascular functioning in anginal patients," *Biological Psychiatry* **12**: 255–266.

Greenberg, I., Kuehnle, J., Medelson, J. H. and Bernstein, J. G. (1976) "Effects of marijuana use on body weight and caloric intake in humans," *Psychopharmacology* **49**: 79–84.

Halibo, J. A. *et al.* (1971) "Marijuana effects, a survey of repeat users," *The Journal of American Medical Association* **217**: 692–694.

Hollister, L. E. (1970) "Hunger and appetite after single doses of marijuana, alcohol and dextroamphetamine," *Clinical Pharmacology and Therapeutics* **12**: 44–49.

Hollister, L. E. (1988) "Cannabis – 1988," *Acta Psychiatrica Scandinavica* (Suppl.) **345**: 108–118.

Hollister, L. E. (1998) "Health aspects of cannabis; revisited," *International Journal of Neuropsychopharm* **1**: 71–80.

Hubbard, J. R., Franco, S. E. and Onaivi, E. S. (1999) "Marijuana: Medical implications," *American Family Physician* **60**: 2583–2588.

Hubbard, J. R., Workman, E., Marcus, L., Felker, B., Capell, L., Smith, J. *et al.* (1993) "Differences in marijuana use across psychiatric diagnosis. Reasons they use, the side effects they experience," *Poster B37, The Future of VA Mental Health Research*. National Foundation for Brain Research, Washington, DC.

Janowsky, D. S., Meacham, M. P., Blaine, J. D. *et al.* (1976) "Marijuana effects on simulated flying ability," *The American Journal of Psychiatry* **133**: 383–388.

Johnson, M. K., Smith, R. P., Morrison, D., Laszlo, G. and White, R. J. (2000) "Large lung bullae in marijuana smokers," *Thorax* **55**: 340–342.

Kandel, D. B., Murphy, D. and Karus, D. (1985) "Cocaine use in young adulthood: patterns of use and psychosocial correlates," *NIDA research monograph* **61**: 76–110.

Kaslow, R. A., Blackwelder, W. C. and Ostrow, D. G. (1989) "No evidence of a role of alcohol or other psychoactive drugs in accelerating immunodeficiency in HIV-positive individuals," *The Journal of American Medical Association* **261**: 3424–3429.

Klonoff, H. (1974) "Marijuana and driving in real life situations," *Service* **1986**: 317–324.

Leirer, V. V. Yesavag, J. A (1991) "Marijuana carry-over effect on aircraft pilot performance," *Aviation, Space and Environmental Medicine* **62**: 221–227.

Lester, B. M. and Dreher, R. (1989) "Effects of marijuana use during pregnancy in newborn cry," *Child development* **60**: 765–771.

Losken, A., Maviglia, S. and Friedman, L. S. (1996) "Marijuana," in *Source book of substance abuse and addiction*, edited by L. S. Friedman *et al.*, pp. 179–187. Baltimore, MD: Williams & Wilkins.

Lu, R., Hubbard, J. R., Martin, B. R. and Kalimi, M. Y. (1993) "Roles of sulfhydryl and disulfide groups in the binding of CP-55,940 to rat brain cannabinoid receptor," *Molecular and cellular biochemistry* **121**: 119–126.

Martin, P. R. and Hubbard, J. R. (2000) "Substance-related disorders," in *Current diagnosis and treatment in Psychiatry*, edited by M. H. Ebert, P. T. Loosen and B. Nuscombe, pp. 233–259. Lang medical Books.

Nahas, G. and Latour, C. (1992) "The human toxicity of marijuana," *The Medical Journal of Australia* **156**: 495–497.

Nahas, G. (1977) "Biomedical aspects of cannabis usage," *Bull Narc* **29**: 13–27.

Page, J. B., Fletcher, J. and True, W. R. (1988) "Psychosociocultural perspectives on chronic cannabis use: the Costa Rican follow-up," *Journal of Psychoactive Drugs* **20**: 57–65.

Peck, R. C., Biasotti, A., Boland, P. M., Mallory, C. and Reeve, V. (1986) "The effects of marijuana and alcohol on actual driving performance," *Alcohol, drugs, and driving* **2**: 125–154.

Robinson, I. I., Buckley, J. D., Daigle, A. E. *et al.* (1989) "Maternal drug use and risk of childhood nonlymphoblastic leukemia among offspring. An epidomilogic investigation implementing marijuana," *Cancer* **63**: 1909–1910.

Schwartz, R. H., Greenwald, P. J., Klitzner, M. and Fedio, P. (1989) "Short-term memory impairment on cannabis-dependent adolescents," *Am. J. Dis. Child* **143**: 1214–1219.

Schuckit, M. A. (1989) "Cannabinols," in *Drug and alcohol abuse: a clinical guide to diagnosis and treatment*, 3rd ed., pp. 143–157. New York: Plenum Medical.

Smart, R. G, and Adlaf, E. M. (1982) "Adverse reactions and seeking medical treatment among student cannabis users," *Drug Alcohol Depend* **9**: 201–211.

Smiley, A.M. (1986), "Marijuana on road and driving simulations studies," *Alcohol, drugs, driving* **2**: 121–134.

Tennani, F. S. and Guerny, R. L. (1980) "Histopathologic and clinical abnormalities of the respiratory system in chronic hashish smokers," *Subst Alcohol Actions Misuse* **13**: 93–100.

Tunving, K. (1985) "Psychiatric effects of cannabis use," *Acta Psychiatrica Scandinavica* **72**: 209–217.

Walker, A., Rosenberg, M. and Balaban-Gil, K. (1999) *Child and Adolescent Psychiatric Clinics of North America* **8**: 845–867.

Weil, A. T. (1970) "Adverse reactions to marijuana. Classification and suggested treatment," *The New England Journal of Medicine* **282**: 997–1000.

Witorsch, R. J., Hubbard, J. R. and Kalimi, M. Y. (1995) "Reproductive toxic effects of alcohol, tobacco, and substances of abuse," in *Reproduction toxicology*, 2nd ed., edited by R. J. Witorsch, pp. 283–318. New York: Raven Press.

Wu, T. C., Taskkbin, D. P., Jiahed, B. and Rou, J. E. (1988) "Pulmonary hazards of smoking marijuana as compared with tobacco," *The New England Journal of Medicine* **318**: 347–351.

Zuckerman, B., Frank, D., Higson, R. *et al.* (1989) "Effects of maternal marijuana and cocaine on fetal growth," *The New England Journal of Medicine* 320–762.

Index

Milton Keynes UK
Ingram Content Group UK Ltd.
UKHW050458071024
449327UK00015B/424